W9-CNJ-202

Organic Reactions

VOLUME 36

JOHN WILEY & SONS, INC.

New York · Chichester · Brisbane · Toronto · Singapore

Published by John Wiley & Sons, Inc.

Library of Congress Catalog Card Number 42–20265

ISBN 0-471-85748-3

Printed in the United States of America

10 9 8 7 6 5 4 3 2 1

PREFACE TO THE SERIES

In the course of nearly every program of research in organic chemistry the investigator finds it necessary to use several of the better-known synthetic reactions. To discover the optimum conditions for the application of even the most familiar one to a compound not previously subjected to the reaction often requires an extensive search of the literature; even then a series of experiments may be necessary. When the results of the investigation are published, the synthesis, which may have required months of work, is usually described without comment. The background of knowledge and experience gained in the literature search and experimentation is thus lost to those who subsequently have occasion to apply the general method. The student of preparative organic chemistry faces similar difficulties. The textbooks and laboratory manuals furnish numerous examples of the application of various syntheses, but only rarely do they convey an accurate conception of the scope and usefulness of the processes.

For many years American organic chemists have discussed these problems. The plan of compiling critical discussions of the more important reactions thus was evolved. The volumes of *Organic Reactions* are collections of chapters each devoted to a single reaction, or a definite phase of a reaction, of wide applicability. The authors have had experience with the processes surveyed. The subjects are presented from the preparative viewpoint, and particular attention is given to limitations, interfering influences, effects of structure, and the selection of experimental techniques. Each chapter includes several detailed procedures illustrating the significant modifications of the method. Most of these procedures have been found satisfactory by the author or one of the editors, but unlike those in *Organic Syntheses* they have not been subjected to careful testing in two or more laboratories.

Each chapter contains tables that include all the examples of the reaction under consideration that the author has been able to find. It is inevitable, however, that in the search of the literature some examples will be missed, especially when the reaction is used as one step in an extended synthesis. Nevertheless, the investigator will be able to use the tables and their accompanying bibliographies in place of most or all of the literature search so often required.

Because of the systematic arrangement of the material in the chapters and the entries in the table, users of the books will be able to find information desired by reference to the table of contents of the appropriate chapter. In the interest of economy the entries in the indices have been kept to a minimum, and, in particular, the compounds listed in the tables are not repeated in the indices.

The success of this publication, which will appear periodically, depends upon the cooperation of organic chemists and their willingness to devote time and effort to the preparation of the chapters. They have manifested their interest already by the almost unanimous acceptance of invitations to contribute to the work. The editors will welcome their continued interest and their suggestions for improvements in *Organic Reactions.*

Chemists who are considering the preparation of a manuscript for submission to *Organic Reactions* are urged to write either secretary before they begin work.

CUMULATIVE CHAPTER TITLES BY VOLUME

6. **The Preparation of Unsymmetrical Biaryls by the Diazo Reaction and the Nitrosoacetylamine Reaction,** Werner E. Bachmann and Roger A. Hoffman

7. **Replacement of the Aromatic Primary Amino Group by Hydrogen,** Nathan Kornblum

8. **Periodic Acid Oxidation,** Ernest L. Jackson

9. **The Resolution of Alcohols,** A. W. Ingersoll

10. **The Preparation of Aromatic Arsonic and Arsinic Acids by the Bart, Béchamp, and Rosenmund Reactions,** Cliff S. Hamilton and Jack F. Morgan

Volume 3 (*1946*)

1. **The Alkylation of Aromatic Compounds by the Friedel-Crafts Method,** Charles C. Price

2. **The Willgerodt Reaction,** Marvin Carmack and M. A. Spielman

3. **Preparation of Ketenes and Ketene Dimers,** W. E. Hanford and John C. Sauer

4. **Direct Sulfonation of Aromatic Hydrocarbons and Their Halogen Derivatives,** C. M. Suter and Arthur W. Weston

5. **Azlactones,** H. E. Carter

6. **Substitution and Addition Reactions of Thiocyanogen,** John L. Wood

7. **The Hofmann Reaction,** Everett S. Wallis and John F. Lane

8. **The Schmidt Reaction,** Hans Solff

9. **The Curtius Reaction,** Peter A. S. Smith

Volume 4 (*1948*)

1. **The Diels-Alder Reaction with Maleic Anhydride,** Milton C. Kloetzel

2. **The Diels-Alder Reaction: Ethylenic and Acetylenic Dienophiles,** H. L. Holmes

3. **The Preparation of Amines by Reductive Alkylation,** William S. Emerson

4. **The Acyloins,** S. M. McElvain

5. **The Synthesis of Benzoins,** Walter S. Ide and Johannes S. Buck

6. **Synthesis of Benzoquinones by Oxidation,** James Cason

7. **The Rosenmund Reduction of Acid Chlorides to Aldehydes,** Erich Mosettig and Ralph Mozingo

8. **The Wolff-Kishner Reduction,** David Todd

Volume 5 (1949)

1. **The Synthesis of Acetylenes,** Thomas L. Jacobs
2. **Cyanoethylation,** H. A. Bruson
3. **The Diels-Alder Reaction: Quinones and Other Cyclenones,** Lewis W. Butz and Anton W. Rytina
4. **Preparation of Aromatic Fluorine Compounds from Diazonium Fluoborates: The Schiemann Reaction,** Arthur Roe
5. **The Friedel and Crafts Reaction with Aliphatic Dibasic Acid Anhydrides,** Ernst Berliner
6. **The Gattermann-Koch Reaction,** Nathan N. Crouse
7. **The Leuckart Reaction,** Maurice L. Moore
8. **Selenium Dioxide Oxidation,** Norman Rabjohn
9. **The Hoesch Synthesis,** Paul E. Spoerri and Adrien S. DuBois
10. **The Darzens Glycidic Ester Condensation,** Melvin S. Newman and Barney J. Magerlein

Volume 6 (1951)

1. **The Stobbe Condensation,** William S. Johnson and Guido H. Daub
2. **The Preparation of 3,4-Dihydroisoquinolines and Related Compounds by the Bischler-Napieralski Reaction,** Wilson M. Whaley and Tuticorin R. Govindachari
3. **The Pictet-Spengler Synthesis of Tetrahydroisoquinolines and Related Compounds,** Wilson M. Whaley and Tuticorin R. Govindachari
4. **The Synthesis of Isoquinolines by the Pomeranz-Fritsch Reaction,** Walter J. Gensler
5. **The Oppenauer Oxidation,** Carl Djerassi
6. **The Synthesis of Phosphonic and Phosphinic Acids,** Gennady M. Kosolapoff
7. **The Halogen-Metal Interconversion Reaction with Organolithium Compounds,** Reuben G. Jones and Henry Gilman
8. **The Preparation of Thiazoles,** Richard H. Wiley, D. C. England, and Lyell C. Behr
9. **The Preparation of Thiophenes and Tetrahydrothiophenes,** Donald E. Wolf and Karl Folkers
10. **Reductions by Lithium Aluminum Hydride,** Weldon G. Brown

3. **The Thiele-Winter Acetoxylation of Quinones,** J. F. W. McOmie and J. M. Blatchly

4. **Oxidative Decarboxylation of Acids by Lead Tetraacetate,** Roger A. Sheldon and Jay K. Kochi

Volume 20 (*1973*)

1. **Cyclopropanes from Unsaturated Compounds, Methylene Iodide, and Zinc-Copper Couple,** H. E. Simmons, T. L. Cairns, Susan A. Vladuchick, and Connie M. Hoiness

2. **Sensitized Photooxygenation of Olefins,** R. W. Denny and A. Nickon

3. **The Synthesis of 5-Hydroxyindoles by the Nenitzescu Reaction,** George R. Allen, Jr.

4. **The Zinin Reduction of Nitroarenes,** H. K. Porter

Volume 21 (*1974*)

1. **Fluorination with Sulfur Tetrafluoride,** G. A. Boswell, Jr., W. C. Ripka, R. M. Scribner, and C. W. Tullock

2. **Modern Methods to Prepare Monofluoroaliphatic Compounds,** William A. Sheppard

Volume 22 (*1975*)

1. **The Claisen and Cope Rearrangements,** Sara Jane Rhoads and N. Rebecca Raulins

2. **Substitution Reactions Using Organocopper Reagents,** Gary H. Posner

3. **Clemmensen Reduction of Ketones in Anhydrous Organic Solvents,** E. Vedejs

4. **The Reformatsky Reaction,** Michael W. Rathke

Volume 23 (*1976*)

1. **Reduction and Related Reactions of α,β-Unsaturated Compounds with Metals in Liquid Ammonia,** Drury Caine

2. **The Acyloin Condensation,** Jordan J. Bloomfield, Dennis C. Owsley, and Janice M. Nelke

3. **Alkenes from Tosylhydrazones,** Robert H. Shapiro

Volume 24 (*1976*)

1. **Homogeneous Hydrogenation Catalysts in Organic Synthesis,** Arthur J. Birch and David H. Williamson

2. **Ester Cleavages via S_N2-Type Dealkylation,** John E. McMurry

3. **Arylation of Unsaturated Compounds by Diazonium Salts (The Meerwein Arylation Reaction),** Christian S. Rondestvedt, Jr.

4. **Selenium Dioxide Oxidation,** Norman Rabjohn

Volume 25 (*1977*)

1. **The Ramberg-Bäcklund Rearrangement,** Leo A. Paquette

2. **Synthetic Applications of Phosphoryl-Stabilized Anions,** William S. Wadsworth, Jr.

3. **Hydrocyanation of Conjugated Carbonyl Compounds,** Wataru Nagata and Mitsuru Yoshioka

Volume 26 (*1979*)

1. **Heteroatom-Facilitated Lithiations,** Heinz W. Gschwend and Herman R. Rodríguez

2. **Intramolecular Reactions of Diazocarbonyl Compounds,** Steven D. Burke and Paul A. Grieco

Volume 27 (*1982*)

1. **Allylic and Benzylic Carbanions Substituted by Heteroatoms,** Jean-François Biellmann and Jean-Bernard Ducep

2. **Palladium-Catalyzed Vinylation of Organic Halides,** Richard F. Heck

Volume 28 (*1982*)

1. **The Reimer-Tiemann Reaction,** Hans Wynberg and Egbert W. Meijer

2. **The Friedländer Synthesis of Quinolines,** Chia-Chung Cheng and Shou-Jen Yan

3. **The Directed Aldol Reaction,** Teruaki Mukaiyama

Volume 29 (*1983*)

1. **Replacement of Alcoholic Hydroxy Groups by Halogens and Other Nucleophiles via Oxyphosphonium Intermediates,** Bertrand R. Castro

2. **Reductive Dehalogenation of Polyhalo Ketones with Low-Valent Metals and Related Reducing Agents,** Ryoji Noyori and Yoshihiro Hayakawa

3. **Base-Promoted Isomerizations of Epoxides,** Jack K. Crandall and Marcel Apparu

Volume 30 (1984)

1. **Photocyclization of Stilbenes and Related Molecules,** Frank B. Mallory and Clelia W. Mallory

2. **Olefin Synthesis via Deoxygenation of Vicinal Diols,** Eric Block

Volume 31 (1984)

1. **Addition and Substitution Reactions of Nitrile-Stabilized Carbanions,** Siméon Arseniyadis, Keith S. Kyler, and David S. Watt

Volume 32 (1984)

1. **The Intramolecular Diels-Alder Reaction,** Engelbert Ciganek

2. **Synthesis Using Alkyne-Derived Alkenyl- and Alkynylaluminum Compounds,** George Zweifel and Joseph A. Miller

Volume 33 (1985)

1. **Formation of Carbon-Carbon and Carbon-Heteroatom Bonds via Organoboranes and Organoborates,** Ei-Ichi Negishi and Michael J. Idacavage

2. **The Vinylcyclopropane-Cyclopentene Rearrangement,** Tomáš Hudlický, Toni M. Kutchan, and Saiyid M. Naqvi

Volume 34 (1985)

1. **Reductions by Metal Alkoxyaluminum Hydrides,** Jaroslav Málek

2. **Fluorination by Sulfur Tetrafluoride,** Chia-Lin J. Wang

Volume 35 (1987)

1. **The Beckmann Reactions: Rearrangements, Elimination-Additions, Fragmentations, and Rearrangement-Cyclizations,** Robert E. Gawley

2. **The Persulfate Oxidation of Phenols and Arylamines (The Elbs and The Boyland-Sims Oxidations),** E. J. Behrman

3. **Fluorination with Diethylaminosulfur Trifluoride and Related Aminofluorosulfuranes,** Miloš Hudlický

CONTENTS

Organic Reactions

CHAPTER 1

THE [3 + 2] NITRONE–OLEFIN CYCLOADDITION REACTION

Pat N. Confalone and Edward M. Huie

Central Research and Development Department, Experimental Station, E. I. du Pont de Nemours and Company, Wilmington, Delaware

CONTENTS

ACKNOWLEDGMENTS

We wish to thank Dr. Karin J. Karel and Mr. Thomas C. Johns of the Du Pont Company's Lavoisier Library, Wilmington, Delaware, for assistance in the literature searches. In addition, we would like to acknowledge the significant contributions of Mrs. Theresa Bonnes of Du Pont's Central Research Department (Life Sciences Division) for her skills in the preparation of this chapter.

INTRODUCTION

Early researchers studying the condensation product of carbonyl compounds with *N*-substituted hydroxylamines elected to coin the term "nitrone" as a combination of the words "nitrogen" and "ketone."[1] This was done to emphasize the parallel between this newly discovered functionality and the already rich chemistry of the carbonyl group.

$$R^1NHOH + R^2R^3C{=}O \longrightarrow R^2R^3C{=}N^+(O^-)R^1$$

Nitrone

For example, nitrones are capable of reacting with carbanions of various types, a consequence of the iminium species embedded in the nitrone that renders the functionality susceptible to nucleophilic attack. Thus *C*-phenyl-*N*-methylnitrone undergoes a Reformatsky reaction with ethyl bromoacetate in complete analogy with benzaldehyde.[2] The intermediate zinc alkoxide cyclizes to 2-methyl-3-phenylisoxazolidin-5-one, a type of compound that can also be prepared by the related nucleophilic addition of dialkyl malonates to nitrones.[3]

$$C_6H_5CH{=}N^+(O^-)CH_3 + BrCH_2CO_2C_2H_5 \xrightarrow[THF]{Zn} \left[C_6H_5CH(CH_2CO_2C_2H_5)N(CH_3)OZnBr\right]$$

$$\xrightarrow{-ZnBrOC_2H_5} \text{2-methyl-3-phenylisoxazolidin-5-one}$$

This chapter deals with a unique property of nitrones not shared by the corresponding carbonyl compounds, namely, a marked ability to undergo a

[3 + 2] cycloaddition reaction in the presence of a dipolarophile. We address the reaction of nitrones of general structure **1** with substituted olefins, both intermolecular and intramolecular. This process yields isoxazolidines directly, affording products related to those obtained in the Reformatsky reaction but arising by a different reaction mode.

R^3, R^1, N^+, O^-, R^2 + R^4, R^6, R^5, R^7 ⟶ R^1, R^2, R^4, R^5, R^7, R^3—N, O, R^6

1

The earliest description of a [3 + 2] cycloaddition involving a nitrone is that of Beckmann in 1890, who reported that a 1:1 adduct is formed when an aryl isocyanate is heated in the presence of a nitrone.[4] This observation was ignored until the systematic studies of Huisgen in the 1960s established the scope and applications of dipolar cycloadditions as a route to a great variety of heterocycles.[5–7] In 1960 LeBel described an intramolecular nitrone–olefin dipolar cycloaddition that occurs when melonal (**2**) is condensed with *N*-methylhydroxylamine.[8] The cyclopentane derivative **3** is generated with complete stereochemical control over the newly formed asymmetric centers. This prototypical reaction proceeds via the intermediate nitrone olefin. In spite of the obvious power of this reaction in forming carbocyclic rings with high regio- and stereochemical selectivities, applications of this methodology, particularly in natural product total synthesis, have not appeared until recently. Regio- and stereocontrol in the intermolecular version of the reaction is much more complicated and is covered later in this chapter.

OHC, H, **2** + CH_3NHOH ⟶ [^-O, $\overset{+}{N}$, CH_3, H] ⟶ H, O, N, H, H, CH_3, **3**

The intent of the review presented in this chapter is to provide the reader with a thorough understanding of the nitrone–olefin [3 + 2] cycloaddition reaction and to illustrate its power by describing some significant applications to complex synthetic problems. Various aspects have been reviewed.[1,9] This documentation of the nitrone–olefin cycloaddition reaction begins with the preparation and stability of the nitrone component and is followed by mechanistic considerations. A presentation of the dipolarophile syntheses is beyond the scope of this chapter; however, the tabular survey provides leading references to specific examples. The important concepts of regio- and stereoselectivities are introduced next. Since the general rules of regiochemistry that apply in the intermolecular version of the reaction are often reversed in the intramolecular version, the latter are dealt with separately. Finally, important applications to the total synthesis of natural products are presented. The versatile utility of the nitrone–olefin cycloaddition reaction in the synthesis of natural products has been the major driving force in the development of this long-neglected chemistry. The authors believe that an in-depth understanding of the key transformations of the isoxazolidines afforded by the reaction will place this chemistry firmly within the arsenal of organic reactions.

PREPARATION OF THE NITRONE COMPONENT

Oxidation of *N*,*N*-Disubstituted Hydroxylamines

The conversion of *N*,*N*-disubstituted hydroxylamines to cyclic and acyclic nitrones can be accomplished by oxidation with molecular oxygen,[10] yellow mercuric oxide,[11] "active" lead oxide,[12] potassium ferricyanide,[13–16] potassium permanganate,[17] *tert*-butyl hydroperoxide,[18] or hydrogen peroxide.[9]

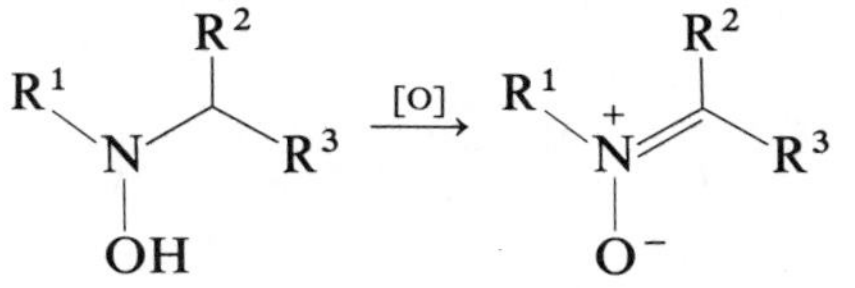

$R^1N(OH)CH(R^2)R^3 \xrightarrow{[O]} R^1\overset{+}{N}(O^-)=C(R^2)R^3$

For example, 5-ethyl-1-hydroxy-2,2-dimethylpyrrolidine yields the cyclic nitrone **4** when air is bubbled into an ethanol solution containing cupric acetate and ammonia.[11] The most important oxidant with wide generality is yellow mercuric oxide, which produces the stabilized nitrone **5** from *N*-hydroxy-1,2,3,4-tetrahydroisoquinoline.[19,20]

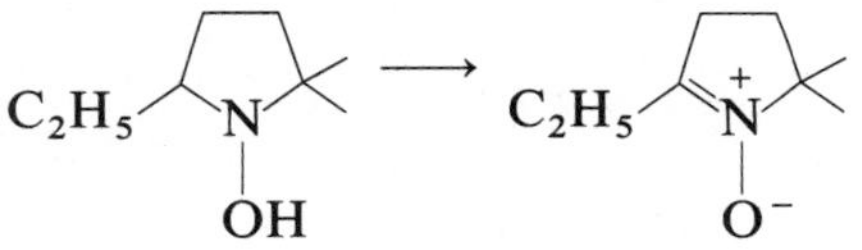

C_2H_5 (N–OH pyrrolidine) ⟶ C_2H_5 ($\overset{+}{N}$–O^- nitrone)

4

(N–OH tetrahydroisoquinoline) $\xrightarrow{HgO}$ (N^+–O^- nitrone)

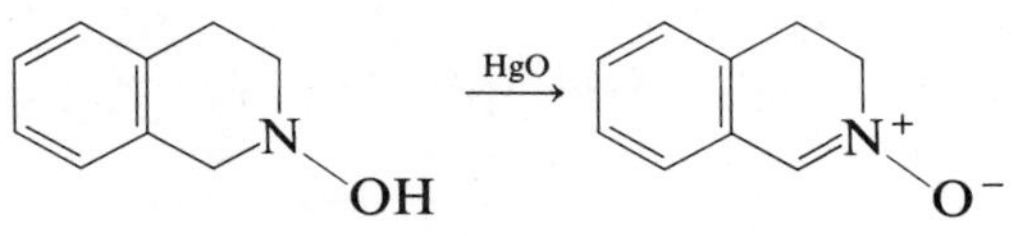

5

Reactive cyclic nitrones such as 1-pyrroline-1-oxide and 2,3,4,5-tetrahydropyridine-1-oxide oligomerize on attempted isolation.[21] These nitrones are therefore generated *in situ* and used directly in solution. The unsubstituted five- and six-membered ring nitrones **8** and **9** are usually generated this way because there are no regiochemical consequences. With unsymmetrical cyclic hydroxylamines, regiochemical control is limited and not readily predictable. For example, the phenyl-substituted compound **6** yields the regioisomers **10** and **11**[22] in a ratio of 3:7, whereas the lower homolog **7** affords only the nitrone **12**.[20]

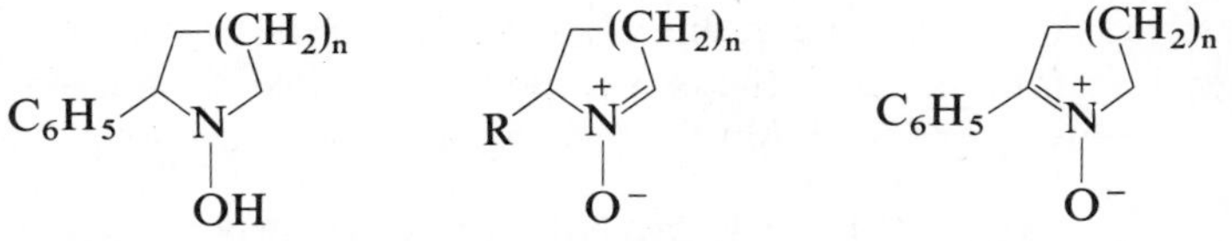

6 n = 2
7 n = 1

8 R = H, n = 1
9 R = H, n = 2
10 R = C_6H_5, n = 2

11 n = 2
12 n = 1

Because of these uncertainties, oxidation of unsymmetrical cyclic hydroxylamines is of limited use unless the regiochemistry is forced as in the conversion of 2,2-dimethyl-1-hydroxypyrrolidine to 5,5-dimethyl-1-pyrroline-1-oxide.[11]

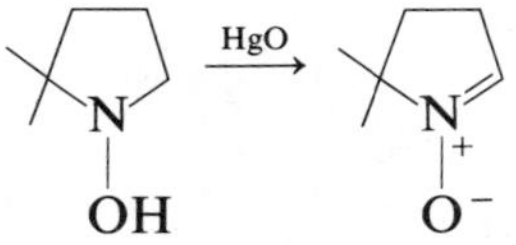

HgO

Condensation Reactions with Carbonyl Compounds

An excellent method for the preparation of nitrones is the reaction of an *N*-substituted hydroxylamine with an aldehyde or ketone.[9,23]

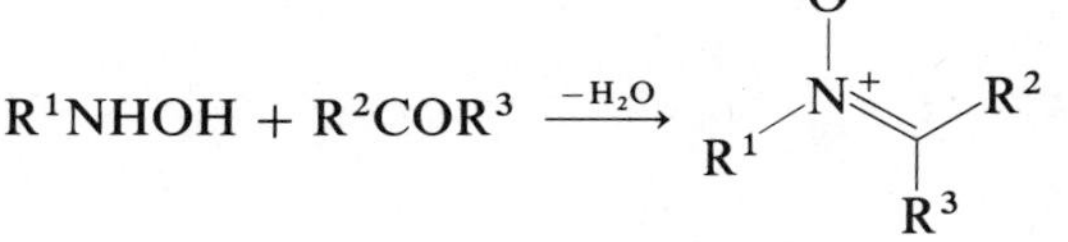

$$R^1NHOH + R^2COR^3 \xrightarrow{-H_2O} R^1N^+(O^-)=CR^2R^3$$

This reaction proceeds smoothly and in high yield when R^1 is an alkyl or aryl group and R^2 and R^3 are not sterically demanding.[24] For example, *N*-phenylhydroxylamine and *n*-butyraldehyde afford an 80% yield of *N*-phenyl-*C*-(*n*-propyl)nitrone (**13**).[13]

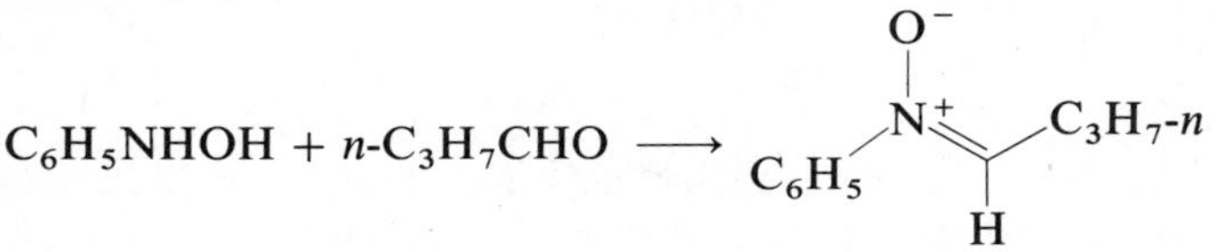

$$C_6H_5NHOH + n\text{-}C_3H_7CHO \longrightarrow C_6H_5N^+(O^-)=CH(C_3H_7\text{-}n)$$

13 (80%)

The extremely hygroscopic nitrone hydrochloride **14** is produced from *N*-methylhydroxylamine hydrochloride and cyclohexanone.[23] With some less reactive ketones, the dialkyl ketal derivatives are used to activate the system for nitrone formation.

$$CH_3NHOH\cdot HCl + \text{cyclohexanone} \xrightarrow[3\,h]{70^\circ} \text{cyclohexylidene}{=}\overset{+}{N}(O^-)CH_3$$

14 (61%)

For example, benzophenone diethyl ketal and *N*-methylhydroxylamine hydrochloride readily afford *C*-diphenyl-*N*-methylnitrone in 84% yield.[23]

$$CH_3NHOH\cdot HCl + (C_6H_5)_2CH(OC_2H_5)_2 \longrightarrow CH_3\overset{+}{N}(O^-){=}C(C_6H_5)_2$$

(84%)

Alternatively, the bisulfite addition compounds of aldehydes and ketones can be used with *N*-substituted hydroxylamines to yield the nitrones, often quantitatively.[15]

Miscellaneous Methods

Five-membered cyclic nitrones are obtained by reductive cyclization of γ-nitroketones in yields of 50–80%, using zinc dust in aqueous ammonium chloride.

$$O_2N\text{–}CH_2CH_2CH_2\text{–}C(=O)\text{–}R \xrightarrow[NH_4Cl]{Zn} \text{1-pyrroline } N\text{-oxide }(\overset{+}{N}\text{–}O^-)\text{, 2-}R$$

An intermolecular variation of this method is the zinc dust reduction of nitrobenzene in the presence of benzaldehyde at -8° for 2 hours to give a 90% yield of *C*,*N*-diphenylnitrone[25–29].

An elegant Grob-type fragmentation of the cyclic hydroxyamine tosylate **15** generates the nitrone–olefin *in situ* and affords the tricyclic adduct **16**.[30] The addition of 1.05 equivalents of potassium *tert*-butoxide is critical since its omission leads only to elimination of *p*-toluenesulfonic acid, affording a low yield of *N*-hydroxy-4a,5-octahydroquinoline. Apparently, the *N*-hydroxylamino nitrogen is too weakly basic to initiate the concerted fragmentation pathway, so the competing E_1 elimination dominates. Formation of the potassium salt of the substrate increases this basicity and facilitates the desired fragmentation.

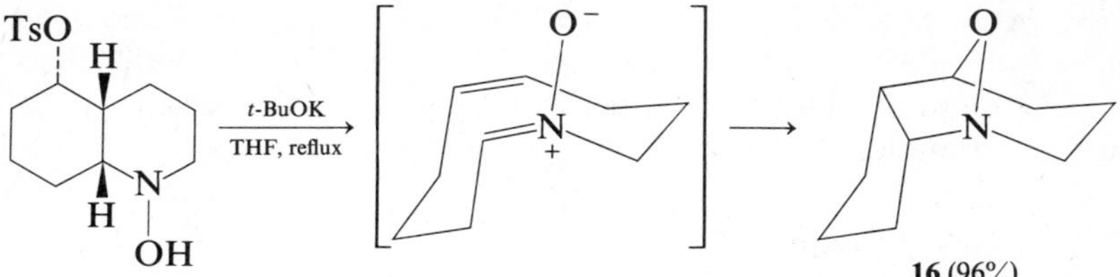

TsO H N H OH **15** $\xrightarrow[\text{THF, reflux}]{t\text{-BuOK}}$ [O^- N^+] → O N **16** (96%)

Nitrones can also be prepared by *N*-alkylation of oximes,[1,31] thermal rearrangement of oxaziranes,[32,33] and a variety of other routes.[34,35] These alternatives have been reviewed;[9] they are usually of limited synthetic utility and often lack generality. Finally, if a nitrone is relatively unstable, generation in solution and immediate reaction with the olefin component will usually overcome difficulties associated with isolation of a reactive nitrone. A recent report,[36] however, describes the preparation of the reactive cyclic nitrone **9** by palladium-catalyzed dehydrogenation of **6** in water at 80° in 56% yield. The generality of this new method has not yet been examined. A simple synthesis of nitrones by oxidation of secondary amines with hydrogen peroxide has appeared recently.[36a] Additional references to the preparation and use of other specific nitrones can be obtained from the tabular survey.

STABILITY OF THE NITRONE COMPONENT

Although nitrones are hydrolytically unstable, reverting to the hydroxylamine and carbonyl precursors (particularly with ketone adducts), a facile dimerization is the chief mode of reactivity for most nitrones. Oxidation of *N*-hydroxypiperidine yields the dimer on attempted isolation.[20] The monomeric cyclic nitrone 2,3,4,5-tetrahydropyridine-1-oxide (THPO) can be trapped in solution with an olefin and is thus available for cycloaddition chemistry. The dimer can also be thermally cracked to yield the monomer, but the former method is preferred.

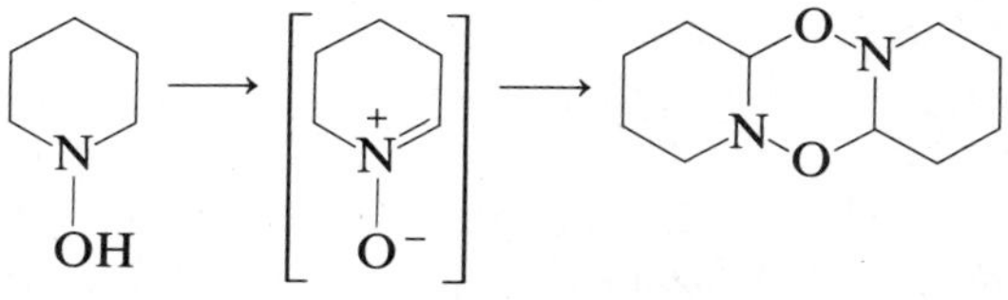

N OH → [N^+ O^-] → O N N O

The structure of the nitrone dimer derived from acetone and *N*-phenylhydroxylamine was finally solved by X-ray diffraction after a number of incorrect assignments had been made.[37,38]

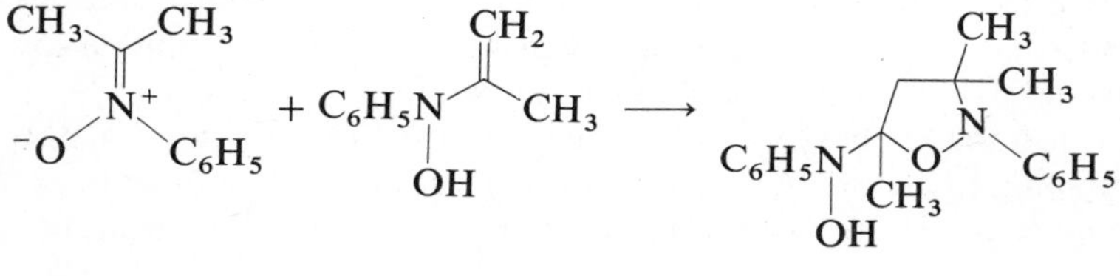

CH_3 CH_3 N^+ ^-O C_6H_5 + $C_6H_5N(OH)C(=CH_2)CH_3$ → CH_3 CH_3 N O C_6H_5 C_6H_5N OH CH_3

This adduct presumably arises from a [3 + 2] cycloaddition of *C*-dimethyl-*N*-phenylnitrone to its tautomer. In general, therefore, the nitrones derived from condensation reactions are frequently prepared in the presence of the olefin component in order to minimize the dimerization reaction. Otherwise, use of an excess of the nitrone component is advisable. The cyclic nitrones, usually prepared by oxidation of a cyclic hydroxylamine precursor, are not isolated in practice but are used immediately in solution.

MECHANISM

The [3 + 2] dipolar cycloaddition reaction of a nitrone to an olefin is an orbital symmetry allowed thermal $[\pi^2 s + \pi^4 s]$ process and is a specific example of the general reaction of a 1,3-dipole adding across a dipolarophile, which is virtually any double or triple bond. The resulting adduct is a five-membered ring heterocycle of general structure **17**, a synthetic concept well exploited by the work of Huisgen and co-workers.[5,39,40]

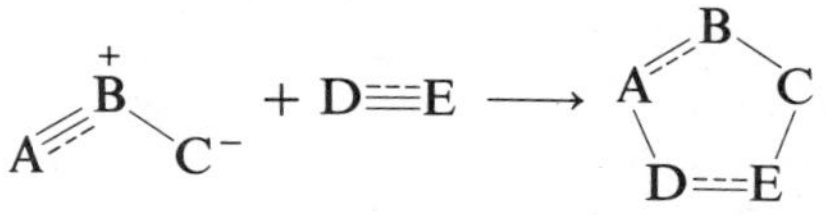

$$\mathrm{A{\equiv}\overset{+}{B}{-}C^-} + \mathrm{D{\equiv}E} \longrightarrow \mathbf{17}$$

17

The cycloaddition reaction involves a nitrone as the 1,3-dipole and olefins or acetylenes as the dipolarophiles and leads to isoxazolidines and isoxazolines, respectively, as the cycloadducts. It was recognized early that a nitrone possesses two all-octet resonance forms.

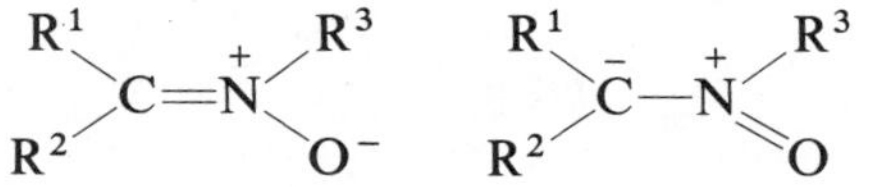

$$\mathrm{R^1R^2C{=}\overset{+}{N}(R^3){-}O^-} \qquad \mathrm{R^1R^2\overset{-}{C}{-}\overset{+}{N}(R^3){=}O}$$

This dipolar character, described in terms of valence-bond theory, possesses proximate nucleophilic and electrophilic centers, which readily explains the facile [3 + 2] mode of cycloaddition characteristic of the nitrone group. In general, the additions of nitrones to dipolarophiles are stereospecifically suprafacial, exhibiting small activation energies and large negative activation entropies. These facts, together with small solvent polarity effects and considerations of reactivity and regioselectivity, all require a highly ordered transition state. Accordingly, some workers have concluded that the dipolar cycloaddition proceeds via a concerted four-center mechanism,[40,41] passing through the transition state **18**.

The principles of orbital symmetry conservation provide a permissive, although not obligatory, theoretical basis for this concerted mechanism,[42,43] as do the observations of $[\pi^4 s + \pi^6 s]$ to the exclusion of $[\pi^4 s + \pi^4 s]$ dipolar cycloadditions in the reactions of 1,3-dipoles with trienes.[44,45]

Nitrone–olefin cycloadditions are minimally influenced by solvent polarity. Thus the rate of the cycloaddition of *N*-methyl-*C*-phenylnitrone to ethyl acrylate increases by a factor of only 2–6 when the solvent is changed from toluene to dimethyl sulfoxide, in spite of a change in dielectric constant by a factor of over 20. These observations have been used to rule out a third mechanism,[46,47] namely, a process involving a zwitterionic species such as **19**, since cycloadditions known to proceed via such species exhibit a marked effect of solvent polarity on rate.[48] In fact, this lack of solvent effect as well as the production of certain byproducts in selected dipolar cycloadditions,[49] have led others to conclude that nitrone–olefin cycloadditions proceed through the spin-paired diradical **20**.[50–52]

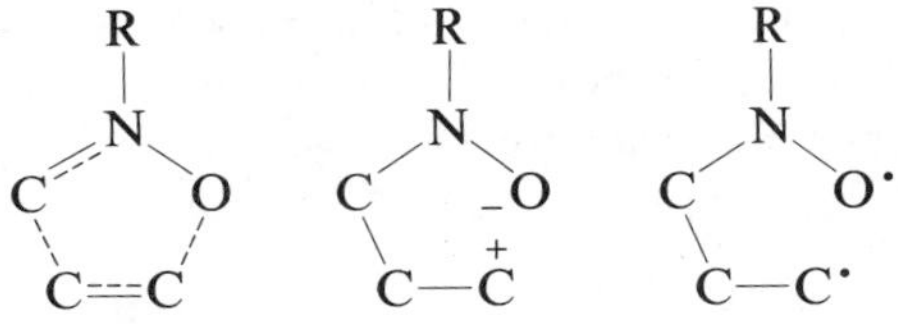

18 **19** **20**

Proponents of the concerted mechanism, however, argue that the solvent polarity data merely indicate an early and highly ordered transition state. This viewpoint is supported by *ab initio* calculations.[53–55]

Perhaps the strongest body of evidence for the concerted mechanism is the marked stereospecific bond formation of nitrone–olefin cycloadditions, which is an absolute consequence of a concerted mechanism. This fact, however, can still be consistent with a diradical mechanism if bond rotations in the intermediate are slower than ring closure to the isoxazolidine. A rate of bond rotation comparable to ring closure would lead to a loss of the observed stereoselectivity. However, the diradical mechanism fails to rationalize adequately the observed regiochemistry of intermolecular nitrone–olefin cycloadditions, particularly of electron-rich monosubstituted olefins compared with their electron-deficient counterparts.

It was only through the application of frontier molecular orbital (FMO) theory to this problem that a coherent picture of a mechanism based on a concerted nonsynchronous process emerged. This theoretical solution to the mechanistic dilemma[48,56–59] went far beyond the usual rationalization of steric and electronic effects in the transition state of a concerted process.[60,61] Thus the reactivity and regiochemical control in the [3 + 2] dipolar cycloaddition of a nitrone to a substituted olefin can now be understood by consideration of the interactions of the corresponding frontier molecular orbitals.

Perturbation theory has been applied to a variety of chemical phenomena and describes the interaction of two molecular orbitals that give rise to a new set of orbitals. Fukui first advocated the FMO approximation of the second-order perturbation expression that governs this phenomenon.[51] Simply stated, this theory holds that the orbitals that overlap best and are closest in energy will interact the most. If this interaction is between orbitals containing two or three electrons, stabilization occurs since electronic energy is lowered on balance. However, the presence of four electrons leads to destabilization as a result of the resulting closed-shell repulsions.[62,63] The key approximation of FMO theory is to include in the stabilizing terms of the perturbation expression only energy changes arising from interactions of the highest occupied molecular orbital (HOMO) of one of the reactants and the lowest unoccupied molecular orbital (LUMO) of the other and vice versa. Sustmann[64] first applied FMO theory to 1,3-dipolar cycloadditions, while Houk[59] and Bastide and Henri-Rousseau[65] used it to explain reactivity and regioselectivity of the reaction.

Consider the reactivity of *N*-methyl-*C*-phenylnitrone with four monosubstituted olefins of differing electronic character ranging from electron-deficient to electron-rich. The energies of the HOMO and LUMO for these reactants, determined by photoelectron spectroscopy and electron affinity data, respectively, are listed in Table A in order of increasing HOMO energies.[57,58,66]

TABLE A

Reactant	*E* (eV) LUMO	*E* (eV) HOMO
N-Methyl-*C*-phenylnitrone	−0.4	−8.58
Nitroethylene	−0.7	−11.40
Methyl acrylate	0.0	−10.72
Propene	+1.8	−9.88
Methyl vinyl ether	+2.0	−9.05

The order of reactivity of the nitrone with these diverse dipolarophiles can be predicted by comparison of the relative energy differences for the dominant HOMO–LUMO interaction for each pairwise reaction. For example, the dominant interaction for the reaction of *N*-methyl-*C*-phenylnitrone and nitroethylene is between the LUMO of the dipolarophile and the HOMO of the dipole since this represents an energy of 7.88 eV, whereas the alternative HOMO dipolarophile/LUMO dipole for this reaction requires an energy of 11.0 eV. Similar calculations for the other three examples lead to the values summarized in Table B, listed in order of increasing energy differences.

TABLE B

Monosubstituted olefin	Energy of Dominant FMO Interation (eV)
Nitroethylene	7.88
Methyl acrylate	8.58
Methyl vinyl ether	8.65
Propene	9.48

Thus the order of reactivity for nitrone [3 + 2] cycloaddition to these monosubstituted olefins is nitro > carbomethoxy > methoxy > methyl, in complete accord with the experimental facts.[67,68] The success of this theoretical approach is quite remarkable since the frontier orbital approximation totally ignores interactions of extra frontier orbitals, closed-shell repulsions, and coulombic terms.[69]

Extension of FMO theory to the vexing problem of nitrone olefin regioselectivity is accomplished by the observation that the LUMO dipole–HOMO dipolarophile interaction dominates in the case of electron-rich olefins. In the alternative very electron-deficient case, the HOMO dipole-LUMO dipolarophile dominates. This switchover clearly complicates intermediate examples and explains the regiochemical mixtures often encountered in certain dipolar cycloadditions of these compounds.

Although reactivity is rationalized in terms of relative energy differences of the reactant frontier orbitals, regiochemistry is correlated with the coefficients associated with the atomic orbitals in each FMO. The dominant stabilizing interaction in the transition state is between atomic orbitals of interacting atoms with the largest coefficients. Consider the general cycloaddition of nitrones to propene, where the dominant interaction is that of the LUMO of the nitrone with the HOMO of propene. It is known that the atomic orbital coefficient of the nitrone LUMO is larger at carbon than at oxygen whereas the coefficient of the terminal carbon of the propene double bond is larger than that of the central carbon in the HOMO of this substrate. Therefore, the transition state for this reaction favors bonding of the nitrone carbon to the terminal olefinic carbon, thus affording the 5-methyl-substituted isoxazolidine as the major regiochemical adduct, in agreement with experimental observations. However, nitrone addition to nitroethylene involves the nitrone HOMO and the LUMO of the substrate. Thus the lower energy transition state should now lead to 4-substituted isoxazolidines, again in accord with the actual results.

A final example is the reaction of *N-tert*-butylnitrone with the dipolarophiles acrylonitrile and cyanoacetylene. The HOMO dipole–LUMO interaction is dominant for the reaction with acrylonitrile since this interaction has a lower relative energy difference ($E = 8.62$ vs. $E = 11.42$ eV) for the alternative LUMO–HOMO couple. However, the coefficients at the termini of the nitrone HOMO are nearly equal, so no particular regioisomer is favored by analysis of this interaction. The alternative higher-energy interaction of the nitrone LUMO–acrylonitrile HOMO couple favors the transition state leading to the 5-cyano-substituted isoxazolidine. Therefore, it is argued that this latter interaction leads to regioisomeric control even though it contributes less to the total energy stabilization of the transition state itself. With cyanoacetylene, a decrease in the nitrone LUMO–cyanoacetylene HOMO interaction practically eliminates this control element, leading to prediction of a mixture of regioisomers. Experimentally, *N-tert*-butylnitrone affords a single 5-substituted cycloadduct with acrylonitrile and a 50:50 mixture of both 4- and 5-substituted isoxazolines on reaction with cyanoacetylene.[48] Thus FMO theory offers a sophisticated theoretical basis for both reactivity and regioselectivity in nitrone–olefin [3 + 2] dipolar cycloadditions. From these considerations a unifying view of the mechanism has emerged that involves an early and highly ordered transition state in which both reactant reactivity and product regiochemistry can be rationalized.

REGIO- AND STEREOSELECTIVITY OF THE INTERMOLECULAR REACTION

As a general rule, the [3 + 2] cycloaddition reaction of nitrones to olefins follows the predictions of frontier orbital calculations; thus most dipolarophiles yield C-5-substituted isoxazolidines as a result of maximum overlap of the LUMO of the nitrone with the HOMO of the olefinic partner. Thus it is predicted and experimentally verified that an increased tendency to form C-4-substituted isoxazolidines will be observed as the electron affinity of the dipolarophile increases. These considerations dominate the usual free-energy arguments based on repulsive steric interactions and attractive van der Waals forces. These latter elements play a major role in transition-state energies and are more influential in determination of stereochemistry rather than regiochemistry.

Aryl Monosubstituted Olefins

The regiochemistry of the [3 + 2] cycloaddition of nitrones to monosubstituted olefins correlates with the electronic nature of the olefinic substituents and their substitution pattern and to a lesser extent their steric requirements. The regiochemistry is primarily a reflection of the HOMO–LUMO considerations outlined in the section on mechanism. For example, the reaction of *C*-cyclopropyl-*N*-methylnitrone with styrene gives the 5-substituted isoxazolidine **21**. The alternative 4-substituted regioisomer **22** is not observed.[70]

O^- N^+ CH_3 + C_6H_5 ⟶ CH_3 N O C_6H_5 **21** (CH_3 N O C_6H_5) **22**

The cycloaddition of THPO to 3,4-dimethoxystyrene in refluxing toluene affords only the adduct **23**.[71]

N^+ O^- + CH_3O CH_3O ⟶ CH_3O CH_3O O–N **23** (93%)

A similar result is obtained when the nitrone **24** is heated at 90° in neat styrene to produce the cycloadduct **25** without contamination by the regioisomer **26**.[72]

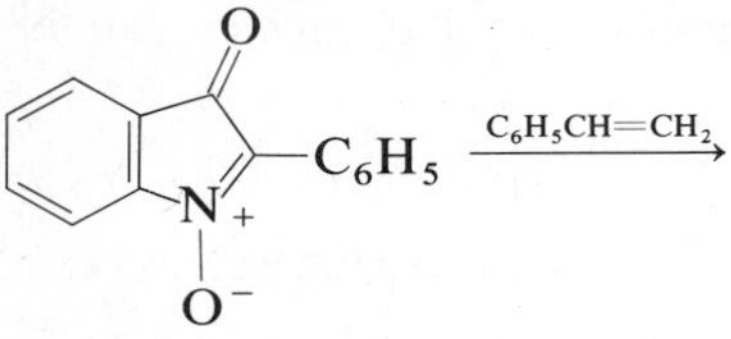

$C_6H_5CH{=}CH_2$

24

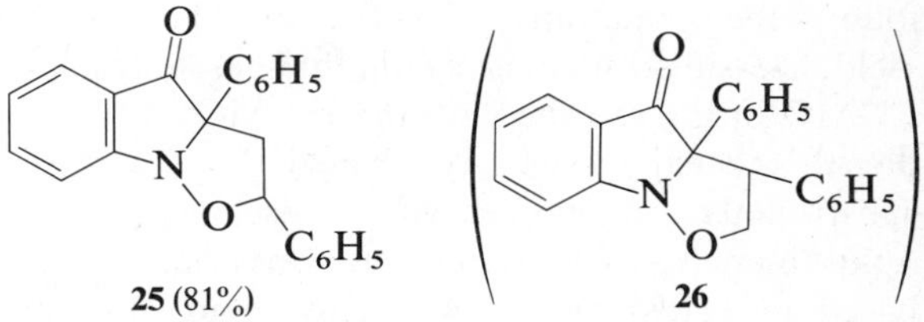

25 (81%)

26

The marked tendency of the C—O bond generated during the cycloaddition to become attached to the benzylic carbon is also found even in simple nitrones such as *N*-methylnitrone, derived from formaldehyde and *N*-methylhydroxylamine, leading to 2-methyl-5-phenylisoxazolidine on reaction with styrene.[73]

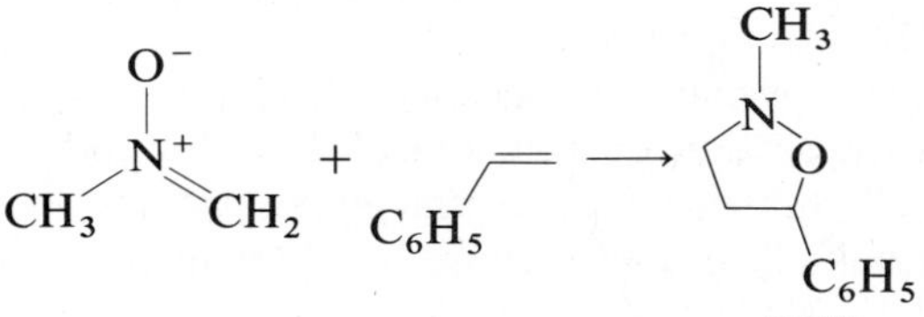

(74%)

The reaction of the nitrofurfural-derived nitrone **27** with 4-vinylpyridine leads regiospecifically to the 5-substituted isoxazolidine **28**.[74]

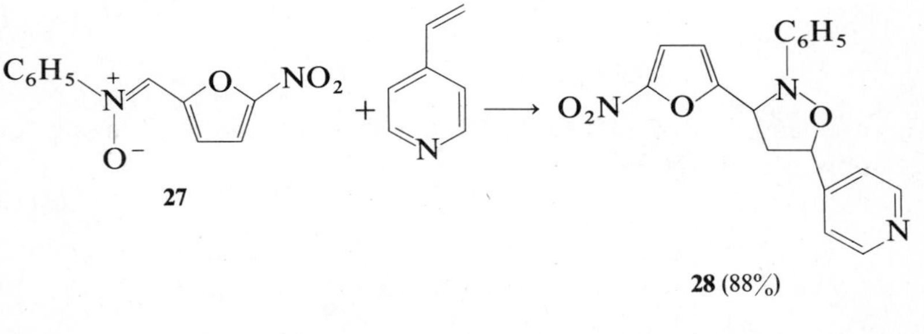

27

28 (88%)

The functionalized cyanonitrone **29** reacts with styrene at 100° to produce the expected 5-substituted product **30**.[75]

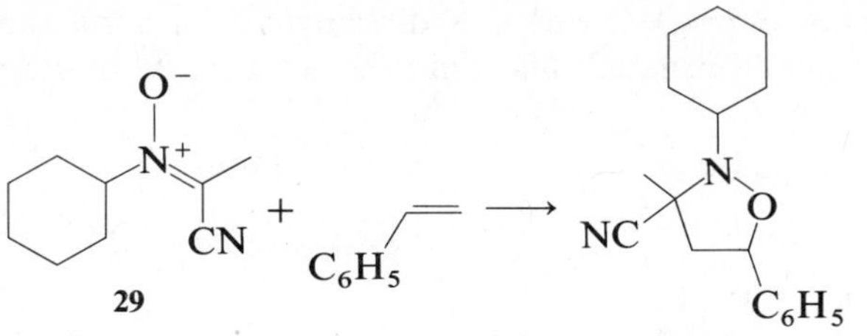

29 + $C_6H_5CH=CH_2$ ⟶ 30 (79%)

Finally, 1-(3,4-dimethoxyphenyl)butadiene reacts exclusively with THPO at the terminal double bond to yield regioisomer **31** as a mixture of diastereomers.[76]

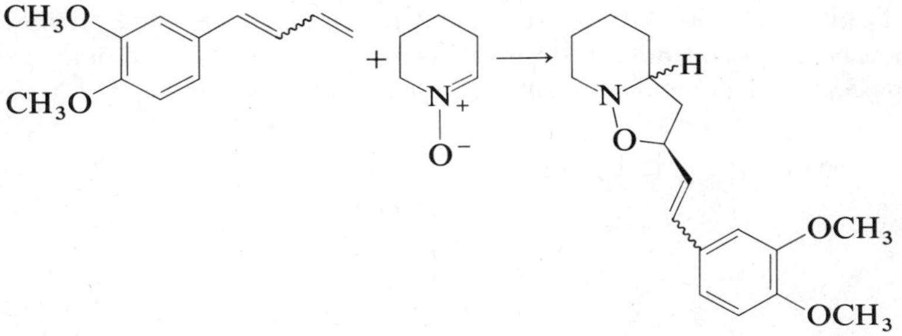

31

Alkyl Monosubstituted Olefins

Monosubstituted olefins bearing a simple aliphatic group behave like the monoaryl analogs, as can be seen from the reactions of *N*-methyl-*C*-phenylnitrone and 1-hexene to yield the 5-substituted product 2-methyl-3-phenyl-5-*n*-butylisoxazolidine exclusively.[77]

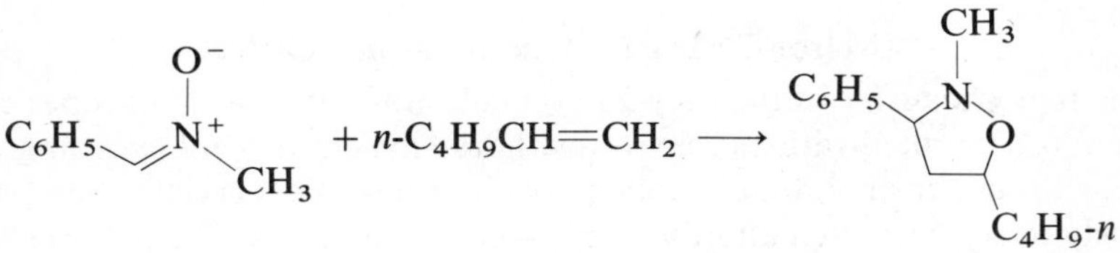

$C_6H_5CH=N^+(O^-)CH_3$ + n-$C_4H_9CH=CH_2$ ⟶

A second example is the reaction of the nitrone **32** with 3-chloropropene to furnish the regioisomer **33**.[78]

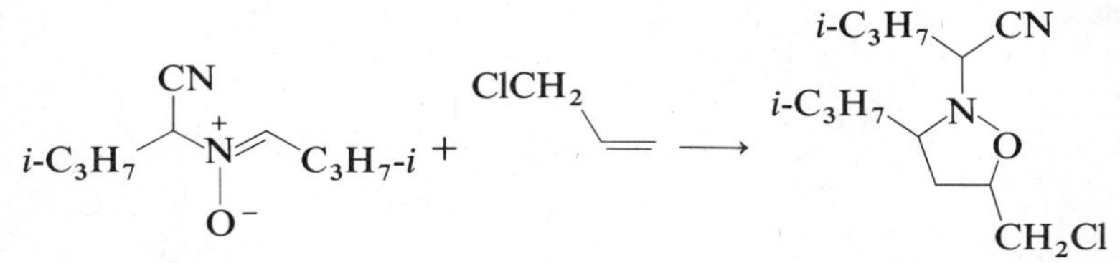

32 + $ClCH_2CH=CH_2$ ⟶ 33 (93%)

In the reaction of propene with *C,N*-diphenylnitrone a trace amount of regioisomer was actually isolated, affording the adducts **34** and **35** in a ratio of 98:2.[79]

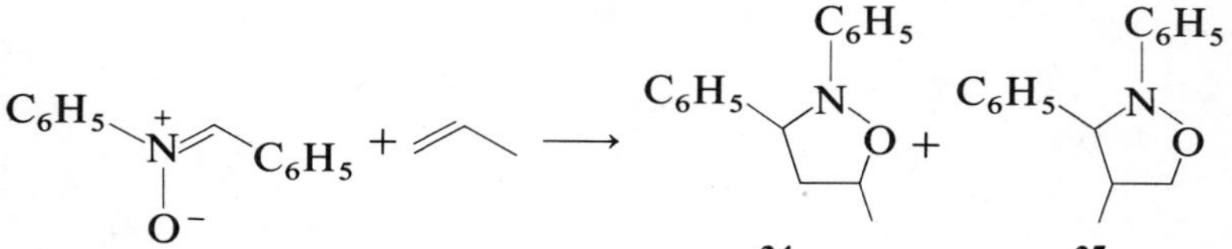

34 **35**

(98:2)

This result emphasizes the marked tendency to form the C-5 regioisomer in [3 + 2] nitrone cycloadditions to unactivated monosubstituted olefins. Cyclic nitrones behave in a similar fashion, as seen in the cycloaddition of the silyloxynitrone **36** and 1-pentene to furnish the bicyclic adduct **37**.[80]

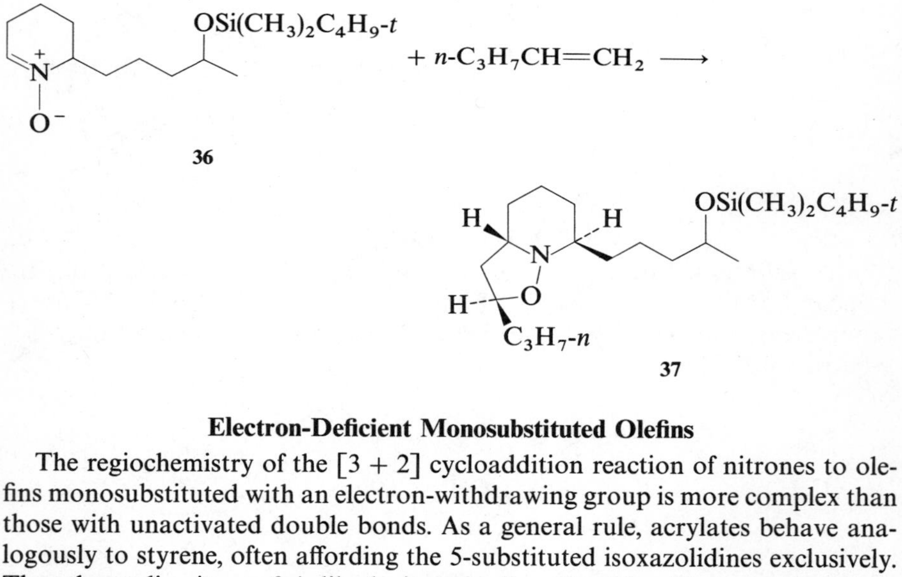

$+ \ n\text{-}C_3H_7CH{=}CH_2 \longrightarrow$

36

37

Electron-Deficient Monosubstituted Olefins

The regiochemistry of the [3 + 2] cycloaddition reaction of nitrones to olefins monosubstituted with an electron-withdrawing group is more complex than those with unactivated double bonds. As a general rule, acrylates behave analogously to styrene, often affording the 5-substituted isoxazolidines exclusively. Thus the cyclic nitrone 3,4-dihydroisoquinoline-*N*-oxide adds to ethyl acrylate to yield only the regioadduct **38**.[81]

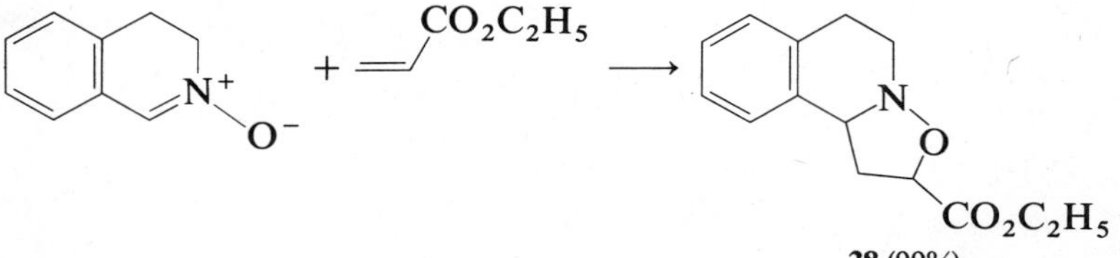

38 (99%)

N-tert-Butylnitrone reacts similarly with methyl acrylate to afford only 2-*tert*-butyl-5-carbomethoxyisoxazolidine.[68]

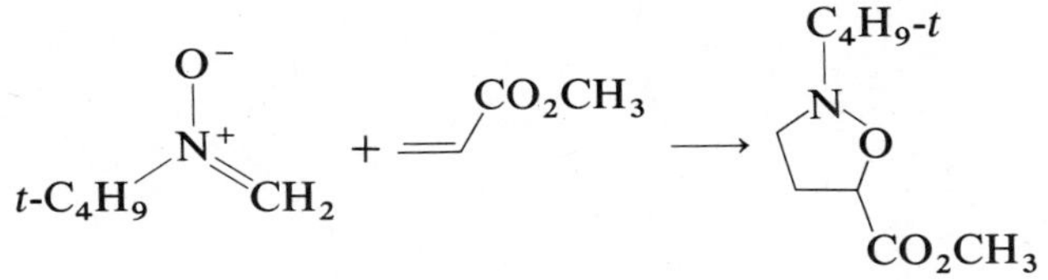

However, there are many examples in which amounts of the 4-substituted isoxazolidines are isolated in the reactions of acrylates. *C*-Cyclopropyl-*N*-methylnitrone adds to methyl acrylate to produce a 70:30 mixture of the 4- and 5-substituted adducts **39** and **40**, respectively.[68]

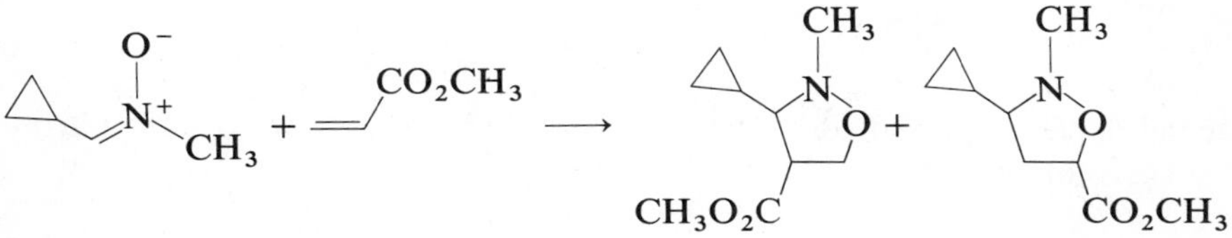

39 **40**

(70:30)

The cycloaddition of *C,N*-diphenylnitrone to methyl acrylate affords a 2:1 ratio of cycloadducts **41** (as a mixture of diastereomers) and **42**.[82]

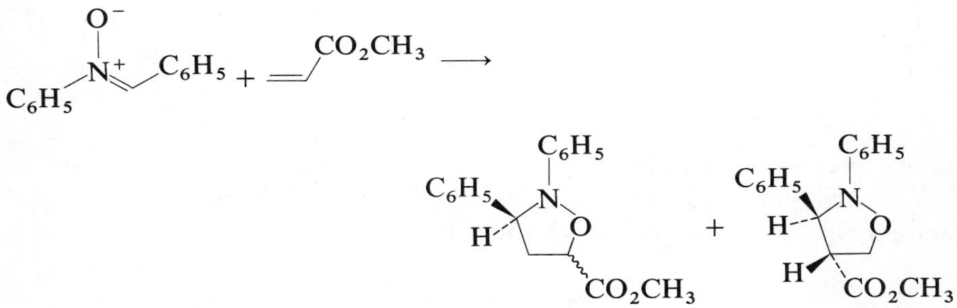

41 **42**

(2:1)

Because of their electronic similarities, acrylonitriles and acrylamides behave analogously to the acrylates. For example, the nitrones **43** and **45** readily add to acrylonitrile to yield the 5-substituted adducts **44**[83] and **46**,[84] respectively.

$4\text{-HOC}_6\text{H}_4\text{NHCO}$ $\text{N}^+(\text{O}^-)$ C_6H_5 + CN ⟶ $4\text{-HOC}_6\text{H}_4\text{NHCO}$ C_6H_5 N O CN

43 **44** (81%)

C_6H_5 $\text{N}^+(\text{O}^-)$ $\text{CONHC}_6\text{H}_4\text{COCH}_3\text{-4}$ + CN ⟶

45

$4\text{-CH}_3\text{COC}_6\text{H}_4\text{NHCO}$ C_6H_5 N O CN

46 (85%)

One of the few reported additions of acrylamide is its reaction with the nitrone **27** to give the product **47**.[74]

C_6H_5 N^+ O^- O NO_2 + CONH_2 ⟶

27

C_6H_5 O_2N O N O CONH_2

47

Severe perturbations in the nitrone structure can result in the production of only 4-substituted isoxazolidines as in the generation of the adduct **49** from the anthraquinone-derived nitrone **48** and acrylonitrile.[85]

O N^+ C_6H_5 O^- + CN ⟶ O NC $\text{N—C}_6\text{H}_5$ O

48 **49** (80%)

There is an excellent correlation between increased production of the 4-substituted regioisomers with increasing strength of the electron-withdrawing substituents. Thus *C*-phenyl-*N*-methylnitrone reacts with nitroethylene to yield the C-4 product exclusively or with phenyl vinyl sulfone to yield a mixture of regioisomers.[68,86]

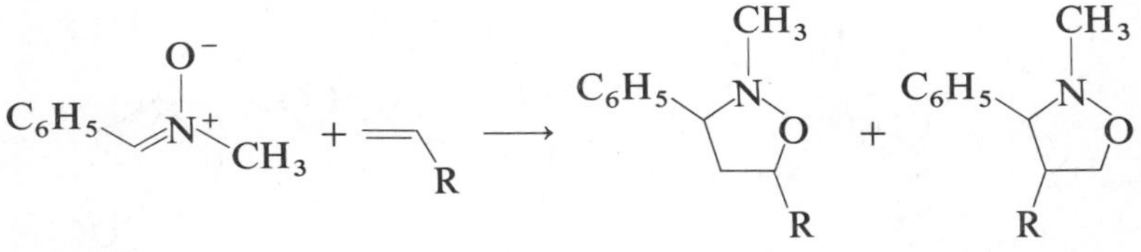

R	
$R = NO_2$	0:100
$= SO_2C_6H_5$	32:68
$= CO_2CH_3$	30:70
$= C_6H_5$	100:0

This same nitrone reacts with methyl acrylate to afford a 70:30 mixture of regioadducts, C-4-substituted predominating, and with styrene to yield 100% of the C-5 product. An apparently anomalous behavior of the nitrone derived from *tert*-butylhydroxylamine and formaldehyde toward nitroethylene is reported,[48] in which only the unexpected 5-substituted product 2-*tert*-butyl-5-nitroisoxazolidine is formed.

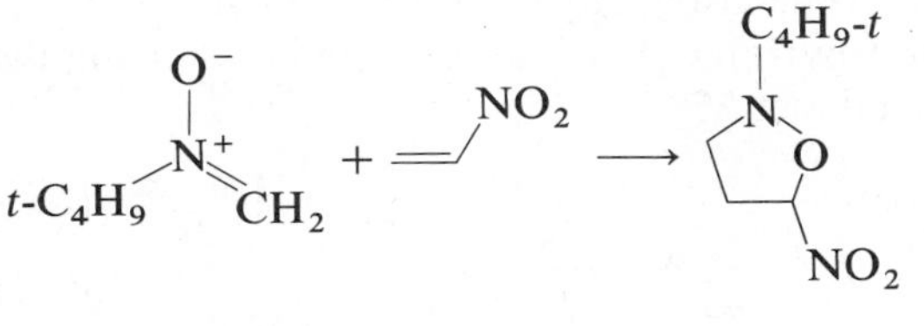

To further complicate theoretical considerations, the cycloheptatrienone-derived nitrone **50** on treatment with phenyl vinyl sulfone produces only the thermally labile C-4 regioisomer **51**, which subsequently rearranges to the fused bicyclic adduct **52**.[87]

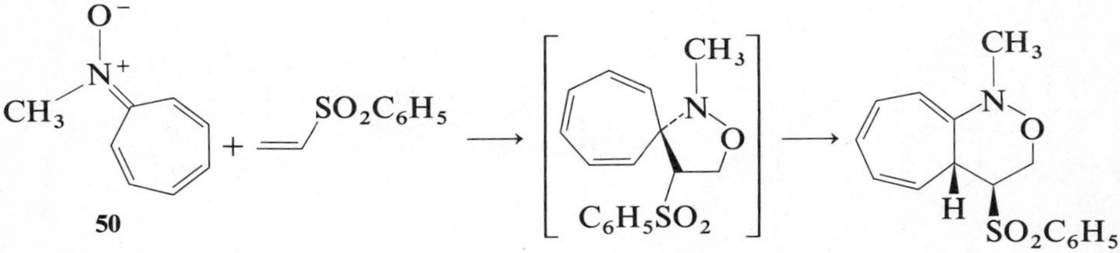

50 **51** **52**

Electron-Rich Monosubstituted Olefins

Significantly fewer examples of nitrone [3 + 2] cycloadditions to electron-rich systems are reported. However, it is quite clear that the reaction is regiospecific, yielding C-5-substituted adducts. Thus reaction of *C*-ethoxycarbonyl-*N*-phenylnitrone with ethyl vinyl ether at 80° produces the product **53** as the sole regioisomer.[88]

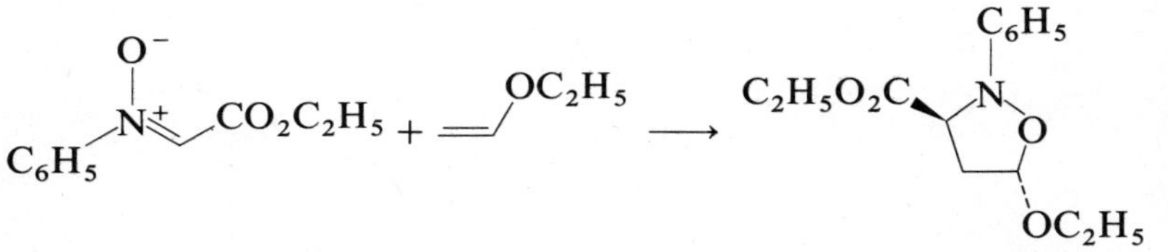

53

A related example is the reaction of *C*,*N*-diphenylnitrone with vinyl acetate to afford the cycloadduct **54**.[89]

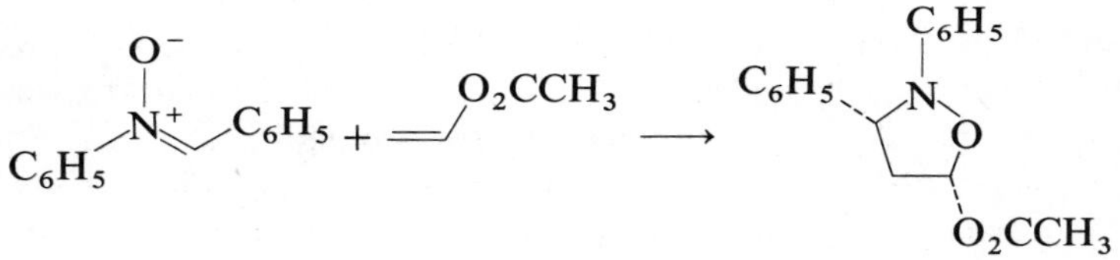

54

The differing stereochemistry at C-3 between **53** and **54** is noteworthy. Cyclic nitrones such as 1-pyrroline-1-oxide react similarly, giving the expected regiochemistry in the adduct **55**.[90]

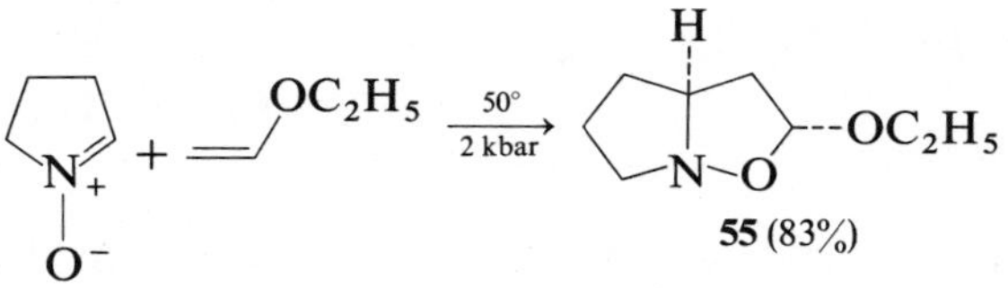

50°, 2 kbar

55 (83%)

Reaction of the complex nitrone **56** with ethyl vinyl ether proceeds to give the C-5 regioisomer **57**,[90] unlike the reaction of the related nitrone **48** with electron-deficient monosubstituted olefins which affords a C-4 regioisomer.

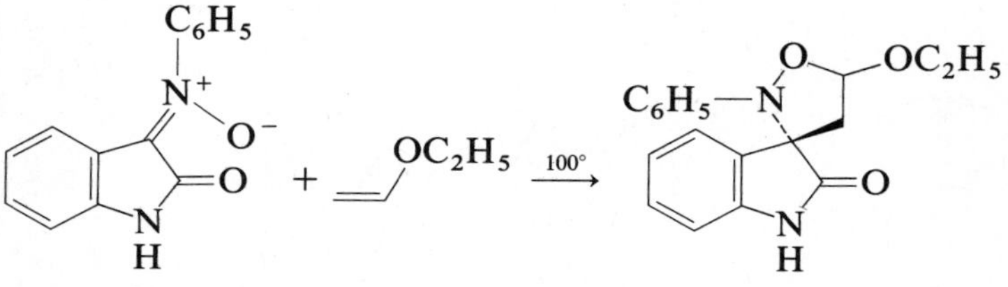

100°

56 **57** (99%)

1,1-Disubstituted Olefins

In general, the [3 + 2] cycloaddition of nitrones to 1,1-disubstituted olefins yields predominantly isoxazolidines disubstituted at C-5. In contrast to monosubstituted olefins, the regiochemical outcome of these reactions is influenced more by the substitution pattern on the olefin than by the electronic nature of the substituents. Thus the reaction of *C,N*-diphenylnitrone with 1-*tert*-butylthio-1-cyanoethylene affords a mixture of products.[91]

SC_4H_9-t, CN; C_6H_5; (76%) (8%) (1%); t-C_4H_9S; NC; CN

N-Methylnitrone adds to ethyl 1-phenylacrylate to yield the adduct **58** as the only regioisomer.[73]

CH_3; CH_2; C_6H_5; $CO_2C_2H_5$; **58**

Even the enamine derived from pyrrolidine and acetophenone yields the related nitrone adduct **59** on treatment with *C*-4-chlorophenyl-*N*-phenylnitrone.[92]

C_6H_4Cl-4; 4-ClC_6H_4; C_6H_5; **59**

In a related example *C,N*-diphenylnitrone reacts with ethyl 1-acetamidoacrylate to furnish the 5,5-disubstituted product **60**.[93] Finally, the exocyclic olefins **61** and **62** derived from norbornane and adamantane react normally with *C,N*-diphenylnitrone to give the expected adducts **63** and **64**, in spite of steric crowding at the 1,1-disubstituted carbons of the double bonds.[94,95]

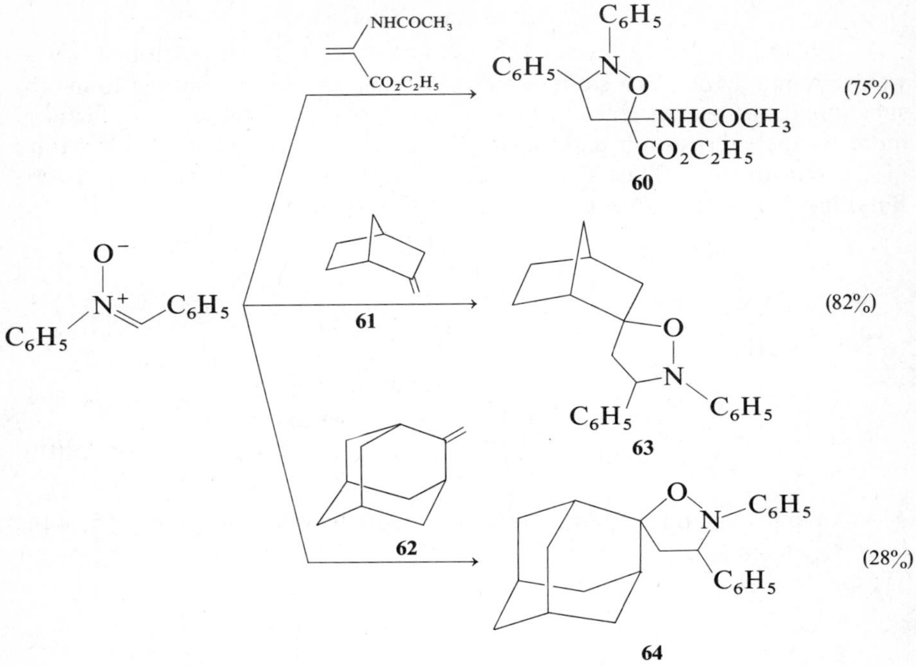

(75%)

61

(82%)

63

62

(28%)

64

The reaction of *C*-benzoyl-*N*-phenylnitrone with diketene is interesting because the expected 5,5-disubstituted regioisomer **65** decarboxylates to yield the "allene" cycloadduct **66**.[96]

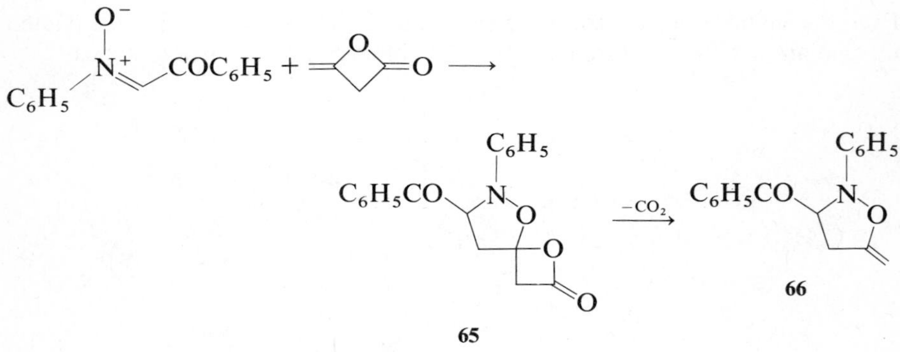

$\xrightarrow{-CO_2}$

66

65

In fact, the only examples of the production of 4,4-disubstituted isoxazolidines from 1,1-disubstituted olefins involve either a single powerful electron-withdrawing substituent (e.g., nitro) or two moderately electron-withdrawing substituents. An illustration of the latter is the reaction of THPO with diethyl methylenemalonate to yield the 4,4-disubstituted bicyclic adduct **67**.[79]

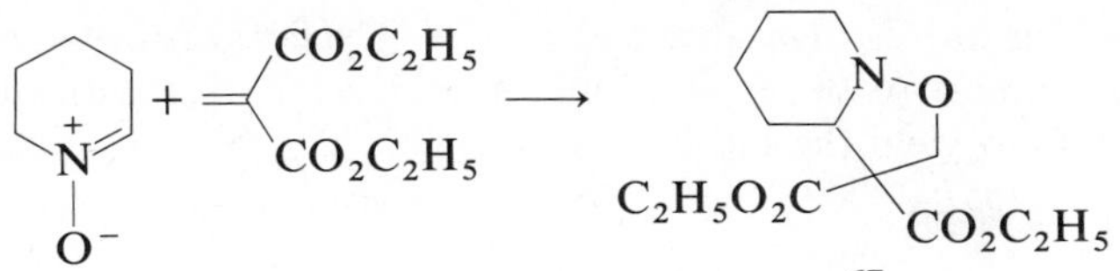

67

1,2-Disubstituted Olefins

The selectivity of symmetrical 1,2-disubstituted olefins in nitrone cycloaddition is obviously uncomplicated by regiochemical considerations, as in the combination of *C,N*-diphenylnitrone and dimethyl fumarate to yield the *trans* adduct **68**.[97]

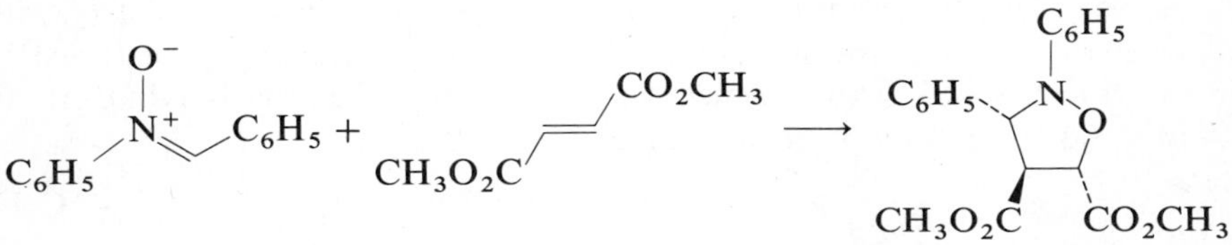

68

Of greater interest is the regioselectivity of the reaction of nitrones with unsymmetrical 1,2-disubstituted olefins. Such substitution patterns may seem to augur poorly for regiochemical control, since the presence of directing groups on opposite carbons will often work at cross purposes. In practice, however, this class of olefins often exhibits an exceptional degree of regioselectivity. For example, the cycloaddition of *C*-benzoyl-*N*-phenylnitrone to *β*-nitrostyrene produces only the regioisomer **69**, in which the more electron-withdrawing nitro substituent occupies the C-4 position.[97]

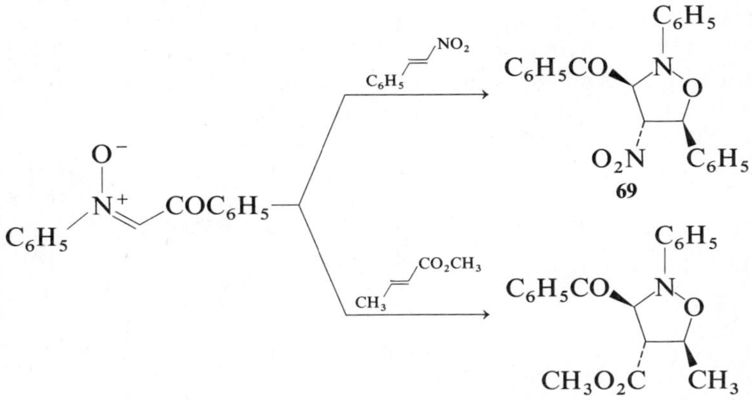

69

70

A related result is shown by the reaction of the same nitrone with methyl crotonate to afford the regioisomer **70**, in which the more electron withdrawing carbomethoxy group is also attached to C-4. Cyclic nitrones such as 5,5-dimethyl-1-pyrroline-1-oxide exhibit similar regioselectivity with, for example, ethyl crotonate to yield the bicyclic product **71**.[98]

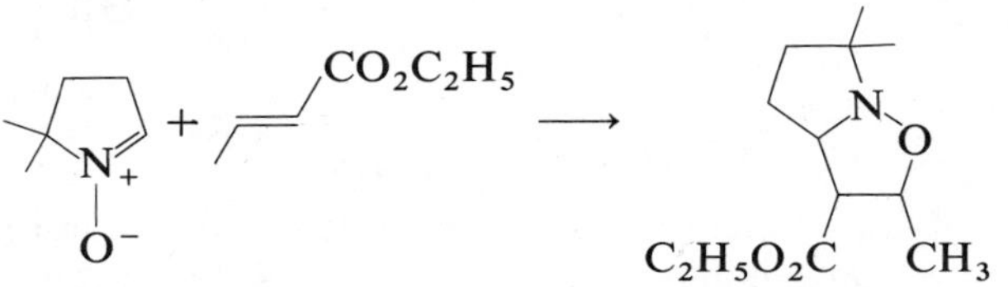

$CO_2C_2H_5$ → $C_2H_5O_2C$ CH_3

71 (78%)

Regiochemical complications arise when the olefin contains substituents of similar electron-withdrawing abilities; *trans*-1-phenylpropene reacts with nitrones to give mixtures of regioisomers.[99] A similar outcome is observed[100] in the reaction of 3-nitropropenenitrile with *C*-phenyl-*N*-*tert*-butylnitrone to afford a 60:40 mixture of isoxazolidines **72** and **73**.

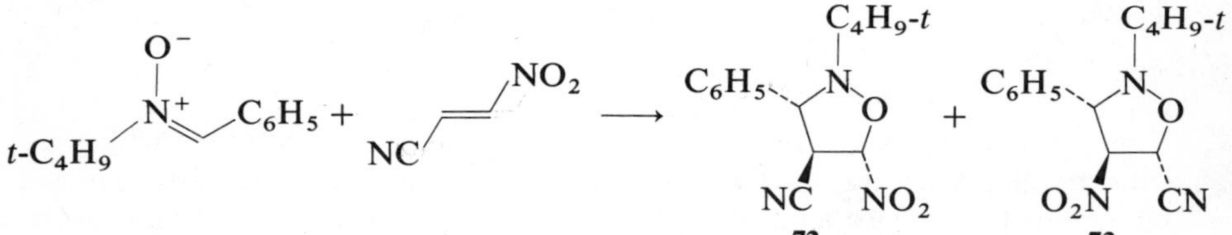

$t\text{-}C_4H_9$, C_6H_5 + NC, NO_2 → **72** + **73** (60:40)

Cyclic Olefins

A special class of 1,2-disubstituted olefins is cyclic olefins, many of which undergo cycloaddition reactions with nitrones. Regiochemical concerns are absent in a symmetrical cyclic olefin such as 1,4-cyclohexadiene, whose monoadduct with *N*-methylnitrone can only be the bicyclic compound **74**.[101] Although 1,3-cyclohexadiene is capable of regiochemical complications, its monoadduct **75** is formed regioselectively in 78% yield.[101]

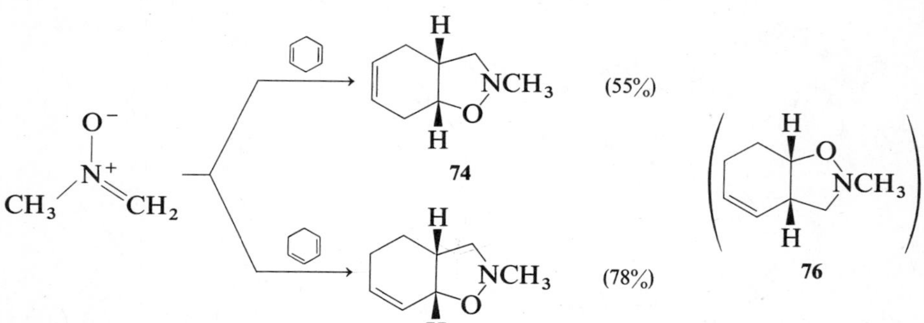

CH_3–$N^+(O^-)$=CH_2 → **74** (55%); **75** (78%); (**76**)

No trace of the isomeric adduct **76** was detected. Addition of 1-pyrroline-1-oxide to 2,3-dihydrofuran yields the tricyclic compounds **77** and **78**.

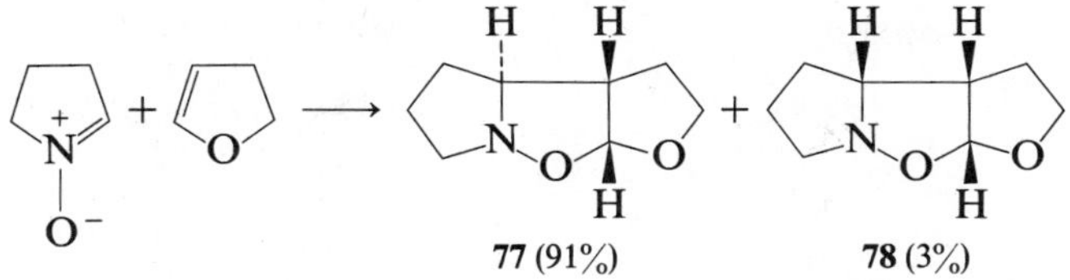

77 (91%) **78** (3%)

The ratio of products is rationalized on the basis of the *exo* mode of cycloaddition as depicted in **79** rather than the *endo* mode **80**, which incorporates significant steric interactions not present in **79**.[102] Both products arise from attachment of the nitrone oxygen to the olefinic carbon bearing the heteroatom. If the heteroatom is electron withdrawing, as in benzo[*b*]thiophene *S*,*S*-dioxide, the reaction with *C*,*N*-diphenylnitrone yields the regioisomeric analog **81**.[103]

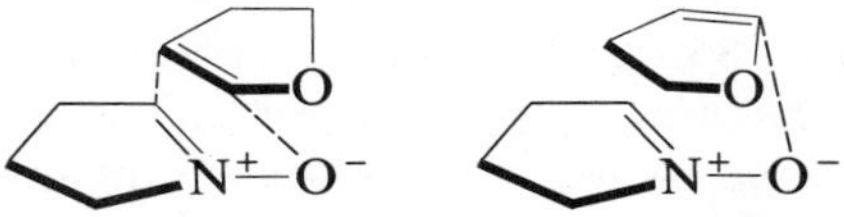

79 **80**

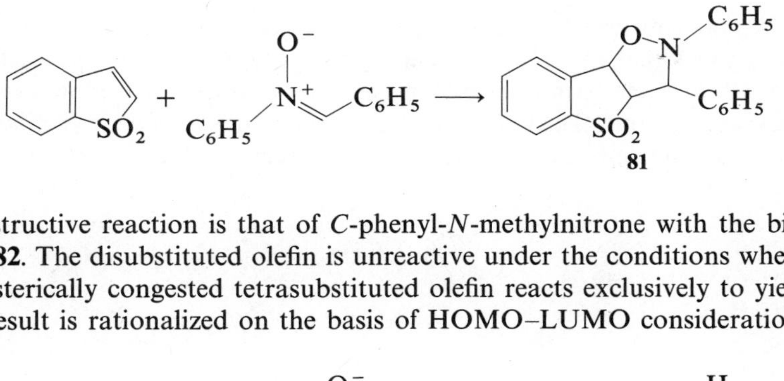

81

An instructive reaction is that of *C*-phenyl-*N*-methylnitrone with the bicyclic diene **82**. The disubstituted olefin is unreactive under the conditions where the more sterically congested tetrasubstituted olefin reacts exclusively to yield **83**. This result is rationalized on the basis of HOMO–LUMO considerations.[104]

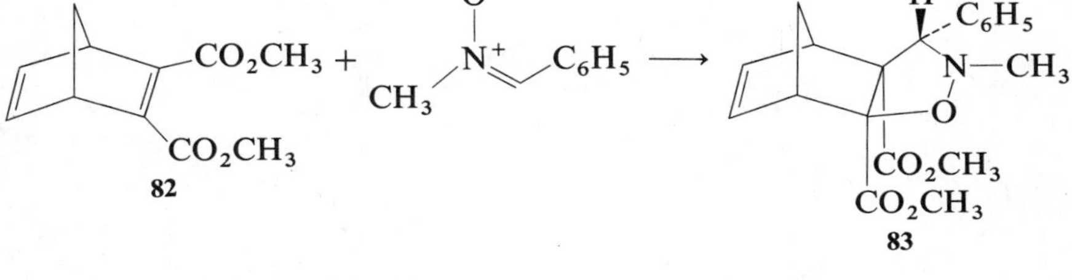

82 **83**

The reaction of nitrones with indenes differs significantly from that with dihydronaphthalenes in regiochemical outcome.

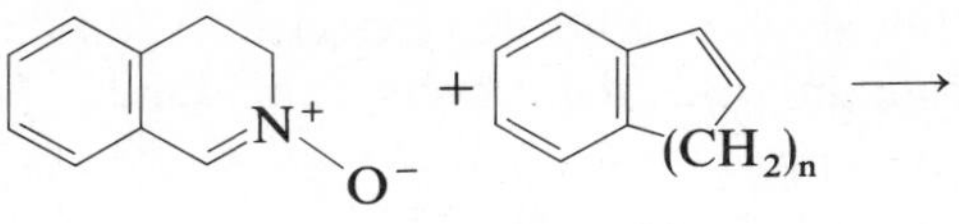

84

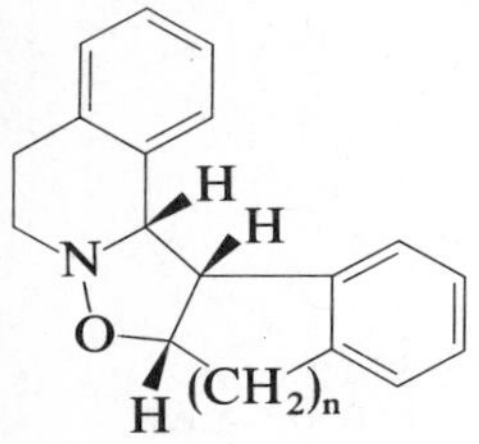

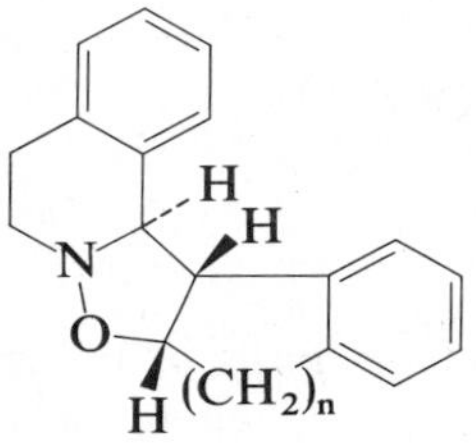

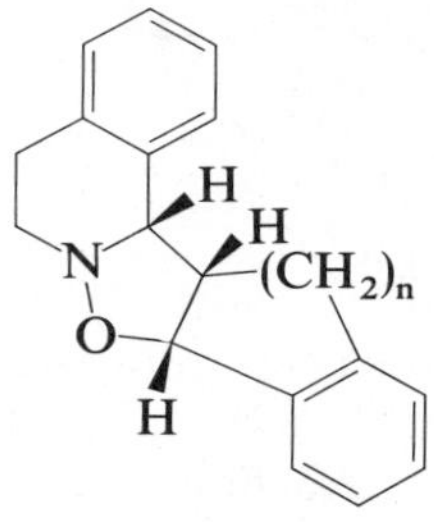

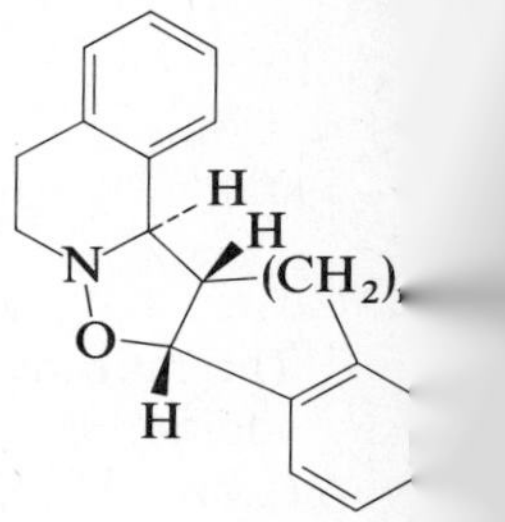

85a
n = 1 (32.4%)
n = 2 (83.7%)

85b
(3.6%)
(0%)

86a
(50.4%)
(1.8%)

86b
(3.6%)
(0%)

Reaction of 3,4-dihydroisoquinoline-*N*-oxide with indene (**84**, n = 1) or with 1,2-dihydronaphthalene (**84**, n = 2) affords the product mixtures indicated. Clearly, the former reaction is only somewhat regioselective for the C-4-substituted tricyclic system **85**, whereas the latter is highly regioselective for the C-5-substituted system **86**.[99]

Acyclic Trisubstituted Olefins

The regiochemistry of the [3 + 2] cycloaddition of nitrones to trisubstituted olefins parallels that of the corresponding mono- and disubstituted olefins. In general, the more electron-poor olefinic carbon becomes attached at C-4 in the product isoxazolidine. Thus *C,N*-diphenylnitrone adds to dimethyl ethylidenemalonate to provide the adduct **87** as the sole regioisomer.[105]

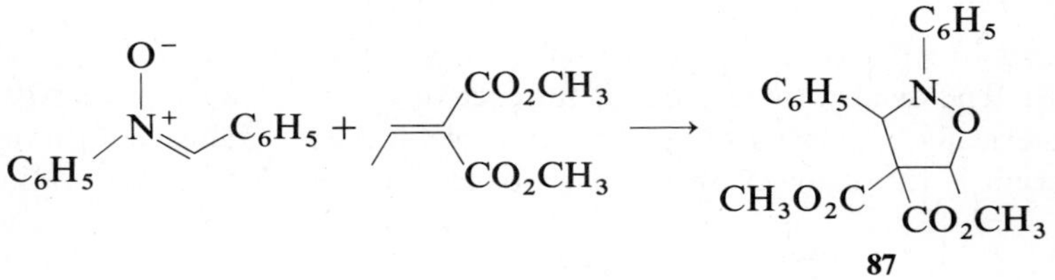

87

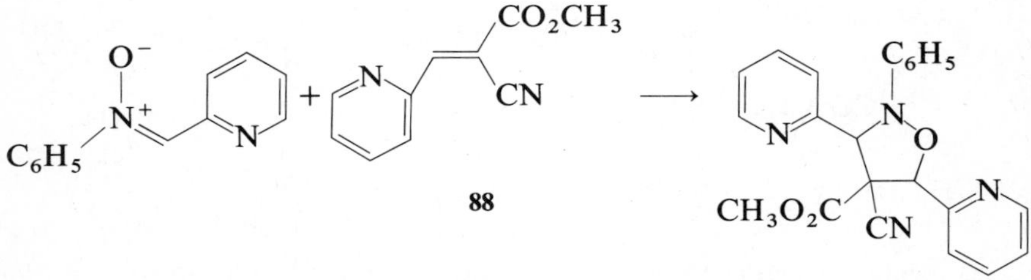

88

89

Similarly, the reaction of *C*-(2-pyridyl)-*N*-phenylnitrone and the trisubstituted olefin **88** affords the cycloadduct **89**.[106]

These generalities are reversed with *C*-acylnitrones. Thus reaction of the dipolarophile tricarbomethoxyethylene with either *C*-phenyl-*N*-methylnitrone or *C*,*N*-diphenylnitrone leads to the expected predominance of the regioisomers **A**. However, under identical reaction conditions, *C*-benzoyl-*N*-phenylnitrone affords the regioisomer **B** as the major product.[107] This reversal of the regiochemical outcome in the case of *C*-acylnitrones is expected to be general.

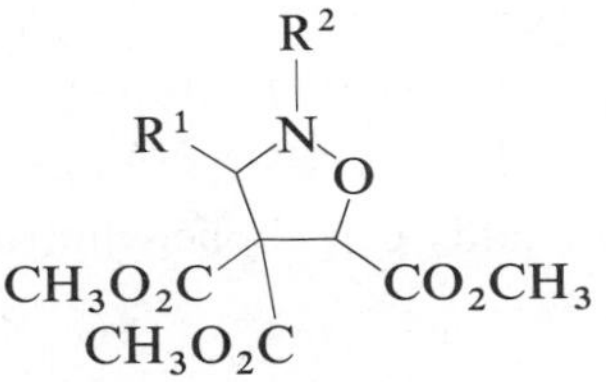

A

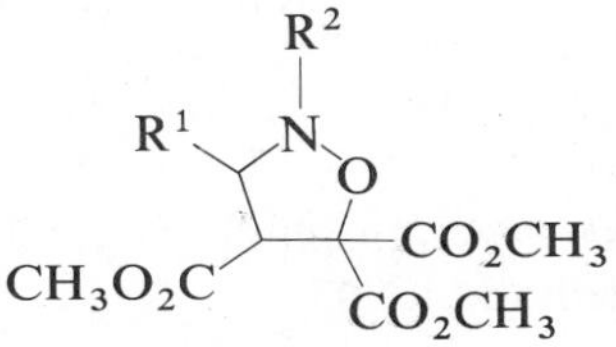

B

Nitrone	Ratio-Regioisomers **A**:**B**	
C-Phenyl-*N*-methylnitrone	($R^1 = C_6H_5$, $R^2 = CH_3$)	99:1
C,*N*-Diphenylnitrone	(R^1, $R^2 = C_6H_5$)	82:8
C-Benzoyl-*N*-phenylnitrone	($R^1 = COC_6H_5$, $R^2 = C_6H_5$)	25:75

Cyclic Trisubstituted Olefins

C-Phenyl-*N*-methylnitrone adds to carvone to yield the cycloadduct **90**,[81] whereas *C*-(2-nitrostyryl)-*N*-phenylnitrone adds to the enamine *1*-(*N*-morpholinyl)cyclohexene to furnish the expected product.[108]

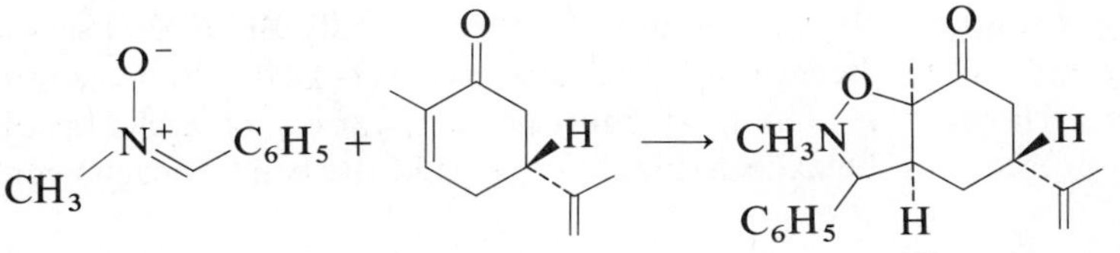

90

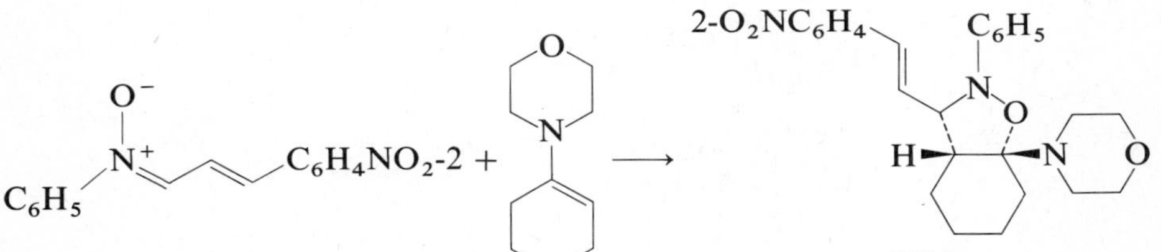

(25%)

The reaction of *C*-4-chlorophenyl-*N*-phenylnitrone with the pyrrolidine-derived enamine **91** proceeds anomalously to afford the bicyclic regioisomer **92**.[109]

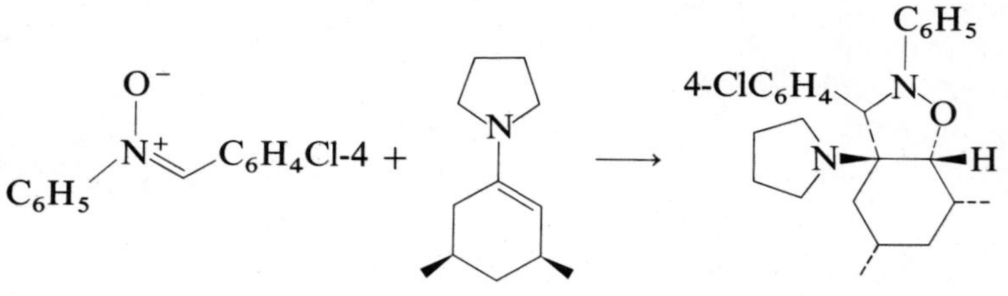

91

92 (27%)

In contrast, 1-(*N*-morpholinyl)cyclohexene adds *C*,*N*-diphenylnitrone to yield the expected adduct **93**.[110]

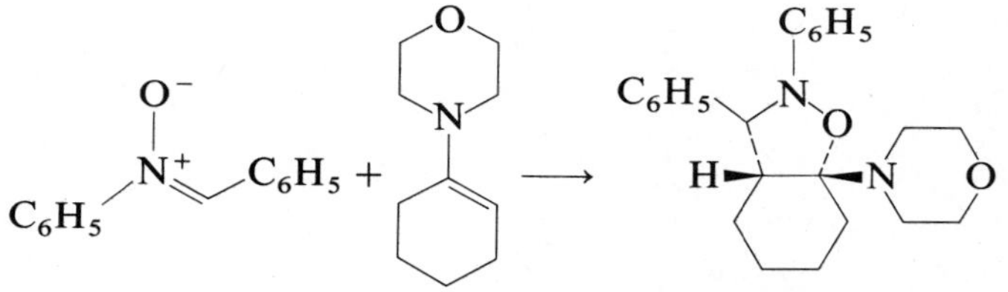

93 (7%)

These contradictory results are rationalized on the basis of differing stabilities of the regioisomeric cycloadducts under the reaction conditions and varying degrees of steric crowding in cyclic enamines. Moreover, the reactions of cyclic enamines with nitrones generally do not give very good yields, rendering product analysis difficult and firm conclusions questionable. However, the regiochemistry found in the production of **93** is generally the one observed and is the regioadduct predicted on the basis of HOMO–LUMO considerations.[92]

The cycloaddition of *N*-methylnitrone to a variety of cyclic trisubstituted olefins proceeds in the expected manner to yield the benzotricyclic compounds **94**.[111]

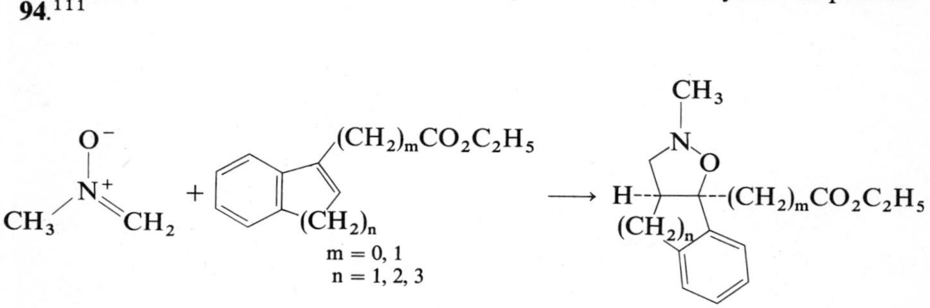

m = 0, 1
n = 1, 2, 3

94

Finally, the strained trisubstituted olefin **95** reacts with *C,N*-diphenylnitrone to afford the complex polycyclic compound **96** as the only reported regioisomer in undisclosed yield.[112]

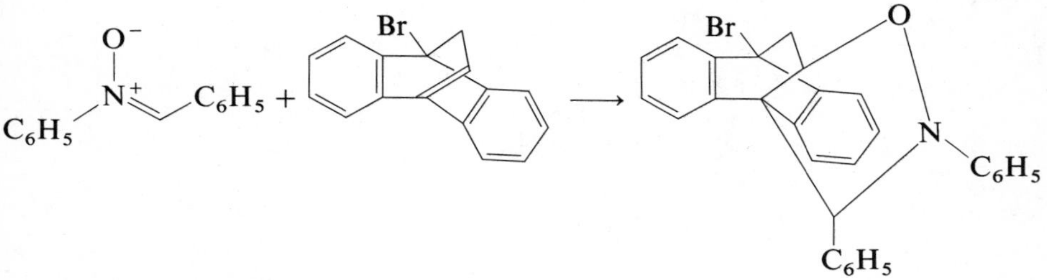

95 **96**

Tetrasubstituted Olefins

Although there are very few examples of the cycloaddition of nitrones to tetrasubstituted olefins, it is expected that the regiochemistry of the products can be predicted by extrapolation of the rules for less-substituted analogs. Thus the reaction of *C,N*-diphenylnitrone with dimethylketene dimethyl acetal affords the regioisomer **97**.[81] As expected, the oxazolidine oxygen becomes attached to the more electron-rich olefinic carbon.

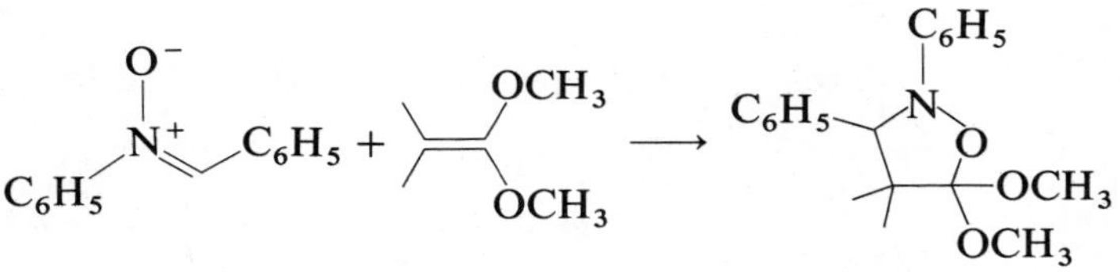

97

Monosubstituted Acetylenes

Terminal acetylenes undergo a [3 + 2] cycloaddition reaction with nitrones to afford the corresponding isoxazolines. Thus *C*-phenyl-*N*-methylnitrone adds to methyl propiolate to yield two regioisomers in a ratio of 58:42.[67]

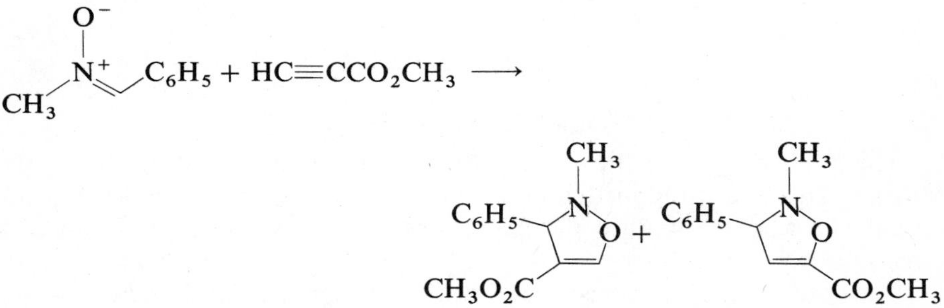

$HC{\equiv}CCO_2CH_3$ ⟶

(58:42)

However, this lack of regioselectivity is seldom observed with electron-deficient monosubstituted acetylenes. The major and often exclusive product is the 4-substituted-4-isoxazoline, except when *N-tert*-butylnitrone is employed, the latter yielding the 5-substituted regioisomer.

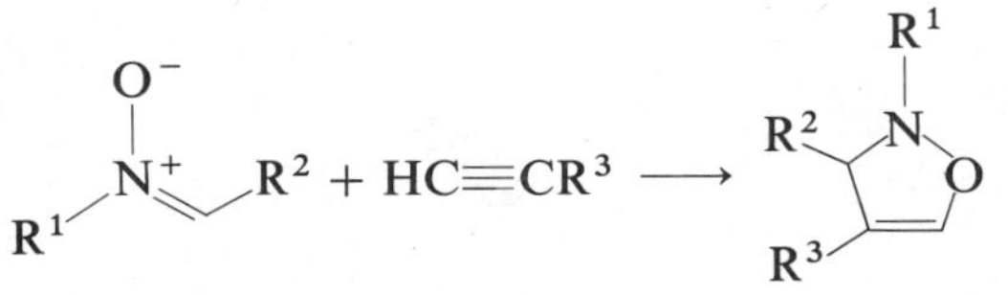

R^3 = electron-withdrawing group

For example, the nitrone **98** reacts with methyl propiolate to furnish the 4-substituted adduct **99**.[113]

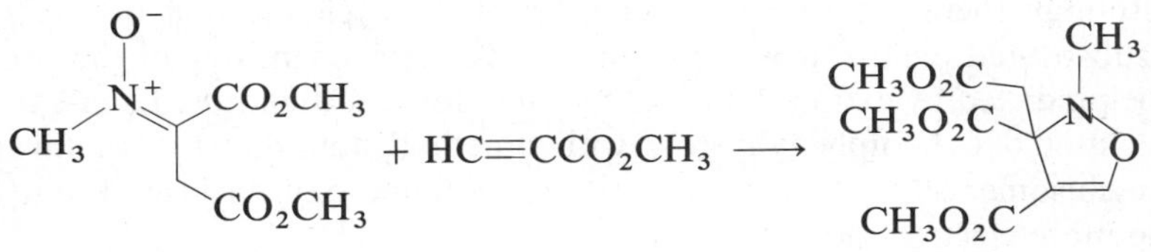

98 **99**

9-Ethynylacridine affords a related product **100** when treated with *C,N*-diphenylnitrone.[114]

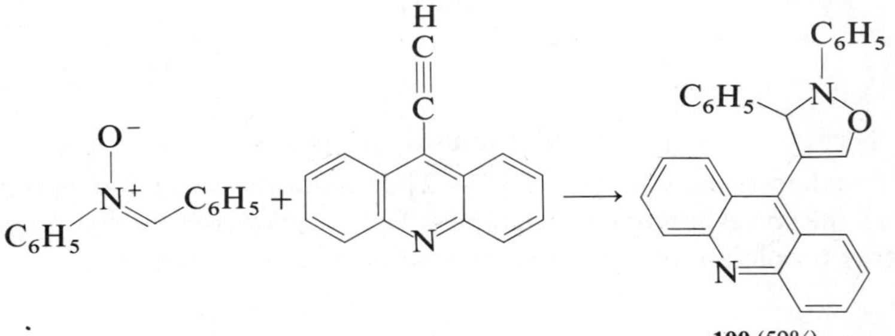

100 (59%)

Disubstituted Acetylenes

The *N*-phenylnitrone **101**, derived from anthraquinone, and the *N*-methylnitrone **102** add to dimethyl acetylenedicarboxylate and phenylacetylene carboxylic acid to produce the 4,5-disubstituted isoxazolines **103** and **104**, respectively.[85]

The latter example demonstrates the principle paralleled in 1,2-disubstituted olefins, that the more electron-withdrawing substituent becomes attached to C-4.

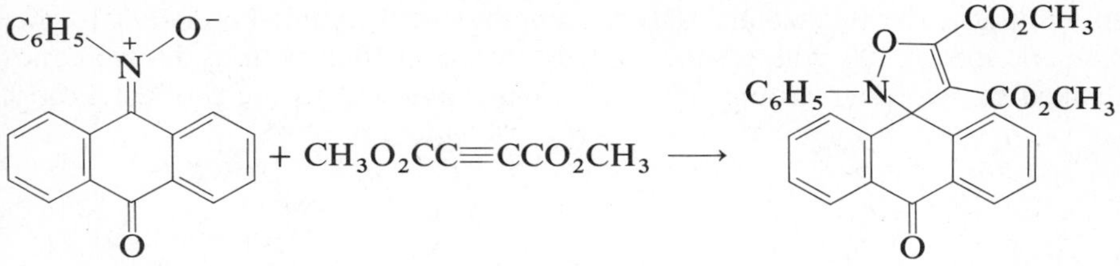

C_6H_5 $\overset{+}{N}$ O^- + $CH_3O_2CC{\equiv}CCO_2CH_3$ ⟶ CO_2CH_3 O C_6H_5—N CO_2CH_3 O

101 **103** (85%)

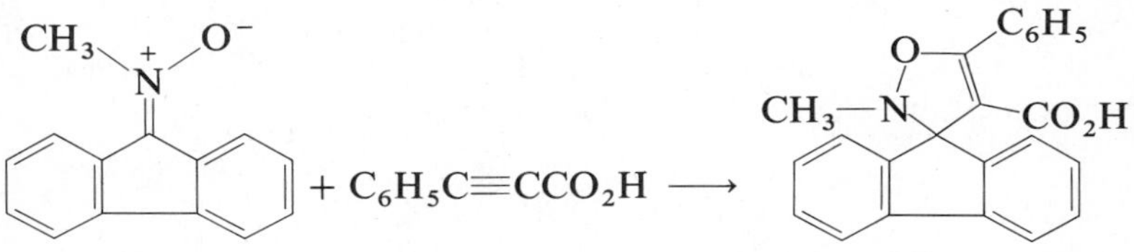

CH_3 $\overset{+}{N}$ O^- + $C_6H_5C{\equiv}CCO_2H$ ⟶ C_6H_5 O CH_3—N CO_2H

102 **104** (80%)

Finally, the trapping of benzyne by *C*-phenyl-*N*-methylnitrone affords the bicyclic adduct **105**.[67,115–117]

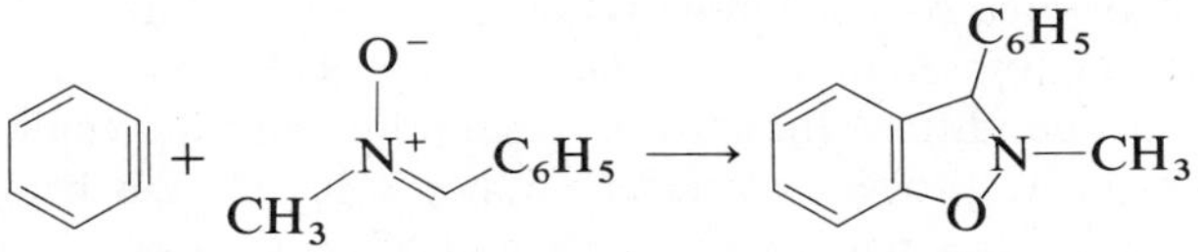

+ O^- N^+ C_6H_5 CH_3 ⟶ C_6H_5 N—CH_3 O

105

Stability of 4-Isoxazolines

A number of facile thermal rearrangements of 4-isoxazolines are reported, indicating a degree of instability in these nitrone–acetylene adducts. For example, the adduct **106**, derived from the cycloaddition of dimethyl acetylenedicarboxylate and *N*-*tert*-butylnitrone, rearranges on heating at 80° to the 4-oxazoline isomer **107**. The intermediacy of the aziridine has been implicated.[118]

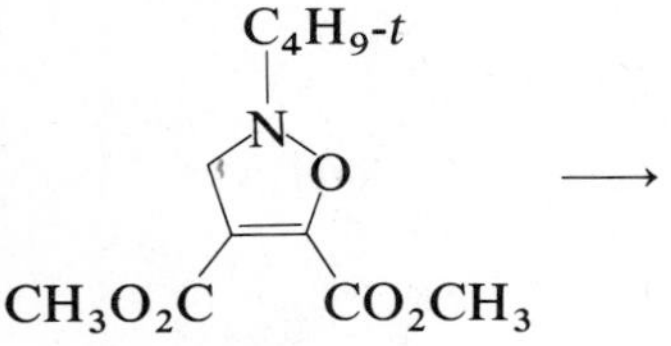

C_4H_9-*t* N O CH_3O_2C CO_2CH_3 ⟶

106

[C_4H_9-*t* N CO_2CH_3 CH_3O_2C O] ⟶ *t*-C_4H_9 N O CH_3O_2C CO_2CH_3

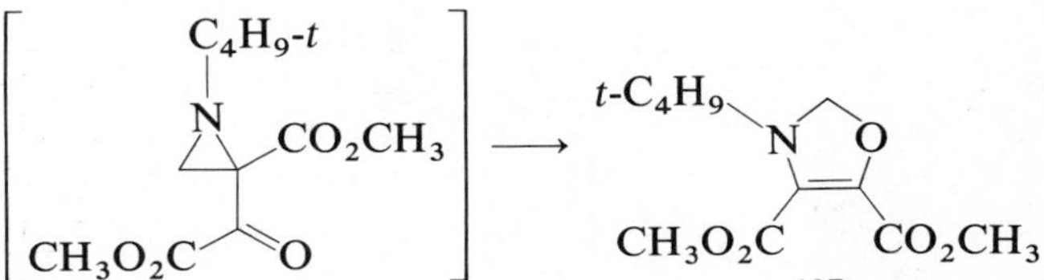

107

Interestingly, the isoxazoline **108a** is stabilized by a double-bond shift to the isomeric species **109**, whereas the related compound **108b**, with its double bond flanked by two ester groups, undergoes rearrangement to the product **110**.[67]

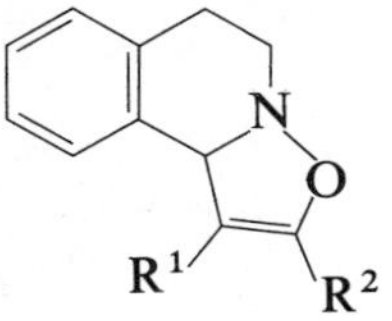

R^1 R^2

108

a $R^1 = CO_2C_2H_5$, $R^2 = C_6H_5$
b $R^1 = R^2 = CO_2CH_3$

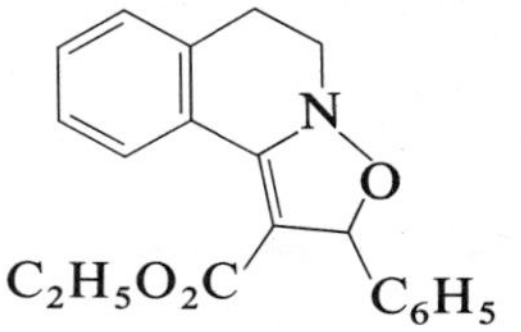

$C_2H_5O_2C$ C_6H_5

109

CH_3O_2C

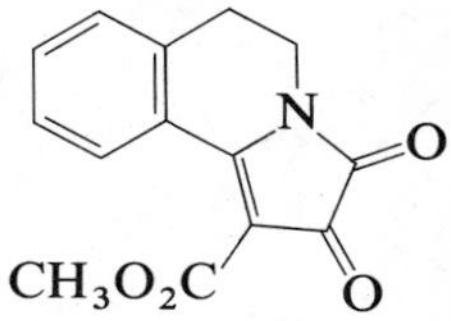

110

REGIOSELECTIVITY AND STEREOSELECTIVITY OF THE INTRAMOLECULAR REACTION

C-Alkenylnitrones

Olefin Aldehydes and *N*-Substituted Hydroxylamines. Entropic factors allow the intramolecular version of the nitrone–olefin cycloaddition reaction to proceed under milder conditions than the corresponding intermolecular equivalent. These intramolecular processes produce either a fused or a bridged bicyclic isoxazolidine when the reactive nitrone and olefinic partners are suitably disposed in a single molecule. This tremendously useful concept was pioneered by LeBel employing substrates in which the nitrone moiety was separated from

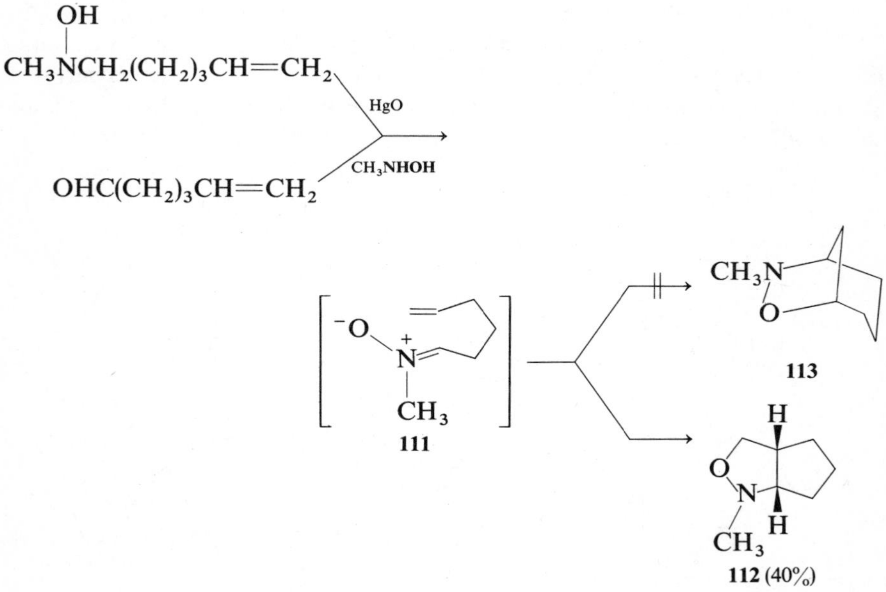

$CH_3N(OH)CH_2(CH_2)_3CH{=}CH_2$ (HgO)

$OHC(CH_2)_3CH{=}CH_2$ (CH_3NHOH)

111

113

112 (40%)

the olefin by a propylene or butylene linkage.[119] Thus the olefinic nitrone **111**, generated by either oxidation of an *N*-alkenylhydroxylamine or condensation of an unsaturated aldehyde with *N*-methylhydroxylamine, is not isolable but cyclizes under the reaction conditions to the fused bicyclic system **112**. None of the isomeric bridged product **113** is produced despite the preference for that regiochemistry in the intermolecular reaction.[119]

However, bridged bicyclic adducts can be formed in selected cases in which steric interactions in transition states leading to the fused systems raise the energetics sufficiently to allow formation of the bridged adduct to compete. Thus the nitrone **114** cyclizes to both the fused adduct and the bridged bicyclic product in a ratio of 55:45.[120]

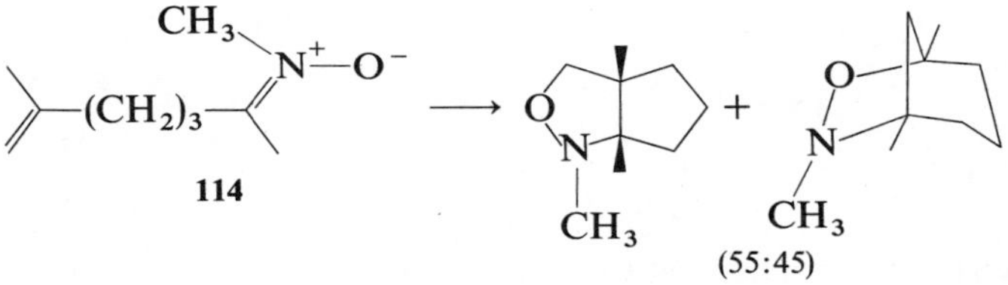

CH_3 $\overset{+}{N}$–O^- $(CH_2)_3$ **114** → O N CH_3 + O N CH_3 (55:45)

The intramolecular cycloaddition of the nitrone olefin derived from *cis*- and *trans*-heptenal illustrates the stereospecificity of the reaction. Thus the *cis* isomer affords only the *trans* methyl product, whereas the corresponding *trans* olefin leads to the epimeric adduct.[8]

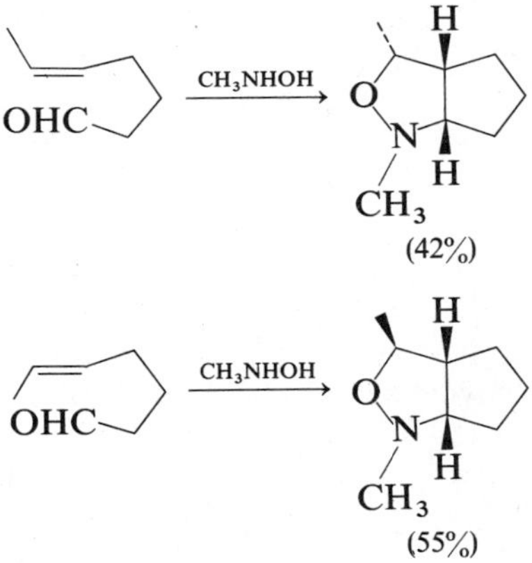

OHC $\xrightarrow{CH_3NHOH}$ H O N H CH_3 (42%)

OHC $\xrightarrow{CH_3NHOH}$ H O N H CH_3 (55%)

When these reactions are extended from *C*-(4-pentenyl)nitrones to the homologous *C*-(5-hexenyl) series, a third stereoisomer is often observed. The cycloaddition of *N*-methyl-*C*-(5-hexenyl)nitrone affords three cycloadducts. The additional product, the *trans*-fused bicyclic adduct, is apparently formed under thermodynamic control.[8]

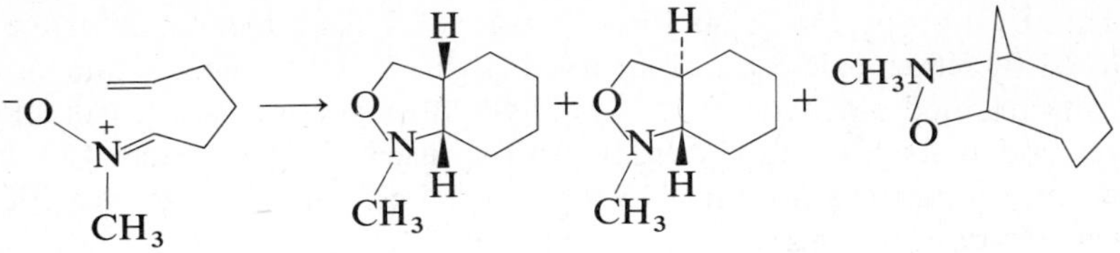

In contrast, the reaction of (+)-citronellal with *N*-substituted hydroxylamines affords predominantly the *trans*-fused product, which isomerizes to the *cis*-fused adduct at 300°.

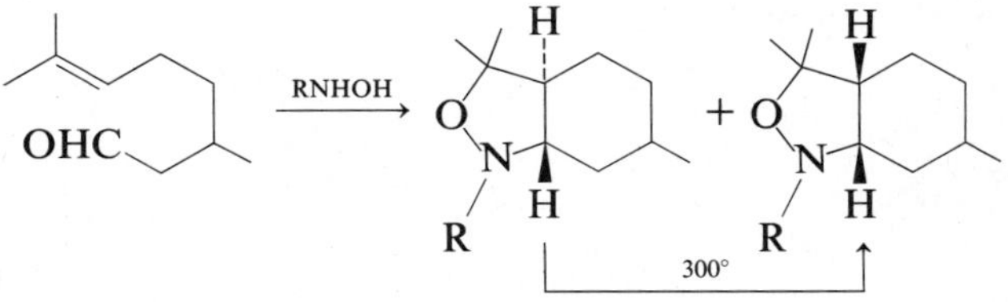

The transition state leading to *cis*-fused adducts involves a twist conformation and allows favorable orbital overlap during cycloaddition for both *syn* (**115**) and *anti* (**116**) nitrone configurations. *Trans*-fused products, on the other hand, are formed from a slightly deformed chair conformation. In this instance, effective orbital overlap is impossible in the *anti* nitrone and exists only for the *syn* configuration **117**.

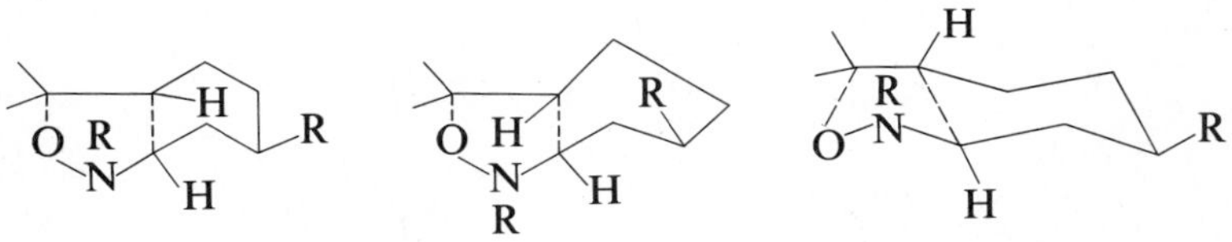

115 **116** **117**

Product distribution can be rationalized on the basis of the relative energy barrier for the *syn* ↔ *anti* interconversion versus that for the cycloaddition itself. In cases where the barrier is high, *cis*-fused adducts are favored.

Reaction of the aldehyde olefin **118** with *N*-phenylhydroxylamine yields only the *cis*-fused adduct **119**.[121] Neither the *trans*-fused isomer nor the bridged regioisomer **120** is detected.[121]

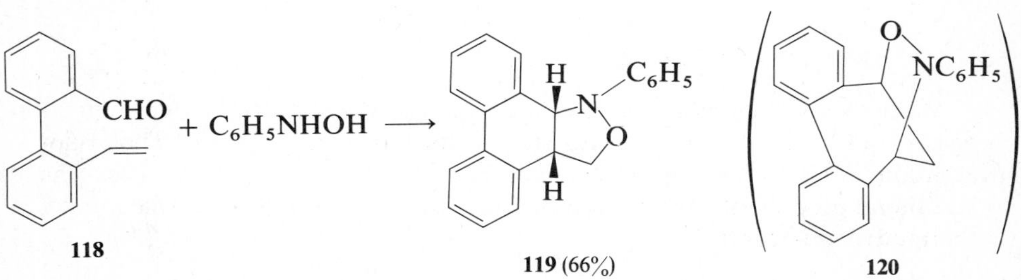

$$\mathbf{118} + C_6H_5NHOH \longrightarrow \mathbf{119}\ (66\%)\quad (\mathbf{120})$$

A useful synthetic entry into morphine-related analgesics employs the cycloaddition of the nitrone olefin derived from the aldehyde **121** and *N*-methylhydroxylamine to furnish the tetracyclic product **122**.[122]

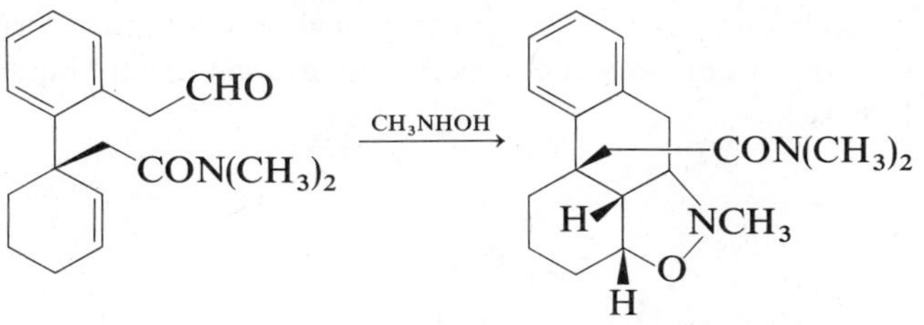

CHO
$CON(CH_3)_2$
CH_3NHOH
$CON(CH_3)_2$
H
NCH_3
O
H

121 **122** (72%)

A [3 + 2] cycloaddition route to carbon-bridged dibenzocycloheptanes involves cyclization of the nitrone olefins **123a** and **123b** to the polycyclic adducts **124a** and **124b**, respectively. The nitrone **123a**, a mixture of *syn* and *anti* isomers, is a stable crystalline product at room temperature, undergoing intramolecular cycloaddition above 110°, whereas the homolog **123b** cyclizes spontaneously. These results are explained by the low reactivity of the stilbene-type olefin and the greater strain present in the isomer **124a** bridged by only a single carbon atom.[123]

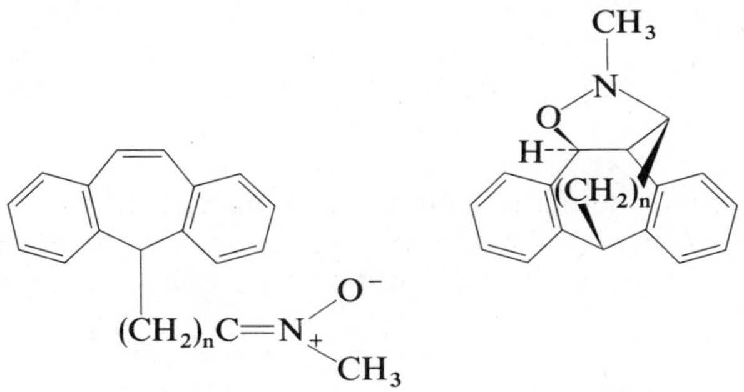

CH_3
N
O
H
$(CH_2)_n$
O^-
$(CH_2)_nC{=}N^+$
CH_3

123
a, n = 0
b, n = 1

124
a, n = 0
b, n = 1

The parent example of this series involves treating 5-cycloheptenecarboxaldehyde with *N*-methylhydroxylamine to furnish a single tricyclic adduct.[124]

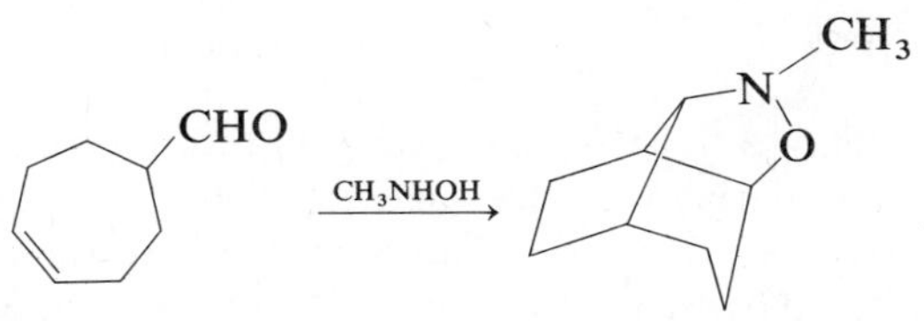

CH_3
CHO
N
O
CH_3NHOH

(60%)

Control of stereochemical centers around a cyclopentane ring is found when the aldehyde olefins **125** and **126** react with *N*-methylhydroxylamine to afford the cycloadducts **127** and **128**, respectively.[125] The three contiguous asymmetric centers are generated stereospecifically in a single step from an acyclic precursor. In complete analogy with the prototypical reaction, the thermodynamic product has the more sterically encumbered substituent occupying the less congested *exo* orientation.

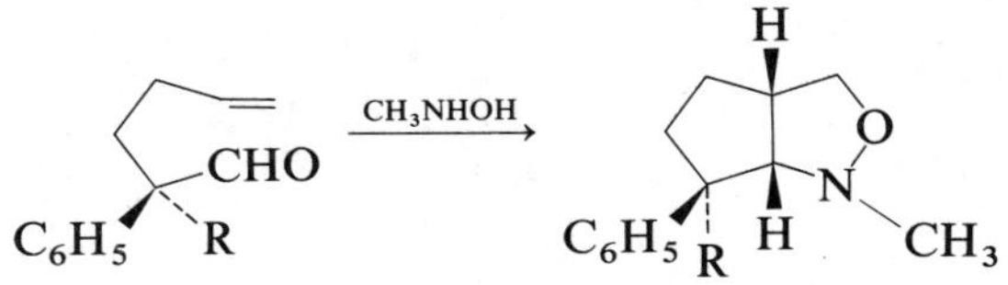

CH_3NHOH

125, R = CH_3
126, R = SCH_3

127, R = CH_3
128, R = SCH_3

The optically active aldehyde olefin **129**, obtained by reductive fragmentation of 5-bromo-5-deoxyglucosides, affords a single stereoisomer **130** on treatment with *N*-methylhydroxylamine.[126]

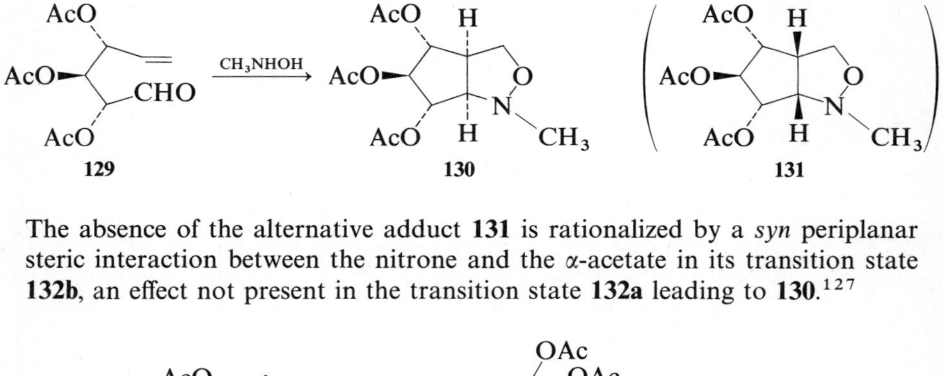

CH_3NHOH

129 **130** **131**

The absence of the alternative adduct **131** is rationalized by a *syn* periplanar steric interaction between the nitrone and the α-acetate in its transition state **132b**, an effect not present in the transition state **132a** leading to **130**.[127]

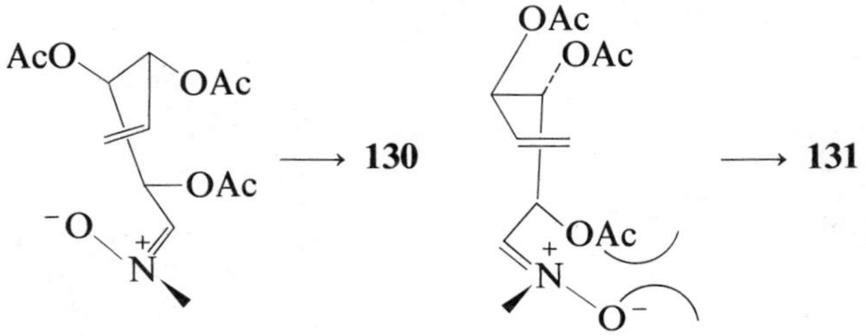

132a → **130**

132b → **131**

The intramolecular [3 + 2] dipolar cycloaddition reaction affords a ready entry to adamantane derivatives. The bicyclic olefin aldehyde **133** yields the protoadamantane system and the adamantane analog on treatment with a variety of *N*-substituted hydroxylamines.[128,129] Although the protoadamantane

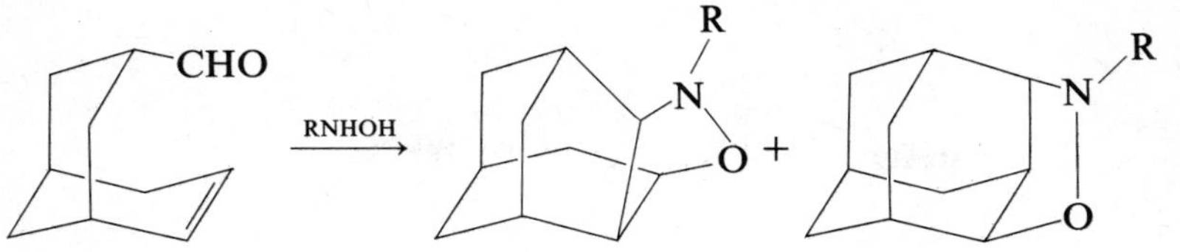

CHO

RNHOH

R N O +

R N O

133

product always predominates, the product ratio is markedly influenced by reaction conditions, especially solvent.

Related reactions yield other classes of polycyclics. For example, aldehydes **134**[130] and **135**[124] lead to the indicated adducts.

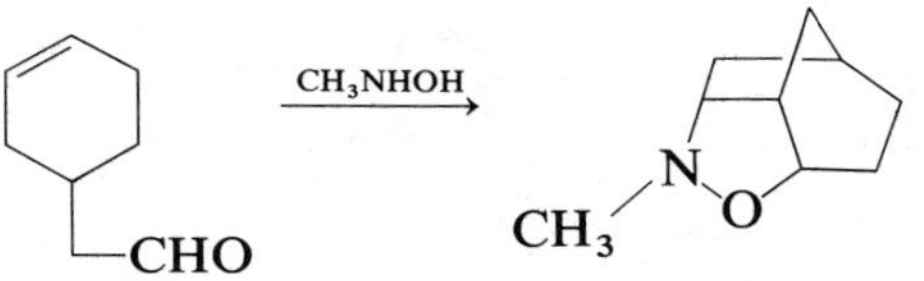

CH_3NHOH

CHO

134

N O CH_3

(55%)

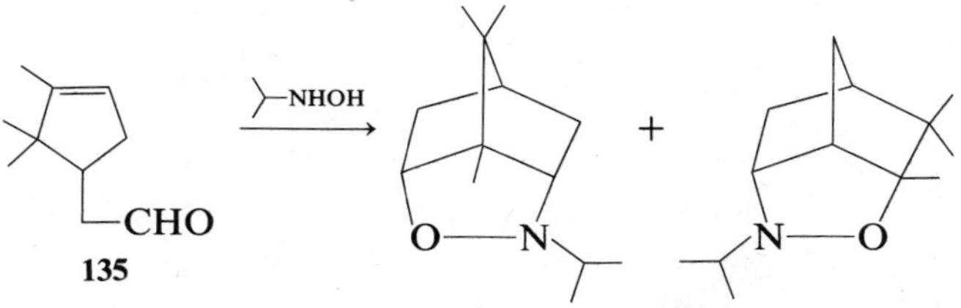

NHOH

CHO

135

O—N

+

N—O

(2:3)

The tricyclic adducts **136** and **137** are produced on cyclization of the cyclopentenyl- and cyclohexenylnitrones. Only the former reaction proceeds stereospecifically; the latter leads to a mixture of diastereomers.[120]

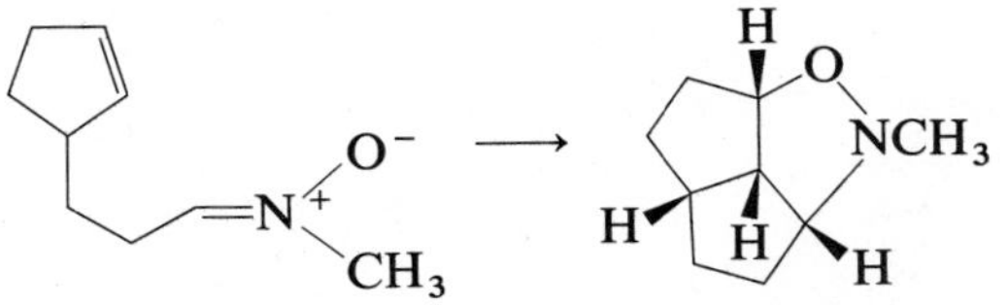

O^- N^+ CH_3 →

H O NCH_3 H H H

136

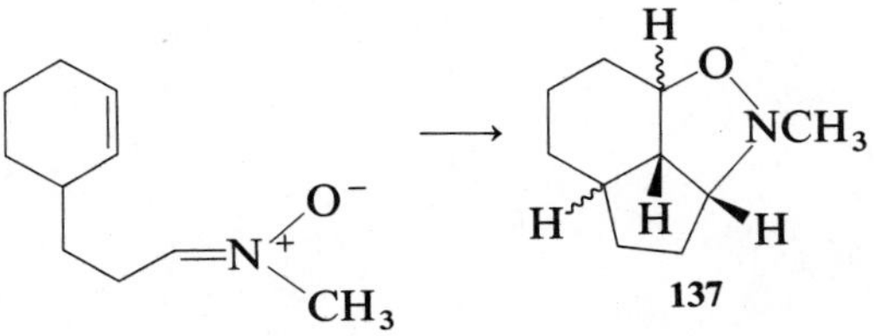

→ O^- N^+ CH_3

H O NCH_3 H H H

137

A convenient route to 2,4-disubstituted noradamantane analogs proceeds via cycloaddition of *C*-bicycloalkenylnitrones such as **138** and **139**. The products are produced under quite mild conditions in high yield, a result of favorable entropic factors in these rigid substrates.[128,131]

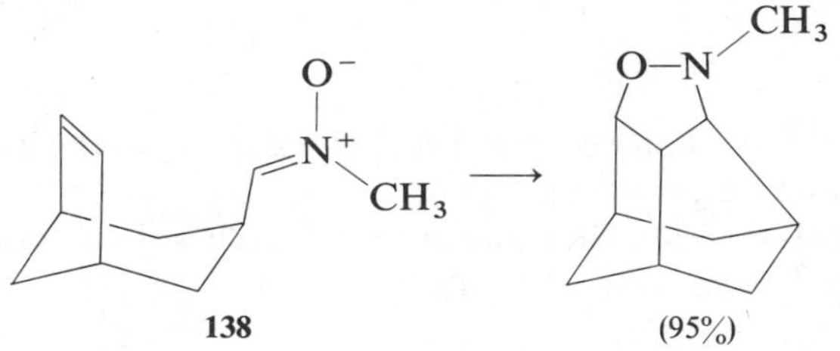

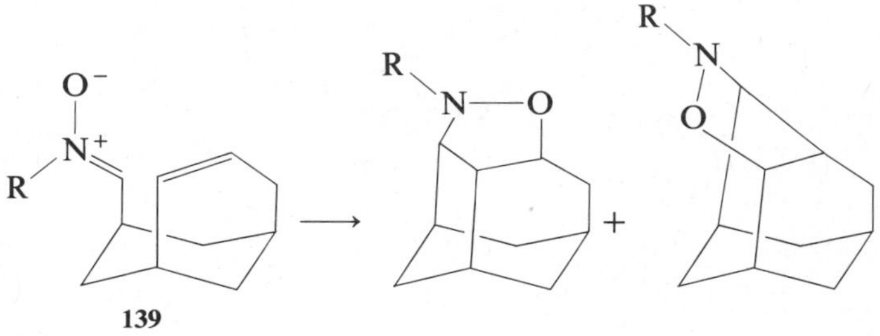

138 (95%)

139

A novel synthesis of unusual azabicyclic diols relies on nitrone–olefin cycloaddition reactions. Thermolysis of the *N*-hydroxyoxazine **140** produces the fragmentation products **141** and **142**, which then cyclize stereospecifically to the observed products **143–145.**

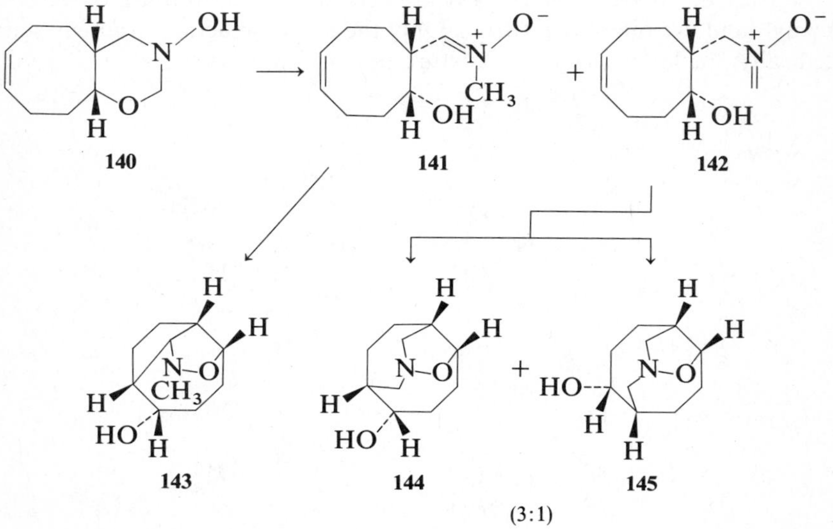

140 **141** **142**

143 **144** **145**

(3:1)

The product ratios are rationalized on the basis of severe 1,5-transannular nonbonded interactions in the transition states of these kinetically controlled reactions.[101]

Heteroatom-Linked Olefin Aldehydes and *N*-Substituted Hydroxylamines. The intramolecular nitrone–olefin cycloaddition reaction includes heteroatom-linked olefinic aldehydes. For example, treatment of the formamido derivative **146** with *N*-methylhydroxylamine yields the *cis*-fused tricyclic adduct in 80% yield.[132]

CHO N CHO **146** $\xrightarrow[110°]{CH_3NHOH}$ CH_3 N–O H H N CHO (80%)

An oxa-linked example is the cycloaddition of *O*-vinylsalicylaldehyde to afford the fused and bridged products in a ratio of 1:2.

CHO O $\xrightarrow{CH_3NHOH}$ H N CH_3 O O H (10%) + CH_3 N–O O (21%)

In most examples studied, the predominance of *cis*-fused products over the *trans*-fused isomers is observed. For example, the ratio of tetracyclic adducts **147** and **148** is 4:1, favoring *cis*-fusion.[132]

CHO N CHO $\xrightarrow{CH_3NHOH}$ CH_3 N–O H H H N H CHO **147** (64%) + CH_3 N–O H H H N H CHO **148** (17%)

An interesting and potentially important variation is the *in situ* generation of a heteroatom-linked system in the reaction of keto nitrone **149** with allylamine to produce the cycloadduct **150**.[46]

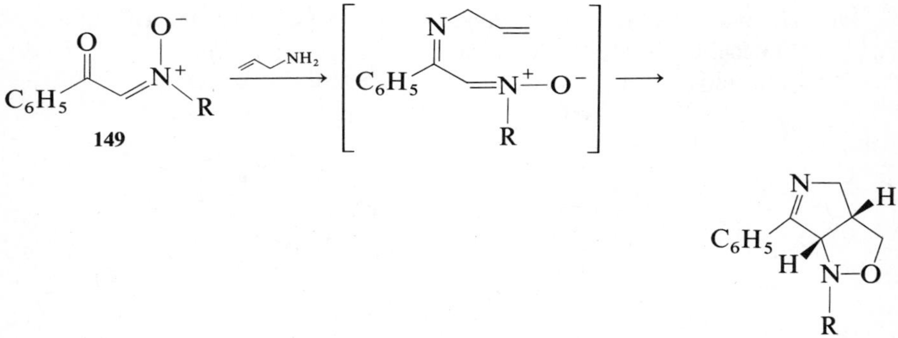

149

150

a, R = n-C_6H_{13}
b, R = C_6H_5
c, R = CH_3

Olefin Ketones and *N*-Substituted Hydroxylamines. The earliest report of a nitrone–olefin cycloaddition derived from an olefinic ketone is that in which 7-octen-2-one is treated with *N*-methylhydroxylamine to yield a stereoisomeric mixture of fused and bridged bicyclic adducts in a ratio of 2:1, respectively.[133]

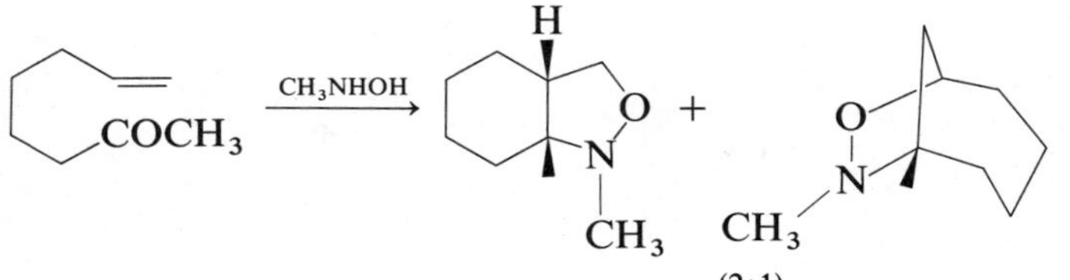

CH_3NHOH (2:1) (69%)

In contrast, 6-hepten-2-one affords only a *cis*-fused bicyclic compound.[8,133]

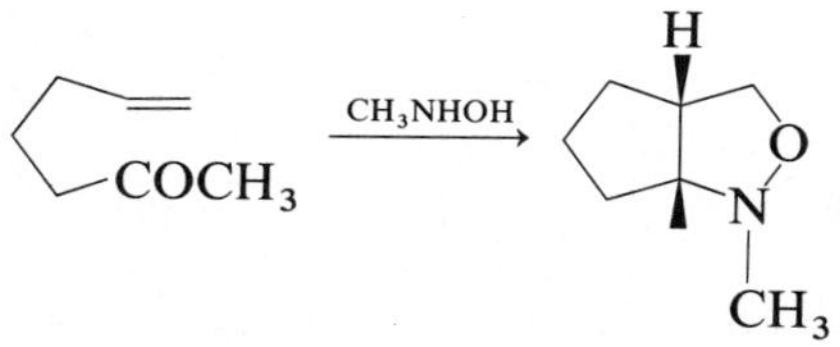

CH_3NHOH (80%)

In a related reaction, 5,6-heptadien-2-one yields the unsaturated analog **151**. This result is a consequence of reaction solely at the terminal double bond.

However, 6,7-octadien-2-one affords a mixture of three stereoadducts in a ratio of 44:16:40.[134] This is a result of cycloaddition to both allenic double bonds followed by acid-catalyzed addition of ethanol to the regioisomer derived from addition to the internal double bond.

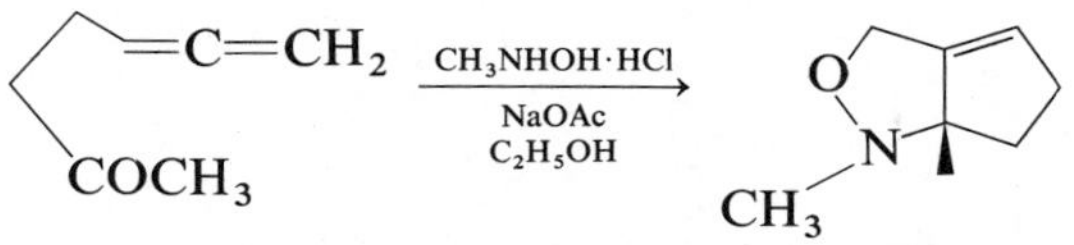

$=C=CH_2$ / $COCH_3$ —($CH_3NHOH \cdot HCl$, NaOAc, C_2H_5OH)→ O, N, CH_3

151 (45%)

$=C=CH_2$ / $COCH_3$ —($CH_3NHOH \cdot HCl$, NaOAc, C_2H_5OH)→ C_2H_5O, H, O, N, CH_3 + OC_2H_5, H, O, N, CH_3 + O, N, CH_3

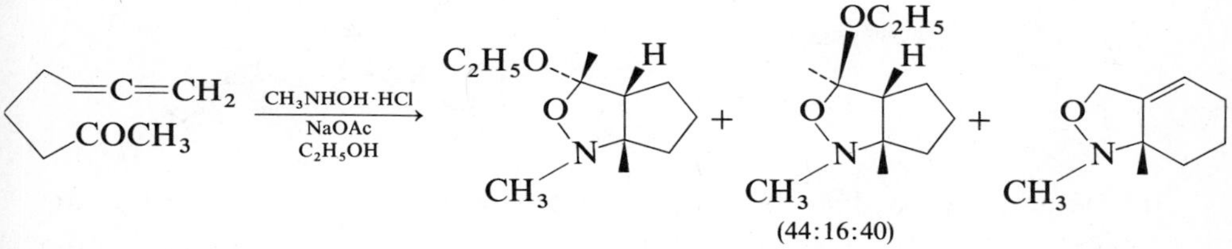

(44:16:40)

A transannular nitrone cycloaddition is the reaction of $\Delta^{1,10}$-*trans*-5-oxo-5,10-secosteroid **152** with *N*-methylhydroxylamine to yield the complex steroid analog **153a**. An analogous reaction also occurs with hydroxylamine itself to yield the unsubstituted derivative **153b**.

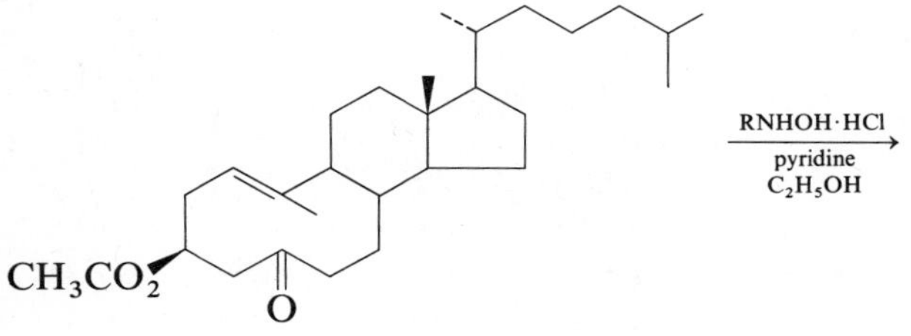

CH_3CO_2, O

152

—(RNHOH·HCl, pyridine, C_2H_5OH)→

CH_3CO_2, O, N, R

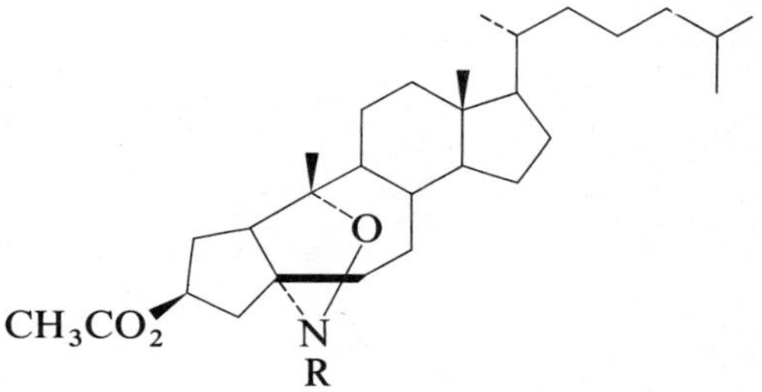

153

a, R = CH_3 (68%)
b, R = H (100%)

The *cis* isomer of **152** fails to react, a consequence of the greater distance between the reactive centers compared to those in the *trans* isomer.[135]

The functionalized nitrones **154a,b** derived from keto olefins undergo cycloaddition to yield the products **155a,b**.

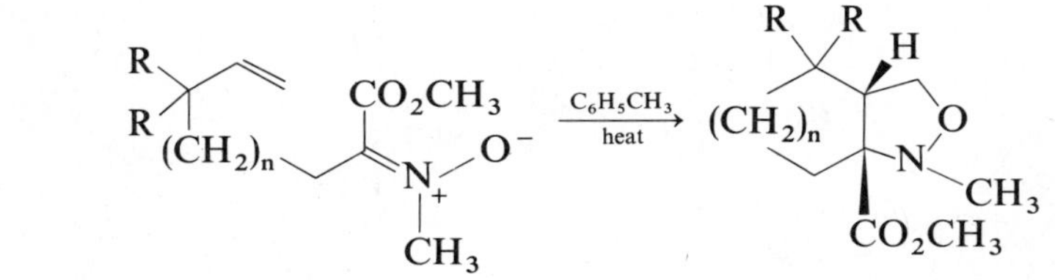

154a,b $\xrightarrow[\text{heat}]{C_6H_5CH_3}$ **155**

a, n = 1, R = CH_3 (83%)
b, n = 2, R = H (78%)

It is noteworthy that **154b** affords a single stereoisomeric product in contrast to its descarbomethoxy analog, which produces a mixture.[133] A useful synthesis of the functionalized bicyclo[$x \cdot 1 \cdot y$] system **156** relies on nitrone–olefin cycloaddition reactions of the general structure **157**. Where m = 1, n = 3, a bridged regioisomer **158** is obtained along with an equal amount of the expected product **156** (m = 1, n = 3).[136,137]

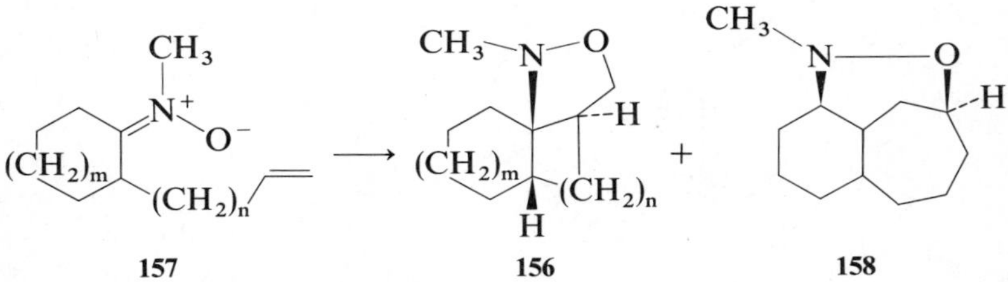

157 → **156** + **158**

The isolation of only *cis*-fused adduct where n = 2 is attributable to the transition state that permits maximum orbital overlap of the nitrone and olefinic centers.

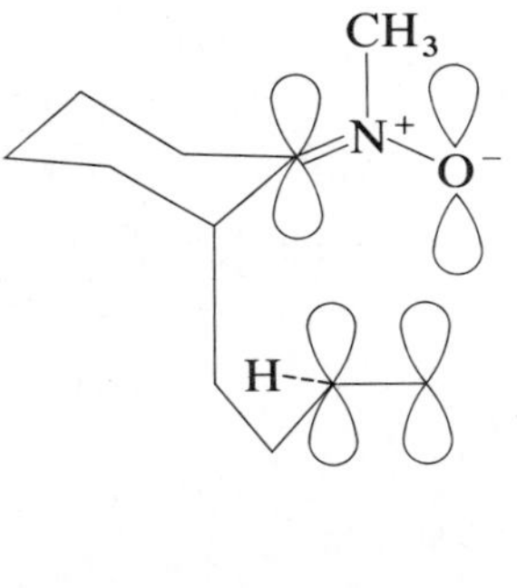

Cycloalkanones such as **159** and **160** react with *N*-substituted hydroxylamines to give the isoxazolidines via the intermediate nitrones. Only a single polycyclic diastereomer is formed by the cycloaddition reaction, affording a novel synthesis of functionalized bridged bicycloalkanes of general structure **161**.[138]

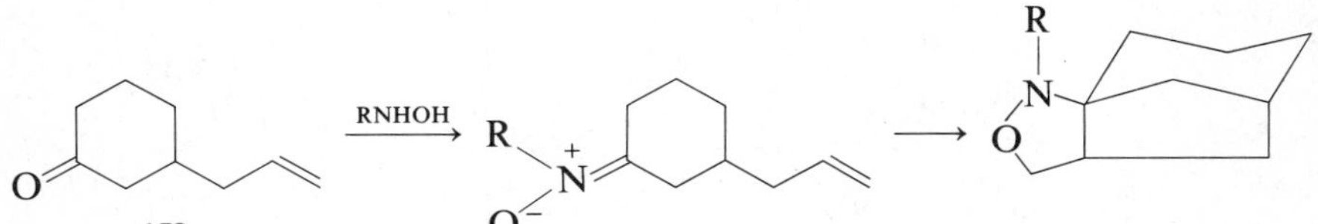
RNHOH
R
N
O
R
N
O

159

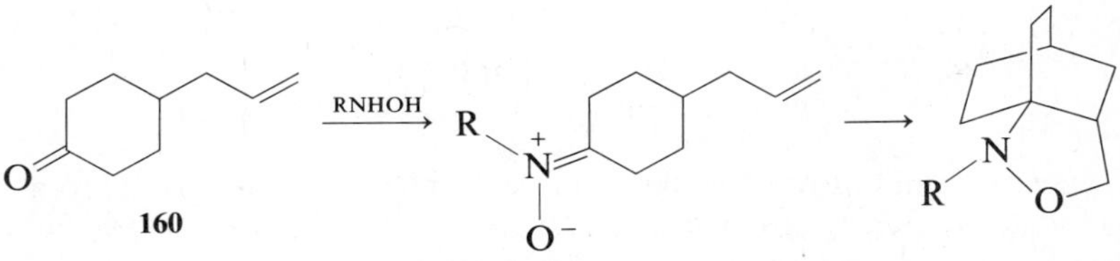
RNHOH
R
N
O
N
R
O

160

$R = CH_3, C_6H_5, C_6H_5CH_2$

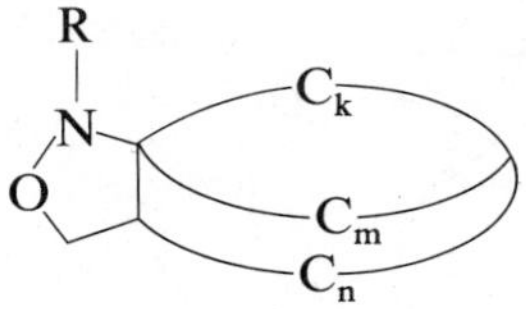
R
N
O

161 k = 1, 2
m = 1, 2, 3
n = 1, 2

A related reaction affords the tetracyclic linearly fused cyclopentanoid framework on treatment of the nitrone precursor **162** with *N*-methylhydroxylamine.

Inclusion of sodium ethoxide allows both stereoisomers of **162** to be converted to the same cycloadduct. However, in the absence of sodium ethoxide, only the stereoisomer directly leading to nitrone **162a** undergoes cyclization; the other leads to intractable material. This is a result of base-catalyzed epimerization affording the desired intermediate stereoisomer.[139]

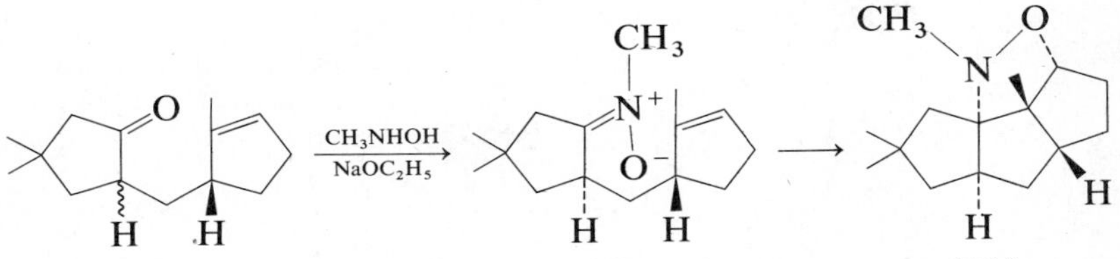
O
H H
CH3NHOH
NaOC2H5
CH3
N+
O−
H H
CH3
O
N
H
H

162 **162a** (75%)

Heteroatom-Linked Olefin Ketones and *N*-Substituted Hydroxylamines. One of the few examples of this class of intramolecular [3 + 2] nitrone olefin dipolar cycloadditions is the reaction of the keto amide olefin **163** with *N*-methylhydroxylamine to yield the isomeric tetracyclic adducts **164** and **165** in a ratio of 1:13.[140,141]

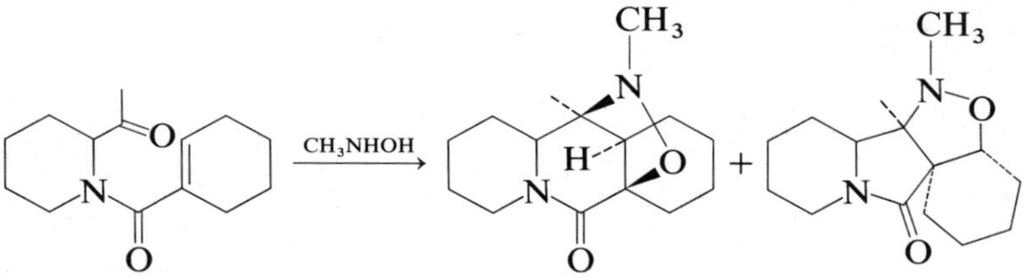

CH_3NHOH

163 **164** **165**

1:13

A highly unusual intramolecular cycloaddition of a cyclic nitrone to a cyclic olefin linked by two heteroatoms is that of the tricyclic nitrone **166**, which is thermally transformed into an exotic pentacyclic trioxadiaza cage compound.[142]

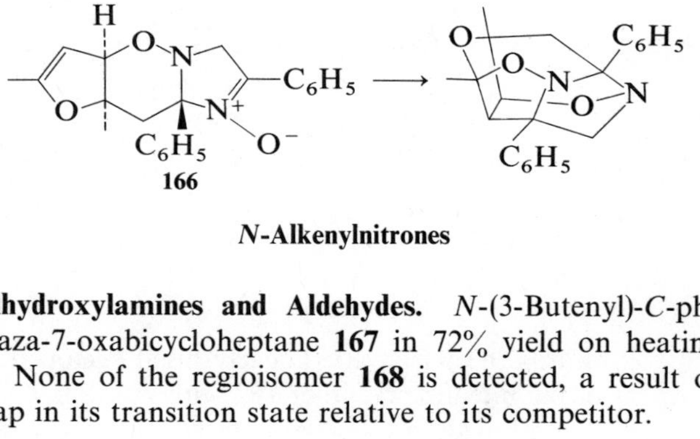

166

N-Alkenylnitrones

***N*-Alkenylhydroxylamines and Aldehydes.** *N*-(3-Butenyl)-*C*-phenylnitrone yields the 1-aza-7-oxabicycloheptane **167** in 72% yield on heating in xylene under reflux. None of the regioisomer **168** is detected, a result of a lack of orbital overlap in its transition state relative to its competitor.

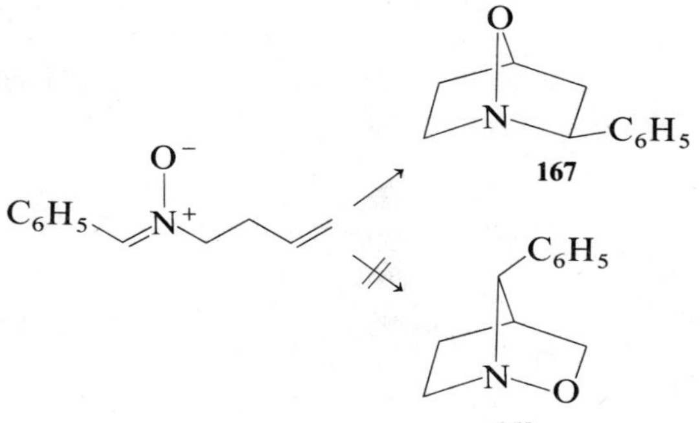

167

168

This preference may also be rationalized on electronic considerations since the production of **167** is the expected mode of addition of a nitrone to a terminal double bond.[120]

Product regiochemistry in these cyclizations is markedly influenced by the number of carbon atoms in the spacer that links the nitrone nitrogen and the double bond. Thus the series of homologous *N*-alkenylnitrones **169** yield cycloadducts of differing regiochemistry. The nitrone **169a** affords only the regioisomer **170a**, whereas its homolog **169b** furnishes only **171b**, a product with the opposite sense of regiochemical cycloaddition. The bis-homo analog **169c** is converted to a 1:3 mixture of cycloadducts **170c** and **171c**.

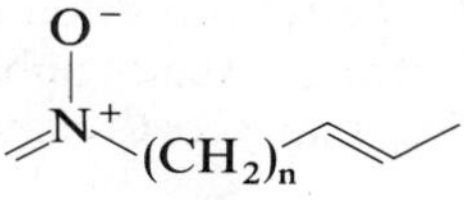

169 **a**, n = 1 **b**, n = 2 **c**, n = 3

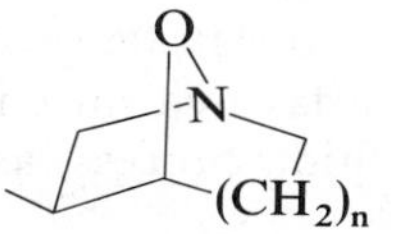

170 **a**, n = 0 **c**, n = 2

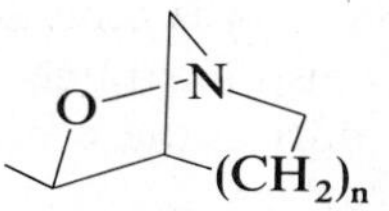

171 **b**, n = 1 **c**, n = 2

A similar reaction is observed with *C*-4-nitrophenylnitrones **172**, yielding exclusively **173** and **174**, respectively.[143]

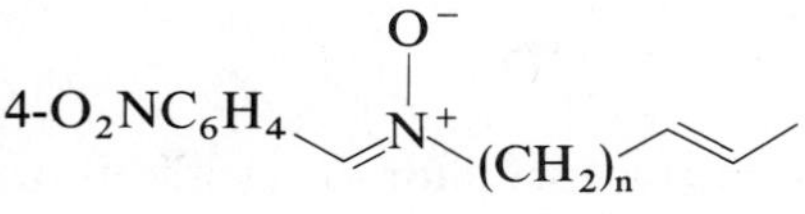

172 **a**, n = 2 **b**, n = 3

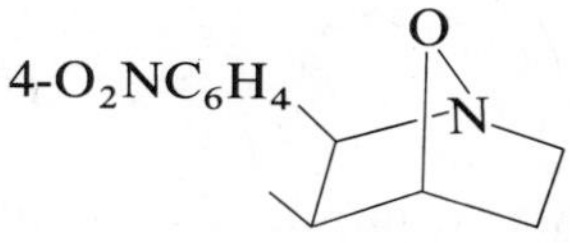

173

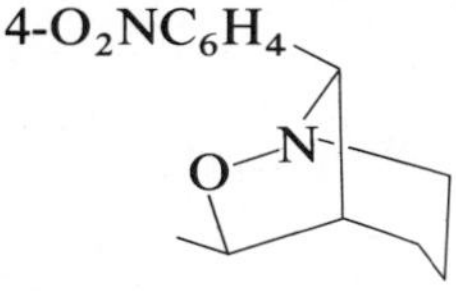

174

These diverse results are rationalized by assuming that C—C bond formation is more advanced than C—O bond formation in the transition state of the cycloaddition reaction. This assumption is readily supported for dipolar cycloadditions in which the LUMO (dipole)–HOMO (dipolarophile) interaction is the controlling factor. Thus if a two-carbon spacer is placed between the nitrone nitrogen and the olefin, the transition state **175** is favored compared with the highly strained alternative **176**.

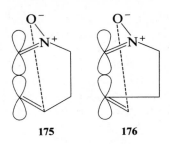

175 176

Since the higher energy of **176** arises from angle strain, which is relieved in the homologous transition state **177**, it is not unexpected that the opposite regiochemistry is generated. In this instance, the alternative transition state **178**, which is also unstrained, is disfavored since it requires a seven-membered ring closure during the cycloaddition process, assuming that C—C bond formation is early.

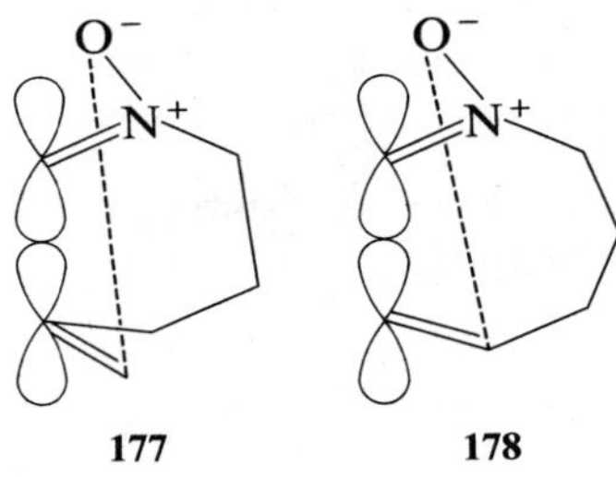

177 178

N-Alkenylnitrone cycloadditions are also useful for the synthesis of bridged polycyclics, as exemplified in the cyclization of the bicyclic nitrone **179** to the adduct **180**.[131]

$C_6H_5CH_3$, 125°

179 → 180

APPLICATIONS TO NATURAL PRODUCT TOTAL SYNTHESIS

Introduction

The utility of nitrone–olefin [3 + 2] cycloadditions in the total synthesis of natural products arises not from the presence of the isoxazolidine moiety in naturally occurring substances but from the various transformation products of this saturated heterocycle. The most important of these is the conversion of the isoxazolidine ring system to an *N*-substituted 1,3-amino alcohol by a reduction step.

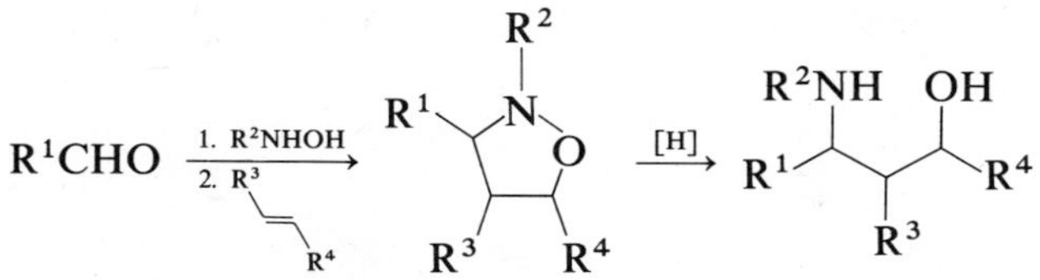

The attachment to the aldehyde of two additional carbon atoms laden with the useful functionality present in the 1,3-amino alcohol is very powerful methodology. Nevertheless, applications to natural product synthesis did not appear in force until the mid-1970s.[144,145] Since this transformation not only yields new carbon–carbon and carbon–oxygen bonds but also incorporates a nitrogen atom in the molecule, the obvious synthetic targets are nitrogenous substances, particularly alkaloids. The reaction is not limited to the synthesis of nitrogen-containing compounds, however, and examples of applications to non-nitrogenous targets have been reported.

Intramolecular Cycloadditions

Biotin (Vitamin H). The use of the [3 + 2] nitrone–olefin cycloaddition reaction requires the identification of an actual or latent 1,3-amino alcohol in the target structure. The latter is exemplified in the total synthesis of the important vitamin biotin (**181**).[144,145] Although there is no 1,3-amino alcohol present in biotin, the diamino precursor **182** can be derived in principle from the 1,3-amino alcohol starred in **183**.

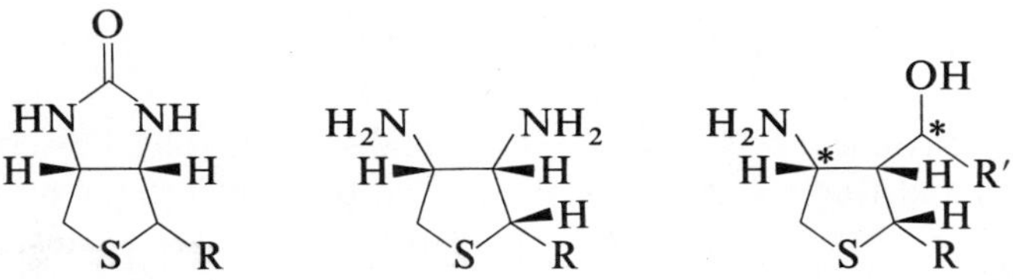

181 **182** **183**

$R = (CH_2)_4CO_2H$

This transformation could be accomplished in theory by oxidation of **183** to an amino ketone followed by a Beckmann rearrangement and hydrolysis to yield **182**, thus illustrating the identification of a *latent* 1,3-amino alcohol in the target molecule. The actual synthesis involves the intramolecular cycloaddition of the olefinic nitrone **184** to yield the tricyclic isoxazolidine **185**. This serves to fix the three contiguous asymmetric centers on the tetrahydrothiophene ring in the required all-*cis* configuration of biotin. Reduction of the N—O bond of **185** with zinc and acid yields the desired *N*-benzylamino alcohol **186**, which is converted to biotin in two steps after oxidation and a Beckmann rearrangement. The formation of **185** is achieved with complete stereospecificity and is an excellent example of the ability of the reaction to proceed in a stereocontrolled manner.

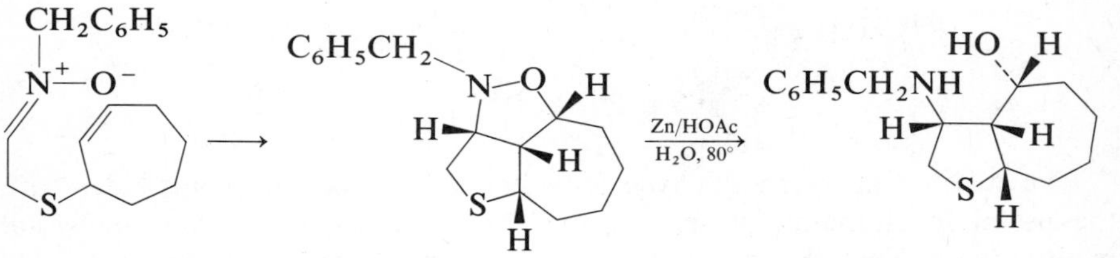

184 (66%) **185** **186**

An alternative synthesis of biotin[146] employs a [3 + 2] cycloaddition of the nitrone enol sulfide **187**, derived from L-(+)-cysteine. This substance undergoes spontaneous cyclization at 25° to yield a mixture of adducts **188** and **189** in a ratio of 9:1.

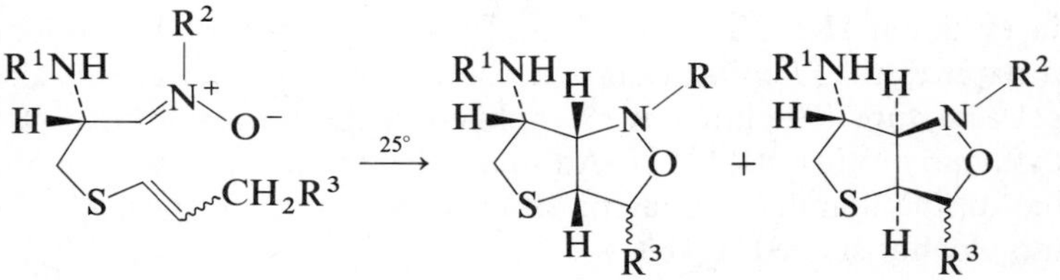

187 **188** (9:1) **189**

The lack of stereochemical control in this initial study is solved by constructing the 10-membered ring compound **190**, inspired by deliberations on the transition states leading to **188** and **189**. This substrate cyclizes exclusively to the desired tricyclic structure, which is then converted into biotin in six steps.

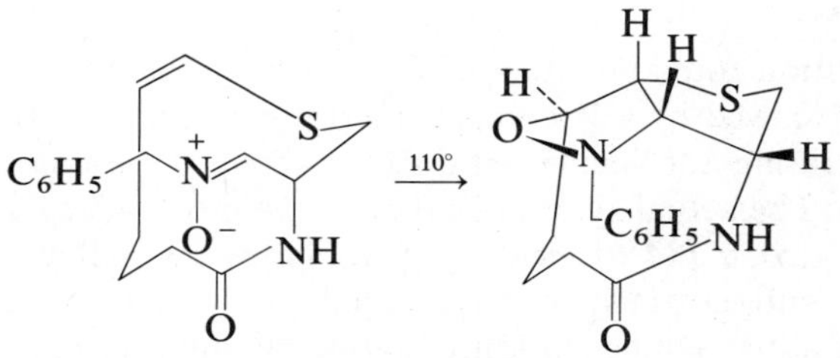

190 (63%)

Amino Glycosides. A chiral synthesis[147] of L-daunosamine, the sugar moiety of the antitumor agent adriamycin, involves reaction of the masked aldehyde **191** with (*S*)-(−)-*N*-hydroxymethylbenzenemethanamine to yield the nitrone **192**, which cyclizes in refluxing xylene to produce the diastereomers **193** and **194** as an 82:18 mixture. The desired major product **193** is separated in optically pure form and carried forward in the synthesis. The ability of the intramolecular version of this reaction to reverse the normal mode of cycloaddition to an enol ester system is noteworthy. The driving force to form the fused 5,5-bicyclic nucleus overcomes other energetics usually at play in the intermolecular reaction. Thus the nitrone carbon of **192** becomes exclusively bound to the oxygen-bearing carbon of the enol acetate unit in **191**, an orientation opposite to that found in the intermolecular variant. The application of the nitrone–olefin reaction in this instance is predicated on an acid-catalyzed transformation of the lactol **195**, prepared in three steps from **193**, to the pyranose anomers **196**.

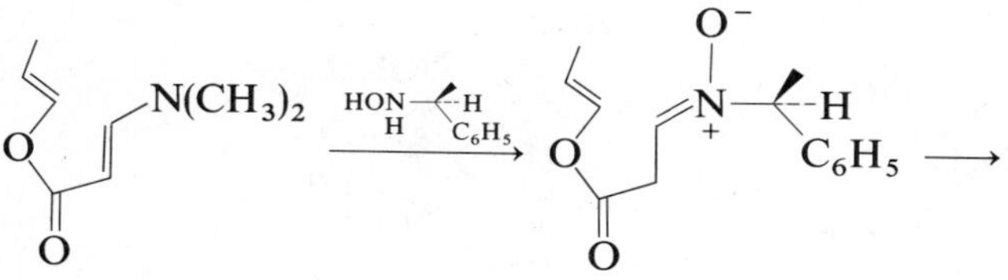

$N(CH_3)_2$ HON(H)–CH(C_6H_5) → O^- N⁺–C(H)(C_6H_5) →

191 **192**

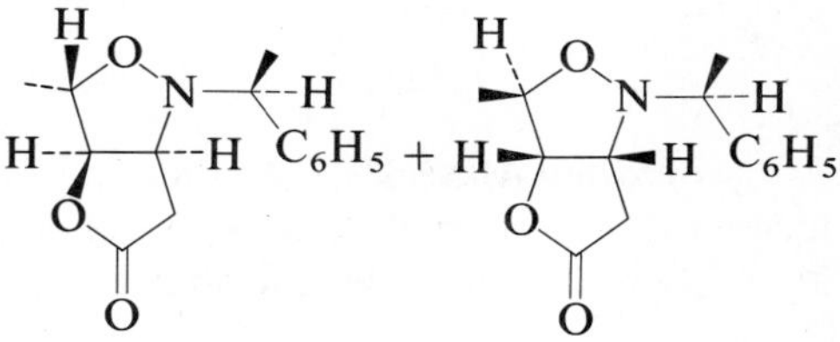

193 (56%) + **194** (12%)

CO_2CH_3 OH H C_6H_5 → OCH_3 CO_2CH_3 OH C_6H_5

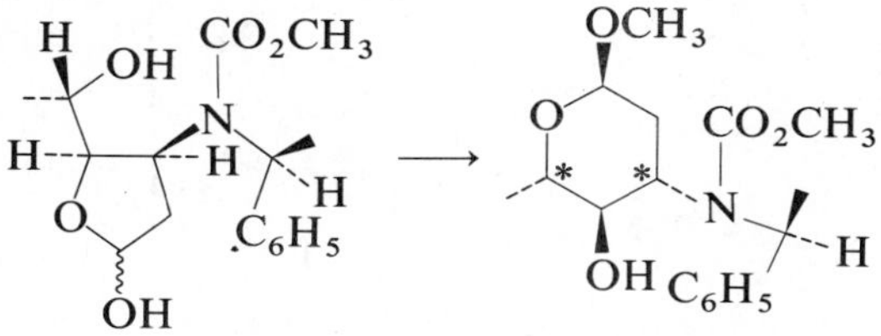

195 **196**

The latent 1,3-amino alcohol in the target is starred; the alcohol portion is masked by its involvement in the hemiacetal.

Ergot Alkaloids. The synthesis[148] of the ergot alkaloids (+)-chanoclavine I and (+)-isochanoclavine utilizes an intramolecular [3 + 2] cycloaddition of the indolic nitrone **197** to furnish the *cis*-fused tetracyclic isoxazolidine **198**. The transient nitrone **197** is prepared and cyclized in refluxing benzene–methanol with azeotropic removal of water.

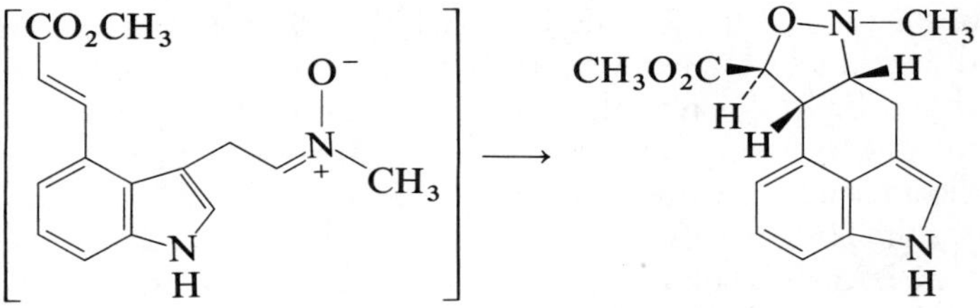

197 → **198**

Conversion of **198** to the diol reveals the 1,3-amino alcohol (starred) in this key intermediate, which is converted into (+)-chanoclavine I in five facile steps.

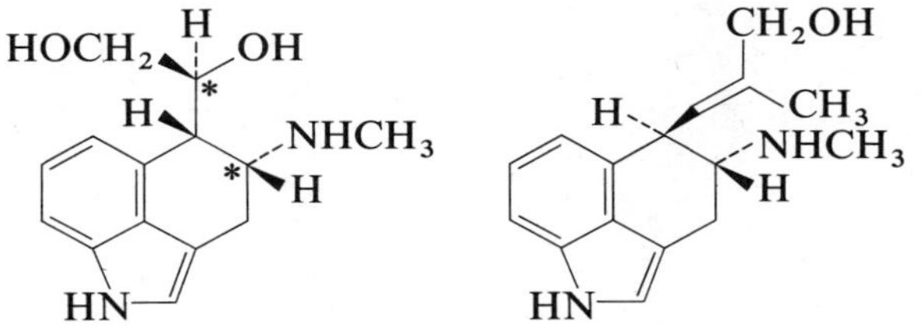

(+)-chanoclavine I

Lycopodium Alkaloids. An efficient synthesis[149] of the lycopodium alkaloid luciduline centers on the thermolysis of the easily available hydroxylamine derivative **199** in the presence of formaldehyde. The transient nitrone is presumably

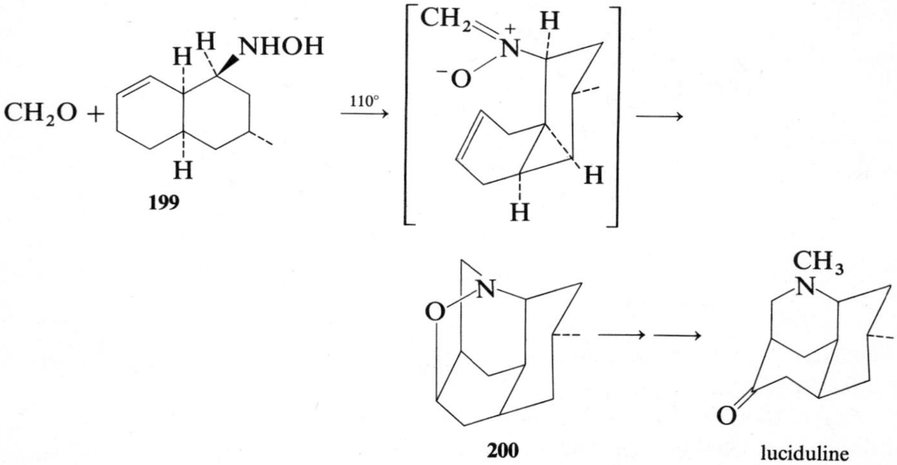

CH_2O + **199** $\xrightarrow{110°}$ → **200** → → luciduline

generated and spontaneously cyclizes to yield the desired isoxazolidine **200**. This product is readily converted to luciduline by the sequence of methylation, reduction, and oxidation. The overall synthesis is far superior to all previously reported multistep preparations of luciduline and was the first comparative demonstration of the prowess of nitrone–olefin cycloadditions in the total synthesis of natural products.

Tropane Alkaloids. The cycloaddition of the 1-pyrroline-1-oxide derivative **201** provides a route to tropane alkaloids. An aza-Cope rearrangement of **201** affords the isomeric **202**, whereas the desired crystalline cycloadduct **203** is obtained exclusively at higher temperatures. Alkylation of **203** and hydrogenation yields the bicyclic compound **204**, an analog of pseudotropine. This pathway represents a substantial improvement over more classical routes to substituted tropanes.[150]

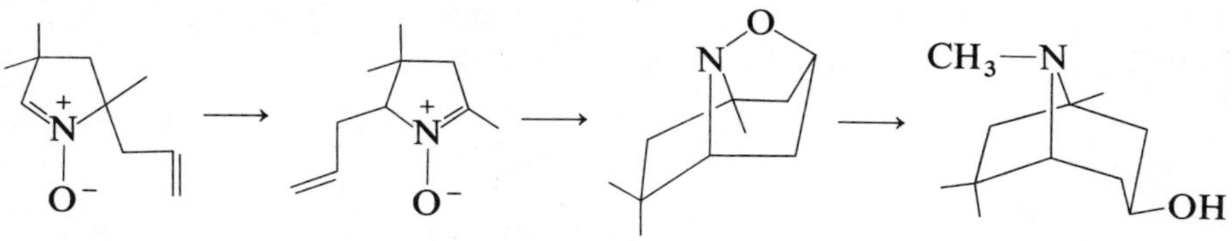

201 **202** **203** (20%) **204**

The parent tropane system can be prepared by thermolysis of the unsubstituted precursor **205**, generated *in situ* from the nitro acetal.[151] The cycloadduct **206** is sublimed and hydrogenated to yield norpseudotropine (**207a**), convertible to pseudotropine (**207b**) by an Eschweiler–Clarke procedure.

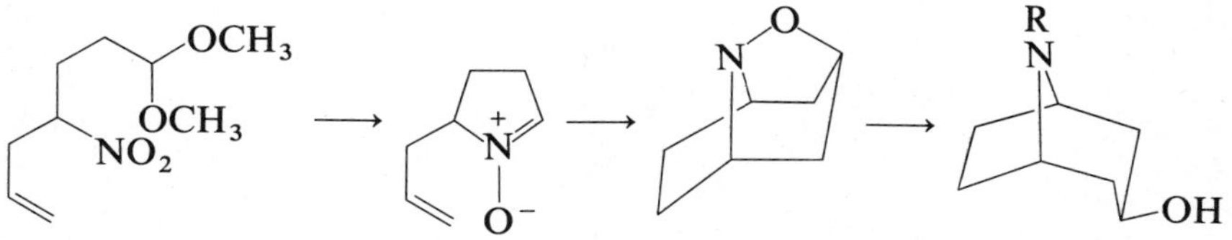

205 **206** **207**

a, R = H
b, R = CH_3

These results form the basis for two syntheses of cocaine[152,153] and are particularly instructive for illustrating various features of nitrone–olefin chemistry. Thus the protected aldehyde **208** can be hydrolyzed to generate the desired nitrone ester **209**, which cyclizes to the tricyclic isoxazolidine **210** with complete regio- and stereospecificity. *N*-Methylation, reduction of the N—O bond, and benzoylation serve to convert the isoxazolidine into (+)-cocaine. The derived 1,3-amino alcohol is starred.

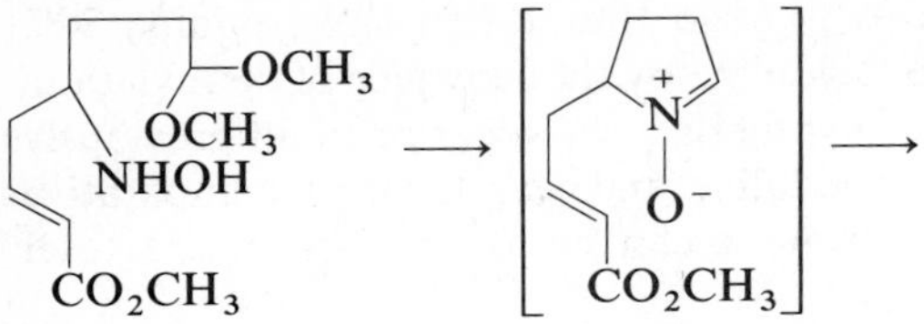

208 209

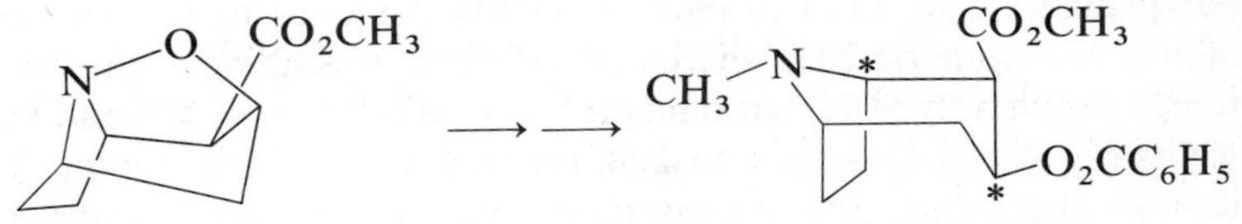

210 (+)-cocaine

An improved version[152] involves a cascade of nitrone–olefin cycloadditions. Reaction of 1-pyrroline-1-oxide with methyl 3-butenoate in refluxing toluene for 15 hours yields the desired adduct regiochemically pure.

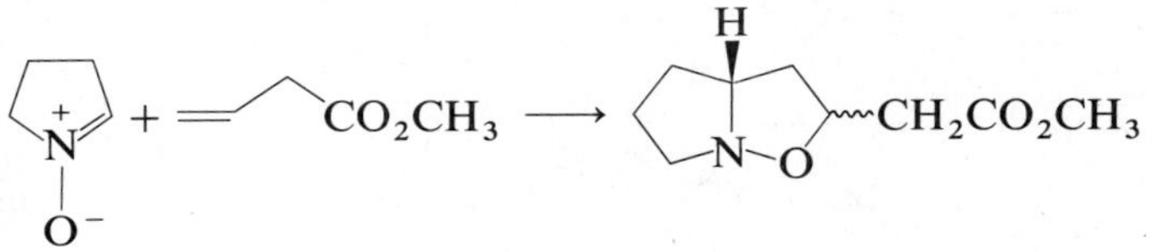

Oxidation of this adduct with 1 equivalent of *m*-chloroperbenzoic acid (MCPBA) produces the hydroxy nitrone **211**. Since the nitrone group complicates the required dehydration of **211**, a novel protecting group is introduced. Thus treatment of **211** with methyl acrylate yields the isoxazolidine **212**, which can be readily dehydrated to the required olefin **213**.

Heating of **213** in refluxing xylene cleaves the protected nitrone by a [3 + 2] cycloreversion, expelling methyl acrylate in the process. The newly unveiled

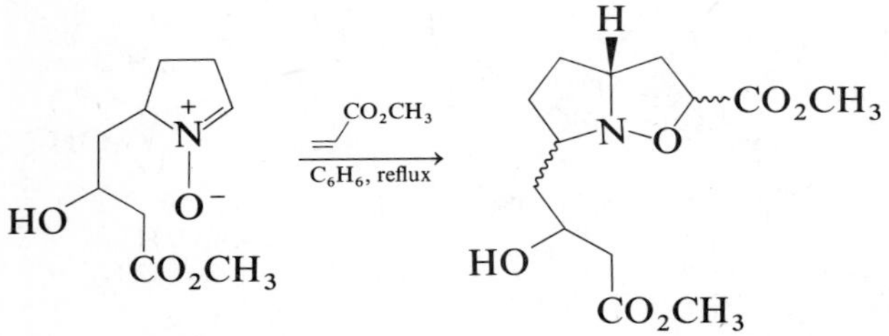

211 212

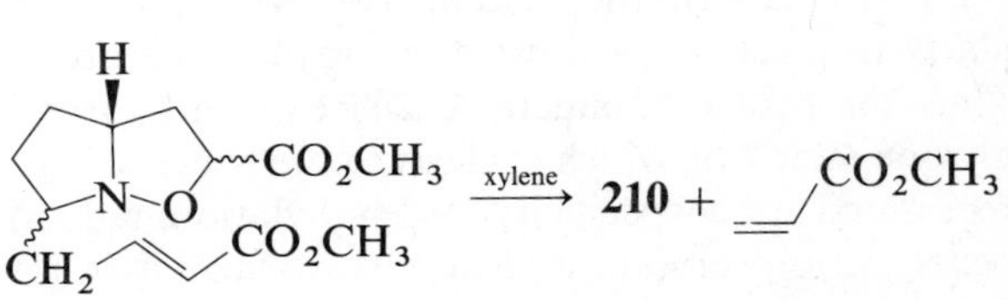

213

nitrone (identical to **209**) spontaneously cyclizes to afford the desired **210**, taking advantage of the favored mode for intramolecular reaction, leaving methyl acrylate unreacted.

(+)-Adaline. The total synthesis of (+)-adaline is achieved by a sequence incorporating a key nitrone–olefin cycloaddition of the intermediate **214**.[154] The tricyclic adduct is converted into the target (+)-adaline by hydrogenation over Raney nickel, followed by oxidation with pyridinium chlorochromate.

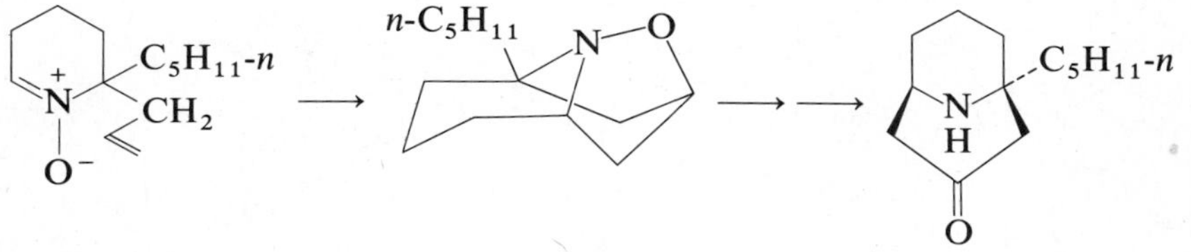

214 (66%) (+)-adaline

Nitrogen-Free Systems. The nitrone–olefin reaction is used in a chiral preparation of the epoxylactone **217**, a prostanoid intermediate, from D-glucose.[127] Thus the aldehyde **215** is treated with *N*-methylhydroxylamine hydrochloride to yield the bicyclic isoxazolidine. Extrusion of the unwanted nitrogen is accomplished by reduction to give the aziridine, which is converted into the required hydroxy olefin **216** by oxidation. Transformation of **216** to the epoxide **217** over several steps yields optically pure material.

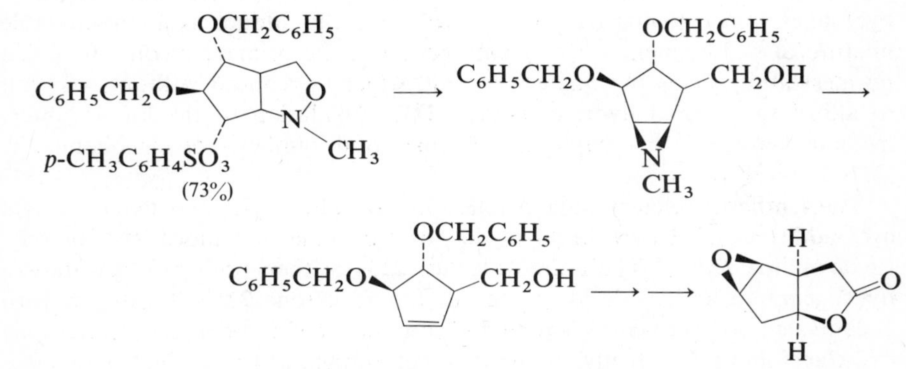

D-glucose → **215** → (73%) → → **216** → → **217**

(−)-Bisabolol can be synthesized by thermolysis of the nitrone **218** derived from farnesal.[155,156] The product isoxazolidines are alkylated to the methiodides and reduced with lithium aluminum hydride to yield the *N,N*-dimethylamino alcohols **219**. Methylation on nitrogen and sodium–ammonia reduction of the resulting quaternary salts gives (−)-bisabolol, illustrating a more complex excision of nitrogen.

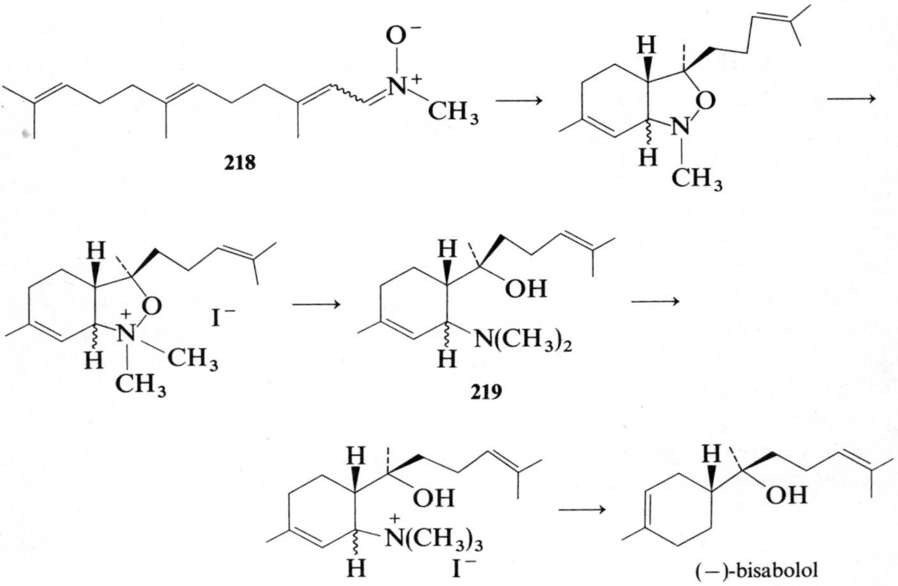

O^- N^+ CH_3 H O N H CH_3

218

H O + N I⁻ H CH_3 CH_3 H OH $N(CH_3)_2$ H

219

H OH + $N(CH_3)_3$ H I^- H OH

(−)-bisabolol

Intermolecular Cycloaddition Syntheses Based on 1-Pyrroline-1-oxide

Elaeocarpus Alkaloids. The elaeocarpus alkaloids constitute a diverse group of naturally occurring alkaloids derived from the plant family Elaeocarpaceae found primarily in New Guinea and India. The cycloaddition of 1-pyrroline-1-oxide to the *trans*-enone **220** gives the adduct as an inseparable mixture of C-3 epimers.[157,158] Deprotection of the primary alcohol followed by mesylation yields the quaternary salt, which is reduced without isolation to afford the desired hydroxy ketone **221a**. Oxidation to the known intermediate ketone **221b** completes a formal total synthesis of elaeokanine C (**221c**).

The synthesis of elaeokanine A relies on the highly regio- and stereoselective cycloaddition of 1-pyrroline-1-oxide to 1-pentene to afford the bicyclic isoxazolidine **222**.[159] The amino alcohol **223a** is then prepared by catalytic hydrogenation and oxidized to the key amino ketone **223b**. Treatment with 1 equivalent of acrolein followed by concentrated hydrochloric acid yields (+)-elaeokanine C directly. If the acrolein adduct **224** is treated with base, (+)-elaeokanine A (**225**) is produced instead.

This approach can be extended[160] to the synthesis of aromatic elaeocarpus alkaloids by starting with 2-methoxy-6-methylstyrene. Reaction with 1-pyrroline-1-oxide affords the adduct highly selectively. Alkylation with 3-bromo-1-propanol yields the quaternary salt **226**, which undergoes a dramatic series of reactions when exposed to potassium *tert*-butoxide and benzophenone at reflux temperature. Under these conditions, **226** undergoes a base-induced isoxazolidine ring opening, followed by a modified Oppenauer oxidation of the primary alcohol and aldol closure to the desired enone. This sequence further exemplifies the ability of the isoxazolidine nucleus to be modified in a variety of useful ways. Conversion of the enone to (+)-elaeocarpine involves demethylation and base-catalyzed cyclization.

The indolizidine alkaloid septicine is synthesized from the cycloadduct **227** arising from the reaction of 1-pyrroline-1-oxide with 2,3-di(3,4-dimethoxyphenyl)butadiene.[161] Elaboration to the target substance is achieved by

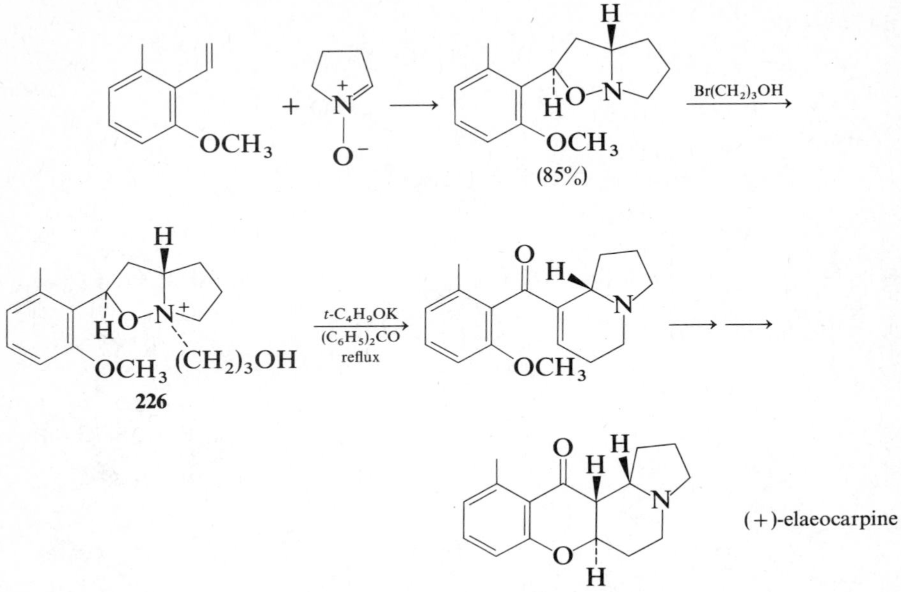

Br(CH$_2$)$_3$OH

(85%)

226

t-C$_4$H$_9$OK
(C$_6$H$_5$)$_2$CO
reflux

(+)-elaeocarpine

alkylative bromination of **227** to the quaternary salt, which is converted to (+)-septicine in three steps.

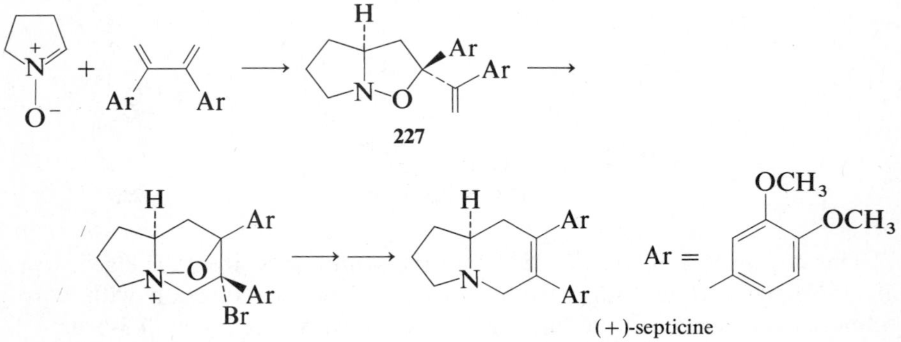

227

(+)-septicine

Ar = 3,4-(OCH$_3$)$_2$C$_6$H$_3$

Pyrrolizidine Alkaloids. The reduced pyrrole nucleus imbedded in this alkaloidal class affords a straightforward application of 1-pyrroline-1-oxide–olefin cycloaddition chemistry. The synthesis of (+)-supinidine, the necine base obtained from supinine and its congeners,[162,163] involves reaction of 1-pyrroline-1-oxide with methyl-γ-hydroxycrotonate to afford the desired regio adduct **228a**. Then, in a procedure that has wide applicability in pyrrolizidine synthesis, **228a** is converted to the mesylate **228b**, which is hydrogenolyzed over palladium on carbon. The transient 1,3-amino alcohol is not isolated since it undergoes facile N-alkylation to the hydroxy ester, which is dehydrated to (+)-supinidine.

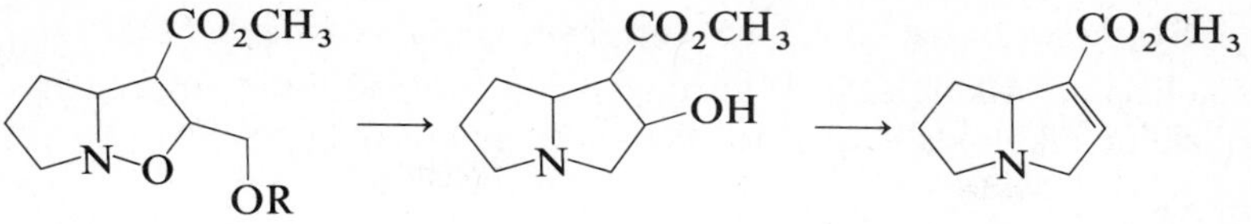

CO_2CH_3 CO_2CH_3 CO_2CH_3

OH

N–O N N

OR

228 (+)-supinidine

a, R = H
b, R = SO_2CH_3

The total synthesis of (+)-isoretronecanol is achieved starting from the *exo*-cycloaddition product of 1-pyrroline-1-oxide and 2,3-dihydrofuran.[102,155] Reduction with lithium aluminum hydride yields the amine diol **229a**, which is silylated by *N*-(trimethylsilyl)diethylamine at 145° to afford the persilylated derivative **229b**. Conversion to the iodide **229c** occurs selectively with trimethylsilyl iodide. Finally, treatment with fluoride ion yields (+)-isoretronecanol by intramolecular alkylation of the desilylated intermediate. The 1,3-amino alcohol of the target compound is starred.

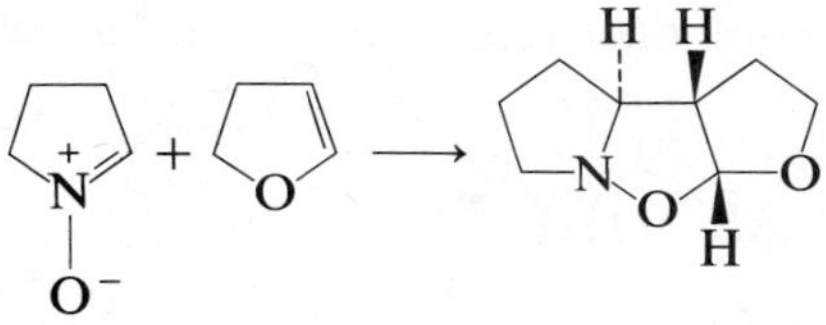

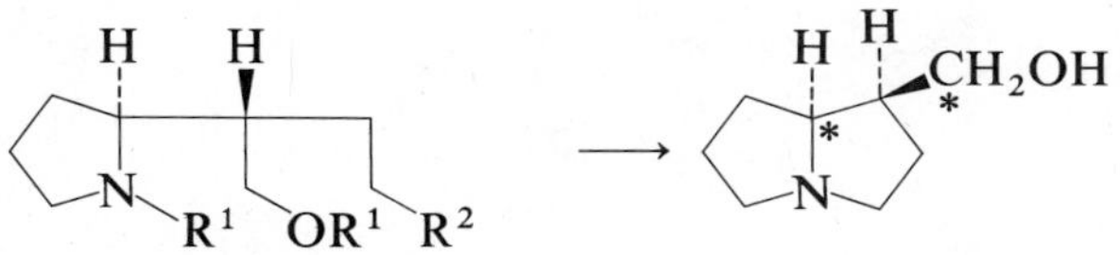

H H

N O O

O^- H

H H H H CH_2OH

N N

R^1 OR^1 R^2

229 (+)-isoretronecanol

a, R^1 = H, R^2 = OH
b, R^1 = TMS, R^2 = OTMS
c, R^1 = TMS, R^2 = I

Finally, a modification of an earlier synthesis of (+)-supinidine utilizing 4,4-dimethoxy-1-pyrroline-1-oxide is reported. The approach also affords a preparation of (+)-retronecine.[164]

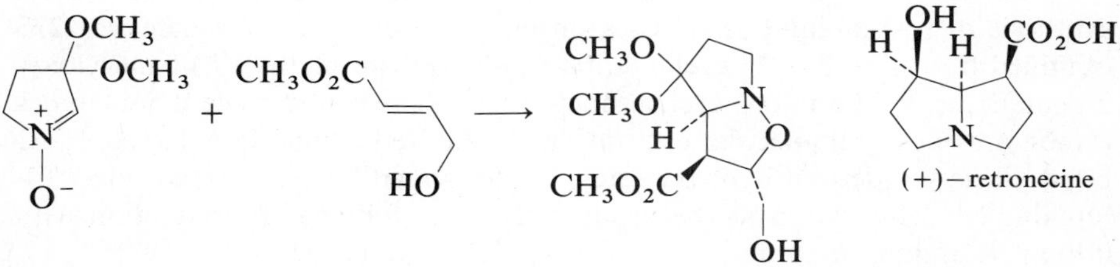

OCH_3 OH
OCH_3 CH_3O_2C CH_3O H H CO_2CH_3
CH_3O N
N H O N
O^- HO CH_3O_2C (+)–retronecine

OH

Syntheses Based on 2,3,4,5-Tetrahydropyridine-1-oxide (THPO)

Quinolizidine Alkaloids. A nitrone–olefin cycloaddition approach to the quinolizidine alkaloids is used to prepare (+)-abresoline.[165,166]

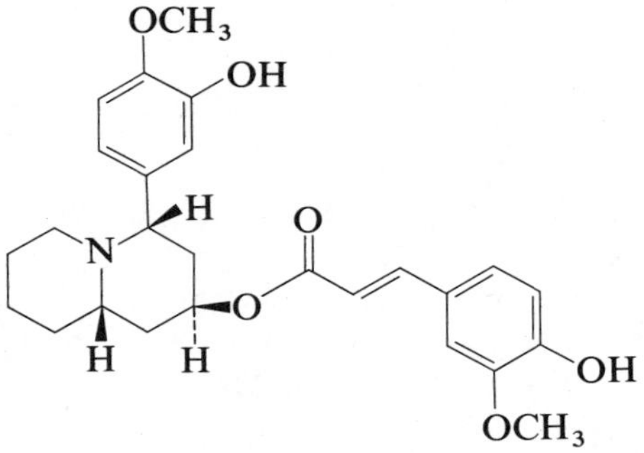

OCH_3 OH H N O O H H OH OCH_3

(+)-abresoline

Reaction of the homoallylic alcohol **230** with THPO affords the desired cycloadduct **231** as a mixture of stereoisomers at the benzylic carbon atom.

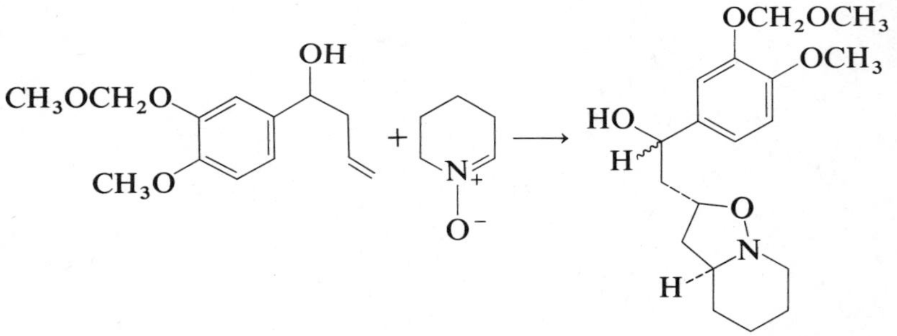

OH CH_3OCH_2O CH_3O + N^+ O^- → OCH_2OCH_3 OCH_3 HO H O N H

230 **231**

Treatment of alcohol **231** with methanesulfonyl chloride in pyridine followed by reduction with zinc in 50% aqueous acetic acid yields the alcohol **232a** via the intermediacy of the quarternary cation. Conversion of **232a** to (+)-abresoline involves separation of the derived acetates **232b**, basic hydrolysis, attachment of the requisite side chain with inversion, and deprotection.

The methodology evolved for synthesis of the pyrrolizidines provides a total synthesis of (+)-lupinine that relies on elaboration of the mesylate salt **233** obtained from the [3 + 2] cycloaddition of THPO to methyl (*E*)-5-mesyloxy-2-pentenoate.[21] The initial cycloadduct cannot be isolated since it undergoes a spontaneous intramolecular cyclization to **233**. Reduction of the labile N—O bond yields the desired hydroxy ester, which is readily transformed into (+)-lupinine by dehydration to the unsaturated ester followed by reduction with lithium aluminum hydride.

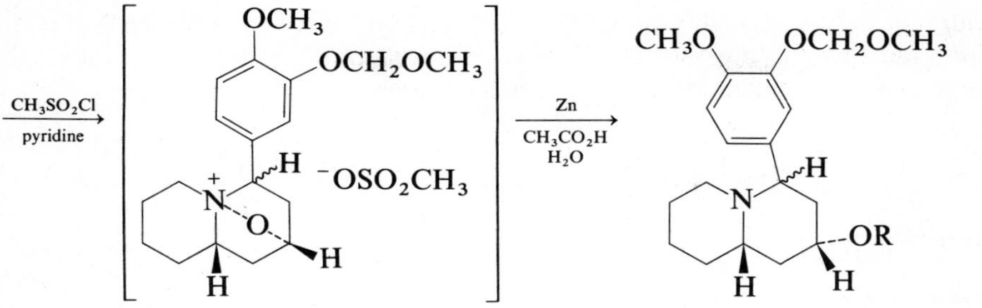

232

a, R = H
b, R = CH_3CO

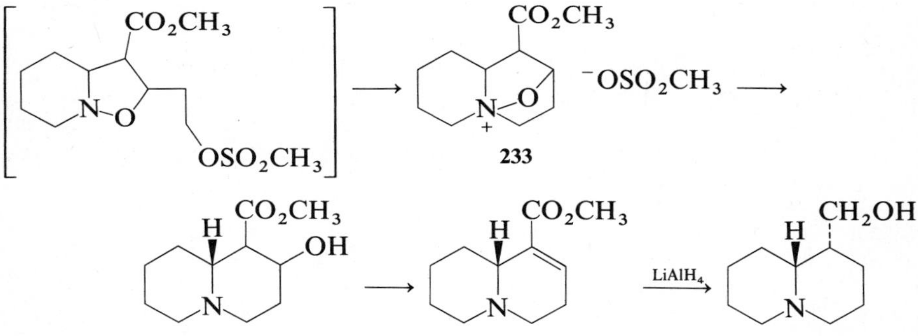

233

(+)-lupinine

Related chemistry conducted on the adduct **234** of THPO and *trans*-piperylene yields the quinolizidine base (+)-myritine.[167] The reaction sequence of N—O bond cleavage, oxidation, and Michael addition of the derived amino enone affords the target substance. The marked efficiency of this synthesis stems from the high chemo- and regioselective controls in the key cycloaddition step that constructs adduct **234**.

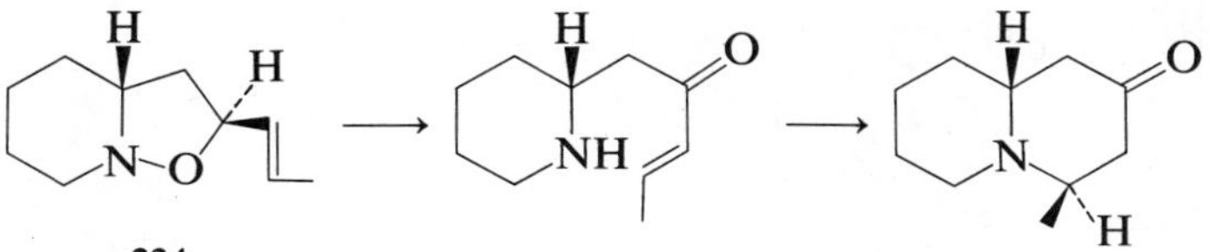

234

(+)-myritine

Miscellaneous Alkaloids. The synthesis of (+)-porantheridine[80] involves the cycloaddition of the substituted nitrone **235** and 1-pentene to yield the key intermediate bicyclic isoxazolidine. This compound is transformed into the transient iminium alcohol by sequential hydrogenation, acid hydrolysis, oxidation, and basic hydrolysis and then cyclization to afford the target alkaloid, shown with its embedded 1,3-amino alcohol starred.

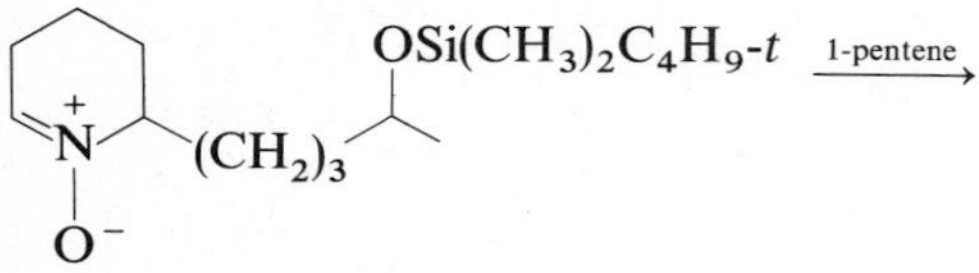

$OSi(CH_3)_2C_4H_9$-*t* 1-pentene →

235

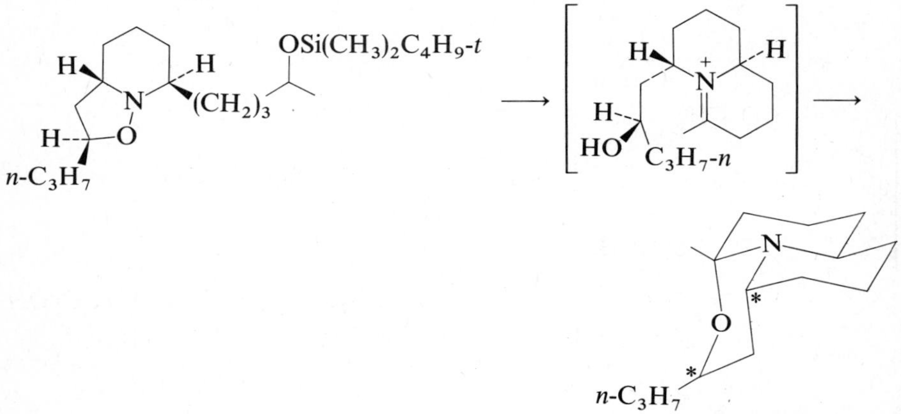

(+)-porantheridine

The anticancer alkaloid (+)-cryptopleurine, a member of a rare class of naturally occurring substances containing a phenanthro[9,10-*b*]quinolizidine nucleus, is prepared by cycloaddition of THPO to 3,4-dimethoxystyrene to afford the bicyclic compound **236**.[71] Reduction of the N—O bond and *N*-acylation to the amido alcohol sets the stage for conversion to (+)-cryptopleurine by oxidation followed by a dehydrative aldol procedure.

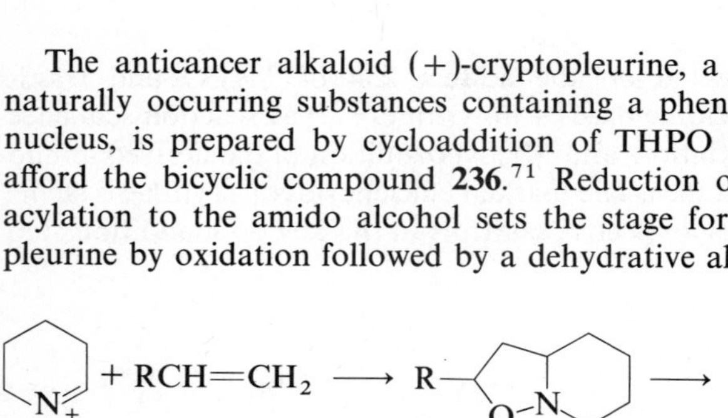

+ $RCH{=}CH_2$ →

236

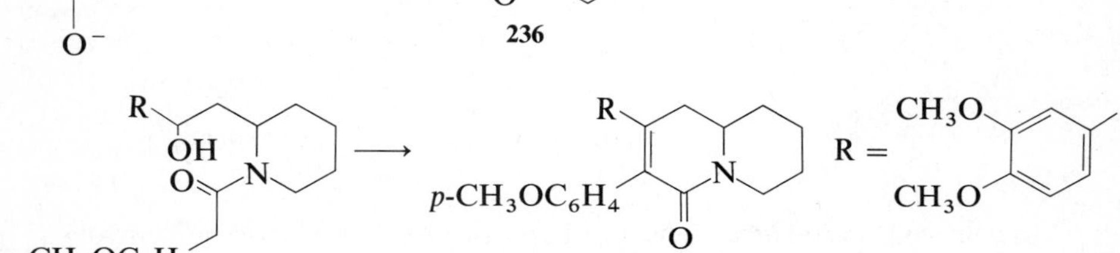

(+)-crytopleurine

R = 3,4-$(CH_3O)_2C_6H_3$

A simple example of the utility of the nitrone–olefin reaction is afforded by the synthesis of (±)-sedridine,[168] which takes advantage of the regio- and stereospecificity of the reaction of THPO and propene. Reduction of the adduct leads exclusively to (±)-sedridine.

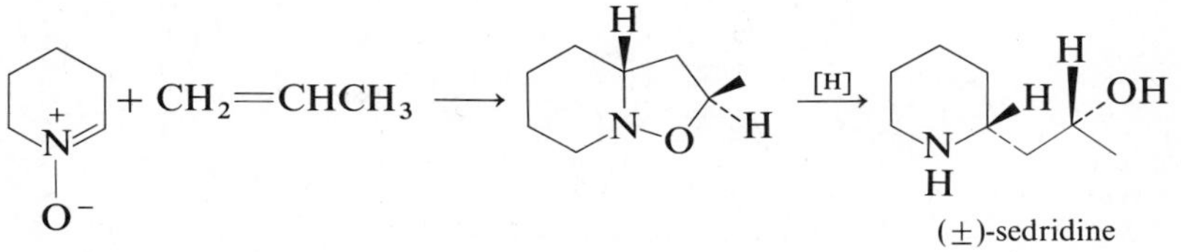

+ $CH_2{=}CHCH_3$ → [H] →

(±)-sedridine

Miscellaneous Intermolecular Syntheses

The structure of hydroxycotinine, the mammalian metabolite of nicotine isolated from the urine of smokers, has been established as *trans*-1-methyl-3-(*R*)-hydroxy-5-(*S*)-(3-pyridyl)-2-pyrrolidinone (**237**) by the following synthesis.[169] The pyridyl nitrone **238** reacts with methyl acrylate to afford predominantly the desired isoxazolidine **239**. Conversion to both the natural product **237** and its 3-hydroxy epimer **240** confirms the structural assignment.

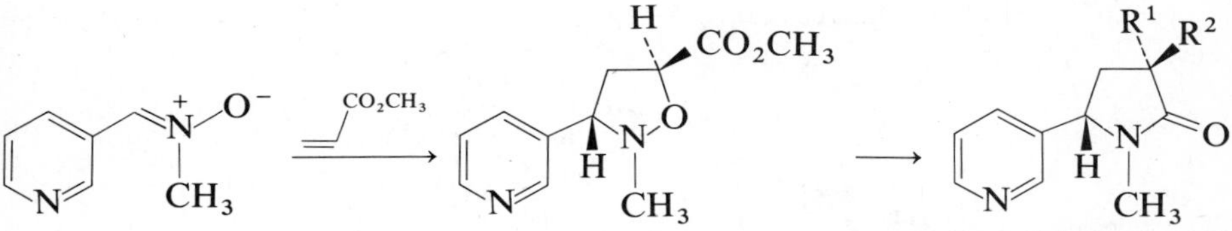

238 **239**

237 $R^1 = H, R^2 = OH$
240 $R^1 = OH, R^2 = H$

Amino Acids. A short synthesis of 4-hydroxyprolines involves the cycloaddition of *N*-benzyl-α-methoxycarbonylmethanimine-*N*-oxide to acrolein.[170] The cycloadduct is hydrogenolyzed to yield a mixture of epimeric 4-hydroxyproline methyl esters via the intermediate imine. The mixture is hydrolyzed to the parent amino acids.

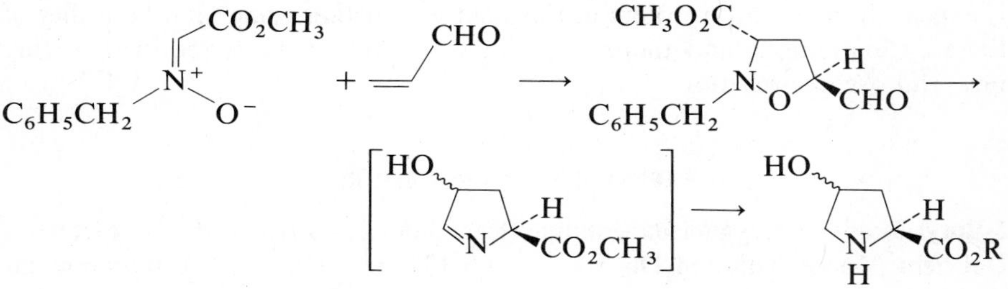

β-Lactams. A three-step procedure for the synthesis of *trans*-3,4-disubstituted azetidin-2-ones is used for the preparation of intermediates to thienamycin, an exceptionally potent β-lactam antibiotic.[171]

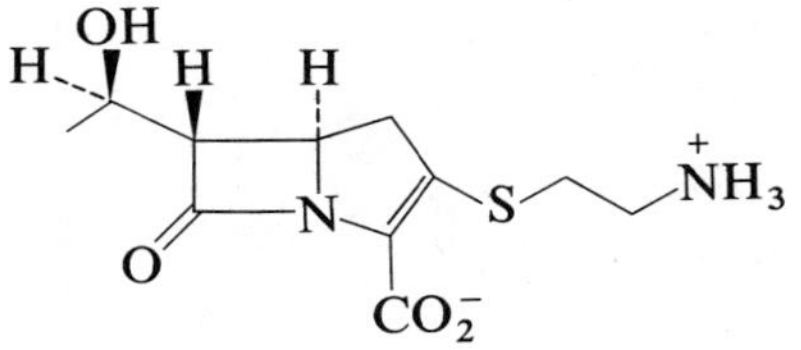

thienamycin

Thus *C*-chloro-*N*-benzylnitrone undergoes cycloaddition with benzyl crotonate to afford the required isoxazolidines **241a** and **241b** in a ratio of 1:5. Hydrogenolysis leads to the amino acids, which are cyclized with dicyclohexylcarbodiimide in acetonitrile to the desired azetidinone **242** and its C-4 epimer in overall yields of 30 and 5%, respectively. Similar thienamycin intermediates can be synthesized by nitrone–olefin methodology.[172]

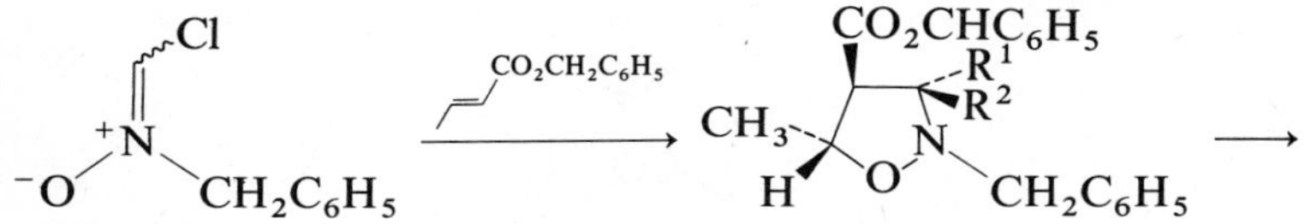

241 **a**, $R^1 = Cl$, $R^2 = H$
b, $R^1 = H$, $R^2 = Cl$

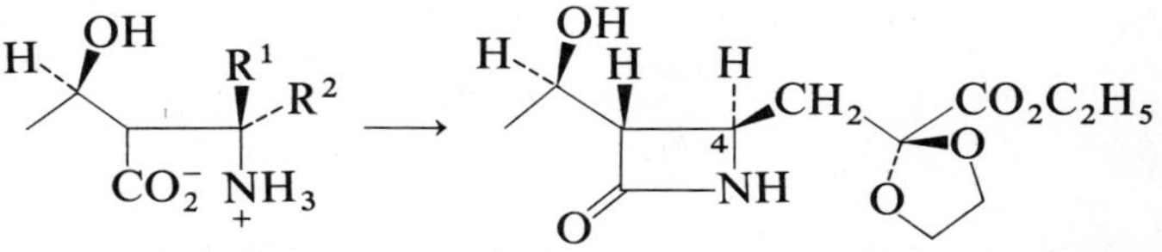

$R^1 = Cl$, $R^2 = H$
$R^1 = H$, $R^2 = Cl$

242

In conclusion, this survey of applications of the nitrone–olefin [3 + 2] cycloaddition reaction to the synthesis of natural products chronicles the elegance and efficiency of the title reaction for the preparation of complex structures. It is expected that future workers in this field will further exemplify the utility of these cycloadditions and thereby continue to expand the scope of this truly powerful organic reaction.

EXPERIMENTAL PROCEDURES

2-Butyl-3-methyl-5-cyanoisoxazolidine (Reaction of a Nitrone with an Electron-Deficient Monosubstituted Olefin).[173] To 17.0 g of *C*-methyl-*N*-butylnitrone was added 30 mL of acrylonitrile. The mixture was heated at 60° for 10 hours; after cooling, the excess acrylonitrile was evaporated. The residue was distilled to afford the product as a colorless oil, bp 93°/1 mm, in 58% yield. Anal. Calcd. for $C_9H_{16}N_2O$: C, 64.28; H, 9.53; N, 16.66. Found: C, 63.94; H, 9.54; N, 16.48.

5-Ethoxy-2-methyl-3-phenylisoxazolidine (Reaction of a Nitrone with an Electron-Rich Monosubstituted Olefin).[89] *C*-Phenyl-*N*-methylnitrone (1.00 g, 7.4 mmol) was dissolved in an excess of freshly distilled ethyl vinyl ether (25.0 mL, 161 mmol), and the solution was sealed in a thick-walled glass reaction tube. The tube was heated at 80° for 72 hours and then cooled, and the excess ethyl vinyl ether was removed by evaporation under high vacuum. The residue was chromatographed (CH_2Cl_2) to remove unreacted nitrone followed by bulb-to-bulb distillation at 145° (1 torr) to give a yellow oil (1.19 g, 78%). As evidenced by GLC (SE-30 column/140°) and NMR analysis, the reaction mixture contained a 50:50 mixture of *cis* and *trans* isomers (retention times 4.0 and 5.4 minutes, respectively). The isomers were separated by column chromatography (120 g; 10:1 hexane:ethyl acetate). The less polar isomer was the *cis* isomer, while the more polar compound was the *trans* isomer.

Cis isomer (oil): ^1H NMR ($CDCl_3$, 200 MHz) δ:1.25 (t, $J = 7$ Hz, 3*H*), 2.32 (ddd, $J = 3$, 10, and 13 Hz, 1*H*), 2.55 (s, 3*H*), 2.86 (ddd, $J = 6$, 10, and 13 Hz, 1*H*), 3.34 (t, $J = 10$ Hz, 1*H*), 3.45 (dq, $J = 7$ and 14 Hz, 1*H*), 3.93 (dq, $J = 7$ and 14 Hz, 1*H*), 5.15 (dd, $J = 3$ and 6 Hz, 1*H*), 7.33 (m, 5*H*); IR (CCl_4) 3100–3000 (m), 3000–2800 (vs), 1455 (s), 1370 (s), and 1100 (vs) cm^{-1}; mass spectrum (EI), m/z (relative intensity) 207 (M^+, 4), 161 (36), 118 (100), 77 (75). Anal. Calcd. for $C_{12}H_{17}NO_2$: C, 69.54; H, 8.27; N, 6.76. Found: C, 69.44; H, 8.10; N, 6.69.

Trans isomer (oil): ^1H NMR ($CDCl_3$, 200 MHz) δ: 1.23 (t, $J = 7$ Hz, 3*H*), 2.41 (ddd, $J = 5$, 9, and 13 Hz, 1*H*), 2.57 (dd, $J = 6$ and 13 Hz, 1*H*), 2.78 (s, 3*H*), 3.48 (dq, $J = 7$ and 10 Hz, 1*H*), 3.86 (dq, $J = 7$ and 10 Hz, 1*H*), 4.03 (dd, $J = 6$ and 9 Hz, 1*H*), 5.16 (d, $J = 5$ Hz, 1*H*), 7.30 (m, 5*H*); IR (CCl_4) 3100–3000 (m), 3000–2800 (vs), 1455 (s), 1370 (s), and 1100 (vs) cm^{-1}; mass spectrum (EI), m/z (relative intensity): 207 (M^+, 9), 161 (100), 134 (63), 118 (53), 77 (48). Anal. Calcd. for $C_{12}H_{17}NO_2$: C, 69.54; H, 8.27; N, 6.76. Found: C, 69.37; H, 8.10; N, 6.77.

2-Phenyl-3-*n*-propylisoxazolidine-4,5-*cis*-dicarboxylic Acid *N*-Phenylimide (Reaction of a Nitrone with a Cyclic Disubstituted Olefin).[174] *N*-Phenylhydroxylamine (11 g, 0.10 mol) and *N*-phenylmaleimide (17.4 g, 0.10 mol) were suspended in 40 mL of ethanol contained in a 200-mL Erlenmeyer flask. To the mixture was added immediately 8.98 g (11.2 mL, 0.124 mol) of freshly distilled *n*-butyraldehyde. An exothermic reaction ensued, and the mixture spontaneously heated to the boiling point. A clear slightly yellow solution resulted, which, on cooling, deposited an almost colorless crystalline cake. The mixture was cooled and filtered to afford 26–27 g (92%) of pure product, mp 106–107°, after a single recrystallization from ethanol.

Ethyl 4-Cyano-2-phenyl-3,5-di(2′-pyridyl)isoxazolidine-4-carboxylate (Reaction of a Nitrone with an Acyclic Trisubstituted Olefin).[106] Ethyl (2′-pyridylmethylene)cyanoacetate (1.0 g, 5 mmol) and *N*-(2′-pyridylmethylene)aniline-*N*-oxide (0.93 g, 5 mmol) were heated under reflux in dry benzene for 72 hours. The benzene was evaporated under reduced pressure, and the crystalline product was recrystallized from ethanol to afford the adduct as white needles (1.7 g, 91%), mp 124°. IR 2275, 1743, 1593, 1573, 1468, 1377, 1333, 860, 757, and 696 cm^{-1}; ^1H NMR δ: 1.35 (t, $CO_2C_2H_5$), 4.47 (q, $CO_2C_2H_5$), 5.78, (s, H3), 5.82 (s, H5), 6.9–8.0 (m, aryl), 8.54 (m, pyridyl); mass spectrum m/z (relative intensity) 401 (M^+ + 1, 20), 181 (21), 158 (52), 157 (25), 130 (100), 129 (21), 94 (22), 91 (32), 79 (52), 78 (74), 77 (42), 52 (25), 51 (57), 31 (24). Anal. Calcd. for $C_{23}H_{20}N_4O_3$: C, 69.0; H, 5.0; N, 14.0. Found: C, 68.7; H, 5.2; N, 13.9.

4-(9-Acridinyl)-2,3-diphenyl-4-isoxazoline (100) (Reaction of a Nitrone with an Acetylene).[114] A solution of 1.0 g of 9-ethynylacridine and 1.09 g of *C,N*-diphenylnitrone in 20 mL of ethanol containing 0.02 mL of concentrated HCl was stirred at 25° for 10 minutes. Filtration afforded a 59% yield of the cycloadduct as yellow crystals, mp 115–116° (ethanol). ^1H NMR (CS_2) δ:5.83 (d, $J = 2.0$ *Hz*, 1*H*), 6.83 (d, $J = 2.0$ Hz, 1*H*), 7.0–8.1 (m, 18*H*); mass spectrum m/z 400 (M^+). Anal. Calcd. for $C_{28}H_{20}N_2O$: C, 83.97; H, 5.03; N, 7.00%; Found: C, 83.77; H, 5.24; N, 6.71%

13-Methyl-12-oxa-13-azatricyclo[8.3.0.0^{1,7}]tridecane (156, *m* = 2, *n* = 1) (*C*-Alkenylnitrone Cycloaddition of a Keto Olefin and an *N*-Substituted Hydroxylamine).[137] A solution of 2-(3-butenyl)cycloheptanone (520 mg, 3.13 mmol), *N*-methylhydroxylamine hydrochloride (449 mg, 3.74 mmol), and potassium hydroxide (238 mg, 4.24 mmol) in methanol (20 mL) was heated at reflux for 2 days, cooled, and diluted with diethyl ether. The precipitate was filtered, and the filtrate was concentrated. The residual oil was distilled with a Kugelrohr apparatus, affording the product (514 mg, 84%). ^1H NMR ($CDCl_3$) δ: 3.82 (t, $J = 8.0$ Hz, O-$CH_AH_BCH_x$, 1*H*), 3.23 (dd, $J = 4.5$ and 8.0 Hz, $OCH_AH_BCH_x$, 1*H*), 2.52 (s, CH_3N, 3*H*). An analytical sample was obtained by short-path distillation at 140°/17 torr. Anal. Calcd. for $C_{12}H_{21}NO$: C, 73.79; H, 10.83; N, 7.17. Found: C, 73.65; H, 10.92; N, 7.00.

1,11β-Dimethyl-2aβ,3,4,5,5aβ,8,9,10,11aα,11b,11cβ-dodecahydro-2*H*-isoxazolo[5,4,3-*k,l*]benzo[*b*]quinolizin-6-one (Heteroatom-Linked *C*-Alkenylnitrone Cycloaddition of a Keto Olefin and an *N*-Substituted Hydroxylamine).[140] A solution of 2-acetyl-1-(cyclohexenylcarbonyl)piperidine (24.3 g), *N*-methylhydroxylamine hydrochlorine (17.2 g), diisopropylethylamine (36.0 mL), and ethanol (115 mL) was refluxed for 34 hours, evaporated, diluted with water (450 mL), and extracted ($CHCl_3$). Chromatography of the extract (Al_2O_3, elution with diethyl ether, or silica gel and elution with 1% CH_3OH–$CHCl_3$ followed by crystallization (diethyl ether) gave 13.94 g (51%) of product as colorless prisms, mp 109.5–111.5° (Et_2O–hexanes). IR 1640, 1460, and 1440 cm^{-1}; ^1H NMR δ: 4.54 (br d, $J = 13$ Hz, H_{8eq}) and 4.28 (br, H_{2a}, $2H$), 3.32 (br dd, $J = 10$ and 2 Hz, H_{11a}), 2.84–2.25 and 2.64 (s, NCH_3, $7H$), 2.08–1.06 and 1.24 (s, CCH_3, $14H$). Anal. Calcd. for $C_{15}H_{24}N_2O_2$: C, 68.15; H, 9.15; N, 10.60). Found: C, 68.10; H, 9.43; N, 10.62.

1,3,3a,11b-Tetrahydro-1-phenylphenanthro[9,10-*c*]isoxazole (119)(*C*-Alkenylnitrone Cycloaddition of an Aldehyde Olefin and an *N*-Substituted Hydroxylamine).[121] A solution containing 550 mg of *N*-phenylhydroxylamine and 1.04 g of 2′-vinyl-2-biphenylcarboxaldehyde in 5 mL of ethanol was allowed to stand at room temperature for 4 hours, at which time a pale-yellow oil had separated and solidified. Recrystallization ($CHCl_3$-hexane) afforded 998 mg (66%) of pure adduct, mp 147–149°. IR (KBr) 6.28, 6.77, 7.91, 8.26, 8.84, 9.26, 9.81, 10.25, 10.64, 11.13, 13.15, 13.61, and 14.34 μm; UV, nm max (ε) (methanol) 266 (17,000) and 301 (1710); ^1H NMR (100 MHz, $CDCl_3$) δ:6.12–6.61 (m, $2H$), 5.50 (dd, $J = 16.0$ and 12.0 Hz, $1H$), 4.86 (d, $J = 6.0$ Hz, $1H$), 2.52–3.11 (m, $11H$), and 2.13–2.36 (m, $2H$); mass spectrum m/z 299 (M^+), 269, 206, 205, 191, 179, 178 (base), 177, 176, and 93. Anal. Calcd. for $C_{21}H_{17}NO$: C, 84.24; H, 5.72; N, 4.68. Found: C, 84.24; H, 5.76; N, 4.68.

2aβ,4aβ,5,6,7,8,8aβ,8bβ-Octahydro-2-benzyl-2*H*,3*H*-thieno[3′,4′,5′:3,3a,4]-cyclohept[*d*]isoxazole (185)(Heteroatom-Linked *C*-Alkenylnitrone Cycloaddition of an Aldehyde Olefin and an *N*-Substituted Hydroxylamine).[145] A solution of 21.30 g (0.125 mol) of 2-[(1-cyclohepten-3-yl)thio]acetaldehyde in 150 mL of acetonitrile was treated with 15.3 g (0.125 mol) of *N*-benzylhydroxylamine and 1 mL of triethylamine. The reaction mixture was heated under reflux for 2 hours, cooled, and evaporated to dryness. The residue was triturated with benzene: ethyl acetate (98:2) to dissolve the product. An insoluble impurity was filtered, and the filtrate was concentrated and chromatographed over silica, using the same solvent system for elution. The product was eluted after a less polar byproduct and yielded 22.80 g (0.083 mol, 66%) of a white solid, mp 56–57° (petroleum ether); IR ($CHCl_3$) 3010, 2920 (CH), 1605, 1500 (aryl), 1228 (C—O), and 700 cm^{-1}; ^1H NMR ($CDCl_3$) δ:7.4–7.2 (m, $5H$), 4.33 (septet, $1H$, CHO), 4.1–3.3 (m, $5H$), 2.91 (d, $2H$, CH_2S), 2.7–1.1 (m, $8H$); mass spectrum m/z 275 (M^+), 288, 91 (base).

12-*exo*-Phenyl-4-oxa-5-azatetracyclo[5.3.1.1.^{2,5}0^{3,9}]dodecane(Table XX, p. 143)(*N*-Alkenylnitrone Cycloaddition of an *N*-Alkenylhydroxylamine and an Aldehyde).[131] A mixture of *N*-(*endo*-bicyclo[3.2.1]oct-6-en-3-ylmethyl)hydroxylamine (153 mg, 1 mmol), benzaldehyde (153 mg, 1.44 mmol), and 4Å molecular sieves (0.6 g) in xylene (3 mL) was heated under argon at reflux for 11 hours. The solvent was removed under vacuum, and the residue was chromatographed on a silica gel column (diethyl ether–hexane) to afford the cycloadduct as 229 mg (95%) of colorless crystals, mp 85–86° (methylene chloride–hexane). IR

(KBr) 3050, 1600, 1500, 1450, 1350, 1310, 1250, 1050, 1010, 960, 920, 780, 740, and 690 cm^{-1}; ^{1}H NMR ($CDCl_3$) δ: 7.6–7.1 (m, 5*H*), 4.90 (dd; $J = 9.6$, 1 Hz), 4.20 (s, 1*H*), 3.6–2.6 (m, 3*H*), 2.6–1.4 (m, 9*H*); mass spectrum m/z 242 (25), 241 (75) (M^+), 213 (27), 212 (71), 170 (33), 118 (46), 117 (41), 104 (37), 91 (100).

TABULAR SURVEY

The contents of the tables were derived by searching *Chemical Abstracts.* An extensive computer search from 1967 to May 1985 was performed on the CAS Registry File, using the isoxazolidine substructure. The previous literature was searched by hand.

The tabular organization of the intermolecular nitrone–olefin cycloaddition is first divided into acyclic and cyclic nitrones. In the former, the class is further broken down into aldehyde-derived and ketone-derived nitrones. Since the reactions of aldehyde-derived acyclic nitrones are so extensive, additional organization is based on the degree and nature of the substituents on the olefin partner of acyclic olefins. In addition, a separate table is used for nitrone–acetylenes.

For aldehyde-derived nitrones and cyclic olefins, the tables were divided into exo- and endocyclic olefins, as well as by the numbers of rings and the nature of the ring junction (bridged or fused). The addition of specialized tables in this context, such as aldehyde-derived nitrones and endocyclic olefins, is incorporated.

The cyclic nitrones are divided into exo- and endocyclic classes and further separated depending on whether they reacted with a cyclic or acyclic olefin. Finally, a miscellaneous table of reactions is used to describe certain examples that do not fit rationally into the existing tables.

The individual entries are ordered based on increasing carbon number of the *C*-nitrone substituent. In cases of equal carbon number, the order is by increasing hydrogen number. For examples of identical nitrones, the total carbon number of the olefin partner is used as the basis for the ordering. In the event of identical nitrones with olefins of equal carbon number, the number of hydrogens is employed to determine the sequence.

The intramolecular cycloaddition reactions are grouped into tables according to how the unsaturation is linked to the nitrone. Entries with the chain containing the unsaturation attached to the nitrone nitrogen are grouped in Table XX; accordingly, Table XXII lists those that are attached to the nitrone carbon. Cyclic nitrones are organized separately in Table XXI since they could be considered in either of the two preceding tables. The entries in these tables are listed in order of increasing number of: (1) atoms between the participating unsaturation and carbonyl groups, (2) rings in the reactant, (3) carbons and heteroatoms in the rings, and (4) unsaturations in the rings.

In the yield column, numbers in parentheses represent isolated yield, whereas numbers separated by a colon are ratios. A dash in the tables indicates that the data were not provided.

The following abbreviations are used in the tables:

Ac	acetyl
BDMS	*tert*-butyldimethylsilyl
CBz	carbobenzyloxy
DMF	*N,N*-dimethylformamide
DMSO	dimethyl sulfoxide
Ether	diethyl ether
MEM	methoxyethoxymethyl
THP	tetrahydropyranyl
TMS	trimethylsilyl

TABLE I. Aldehyde-Derived Nitrones and Monosubstituted Olefins

A.

R^1	R^2	R^3	Reaction Conditions	Product(s) and Yield(s) (%)	Ref.
H	CH_3	CN	C_2H_5OH, 25°, 18 h	**A** (63)	73
"	"	CO_2CH_3	C_2H_5OH, 25°, 18 h	**A** (85)	73
"	"	C_6H_5	C_2H_5OH, 25°, 18 h	**A** (74)	73
"	"	1-Naphthyl	C_2H_5OH, 25°, 18 h	**A** (49)	73
"	t-C_4H_9	NO_2	25°	**A** (—)	48
"	"	CN	—	**A** (—)	68
"	"	CO_2CH_3	—	**A** (—)	68
"	"	$C_6H_5SO_2$	—	**A**:**C** = 70:30	68
"	"	$C_6H_5SO_2$	$CHCl_3$, 25°	**A**:**C** = 70:30	48
"		$CO_2C_2H_5$	Neat, reflux, 1.5 d	**A** (84)	175
"	"	n-C_4H_9	Neat, reflux, 1.5 d	**A** (78)	175
"	"	C_6H_5	Neat, reflux, 1.5 d	**A** (89)	175
"	"	$CBzNHCHCO_2CH_3$	Neat, reflux, 1.5 d	**A** (90)	175
"		C_6H_5	—	**A** (—)	176

”	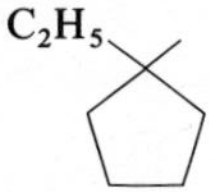 C_2H_5	”	—	**A** (—)	176
”	$C_6H_5CHCH_3$	CH_2OH	C_6H_6, reflux, 1.5 d	**A** (86)[a]	177
”	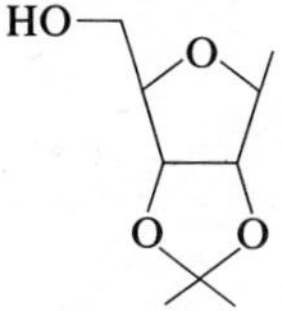 HO—	$n\text{-}C_6H_{13}$	Neat, 85°, 45 min	**A, B** (82)[a]	178
”	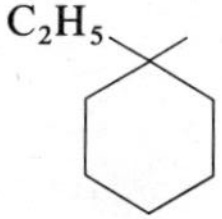 C_2H_5	C_6H_5	C_6H_6, 25°, 2 h	**A** (95)	176
”	$t\text{-}C_4H_9C(CH_3)_2$	C_6H_5	—	**A** (—)	176
”	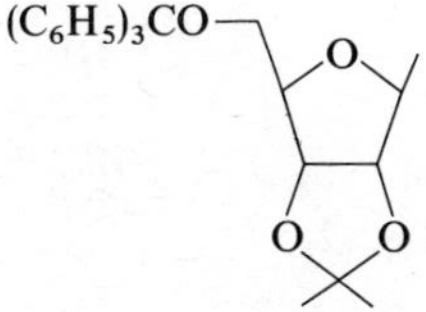 $(C_6H_5)_3CO$—	CO_2CH_3	$CHCl_3$, 70°, 0.5 h	**A, B** (58, 34)[a]	176
$CONH_2$	C_6H_5	CH_2Cl	—	**A, B** (35)	179
”	”	CH_2Br	—	**A, B** (59)	179
”	”	CO_2CH_3	—	**A, B** (67)	179
”	”	CH_2CN	—	**A, B** (78)	179
”	”	C_6H_5	—	**A, B** (68)	179

TABLE I. ALDEHYDE-DERIVED NITRONES AND MONOSUBSTITUTED OLEFINS (*Continued*)

R^1	R^2	R^3	Reaction Conditions	Product(s) and Yield(s) (%)	Ref.
$CONH_2$	C_6H_5	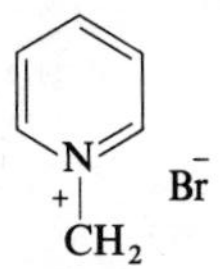	—	**A, B** (26)	179
CH_3	C_2H_5	C_2H_5O	Neat, 120°, 96 h	**A, B** (42)	88
"	n-C_4H_9	CN	—	**A, B** (58)	173
"	"	CO_2CH_3	—	**A, B** (64)	173
"	C_6H_5	C_6H_5	C_2H_5OH, 60°, 45 h	**A, B** (66)	77
CO_2CH_3	CH_3	CHO	C_6H_6, 25°, 24 h	**A, B** (—)	170
"	$C_6H_5CH_2$	"	C_6H_6, 25°, 24 h	**A, B** (61, 39)	170
"	$C_6H_5CHCH_3$	C_6H_5	C_6H_6, 25°, 1 h	**A, B** (95)	180
"	$(C_6H_5)_2CH$	CHO	C_6H_6, 25°, 24 h	**A, B** (81, 19)	170
$Si(CH_3)_2OTMS$	C_6H_5	CN	Neat, 25°, 3 weeks	**A, B** (—)	181
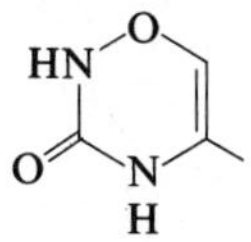	"	$CONH_2$	—	**A, B** (—)	182
"	"	CH_2OH	DMF, 85°, 12 h	**A, B** (100)	182
"	"	CO_2CH_3	—	**A, B** (—)	182
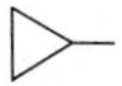	CH_3	CN	$CHCl_3$, 63°	**A** + **B**:**C** + **D** = 2:1	70
"	"	CO_2CH_3	$CHCl_3$, 63°	**A** + **B**:**C** + **D** = 4:1	70
"	"	C_6H_5	$CHCl_3$, 63°	**A, B** (—)	70
"	"	$C_6H_5SO_2$	$CHCl_3$, 63°	**A** + **B**:**C** + **D** = 1:1.6	70
$CO_2C_2H_5$	"	C_2H_5O	Neat, 80°, 72 h	**A** (92)	89

"	"	TMS	C_6H_6, 25°, 24 h	**A, B** (85)	183
"	$C_6H_5CH_2$	CO_2CH_3	C_6H_6, 25°, 36 h	**A, B** (67, 17)	60
n-C_3H_7	C_2H_5	CN	—	**A, B** (48)	173
"	$(CH_2)_2OH$	CO_2CH_3	—	**A, B** (45)	173
"	n-C_3H_7	CN	—	**A, B** (56)	173
"	"	C_6H_5	—	**A, B** (46)	173
"	C_6H_5	$CO_2C_2H_5$	C_2H_5OH, 85°, 60 h	**A, B** (96)	77
"	"	C_6H_5	C_2H_5OH, 65°, 43 h	**A, B** (99)	77
"	C_6H_{11}	CO_2CH_3	Neat, 85°, 24 h	**A, B** (85)	77
"	"	C_6H_5	Neat, 98°, 22 h	**A, B** (98)	77
i-C_3H_7	i-C_3H_7CHCN	CN	" " 20 h	**A, B** (91)	78
"	"	CH_2Cl	" " 18 h	**A, B** (93)	78
"	"	$CONH_2$	" " 20 h	**A, B** (92)	78
"	"	CO_2CH_3	" " 7 h	**A, B** (93)	78
"	"	$CO_2C_2H_5$	" " —	**A, B** (97)	78
"	"	n-C_4H_9	" " 65 h	**A, B** (64)	78
"	"	n-C_4H_9O	" " 25 h	**A, B** (97)	78
"	"	4-ClC_6H_4	" " 200 h	**A, B** (—)	78
"	"	C_6H_5	" " 11 h	**A, B** (80)	78
"	"	$C_6H_5CH_2$	" " 54 h	**A, B** (—)	78
"	"	$C_6H_5CH_2$	" " 54 h	**A, B** (—)	78
2-Nitro-5-furyl	CH_3	4-Pyridyl	C_2H_5OH, reflux, 30 h	**A, B** (26)	74
"	"	2-Propenyl	C_2H_5OH, reflux, 24 h	**A, B** (31)	74
"	C_6H_5	C_6H_5	C_2H_5OH, reflux, 5 h	**A, B** (41)	74
"	"	"	Toluene, reflux, 2 h	**A, B** (98)	74
"	"	4-Pyridyl	Toluene, reflux, 6 h	**A, B** (88)	74
"	"	CH_2OH	C_2H_5OH, reflux, 72 h	**A, B** (16)	74
"	"	"	Toluene, reflux, 3.5 h	**A, B** (95)	74
	"	CH_2Br	Toluene, reflux, 2 h	**A, B** (66)	74
"	"	$CONH_2$	Toluene, reflux, 4 h	**A, B** (77)	74
2-Furyl	CH_3	TMS	C_6H_6, reflux, 24 h	**A, B** (89)	183
"	C_6H_5	CN	—	**A, B** (65)	184
"	"	C_6H_5	—	**A, B** (65)	184
"	"	C_2H_5O	—	**A, B** (—)	185

TABLE I. ALDEHYDE-DERIVED NITRONES AND MONOSUBSTITUTED OLEFINS (*Continued*)

R^1	R^2	R^3	Reaction Conditions	Product(s) and Yield(s) (%)	Ref.
H N O NH O	C_6H_5	CO_2CH_3	DMF, 85°, 5 h	**A, B** (100)	186
"	"	CH_2OH	DMF, 85°, 12 h	**A, B** (100)	186
"	"	$CONH_2$	DMF, 85°, 7 h	**A, B** (100)	186
"	"	C_6H_5	DMF, 85°, 9 h	**A, B** (100)	186
"	"	Pyridyl[a]	DMF, 85°, 8 h	**A, B** (100)	186
t-C_4H_9	$C_6H_5CHCH_3$	C_6H_5	15 h	**A, B** (48, 42)[a]	180
2-FurylCH=CH	C_6H_5	C_6H_5	CH_3OH, reflux, 30 h	**A, B** (57)	187
2-Pyridyl	"	C_2H_5O	—	**A, B** (—)	185
3-Pyridyl	CH_3	CO_2CH_3	Neat, reflux, 2 h	**A**:**B**:(**C** + **D**) = 8:1:1	169
"	C_6H_5	C_2H_5O	—	**A, B** (—)	185
4-Pyridyl	"	"	—	**A, B** (—)	185
$CO_2C_4H_9$-n	$C_6H_5CHCH_3$	C_6H_5	C_6H_6, reflux, 1 h	**A, B** (95)[b]	180
$Si(CH_3)_2OTMS$	C_6H_5	CN	Neat, 25°, 3 weeks	**A, B** (25)	181
4-ClC_6H_4	CH_3	C_6H_5	Neat, 25°, 80 h	**A, B** (99)	77
3-$O_2NC_6H_4$	C_6H_5	n-$C_{14}H_{29}$	—	**A, B** (67)	188
4-$O_2NC_6H_4$	"	C_2H_5O	—	**A, B** (—)	185
"	"	$C_6H_5SO_2$	C_6H_6, reflux, 30 h	**C** (75)	189
C_6H_5	CH_3	NO_2	—	**C**:**D** = 1:2	86
"	"	"	60°	**C, D** (—)	48
"	"	CN	Neat, 95°, 16 h	**A, B** (91)	81
"	"	"	—	**A**:**B** = 77:23	86
"	"	CO_2CH_3	—	**A**:**B**:**C**:**D** (4:77:15:4)	86
"	"	C_2H_5O	Neat, 80°, 72 h	**A**:**B** (39:39)	190
"	"	$CO_2C_2H_5$	Neat, 110°, 21 h	**A, B** (99)	81
"	"	TMS	C_6H_6, reflux, 24 h	**A, B** (95)	183
"	"	n-C_4H_9	—	**A, B** (—)	191

"	"	"	Neat, 70°, 48 h	**A, B** (83)	192
"	"	2-Pyridyl	Toluene, 110°	**A, B** (100)	192
"	"	C_6H_5	Neat, 85°, 15.5 h	**A, B** (32, 63)	192
"	"	$C_6H_5SO_2$	—	(**A** + **B**):(**C** + **D**) = 32:68	68
"	"	p-$CH_3C_6H_4SO$	C_6H_6, reflux, 20 h	**C** (49)[c]	193
"	"	n-C_8H_{17}	—	**A, B** (—)	191
"	"	n-$C_{10}H_{21}$	—	**A, B** (—)	191
"	"	n-$C_{12}H_{25}$	—	**A, B** (—)	191
"	"	n-$C_{14}H_{29}$	—	**A, B** (—)	191
"	t-C_4H_9	NO_2	—	**C, D** (50)	86
"	"	CN	—	**A, B** (—)	86
"	"	CO_2CH_3	—	**A** (91)	86
"	"	CO_2[(−)-menthyl]	Neat, 80°, 1 h	**A**:**B** = 4:1	194
"	$CH_3CHCO_2C_2H_5$	CO_2CH_3	3 h	**A, B** (95)	180
"	C_6H_5	CN	Neat, 20°, 48 h	**A** (100)	81
"	"	CN	—	(**A** + **B**):(**C** + **D**) = 91:9	79
"	"	CHO	—	(**A** + **B**):(**C** + **D**) = 82:18	79
"	"	CH_3	—	(**A** + **B**):(**C** + **D**) = 92:2	79
"	"	"	—	(**A** + **B**):(**C** + **D**) = 83:17	79
"	"	CO_2CH_3	Toluene, 110°	**A, B, C** (97)	82, 107
"	"	"	Neat, 20°, 8 h	**A** (100)	81
"	"	CH_3CO_2	80°, 72 h, (dark)	**B** (70)	89
"	"	C_2H_5O	—	**A, B** (—)	185
"	"	Imidazol-1-yl	C_6H_6, 130°, 48 h	**A, B** (62)	195
"	"	$CO_2C_2H_5$	—	(**A** + **B**):(**C** + **D**) = 70:30	79
"	"	"	Toluene, 4 h	**A, B** (100)	81
"	"	C_6H_5	—	**B** (100)	79
"	"	"	Neat, 60°, 40 h	**A**:**B** = 1:9 (95)	77
"	"	"	C_6H_6, 60°, 40 h	**A, B** (82)	195
"	"	$C_6H_5SO_2$	C_6H_6, reflux, 24 h	**C** (80)	189
"	"	p-$CH_3C_6H_4SO$	C_6H_6, reflux, 20 h	**C** (57)[d]	193
"		"	—	**A** (low)	196
"	"	$C_2H_5O_2C(CH_2)_8$	—	**A, B** (30)	188

TABLE I. ALDEHYDE-DERIVED NITRONES AND MONOSUBSTITUTED OLEFINS (*Continued*)

R^1	R^2	R^3	Reaction Conditions	Product(s) and Yield(s) (%)	Ref.
C_6H_5	CH_3	9-Acridinyl	Xylene, reflux, 1.5 h	**A, B** (11)	114
"	"	n-$C_{14}H_{29}$	Xylene, reflux, 24 h	**A, B** (67)	188
"	p-$CH_3C_6H_4$	CO_2[(−)-menthyl]	Neat, 80°, 1 h	**A, B** (—)	194
"	$C_6H_5CH_2$	CO_2[(−)-menthyl]	Neat, 80°, 1 h	**A, B** (—)	194
"	$C_6H_5CHCH_3$	CO_2CH_3	C_6H_6, 4 h	**A, B** (29, 40)[b]	180
"	"	C_6H_5	C_6H_6, 85°, 15 h	**A, B** (11, 78)[b]	197
2-HOC_6H_4	C_6H_5	$CH_3O_2C(CH_2)_8$	Xylene, reflux, 12–24 h	**A, B** (50)	188
3-HOC_6H_4	"	"	Xylene, reflux, 12–24 h	**A, B** (50)	188
3-HOC_6H_4	"	n-$C_{14}H_{29}$	Xylene, reflux, 12–24 h	**A, B** (18)	188
4-HOC_6H_4	"	$CH_3O_2C(CH_2)_8$	Xylene, reflux, 12–24 h	**A, B** (50)	188
[structure]	CH_3	TMS	C_6H_6, reflux, 24 h	**A, B** (90)	198
	$C_6H_5CH_2$	CH_3CO_2	Neat, reflux, 36 h, dark	**A, B** (14, 56)	199
	"	OC_2H_5	Neat, 35°, 72 h	**A** (93)	199a
"	"	"	Neat, reflux, 72 h	**A** (93)	199
n-C_6H_{13}	CH_3	TMS	Benzene, reflux, 24 h	**A, B** (78)	183
"	C_6H_5	$CH_3O_2C(CH_2)_8$	Xylene, reflux, 12–24 h	**A, B** (60)	188
4-$O_2NC_6H_4$	CH_3	CN	$CHCl_3$–C_6H_6, 20°, 1 d	**A, B** (75)	72
"	C_6H_5	CO_2CH_3	$CHCl_3$–C_6H_6, 20°, 10 min	**A, B** (100)	72
"	"	n-C_4H_9	$CHCl_3$–C_6H_6, 20°, 48 h	**A, B** (82)	72
"	"	C_6H_5	$CHCl_3$–C_6H_6, 20°, 24 h	**A, B** (99)	72
C_6H_5CO	"	CN	$CHCl_3$–C_6H_6, 20°, 1 d	**A, B** (78)	72
"	"	CO_2H	$CHCl_3$–C_6H_6, 20°, 2 h	**A, B** (76)	72
"	"	CH_3	$CHCl_3$–C_6H_6, 20°, 18 h	**A, B** (99)	72

”	”	CH_2OH	$CHCl_3$–C_6H_6, 20°, 24 h	**A, B** (84)	72
”	”	$CH{=}CH_2$	$CHCl_3$–C_6H_6, 20°, 20 h	**A, B** (100)	72
”	”	CO_2CH_3	$CHCl_3$–C_6H_6, 20°, 15 min	**A, B** (100)	72
”	”	*n*-C_4H_9	$CHCl_3$–C_6H_6, 20°, 1 d	**A, B** (93)	72
”	”	C_6H_5	$CHCl_3$–C_6H_6, 20°, 30 min	**A, B** (73)	72
”	”	2-Pyridyl	Ether, reflux, 6.5 h	**A, B** (83)	46
”	”	$CH_2NHC_6H_5$	Ether, reflux, 14 h	**A, B** (89)	46
”	C_6H_{11}	2-Pyridyl	Ether, reflux, 24 h	**A, B** (45)	46
N, N H	”	CO_2CH_3	Dioxane, 90°, 27 h	**A, B** (50)	200
”	”	C_6H_5	Dioxane, 90°, 27 h	**A, B** (40)	200
4-$CH_3C_6H_4$	C_6H_5	Imidazol-1-yl	C_6H_6, 130°, 48 h	**A, B** (58)	195
3-$CH_3OC_6H_4$	”	*n*-$C_{14}H_{29}$	Xylene, reflux, 12–24 h	**A, B** (22)	188
4-$CH_3OC_6H_4$	”	Imidazol-1-yl	C_6H_6, 130°, 48 h	**A, B** (50)	195
”	”	$C_6H_5SO_2$	C_6H_6, reflux, 24 h	**C** (95)	189
”	”	CO_2[(−)-menthyl]	Neat, 80°, 1 hr	**A**:**B** = 70:30	194
”	4-$CH_3OC_6H_4$	9-Acridinyl	Xylene, reflux, 1.5 h	**A, B** (11)	114
t-C_4H_9, CO, HO, C_4H_9-*t*	C_6H_5	CN	—	**A, B** (61)	190
”	”	C_6H_5	—	**A, B** (76)	190

TABLE I. ALDEHYDE-DERIVED NITRONES AND MONOSUBSTITUTED OLEFINS (*Continued*)

R^1	R^2	R^3	Reaction Conditions	Product(s) and Yield(s) (%)	Ref.
t-C_4H_9, NHCO, HO, C_4H_9-t	C_6H_5	CN	—	**A, B** (75)	190
"	"	C_6H_5	—	**A, B** (65)	190
"	"	CN	—	**A, B** (63)	190
"	"	C_6H_5	—	**A, B** (56)	190
$C_6H_5CH{=}CH$	"	CN	Neat, 3 d	**A, B** (66)	187
"	"	CH_2Cl	Toluene, reflux, 20 h	**A, B** (65)	187
"	"	C_6H_5	Toluene, reflux, 26 h	**A, B** (30)	187
2,4,6-$(CH_3)_3C_6H_2$	CH_3	NO_2	—	**C, D** (—)	68
"	"	CN	—	(**A** + **B**):(**C** + **D**) = 15:85	68
"	"	CO_2CH_3	—	(**A** + **B**):(**C** + **D**) = 1:1	68
2,4,6-$(CH_3)_3C_6H_2$	CH_3	$C_6H_5SO_2$	—	**C, D** (—)	68
CO_2[(−)-menthyl]	"	C_6H_5	C_6H_6, reflux, 3 h	**A**:**B** = 4:1 (60)	194
"	t-C_4H_9	"	C_6H_6, reflux, 3 h	**A**:**B** = 38:62 (96)	194
"	$C_6H_5CH_2$	"	C_6H_6, reflux, 3 h	**A**:**B** = 67:33 (85)	194
"	$(C_6H_5)_2CH$	"	C_6H_6, reflux, 1.5 h	**A**:**B** = 83:17 (87)	194
9-Acridinyl	C_6H_5	4-ClC_6H_4	Xylene, reflux, 1.5 h	**A, B** (41)	114
"	"	C_6H_5	Xylene, reflux, 1.5 h	**A, B** (37)	114
S, O, N, H, H H, O^-, S^+, N, O, $(C_6H_5)_2CHO_2C$	CH_3	$CO_2C_2H_5$	Neat, 80°, 1 h	**C, D** (52)	201

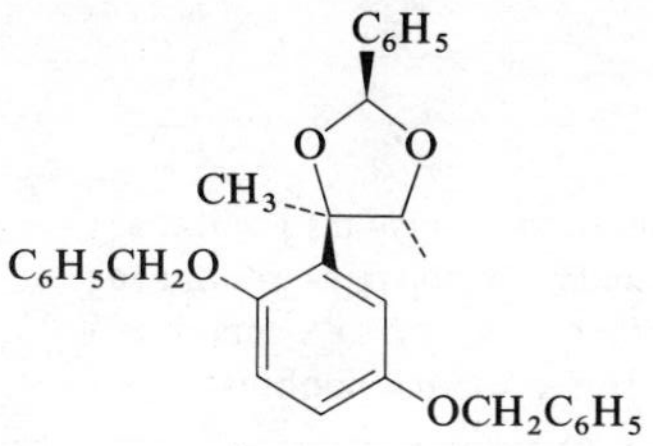 $C_6H_5CH_2O$ / $OCH_2C_6H_5$ (structure, C_6H_5, CH_3)	"	TMS	80°, 24 h	**A, B** (88)	183

B.

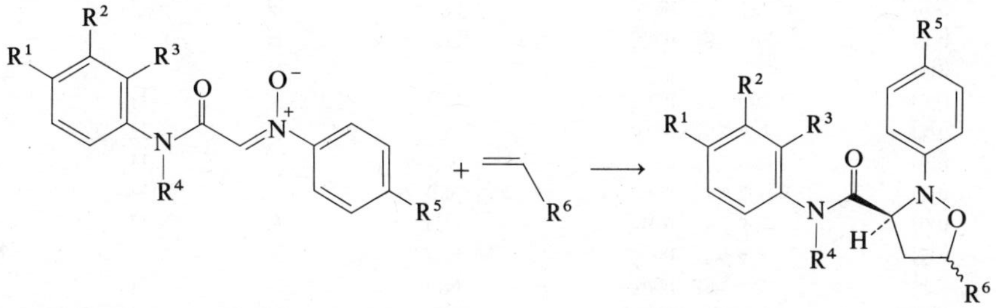

R^1	R^2	R^3	R^4	R^5	R^6	Reaction Conditions	Product and Yield (%)	Ref.
H	H	NO_2	H	H	CN	Neat, 25°, 10 h	(90)	84
NO_2	H	H	H	H	CN	Neat, 25°, 10 h	(88)	84
H	H	H	H	H	CN	Neat, 25°, 3 h	(86)	84
H	H	H	H	H	CH_2Br	—	(66)	179
H	H	H	H	H	CH_2NCS	—	(98)	179
H	H	H	H	H	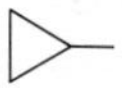 cyclopropyl	—	(—)	202

TABLE I. ALDEHYDE-DERIVED NITRONES AND MONOSUBSTITUTED OLEFINS (*Continued*)

R^1			R^2		R^3	Reaction Conditions	Product(s) and Yield(s) (%)	Ref.
H	H	H	H	H	N^+–CH_2 Br^-	—	(83)	179
H	H	H	H	$(CH_3)_2N$	CN	Neat, 25°, 400 h	(26)	84
H	OH	H	H	H	CN	Neat, 25°, 4 h	(81)	83
OH	H	H	H	H	CN	Neat, 25°, 4 h	(44)	83
H	H	CH_3	H	H	CN	Neat, 25°, 4 h	(50)	84
CH_3	H	H	H	H	CN	Neat, 25°, 3 h	(60)	84
H	H	CH_3	H	$(CH_3)_2N$	CN	Neat, 25°, 400 h	(10)	84
CH_3	H	H	H	$(CH_3)_2N$	CN	Neat, 25°, 360 h	(36)	84
CH_3O	H	H	H	H	CN	Neat, 25°, 4 h	(56)	84
CH_3O	H	H	H	$(CH_3)_2N$	CN	Neat, 25°, 400 h	(37)	84
H	H	CH_3CO_2	H	H	CN	—	(60)	203
CH_3CO	H	H	H	H	CN	Neat, 25°, 10 h	(98)	84
H	H	H	C_6H_5	H	CN	Neat, 25°, 18 s	(92)	84

[a] The substitution pattern of this vinylpyridine is not clearly stated in the paper.
[b] The products are mixtures of diastereomers arising from optically active R^2 group.
[c] The ratio of sulfoxide diastereomers is 9:1.
[d] The ratio of sulfoxide diasteromers is 18:1.

TABLE II. ALDEHYDE-DERIVED NITRONES AND 1,1-DISUBSTITUTED OLEFINS

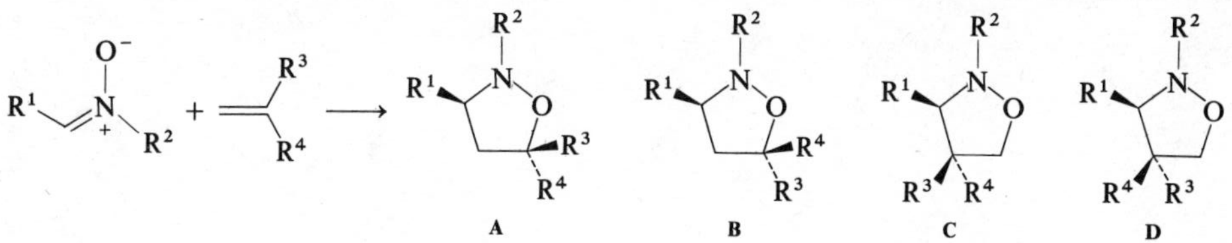

A B C D

R^1	R^2	R^3	R^4	Reaction Conditions	Product(s) and Yield(s) (%)	Refs.
H	CH_3	CH_3	CO_2CH_3	C_2H_5OH, 25°, 18 h	**A, B** (52)	73
H	"	$CO_2C_2H_5$	$CO_2C_2H_5$	—	**A** (—)	79
H	"	C_6H_5	CH_3	C_2H_5OH, 25°, 18 h	**A, B** (45)	73
H	"	"	CO_2CH_3	C_2H_5OH, 25°, 18 h	**A, B** (79)	73
H	"	"	$CO_2C_2H_5$	C_2H_5OH, 25°, 18 h	**A, B** (90)	73
H	"	"	C_6H_5	C_2H_5OH, 25°, 18 h	**A, B** (9)	73
H	"	"	C_6H_5CO	C_2H_5OH, 25°, 18 h	**A, B** (86)	73
H	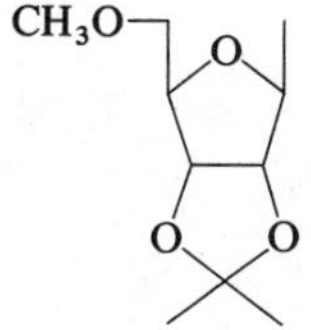 CH_3O— (structure)	CH_3	CO_2CH_3	Neat, 60°, 3 h	**A, B** (82)[a]	178
H	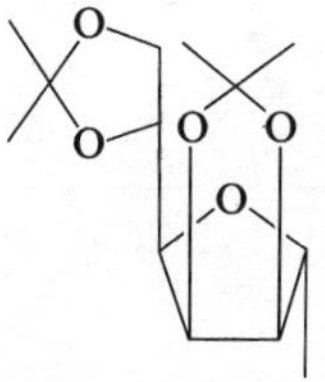 (structure)	"	"	Neat, 65°, 3 h	**A, B** (98)[a]	178

TABLE II. ALDEHYDE-DERIVED NITRONES AND 1,1-DISUBSTITUTED OLEFINS (*Continued*)

R^1	R^2	R^3	R^4	Reaction Conditions	Product(s) and Yield(s) (%)	Refs.
H	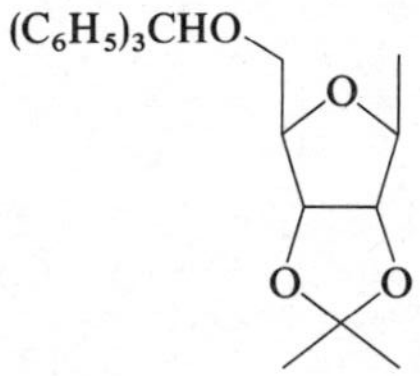 $(C_6H_5)_3CHO$–[structure]	CH_3	CO_2CH_3	Neat, 65°, 2.5 h	**A, B** (94)[a]	204
CH_3	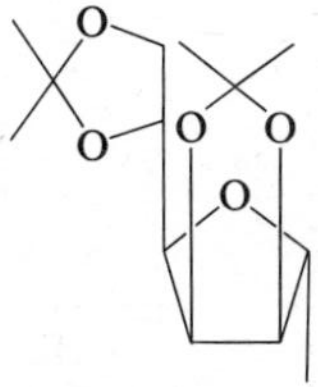 [structure]	"	"	Neat, 65°, 3 h	**A, B** (—)[a]	178
"	"	"	"	Neat, 60°, 16 h	**A, B** (—)[a]	178
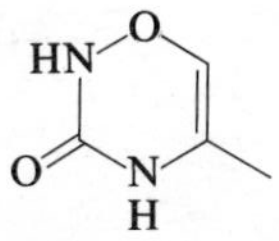 [structure]	C_6H_5	"	"	—	**A, B** (—)	182
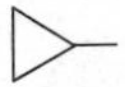 [structure]	CH_3	"	"	$CHCl_3$, 63°	(**A** + **B**):(**C** + **D**) = 1:4	70
$CO_2C_2H_5$	$C_6H_5CH_2$	"	CH_3	Neat, 80°, 3 d	**A** (—)	60
n-C_3H_7	C_6H_{11}	"	CO_2CH_3	Toluene, 60°, 17 h	**A, B** (96)	81

i-C_3H_7	i-C_3H_7CHCN	"	CN	C_6H_6, reflux, 40 h	**A, B** (80)	78
"	"	"	CO_2CH_3	C_6H_6, reflux, 9 h	**A, B** (90)	78
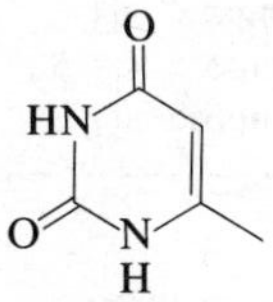	C_6H_5	"	"	DMF, 85°, 5 h	**A, B** (86)	186
5-Nitro-2-furyl	C_6H_5	CH_3	CH_2Cl	Toluene, reflux, 12 h	**A, B** (44)	74
4-ClC_6H_4	"	C_6H_5	1-Pyrrolidinyl	C_6H_6, 60°, 1.5 h	**A, B** (62)	92
4-$O_2NC_6H_4$	"	"	"	"	**A, B** (49)	92
C_6H_5	CH_3	CH_3	CO_2CH_3	Toluene, 105°, 24 h	**A, B** (97)	81
"	"	"	C_6H_5	Neat, 85°, 64 h	**B** (85)	192
"	"	"	"	—	**A**:**B** = 55:45	168
"	"	C_6H_5	"	Neat, 120°, 10 d	**A** (65)	192
"	C_6H_5	CH_3	CO_2CH_3	Neat, 70°, 23 h	**A, B** (97)	81
"	"	C_2H_5O	C_2H_5O	Toluene, 100°, 19 h	**A** (98)	192
"	"	$CO_2C_2H_5$	CH_3CONH	THF, 60°, 26 h	**A, B** (75)	93
"	"	SC_4H_9-t	CN	C_6H_6, 80°, 1 h	**A** + **B, C, D** (85, 1, 8)	91
"	"	$CO_2C_2H_5$	$CO_2C_2H_5$	—	**C, D** (—)	79
"	"	C_6H_5	CH_3	—	**A**:**B** = 1:1	192
"	"	"	1-Pyrrolidinyl	C_6H_6, 60°, 1 h	**A, B** (48)	195, 92
"	"	"	C_6H_5	Neat, 85°, 24 h	**A** (86)	192
"	$C_6H_5CHCH_3$	CH_3	CO_2CH_3	Neat, 70°, 12 h	**A, B** (90)[a]	180
C_6H_5CO	C_6H_5	"	"	$CHCl_3$–C_6H_6, 20°, 10 min	**A, B** (96)	72
"	"	"	i-$C_3H_7N{=}N$	C_6H_6, 25°, 5 d	**A, B** (49)	205
"	"	"	C_6H_5	$CHCl_3$–C_6H_6, 100°	**A, B** (85)	205
C_6H_5NHCO	"	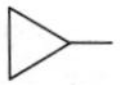	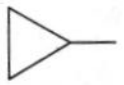	—	**A** (—)	202
$C_6H_5CH{=}CH$	"	CH_3	CO_2CH_3	Toluene, reflux, 10 h	**A, B** (61)	187
2,5$(CH_3O)_2C_6H_3$	"	"	"	$CHCl_3$, reflux	**B** (87)	206

TABLE II. ALDEHYDE-DERIVED NITRONES AND 1,1-DISUBSTITUTED OLEFINS (*Continued*)

R^1	R^2	R^3	R^4	Reaction Conditions	Product(s) and Yield(s) (%)	Refs.
CO_2[(−)-menthyl]	CH_3	C_6H_5	C_6H_5	Toluene, reflux, 5 h	**A** (—)[b]	194
"	t-C_4H_9	"	"	Toluene, reflux, 5 h	**A** (—)[b]	194
"	$C_6H_5CH_2$	"	"	Toluene, reflux, 5 h	**A** (—)[b]	194
"	$(C_6H_5)_2CH$	"	"	Toluene, reflux, 5 h	**A** (—)[b]	194
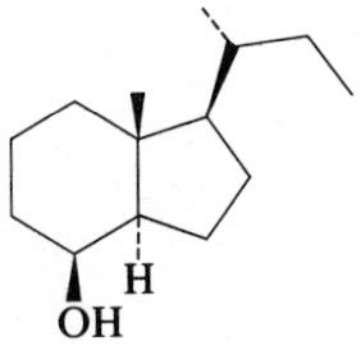 [structure: hydrindane with OH, H]	CH_3	CH_3	CO_2CH_3	Toluene, 90°, 2 h	**A, B** (19, 81)[b]	207
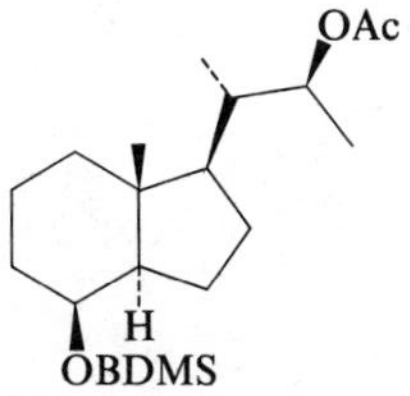 [structure: hydrindane with OAc side chain, H, OBDMS]	t-C_4H_9	"	"	Neat, 50°, 42 h	**A** (100)[c]	207
"	$C_6H_5CH_2$	"	"	Neat, 25°, 15 h	**A** (92)[d]	207

[a] The product is a mixture of diastereomers arising from the optically active R^2 group.
[b] The product is a mixture of diastereomers arising from the optically active R^1 group.
[c] The product is a mixture of diastereomers arising from the optically active R^1 group (ratio = 81:18).
[d] The product is a mixture of diastereomers arising from the optically active R^1 group (ratio = 82:10).

TABLE III. ALDEHYDE-DERIVED NITRONES AND 1,2-DISUBSTITUTED *cis*-OLEFINS

R^1	R^2	R^3	R^4	Reaction Conditions	Product (s) and Yield(s) (%)	Refs.
i-C_3H_7	i-C_3H_7CHCN	$CO_2C_2H_5$	$CO_2C_2H_5$	C_6H_6, reflux, 16 h	**A, B** (50)	78
C_6H_5	CH_3	CO_2CH_3	CO_2CH_3	Toluene, 63°, 70 h	**A** (91)	208
"	"	C_6H_5CO	C_6H_5CO	Toluene, 85°, 3 h	**A** (81)	208
"	C_6H_5	CO_2CH_3	CO_2CH_3	C_6H_6, reflux, 48 h	**A, B** (90, 10)	97
"	"	$CO_2C_2H_5$	$CO_2C_2H_5$	$CHCl_3$–C_6H_6, reflux	**A, B** (—)	206
"	"	C_6H_5	CN	C_6H_6, reflux, 9 d	**A, B** (44, 26)	97
"	"	"	CO_2CH_3	C_6H_6, reflux, 48 h	**A, B** (35, 35)	97, 209
"	"	n-C_5H_{11}	$C_{10}H_{18}CO_2CH_3$	Xylene, reflux, 12–24 h	**A, B** (41)[a]	188
"	"	n-C_8H_{17}	$CH_3O_2C(CH_2)_7$	Xylene, reflux, 12–24 h	**A, B** (16)[a]	188
2,3-$(CH_3O)_2C_6H_3$	"	$CO_2C_2H_5$	$CO_2C_2H_5$	$CHCl_3$–C_6H_6, reflux, 4.5 d	**B** (—)	206
C_6H_5CO	"	CO_2CH_3	CO_2CH_3	—	**B** (100)	97, 97
"	"	"	"	$CHCl_3$–C_6H_6, 20°, 4 d	**A** (99)	72
"	"	C_6H_5	CN	C_6H_6, reflux, 9 d	**B** (10)	97
"	"	"	CO_2CH_3	C_6H_6, reflux, 4 d	**B** (70)	97
"	"	"	"	—	**B** (100)	209

[a] The product is a complex mixture of stereoisomers.

TABLE IV. ALDEHYDE-DERIVED NITRONES AND 1,2-DISUBSTITUTED *trans*-OLEFINS

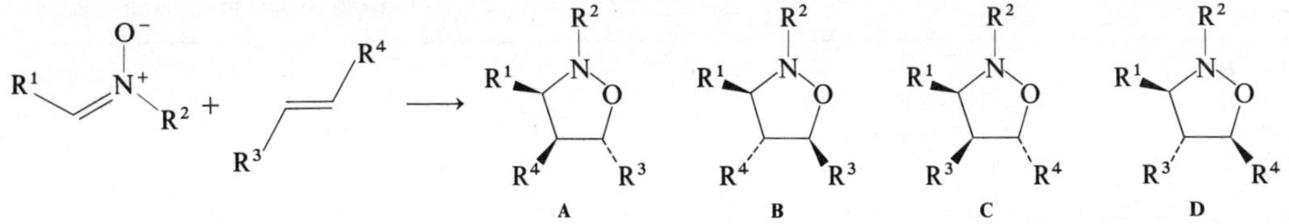

R^1	R^2	R^3	R^4	Reaction Conditions	Product(s) and Yield(s) (%)	Refs.
H	CH_3	n-$C_6F_{13}O_2S$	2-Furyl	C_2H_5OH–H_2O, 60°, 72 h	**A** (22)	210
"	"	"	2-Thienyl	C_2H_5OH–H_2O, 60°, 72 h	**A** (25)	210
"	"	"	4-$CH_3OC_6H_4$	C_2H_5OH–H_2O, 60°, 72 h	**A** (30)	210
$CO_2C_2H_5$	$C_6H_5CH_2$	CH_3	CO_2CH_3	C_6H_6, 25°, 3 d	**A, B** (68, 15)	60
i-C_3H_7	i-C_3H_7CHCN	"	CN	C_6H_6, reflux, 40 h	**A, B** (62)	78
"	"	"	CO_2CH_3	C_6H_6, reflux, 9 h	**A, B** (86)	78
"	"	CO_2CH_3	"	C_6H_6, reflux, 54 h	**A, B** (54)	78
"	"	$CO_2C_2H_5$	CH_3	C_6H_6, reflux, 12 h	**C, D** (63)	78
"	"	"	$CO_2C_2H_5$	C_6H_6, reflux, 24 h	**A, B** (75)	74
4-ClC_6H_4	"	4-ClC_6H_4CO	4-ClC_6H_4CO	—	**A**:**B** = 1:4.7	211
"	"	C_6H_5CO	C_6H_5CO	—	**A**:**B** = 1:5	211
"	"	4-$CH_3OC_6H_4CO$	4-$CH_3OC_6H_4CO$	—	**A**:**B** = 1:4.5	211
4-$O_2NC_6H_4$	CH_3	CH_3	CO_2CH_3	—	**A**:**B** = 1:4	212
"	C_6H_5	"	"	—	**B** (–)	212
"	"	"	$C_6H_5SO_2$	C_6H_6, reflux, 72 h	**B** (61)	189
"	"	4-ClC_6H_4	4-ClC_6H_4	—	**A**:**B** = 1:4.5	211
"	"	C_6H_5	C_6H_5	—	**A**:**B** = 1:4.7	211
"	"	4-$CH_3OC_6H_4CO$	4-$CH_3OC_6H_4CO$	—	**A**:**B** = 1:6.3	211
C_6H_5	CH_3	Cl	NO_2	C_6H_6, 25°, 18 h	**B** (41)[a]	86
"	"	CN	"	C_6H_6, 25°, 24 h	**B**:**C**:**D** = 29:60:11	86
"	"	CO_2CH_3	"	C_6H_6, 25°, 3 h	**A, B, C, D** (30, 30, 25, 15)	86
"	"	"	CH_3	—	**C**:**D** = 1:1	212
"	"	"	CO_2CH_3	$CHCl_3$, 25°, 13 d	**A, B** (90)	208

”	”	$CO_2C_2H_5$	CH_3	Neat, 80°, 36 h	**C, D** (95)	81
”	”	C_6H_5CO	C_6H_5CO	Toluene, 85°, 3 h	**A, B** (85)	208
”	”	$C_6H_5SO_2$	CH_3	C_6H_6, reflux, 24 h	**C**:**D** = 1:16	213
”	t-C_4H_9	CN	NO_2	C_6H_6, reflux, 24 h	**B, D** (57, 43)	86
”	”	CO_2CH_3	”	C_6H_6, reflux, 24 h	**A, B, C, D** (50, 5, 13, 7)	86
”	4-ClC_6H_4	4-ClC_6H_4CO	4-ClC_6H_4CO	—	**A**:**B** = 1:5.2	211
”	C_6H_5	CN	CN	C_6H_6, reflux, 4 h	**B** (100)	97
”	”	CH_3	CH_2OH	—	**A, B** (—)	81
	”	”	CO_2CH_3	—	**A**:**B** = 13:87	212, 97
”	”	”	”	Neat, 100°	**A, B** (92)	81
”	”	CO_2CH_3	CO_2CH_3	C_6H_6, 25°, 2 d	**A, B** (20, 80)	97
”	”	$CO_2C_2H_5$	CH_3	Neat, 100°, 24 h	**C, D** (96)	81
”	”	Imidazol-1-yl	CO_2CH_3	C_6H_6, 130°, 60 h	**A, B** (39)	195
”	”	C_6H_5	NO_2	C_6H_6, reflux, 48 h	**A, B** (—)	97
”	”	”	CN	C_6H_6, reflux, 48 h	**A, B** (4, 74)	97
”	”	$C_6H_5SO_2$	CH_3	C_6H_6, reflux, 24 h	**D** (53)	189
”	”	C_6H_5	CO_2CH_3	C_6H_6, reflux, 48 h	**A, B** (5, 95)	97, 195
”	”	4-$O_2NC_6H_4$	$CO_2C_2H_5$	C_6H_6, 100°, 40 h	**A, B** (90)	81 189
”	”	C_6H_5	$C_6H_5SO_2$	C_6H_6, reflux, 72 h	**B** (60)	195, 81
”	”	”	4-$CH_3C_6H_4SO_2$	C_6H_6, reflux, 40 h	**C, D** (68)[b]	214
”	”	4-ClC_6H_4CO	4-ClC_6H_4CO	—	**A**:**B** = 1:4	211
”	”	C_6H_5CO	C_6H_5CO	C_6H_6, reflux	**A**:**B** = 1:4.2	211
”	”	”	$C_6H_5SO_2$	C_6H_6, reflux, 4 h	**B** (78)	189
”	”	4-$CH_3OC_6H_4CO$	4-$CH_3OC_6H_4CO$	—	**A**:**B** = 1:4.4	211
”	4-$CH_3C_6H_4$	4-ClC_6H_4CO	4-ClC_6H_4CO	—	**A**:**B** = 1:3.9	211
”	”	C_6H_5CO	C_6H_5CO	—	**A**:**B** = 1:3	211
”	4-$CH_3OC_6H_4$	CH_3	CO_2CH_3	—	**A**:**B** = 18:82	212
4-ClC_6H_4CO	4-$(CH_3)_2NC_6H_4$	C_6H_5CO	C_6H_5CO	C_6H_6, 25°, 18 h	**A** (67)	215
4-$O_2NC_6H_4CO$	C_6H_5	CO_2CH_3	CO_2CH_3	Neat, 20°, 1 d	**A, B** (89)	72
C_6H_5CO	”	CN	CN	C_6H_6, reflux, 1 h	**B** (100)	97
”	”	CH_3	CO_2CH_3	$CHCl_3$–C_6H_6, 100°, 3 h	**A, B** (92)	72

TABLE IV. Aldehyde-Derived Nitrones and 1,2-Disubstituted *trans*-Olefins (*Continued*)

R^1	R^2	R^3	R^4	Reaction Conditions	Product(s) and Yield(s) (%)	Refs.
C_6H_5CO	C_6H_5	CO_2CH_3	CO_2CH_3	$CHCl_3$, 20°, 1 h	**B** (70)	72
"	"	"	"	C_6H_6, reflux, 1 h	**B** (100)	97
"	"	C_6H_5	NO_2	C_6H_6, reflux, 6 h	**A, B** (—)	97
"	"	"	CN	C_6H_6, reflux, 48 h	**B** (40)	97
"	"	"	CO_2CH_3	C_6H_6, reflux, 24 h	**B** (100)	209, 97
4-$CH_3OC_6H_4$	CH_3	CH_3	CO_2CH_3	—	**A**:**B** = 3:2	212
"	C_6H_5	"	"	—	**A**:**B** = 3:7	212
"	"	$C_6H_5SO_2$	$C_6H_5SO_2$	C_6H_6, reflux, 8 h	**B** (51)	189
"	"	4-ClC_6H_4CO	4-ClC_6H_4CO	—	**A**:**B** = 1:4.5	211
2,3-$(CH_3O)_2C_6H_3$	"	CO_2CH_3	CO_2CH_3	Neat, 25°, 16 d	**A**:**B** = 8:92	206
2,5-$(CH_3O)_2C_6H_3$	"	"	"	$CHCl_3$, reflux, 40 h	**A, B** (—)	206
$C_2H_5O_2C$ (O O)	$C_6H_5CH_2$	CH_3	$CO_2CH_2C_6H_5$	C_6H_6, reflux, 6 h	**A** (84)	172
"	"	"	"	Toluene, 100°, 5 h	**A, B** (71, 14)	171
CH_3O_2C– Fe $(CO)_3$	CH_3	"	CO_2CH_3	C_6H_6, 60°	**A** or **B** (100)[c]	216
"	"	CO_2CH_3	"	C_6H_6, reflux, 14 h	**A**:**B** = 9:1	216
4-(*i*-C_3H_7)C_6H_4	C_6H_5	CH_3	"	—	**B** (100)	212
"	CO_2CH_3	"	"	—	**A, B** (95)	190

[a] The addition of 1 equivalent of strontium carbonate raises this yield to 85%.
[b] The regiochemical assignment of the stated product presumably should be a mixture of **C** and **D**.
[c] A single diastereomer was obtained, but no specific structure was assigned.

TABLE V. ALDEHYDE-DERIVED NITRONES AND TRISUBSTITUTED OLEFINS

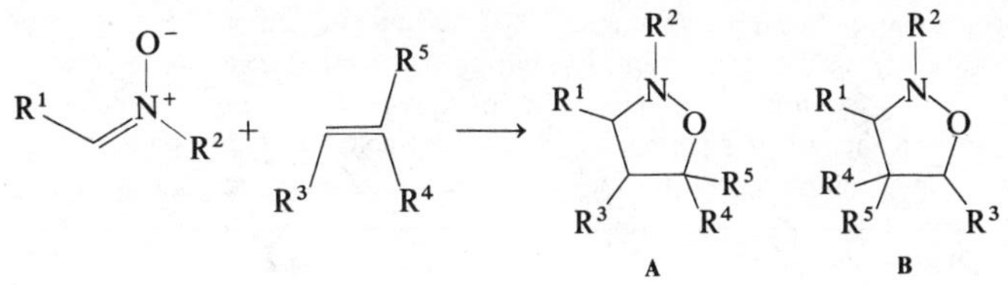

R^1	R^2	R^3	R^4	R^5	Reaction Conditions	Product(s) and Yield(s) (%)	Ref.
2-Pyridyl	C_6H_5	2-Pyridyl	CN	$CO_2C_2H_5$	C_6H_6, reflux, 72 h	**B** (91)	106
"	"	C_6H_5	"	"	Acetone, 25°, 4 d	**B** (70)	106
"	"	2-$O_2NC_6H_4$	"	"	C_6H_6, 25°, 5 d	**B** (50)	106
"	"	4-$O_2NC_6H_4$	"	"	C_6H_6, 25°, 60 h	**B** (90)	106
"	"	4-$CH_3OC_6H_4$	"	"	DMSO, 25°, 5 d	**B** (23)	106
"	"	2-Pyridyl	$CO_2C_2H_5$	"	C_6H_6, reflux, 22 h	**B** (18)	106
"	"	2-$O_2NC_6H_4$	"	"	DMSO, 25°, 10 d	**B** (40)	106
"	"	4-$O_2NC_6H_4$	"	"	DMSO, 25°, 10 d	**B** (40)	106
4-$O_2NC_6H_4$	"	$CO_2C_2H_5$	CH_3	$C_6H_5N{=}N$	C_6H_6, reflux, 24 h	**A** (43)	205
C_6H_5	CH_3	CH_3CO	"	CH_3	Neat, 70°, 69 h	**A** (26)	81
"	"	CO_2CH_3	CO_2CH_3	"	Neat, 85°, 2 d	**A** (86)	208
"	"	"	CH_3	CO_2CH_3	Neat, 90°, 24 h	**A** (100)	208
"	"	"	"	CH_3	Neat, 100°, 48 h	**A** (81)	81
"	C_6H_5	CH_3	"	"	Toluene, 110°	**A** (11)	217
"	"	"	CN	CO_2CH_3	C_6H_6, 25°, 4 d	**B** (70)[a]	218
"	"	CO_2CH_3	CH_3	CH_3	Neat, 100°, 20 h	**A** (97)	81
"	"	CH_3	CO_2CH_3	CO_2CH_3	C_6H_6, 25°, 10 d	**B** (85)[b]	218
"	"	$CO_2C_2H_5$	CH_3	CH_3	Toluene, 100°, 93 h	**A** (92)	81
"	"	CO_2CH_3	CO_2CH_3	CO_2CH_3	C_6H_6, 25°	**B** (100)	107
"	"	"	"	"	C_6H_6, 80°	**A, B** (8, 92)	107

TABLE V. ALDEHYDE-DERIVED NITRONES AND TRISUBSTITUTED OLEFINS (*Continued*)

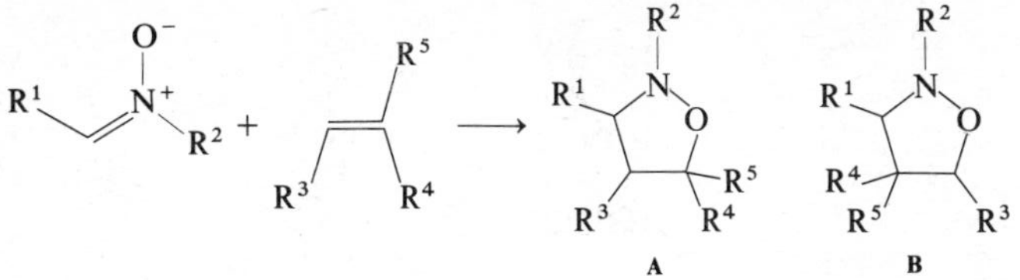

R^1	R^2	R^3	R^4	R^5	Reaction Conditions	Product(s) and Yield(s) (%)	Ref.
C_6H_5	C_6H_5	C_6H_5	CN	CO_2CH_3	C_6H_6, 25°, 4 d	**B** (100)[c]	218
"	"	2-Pyridyl	"	$CO_2C_2H_5$	Ether, reflux, 72 h	**B** (23)	106
"	"	2-$O_2NC_6H_4$	"	"	C_6H_6, 25°, 3 d	**B** (50)	106
"	"	C_6H_5	"	"	DMSO, 25°, 24 h	**B** (25)	106
"	"	"	CO_2CH_3	CO_2CH_3	C_6H_6, 25°, 4 months	**B** (80)	218
"	"	$CO_2C_2H_5$	CH_3	C_6H_5N N	C_6H_6, reflux, 24 h	**A** (24)	205
C_6H_5CO	CH_3	CO_2CH_3	CO_2CH_3	CO_2CH_3	C_6H_6, 25°	**B** (100)	107
"	C_6H_5	CH_3	CN	"	C_6H_6, 25°, 4 d	**B** (65)[d]	218
"	"	"	CO_2CH_3	"	C_6H_6, 25°, 10 d	**B** (83)[e]	218
"	"	CO_2CH_3	"	"	C_6H_6, 25°, 8 h	**A**:**B** = 84:16[f]	107
"	"	C_6H_5	CN	"	C_6H_6, 25°, 19 h	**B** (100)[g]	218
"	"	4-$O_2NC_6H_4$	CN	$CO_2C_2H_5$	C_6H_6, 25°, 14 h	**B** (50)	106
"	"	C_6H_5	CO_2CH_3	CO_2CH_3	C_6H_6, 25°	**B** (20)[h]	187
$C_6H_5CH{=}CH$	"	CH_3CO	CH_3	CH_3	Toluene, reflux, 2 d	**B** (45)	187

[a] The product was obtained as essentially one isomer and was assigned a *cis* relationship between the phenyl and cyano substituents.
[b] The product obtained is essentially a single stereoisomer in which the phenyl and methyl substituents are *trans*.
[c] The product is a 4:1 mixture of stereoisomers with the *cis* phenyl-substituted product predominating.
[d] The product is essentially only the stereoisomer in which the benzoyl and cyano substituents are *cis*. A trace of the *trans* isomer was reported.
[e] The product is primarily the stereoisomer in which the benzoyl and methyl substituents are *trans*. Only a trace of the *cis* isomer was detected.
[f] The **A**:**B** ratio is 75:25 at 80° in benzene after 8 hours.
[g] The ratio of benzoyl-cyano *cis* isomer to the corresponding *trans* isomer is 82:18.
[h] The only product reported has a benzoyl group *trans* to the phenyl substituent.

TABLE VI. Aldehyde-Derived Nitrones and Acetylenes

$R^1CH{=}N^+(O^-)R^2 + R^3C{\equiv}CR^4 \longrightarrow$ **A** + **B**

R^1	R^2	R^3	R^4	Reaction Conditions	Product(s) and Yield(s) (%)	Ref.
H	t-C_4H_9	H	CN	C_6H_6, 25°	**A**:**B** = 1:1	48
"	"	"	CO_2CH_3	—	**A**:**B** = 7:3	68
"	"	"	$CO_2C_2H_5$	CCl_4, 25°	**A**:**B** = 7:3	48
"	"	CO_2CH_3	CO_2CH_3	Neat, 0°	**A** (100)	118
"	"	$(CH_3)_2COH$	H	74°, 10 min	**B** (—)	118
CO_2CH_3	CH_3	H	CO_2CH_3	Ether, 25°, 5 d	**A**, **B** (29, 30)[a]	113
"	"	CO_2CH_3	"	Ether, 25°, 24 h	**A** (70)	113
Cyclopropyl	"	H	$CO_2C_2H_5$	$CHCl_3$, 63°	**A**:**B** = 1:4	70
$CH_3O_2CCH_2$	"	"	CO_2CH_3	—	**A**, **B** (—)	219

TABLE VI. ALDEHYDE-DERIVED NITRONES AND ACETYLENES (*Continued*)

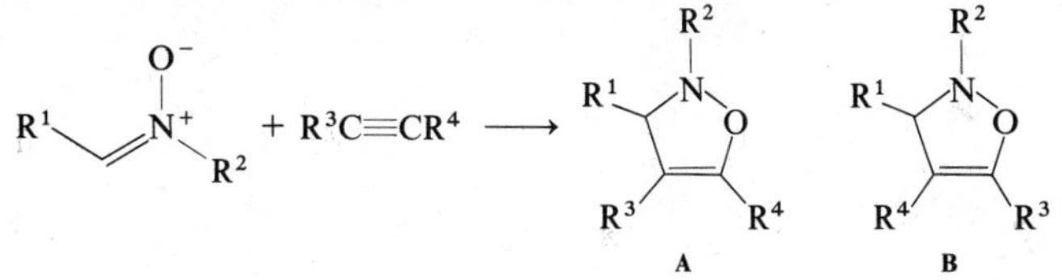

R^1	R^2	R^3	R^4	Reaction Conditions	Product(s) and Yield(s) (%)	Ref.
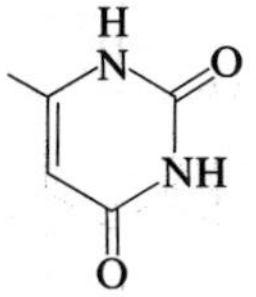	CH_3	H	C_6H_5	Dioxane, 100°, 48 h	**A** (61)	186
"	C_6H_5	"	CN	Dioxane, 100°, 24 h	**A** (48)	186
"	"	"	CH_2OH	Dioxane, 100°, 24 hr	**A** (62)	186
4-ClC_6H_4	4-ClC_6H_4	"	9-Acridinyl	C_2H_5OH/HCl, 25°	**B** (41)	21
C_6H_5	CH_3	"	CN	C_6H_6, 25°	**B** (100)	48
"	"	"	CO_2CH_3	DMF, 85°	**A**:**B** = 42:58	48
"	C_6H_5	"	9-Acridinyl	C_2H_5OH/HCl, 25°	**B** (59)	21
4-$CH_3OC_6H_4$	4-$CH_3OC_6H_4$	"	"	C_2H_5OH/HCl, 25°	**B** (73)	21
2,4,6-$(CH_3)_3C_6H_2$	CH_3	"	CN	—	**B** (—)	68
"	"	"	CO_2CH_3	—	**B** (—)	68
2-($CH_3CH{=}CHCH_2O$)C_6H_4	"	CO_2CH_3	"	Toluene, 25°	**A** (85)	220
2-(2-FurylCH_2O)C_6H_4	"	"	"	C_6H_6, 25°	**A** (100)	220

[a] Product **B** was isolated as a 2:1 adduct with the nitrone adding again to the double bond of the initially formed product (**B**).

TABLE VII. ALDEHYDE-DERIVED NITRONES AND ENDOCYCLIC ENAMINES

A B

R^1	R^2	R^3	R^4	Reaction Conditions	Product(s) and Yield(s) (%)	Ref.
(uracil structure: H, O, N, HN, O)	CH_3	1-Morpholino	H	Dioxane, 85°, 3 d	**A** (72)	186
4-ClC_6H_4	C_6H_5	1-Pyrrolidino	"	C_6H_6, reflux, 4.5 h	**A** (69)	92
"	"	1-Morpholino	"	C_6H_6, reflux, 3 h	**A** (3)	92
"	"	1-Pyrrolidino	CH_3	DMF, 25°, 48 h	**B** (27)	109
3-$O_2NC_6H_4$	"	"	"	DMF, 25°, 48 h	**B** (66)	109
4-$O_2NC_6H_4$	"	"	H	C_6H_6, 65°, 4.5 h	**A** (60)	92
C_6H_5	"	"	"	C_6H_6, 65°, 4.5 h	**A** (61)	92
"	"	"	"	DMF, 25°, 24 h	**A** (55)	110
"	"	1-Morpholino	"	DMF, 25°, 24 h	**A** (7)	110
4-$CH_3OC_6H_4$	"	1-Pyrrolidino	"	C_6H_6, 65°, 4.5 h	**A** (50)	92
2-$O_2NC_6H_4$(H)C=C(H)–	"	1-Morpholino	"	C_6H_6, reflux, 3 h	**A** (25)	108

TABLE VIII. ALDEHYDE-DERIVED NITRONES AND MONOCYCLIC ENDOCYCLIC OLEFINS

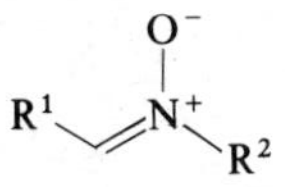

$R^1CH{=}N^+(O^-)R^2$

R^1	R^2	Olefin	Reaction Conditions	Product and Yield (%)	Ref.
H	CH_3	1,3-Cyclohexadiene	C_2H_5OH, reflux, 2 d	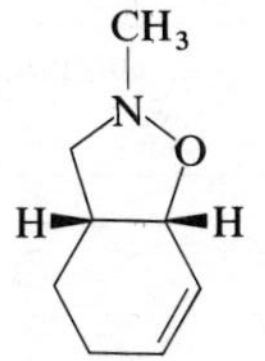 (structure: N-CH$_3$ isoxazolidine, H, H) (78)	101
"	"	1,4-Cyclohexadiene	C_2H_5OH, reflux, 2 d	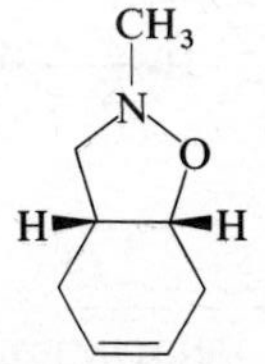 (structure: N-CH$_3$ isoxazolidine, H, H) (55)	101

”	”	1,5-Cyclooctadiene	C_2H_5OH, reflux, 2 d	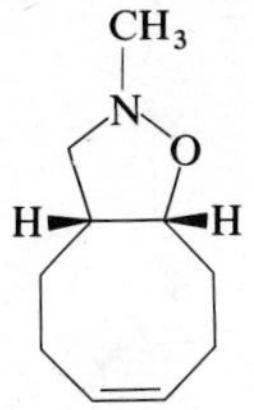 (61)	101
”	C_6H_5	”	C_2H_5OH, reflux, 2 d	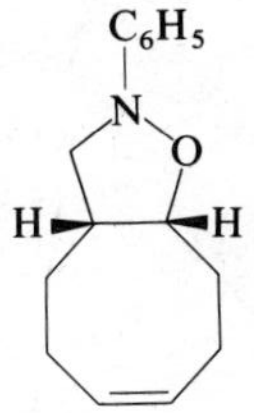 (12)	101
5-Nitro-2-furyl	”	”	Toluene, reflux, 9.5 h	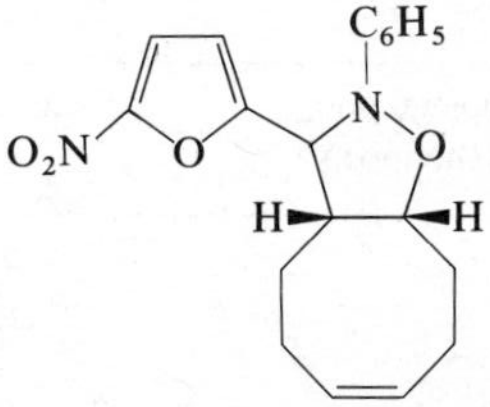 (61)	74

TABLE VIII. ALDEHYDE-DERIVED NITRONES AND MONOCYCLIC ENDOCYCLIC OLEFINS (*Continued*)

A. $R^1CH{=}N^+(O^-)R^2$ + (CH$_2$)$_n$ / X cycloalkene ⟶ **A** + **B**

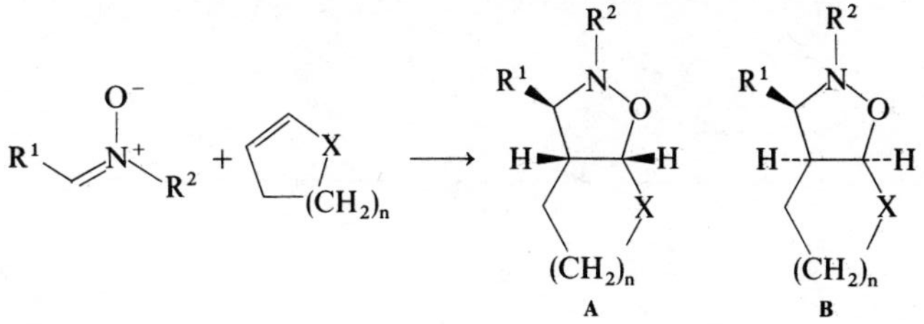

R^1	R^2	*n*	X	Reaction Conditions	Product(s) and Yield(s) (%)	Ref.
H	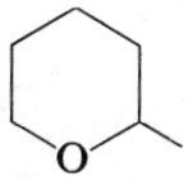	1	CH_2	Neat, reflux, 1.5 d	**A** (59)	175
"	"	2	"	Neat, reflux, 1.5 d	**A** (72)	175
"	"	3	"	Neat, reflux, 1.5 d	**A** (81)	175
"	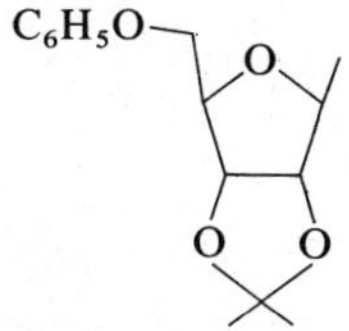 C_6H_5O—	2	"	Neat, 100°, 3 h	**A** (71)	204
5-Nitro-2-furyl	C_6H_5	1	O	—	**A, B** (—)	185
2-Pyridyl	"	"	"	Neat, 5 d	**A, B** (70)	185
3-Pyridyl	"	"	"	—	**A, B** (—)	185
4-Pyridyl	"	"	"	—	**A, B** (—)	185

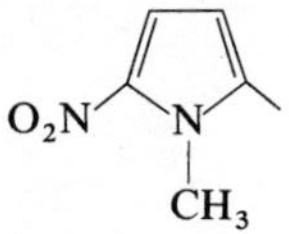 1-methyl-5-nitropyrrol-2-yl (O_2N, N–CH_3)	CH_3	4	CH_2	Toluene, reflux, 16 h	**A, B** (—)	221
4-$O_2NC_6H_4$	C_6H_5	1	O	—	**A, B** (—)	185
C_6H_5	CH_3	"	CH_2	—	**A, B** (72, 17)	222
"	"	2	"	—	**A, B** (68, 10)	222
"	C_6H_5	1	O	Neat, 70°, 24 h	**A, B** (18, 71)	223
C_6H_5CO	"	"	"	Neat, 40°, 8 h	**A, B** (47, 51)	223
"	"	"	S	Toluene, 40°, 15 h	**A, B** (76, 23)	224
"	"	"	CH_2	$CHCl_3$-C_6H_6, 62°, 3 d	**A, B** (49)	72

B.

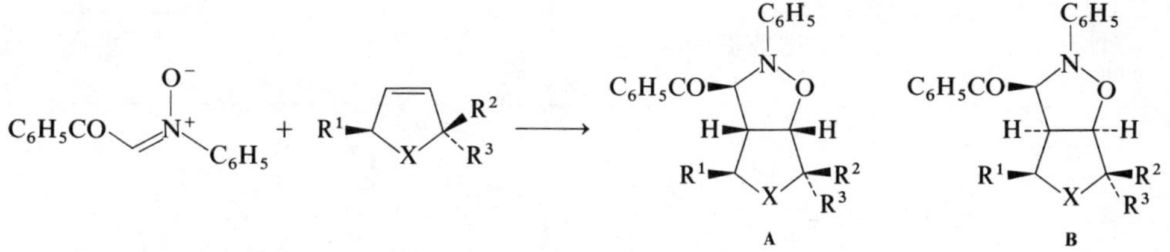

R^1	R^2	R^3	X	Reaction Conditions	Product and Yield(s) (%)	Ref.
H	H	H	O	—	**A, B** (17, 57)	223
"	"	"	S	Toluene, 40°, 15 h	**A**:**B** = 66:34	224
CH_3O	CH_3O	"	O	Neat, 55°, 30 h	**A, B** (17, 16)	223
"	H	CH_3O	"	Neat, 55°, 30 h	**A** (48)	223
CH_3CO_2	CH_3CO_2	H	"	Neat, 40°, 40 h	**A, B** (42, 18)	223

TABLE IX. ALDEHYDE-DERIVED NITRONES AND *N*-PHENYLMALEIMIDES

R^1	R^2	R^3	Reaction Conditions	Product(s) and Yield(s) (%)	Refs.
n-C_3H_7	C_6H_5	C_6H_5	C_2H_5OH, 0°, 1 d	**A**, **B** (92)	174
2-Furyl	"	"	THF, 100°, 10 h	**A**, **B** (—)	225
2-Thienyl	"	"	THF, 100°, 10 h	**A**, **B** (—)	225
4-BrC_6H_4	"	"	Toluene, 80°, 5 h	**A**, **B** (70)	226
4-$O_2NC_6H_4$	"	"	Toluene, 80°, 5 h	**A**:**B** = 2.6:1	226, 227
C_6H_5	CH_3	"	C_6H_6, reflux, 4 d	**A**, **B** (95)	208, 228
"	4-ClC_6H_4	4-$O_2NC_6H_4$	THF, 100°, 10 h	**A**:**B** = 1.6:1 (100)	227
"	"	C_6H_5	THF, 100°, 10 h	**A**:**B** = 3.1:1 (100)	227
"	"	4-$CH_3OC_6H_4$	THF, 100°, 10 h	**A**:**B** = 3.6:1 (100)	227
"	C_6H_5	4-$O_2NC_6H_4$	THF, 100°, 10 h	**A**:**B** = 2.1:1 (100)	227
"	"	C_6H_5	Toluene, 80°, 5 h	**A**:**B** = 1.7:1 (70)	227, 208

"	"	4-$CH_3OC_6H_4$	THF, 100°, 10 h	**A**:**B** = 1.5:1 (100)	227
"	4-$CH_3C_6H_4$	4-$O_2NC_6H_4$	THF, 100°, 10 h	**A**:**B** = 3:1 (100)	227
"	"	C_6H_5	THF, 100°, 10 h	**A**:**B** = 2.9:1 (100)	227
"	"	4-$CH_3OC_6H_4$	THF, 100°, 10 h	**A**:**B** = 2.4:1 (100)	227
"	4-t-$C_4H_9C_6H_4$	4-$O_2NC_6H_4$	THF, 100°, 10 h	**A**:**B** = 3.9:1 (100)	227
"	"	C_6H_5	THF, 100°, 10 h	**A**:**B** = 2.1:1 (100)	227
"	"	4-$CH_3OC_6H_4$	THF, 100°, 10 h	**A**:**B** = 2.1:1 (100)	227
2-HOC_6H_4	C_6H_5	4-$O_2NC_6H_4$	C_6H_6, reflux, 0.5 h	**A**, **B** (31, 46)	229
"	"	C_6H_5	C_6H_6, reflux, 0.5 h	**A**, **B** (22,40)	229
"	"	4-$CH_3C_6H_4$	C_6H_6, reflux, 0.5 h	**A**, **B** (26, 48)	229
3-HOC_6H_4	"	4-$O_2NC_6H_4$	C_6H_6, reflux, 0.5 h	**A**, **B** (22, 34)	229
"	"	C_6H_5	C_6H_6, reflux, 0.5 h	**A**, **B** (22, 34)	229
"	"	4-$CH_3C_6H_4$	C_6H_6, reflux, 0.5 h	**A**, **B** (22, 40)	229
4-HOC_6H_4	"	4-$O_2NC_6H_4$	C_6H_6, reflux, 1.5 h	**A**, **B** (30, 46)	229
"	"	C_6H_5	C_6H_6, reflux, 0.75 h	**A**, **B** (25, 41)	229
"	"	4-$CH_3C_6H_4$	C_6H_6, reflux, 2 h	**A**, **B** (27, 51)	229
C_6H_5CO	"	C_6H_5	CH_2Cl_2–C_5H_6, 20°, 14 h	**A**, **B** (92)	72
4-$CH_3OC_6H_4$	"	"	Toluene, 80°, 5 h	**A**:**B** = 1.4:1 (70)	227, 226
4-HO-3-$CH_3OC_6H_3$	"	4-$O_2NC_6H_4$	C_6H_6, reflux, 0.5 h	**A**, **B** (32, 47)	229
"	"	C_6H_5	C_6H_6, reflux, 1 h	**A**, **B** (23, 40)	229
"	"	4-$CH_3C_6H_4$	C_6H_6, reflux, 2.5 h	**A**, **B** (29, 47)	229
3,4-$(CH_3O)_2C_6H_3$	"	4-$O_2NC_6H_4$	C_6H_6, reflux, 0.5 h	**A**, **B** (29, 47)	229
"	"	C_6H_5	C_6H_6, reflux, 0.5 h	**A**, **B** (23, 42)	229
"	"	4-$CH_3C_6H_4$	C_6H_6, reflux, 0.5 h	**A**, **B** (33, 49)	229
4-$(CH_3)_2NC_6H_4$	"	C_6H_5	Toluene, 80°, 5 h	**A**, **B** (70)	226

TABLE X. Aldehyde-Derived Nitrones and Benzofused Bi- and Tricyclic Olefins

A.

R^1	R^2	R^3	R^4	R^5	X	Reaction Conditions	Product(s) and Yield(s) (%)	Ref.
H	CH_3	H	$CO_2C_2H_5$	H	$(CH_2)_2$	—	**A** (—)	111
"	"	"	$C_2H_5O_2CCH_2$	"	CH_2	—	**A** (—)	111
"	"	"	"	"	$(CH_2)_2$	—	**A** (—)	111
"	"	"	"	"	$(CH_2)_3$	—	**A** (—)	111
C_6H_5	"	"	H	"	CH_2	Neat, 90°, 65 h	**A, B, C, D** (66, 10, 11, 3)	222, 192
"	"	"	"	"	$(CH_2)_2$	Neat, 100°, 139 h	**A, B, C, D** (25, 4, 57, 11)	222, 192
"	C_6H_5	"	"	Cl	SO_2	C_6H_6, reflux, 15 d	**A, B** (5)	230
"	"	"	"	Br	SO	$CHCl_3$, reflux, 15 d	**A, B** (16)	230
"	"	"	"	"	SO_2	$CHCl_3$, reflux, 15 d	**A, B** (30)	230
"	"	"	"	H	"	C_6H_6, reflux, 30 h	**B** (85)	230
"	"	NO_2	CH_3	"	"	Toluene, reflux, 24 h	**A, B** (68)	103
"	"	H	H	CH_3	SO	C_6H_6, reflux, 15 d	(**A** + **B**), (**C** + **D**) (4, 31)	230
"	"	"	"	"	SO_2	C_6H_6, reflux, 13 h	(**A** + **B**), (**C** + **D**) (35, 3)	230

”	”	”	CH_3	H	SO	C_6H_6, reflux, 27 h	**A, B** (71)	230
”	”	”	”	”	SO_2	Toluene, reflux, 22 h	**A, B** (51)	230, 103
”	”	”	CH_3CO	”	”	C_6H_6, reflux, 2 d	**A, B** (16)	230
”	”	”	CH_3	CH_3	SO	C_6H_6, reflux, 15 d	**A, B** (4)	230
”	”	”	”	”	SO_2	$CHCl_3$, reflux, 15 d	**A, B** (7)	230
”	”	”	1-Pyrrolidino	H	”	$Cl_2C{=}CHCl$, reflux, 2 d	**A, B** (85)	230
”	”	”	C_6H_5	”	SO	$Cl_2C{=}CHCl$, reflux, 10 d	**A, B** (58, 4)	230
”	”	”	”	”	SO_2	C_6H_6, reflux, 2 d	**A, B** (60)	230
4-HOC_6H_4	CH_3	”	H	”	”	Toluene, reflux, 20 h	**A, B** (51)	103
C_6H_5CO	C_6H_5	”	”	”	CH_2	CH_2Cl_2–C_6H_6, 40°, 9.5 h	**A, B** (84)	72
$C_6H_5O_2C$	”	”	”	”	O	C_6H_6, 55°, 40 h	**A, B** (35)	231
4-$CH_3OC_6H_4$	”	”	”	”	SO_2	Toluene, reflux, 16 h	**A, B** (86)	103
”	”	”	CH_3	”	”	Toluene, reflux, 22 h	**A, B** (63)	103

B.

R^1	R^2	Reaction Conditions	Yield (%)	Ref.
n-C_3H_7	C_6H_5	C_2H_5OH, reflux, 40 h	(73)	192
C_6H_5	CH_3	C_6H_6, 80°, 48 h	(78)	192
”	C_6H_5	C_6H_6, 80°, 42 h	(94)	192
C_6H_5CO	”	CH_2Cl_2–C_6H_6, 50°, 2 h	(71)	72

TABLE XI. Aldehyde-Derived Nitrones and Bridged Bi- and Tricyclic Olefins

A.

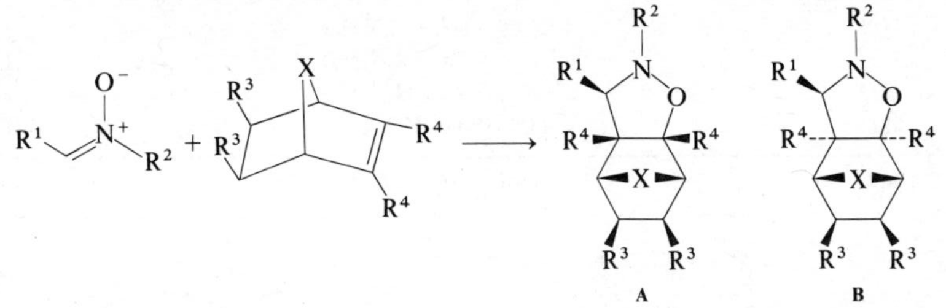

R^1	R^2	R^3	R^4	X	Reaction Conditions	Product(s) and Yield(s) (%)	Ref.
C_6H_5	CH_3	H	H	CH_2	Toluene, reflux, 12 h	**A, B** (79, 9)	232
"	"	"	CO_2CH_3	O	C_6H_6, reflux, 10 d	**A, B** (72, 8)	104
"	"	"	"	CH_2	C_6H_6, reflux, 10 d	**A** (70)	104
"	"	"	"	$(CH_3)_2C{=}C$	C_6H_6, reflux, 10 d	**A** (77)	104
"	C_6H_5	CO_2CH_3	H	O	C_6H_6, 30°, 6 h	**B** (71)	233
"	"	H	CO_2CH_3	"	Ether, 60°, 30 h	**A** (81)	233
C_6H_5CO	"	"	H	CH_2	Neat, 20°, 15 d	**A, B** (35)	72
"	"	CO_2CH_3	"	O	C_6H_6, 40°, 0.5 h	**B** (62)	233
"	"	H	CO_2CH_3	"	C_6H_6, 40°, 30 h	**A** (66)	233

B.

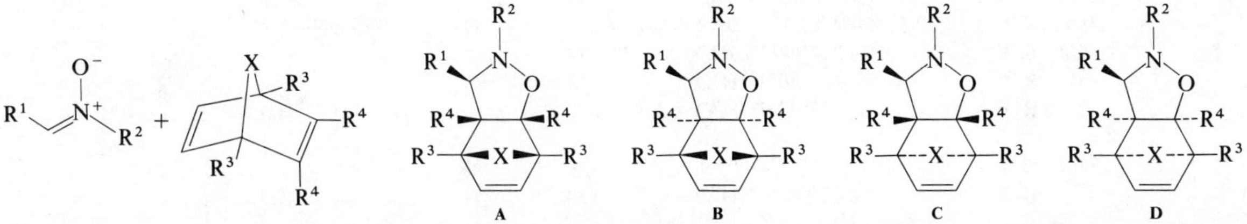

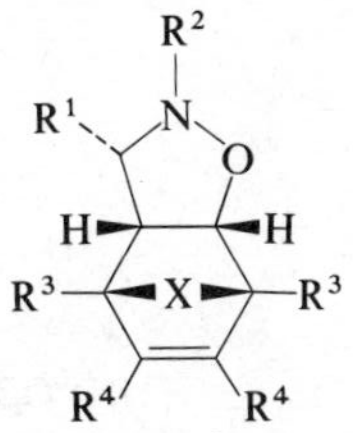

E 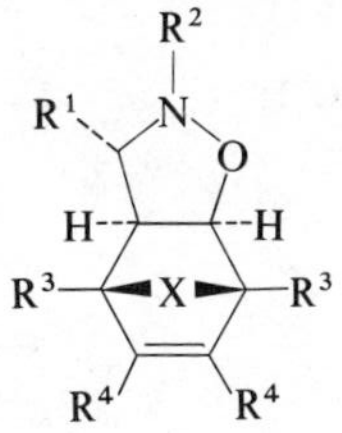 F 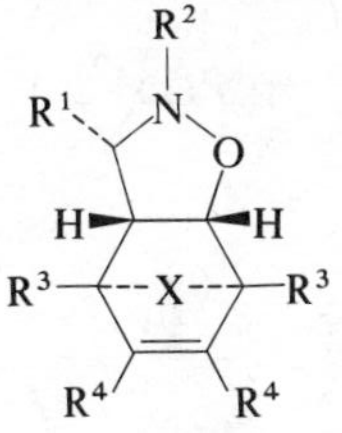 G 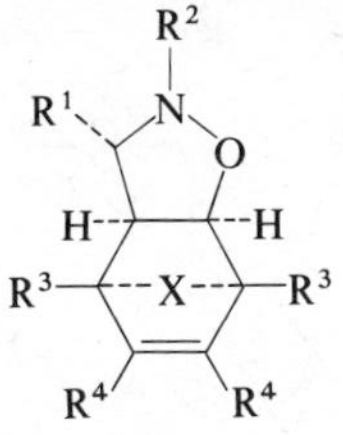 H

R^1	R^2	R^3	R^4	X	Reaction Conditions	Product(s) and Yield(s) (%)	Ref.
4-ClC_6H_4	CH_3	H	CO_2CH_3	O	—	**B**, **C** (100)	234
"	"	"	"	$C_2H_5O_2CN$	—	**B**, **C** (100)	234
C_6H_5	"	"	H	CH_2	Neat, reflux, 4 h	**C** (82)	232
"	"	"	CO_2CH_3	O	C_6H_6, reflux, 10 d	**B**, **C** (25, 60)	104
"	"	"	"	CH_2	C_6H_6, reflux, 10 d	**C** (73)	104
"	"	"	"	$(CH_3)_2C{=}C$	C_6H_6, reflux, 10 d	**C** (76)	104
"	"	Cl	Cl	Cl_2C	Ether, reflux, 5 d	**G** (92)	235
"	"	"	"	ClHC	Ether, reflux, 5 d	**G** (75)[a]	235
"	"	"	"	"	Ether, reflux, 5 d	**E**, **G** (39, 49)[b]	235
"	"	"	"	CH_2	Ether, reflux, 5 d	**E** (90)	235
4-$O_2NC_6H_4$	C_6H_5	H	H	"	C_6H_6, reflux, 18 h	**A**, **B**, **C** (3, 51, 22)	236
C_6H_5	"	"	"	"	Toluene, reflux, 4 h	(**A** + **B**), (**C** + **D**) (32, 41)	236
"	"	"	CO_2CH_3	O	Ether, 40°, 2 h	**C**, **F**, **G** (29, 15, 3)	233
C_6H_5CO	"	"	"	"	C_6H_6, 40°, 10 h	**B**, **C**, **F** (20, 32, 36)	233
$(CH_3)_5C_6$	CH_3	Cl	Cl	Cl_2C	Ether, room temp, 20 d	**G** (90)	235
"	"	"	"	ClHC	Ether, room temp, 20 d	**E**, **G** (36, 54)[a]	235
"	"	"	"	ClHC	Ether, room temp, 20 d	**E** (88)[b]	235
"	"	"	"	CH_2	Ether, room temp, 20 d	**G** (92)	235

TABLE XI. ALDEHYDE-DERIVED NITRONES AND BRIDGED BI- AND TRICYCLIC OLEFINS (*Continued*)

Nitrone	Olefin	Reaction Conditions	Product(s) and Yield(s)(%)	Ref.
C.				
O_2N, O^-, N^+, C_6H_5		Toluene, reflux, 4 h	O_2N, C_6H_5, N, O (47)[c]	74
D. 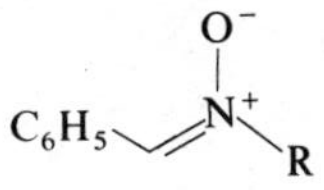 C_6H_5, O^-, N^+, R +	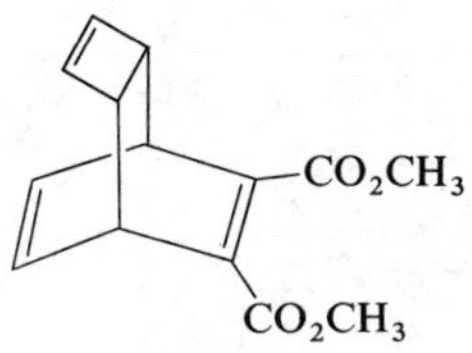 CO_2CH_3, CO_2CH_3	⟶	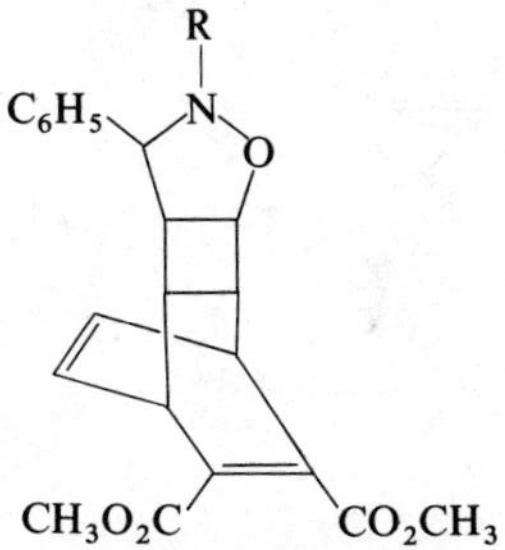 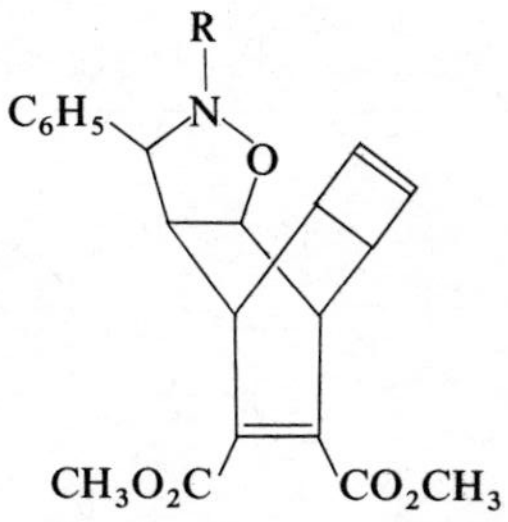 R, C_6H_5, N, O, CH_3O_2C, CO_2CH_3 **A** + R, C_6H_5, N, O, CH_3O_2C, CO_2CH_3 **B**	237

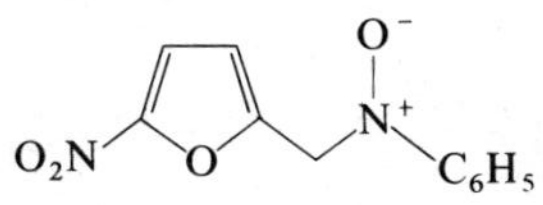

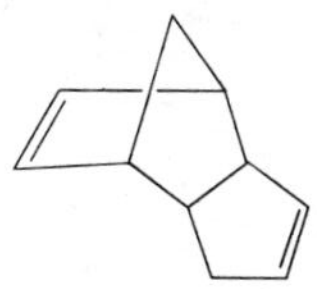

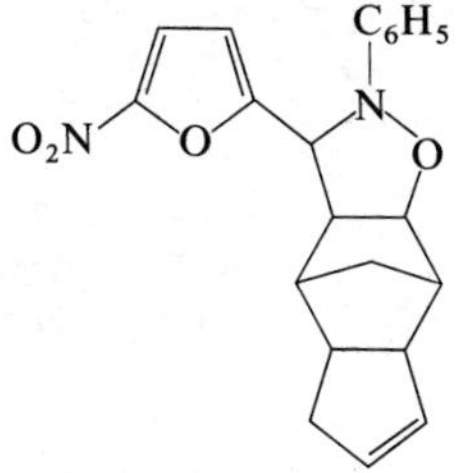

R	Reaction Conditions	Product(s) and Yield(s) (%)	Ref.
CH_3	C_6H_6, reflux, 150 h	**A, B** (61, 22)	237
p-ClC_6H_4	C_6H_6, reflux, 150 h	**A** (82)	237
C_6H_5	C_6H_6, reflux, 150 h	**A, B** (81, 6)	237

E.

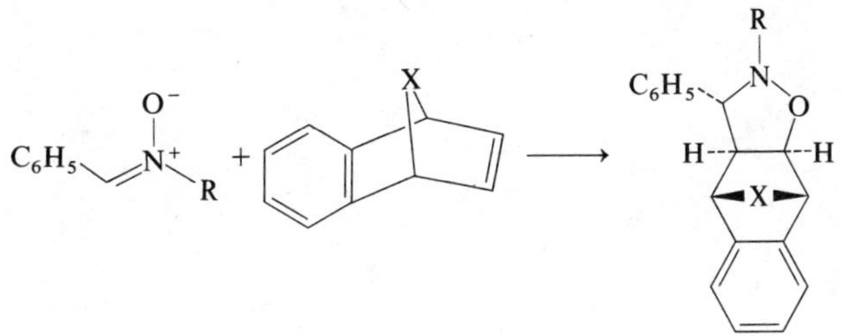

R	X	Reaction Conditions	Product(s) and Yield(s) (%)	Ref.
C_6H_5	t-$C_4H_9O_2CN$	Toluene, reflux, 9 h	(75)	238
$C_6H_5CH_2$	$(CH_3)_2C{=}C$	Toluene, reflux, 15 h	(75)	239
"	t-$C_4H_9O_2CN$	Toluene, reflux, 9 h	(65)	238

[a] The hydrogen of the bridging CHCl group is *syn* to the dichloroolefin moiety.
[b] The hydrogen of the bridging CHCl group is *anti* to the dichloroolefin moiety.
[c] The products are presumably a mixture of *endo* and *exo* adducts.

TABLE XII. ALDEHYDE-DERIVED NITRONES AND EXOCYCLIC OLEFINS

	Nitrone	Olefin	Reaction Conditions	Product(s) and Yield(s) (%)	Ref.
A	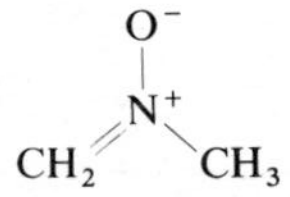 O^-, N^+, CH_2, CH_3	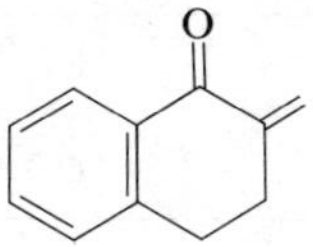 O	C_2H_5OH, 25°, 18 h	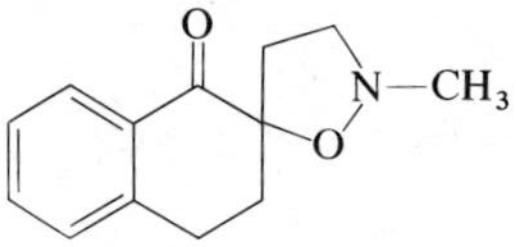 O, N—CH_3, O (63)	73
	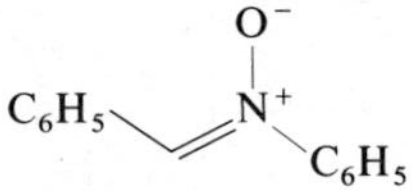 O^-, C_6H_5, N^+, C_6H_5	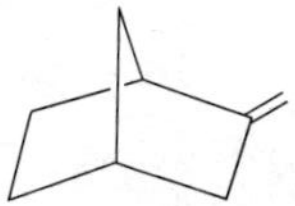	Toluene, reflux, 24 h	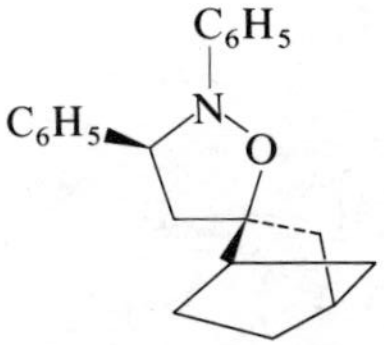 C_6H_5, C_6H_5, N, O (41) 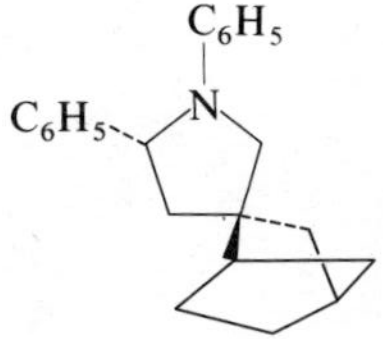 + C_6H_5, C_6H_5, N (41)	94
		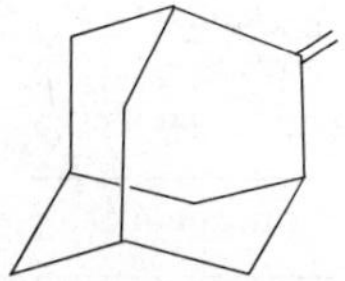	Toluene, reflux, 28 h	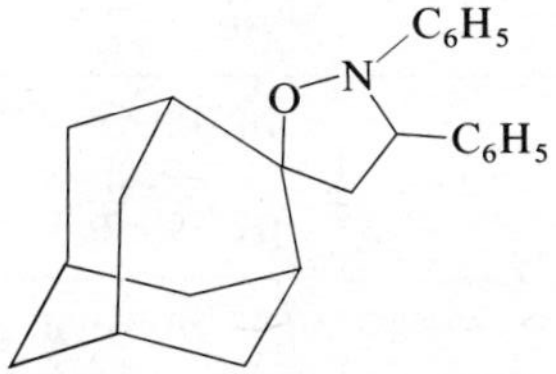 C_6H_5, O—N, C_6H_5 (28)[a]	95

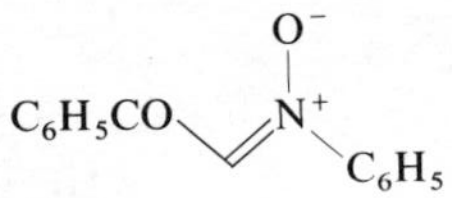

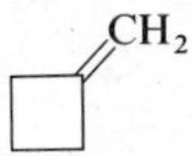

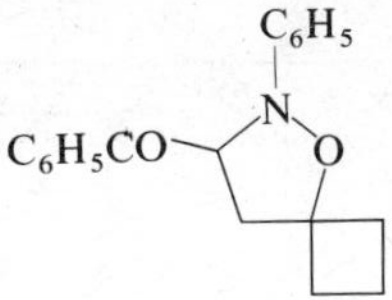

C_6H_5CO–CH=$N^+(O^-)$–C_6H_5	CH_2	—	C_6H_5, N, O, C_6H_5CO (—)	240
	C_6H_5, O, O	$ClCH_2CH_2Cl$, 4 h	C_6H_5, N, O, C_6H_5CO, O, O, C_6H_5 (53)	241
C_6H_5NHCO–CH=$N^+(O^-)$–C_6H_5	CH_2	—	C_6H_5, N, O, C_6H_5NHCO (92)	202
		—	C_6H_5, N, O, C_6H_5NHCO (–)	240

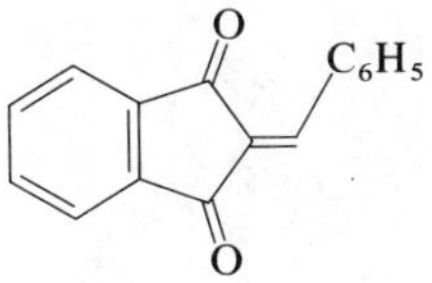

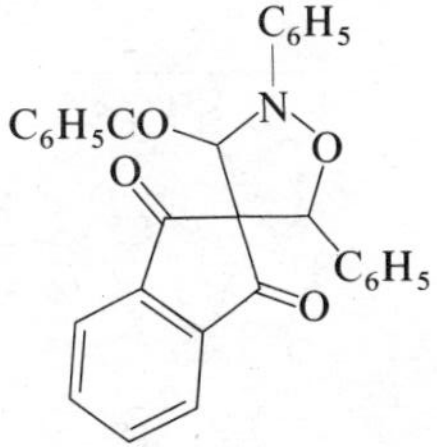

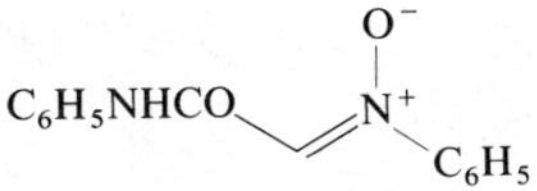

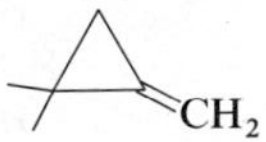

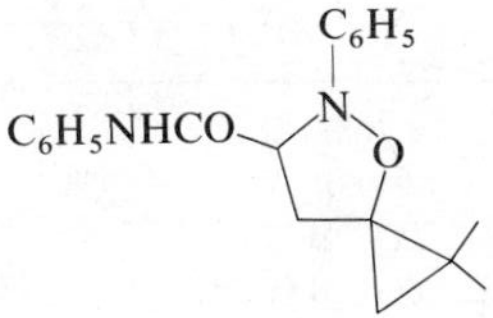

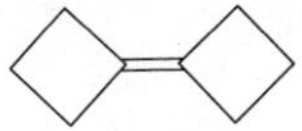

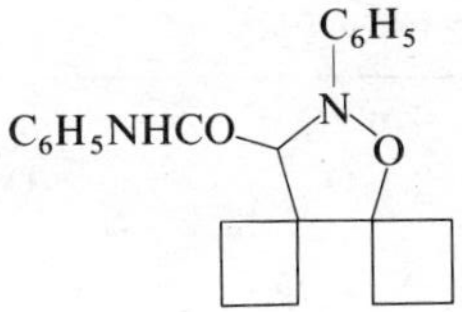

TABLE XII. ALDEHYDE-DERIVED NITRONES AND EXOCYCLIC OLEFINS (*Continued*)

B.

R^1	R^2	Reaction Conditions	Product(s) and Yield(s) (%)	Ref.
C_6H_5	H	Neat, 110°, 3 h	**A, B** (22, 52)	242
"	CH_3	Neat, 100°, 3 h	**A, B** (37, 30)	242
"	H	C_6H_6, 60°, 4 h	**A** (80)	242
2,4,6-$(CH_3)_3C_6H_2$	"	Neat, 100°, 100 h	**B** (20)	242

C.

R^1	R^2	Reaction Conditions	Yield (%)	Ref.
4-ClC_6H_4	4-ClC_6H_4	Xylene, reflux, 10 h	(33)	243
"	C_6H_5	Xylene, reflux, 10 h	(37)	243
C_6H_5	4-ClC_6H_4	Xylene, reflux, 5 h	(51)	243
"	C_6H_5	Xylene, reflux, 18 h	(48)	243

[a] A corrected yield of 43% was also reported since the starting olefin was recovered in part.

TABLE XIII. ALDEHYDE-DERIVED NITRONES AND ENDOCYCLIC OLEFINS

Nitrone	Olefin	Reaction Conditions	Product(s) and Yield(s) (%)	Ref.
A.				
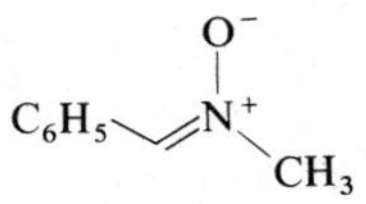 $C_6H_5CH{=}N^+(O^-)CH_3$	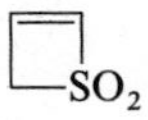 SO_2	Ether, 25°, 1 month	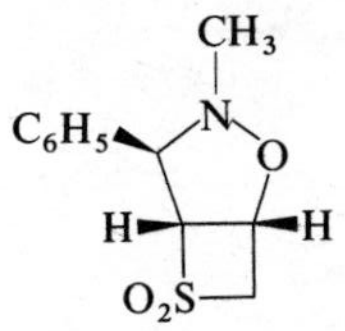 CH_3, N, O, C_6H_5, H, H, O_2S (73)	213
	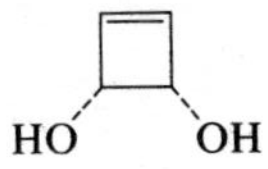 HO, OH	C_6H_6, reflux	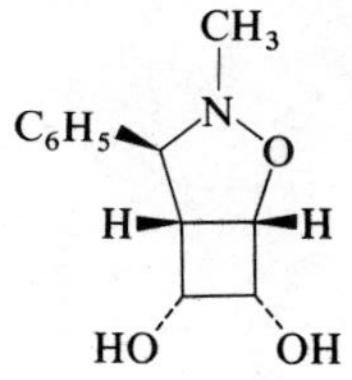 CH_3, N, O, C_6H_5, H, H, HO, OH (38)	244
	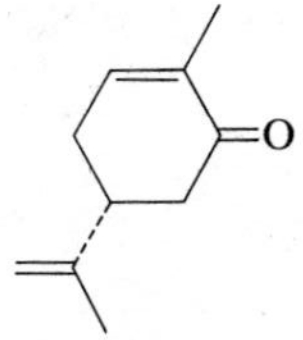 O	Neat, 110°, 46 h	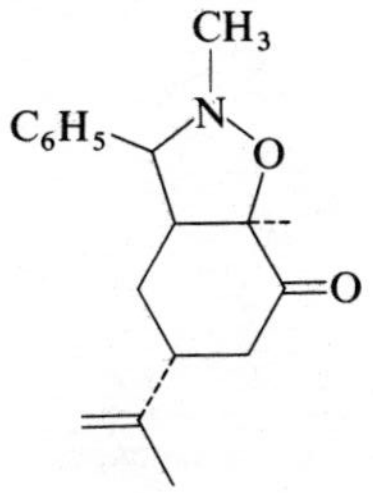 CH_3, N, O, C_6H_5, O (85)	81
	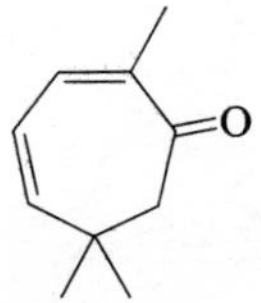 O	Neat, 115°, 128 h	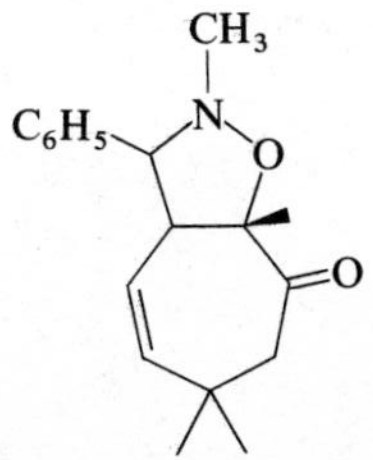 CH_3, N, O, C_6H_5, O (68)	81

TABLE XIII. ALDEHYDE-DERIVED NITRONES AND ENDOCYCLIC OLEFINS (*Continued*)

Nitrone	Olefin	Reaction Conditions	Product(s) and Yield(s) (%)	Ref.
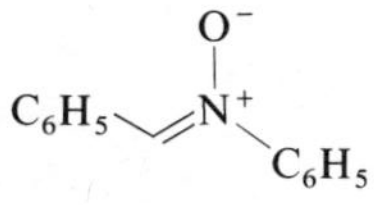	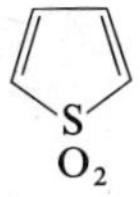	$CHCl_3$, reflux, 4 h	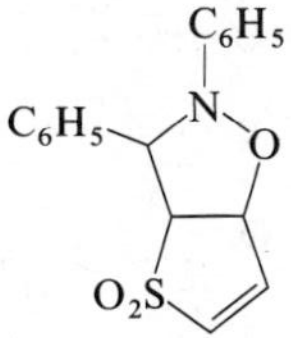 (70)	245
	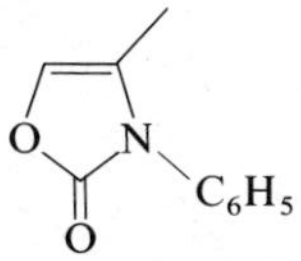	C_6H_6, 60°, 4 h	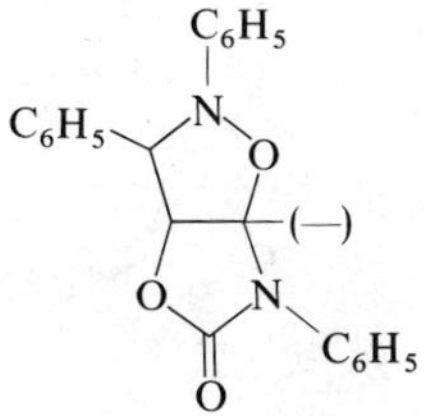 (—)	242
	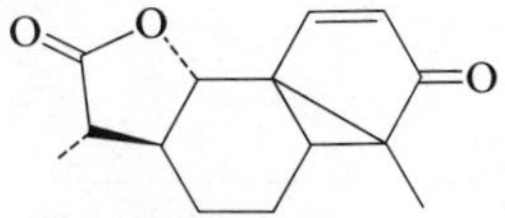	C_6H_6, reflux, 15 h	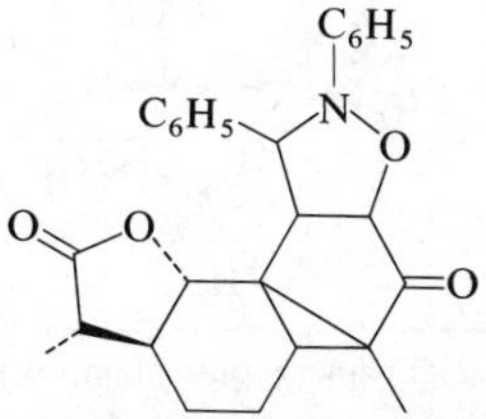 (8)	246

Reactant	Conditions	Product(s) and Yield(s) (%)	Refs.
		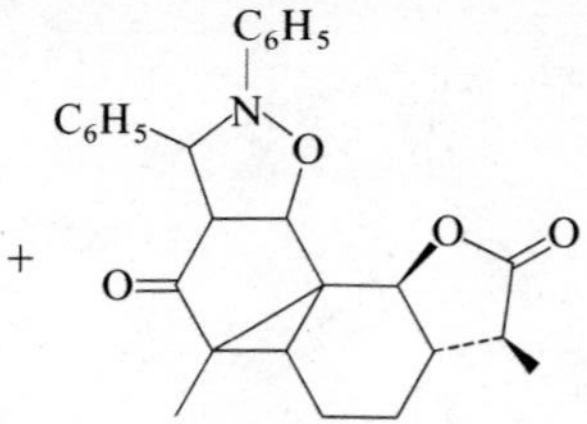 + C_6H_5, C_6H_5, N, O, O, O, O (24)	
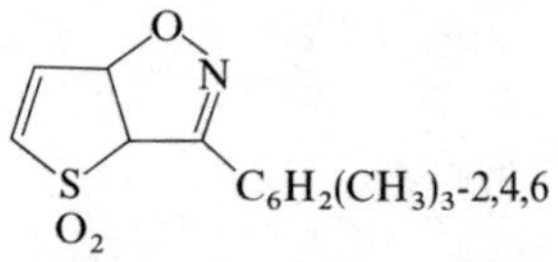 O, N, S, O_2, $C_6H_2(CH_3)_3$-2,4,6	$CHCl_3$, reflux, 3 d	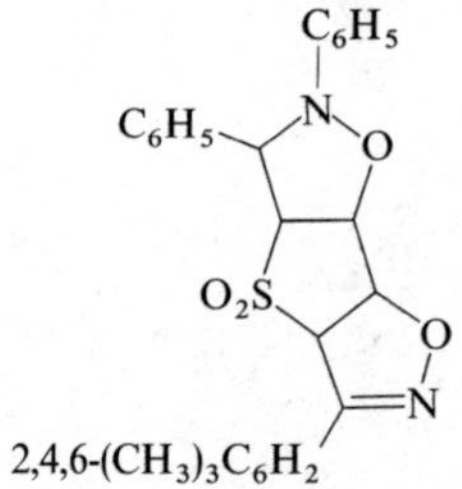 C_6H_5, C_6H_5, N, O, O_2S, O, N, 2,4,6-$(CH_3)_3C_6H_2$ (75)	245
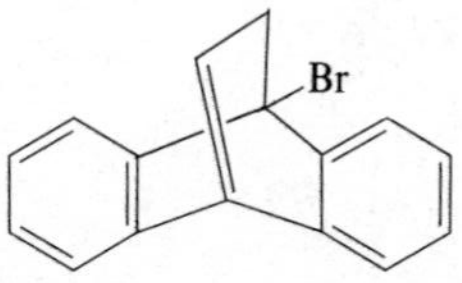 Br	THF, reflux, 5 d	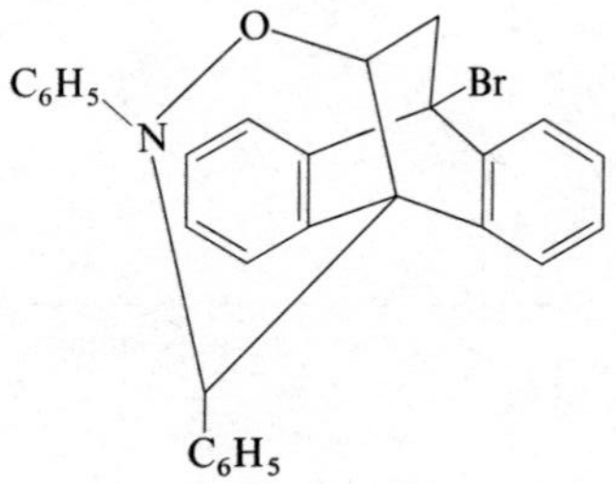 O, C_6H_5, N, Br, C_6H_5 (—)	112
 $P(O)(C_6H_5)_2$	C_6H_6, reflux, 2 h	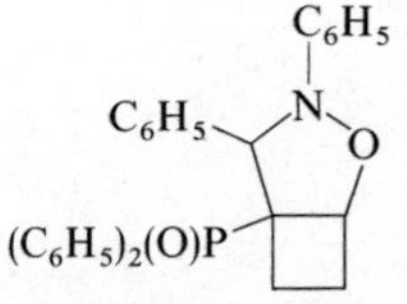 C_6H_5, C_6H_5, N, O, $(C_6H_5)_2(O)P$ (81)	247

TABLE XIII. ALDEHYDE-DERIVED NITRONES AND ENDOCYCLIC OLEFINS (*Continued*)

Nitrone	Olefin	Reaction Conditions	Product(s) and Yield(s) (%)	Ref.
	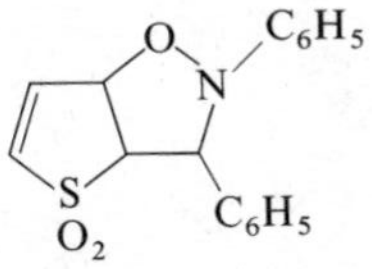 [structure: C_6H_5, C_6H_5, O_2]	$CHCl_3$, reflux, 2 d	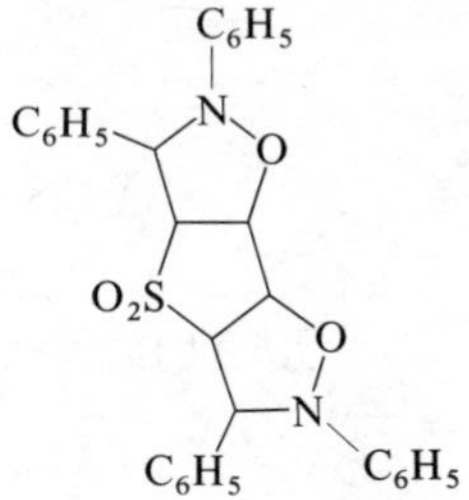 [structure: C_6H_5, O_2S] (70)	245
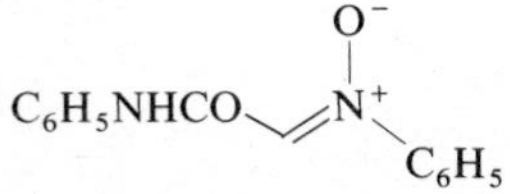 C_6H_5NHCO–CH=$N^+(O^-)$–C_6H_5	 [structure]	—	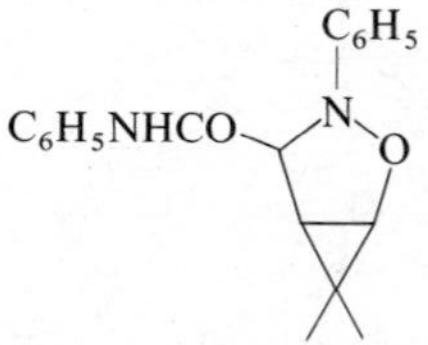 [structure: C_6H_5NHCO, C_6H_5] (96)	202
	 [structure]	—	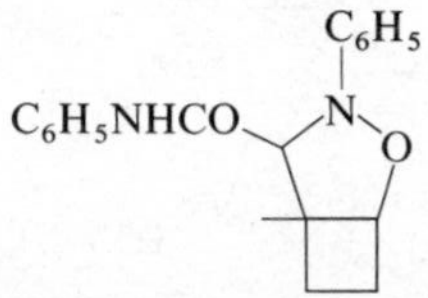 [structure: C_6H_5NHCO, C_6H_5] (90)	240

B.

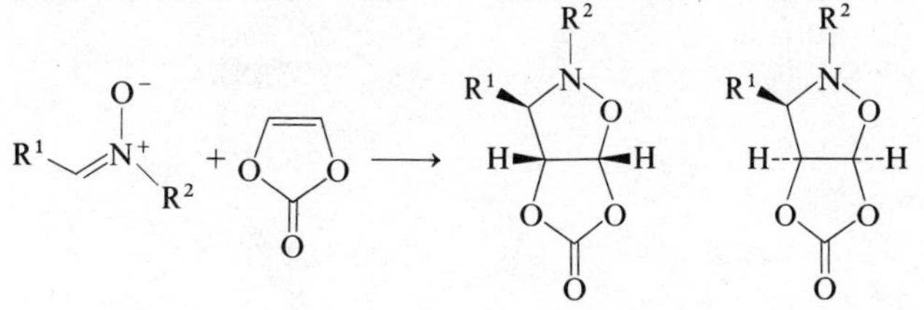

R[1]CH=N[+](O[−])R[2] + vinylene carbonate ⟶ A + B

R^1	R^2	Reaction Conditions	Product(s) and Yield(s) (%)	Ref.
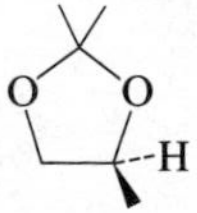 [structure]	CH_3	Neat, 45°, 48 h	**A, B** (42, 42)	199
"	$C_6H_5CH_2$	Neat, 90°, 72 h	**A, B** (27, 54)	199
C_6H_5	CH_3	Neat, 115°, 72 h	**A, B** (51)	88
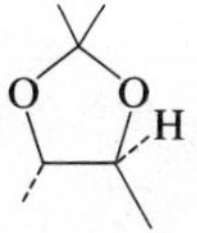 [structure]	$C_6H_5CH_2$	Neat, 95°, 72 h	**A, B** (8, 72)	199

TABLE XIII. ALDEHYDE-DERIVED NITRONES AND ENDOCYCLIC OLEFINS (*Continued*)

C.

$R^1CH{=}N^+(O^-)R^2$ + maleic anhydride ⟶ A + B

R^1	R^2	Reaction Conditions	Product(s) and Yield(s) (%) (**A**, **B**)	Refs
H	1-(C_2H_5)cyclohexyl	C_6H_6, 25°, 3 h	**A** (—)[a]	176
C_6H_5	CH_3	C_6H_6, 25°, 1 h	**A**, **B** (49)	208, 228
"	C_6H_5	$CDCl_3$, 25°, 2 h	**A**, **B** (55, 45)	97
C_6H_5CO	"	$CDCl_3$, 25°, 2 h	**B** (100)	97

D.

R^1	R^2	R^3	Reaction Conditions	Product(s) and Yield(s) (%)	Ref.
4-BrC_6H_4	C_6H_5	H	Toluene, 60°, 18 h	**A** (16)	248
C_6H_5CO	"	"	Toluene, 60°, 18 h	**A, B, C** (26, 15, 24)	248
"	"	CH_3	—	**A, C** (37, 14)	249
"	"	CH_2OH	Toluene, 60°, 15 h	**A, C** (21, 54)	250
"	"	CH_2SH	Toluene, 60°, 15 h	**A** (64)	250
"	"	C_2H_5	—	**A** (47)	249
"	"	$CH_2O_2CCH_3$	Toluene, 60°, 15 h	**A, C** (71, 12)	250
"	"	C_6H_5	Toluene, 60°, 15 h	**A, C** (38, 40)	250
[structure]	$CO_2C_4H_9$-*t*	H	$CHCl_3$, 80°, 17 h	**A** (38)[b]	250

TABLE XIII. ALDEHYDE-DERIVED NITRONES AND ENDOCYCLIC OLEFINS (*Continued*)

E.

R^1	R^2	Reaction Conditions	Yield	Ref.
CH_3	F	C_2H_5OH, reflux, 2 d	(35)	251
"	H	C_2H_5OH, reflux, 2 d	(40)	251
$C_6H_5CH_2$	F	C_2H_5OH, reflux, 2 d	(24)	251
"	H	C_2H_5OH, reflux, 2 d	(33)	251

[a] The product was isolated as the corresponding diacid.

TABLE XIV. KETONE-DERIVED ACYCLIC NITRONES AND ACETYLENES OR OLEFINS

A.

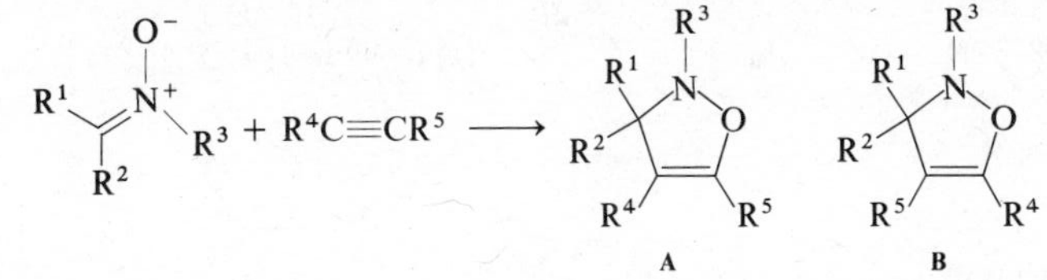

$R^1R^2C{=}N^+(O^-)R^3 + R^4C{\equiv}CR^5 \longrightarrow$ **A** + **B**

R^1	R^2	R^3	R^4	R^5	Reaction Conditions	Product(s) and Yield(s) (%)	Refs.
CO_2CH_3	$CH_3O_2CCH_2$	CH_3	H	CO_2CH_3	Ether, 25°, 14 d	**A**, **B** (41)	219, 113
"	"	"	CO_2CH_3	"	Ether, 0°, 12 h	**A** (58)	113
"	"	C_6H_5	"	"	—	**A** (—)	113
"	"	CH_3	H	$CO_2C_2H_5$	$CHCl_3$, 63°	**B** (—)	70
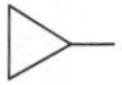 cyclopropyl	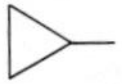 cyclopropyl	CH_3	H	"	$CHCl_3$, 63°	**B** (—)	70

B.

$R^1R^2C{=}N^+(O^-)R^3 + R^4CH{=}CR^5R^6 \longrightarrow$ **A** + **B**

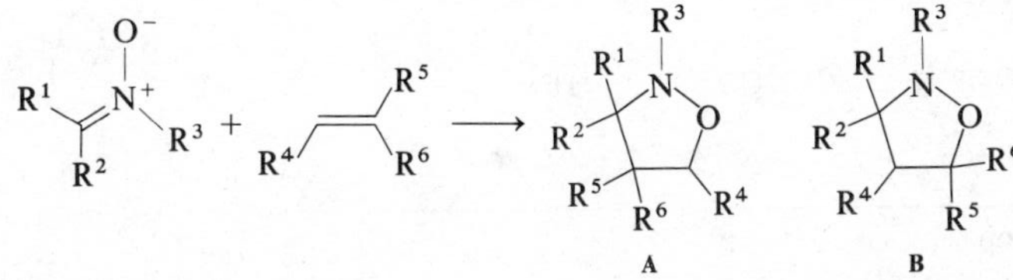

R^1	R^2	R^3	R^4	R^5	R^6	Reaction Conditions	Product(s) (**A**, **B**) and Yield(s) (%)	Ref.
CH_3	CN	C_6H_{11}	C_6H_5	H	H	Neat, 100°, 12 h	**A** (79)	75
"	CH_3	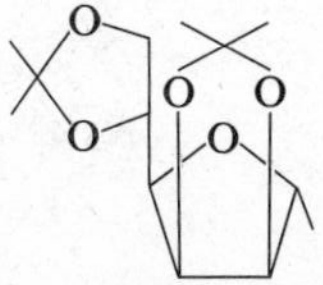 [di-O-isopropylidene sugar structure]	H	CH_3	CO_2CH_3	Acetone, reflux, 48 h	**B** (76)[a]	178

TABLE XIV. KETONE-DERIVED ACYCLIC NITRONES AND ACETYLENES OR OLEFINS (*Continued*)

R^1	R^2	R^3	R^4	R^5	R^6	Reaction Conditions	Product(s) (**A**, **B**) and Yield(s) (%)	Ref.
CH_3	CH_3	$(C_6H_5)_3CO$— [structure]	H	CH_3	CO_2CH_3	Neat, reflux, 130 h	**B** (91)[a]	204
CO_2CH_3	CO_2CH_3	CH_3	CO_2CH_3	H	"	—	**A** (—)	252
"	"	"	"	CO_2CH_3	H	—	**A** (—)	252
[structure]	[structure]	"	CN	H	"	$CHCl_3$, 63°	**A**:**B** = 1:3	70
"	"	"	CO_2CH_3	"	"	$CHCl_3$, 63°	**A**:**B** = 1:1	70
"	"	"	H	CH_3	CO_2CH_3	$CHCl_3$, 63°	**A**:**B** = 1:1	70
"	"	"	C_6H_5	H	H	$CHCl_3$, 63°	**A** (—)	70
"	"	"	$C_6H_5SO_2$	"	"	$CHCl_3$, 63°	**B** (—)	70
C_6H_5	C_6H_5	"	CH_3CO	"	"	C_6H_6, reflux, 20	**A** (75)	253
"	"	"	[structure]	"	"	—	**A** (—)	254
"	"	"	CO_2[(−)-menthyl]	"	"	Neat, 80°, 10 h	**A** (—)	194
"	"	CH_3SCH_2	CO_2CH_3	"	"	Neat, 85°, 4 h	**A** (52)	255

"	"	$CH_3CO(CH_2)_2$	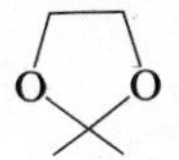 [structure]	"	"	—	**A** (—)	254
"	"	$CH_3O_2C(CH_2)_2$	CH_3CO	"	"	—	**A** (—)	254
"	"	C_6H_5	CO_2CH_3	"	CO_2CH_3	Toluene, 100°, 23 h	**A** (54)	208
"	"	"	$CO_2C_2H_5$	CH_3	H	Neat, 90°	**B** (80)[b]	81
"	"	"	C_6H_5	H	"	Neat, 100°, 41 h	**A** (86)	254
"	"	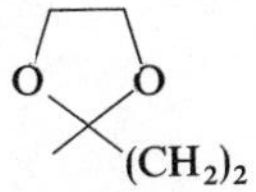 [structure] $(CH_2)_2$	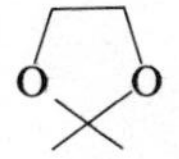 [structure]	"	"	—	**A** (—)	254
"	"	$C_6H_5CH_2$	$CO_2[(-)\text{-menthyl}]$	"	"	Neat, reflux, 10 h	**A** (—)	194

	Nitrone	Olefin	Reaction Conditions	Product and Yield (%)	Ref.
C.	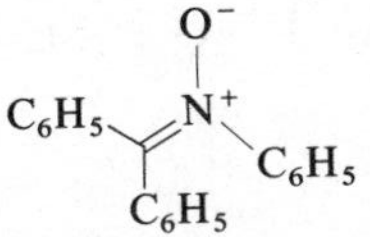 [structure]	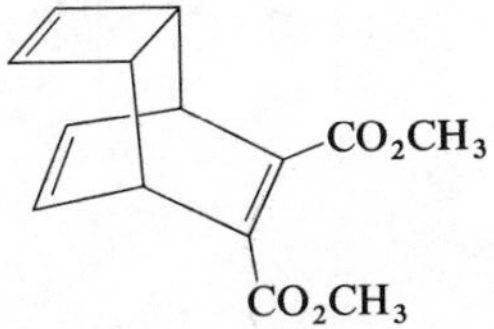 [structure]	Toluene, reflux, 10 d	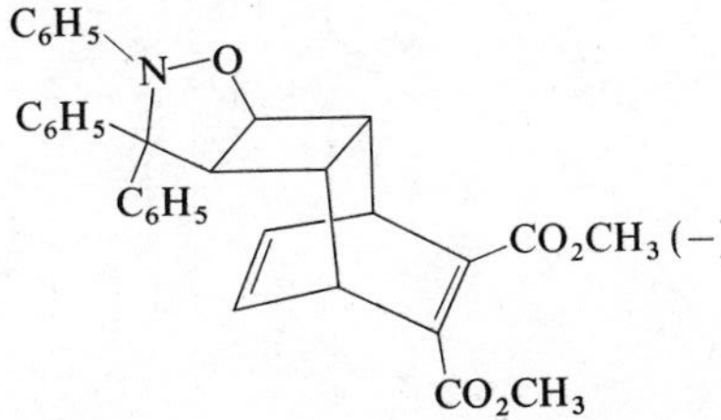 [structure] (—)	237

[a] The product is a mixture of diastereomers arising from the optically active R^2 group.
[b] The carbomethoxy group and the C-5 methyl are *trans* in the product.

TABLE XV. Endocyclic Nitrones

A.

O^- R^1 N^+ R^2 R^3 X + R^5 R^4 R^6 ⟶ R^2 R^3 X N O R^1 R^5 R^6 R^4

A

R^1	R^2	R^3	R^4	R^5
H	H	H	C_2H_5O	H
”	”	”	CH_2OH	CH_3O_2C
”	”	”	$C_2H_5O_2C$	H
”	”	”	$n\text{-}C_3H_7$	”
”	”	”	CO_2CH_3	CO_2CH_3
”	”	”	CH_3CHOH	$CH_2{=}CH$
”	”	”	CO_2CH_3	CH_2OH
”	”	”	C_6H_5	H
”	”	”	$CO_2C_2H_5$	”
”	”	”	”	$CO_2C_2H_5$
”	”	”	$4\text{-}CH_3OC_6H_4CH_2$	H
”	”	”	$2\text{-}CH_3\text{-}6\text{-}CH_3OC_6H_3$	”
”	”	”	$3,4\text{-}(CH_3O)_2C_6H_3$	”
”	”	”	$CO_2C_3H_7\text{-}n$	$THPO(CH_2)_2$
”	”	”	$n\text{-}C_3H_7CO$	”
”	”	”	”	H
”	”	”	3,4-$(CH_3O)_2C_6H_3$–C$=CH_2$ (structure: CH_3O, CH_3O, C$=CH_2$)	”
”	”	”	CH_3	”
”	”	”	CH_2OH	”
”	”	”	$COCH_3$	”
”	”	”	CO_2CH_3	”
”	”	”	”	”
”	”	”	$CO_2C_2H_5$	”
”	”	”	”	CH_3
”	”	”	$n\text{-}C_4H_9O$	H
”	”	”	CO_2CH_3	$CH_3SO_3(CH_2)_2$
”	”	”	C_6H_5	H
”	”	”	H	$CO_2C_2H_5$
”	”	”	$THPOCH_2$	H
”	”	”	$C_6H_5CH{=}CH$	”
”	”	”	$3,4\text{-}(CH_3O)_2C_6H_3$	”
”	”	”	$3,4(CH_3O)_2C_6H_3CH{=}CH_2$	”

AND ACYCLIC OLEFINS

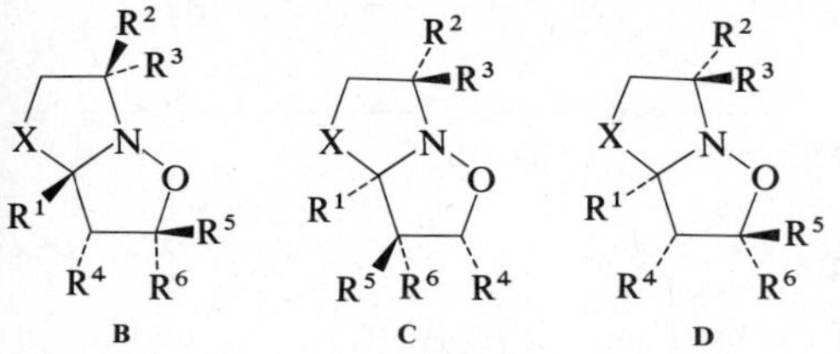

B C D

R^6	X	Reaction Conditions	Product(s) (**A**, **B**, **C**, **D**) and Yield(s) (%)	Ref.
H	CH_2	50°, 23 h, 2 kbar	**C** (83)	98
"	"	$CHCl_3$, reflux, 12 h	**A, C** (80)	163
"	"	Neat, 100°, 12 h	**D** (59)	256
"	"	Neat, 110°	**C** (72)	159
"	"	$CHCl_3$, reflux, 1 d	**A, C** (62)	163
"	"	C_6H_6, reflux	**C** (70)	257
CH_3	"	$CHCl_3$, reflux, 12 h	**B, D** (32)	258
H	"	Neat, 100°, 12 h	**C** (73)	256
$CO_2C_2H_5$	"	$CHCl_3$, reflux, 1 h	**A, C** (16, 8)	163
H	"	$CHCl_3$, reflux, 1 d	**A, C** (90)	163
"	"	Toluene, reflux	**C** (70)	259
"	"	Toluene, 95°	**C** (85)	160
"	"	Toluene, reflux, 3 h	**A, C** (25, 63)	260
"	"	Toluene, reflux, 3 h	**B, D** (42, 28)	157
"	"	$CHCl_3$, reflux	**B, D** (88)	157
$THPO(CH_2)_2$	"	$CHCl_3$, reflux, 14 h	**B, D** (90)	157
3,4-$(CH_3O)_2C_6H_3$	"	—	**B, D** (—)	161
H	$(CH_2)_2$	Toluene, 110°	**C** (53)	168 79
"	"	$CHCl_3$	**A, C** (5, 88)	168
"	"	—	**C** (100)	79
"	"	—	**A**:**C** = 18:22	168
"	"	—	**C**:**D** = 84:16	79
"	"	Neat, 100°, 24 h	**D** (73)	256
"	"	Neat, 100°, 12 h	**D** (85)	256
"	"	Neat, 100°	**C** (64)	256
"	"	Toluene, 5°, 60 h	**B, D** (74)[a]	21
"	"	Toluene, reflux	**A, C** (20, 71)	168
$CO_2C_2H_5$	"	—	**A, C** (—)	79
H	"	110°	**C** (66)	256
"	"	Toluene, reflux	**A, C** (71)	261
"	"	Toluene, reflux, 5 h	**A, C** (4, 89)	260
"	"	Toluene, reflux, 4 h	**A, C** (71)[b]	76

TABLE XV. Endocyclic Nitrones

R^1	R^2	R^3	R^4	R^5
H	H	H	MEMO, $CHOHCH_2$ CH_3O	H
"	"	"	$C_6H_5CH_2O$, $CHOHCH_2$ CH_3O	"
"	"	"	$C_6H_5CH_2O$, $CH(OAc)CH_2$ CH_3O	"
"	"	"	CH_3O, $CHOHCH_2$ CH_3O, O $(CH_2)_2CO_2CH_3$	"
"	CO_2CH_3	"	CO_2CH_3	"
"	"	"	C_6H_5	"
"	CH_3	CH_3	$CONH_2$	"
"	"	"	H	CO_2CH_3
"	"	"	$CO_2C_2H_5$	CH_3
"	"	"	CH_3CO	"
"	"	"	t-C_4H_9NHCO	H
"	"	"	C_6H_5	CO_2CH_3
"	H	H	CH_2OH	"
CH_3	CH_3	CH_3	CN	H
"	"	"	CO_2CH_3	"
"	"	"	H	CN
"	"	"	CN	H
"	"	"	CO_2CH_3	"
H	$(CH_2)_3CH(OBDMS)CH_3$	H	n-C_3H_7	"

AND ACYCLIC OLEFINS (*Continued*)

R^6	X	Reaction Conditions	Product(s) (**A**, **B**, **C**, **D**) and Yield(s) (%)	Ref.
H	$(CH_2)_2$	Toluene, reflux, 2.5 h	**C** (99)	165
"	"	Toluene, reflux, 4 h	**C** (100)	166
"	"	Toluene, reflux, 4 h	**C** (100)	166
"	"	Toluene, reflux	**C** (99)	262
"	CH_2	CH_2Cl_2, 25°, 2 h	**A**:**C** = 85:15	263
"	$(CH_2)_3$	Neat, 110°, 12 h	**C** (61)	256
"	CH_2	Neat, 25°, 2 d	**A**, **C** (89)	98
CH_3	"	Neat, 25°, 2 d	**B**, **D** (94)	98
H	"	Neat, 100°, 6 h	**B**, **D** (78)	98
CH_3	"	Neat, 100°, 12 h	**B**, **D** (63)	98
H	"	Neat, 25°, 2 d	**A**, **C** (86)	98
"	"	Neat, 100°, 3 h	**A**, **C** (84)	98
"	$(CH_3O)_2C$	$CHCl_3$, 45°	**C** (86)	164
"	O	CH_2Cl_2, 25°, 24 h	(**A** + **C**), (**B** + **D**) (20, 61)	264
"	"	CH_2Cl_2, 25°, 16 h	(**A** + **C**), **B**, **D** (39, 32, 16)	264
Cl	CH_2	Hexane, reflux, 2.5 d	**B**, **D** (52)	265
H	"	Neat, 25°, 2.25 d	**A**, **C** (49)	264
"	"	Neat, reflux, 3.3 d	(**A** + **C**), (**B** + **D**) (45, 27)	264
"	$(CH_2)_2$	$CHCl_3$, 48°, 135 h	**C** (—)	80

TABLE XV. ENDOCYCLIC NITRONES

B. [3,4-dihydroisoquinoline N-oxide] + $R^1CH{=}CR^2R^3$ ⟶

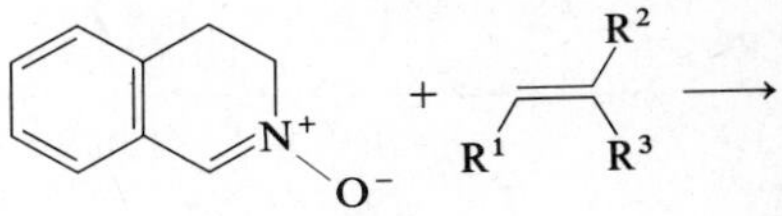

R^1	R^2	R^3
CN	H	H
CO_2CH_3	"	"
$CO_2C_2H_5$	"	"
H	CH_3	CO_2CH_3
CO_2CH_3	H	"
"	CO_2CH_3	H
$CO_2C_2H_5$	CH_3	"
H	C_2H_5O	C_2H_5O
$CO_2C_2H_5$	CH_3	CH_3
C_6H_5	H	H
"	C_6H_5	C_6H_5

C. [2-R^1-5-R^2-benzoxazole/benzothiazole N-oxide (X)] + $R^3CH{=}CR^4R^5$ ⟶ A

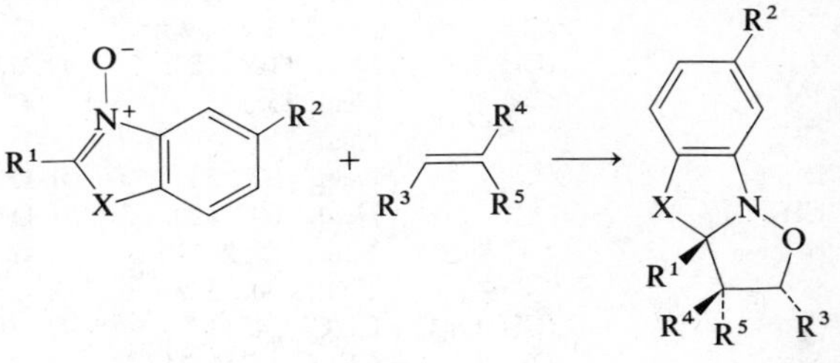

R^1	R^2	R^3	R^4	R^5
C_6H_5	H	$CO_2C_2H_5$	H	H
"	"	C_6H_5	"	"
"	"	n-C_4H_9	"	"
"	"	$CO_2C_2H_5$	$CO_2C_2H_5$	"
H	t-C_4H_9	CN	CN	"
"	"	CO_2CH_3	H	CO_2CH_3
"	"	"	CO_2CH_3	H

AND ACYCLIC OLEFINS (*Continued*)

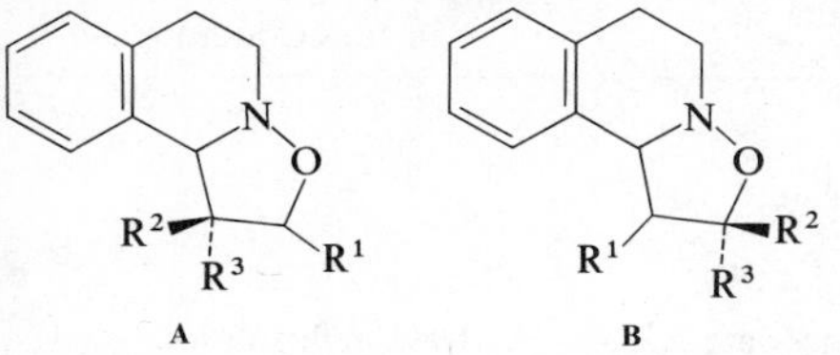

Reaction Conditions	Product(s) and Yield(s) (%)	Refs.
—	**A**:**B** = 20:80	68
Neat, 0°	**A** (—)	81, 68
Neat, 4 h	**A** (99)	81
Neat, 80°, 24 h	**B** (100)	81
C_6H_6, 25°, 24 h	**A** (96)[c]	208
$CHCl_3$, 25°, 46 h	**A** (97)[d]	208
Toluene, 85°, 3 h	**B** (79)[e]	81
Toluene, 100°, 15 h	**B** (98)	192
Neat, 100°, 4 d	**B** (93)	81
Neat, 95°, 22 h	**A** (82)	256
		192
Neat, 98°, 21 h	**B** (86)	77

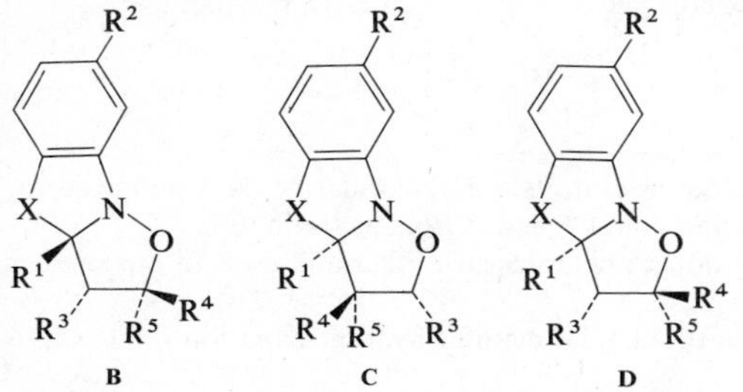

X	Reaction Conditions	Product(s) and Yield(s) (%)	Ref.
CO	Neat, 99°, 4 h	**A, C** (93)	72
"	Neat, 90°, 15 h	**A, C** (81)	72
"	Neat, 90°, 21 h	**A, C** (42)	72
"	Neat, 90°, 1.6 d	**A, C** (86)	72
$(CH_3)_2C$	C_6H_6, 25°, 2 h	**D** (79)	266
"	C_6H_6, 75°, 0.25 h	**D** (85)	266
"	C_6H_6, 65°, 0.25 h	**D** (90)	266

TABLE XV. ENDOCYCLIC NITRONES

Nitrone	Olefin	Reaction Conditions
D.		
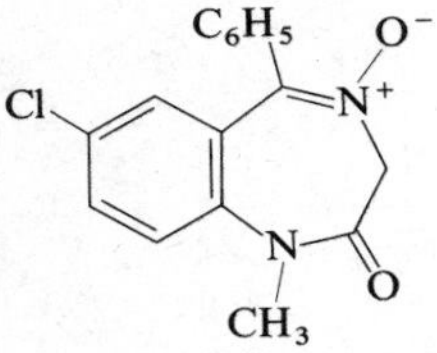 7-Chloro-5-phenyl-1-methyl-1,3-dihydro-2H-1,4-benzodiazepin-2-one 4-oxide (structure: C_6H_5, O^-, N^+, Cl, N, CH_3, O)	Methyl methacrylate	Neat, reflux, 2 h
	Methyl crotonate	Neat, reflux, 2 h
	$CH_2{=}CHCO_2R$ $R{=}CH_3, C_2H_5$	Neat, reflux, 2 h

[a] The isoxazolidine product spontaneously cyclizes to its intramolecular *N*-alkylation product.

[b] The starting olefins and products are a mixture of *E* and *Z* isomers (ratio 9:5).

[c] The two carbomethoxy groups in the product are *cis* to each other and *trans* to the angular methine hydrogen.

[d] The C–4 carbomethoxy group is *trans* to the angular methine hydrogen and *trans* to the C–5 carbomethoxy group.

[e] The C–4 carbethoxy group is *trans* to the angular methine hydrogen and *trans* to the C–5 methyl group.

AND ACYCLIC OLEFINS (*Continued*)

Product(s) and Yield(s) (%)	Ref.
(70)	267
(37) + (37)	267
R=CH_3 (64), R=C_2H_5 (60)	267

TABLE XVI. ENDOCYCLIC NITRONES AND ENDOCYCLIC OLEFINS

Nitrone	Olefin	Reaction Conditions	Product(s) and Yield(s) (%)	Ref.
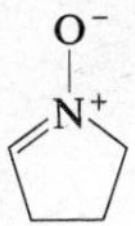 O^- N^+	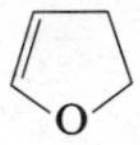 O	Xylene, reflux, 1 h	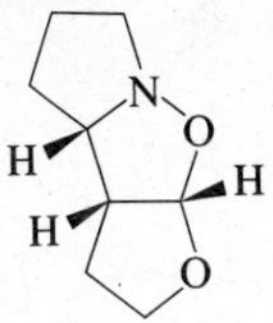 H H N O O H (3) 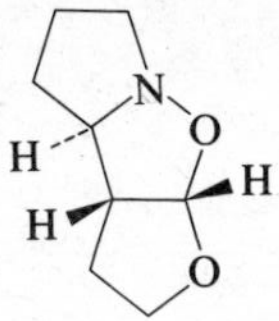 + H H N O O H (92)	268
	 O	C_6H_6, 40°, 1 h	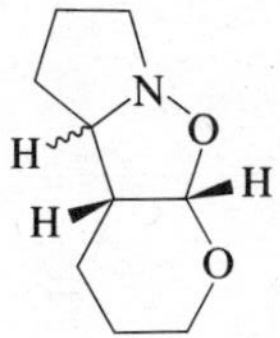 H H N O O H (28)	102 268
	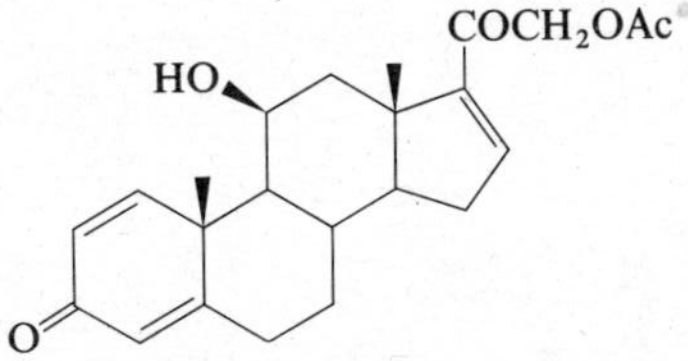 HO COCH$_2$OAc O	C_2H_5OH, reflux, 1 d	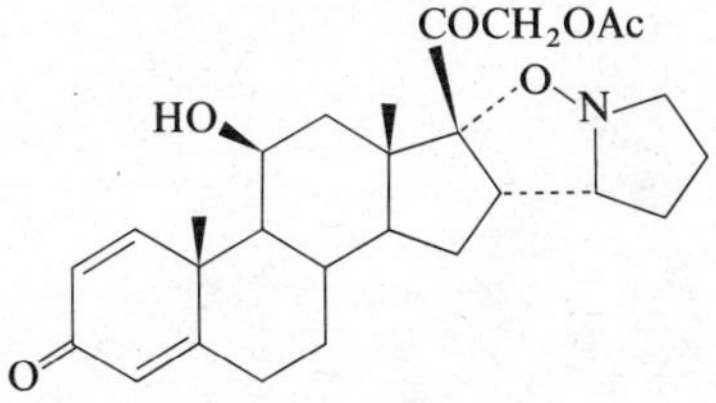 HO COCH$_2$OAc O N O (43)	251
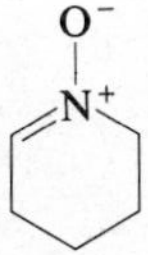 O^- N^+	 R^1 R^2 O $R^1 = R^2 = H$ $R^1, R^2 = C{=}O$	—	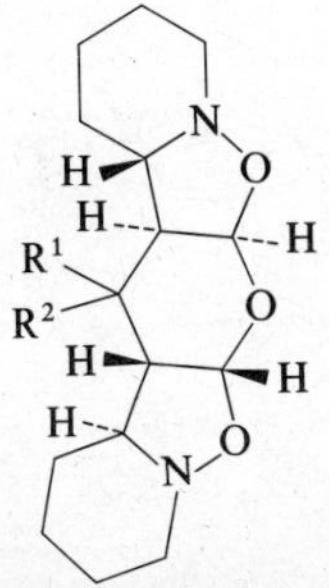 H N O H H R^1 R^2 O H H H O N $R^1 = R^2 = H$ (70) $R^1, R^2 = C{=}O$ (65)	269

Nitrone	Dipolarophile	Conditions	Product(s) and Yield(s) (%)	Refs.
		$CDCl_3$, 23°, 12 h	(—)	270
		Ether, 25°, 2 d	(85)	213
		C_6H_6, 25°	(87)	244
		C_6H_6, 25°	(8) + (65)	244

TABLE XVI. ENDOCYCLIC NITRONES AND ENDOCYCLIC OLEFINS (*Continued*)

Nitrone	Olefin	Reaction Conditions	Product(s) and Yield(s) (%)	Ref.
	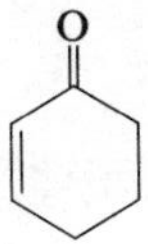 2-Cyclohexen-1-one	Neat, 100°, 6 h	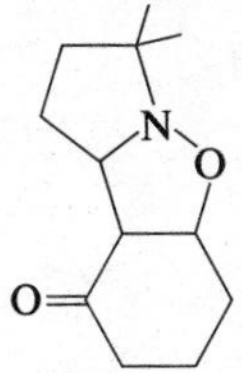 (—)	98
	$R^1 = R^2 = H$	C_6H_6, 25°, 24 h	(87)	235
	$R^1 = H$, $R^2 = Cl$	"	(74)	235
	$R^1 = R^2 = Cl$	"	(90)	235
3,4-Dihydroisoquinoline N-oxide	X = O	C_6H_6, 20°, 15 h	(86)	208
	X = C_6H_5N	C_6H_6, 20°, 15 h	(82)	208

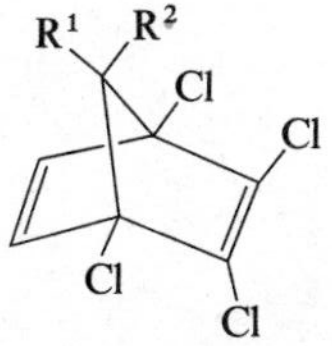

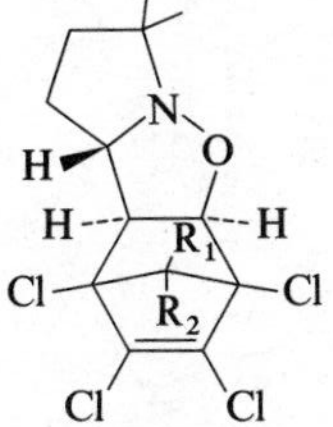

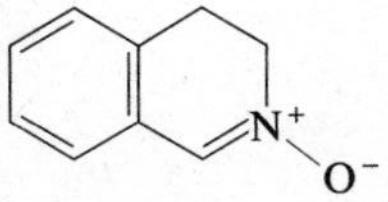

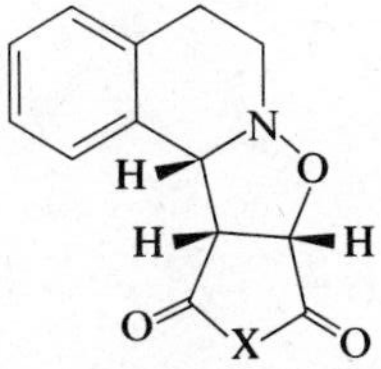

Reactant	Conditions	Product(s) (Yield, %)	Refs.
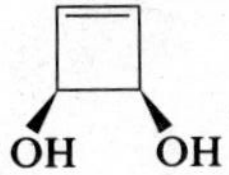	C_6H_6, 25°	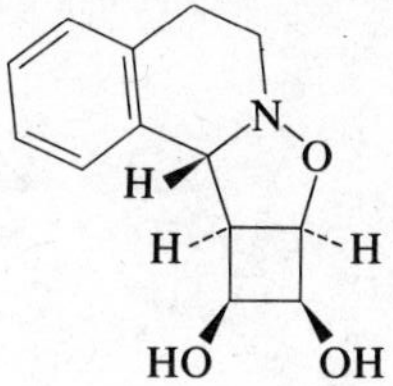 (92)	244
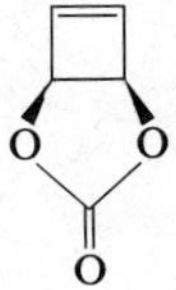	”	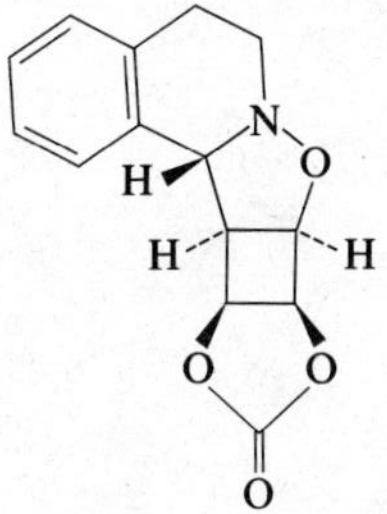 (4) 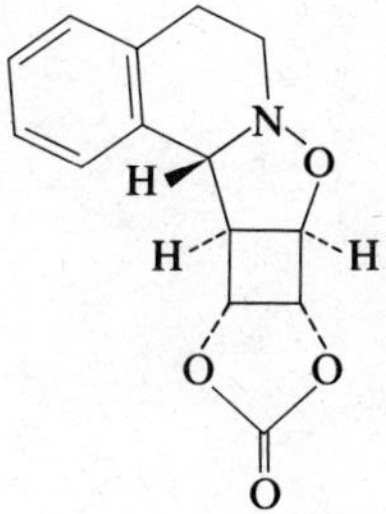 + 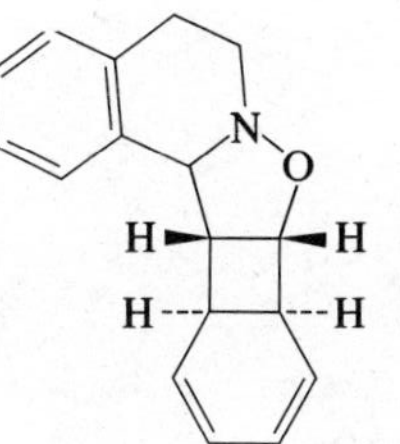 (75)	244
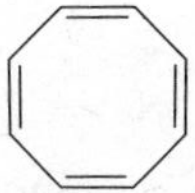	Neat, 65°, 26 h	(50)	237 222

TABLE XVI. ENDOCYCLIC NITRONES AND ENDOCYCLIC OLEFINS (*Continued*)

Nitrone	Olefin	Reaction Conditions	Product(s) and Yield(s) (%)	Ref.
	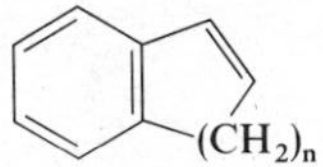 $(CH_2)_n$ n = 1, 2	Neat, 85°, 15 h	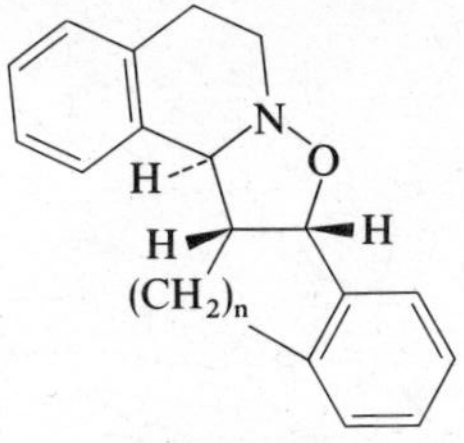 $(CH_2)_n$ n = 1 (50), n = 2 (7) 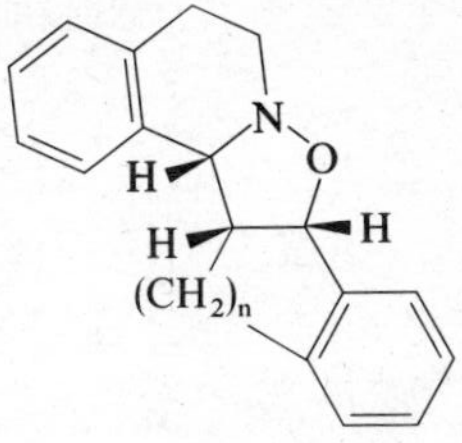 + $(CH_2)_n$ n = 1 (4), n = 2 (0) 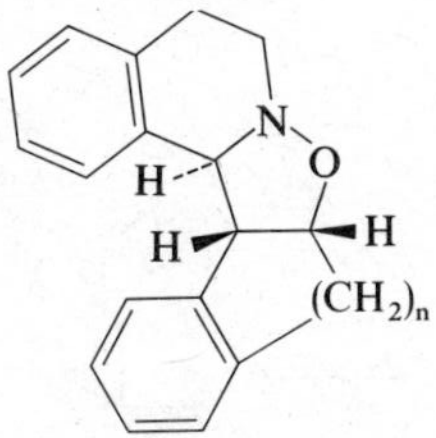 + $(CH_2)_n$ n = 1 (32), n = 2 (84) 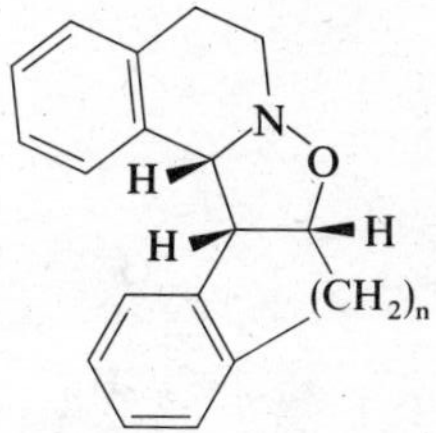 + $(CH_2)_n$ n = 1 (4), n = 2 (0)	222
	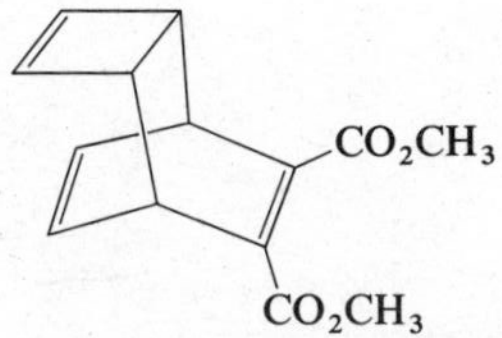 CO_2CH_3, CO_2CH_3	C_6H_6, reflux, 5 h	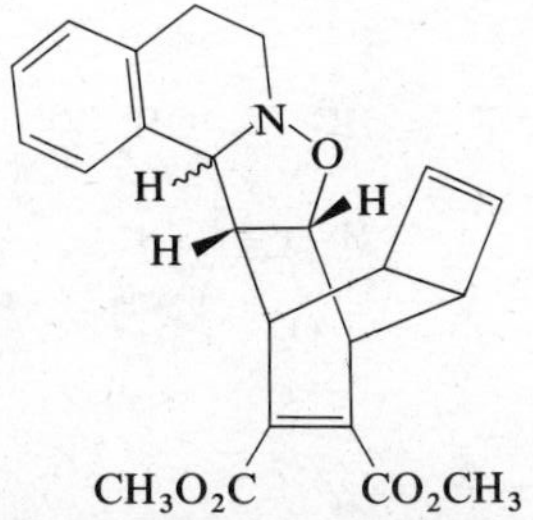 CH_3O_2C, CO_2CH_3 (25)	237

Nitrone	Dipolarophile	Conditions	Product(s) and Yield(s) (%)	Refs.
			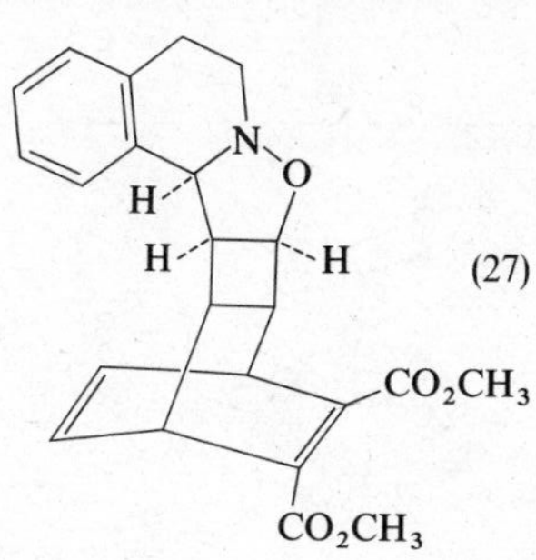 + (27) CO_2CH_3, CO_2CH_3	
	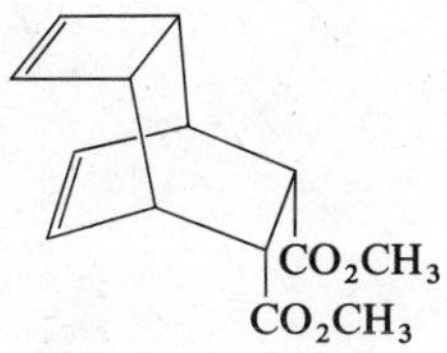 CO_2CH_3, CO_2CH_3	C_6H_6, reflux, 18 h	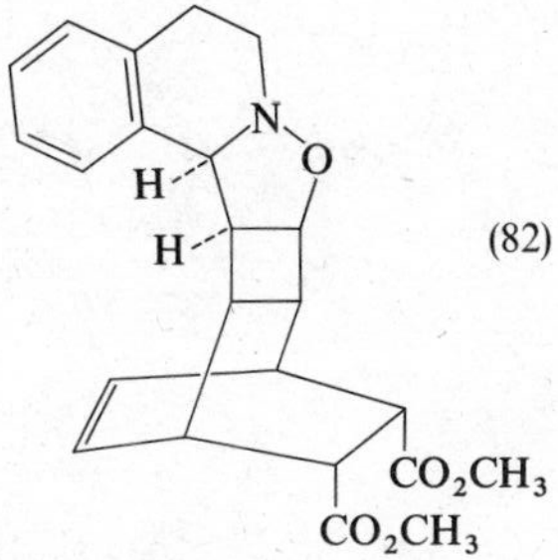 (82) CO_2CH_3, CO_2CH_3	237
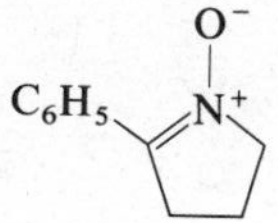 C_6H_5, O^-, N^+	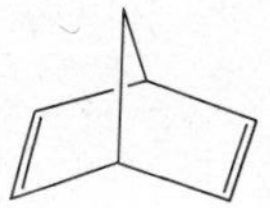	Toluene, reflux, 4 d	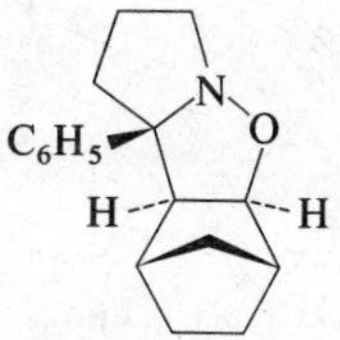 C_6H_5 (75)	235

TABLE XVI. ENDOCYCLIC NITRONES AND ENDOCYCLIC OLEFINS (*Continued*)

Nitrone	Olefin	Reaction Conditions	Product(s) and Yield(s) (%)	Ref.
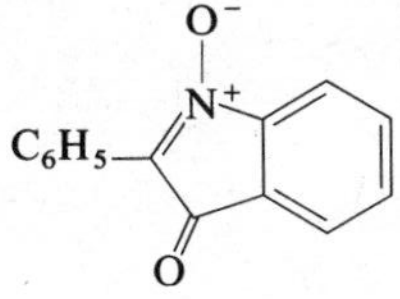 O^-, N^+, C_6H_5, O		Neat, 80°, 40 h	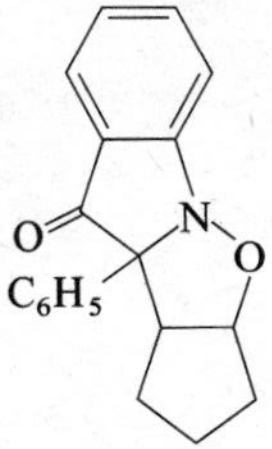 O, N, O, C_6H_5 (65)	72
	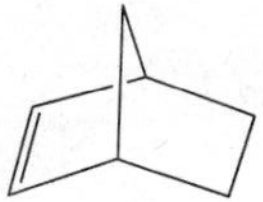	C_6H_6, 80°, 1.75 d	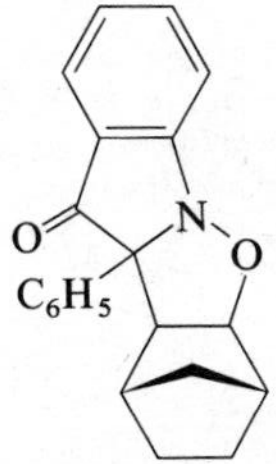 O, N, O, C_6H_5 (81)	72
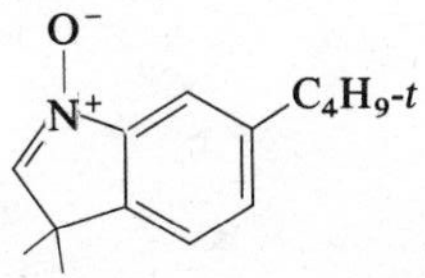 O^-, N^+, C_4H_9-*t*	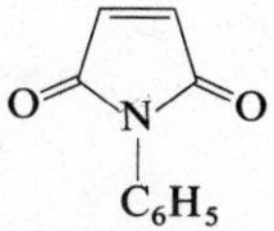 O, N, O, C_6H_5	C_6H_6, 25°, 1 h	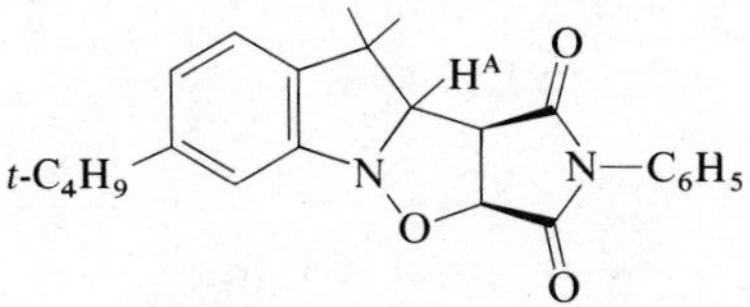 *t*-C_4H_9, N, O, H^A, O, N—C_6H_5, O; H^A-*cis* (70); H^A-*trans* (21)	266

TABLE XVII. Exocyclic Nitrones and Dipolarophiles

Nitrone	Dipolarophile	Reaction Conditions	Product(s) and Yield(s) (%)	Ref.
A.				
O^-, N^+, O, C_6H_5, O, N, C_6H_5	$C_2H_5OCH{=}CH_2$	Neat, reflux, 8 h	C_6H_5, C_6H_5, N, N, O, O, O, O, OC_2H_5 (99)	90
O^-, N^+, R^1, N, N, O, C_6H_5	$C_2H_5OCH{=}CH_2$		R^1, N, N, N, O, C_6H_5, O, OC_2H_5	
R^1				
C_6H_5		Neat, 25°, 48 h	(99)	90
4-BrC_6H_4		Neat, room temp, 24 h	(—)	271
4-ClC_6H_4		Neat, room temp, 24 h	(—)	271
4-FC_6H_4		Neat, room temp, 24 h	(—)	271
4-$O_2NC_6H_4$		Neat, room temp, 75 min	(—)	271
4-$CH_3C_6H_4$		Neat, room temp, 3 d	(—)	271
4-$CH_3CO_2C_6H_4$		Neat, room temp, 24 h	(—)	271
O^-, N^+, CH_3	$CH_2{=}CHSO_2C_6H_5$	$CHCl_3$, 70°, 2 d	CH_3, N, O, $SO_2C_6H_5$ (—)[a]	87

TABLE XVII. EXOCYCLIC NITRONES AND DIPOLAROPHILES (*Continued*)

Nitrone	Dipolarophile	Reaction Conditions	Product(s) and Yield(s) (%)	Ref.
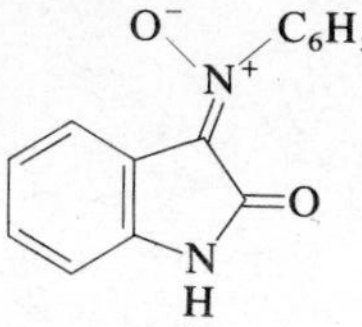 3-(N-oxido-N-phenylimino)indolin-2-one (O^-, N^+, C_6H_5, =O, NH)	$CH_2{=}CHOC_2H_5$	Neat, 100°, 15 h	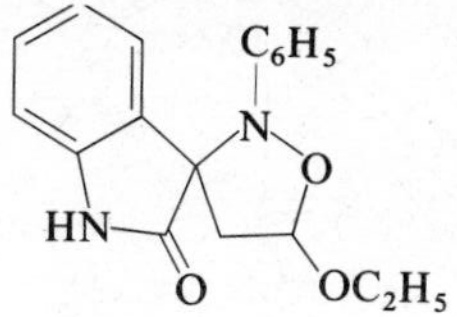 spiro-isoxazolidine (C_6H_5–N–O, HN, O, OC_2H_5) (99)	90

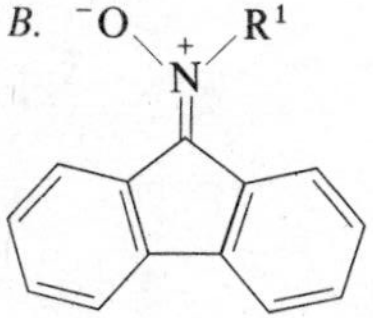

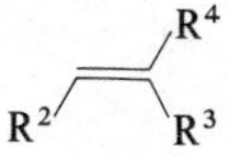

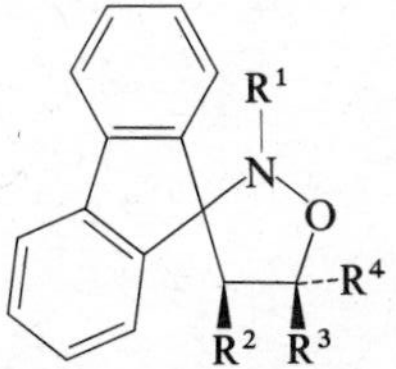

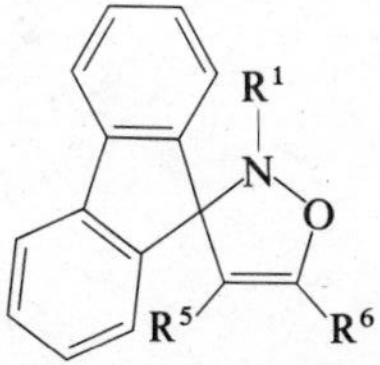

B. Fluorenone nitrone (^-O, $\overset{+}{N}$, R^1) + $R^2CH{=}CR^3R^4$ or $R^5C{\equiv}CR^6$ → **A** (spiro-isoxazolidine: R^1–N–O, R^2, R^3, R^4) or **B** (spiro-isoxazoline: R^1–N–O, R^5, R^6)

R^1	R^2	R^3	R^4	R^5	R^6	Reaction Conditions	Product(s) and Yield(s) (%)	Ref.
CH_3	CN	H	H	—	—	Neat, reflux, 1 h	**A** (50)	85
"	CO_2CH_3	"	"	—	—	C_6H_6, reflux, 2 h	**A** (64)[b]	272
"	"	"	"	—		Neat, reflux, 5.5. h	**A** (41)	85
"	"	CO_2CH_3	"	—	—	Xylene, reflux, 2 d	**A** (42)	272
"	"	H	CO_2CH_3	—	—	Xylene, reflux, 3 d	**A** (18)	272
"	—	—	—	CO_2CH_3	CO_2CH_3	$CHCl_3$, 25°, 17 h	**B** (80)	85
"	—	—	—	CO_2CH_3	CO_2CH_3	C_6H_6, reflux, 2 h	**B** (65)	272
"	—	—	—	CO_2H	C_6H_5	Xylene, reflux, 1 h	**B** (24)	85
"	—$CON(C_6H_5)CO$—		H	—	—	Xylene, reflux, 4 h	**A** (67)	85
C_2H_5	CO_2CH_3	H	H	—	—	C_6H_6, reflux, 2 h	**A** (38)	272
"	—	—	—	CO_2CH_3	CO_2CH_3	C_6H_6, reflux, 2 h	**B** (36)	272
i-C_3H_7	—	—	—	CO_2CH_3	CO_2CH_3	C_6H_6, reflux, 2 h	**B** (81)	272

C. $^{-}O-\overset{+}{N}(C_6H_5)=$ anthrone + $R^1CH=CHR^2$ or $R^3C\equiv CR^3$ ⟶ A + B

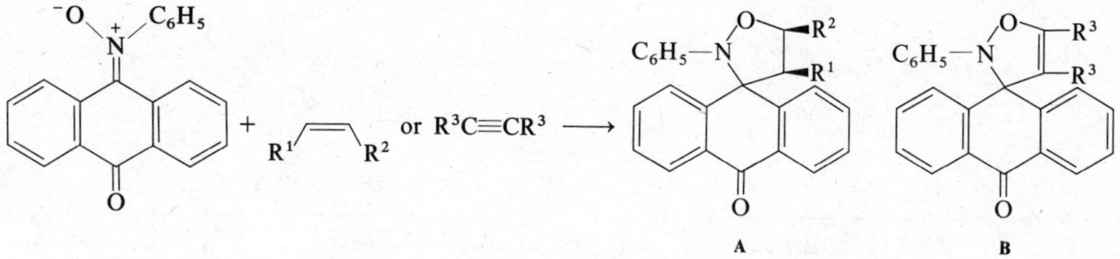

R^1	R^2	R^3	Reaction Conditions	Product(s) and Yield(s) (%)	Ref.
CN	H	—	Neat, reflux, —	**A** (80)	85
CO_2CH_3	H	—	Neat, reflux, —	**A** (60)	85
—	—	CO_2CH_3	$CHCl_3$, 25°, 1 d	**B** (85)	85
—CON(C_6H_5)CO—		—	—	**A** (85)	85

[a] The product was seen by NMR but could not be isolated. Instead a 45% yield of **I** was obtained:

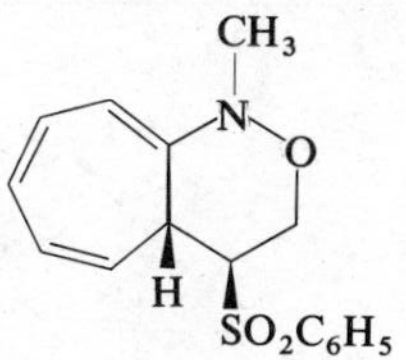

I

[b] The regioisomer was also obtained in 16% yield.

TABLE XVIII. Endocyclic Nitrones and Acetylenes

A. $+ R^6C{\equiv}CR^7 \longrightarrow$ **A** + **B**

R^1	R^2	R^3	R^4	R^5	R^6	R^7	Reaction Conditions	Product(s) and Yield(s) (%)	Ref.
H	CH_3	CH_3	H	H	CO_2CH_3	CO_2CH_3	CH_2Cl_2, 0°, 0.5 h	**A** (80)	273
"	C_6H_5	H	$CON(C_2H_5)_2$	CH_3	H	"	CH_2Cl_2, 25°, 24 h	**A**:**B** = 1:1	274
"	"	"	"	"	CO_2CH_3	"	$CHCl_3$, 25°, 4 h	**A** (69)	274
$R^1, R^2 = (CH_2)_6$		"	"	"	H	"	CH_2Cl_2, reflux, 48 h	**A**:**B** = 13:87	274
"	"	"	"	"	CO_2CH_3	"	$CHCl_3$, 25°, 24 h	**A** (72)	274
CH_3	C_6H_5	"	"	"	H	"	CH_2Cl_2, reflux, 24 h	**A**:**B** = 14:84	274
"	"	"	"	"	CO_2CH_3	"	$CHCl_3$, 25°, 24 h	**A** (82)	274
"	"	"	$CON(CH_2)_4$	C_6H_5	"	"	$CHCl_3$, 25°, 24 h	**A** (75)	274

B. $+ R^2C{\equiv}CR^3 \longrightarrow$ **A** + **B**

R^1	R^2	R^3	X	Reaction Conditions	Product(s) and Yield(s) (%)	Ref.
CH_3	CO_2CH_3	CO_2CH_3	O	CH_2Cl_2, 25°, 5 h	**A** (98)	264
"	C_6H_5	CN	"	CH_2Cl_2, 45°, 24 h	**B** (95)	264
"	"	CO_2CH_3	"	CH_2Cl_2, 35°, 24 h	**B** (74)	264
"	"	CN	CH_2	C_6H_6, reflux, 40 h	**B** (95)	264
"	"	CO_2CH_3	"	C_6H_6, reflux, 32 h	**B** (94)	264
C_2H_5	CO_2CH_3	"	O	CH_2Cl_2, 25°, 5 h	**A** (93)	264
"	C_6H_5	CN	"	CH_2Cl_2, 25°, 48 h	**B** (90)	264
"	"	CO_2CH_3	"	CH_2Cl_2, 25°, 48 h	**B** (73)	264

C. 3,4-dihydroisoquinoline N-oxide + $R^1C{\equiv}CR^2$ ⟶ **A** + **B**

R^1	R^2	Reaction Conditions	Product(s) and Yield(s) (%)	Ref.
H	CN	—	**B** (100)	68
"	CO_2CH_3	—	**B** (100)	68
CH_3	"	20°, 10 d	**B** (—)	275
H	C_6H_5	—	**A** (—)	275

TABLE XVIII. ENDOCYCLIC NITRONES AND ACETYLENES (*Continued*)

Nitrone	Acetylene	Reaction Conditions	Product(s) and Yield(s) (%)	Ref.
D.				
Cl, C_6H_5, $\overset{+}{N}$–O^-, N–CH_3, O	$HC{\equiv}CCO_2C_2H_5$	THF, reflux, 18 h	$CO_2C_2H_5$, C_6H_5, O, N, Cl, N–CH_3, O (32)	276

E. R[1], O^-, N^+, R[4], R[2], R[3], N, $CH_2C_6H_5$ + $CH_3O_2CC{\equiv}CCO_2CH_3$ ⟶ $CH_2C_6H_5$, N, R[4], R[2], R[3], R[1], N, O, CH_3O_2C, CO_2CH_3

R[1]	R[2]	R[3]	R[4]	Reaction Conditions	Yield (%)	Ref.
i-C_3H_7	CH_3	CH_3	H	$CHCl_3$, reflux, 4 h	(79)	277
"	"	"	C_6H_5	$CHCl_3$, reflux, 4 h	(53)	277
$C_6H_5CH_2$	C_6H_5	H	H	$CHCl_3$, reflux, 4 h	(75)	277

TABLE XIX. MISCELLANEOUS REACTIONS OF NITRONES AND DIPOLAROPHILES

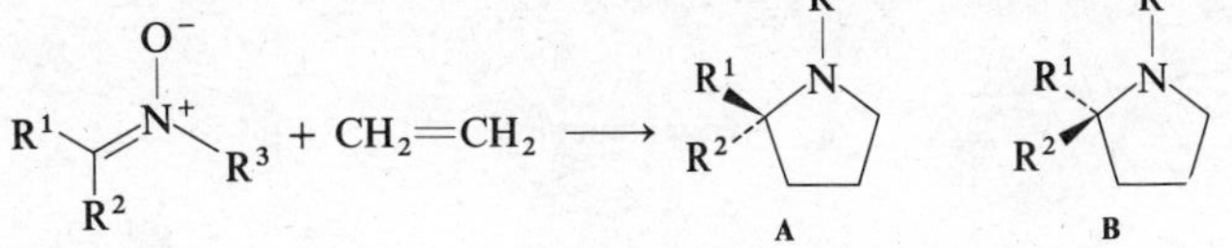

$R^1R^2C{=}N^+(O^-)R^3 + CH_2{=}CH_2 \longrightarrow$ A + B

R^1	R^2	R^3	Reaction Conditions	Product(s) and Yield(s) (%)	Ref.
CO_2CH_3	H	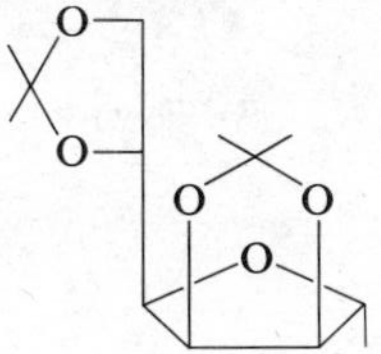 [structure]	$CHCl_3$, 75°, 17 h/65 kbar	**A, B** (92)	278
$(CH_3O)_2PO$	"	"	$CHCl_3$, 75°, 16 h/65 kbar	**A, B** (61, 20)	279
$(C_2H_5O)_2PO$	"	"	$CHCl_3$	**A, B** (60, 20)	279
$CO_2C_4H_9$-*t*	"	"	$CHCl_3$, 75°, 17 h/65 kbar	**A, B** (93)	278
"	"	"	$CHCl_3$, 80°, 17 h	**A, B** (70, 21)	280
$CO_2CH_2C_6H_5$	"	"	$CHCl_3$, 75°, 17 h/65 kbar	**A, B** (86)	278
$CO_2C_4H_9$-*t*	"	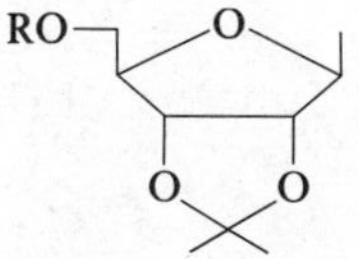 RO– [structure] R = H, $(C_6H_5)_3C$	$CHCl_3$, 80°, 17 h/65 kbar	R = H, **A, B** (25, 62) R = $(C_6H_5)_3C$, **A, B** (78)	280 278
CH_3	CH_3	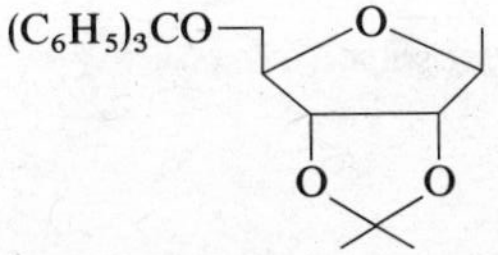 $(C_6H_5)_3CO$– [structure]	—	**A, B** (91)	204
$4\text{-}O_2NC_6H_4CO$	H	C_6H_5	$CHCl_3/C_6H_6$, 20°, 8 h	**A, B** (90)	72
C_6H_5CO	"	"	$CHCl_3/C_6H_6$, 20°, 8 h	**A, B** (79)	72

TABLE XIX. MISCELLANEOUS REACTIONS OF NITRONES AND DIPOLAROPHILES (*Continued*)

Nitrone	Dipolarophile	Reaction Conditions	Product(s) and Yield(s) (%)	Ref.
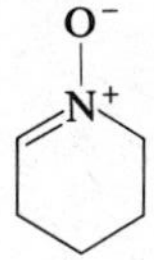 [structure]	$CH_2{=}C{=}CH_2$	—	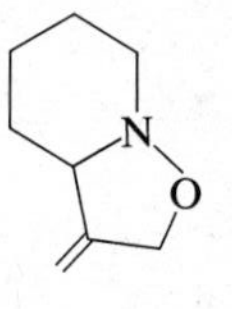 [structure] (—)	79
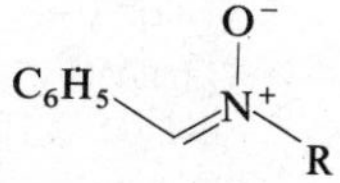 [structure] $R = CH_3, C_6H_5$	"	72°, 60 h	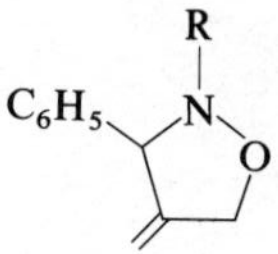 [structure] $R = CH_3$ (60) $R = C_6H_5$ (22)	82
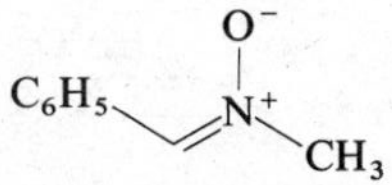 [structure]	1,4-Dibenzyne	—	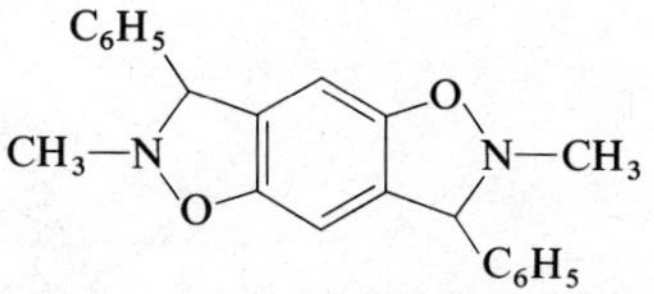 [structure] (91)	281
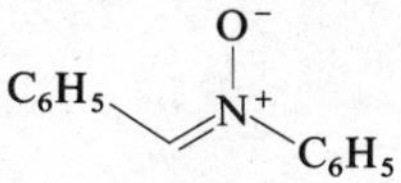 [structure]	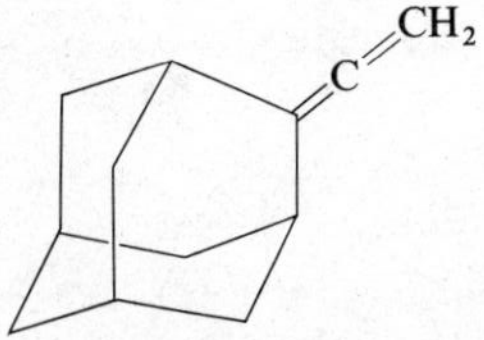 [structure]	Toluene, reflux, 15 h	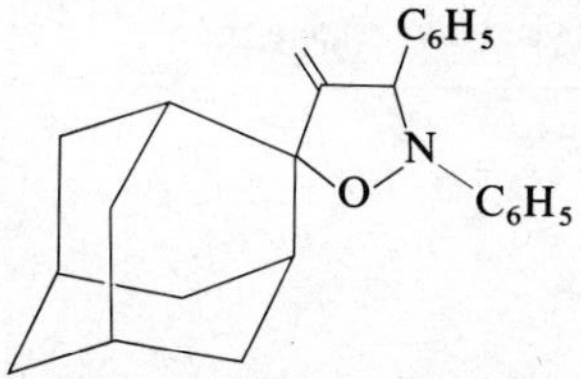 [structure] (16)	282

Nitrone	Dipolarophile	Conditions	Product(s) and Yield(s) (%)	Refs.
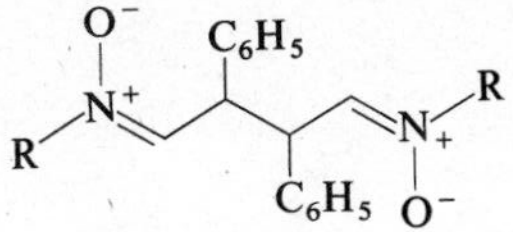 R = CH_3, t-C_4H_9	$CH_3O_2CC{\equiv}CCO_2CH_3$	C_2H_5OH, 60°, 0.5 h (R = CH_3) Neat, 80°, 1 h (R = t-C_4H_9)	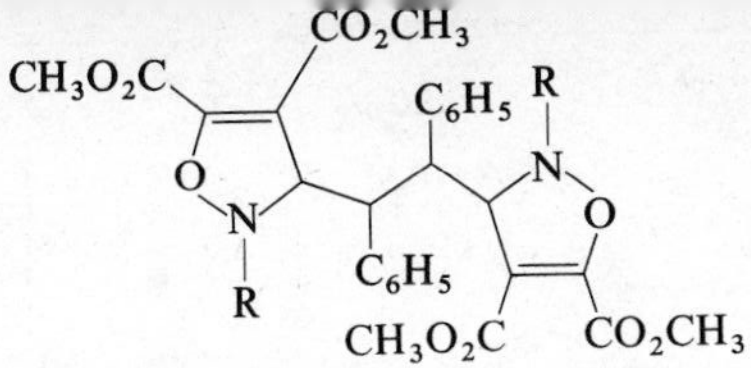 R = CH_3 (90) R = t-C_4H_9 (55)	283
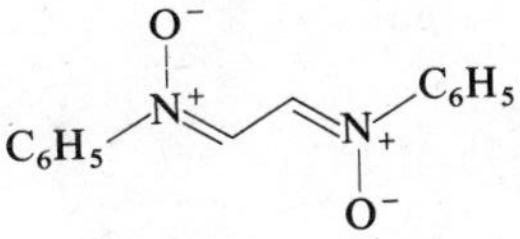	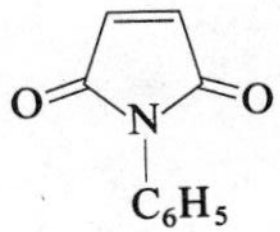	DMF, 110°, 1.5 h	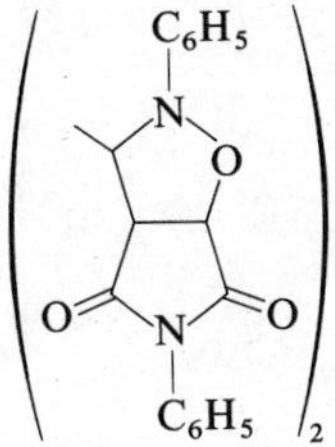 (—)	284
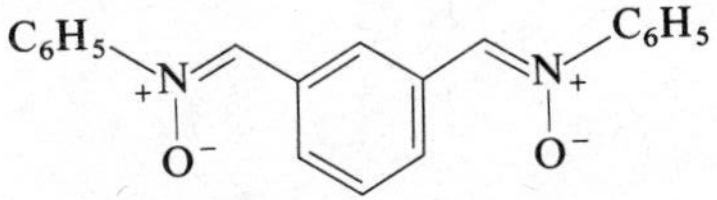	"	DMF, 110°, 1.5 h	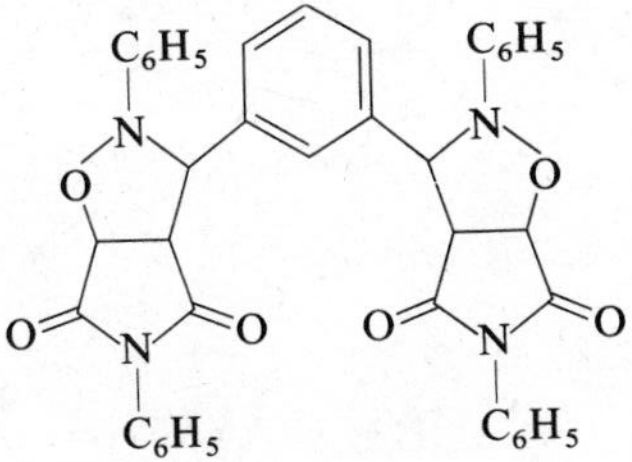 (—)	284
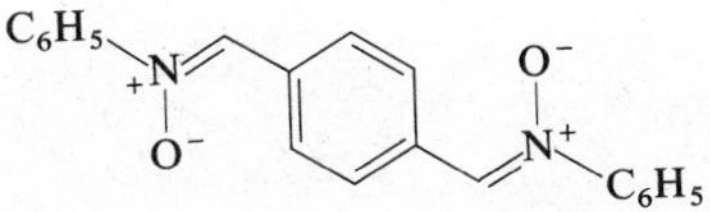	$C_6H_5CH{=}CH_2$	DMF, 110°, 1.5 h	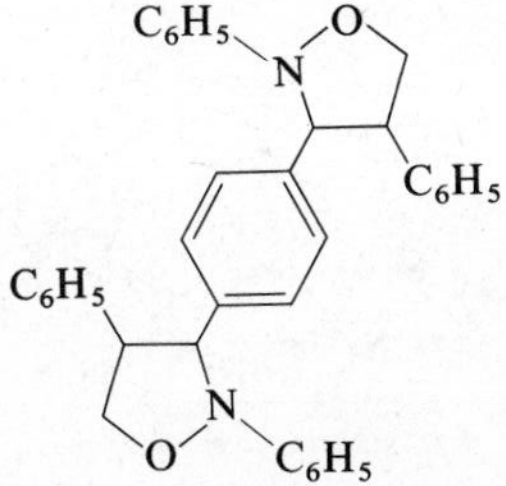 (—)	284

TABLE XIX. MISCELLANEOUS REACTIONS OF NITRONES AND DIPOLAROPHILES (*Continued*)

Nitrone	Dipolarophile	Reaction Conditions	Product(s) and Yield(s) (%)	Ref.
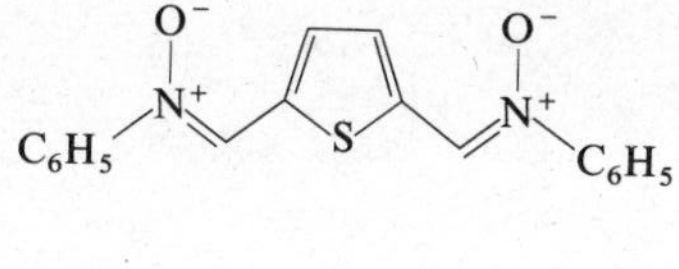 [structure]	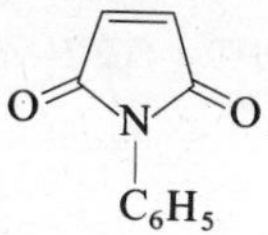 [structure]	THF, heat	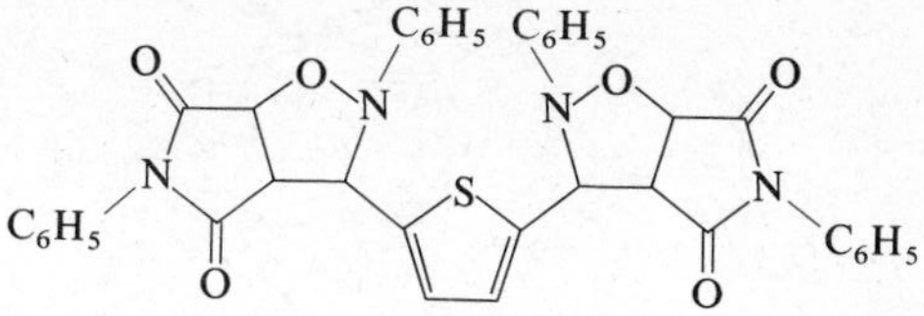 [structure] (—)	285
	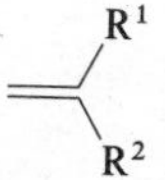 [structure] $R^1 = CH_2OH$, $R^2 = H$; $R^1 = CO_2CH_3$, $R^2 = H$; $R^1 = CO_2CH_3$, $R^2 = CH_3$	"	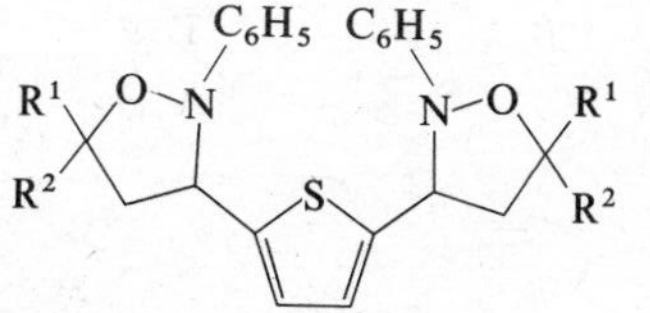 [structure] $R^1 = CH_2OH$, $R^2 = H$ (—); $R^1 = CO_2CH_3$, $R^2 = H$ (—); $R^1 = CO_2CH_3$, $R^2 = CH_3$ (—)	285
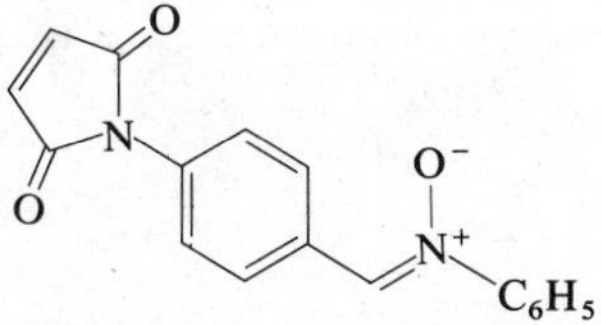 [structure]		DMF, 100°, 10 min	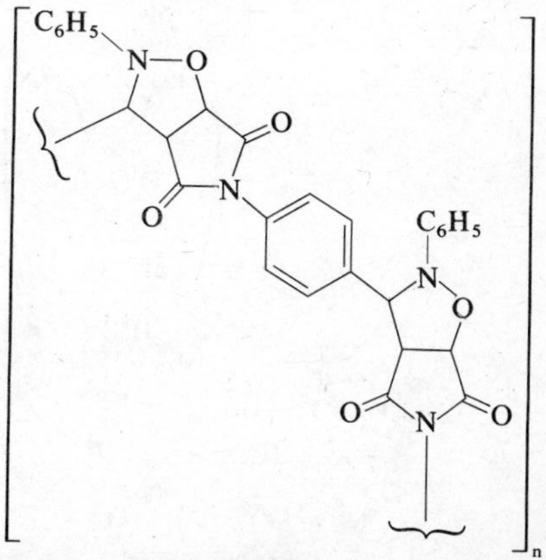 [structure]$_n$ (91)	286

TABLE XX. Intramolecular Cycloadditions of N-Alkenyl Substrates

Hydroxylamine	Aldehyde	Reaction Conditions	Product(s) and Yield(s) (%)		Ref.
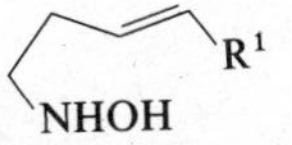 NHOH, R[1] (alkenyl hydroxylamine)	R^2CHO		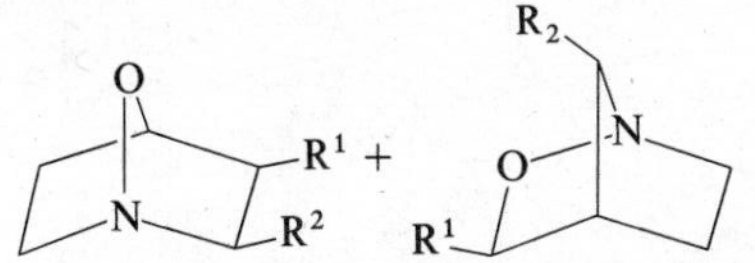 bicyclic N–O product with R^1, R^2 +	tricyclic product with R_2, R^1	
R^1 / R^2					
H / C_6H_5		Xylene, reflux, 24 h	(87)	(—)	120
CH_3 / H		Toluene, 110°, 6 h	(76)	(—)	143
CH_3 / 4-$O_2NC_6H_4$		1,2-$Cl_2C_6H_4$, 180°, 1.5 h	(85)	(—)	143
C_6H_5 / H		Toluene, 110°, 2 h	(56)	(9)	143
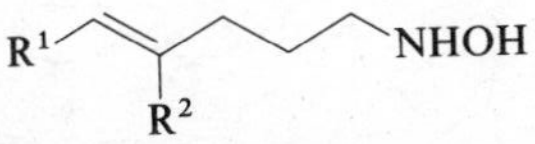 R^1, R^2 alkenyl NHOH	R^3CHO		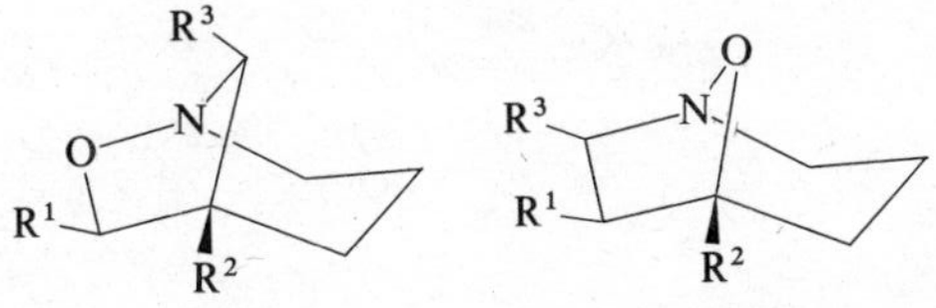 product with R^3, R^1, R^2	product with R^3, R^1, R^2	
R^1 / R^2 / R^3					
H / H / H		Toluene, 110°, 3 h	(47)	(23)	143
H / CH_3 / H		Toluene, 110°, 2.5 h	(10)	(77)	143
H / C_6H_5 / H		Toluene, 110°, 6 h	(0)	(82)	143
CH_3 / H / H		Toluene, 110°, 3 h	(95)	(0)	143
CH_3 / H / 4-$O_2NC_6H_4$		Toluene, 110°, 18 h	(69)	(0)	143
C_6H_5 / H / H		Toluene, 110°, 3 h	(87)	(0)	143
C_6H_5 / H / 4-$O_2NC_6H_5$		Toluene, 110°, 11 h	(95)	(0)	143

TABLE XX. INTRAMOLECULAR CYCLOADDITIONS OF *N*-ALKENYL SUBSTRATES (*Continued*)

Hydroxylamine	Aldehyde	Reaction Conditions	Product(s) and Yield(s) (%)	Ref.
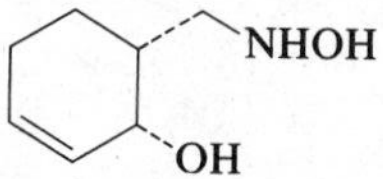 NHOH, OH	CH_2O	Toluene, reflux, 3 d	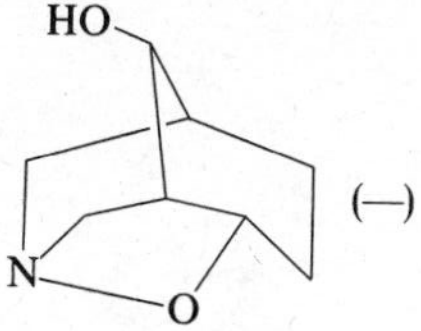 HO, N, O (—)	101
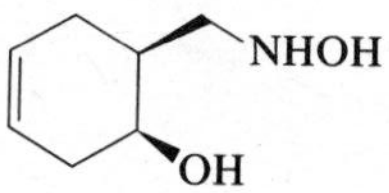 NHOH, OH	CH_2O	Toluene, reflux	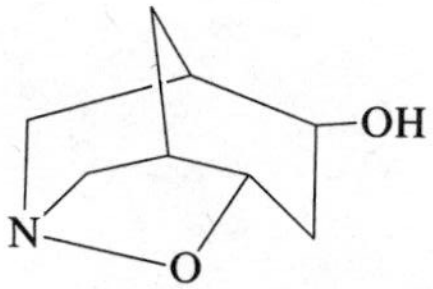 OH, N, O (54)	101
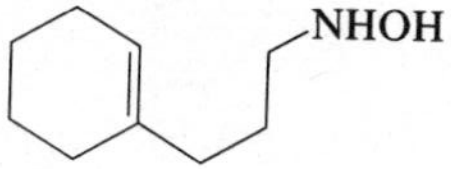 NHOH	CH_2O	1. Toluene, 0° 2. reflux, 24 h	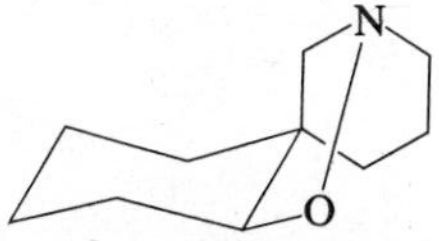 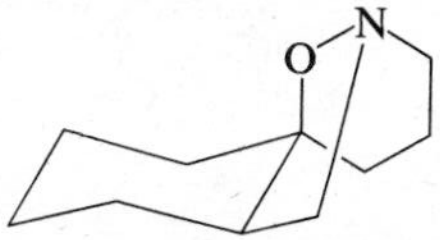 N, O (50)[a]; O, N (20)	287
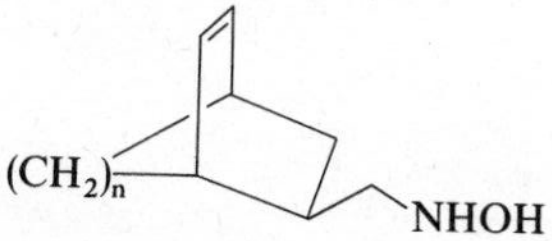 $(CH_2)_n$, NHOH	R^1CHO		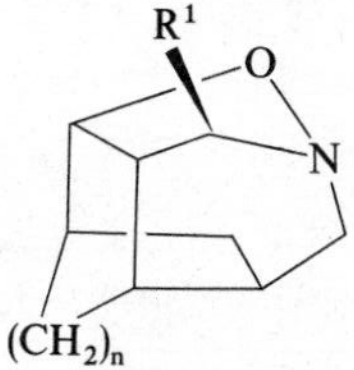 R^1, O, N, $(CH_2)_n$	
R^1 = H, n = 1		C_6H_6, 90°, 4 d	(23)	131
R^1 = C_6H_5, n = 1		Toluene, 125°, 2 d	(67)	131
R^1 = H, n = 2		C_6H_6, 80°, 5 d	(49)	131
R^1 = C_6H_5, n = 2		Toluene, 125°, 2 d	(59)	131

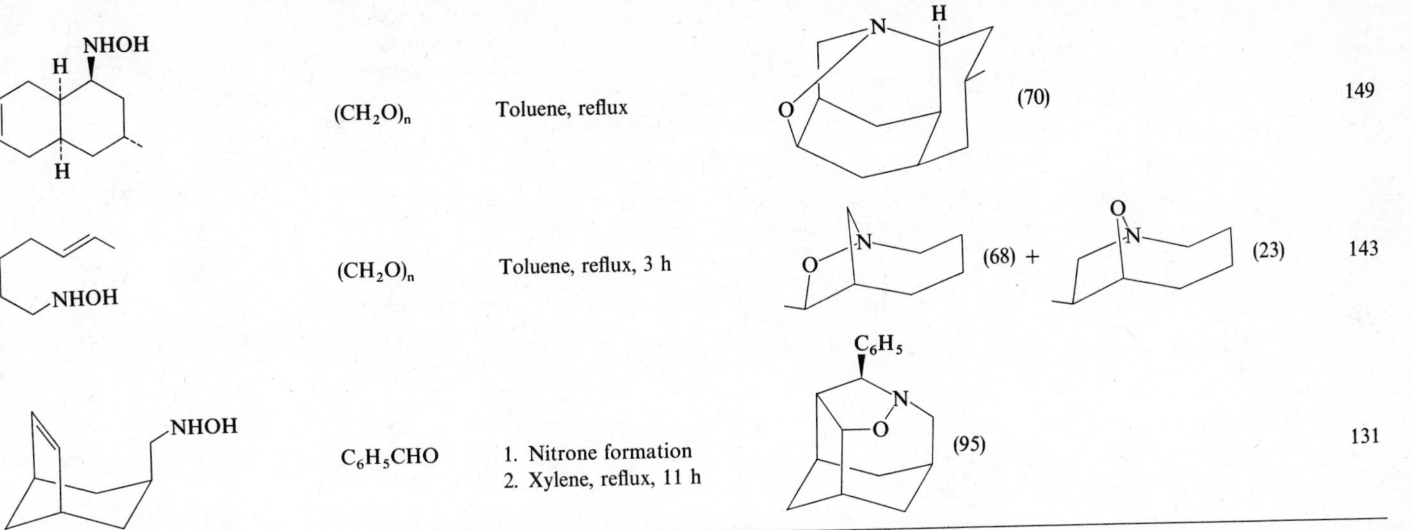

Reactant	Reagent	Conditions	Product(s) and Yield(s) (%)	Refs.
NHOH	$(CH_2O)_n$	Toluene, reflux	(70)	149
NHOH	$(CH_2O)_n$	Toluene, reflux, 3 h	(68) + (23)	143
NHOH	C_6H_5CHO	1. Nitrone formation 2. Xylene, reflux, 11 h	C_6H_5 (95)	131

[a] The yield is based on a three-step sequence starting with sodium cyanoborohydride reduction of the corresponding oxime.

TABLE XXI. Intramolecular Cycloadditions of Cyclic Nitrones

Substrate				Reaction Conditions	Product(s) and Yield(s)	Ref.
R^1	R^2	R^3	R^4			
H	H	H	H	Toluene, reflux[a]	(—)	151
H	H	H	H	Toluene, reflux, 6 h[a]	(—)	152
H	H	H	CO_2CH_3	Toluene, reflux[a]	(4.3)	152
H	H	H	CO_2CH_3	Toluene, reflux[a]	(10)	153
CH_3	CH_3	CH_3	H	Xylene, reflux[b]	(20)	150
CH_3O_2C						
R^1	R^2					
H	CN			Xylene, reflux, 18 h	(56)	152
H	CO_2CH_3			Xylene, reflux	(40)	152
CN	H			Xylene, reflux, 8 h	(—)	152

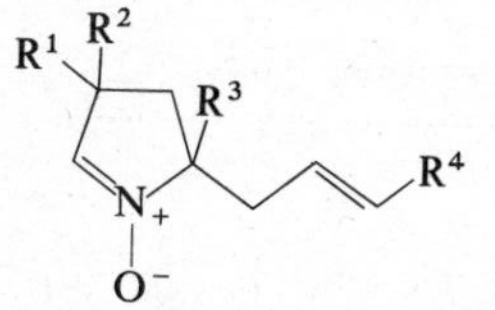

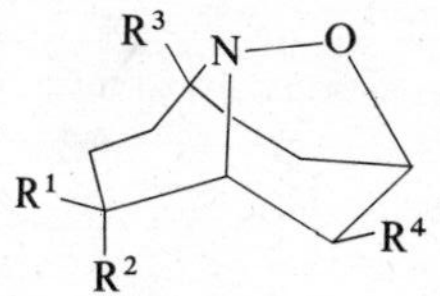

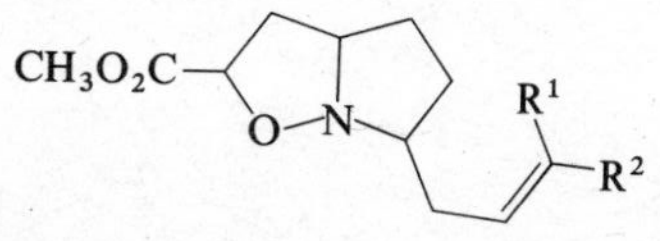

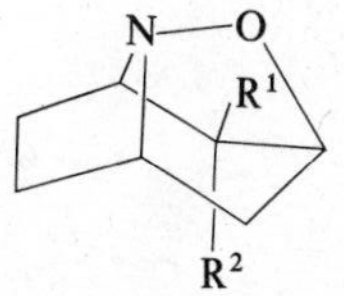

Reactant	Conditions	Product(s) and Yield(s) (%)	Refs.
OH, $COCH_3$, N^+, O^-	C_6H_6, 45°	N, O, HO, $COCH_3$ (71)	257
C_6H_5, N, $\overset{+}{N}$, O^-, C_6H_5	Ether, reflux, 20 h	C_6H_5, C_6H_5, N, O, N (22)	46
R^1, R^2, N^+, O^-		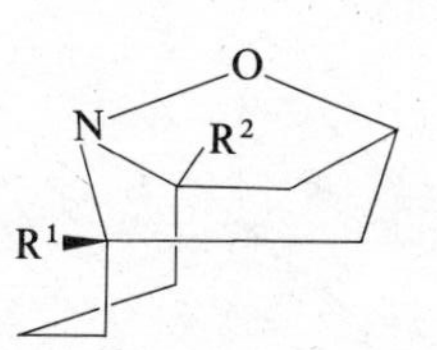	
R^1 / R^2			
H / CH_3	—[c]	(—)	288
H / n-C_5H_{11}	$CHCl_3$, reflux, 1 h	(66)	154
n-C_5H_{11} / H	Toluene, reflux, 14 h	(—)	154

TABLE XXI. INTRAMOLECULAR CYCLOADDITIONS OF CYCLIC NITRONES (*Continued*)

Substrate	Reaction Conditions	Product(s) and Yield(s)	Ref.
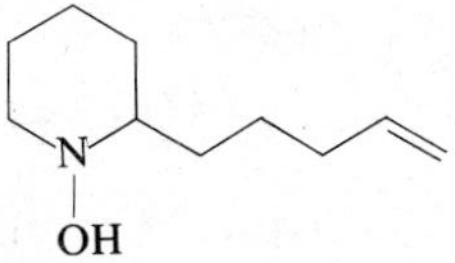 N–OH	1. HgO 2. Toluene, reflux, 3 h	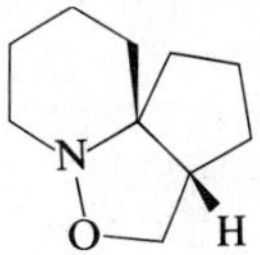 (64) 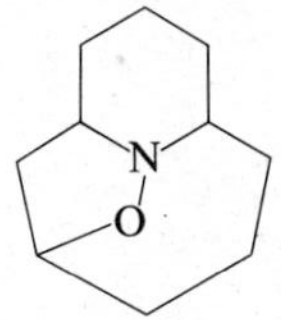 + (7) 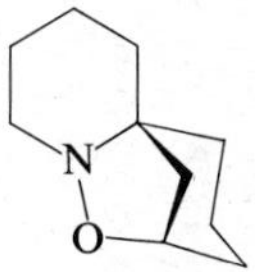 + (1.4)	289
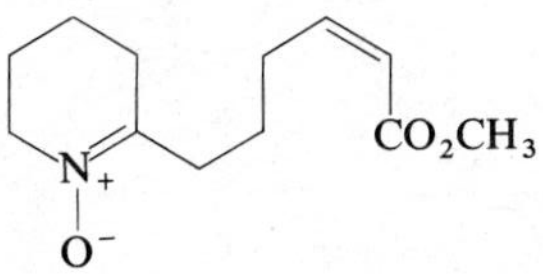 CO_2CH_3, N^+–O^-	CH_3OH, room temp, 4.5 h	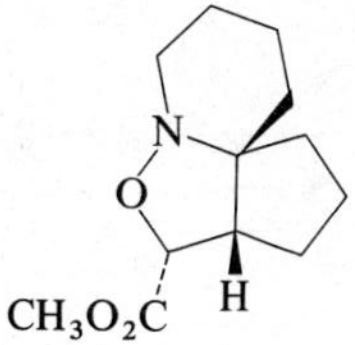 CH_3O_2C, H (51)	290
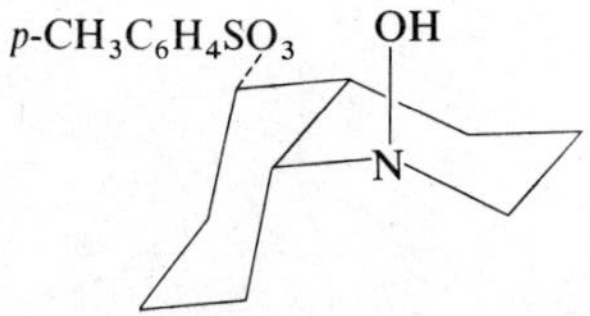 p-$CH_3C_6H_4SO_3$, OH, N	NaH, styrene, THF, heat, 24 h	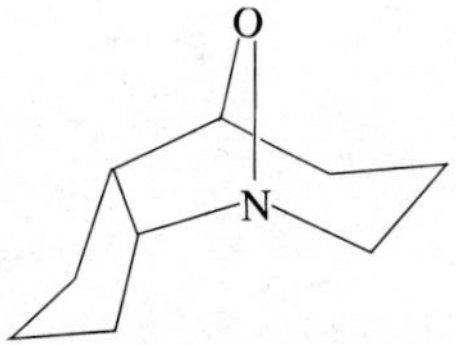 O, N	30

[a] The starting nitrone was formed by reduction and cyclization of the corresponding nitroacetaldehyde.
[b] The starting nitrone was formed by mercuric oxide oxidation followed by a [3, 3] sigmatropic rearrangement.
[c] The starting nitrone was generated by oxidation of the corresponding hydroxylamine with palladium metal.

TABLE XXII. Intramolecular Cycloadditions of C-Alkenyl Substrates

Substrate + Hydroxylamine → Product(s): R^8NHOH

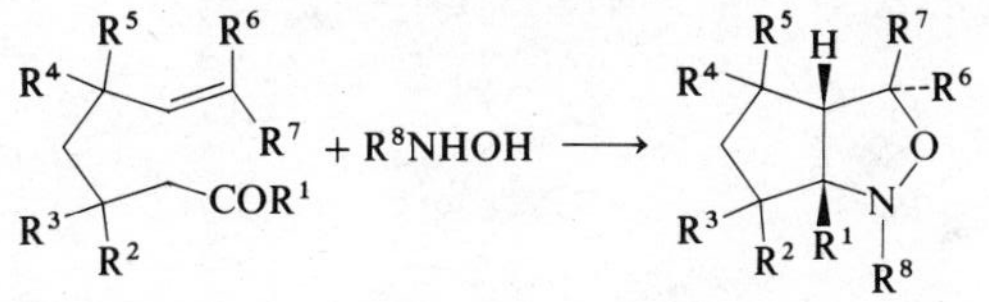

R^1	R^2	R^3	R^4	R^5	R^6	R^7	R^8	Reaction Conditions	Product(s) and Yield(s) (%)	Ref.
H	H	H	H	H	H	H	CH_3	Toluene, reflux	(41)	8
H	H	H	H	H	H	H	CH_3	—	(—)	119
H	H	H	H	H	H	H	C_2H_5	Toluene, reflux	(42)	8
H	H	H	H	H	H	H	i-C_3H_7	Toluene, reflux	(44)	8
H	H	H	H	H	CH_3	H	CH_3	Toluene, reflux	(42)	8
H	H	H	H	H	CH_3	H	CH_3	—	(—)	120
H	H	H	H	H	H	CH_3	CH_3	—	(—)	120
H	H	H	H	H	H	CH_3	CH_3	Toluene, reflux	(55)	8
CH_3	H	H	H	H	H	H	CH_3	Toluene, reflux	(80)	8
H	H	H	H	H	H	CH_3	C_2H_5	Toluene, reflux	(77)	8
CO_2CH_3	H	H	H	H	H	H	CH_3	—	(—)	291
H	CH_3	H	H	H	CH_3	CH_3	CH_3	Toluene, reflux	(55)	8
CO_2CH_3	H	H	CH_3	CH_3	H	H	CH_3	1. CH_2Cl_2 2. Toluene, reflux	(83)	291
H	C_6H_5	CH_3	H	H	H	H	CH_3	C_2H_5OH, pyridine, reflux, 24 h	(57)	125
H	C_6H_5	SCH_3	H	H	H	H	CH_3	C_2H_5OH, pyridine, reflux, 24 h	(62)	125

TABLE XXII. INTRAMOLECULAR CYCLOADDITIONS OF C-ALKENYL SUBSTRATES (*Continued*)

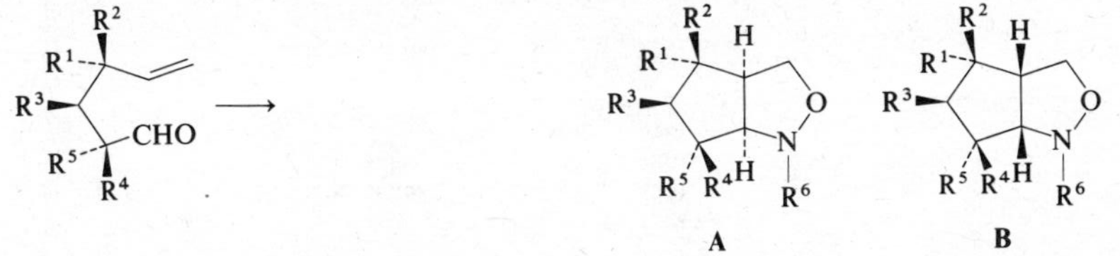

R^1	R^2	R^3	R^4	R^5	R^6	Reaction Conditions	Yield A (%)	Yield B (%)	Ref.
OH	H	—OC(CH$_3$)$_2$O—		H	CH_3	CH_3OH, reflux, 20 min	(—)	(84)	292
OAc	H	OAc	H	OAc	CH_3	C_2H_5OH, pyridine	(50)		126
H	—OC(CH$_3$)$_2$O—		H	OBDMS	CH_3	—	(—)	(—)	293
H	—OC(CH$_3$)$_2$O—		H	$OCH_2C_6H_5$	CH_3	C_2H_5OH, pyridine, 45°	(—)	(—)	127, 293
$O_2CC_6H_5$	H	$O_2CC_6H_5$	H	$OSO_2C_6H_4CH_3$-4	CH_3	—	(73)		127
$OCH_2C_6H_5$	H	$OCH_2C_6H_5$	H	$OCH_2C_6H_5$	CH_3	Propanol, H_2O	(80)		292
$OCH_2C_6H_5$	H	$OCH_2C_6H_5$	H	$OCH_2C_6H_5$	CH_3	—	(—)	(—)	293
"	"	"	$OCH_2C_6H_5$	H	CH_3	—	(—)	(—)	293
"	"	"	H	$OCH_2C_6H_5$	CH_3	$CHCl_3$, reflux, 1 h	(54)		292
"	"	"	"	"	CH_3	$CHCl_3$, reflux, 30 min	(49)		292
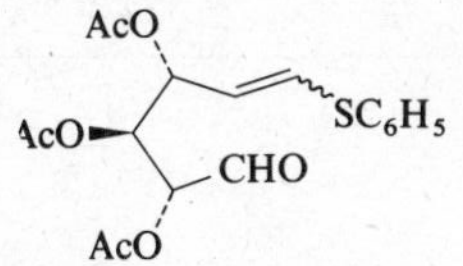			CH_3NHOH			C_2H_5OH, heat	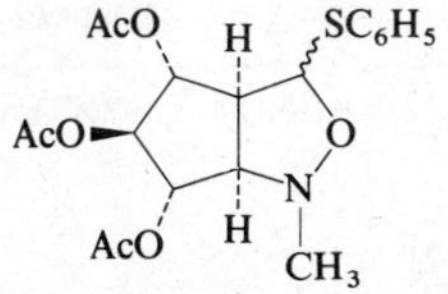	(54)	126

Substrate	Hydroxylamine	Conditions	Product(s) and Yield(s) (%)	Refs.
O=, O, $N(CH_3)_2$	HOHN—C(H)(C_6H_5)	Xylene, reflux	O=, O, H, O, N, H, C_6H_5 H (56) + O=, O, H, O, N, H, C_6H_5 H (12)	147
S, C_4H_9-*n*, CHO, CH_3CO_2NH	$C_6H_5CH_2NHOH$	room temp.	S, H, C_4H_9-*n*, O, N, H, CH_3CO_2NH, $CH_2C_6H_5$ + S, H, C_4H_9-*n*, O, N, H, CH_3CO_2NH, $CH_2C_6H_5$ (1.9:1)	146
C≡CH, $COCH_3$	CH_3NHOH	C_2H_5OH, reflux, 12 h	H, OC_2H_5, H, O, N, CH_3 + H, H, OC_2H_5, O, N, CH_3 (92:8)	134
CH=C=CH_2, $COCH_3$	CH_3NHOH	C_2H_5OH, reflux, 12 h	H, OC_2H_5, O, N, CH_3 (29) + H, OC_2H_5, O, N, CH_3 (10) + O, N, CH_3 (26)	134

TABLE XXII. INTRAMOLECULAR CYCLOADDITIONS OF C-ALKENYL SUBSTRATES (*Continued*)

Substrate	Hydroxylamine	Reaction Conditions	Product(s) and Yield(s) (%)	Ref.
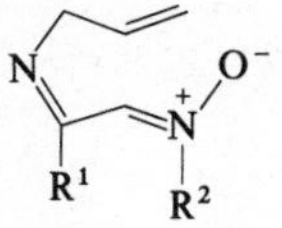 [structure]	(—)		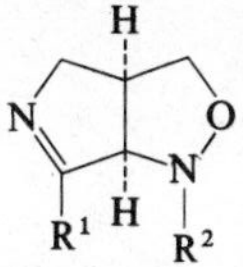 [structure]	
R^1 / R^2				
C_6H_5 / CH_3		Ether, reflux, 2 h	(73)	46
C_6H_5 / C_6H_5		Ether, 30 min	(85)	46
C_6H_5 / C_6H_{11}		Ether, reflux, 14 h	(73)	46
p-BrC_6H_4 / CH_3		Ether, reflux, 4.5 h	(82)	46
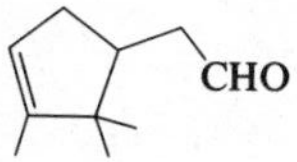 [structure] CHO	R^1NHOH	⟶	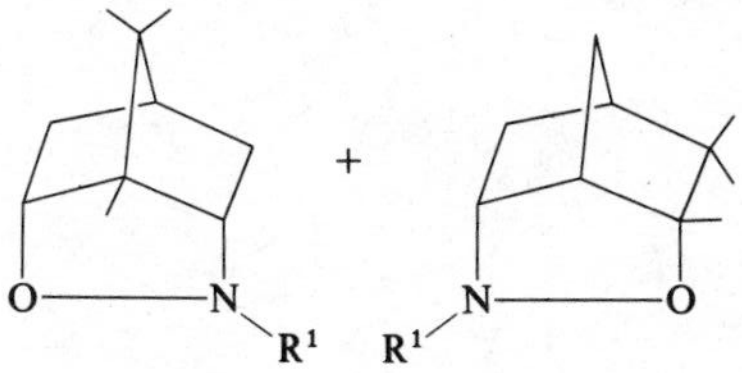 [structure] + [structure]	
	R^1			
	CH_3	—	2:3[b]	124
	i-C_3H_7	—	2:3[b]	124
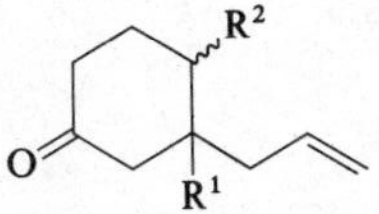 [structure]	R^3NHOH	⟶	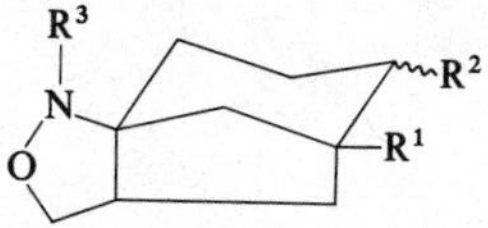 [structure]	
R^1 / R^2 / R^3				
H / H / CH_3		C_6H_6, 80°, 3 h	(84)	138
OH / H / CH_3		C_6H_6, 80°, 3 h	(66)	138
H / CH_2CHCH_2 / CH_3		C_6H_6, 80°, 5 h	(75)	138
H / H / C_6H_5		C_6H_6, 80°, 3 h	(74)	138
H / H / $CH_2C_6H_5$		C_6H_6, 80°, 3 h	(100)	138
OH / H / $CH_2C_6H_5$		C_6H_6, 80°, 3 h	(70)	138
H / CH_2CHCH_2 / "		C_6H_6, 80°, 5 h	(90)	138

Substrate	Reagent	Conditions	Product(s) and Yield(s) (%)	Refs.
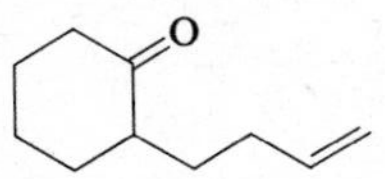	CH_3NHOH	$CDCl_3$, room temp	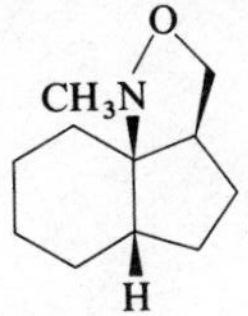 CH_3N, O, H (100)	136
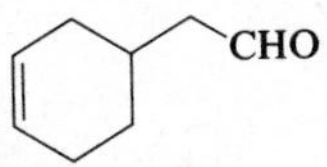 CHO	CH_3NHOH	Toluene, CH_3OH, reflux, 16 h	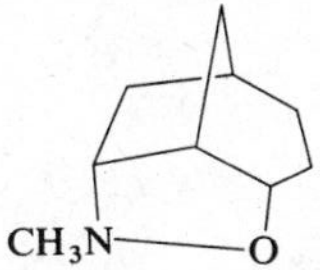 CH_3N, O (55)	130
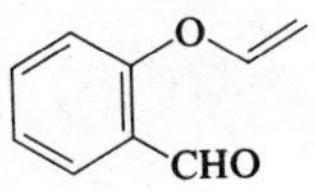 O, CHO	CH_3NHOH	—	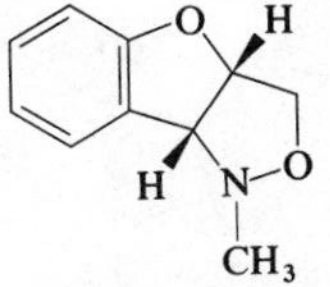 O, H, H, N–O, CH_3 (10) 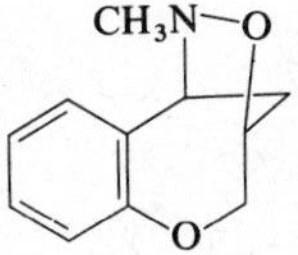 + CH_3N–O, O (21)	132
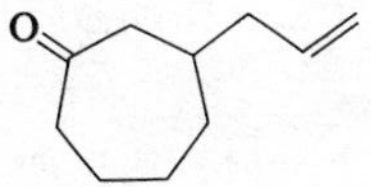 O	R^1NHOH	⟶	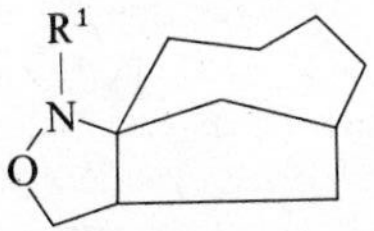 R^1, N, O	
	R^1			
	CH_3	C_6H_6, 80°, 3 h	(84)	138
	$CH_2C_6H_5$	C_6H_6, 80°, 3 h	(82)	138
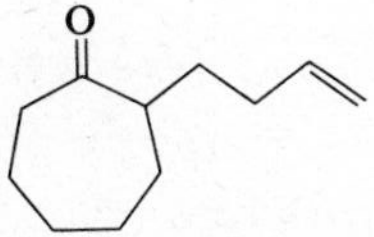 O	CH_3NHOH	CH_3OH, reflux, 48 h	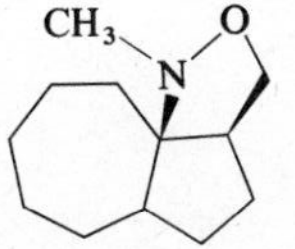 CH_3, N, O (84)	137

TABLE XXII. Intramolecular Cycloadditions of *C*-Alkenyl Substrates (*Continued*)

Substrate	Hydroxylamine	Reaction Conditions	Product(s) and Yield(s) (%)	Ref.
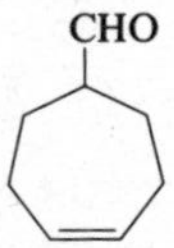 CHO	"	—	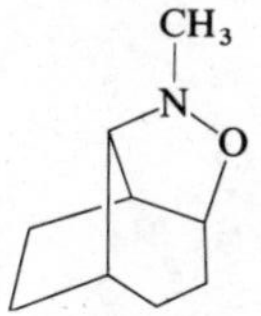 CH_3 N O (60)	124
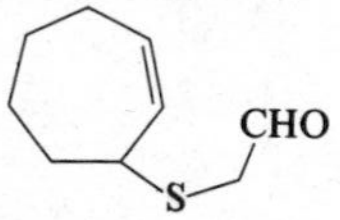 CHO S	R^1NHOH	⟶	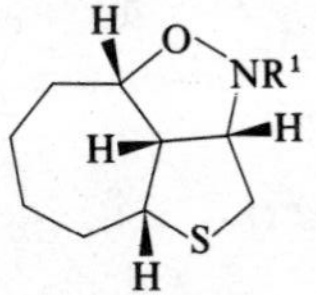 H O NR^1 H H S H	
	R^1			
	CH_3	CH_3CN, reflux, 2 h	(—)	145
	$CH_2C_6H_5$	CH_3CN, reflux, 2 h	(66)	145
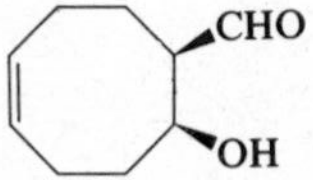 CHO OH	R^1NHOH	⟶	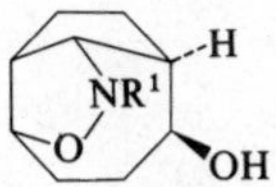 H NR^1 O OH	
	R^1			
	CH_3	2-Methyl-2-butanol, reflux, 3 d	(65)	101
	C_6H_5	2-Methyl-2-butanol, reflux, 3 d	(27)	101

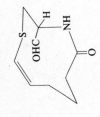

Reactant	Hydroxylamine	Conditions	Product(s) and Yield(s) (%)	Refs.
OHC, H, S, NH, O	$C_6H_5CH_2NHOH$	1. CH_2Cl_2 2. Toluene, reflux	H, O, S, N, $CH_2C_6H_5$, NH, O (63)	146
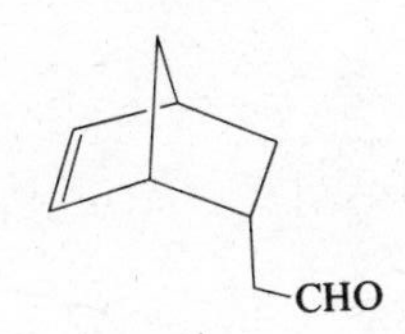	CH_3NHOH	—	O, CH_3N (—)	120
CHO	R^1NHOH	⟶	O, NR^1	
	R^1			
	C_6H_5	C_6H_6, 20°, 2 d	(44)	129
	$CH_2C_6H_5$	C_6H_6, 20°, 2 d	(87)	129
	"	Toluene, reflux, 10 h	(80)	129

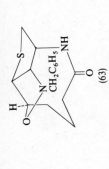

TABLE XXII. INTRAMOLECULAR CYCLOADDITIONS OF C-ALKENYL SUBSTRATES (*Continued*)

Substrate	Hydroxylamine	Reaction Conditions	Product(s) and Yield(s) (%)		Ref.
[structure: CHO-substituted bicyclic alkene]	R'NHOH	⟶	A [structure]	B [structure]	
	R^1				
	CH_3	C_6H_6, 70°, 41 h	(29)	(25)	129
	"	C_2H_5OH, 70°, 24 h	(14)	(20)	129
	"	DMSO, 70°, 10 h	(12)	(31)	129
	"	Toluene, reflux, 11 h	(26)	(7)	128
		C_6H_6, 80°, 15 h	(35)	(14)	129
	"	C_2H_5OH, 80°, 15 h	(10)	(10)	129
	"	DMSO, 80°, 10 h	(18)	(13)	129
	C_6H_5	C_6H_6, room temp, 12 h	(17)	(39)[c]	129
	"	C_2H_5OH, room temp, 10 h	(6)	(37)[c]	129
	"	DMSO, room temp, 20 h	(23)	(40)[c]	129
	$C_6H_5CH_2$	C_6H_6, 70°, 10 h	(38)	(20)[c]	129
	"	C_2H_5OH, 70°, 10 h	(19)	(2)[d]	128
	"	Toluene, reflux, 12 h	(31)	(3)	129
	2-Adamantyl	C_6H_6, 80°, 10 h	(40)	(30)	128
	"	C_2H_5OH, 80°, 15 h	(13)	(30)	129
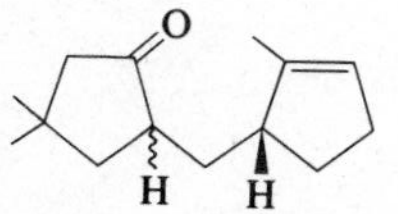 [structure: dimethylcyclopentanone linked to methylcyclopentene; H, H]	CH_3NHOH	Toluene, reflux, 24–36 h	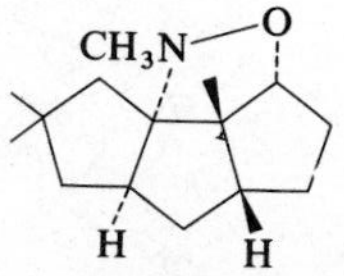 [structure: CH_3N–O tricyclic isoxazolidine; H, H] (75)[e]		139

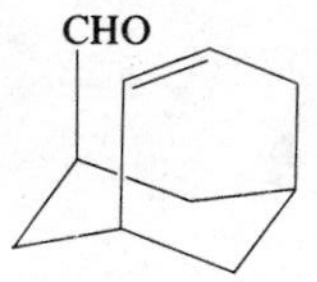

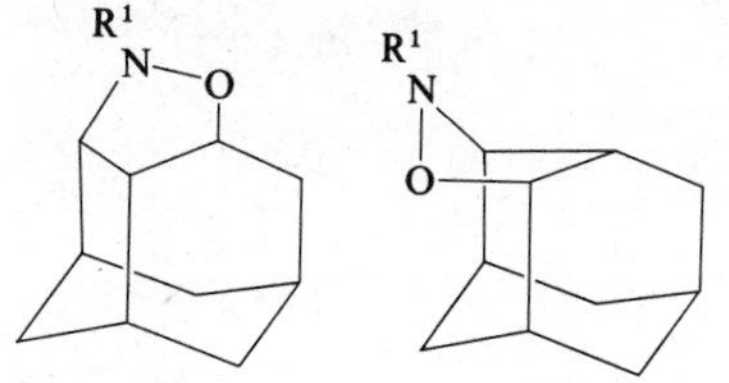

Substrate	Reagent	Conditions	Product(s) and Yield(s) (%)	Refs.
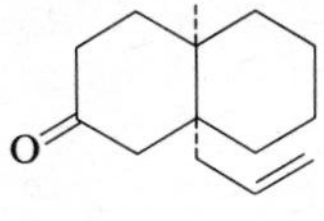	$C_6H_5CH_2NHOH$	C_6H_6, 80°, 2 h 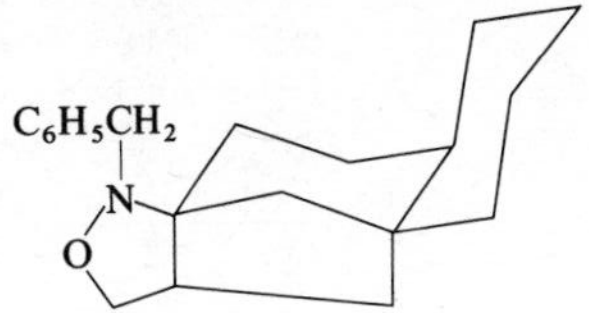	$C_6H_5CH_2$ (94)	138
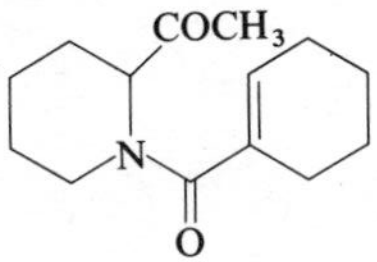 $COCH_3$	CH_3NHOH	C_2H_5OH, reflux, 19 h 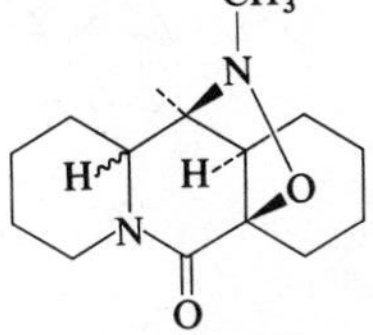	CH_3 (5.3) + CH_3 (64.7)	140
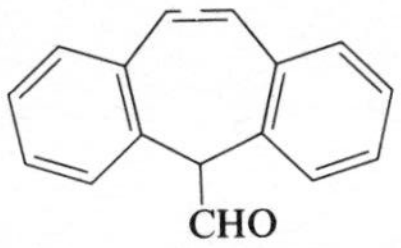 CHO	CH_3NHOH	1. C_2H_5OH 2. Toluene, reflux, 3.5 h 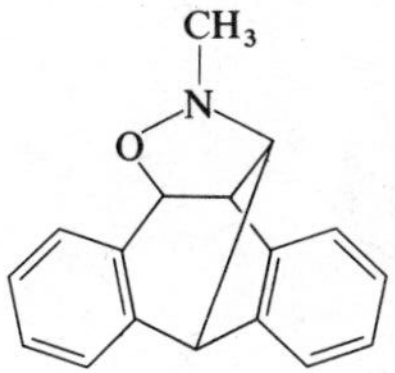	CH_3 (90)	123

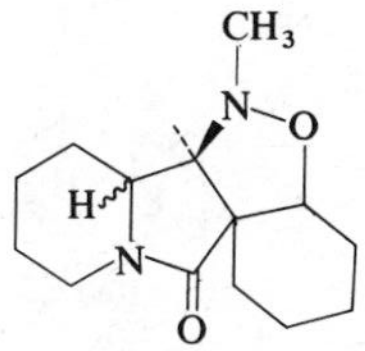

TABLE XXII. INTRAMOLECULAR CYCLOADDITIONS OF C-ALKENYL SUBSTRATES (*Continued*)

Substrate	Hydroxylamine	Reaction Conditions	Product(s) and Yield(s) (%)			Ref.
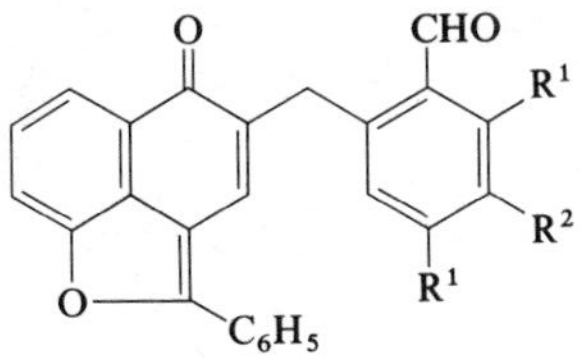 (naphthofuranone–benzaldehyde substrate: R¹, R², CHO, C_6H_5)	C_6H_5NHOH		(bridged isoxazolidine product)	+ (spiro isoxazolidine product)		
R^1 = H, R^2 = H		C_2H_5OH, room temp, 5 d	(87)	(0)		294
R^1 = H, R^2 = H		C_2H_5OH, room temp, 30 h	(80)	(11)		294
R^1 = OCH_3, R^2 = CO_2CH_3		C_2H_5OH, room temp, 6 d	(80)	(0)		294
$R^2CH(CH_2COR^1)CH_2CH_2CH=CR^3R^4$	R^5NHOH	⟶	**A**	+ **B**	+ **C**	

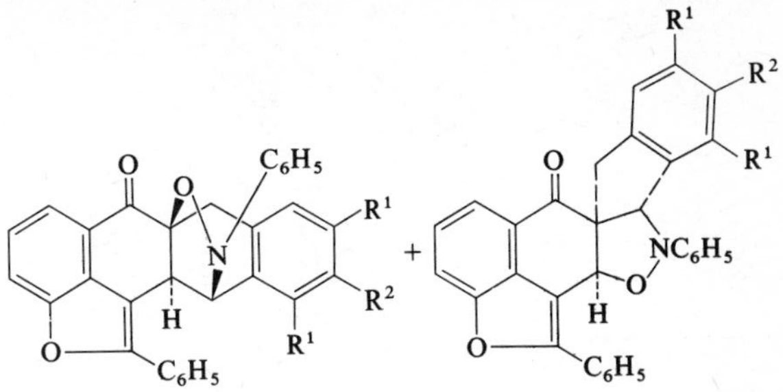

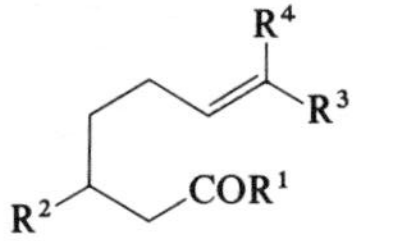

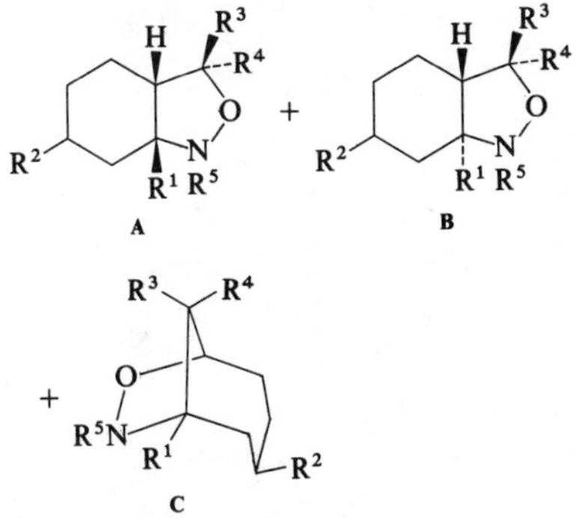

R^1	R^2	R^3	R^4	R^5	Reaction Conditions	A	B	C	Ref.
H	H	H	H	CH_3	—	3	: 1	: 1	133
H	H	H	H	CH_3	Toluene, reflux	(48)[f]	(0)	(0)	8
CH_3	H	H	H	CH_3	—	(—)	(0)	(—)	133
CH_3	H	H	H	CH_3	Toluene, reflux	(46)	(0)	(23)	8
H	CH_3	H	H	CH_3	Toluene, reflux, 19 h	(28)	(49)	(4)	133
H	CH_3	H	H	CH_3	C_2H_5OH	(22)	(74)	(4)	133

R^1	R^2	R^3	R^4	R^5	Conditions	A	B	C	Refs.
H	H	CH_3	CH_3	CH_3	C_2H_5OH	(19.5)	(3.5)	(0)	133
CO_2CH_3	H	H	H	CH_3	1. CH_2Cl_2 2. Toluene, reflux	(—)	(0)	(0)	291
H	CH_3	CH_3	CH_3	CH_3	25°	97 :	3 :	0	133
H	CH_3	CH_3	CH_3	CH_3	138°	87 :	13 :	0	133
H	CH_3	CH_3	CH_3	CH_3	Toluene, reflux	(65)	(0)	(0)	8
H	H	CH_3	H	i-C_3H_7	Toluene, reflux	(45)	(0)	(0)	8
H	CH_3	CH_3	CH_3	C_2H_5	Toluene, reflux	(71)	(0)	(0)	8
H	CH_3	CH_3	CH_3	i-C_3H_7	Toluene, reflux	(74)	(0)	(0)	8
H	CH_3	CH_3	CH_3	C_6H_5	Toluene, reflux	(31)	(0)	(0)	8

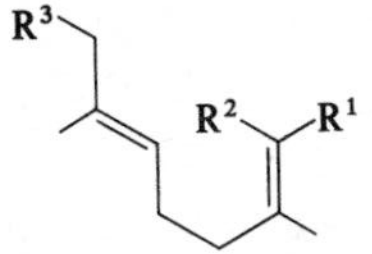

Substrate: R^3, R^2, R^1 (structure) — Reagent: CH_3NHOH — $\longrightarrow$ — Product: H, R^3, O, N, H, CH_3 (structure)

R^1	R^2	R^3	Conditions	Product (yield)	Refs.
CHO	H	H	1. — 2. Xylenes, reflux	(79)	156
H	CHO	H	1. — 2. Xylenes, reflux	(80)[g]	156
(CHO)	(H)[h]	$(CH_3)_2C{=}CHCH_2$	1. — 2. Xylenes, reflux	(—)	155

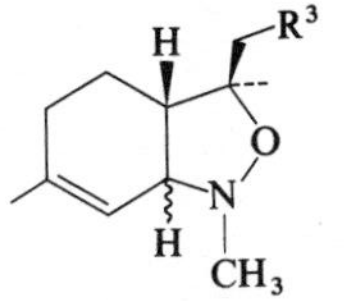

Substrate	Reagent	Conditions	Product (yield)	Refs.
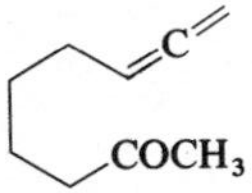 C, $COCH_3$ (structure)	CH_3NHOH	Toluene	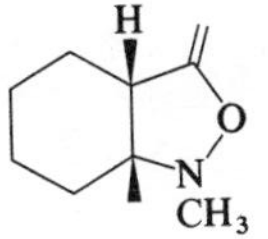 H, O, N, CH_3 (structure) (15)	134
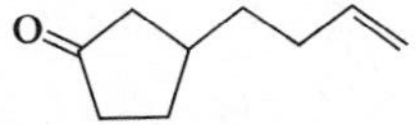 O (structure)	$C_6H_5CH_2NHOH$	Toluene, 111°, 24 h	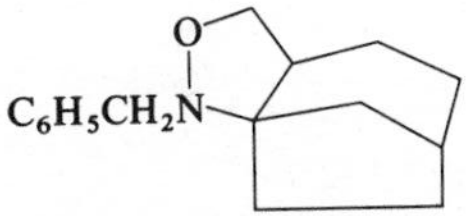 $C_6H_5CH_2N$, O (structure) (50)	138

TABLE XXII. INTRAMOLECULAR CYCLOADDITIONS OF *C*-ALKENYL SUBSTRATES (*Continued*)

Substrate	Hydroxylamine	Reaction Conditions	Product(s) and Yield(s) (%)	Ref.
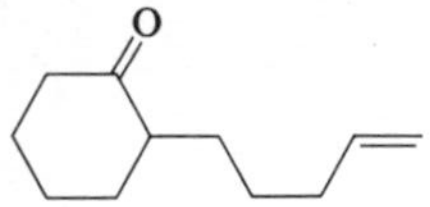	CH_3NHOH	1. — 2. Toluene, reflux, 16 h	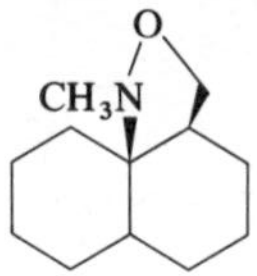 CH_3N (28) 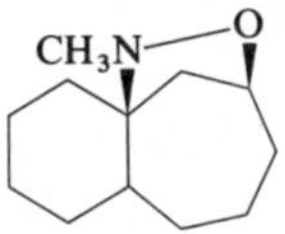 + CH_3N (28)	137
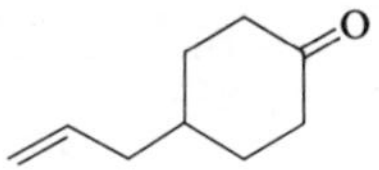	$C_6H_5CH_2NHOH$	Toluene, 111°, 36 h	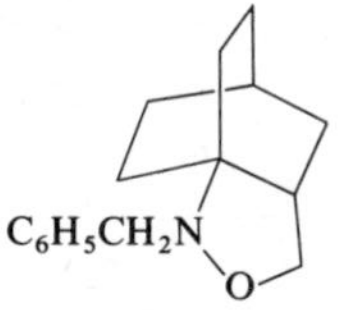 $C_6H_5CH_2N$ (48)	138
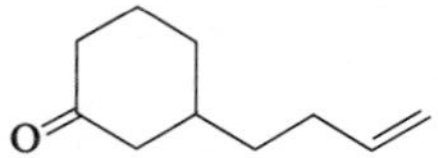	CH_3NHOH	Toluene, reflux, 48 h	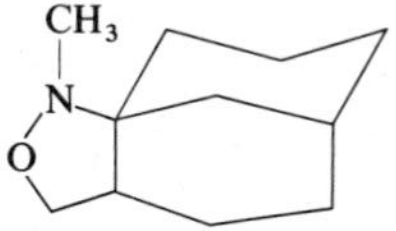 CH_3 N O (46)	138
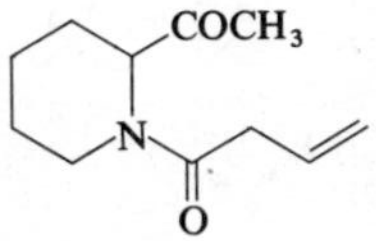 COCH₃ N O	R^1NHOH	⟶	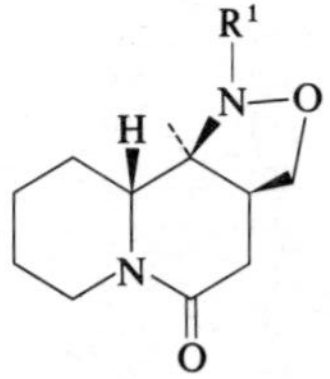 R^1 N O H N O	
	R^1			
	CH_3	C_2H_5OH, reflux, 25 h	(45)	140
	$C_6H_5CH_2$	C_2H_5OH, reflux, 4 d	(21)	140

Reactant	Reagent	Conditions	Products (yield %)		Refs.
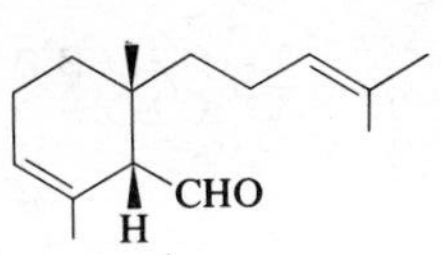 CHO, H	CH_3NHOH	→	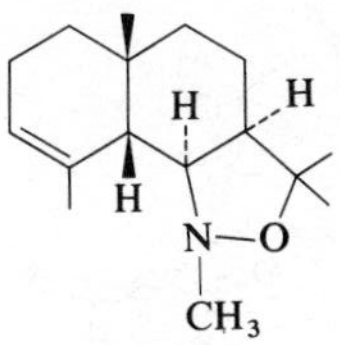 H, H, H, N—O, CH_3	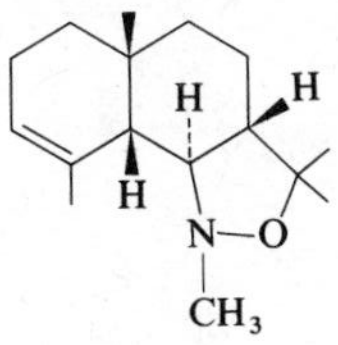 H, H, H, N—O, CH_3	
		1. C_2H_5OH, room temp 2. Toluene, 70°, 4–6 d	(74)	(9)	295
		1. C_2H_5OH, room temp 2. Xylene, 140°, 15–17 h	(31)	(34)	295
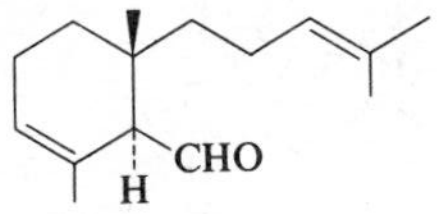 CHO, H	CH_3NHOH	→	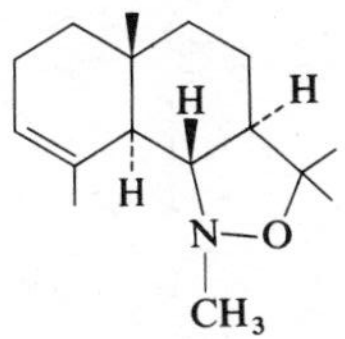 H, H, H, N—O, CH_3	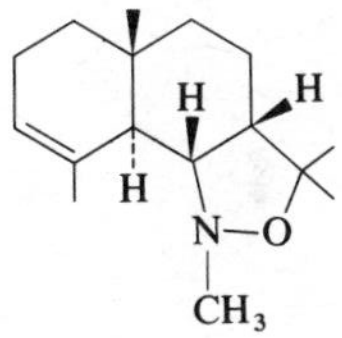 H, H, H, N—O, CH_3	
		1. C_2H_5OH, 25° 2. Toluene, 90°, 4–6 d	(50)	(12)	295
		1. C_2H_5OH 2. Xylene, 140°, 15–17 h	(44)	(20)	295

TABLE XXII. INTRAMOLECULAR CYCLOADDITIONS OF C-ALKENYL SUBSTRATES (*Continued*)

Substrate				Hydroxylamine	Reaction Conditions	Product(s) and Yield(s) (%)	Ref.
2-(R^1(R^2)C=C(R^3)... see structure): 2-CHO-C_6H_4-X-$CH_2C(R^1)$=CR^2R^3				R^4NHOH	→	Tricyclic isoxazolidine (N-R^4; R^1, R^2, R^3; ring X)	
X	R^1	R^2	R^3	R^4			
O	H	H	H	H	Toluene, 110°	(40)	132
O	H	H	H	CH_3	Toluene, reflux	(—)	296
O	Br	H	H	CH_3	Toluene, 110°, 5 h.	(75)	132
O	H	H	CH_2Cl	CH_3	Toluene, reflux	(—)	296
O	H	H	CH_3	CH_3	Toluene, reflux	(—)	296
O	H	CH_3	H	CH_3	Toluene, reflux	(—)	296
O	CH_3	H	H	CH_3	Toluene, 110°, 5 h	(75)	132
NCHO	H	H	H	CH_3	Toluene, 110°	(80)	132
O	H	H	H	i-C_3H_7	Toluene, 110°, 5 h	(77)	132
O	H	H	CO_2CH_3	CH_3	Toluene, reflux	(—)	296
O	$CO_2C_2H_5$	H	H	CH_3	Toluene, 110°, 5 h	(64)[i]	132
O	H	H	C_6H_5	CH_3	Toluene, reflux	(—)	296

Substrate			Hydroxylamine	Reaction Conditions	Product(s) and Yield(s) (%)		Ref.
4-R^1-2-(R^2CH=CH)-C_6H_3-O-CH_2CHO			R^3NHOH	→	Isoxazolidino-chroman (R^1, R^2, NR^3)	+ bridged N-CH_3 isoxazolidine	
R^1	R^2	R^3					
H	H	CH_3		—	(10)	(40)	297
H	$CO_2C_2H_5$	CH_3		—	(58)		297
$C_6H_5CH_2O$	$CO_2C_2H_5$	CH_3		—	(43)		297
$C_6H_5CH_2O$	$CO_2C_2H_5$	i-C_3H_7		—			297

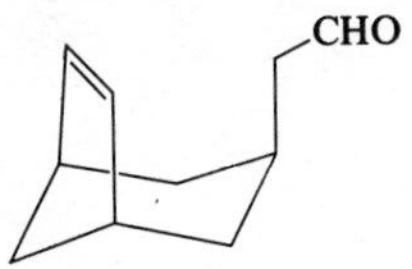

Reactant	Nitrone precursor	Conditions	Product(s) and Yield(s) (%)	Refs.
CHO (bicyclic aldehyde)	$C_6H_5CH_2NHOH$	C_6H_6, reflux, 11 h	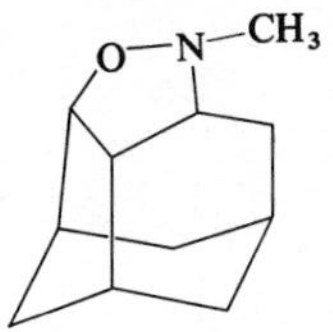 O—N—CH_3 (92)	129

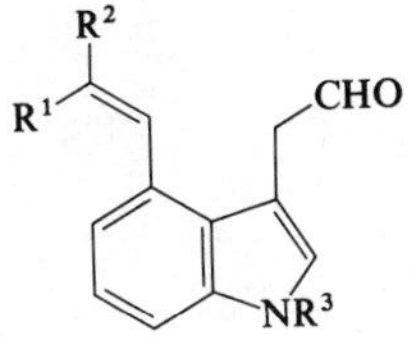

Reactant: 4-(R^1,R^2-vinyl)indole-3-acetaldehyde (CHO, NR^3); nitrone precursor: CH_3NHOH; products: A + B + C (CH_3, O—N, H, R^1, R^2, NR^3)

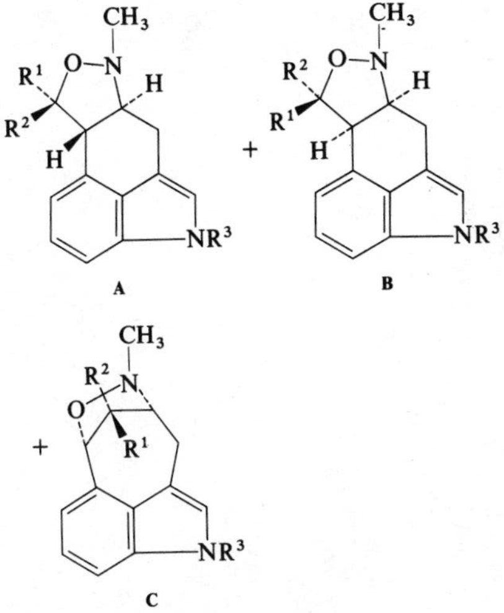

R^1	R^2	R^3	Conditions	A	B	C	Refs.
H	H	H	C_6H_6, CH_3OH, reflux, 15 h	(0)	(0)	(56)[j]	298
OCH_3	H	H	C_6H_6, CH_3OH, reflux, 3 h	(30)[j]	(16)	(0)	298
OCH_3	H	H	C_6H_6, reflux, 12 h	(36)	(20)	(0)	298
H	OCH_3	$CO_2C_2H_5$		(0)	(—)	(0)	299
H	CO_2CH_3	H	C_6H_6, reflux	(0)	(—)	(0)	148
H	CO_2CH_3	H	CH_2Cl_2, CH_3OH, 70°, 2 h	(0)	(43)[j]	(11)	298
H	CO_2CH_3	H	Toluene, CH_3OH, 70°	(0)	(43)	(—)	298

TABLE XXII. INTRAMOLECULAR CYCLOADDITIONS OF *C*-ALKENYL SUBSTRATES (*Continued*)

Substrate	Hydroxylamine	Reaction Conditions	Product(s) and Yield(s) (%)	Ref.
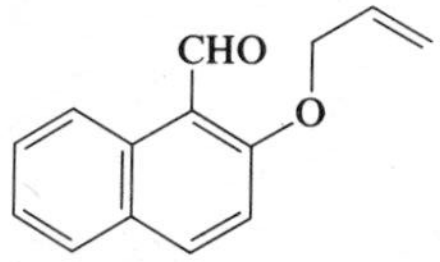 CHO, O	CH_3NHOH	Toluene, 110°	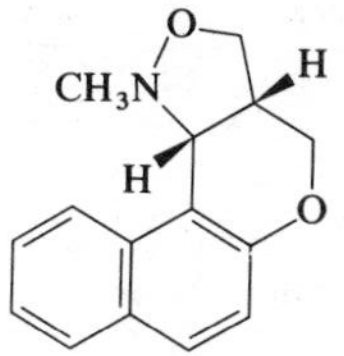 CH_3N, O, H, H, O (79)	132
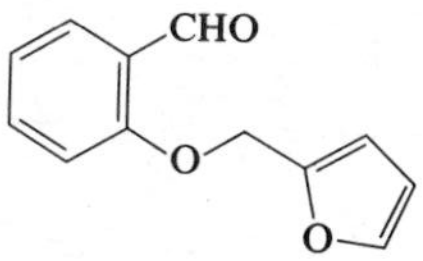 CHO, O, O	C_6H_5NHOH	Toluene, reflux, 5 h	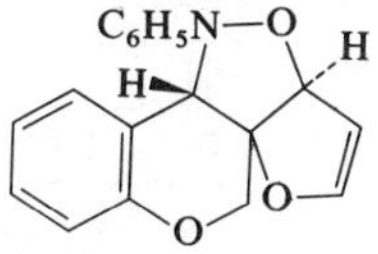 C_6H_5N—O, H, H, O, O (76)	220
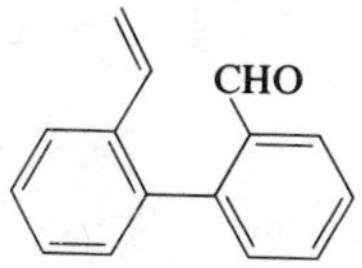 CHO	C_6H_5NHOH	C_2H_5OH, room temp, 4 h	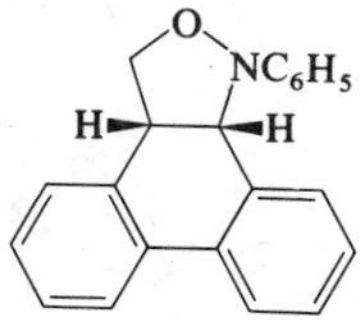 O, NC_6H_5, H, H (66)	121
CHO, X	CH_3NHOH		CH_3N—O, H, H, H, X, H CH_3N—O, H, H, H, X, H	
X: O		Toluene, 110°	(51) (0)	132
X: NCHO		Toluene, 100°	(64) (17)	132

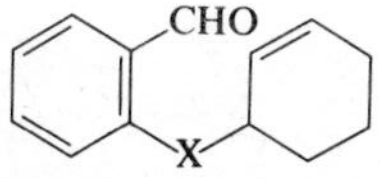

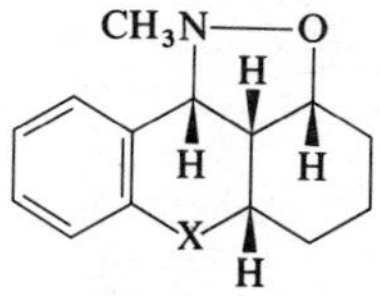

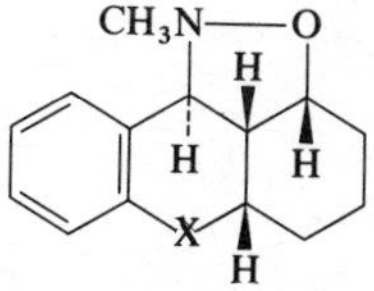

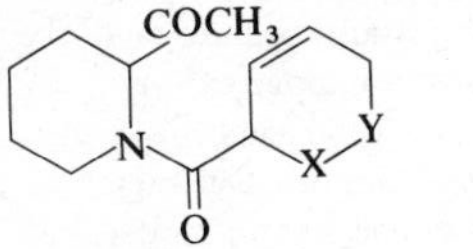

Reactant	Reagent	Conditions	Product(s) and Yield(s) (%)	Refs.
COCH$_3$ … N … X–Y (diene amide)	R^1NHOH		R^1–N–O … H, H, H, H (tetracyclic isoxazolidine)	
X–Y: $-CH_2CH_2-$; R^1: CH_3		C_2H_5OH, reflux, 34 h	(51)	140
X–Y: $-CH_2CH_2-$; R^1: $C_6H_5CH_2$		C_6H_6, reflux, 71 h	(60)	140
X–Y: $-CH{=}CH-$; R^1: CH_3		C_2H_5OH, reflux, 21 h	(3)	140
X–Y: $-CH{=}CH-$; R^1: $C_6H_5CH_2$		C_6H_6, reflux	(32)	140
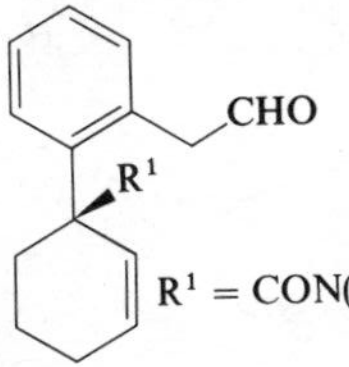 CHO, R^1; $R^1 = CON(CH_3)_2$	CH_3NHOH	C_6H_6, reflux	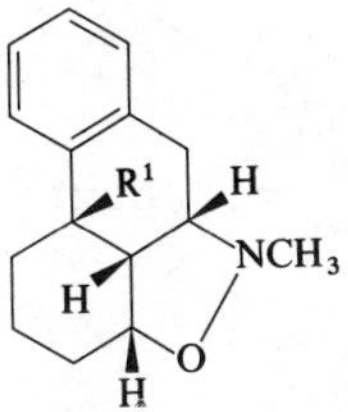 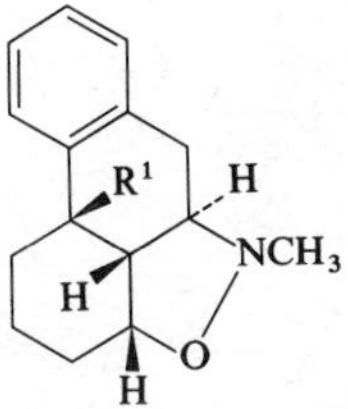 R^1, H, NCH_3, O, H, H (3C) + R^1, H, NCH_3, O, H, H (36)	122
X, CHO (dibenzazepine/dibenzocycloheptene)	CH_3NHOH		CH_3–N–O … X	
X: N		CH_2Cl_2, 25°, 72 h	(70)	123
X: CH		1. CH_2Cl_2 2. Toluene, reflux, 0.5 h	(87)	123

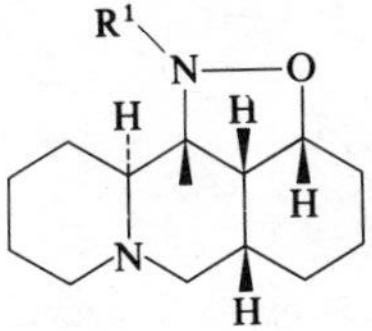

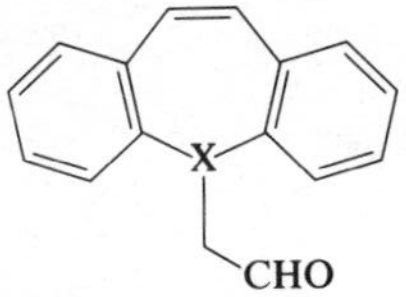

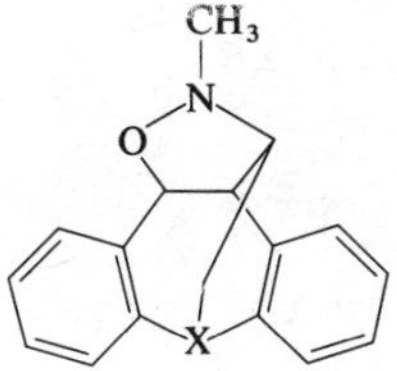

TABLE XXII. INTRAMOLECULAR CYCLOADDITIONS OF *C*-ALKENYL SUBSTRATES (*Continued*)

Substrate	Hydroxylamine	Reaction Conditions	Product(s) and Yield(s) (%)	Ref.
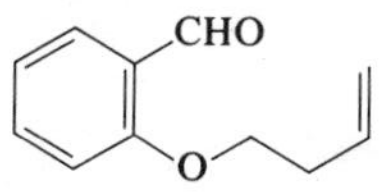 CHO, O (structure)	CH_3NHOH	Toluene, 110°	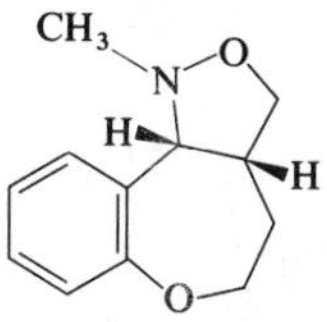 CH_3, N, O, H, H, O (structure) (26)	132
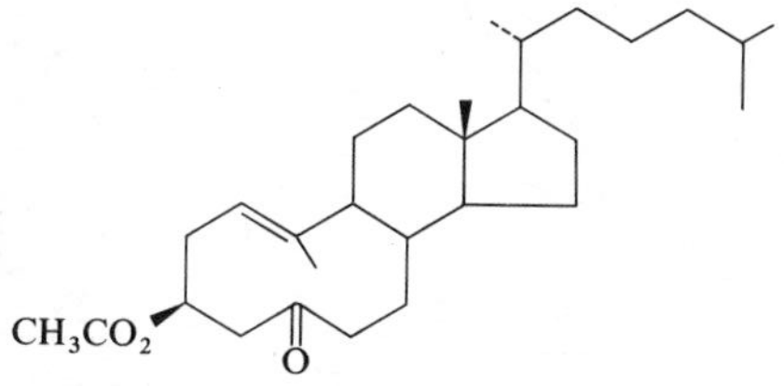 CH_3CO_2, O (structure)	R^1NHOH		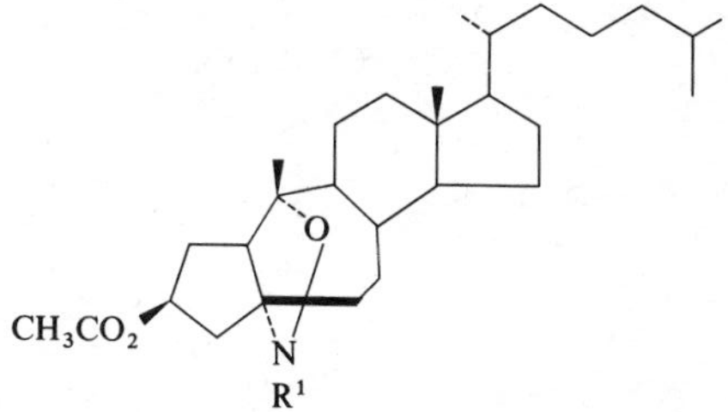 CH_3CO_2, O, N, R^1 (structure)	
R^1 = H		C_2H_5OH, HCl/pyridine, reflux, 7 h	(53)	135
R^1 = CH^3		C_2H_5OH, HCl/pyridine, reflux, 2d	(65)	135

[a] A 3.2% yield of the corresponding regioisomer was also obtained.
[b] The products were obtained in yields ranging from 70 to 80%.
[c] A minor amount of the product corresponding to the reduction of **B** to its amino alcohol was formed.
[d] A 32% yield of the product corresponding to the reduction of **B** to its amino alcohol was formed.
[e] Under these reaction conditions, a *syn* isomer epimerizes to the *anti* isomer, which is the only one that can undergo cycloaddition.
[f] This refers to the combined yield of **A** and **B**.
[g] The isomer ratio at the ring junction is 3:1 (*cis*:*trans*).
[h] The cycloaddition was carried out on the *cis* and *trans* mixture of nitrones.
[i] The bridged product corresponding to cycloaddition of the other regioisomer was formed in 26% yield.
[j] This is the overall yield for a three-step procedure starting from the cyanoolefin.

REFERENCES

[1] L. I. Smith, *Chem. Rev.*, **23**, 193 (1938).
[2] H. Stamm and J. Hoenicke, *Justus Liebigs Ann. Chem.*, **749**, 146 (1971).
[3] H. Steudle and H. Stamm, *Arch. Pharm.*, **309**, 935 (1976).
[4] E. Beckmann, *Ber. Dtsch. Chem. Ges.*, **23**, 1680, 3331 (1890).
[5] R. Huisgen, *Angew. Chem., Int. Ed. Engl.*, **7**, 321 (1968).
[6] R. Huisgen, *J. Org. Chem.*, **33**, 2291 (1968).
[7] R. Huisgen, *Helv. Chim. Acta*, **50**, 2421 (1967).
[8] N. A. LeBel, M. E. Post, and J. J. Whang, *J. Am. Chem. Soc.*, **86**, 3759 (1964).
[9] For a review of these methodologies, see J. Hamer and A. Macaluso, *Chem. Rev.*, **64**, 473 (1964).
[10] D. H. Jonson, M. A. T. Rogers, and G. Trappe, *J. Chem. Soc.*, **1956**, 1093.
[11] R. Bonnett, R. F. C. Brown, V. M. Clark, I. O. Sutherland, and A. Todd, *J. Chem. Soc.*, **1959**, 2094, 2109.
[12] J. Thesing, *Chem. Ber.*, **87**, 507 (1954).
[13] G. Renner, *Anal. Chem.*, **193**, 92 (1963).
[14] G. E. Utzinger and F. A. Regenass, *Helv. Chim. Acta*, **37**, 1892 (1954).
[15] P. Grammaticakis, *Compt. Rend.*, **224**, 1066 (1947).
[16] J. Thesing, A. Muller, and G. Michel, *Chem. Ber.*, **88**, 1030 (1955).
[17] G. E. Utzinger, *Justus Liebigs Ann. Chem.*, **55**, 903 (1961).
[18] H. E. De La Mare and G. M. Coppinger, *J. Org. Chem.*, **28**, 1068 (1963).
[19] J. Thesing and W. Sirrenberg, *Justus Liebigs Ann. Chem.*, **609**, 46 (1957).
[20] J. Thesing and H. Maver, *Chem. Ber.*, **89**, 2159 (1956).
[21] J. J. Tufariello and J. J. Tegeler, *Tetrahedron Lett.*, **1976**, 4037.
[22] J. J. Tufariello and S. A. Ali, State University of New York, Buffalo, unpublished observation.
[23] O. Exner, *Collect. Czech. Chem. Commun.*, **16**, 258 (1951).
[24] J. Meisenheimer and J.-L. Chou, *Justus Liebigs Ann. Chem.*, **539**, 78 (1939).
[25] R. F. C. Brown, V. M. Clark, and A. Todd, *Proc. Chem. Soc.*, **1957**, 97.
[26] G. D. Buckley and N. H. Ray, *J. Chem. Soc.*, **1949**, 1154.
[27] G. R. Delpierre and M. Lamchen, *J. Chem. Soc.*, **1963**, 4693.
[28] M. C. Kloetzel, F. L. Chubb, R. Gobran, and J. L. Pinkus, *J. Am. Chem. Soc.*, **83**, 1128 (1961).
[29] J. Wiemann and C. Glacet, *Bull. Soc. Chim. Fr.*, **17**, 176 (1950).
[30] N. A. LeBel and B. W. Caprathe, *J. Org. Chem.*, **50**, 3938 (1985).
[31] P. A. S. Smith and J. E. Robertson, *J. Am. Chem. Soc.* **84**, 1197 (1962).
[32] W. D. Emmons, *J. Am. Chem. Soc.*, **78**, 6208 (1956).
[33] W. D. Emmons, *J. Am. Chem. Soc.*, **79**, 5739 (1957).
[34] M. F. Hawthorne and R. D. Strahm, *J. Org. Chem.*, **22**, 1263 (1957).
[35] J. S. Splitter and M. Calvin, *J. Org. Chem.*, **23**, 651 (1958).
[36] S. Murahashi, *Tetrahedron Lett.*, **24**, 1049 (1983).
[36a] S. Murahashi and T. Shiota, *Tetrahedron Lett.*, **28**, 2383 (1987).
[37] R. Foster, J. Iball, and R. Nash, *J. Chem. Soc., Chem. Commun.*, **1968**, 1414.
[38] R. Foster, J. Iball, and R. Nash, *J. Chem. Soc., Perkin Trans. 2*, **1974**, 1210.
[39] R. Huisgen, *Angew. Chem., Int. Ed. Engl.*, **2**, 565 (1963).
[40] R. Huisgen, R. Sustmann, and K. Bunge, *Chem. Ber.*, **105**, 1324 (1972).
[41] R. B. Woodward and R. Hoffmann, *The Conservation of Orbital Symmetry*, Academic Press, New York, **1970**, pp. 87–89.
[42] A. Eckell, R. Huisgen, R. Sustmann, Gl. Wallbillich, D. Grashey, and E. Spindler, *Chem. Ber.*, **100**, 2192 (1967).
[43] K. Fukui, *Bull. Chem. Soc. Jpn.*, **39**, 498 (1966).
[44] K. N. Houk and C. R. Watts, *Tetrahedron Lett.*, **1970**, 4025.
[45] K. N. Houk and L. J. Luskus, *Tetrahedron Lett.*, **1970**, 4029.
[46] D. St. C. Black, R. F. Crozier, and I. D. Rae, *Aust. J. Chem.*, **31**, 2239 (1978).
[47] Y.-M. Chang, J. Sims, and K. N. Houk, *Tetrahedron Lett.*, **1975**, 4445.

[48] J. Sims and K. Houk, *J. Am. Chem. Soc.*, **95**, 5798 (1973).
[49] R. A. Firestone, *Tetrahedron*, **33**, 3009 (1977).
[50] R. A. Firestone, *J. Org. Chem.*, **33**, 2285 (1968).
[51] R. A. Firestone, *J. Org. Chem.*, **37**, 2181 (1972).
[52] R. A. Firestone, *J. Chem. Soc.* (*A*), **1970**, 1570.
[53] G. Leroy, M. T. Nguyen, and M. Sana, *Tetrahedron*, **34**, 2459 (1978).
[54] G. Leroy and M. Sana, *Tetrahedron*, **32**, 709 (1976).
[55] G. Leroy and M. Sana, *Tetrahedron*, **31**, 2091 (1975).
[56] K. N. Houk, *J. Am. Chem. Soc.*, **94**, 8953 (1972).
[57] K. N. Houk, J. Sims, R. E. Duke, Jr., K. W. Strozier, and J. K. George, *J. Am. Chem. Soc.*, **95**, 7287 (1973).
[58] K. N. Houk, J. Sims, C. R. Watts, and L. J. Luskus, *J. Am. Chem. Soc.*, **95**, 7301 (1973).
[59] K. N. Houk, *Acc. Chem. Res.*, **8**, 361 (1975).
[60] Y. Inouye, Y. Watanabe, S. Takahashi, and H. Kakisawa, *Bull. Chem. Soc. Jpn.*, **52**, 3763 (1979).
[61] R. Huisgen, *J. Org. Chem.*, **41**, 403 (1976).
[62] L. Salem, *J. Am. Chem. Soc.*, **90**, 553 (1968).
[63] A. Devaquet and L. Salem, *J. Am. Chem. Soc.*, **91**, 3793 (1969).
[64] R. Sustmann, Tetrahedron Lett., **1971**, 2717.
[65] J. Bastide and O. Henri-Rousseau, *Bull. Soc. Chim. Fr.*, **1973**, 2290.
[66] K. N. Houk, Y.-M. Chang, R. W. Strozier, and P. Carmella, *Heterocycles*, **7**, 793 (1977).
[67] R. Huisgen, H. Seidl, and I. Bruning, *Chem. Ber.*, **102**, 1102 (1969).
[68] K. Houk, A. Bimanand, D. Mukherjee, J. Sims, Y.-M. Chang, D. Kaufman, and L. Domelsmith, *Heterocycles*, **7**, 293 (1977).
[69] G. Klopman, *J. Am. Chem. Soc.*, **90**, 223 (1968).
[70] A. Z. Bimanand and K. N. Houk, *Tetrahedron Lett.*, **24**, 435 (1983).
[71] H. Iida and C. Kibayashi, *Tetrahedron Lett.*, **22**, 1913 (1981).
[72] R. Huisgen, H. Hauck, H. Seidl, and M. Burger, *Chem. Ber.*, **102**, 1117 (1969).
[73] E. J. Fornefeld and A. J. Pike, *J. Org. Chem.*, **44**, 835 (1979).
[74] T. Sasaki, T. Yoshioka, and I. Izure, *Bull. Chem. Soc. Jpn.*, **41**, 2964 (1968).
[75] S. Shatzmiller, E. Shabm, R. Lidor, and E. Tartkovski, *Justus Liebigs Ann. Chem.*, **1983**, 906.
[76] H. Iida, M. Tanaka and C. Kibayashi, *J. Org. Chem.*, **49**, 1909 (1984).
[77] R. Huisgen, R. Geashey, H. Hauck, and H. Seidl, *Chem. Ber.*, **101**, 2043 (1968).
[78] M. Masui, K. Suda, M. Yamauchi, and C. Yijima, *Chem. Pharm. Bull.*, **21**, 1605 (1973).
[79] S. A. Ali, P. Senaratne, C. Illig, H. Meckler, and J. Tufariello, *Tetrahedron Lett.*, **1979**, 4167.
[80] E. Gossinger, *Tetrahedron Lett.*, **21**, 2229 (1980).
[81] R. Huisgen, H. Hauck, R. Grashey, and H. Seidl, *Chem. Ber.*, **101**, 2568 (1968).
[82] J. J. Tufariello, S. A. Ali, and H. O. Klingele, *J. Org. Chem.*, **44**, 4213 (1979).
[83] Y. V. Svetkin, N. A. Akmanova, and G. I. Plotnikova, *Zh. Org. Khim.* (*USSR*), **8**, 2475 (1972); *J. Org. Chem. USSR.* (*Engl. Transl.*), **8**, 2475 (1972).
[84] Y. Svetkin, N. Akmanova, and G. Karataeva, *Zh. Org. Khim.*, **8**, 2431 (1972); *J. Org. Chem. USSR* (*Engl. Transl.*) **8**, 2478 (1972).
[85] J. A. Damavandy and R. A. Y. Jones, *J. Chem. Soc. Perkin Trans. 1*, **1981**, 712.
[86] A. Padwa, L. Fisera, K. F. Koehler, A. Rodriguez, and G. S. Wong, *J. Org. Chem.*, **49**, 276 (1984).
[87] D. Mukherjee, L. N. Domelsmith, K. N. Houk, *J. Am. Chem. Soc.*, **100**, 1954 (1978).
[88] C. M. Dicken and P. DeShong, *J. Org. Chem.*, **47**, 2047 (1982).
[89] P. DeShong, C. M. Dicken, R. R. Staib, A. J. Freyer, and S. M. Weinreb, *J. Org. Chem.*, **47**, 4397 (1982).
[90] G. Tacconi, P. P. Righetti, and G. Desimoni, *J. Prakt. Chem.*, **322**, 679 (1980).
[91] D. Dopp and M. Henseleit, *Chem. Ber.*, **115**, 798 (1982).
[92] O. Tsuge, M. Tashiro, and Y. Nishihara, *Tetrahedron Lett.*, **1967**, 3769.
[93] H. Horikawa, T. Nishitani, T. Iwasaki, and I. Inoue, *Tetrahedron Lett.*, **24**, 2193 (1983).
[94] H. Taniguchi, T. Ikeda, and E. Imota, *Bull. Chem. Soc. Jpn.*, **51**, 1859 (1978).
[95] T. Sasaki, S. Eguchi, and Y. Hirako, *Tetrahedron*, **32**, 437 (1976).

[96] M. C. Aversa, G. Cum, G. Stagno d'Alcontres, and N. Vccella, *J. Chem. Soc., Perkin Trans. 1*, **1972**, 222.
[97] M. Joucla, *Tetrahedron*, **29**, 2315 (1973).
[98] B. G. Murray and A. F. Turner, *J. Chem. Soc. (C)*, **1966**, 1338.
[99] G. Bianchi, C. DeMicheli, and R. Gandolfi, *J. Chem. Soc., Perkin Trans. 1*, **1976**, 1518.
[100] A. Padwa, K. F. Koehler, and A. Rodriguez, *J. Am. Chem. Soc.*, **103**, 4974 (1981).
[101] J. T. Bailey, I. Berger, R. Friary, and M. S. Puar, *J. Org. Chem.*, **47**, 858 (1982).
[102] T. Iwashita, T. Kusumi, and H. Kakisawa, *Chem. Lett.*, **1979**, 1337.
[103] F. Santer and G. Buyuk, *Monatsh. Chem.*, **105**, 254 (1974).
[104] D. Cristina, M. DeAmici, C. DeMicheli, and R. Gandolfi, *Tetrahedron*, **37**, 1349 (1981).
[105] M. Jovela, J. Jammolin, and R. Carrie, *Bull. Soc. Chim. Fr.*, **1973**, 3116.
[106] D. St. C. Black, R. F. Crozier, and I. D. Rae, *Aust. J. Chem.*, **31**, 2239 (1978).
[107] M. Joucla and J. Hamelin, *J. Chem. Res. (S)*, **1978**, 276.
[108] N. Singh and K. Krishan, *Indian J. Chem.*, **11**, 1076 (1973).
[109] Y. Nomura, F. Furusaki, and Y. Takeuchi, *Bull. Chem. Soc. Jpn.*, **43**, 1913 (1970).
[110] Y. Nomura, F. Furusaki, and Y. Takeuchi, *Bull. Chem. Soc. Jpn.*, **40**, 1740 (1967).
[111] M. Menard, P. Rivest, L. Morris, J. Meunier, and Y. G. Perron, *Can. J. Chem.*, **52**, 2316 (1974).
[112] T. H. Chan and D. Massuda, *J. Am. Chem. Soc.*, **99**, 936 (1977)
[113] E. Winterfeldt, W. Krohn, and H. Stracke, *Chem. Ber.*, **102**, 2346 (1969).
[114] O. Tsuge and A. Turii, *Bull. Chem. Soc. Jpn.*, **49**, 1138 (1976).
[115] J. C. Mason and G. Tennant, *J. Chem. Soc. Chem. Commun.*, **1972**, 218.
[116] S. Takahashi, S. Hashimoto, and H. Kano, *Chem. Pharm. Bull.* **18**, 1176 (1970).
[117] S. Takahashi and H. Kano, *Chem. Pharm. Bull.*, **16**, 527 (1968).
[118] J. E. Baldwin, R. G. Pudussery, A. L. Oureshi, and B. Sklarz, *J. Am. Chem. Soc.*, **90**, 5325 (1968).
[119] N. A. LeBel and J. J. Whang, *J. Am. Chem. Soc.*, **81**, 6334 (1959).
[120] N. A. LeBel, *Trans. N. Y. Acad. Sci.*, **27**, 858 (1965).
[121] A. Padwa, H. Ku, and A. Mazzu, *J. Org. Chem.*, **43**, 381 (1978).
[122] M. Chandler and P. J. Parsons, *J. Chem. Soc., Chem. Commun.*, **1984**, 322.
[123] P. Confalone and E. Huie, *J. Org. Chem.*, **48**, 2994 (1983).
[124] N. A. LeBel, G. M. J. Slusarczuk, and L. A. Spurlock, *J. Am. Chem. Soc.*, **84**, 4360 (1962).
[125] F. J. Vinick, I. E. Fengler, and H. W. Gschwend, *J. Org. Chem.*, **42**, 2936 (1977).
[126] R. J. Ferrier, R. H. Furneaux, P. Prasit, and P. C. Tyler, *J. Chem. Soc., Perkin Trans. 1*, **1983**, 1621.
[127] R. J. Ferrier and P. Prasit, *J. Chem. Soc., Chem. Commun.*, **1981**, 983.
[128] T. Sasaki, S. Eguchi, and T. Suzuki, *J. Chem. Soc., Chem. Commun.*, **1979**, 506.
[129] T. Sasaki, S. Eguchi, and T. Suzuki, *J. Org. Chem.*, **47**, 5250 (1982).
[130] N. A. LeBel, N. D. Ojha, J. R. Menke, and R. J. Newland, *J. Org. Chem.*, **37**, 2896 (1972).
[131] S. Eguchi, Y. Furukawa, T. Suzuki, K. Kondo, T. Sasaki, M. Honda, C. Katayama, and J. Tanaka, *J. Org. Chem.*, **50**, 1895 (1985).
[132] W. Oppolzer and K. Keller, *Tetrahedron Lett.* **1970**, 1117.
[133] N. A. LeBel and E. G. Banucci, *J. Org. Chem.*, **36**, 2440, (1971).
[134] N. A. LeBel and E. Banucci, *J. Am. Chem. Soc.*, **92**, 5278 (1970).
[135] M. L. Mihailovic, L. Lorenc, Z. Maksimovic, and J. Kalvoda, *Tetrahedron*, **29**, 2683 (1973).
[136] T. Kusumi, S. Takahashi, Y. Sato, and H. Kakisawa, *Heterocycles*, **10**, 257 (1978).
[137] S. Takahashi, T. Kusumi, Y. Sato, Y. Inouye, and H. Kakisawa, *Bull. Chem. Soc. Jpn.*, **54**, 1777 (1981).
[138] R. Funk, L. Horcher, II, J. Daggett, and M. Hansen, *J. Org. Chem.*, **48**, 2632 (1983).
[139] R. Funk and G. Bolton, *J. Org. Chem.*, **49**, 5021 (1984).
[140] R. Brambilla, R. Friary, A. Ganguly, M. S. Puar, B. R. Sunday, J. J. Wright, K. D. Onan, and A. T. McPhail, *Tetrahedron*, **37**, 3615 (1981).
[141] R. Brambilla, R. Friary, A. Ganguly, M. S. Puar, J. G. Topliss, and R. Watkins, *J. Org. Chem.*, **47**, 4137 (1982).
[142] D. Mackay and K. N. Watson, *J. Chem. Soc., Chem. Commun.*, **1982**, 775, 777.

[143] W. Oppolzer, S. Siles, R. Snowden, B. Bakker, and M. Petrzilka, *Tetrahedron Lett.*, **1979**, 4391.
[144] P. N. Confalone, G. Pizzolato, D. L. Confalone, and M. R. Uskokovic, *J. Am. Chem Soc.*, **100**, 6291 (1978).
[145] P. N. Confalone, G. Pizzolato, D. L. Confalone, and M. R. Uskokovic, *J. Am. Chem. Soc.*, **102**, 1954 (1980).
[146] E. G. Baggiolini, H. L. Lee, G. Pizzolato, and M. R. Uskokovic, *J. Am. Chem. Soc.*, **104**, 6460 (1982).
[147] P. M. Wovkulich and M. R. Uskokovic, *J. Am. Chem. Soc.*, **103**, 3956 (1981).
[148] W. Oppolzer and J. I. Grayson, *Helv. Chim. Acta*, **63**, 1706 (1980).
[149] W. Oppolzer and M. Petzilka, *J. Am. Chem. Soc.*, **98**, 6722 (1976).
[150] J. B. Bapat, D. S. Black, R. F. C. Brown, and C. Ichlov, *Aust. J. Chem.*, **25**, 2445 (1972).
[151] J. J. Tufariello and E. J. Trybulski, *J. Chem. Soc., Chem. Commun.*, **1973**, 720.
[152] J. J. Tufariello, G. B. Mullen, J. J. Tegeler, E. J. Trybulski, S. C. Wong, and S. A. Ali, *J. Am. Chem. Soc.*, **101**, 2435 (1979).
[153] J. J. Tufariello, J. J. Tegeler, S. C. Wong, and S. Ali, *Tetrahedron Lett.*, **1978**, 1733.
[154] E. Gossinger and B. Witkop, *Monatsh. Chem.*, **111**, 803 (1980).
[155] T. Iwashita, T. Kusumi, and H. Kakisawa, *Chem. Lett.*, **1979**, 947.
[156] M. A. Schwartz and G. C. Swanson, *J. Org. Chem.*, **44**, 953 (1979).
[157] H. Otomasu, N. Takatsu, T. Honda, and T. Kametani, *Heterocycles*, **19**, 511 (1982).
[158] H. Otomasu, N. Takatsu, T. Honda, and T. Kametani, *Tetrahedron*, **38**, 2627 (1982).
[159] J. J. Tufariello and S. A. Ali, *Tetrahedron Lett.*, **1979**, 4445.
[160] J. J. Tufariello and S. A. Ali, *J. Am. Chem. Soc.*, **101**, 7114 (1979).
[161] T. Iwashita, M. Suzuki, T. Kusumi, and H. Kakisawa, *Chem. Lett.*, **1980**, 383.
[162] J. J. Tufariello and J. P. Tette, *J. Chem Soc., Chem. Commun.*, **1971**, 469.
[163] J. J. Tufariello and J. P. Tette, *J. Org. Chem.*, **40**, 3866 (1975).
[164] J. J. Tufariello and G. E. Lee, *J. Am Chem. Soc.*, **102**, 373 (1980).
[165] A. Takano and K. Shishido, *Heterocycles*, **19**, 1439 (1982).
[166] For preliminary work, see S. Takano and K. Shishido, *J. Chem. Soc., Chem. Commun.*, **1981**, 940.
[167] J. J. Tufariello, *Acc. Chem. Res.*, **12**, 396 (1979).
[168] J. J. Tufariello and S. A. Ali, *Tetrahedron Lett.*, **1978**, 4647.
[169] E. Dagne and N. Castagnoli, Jr., *J. Med. Chem.*, **15**, 356 (1972).
[170] J. Hara, Y. Inouye, and H. Kakisawa, *Bull. Chem. Soc. Jpn.*, **54**, 3871 (1981).
[171] R. V. Stevens and K. Albizati, *J. Chem. Soc., Chem. Commun.*, **1982**, 104.
[172] T. Kametani, S.-P. Huang, A. Nakayama, and T. Honda, *J. Org. Chem.*, **47**, 2328 (1982).
[173] A. D. Nukolaeva and V. S. Perekhod'ko, *J. Org. Chem. USSR* (*Engl. Transl.*), **8**, 2297 (1972).
[174] I. Bruning, R. Grashey, H. Hauck, R. Huisgen, and H. Seidl, *Org. Syn.*, Coll. Vol. 5, J. Wiley, New York, 1973, p. 957.
[175] S. Mzengeza and R. A. Whitney, *J. Chem. Soc., Chem. Commun.*, **1984**, 606.
[176] J. E. Baldwin, A. K. Qureshi, and B. Sklarz, *J. Chem. Soc.* (*C*), **1969**, 1073.
[177] C. M. Tice and B. Ganem, *J. Org. Chem.*, **48**, 5048 (1983).
[178] A. Vasella, *Helv. Chim. Acta*, **60**, 1273 (1977).
[179] N. A. Akmanova and G. K. Kavyeva, *Khimiya Organ. Soedin. Azota.*, **1982**, 94. [*C.A.* **97**, 162875u (1980)].
[180] C. Belzecki and I. Panfil, *J. Org. Chem.*, **44**, 1212 (1979).
[181] Y. V. Svetkin and Y. R. Kolesnik, *Zh. Obshch. Khim.*, **52**, 907 (1982) [*C.A.*, **97**, 72414p (1980)].
[182] T. Sasaki, Japanese Patent 70, 34,586 [*C.A.*, **74**, 76440s (1971)].
[183] P. DeShong and J. M. Leginus, *J. Org. Chem.*, **49**, 3421 (1984).
[184] N. Singh and P. S. Sethi, *Indian J. Chem.*, **13**, 990 (1975).
[185] R. Paul and S. Tchelitsheff, *Bull. Soc. Chim. Fr.*, **1967**, 4180.
[186] T. Sasaki and M. Ando, *Bull. Chem. Soc. Jpn.*, **41**, 2960 (1968).
[187] N. Singh and S. Mohan, *J. Chem. Soc., Chem. Commun.*, **1968**, 787.

[188] H. Basu and H. Schlenk, *Chem. Phys. Lipids*, **6**, 266 (1971).

[189] P. D. Croce, C. La Rosa, R. Stradi, and M. Ballabio, *J. Heterocycl. Chem.*, **20**, 519 (1983).

[190] T. F. Petrushina, V. N. Domrachev, and N. A. Akmanova, *Izv. Vyssh. Uchebn. Zaved., Khim. Khim. Tekhnol.*, **25**, 545 (1982) [*C.A.*, **97**, 127544p (1980)].

[191] I. Ikdea, G. Takemoto, and S. Kemori, *Kogyo Kagaku Zasshi*, **74**, 220 (1971) [*C.A.*, **75**, 5766n (1971)].

[192] R. Huisgen, R. Grashey, H. Seidl, and H. Hauck, *Chem. Ber.*, **101**, 2559 (1968).

[193] T. Koizumi, H. Hirai, and E. Yoshii, *J. Org. Chem.*, **47**, 4005 (1982).

[194] I. Panfil and C. Belzecki, *Pol. J. Chem.*, **55**, 977 (1981) [*C.A.*, **98**, 160619k (1981)].

[195] C. Kashima, A. Tsuzuki, and T. Tajima, *J. Heterocycl. Chem.*, **21**, 201 (1984).

[196] R. A. Jones and M. T. Marriott, *Heterocycles*, **14**, 185 (1980).

[197] C. Belzecki and I. Panfil, *J. Chem. Soc., Chem. Commun.*, **1977**, 303.

[198] P. DeShong and J. M. Leginus, *Tetrahedron Lett.*, **25**, 5355 (1984).

[199] P. DeShong, C. M. Dicken, J. M. Leginus, and R. R. Whittle, *J. Am. Chem. Soc.*, **106**, 5598 (1984).

[199a] P. DeShong, and J. M. Leginus, *J. Am. Chem. Soc.*, **105**, 1687 (1983).

[200] T. Sasaki and T. Ohishi, *Bull. Chem. Soc. Jpn.*, **41**, 3012 (1968).

[201] D. O. Spry, *J. Org. Chem.*, **40**, 2411 (1975).

[202] N. A. Akmanova, K. F. Sagitdinova, and E. S. Balenkova, *Khim. Geterotsikl. Soedin.*, **9**, 1192 (1982) [*C.A.*, **97**, 216064z (1982)].

[203] N. A. Akmanova, V. S. Anisimova, and Y. V. Svetkin, *Zh. Org. Khim.*, **93**, (1976) [*C.A.*, **88**, 152519f (1978)].

[204] A. Vasella, *Helv. Chim. Acta*, **60**, 426 (1977).

[205] L. I. Vasil'eva, G. S. Akimova, and V. N. Chistokletov, *Zh. Org. Khim.*, **20**, 148 (1984) [*C.A.*, 100, 191768x (1983)].

[206] J. Palmer, J. L. Roberts, P. S. Rutledge, and P. D. Woodgate, *Heterocycles*, **5**, 109 (1976).

[207] P. M. Wovkulich, F. Barcelos, A. D. Batcho, J. F. Sereno, E. G. Baggiolini, B. M. Hennessy, and M. R. Uskokovic, *Tetrahedron*, **40**, 2283 (1984).

[208] R. Huisgen, H. Hauck, R. Grashey, and H. Seidl, *Chem. Ber.*, **102**, 736 (1969).

[209] M. M. Joucla and J. Hamelin, *C. R. Acad. Sci. Paris*, **273**, 769 (1971).

[210] E. Abad, J. Fayn, B. Bertaina, and A. Cambon, *J. Fluorine Chem.*, **25**, 453 (1984).

[211] K. Tada, T. Yamada, and F. Toda, *Bull. Chem. Soc. Jpn.*, **51**, 1839 (1978).

[212] M. Joucla, F. Tonnard, D. Gree, and J. Hamelin, *J. Chem. Res.*, **1978**, 240.

[213] P. G. De Benedetti, S. Quartieri, and A. Rastelli, *J. Chem. Soc., Perkin Trans. 2*, **1982**, 95.

[214] S. M. Yarnal and V. V. Badiger, *J. Indian Chem. Soc.*, **48**, 453 (1971).

[215] J. W. Lown and B. E. Landberg, *Can. J. Chem.*, **52**, 798 (1974).

[216] A. Monpert, J. Martelli, and R. Gree, *J. Organomet. Chem.*, **210**, C45 (1981).

[217] G. W. Griffin, D. C. Lankin, N. S. Bhacca, H. Terasawa, and R. K. Sehgal, *Tetrahedron Lett.*, **23**, 2753 (1982).

[218] M. Joucla, J. Hamelin, and R. Carrie, *Bull. Soc. Chim. Fr.*, **1973**, 3116.

[219] G. Schmidt, H. Starcke, and E. Winterfeldt, *Chem. Ber.*, **103**, 3196 (1970).

[220] O. Tsuge, K. Ueno, and S. Kanemasa, *Chem. Lett.*, **5**, 797 (1984).

[221] P. Kulsa and C. Rooney, German Patent 2,100,242 (1971) [*C.A.*, **75**, 110312j (1971)].

[222] G. Biaanchi, C. de Micheli, and R. Gandolfi, *J. Chem. Soc. Perkin Trans. 1*, **1976**, 1518.

[223] L. Fisera, M. Dandarova, J. Kovac, A. Gaplovsky. J. Patus, and I. Goljer, *Collect. Czech. Chem. Commun.*, **47**, 523 (1982).

[224] L. Fisera, J. Kovac, J. Patus, and P. Mesko, *Chem. Zvesti*, **37**, 821 (1983) [*C.A.*, **100**, 156526v (1983)].

[225] Y. Iwakura, K. Uno, and T. Hongu, *Bull. Chem. Soc. Jpn.*, **42**, 2882 (1969).

[226] Y. D. Samuilov, S. E. Solov'eva, T. F. Girutskaya, and A. I. Konovalov, *J. Org. Khim.*, **14**, 1693 (1978); *J. Org. Chem. USSR* (*Engl. Transl.*), **14**, 1579 (1978).

[227] Y. Iwakura, K. Uno, S. Hong, and T. Hongu, *Bull. Chem. Soc. Jpn.*, **45**, 192 (1972).

[228] V. D. Kiselev, D. G. Khuzyasheva, and A. I. Konovalov, *Zh. Org. Khim.*, **19**, 884 (1983); [*C.A.*, **99**, 53651p (1983)].

[229] K. Krishan and B. Singh, *Indian J. Chem.*, **23B**, 620 (1984).

[230] A. Bened, R. Durand, D. Pioch, and P. Geneste, *J. Chem. Soc., Perkin Trans. 2*,**1984**, 1.

[231] L. Fisera, M. Dandarova, J. Kovac, P. Mesko, and A. Krutosikova, *Collect. Czech. Chem. Commun.*, **46**, 2421 (1981).

[232] R. R. Fraser and Y. S. Lin, *Can. J. Chem.*, **46**, 801 (1968).

[233] L. Fisera, J. Kovac, J. Patus, and D. Pavlovic, *Coll. Czech. Chem. Commun.*, **48**, 1048 (1982).

[234] D. N. Reinhoudt and C. G. Kouwenhoven, *Tetrahedron Lett.*, **25**, 2163 (1974).

[235] D. Cristina and C. De Micheli, *J. Chem. Soc., Perkin Trans. 1*, **1979**, 2891.

[236] H. Taniguchi, T. Ikeda, Y. Yoshida, and E. Imoto, *Bull. Chem. Soc. Jpn.*, **50**, 2694 (1977).

[237] G. Bianchi, A. Gamba, and R. Gandolfi, *Tetrahedron*, **28**, 1601 (1972).

[238] T. Sasaki, T. Manabe, and S. Nishida, *J. Org. Chem.*, **45**, 479 (1980).

[239] T. Sasaki, K. Hayakawa, T. Manabe, and S. Nishida, *J. Am. Chem. Soc.*, **103**, 565 (1981).

[240] N. A. Akmanova, Kh. F. Sagitdinova, A. I. Popenova, V. N. Domrachev, and E. S. Balenkova, *Khim. Geterotsikl. Soedin.*, **10**, 1316 (1982) [*C.A.*, **98**, 89222t (1982)].

[241] N. N. Magdesieva and T. A. Sergeeva, *Zh. Org. Khim.*, **20**, 1598 (1984), *J. Org. Chem., USSR (Engl. Transl.)*, **1985**, 1454.

[242] M. C. Aversa, G. Cum, P. D. Giannetto, G. Romeo, and N. Uccella, *J. Chem. Soc., Perkin Trans. 1*, **1974**, 209.

[243] O. Tsuge and I. Shinkai, *Tetrahedron Lett.*, **1970**, 3847.

[244] C. De Michell, A. G. Invernizzi, R. Gandolfi, and L. Scevola, *J. Chem. Soc., Chem. Commun.*, **1976**, 246.

[245] A. Bened, R. Durand, D. Pioch, and P. Geneste, *J. Org. Chem.*, **46**, 3502 (1981).

[246] T. Sasaki and S. Eguchi, *J. Org. Chem.*, **33**, 4389 (1968).

[247] T. Minami, T. Hanamoto, and I. Hirao, *J. Org. Chem.*, **50**, 1278 (1985).

[248] L. Fisera, J. Kovac, J. Poliacikova, and J. Lesko, *Monatsh. Chem.*, **111**, 909 (1980).

[249] L. Fisera, J. Kovac, and J. Poliacikova, *Heterocycles*, **12**, 1005 (1979).

[250] L. Fisera, J. Lesko, M. Dandarova, and J. Kovac, *Collect. Czech. Chem. Commun.*, **45**, 3546 (1980).

[251] M. J. Green, R. L. Tiberi, R. Friary, B. N. Lutsky, J. Berkenkoph, X. Fernandex, and M. Monahan, *J. Med. Chem.*, **25**, 1492 (1982).

[252] A. V. Prosyanik, A. I. Mishchenko, A. B. Zolotoi, and N. L. Zaichenko, *Izv. Akad. Nauk SSSR, Ser. Khim.*, **9**, 2050 (1982) [*C.A.*, **98**, 52774y (1981)].

[253] A. Lablache-Combier and M. L. Villaume, *Tetrahedron*, **24**, 6951 (1968).

[254] A. Lablache-Combier and M. Villaume, *Tetrahedron Lett.*, **49**, 4959 (1967).

[255] D. A. Kerr and D. A. Wilson, *J. Chem. Soc.*, **1970**, 1718.

[256] S. Murahashi, H. Mitsui, T. Watanabe, and S. Zenki, *Tetrahedron Lett.*, **24**, 1049 (1983).

[257] J. J. Tufariello, H. Meckler, K. Pushpananda, and A. Senaratne, *J. Am. Chem. Soc.*, **106**, 7979 (1984).

[258] E. Roeder, H. Wiedenfeld, and E. J. Jost, *Arch. Pharm.*, **317**, 403 (1984).

[259] H. Iida, Y. Watanabe, and C. Kibayashi, *Chem. Lett.*, **1983**, 1195.

[260] H. Iida, Y. Watanabe, M. Tanaka, and C. Kibayashi, *J. Org. Chem.*, **49**, 2412 (1984).

[261] H. Iida, M. Tanaka, and C. Kibayashi, *J. Chem. Soc., Chem. Commun.*, **1983**, 1143.

[262] K. Shishido, K. Tanaka, K. Fukumoto, and T. Kametani, *Tetrahedron Lett.*, **24**, 2783 (1983).

[263] J. E. Baldwin, M. F. Chan, G. Gallacher, P. Monk, and K. Prout, *J. Chem. Soc., Chem. Commun.*, **1983**, 250.

[264] S. P. Ashburn and R. M. Coates, *J. Org. Chem.*, **49**, 3127 (1984).

[265] H. Schneider, *Helv. Chim. Acta*, **65**, 726 (1982).

[266] D. Dopp, C. Kruger, G. Makedakis, and A. M. Nour-el-Din, *Chem. Ber.*, **118**, 510 (1985).

[267] M. C. Aversa, P. Giannetto, and A. Ferlazzo, *J. Chem. Soc., Perkin Trans. 1*, **1982**, 2701.

[268] T. Iwashita, T. Kusumi, and H. Kakisawa, *J. Org. Chem.*, **47**, 230 (1982).

[269] H. Oinuma, S. Dan, and H. Kakisawa, *J. Chem. Soc., Chem. Commun.*, **1983**, 654.

[270] J. B. Hendrickson and D. A. Pearson, *Tetrahedron Lett.*, **24**, 4657 (1983).

[271] R. Commisso, A. C. Coda, G. Desimoni, P. P. Righetti, and G. Tacconi, *Gazz. Chim. Ital.*, **112**, 483 (1982).

[272] A. F. Gettins, D. P. Stokes, G. A. Taylor, and C. B. Judge, *J. Chem. Soc., Perkin Trans. 1*, **1977**, 1849.

[273] M. L. M. Pennings and D. N. Reinhoudt, *J. Org. Chem.*, **48**, 4043 (1983).

[274] M. L. M. Pennings, G. Okay, D. N. Reinhoudt, S. Harkems, and G. J. van Hummel, *J. Org. Chem.*, **47**, 4413 (1982).

[275] R. Huisgen and K. Niklas, *Heterocycles*, **22**, 21 (1984).

[276] J. P. Freeman, D. J. Duchamp, C. G. Chidester, G. Slomp, J. Szmuszkovicz, and M. Raban, *J. Am. Chem. Soc.*, **104**, 1380 (1982).

[277] H. Gnichtel and B. P. Cau, *Justus Liebigs Ann. Chem.*, **1982**, 2223.

[278] A. Vasella and R. Voeffray, *J. Chem. Soc., Chem. Commun.*, **1981**, 97.

[279] A. Vasella and R. Voeffray, *Helv. Chim. Acta*, **65**, 1953 (1982).

[280] A. Vasella, R. Voeffray, J. Pless, and R. Huguenin, *Helv. Chim. Acta*, **66**, 1241 (1983).

[281] H. Hart and D. Ok, *Tetrahedron Lett.*, **25**, 2073 (1984).

[282] T. Sasaki, S. Eguchi, and Y. Hirako, *Tetrahedron Lett.*, **7**, 541 (1976).

[283] F. De Sarlo, A. Brandi, and A. Guarna, *J. Chem. Soc., Perkin Trans. 1*, **1982**, 1395.

[284] V. G. Manecke and J. Klawitter, *Makromol. Chem.*, **108**, 292 (1967).

[285] T. Motomiyz and Y. Iwakura, Japanese Patent **71** 00,025 (1971) [*C.A.*, **75**, 6600r (1971)].

[286] V. G. Manecke and J. Klawitter, *Macromol. Chem.*, **142**, 253 (1971).

[287] B. Snider and C. Cartaya-Marin, *J. Org. Chem.*, **49**, 1688 (1984).

[288] H. Mitsui, S. Zenki, T. Shiota, and S. Murahashi, *J. Chem. Soc., Chem. Commun.*, **1984**, 874.

[289] E. Gossinger, R. Imhof, and H. Wehrli, *Helv. Chim. Acta*, **58**, 96 (1975).

[290] J. J. Tufariello and E. J. Trybulski, *J. Org. Chem.*, **79**, 3378 (1974).

[291] A. Toy and W. J. Thompson, *Tetrahedron Lett.*, **25**, 3533, (1984).

[292] B. Bernet and A. Vasella, *Helv. Chim. Acta*, **62**, 2411 (1979).

[293] B. Bernet and A. Vasella, *Helv. Chim. Acta*, **62**, 1990 (1979).

[294] J. E. Baldwin, D. H. R. Barton, J. J. A. Gutteridge, and R. J. Martin, *J. Chem. Soc. (C)*, **1971**, 2184.

[295] M. Schwartz and A. Willbrand, *J. Org. Chem.*, **50**, 1359 (1985).

[296] W. Oppolzer and H. P. Weber, *Tetrahedron Lett.*, **1970**, 1121.

[297] W. Oppolzer and K. Keller, *Tetrahedron Lett.*, **1970**, 4313.

[298] W. Oppolzer, J. J. Grayson, H. Wegmann, and M. Urrea, *Tetrahedron*, **39**, 3695 (1983).

[299] A. Kozikowski and P. Stein, *J. Am. Chem. Soc.*, **107**, 2569 (1985).

CHAPTER 2

PHOSPHORUS ADDITION AT sp^2 CARBON

ROBERT ENGEL

City University of New York, Queens College, Flushing, New York

CONTENTS

INTRODUCTION

In recent years a renewed interest in the synthesis of organophosphorus compounds has appeared. This has been the result principally of the development of new applications of materials bearing a carbon–phosphorus linkage for chemical synthesis, biological regulation, and a variety of industrial uses.

In chemical syntheses the phosphoryl function has been found to be of great utility as an activator for two major categories of transformations. The first of these is the generation of new carbon–carbon double bonds through the phosphoryl-stabilized anion modifications of the Wittig reaction.[1–7] For these reactions, the phosphoryl function increases the acidity of the hydrogen of the carbon attached to phosphorus and, after addition to a carbonyl linkage, is usually eliminated as a water-soluble byproduct. Alternatively, the initially formed phosphonate anion may be used as a nucleophile in a substitution reaction leading to a derived phosphonate product having its own synthetic utility.[8,9]

In the second category of transformations an appropriately functionalized phosphonate reacts at the carbon attached to phosphorus to give products that can be envisioned as resulting from reaction of an electrophilic substrate with a carbonyl group of inverted polarity.[10–12] The phosphoryl function, which is generally eliminated in the formation of the target material, facilitates these "umpolung" reactions that could not be accomplished directly with a parent material bearing the simple carbon–oxygen double bond.

With regard to biological applications, the synthesis of organophosphorus compounds has been of interest for several reasons. Among these is the discovery in a variety of organisms of naturally occurring compounds bearing the carbon–phosphorus bond. Phosphonomycin[13] and aminoethylphosphonic acid[14] are two well-known examples of such compounds. It has also been recognized that phosphonates may be capable of perturbing metabolic processes normally involving phosphates of similar structure.[15] In addition, materials with other metabolic regulatory capabilities have been found.[16,17] As a result, there have been not only syntheses of many new organophosphorus compounds but also the development of improved methods for such syntheses.

The review presented in this chapter is concerned with one approach to carbon–phosphorus bond formation that is particularly significant for the aforementioned applications, specifically the addition of a (formally) trivalent phosphorus ester or amide to an unsaturated carbon linkage under polar reaction conditions. Reactions of a variety of types of phosphorus reagents with simple carbonyl compounds, imines, and other related polar unsaturated carbon functions, as well as conjugate addition reactions, are reviewed.

MECHANISM

The addition reaction of fully esterified trivalent phosphorus acids at unsaturated carbon necessarily involves at least two steps: (1) formation of the carbon–phosphorus bond and (2) removal of an ester function to generate the phosphonyl linkage. Other nonproductive reactions may, of course, accompany these steps.

With addition to simple carbonyl groups, the first of these steps occurs readily, although reversibly (Eq. 1).[18–19]

$$(RO)_3P \;+\; R'\!-\!\overset{O}{\overset{\|}{C}}\!-\!R'' \;\rightleftharpoons\; R'\!-\!\underset{{}^{+}P(OR)_3}{\underset{|}{\overset{O^-}{\overset{|}{C}}}}\!-\!R'' \qquad \text{(Eq. 1)}$$

Unless stringent conditions are used, however, the second required step does not occur. A favorable trajectory for intramolecular dealkylation cannot be attained, and the starting reagents are regenerated.[20] The isolation and characterization of further adducts of trialkyl phosphites with more than one molecule of reactive aldehyde are indirect evidence for the intermediacy of the tetrahedral zwitterionic adduct. The oxygen anion attacks a second carbonyl carbon leading to either a cyclic phosphorane or some other linear product of dealkylation (Eq. 2).[18,21–29]

$$R'\underset{{}^{+}P(OR)_3}{\underset{|}{\overset{O^-}{\overset{|}{C}H}}} \xrightarrow{R''CHO} R'\underset{{}^{+}P(OR)_3}{\underset{|}{\overset{O-\overset{O^-}{\underset{}{CHR'}}}{\overset{|}{C}H}}} \longrightarrow \begin{cases} \text{cyclic phosphorane: } R'CH{-}O{-}CHR''{-}O{-}P(OR)_3 \text{ (five-membered ring)} \\ \xrightarrow{R''CHO} (RO)_2\overset{O}{\overset{\|}{P}}\underset{R'}{\underset{|}{C}}HO\underset{R''}{\underset{|}{C}}HOCHR''OR \end{cases} \qquad \text{(Eq. 2)}$$

Under conditions of high temperature and pressure, the concentration of the tetrahedral zwitterionic intermediate is sufficiently high that dealkylation can occur in an intermolecular fashion (Eq. 3).[30]

$$(RO)_3\overset{+}{P}-\underset{R''}{\overset{R'}{C}}-O^- + (RO)_3\overset{+}{P}-\underset{R''}{\overset{R'}{C}}-O^- \longrightarrow 2\ (RO)_2\overset{O}{\overset{\|}{P}}-\overset{OR}{\overset{|}{C}}R'R'' \quad \text{(Eq. 3)}$$

With silyl esters of the trivalent phosphorus acids, front-side displacement (S_Ni–Si)[31] is possible for intramolecular desilylation, and the total addition reaction can proceed under relatively mild conditions (Eq. 4).

$$\left[R'_3Si-O-\overset{+}{P}(OR)_2-CR''_2-O^-\right] \longrightarrow (RO)_2\overset{O}{\overset{\|}{P}}-CR''_2-O-SiR'_3 \quad \text{(Eq. 4)}$$

In the presence of a hydroxylic solvent, proton abstraction by the intermediate followed by alkoxide anion dealkylation yields the stable product.

In Michael-type addition reactions with α,β-unsaturated compounds, initial attack by phosphorus is on the β-carbon atom.[32] If the β-carbon site is severely hindered, as in progesterone, no formation of adduct can be observed.[33] Completion of the reaction with phosphonyl bond generation is accomplished by routes entirely analogous to those with addition at the carbonyl carbon.

With phosphorus reagents bearing a single acidic site, other complications in the mechanism of addition might be contemplated. However, the addition to carbonyl carbon appears to be a simple "aldol" type with phosphorus attaching directly to carbon. Isotopic labeling studies indicate that this addition occurs without any exchange of oxygen between carbon and phosphorus.[34]

SCOPE AND LIMITATIONS

In recent years the synthesis of organophosphorus compounds via phosphorus addition reactions at unsaturated carbon atoms has been undertaken for a variety of purposes. In addition to those instances in which the resulting carbon–phosphorus compound (ester or free acid) is the ultimate target, the organophosphorus functionality (carbon–phosphoryl) may also be used as a tool for further synthetic purposes. The phosphoryl entity so introduced is removed in the generation of a further target species. The variety of reactions useful for the preparation of these materials is discussed in this chapter. While the present discussion is concerned primarily with the developments since 1970, some important earlier work is also considered.

Nature of the Phosphorus Reagent

Initial investigations of the addition of trivalent phosphorus to unsaturated carbon were performed with the use of either trialkyl or dialkyl phosphites (esters of phosphorous acid).[30,35] In recent years, however, the use of other derivatives of phosphorous acid, particularly silylated species, has greatly expanded the synthetic utility of the reactions. The variability possible in the structure of the phosphorus reagent permits great versatility in the preparation of particular target species. The utility of each type of reagent is discussed.

Fully Esterified Trivalent Phosphorus Acids. The initial work concerning the addition of a phosphorus triester to a π-bonded electron-deficient carbon involved the treatment of simple aliphatic aldehydes with trialkyl phosphites.[30] Heating of the reaction mixture in a sealed tube at 70–100° for several hours results in the formation of the α-alkoxyalkylphosphonate diesters. In the absence of any other electrophilic species, the anionic oxygen generated from the initial carbonyl function reacts with a quasi-phosphonium site and thus becomes alkylated (Eq. 5).

$$(RO)_3P + R'CHO \longrightarrow (RO)_2P(O)CHR'OR \qquad \text{(Eq. 5)}$$

Significantly milder reaction conditions may be used to generate the α-hydroxyalkylphosphonates from aliphatic aldehydes. Trialkyl phosphite additions to aliphatic aldehydes under milder conditions result in the formation of 2,2,2-trialkoxy-1,4,2-dioxaphospholanes (Eq. 6).[18]

$$2\ RCHO + (R'O)_3P \xrightarrow{20^\circ} \text{[ring: RHC—O—CHR—P(OR')}_3\text{—O—]} \qquad \text{(Eq. 6)}$$

These materials, formed by reaction of a 1:1 adduct (phosphorus initially attacking at carbon) with a second molecule of aldehyde, are rapidly hydrolyzed to generate the α-hydroxyalkylphosphonates. Reactions of trialkyl phosphites with aromatic aldehydes[21] proceed by initial attack of phosphorus on the carbonyl oxygen[36–41] with the formation of 1,3,2-dioxaphospholanes and the generation of a new carbon–carbon bond.[42] The simple α-hydroxyalkylphosphonates are thus not generated with these reagents.

A facile dealkylation of the quasi-phosphonium intermediate from a trialkyl phosphite is observed in the reaction with an oxime (Eq. 7).[43] Here, the phosphonate bearing an ethoxyamino function at the α position is isolated.

$$(C_2H_5O)_3P + \text{2-furyl-}CH{=}NOH \xrightarrow{80^\circ} \text{2-furyl-}CH(NHOC_2H_5)P(O)(OC_2H_5)_2 \quad (39\%) \qquad \text{(Eq. 7)}$$

Reaction between a trialkyl phosphite and thiocarbonyl compounds also results in dealkylation of the quasi-phosphonium intermediate by either displacement or β-elimination routes (Eq. 8).[44]

$$\text{cyclo-}[C(=S)(CH_2)_n] + (RO)_3P \longrightarrow \text{cyclo-}[C(SR)(P(O)(OR)_2)(CH_2)_n]\ (\mathbf{1}) + \text{cyclo-}[C(SH)(P(O)(OR)_2)(CH_2)_n]\ (\mathbf{2}) \qquad \text{(Eq. 8)}$$

When R = methyl, only the displacement product **1** is observed. Use of triethyl phosphite affords mixtures of **1** and **2**, along with ethylene. When triisopropyl phosphite is used, the formation of **1** becomes totally suppressed, and only the free thiol product is observed.

The use of silyl phosphite esters results in rapid "dealkylation" of the quasi-phosphonium intermediate with generation of an *O*-silyl ether. Thus a significant versatility to the reaction is observed. For example, the reaction of tris(trimethylsilyl) phosphite[45,46] with α-haloaldehydes[47] and ketones[47] gives the bis(trimethylsilyl) α-trimethylsiloxyalkylphosphonates in excellent yields (Eqs. 9, 10). When this reaction is followed by the addition of a hydroxylic solvent, the α-hydroxyalkylphosphonic acids can be isolated directly.[48]

$$CH_3COCH_2Cl + [(CH_3)_3SiO]_3P \xrightarrow{THF} CH_3C[OSi(CH_3)_3](CH_2Cl)P(O)[OSi(CH_3)_3]_2 \quad (97\%) \qquad \text{(Eq. 9)}$$

$$(CH_3)_2CClCHO + [(CH_3)_3SiO]_3P \xrightarrow{THF} (CH_3)_2CClCH[OSi(CH_3)_3]P(O)[OSi(CH_3)_3]_2 \quad (92\%) \qquad \text{(Eq. 10)}$$

Similar results are observed with the use of tris(trimethylsilyl) phosphite for conjugate addition reactions (hydrophosphinylation).[49] Initial addition occurs in moderate to good yield (75–84%) with α,β-unsaturated ketones, followed by nearly quantitative hydrolysis to the free γ-ketophosphonic acids (Eq. 11).

$$\mathrm{RCH{=}CHCOR' + [(CH_3)_3SiO]_3P \xrightarrow{C_6H_6} [(CH_3)_3SiO]_2P(O)CHRCH{=}C[OSi(CH_3)_3]R'}$$

$$\xrightarrow{H_2O} \mathrm{H_2O_3PCHRCH_2COR'} \qquad \text{(Eq. 11)}$$

With mixed silyl esters of phosphorous acids, the silyl function is exclusively transferred to the α-oxide anionic site (Eq. 12).[10,50,51]

$$\mathrm{C_6H_5CHO + (CH_3O)_2POSi(CH_3)_3 \xrightarrow{C_6H_6} C_6H_5CH[OSi(CH_3)_3]P(O)(OCH_3)_2} \quad (97\%) \qquad \text{(Eq. 12)}$$

This result is observed when the additional functionality about phosphorus is either ester or amide and presumably occurs by the intramolecular pathway shown in Eq. 13.

$$\mathrm{RR'CO + X_2POSiR''_3 \longrightarrow RR'C(O^-)\text{—}\overset{+}{P}X_2(OSiR''_3) \longrightarrow RR'C(OSiR''_3)\text{—}P(O)X_2} \qquad \text{(Eq. 13)}$$

Ready entrance is thereby gained to either the α-hydroxyalkylphosphonate diester, by simple hydrolysis of the silyl ether, or the α-hydroxyalkylphosphonic acid by further treatment with trimethylsilyl halide.[52,53]

The use of fully esterified phosphorous acid species in which one or more of the linkages is a silyl ester provides chemistry significantly different from that of trialkyl phosphites. Diketones,[54] ketoesters,[54] and trifluoromethyl ketones[54,55] exhibit "anomalous" reaction with trialkyl phosphites but give carbonyl–carbon adducts with the silylated reagents.[54,56–60]

In conjugate addition reactions involving mixed silyl ester reagents (Eq. 14),[51,61,62] it is again only the silyl function that is transferred, although mixtures of conjugate and carbonyl (1,4 and 1,2) products are occasionally obtained.

$$\text{(5,5-dimethylcyclohex-2-enone)} + \mathrm{(C_2H_5O)_2POSi(CH_3)_3 \xrightarrow[180^\circ]{CH_3CN}} \text{(1-OSi(CH}_3)_3\text{-5,5-dimethylcyclohexene-3-P(O)(OC}_2\text{H}_5)_2) \quad (90\%) \qquad \text{(Eq. 14)}$$

For example, in the reaction of phosphonites generated *in situ* from the phosphonous acid with trimethylchlorosilane or *N*, *N*-bis(trimethylsilyl)acetamide, complete 1,2 addition is observed with aldehydes, and complete 1,4 addition with esters and nitriles, but mixtures are obtained with ketones (Eq. 15).[63]

$C_6H_5(CH_2)_4P[OSi(CH_3)_3]_2$

CN → $C_6H_5(CH_2)_4P(O)(OSi(CH_3)_3)CH_2CH_2CN$ (90%)

CHO → $C_6H_5(CH_2)_4P(O)(OSi(CH_3)_3)CH(OH)CH{=}CHCH_3$ (88%)

O → $C_6H_5(CH_2)_4P(O)(OSi(CH_3)_3)C(CH_3)_2CH_2COCH_3$ (37%) + $C_6H_5(CH_2)_4P(O)(OSi(CH_3)_3)C(OH)(CH_3)CH{=}C(CH_3)_2$ (52%) (Eq. 15)

Aromatic esters, such as triphenyl phosphite, do not undergo intramolecular attack to complete the reaction. Instead, a proton source, generally acetic acid, is used in the reaction mixture to generate the α-hydroxyalkylphosphonate diaryl ester. For example, triphenyl phosphite in acetic acid readily undergoes reaction with the imine generated *in situ* from benzyl carbamate and benzyloxyacetaldeyde to form diphenyl 1-benzyloxycarbonylamino-2-benzyloxyethylphosphonate in 51% yield (Eq. 16).[64]

$$C_6H_5CH_2OCH_2CHO + C_6H_5CH_2O_2CNH_2 + (C_6H_5O)_3P \xrightarrow{HOAc}$$

$$C_6H_5CH_2O_2CNHCH(CH_2OCH_2C_6H_5)P(O)(OC_6H_5)_2$$

(51%) (Eq. 16)

This represents a particularly convenient method for generating such aromatic esters. An advantage of the phenyl esters is that they are quite stable to ordinary hydrolytic ester cleavage conditions but are subject to hydrogenolysis readily with platinum catalyst at atmospheric pressure.[65,66]

A convenient approach to the preparation of the dialkyl α-hydroxyalkylphosphonates utilizes the trialkyl phosphite with an intermolecular silylation trap of the intermediate oxide anion (Eq. 17). An equivalent amount of the silyl chloride is used with the aldehyde and trialkyl phosphite.[67]

$$C_2H_5CHO + (CH_3)_2SiCl(OC_2H_5) + (C_2H_5O)_3P \xrightarrow{<25^\circ} C_2H_5CH\left[OSi(OC_2H_5)(CH_3)_2\right]P(O)(OC_2H_5)_2 \quad (96\%) \qquad \text{(Eq. 17)}$$

At the relatively low temperatures used (<25°), reaction of the silyl chloride with the intermediate oxide anion is much faster than with the phosphoryl oxygen of the product phosphonate. The latter would lead to ester cleavage. Moreover, it should be noted that the silyl halide does not undergo reaction with the trialkyl phosphite under these conditions. Such a reaction would generate a mixed silyl phosphite ester capable of yielding the same overall products.[51] With substrates capable of undergoing conjugate addition reactions, either trialkylsilyl chlorides[51,63,68] or *N*,*N*-bis(trimethylsilyl)acetamide[63] can be used as added silyl-trapping reagents. This approach is of particular value for the generation of reagents for reversing the normal polar reactivity characteristics of carbonyl groups (umpolung).[61,69,70]

In conjugate addition reactions using trialkyl phosphites, added hydroxylic reagents are needed both for supplying a proton to the enolate functionality of the intermediate and for performing the displacement at the quasiphosphonium site (Eq. 18). Although ethanol has been used for this purpose,[33] *tert*-butyl alcohol[71,72] and phenol[33] appear to provide better yields of clean product.

$$CH_2{=}CHC(O)CH_3 + (RO)_3P \xrightarrow{R'OH} \left[CH_3C(O^-){=}CHCH_2\overset{+}{P}(OR)_3 \right] \longrightarrow CH_3C(O)CH_2CH_2P(O)(OR)_2 + ROR' \qquad \text{(Eq. 18)}$$

Finally, phosphoryl chlorides in equimolar amount with the trialkyl phosphite in benzene solution can be used to trap the intermediate enolate, generating enol phosphate esters by reaction at oxygen (Eq. 19).[73]

$$(RO)_3P + (R'O)_2P(O)Cl + CH_2{=}CHCHO \xrightarrow{C_6H_6} (RO)_2P(O)CH_2CH{=}CHOP(O)(OR')_2$$

(Eq. 19)

In addition to the phosphites as noted above, mixed ester–amide species can be used in these addition reactions. For example, diethyl anilinophosphinate undergoes addition to *N*-(*p*-chlorophenyl)maleimide.[74] On aqueous workup the phosphonate diethyl ester is isolated, which is identical to that formed with the use of triethyl phosphite. Similarly, fully esterified forms of lower-oxidation-state trivalent phosphorus undergo these reactions. Diethyl phenylphosphonite gives conjugate addition as readily to β-nitrostyrene as do trialkyl phosphites (Eq. 20).[75]

$$C_6H_5P(OC_2H_5)_2 + C_6H_5CH{=}CHNO_2 \xrightarrow[CH_3CN]{50^\circ} C_6H_5P(O)(OC_2H_5)CH(C_6H_5)CH_2NO_2 \quad (61\%)$$

(Eq. 20)

With mixed alkyl esters of the trivalent phosphorus acids, product mixtures may be obtained because of the loss of different alkyl functions from the quasiphosphonium site. The relative tendencies of these alkyl esters to be removed has been measured by competition studies with the conjugate addition reaction of mixed trialkyl phosphites and acrylonitrile.[76] The tendency to be displaced is:

$$C_6H_5CH_2 > t\text{-}C_4H_9 > CH_2{=}CHCH_2 > C_2H_5 > i\text{-}C_3H_7 > n\text{-}C_3H_7$$

Monobasic Phosphorus Acid Esters. The addition of a monobasic trivalent phosphorus acid to a carbonyl carbon atom (or in a conjugate manner to an olefinic carbon atom) is easily accomplished with a wide range of basic reagents. In the earliest work, conjugate addition to several Michael-type acceptors was accomplished with the sodium salts of dialkyl phosphites.[77,78] Reasonable yields can be obtained with either a catalytic or equivalent amount of the salt of the phosphorus acid. With each substrate, yields of the adduct phosphonate are highest with the use of dimethyl phosphite, decreasing slightly with other primary alkyl esters. Significant decreases in yield are found when secondary alkyl phosphorus esters are used.

It was similarly noted in early investigations that α-hydroxyalkylphosphonates can be generated from saturated aliphatic aldehydes or ketones by this approach.[35] High yields of product are reported with the use of either an equivalent amount of the phosphorus acid salt or a catalytic amount of sodium or lithium methoxide with the phosphorus acid. However, stereochemical variations are noted with changing conditions in additions to 1,2:5,6-diisopropylidene-D-glucofuranos-3-ulose.[79]

Sodium salts of the monobasic acids are used quite conveniently in these addition reactions. While di-*n*-butyl phosphite alone undergoes addition to maleate esters poorly, formation of the sodium salt with catalytic amounts of either sodium amide or sodium metal results in formation of the monoadducts in good to excellent yields.[80] Similar results are obtained using dibutyl phosphite with maleate esters. (Eq. 21)[81]

$$\text{(}n\text{-}C_4H_9O_2C\text{)CH=CH(}CO_2C_4H_9\text{-}n\text{)} + (n\text{-}C_4H_9O)_2POH \xrightarrow[NaNH_2]{0.026\ eq} (C_4H_9O)_2P(O)CH(CO_2C_4H_9\text{-}n)CH_2CO_2C_4H_9\text{-}n$$

(85 %)

(Eq. 21)

Sodium diphenylphosphinite, generated by reaction of sodium hydride with diphenylphosphine oxide in tetrahydrofuran, adds readily to the carbonyl carbon of acetone[82] and also participates in Michael-type additions to mesityl oxide[83] or ethyl vinyl ketone.[84] With crotonaldehyde, however, the conjugate addition route is bypassed for addition to the carbonyl carbon atom. The diphenylphosphinite anion reacts in a consistent manner independent of its mode of generation. Equivalent results are obtained in reaction with carbonyl compounds when it is generated by (1) sodium hydride reaction with diphenylphosphine oxide, (2) 3 equivalents of phenyl Grignard in reaction with diethyl phosphite, or (3) sodium iodide action on benzyl diphenylphosphinylformate.[82]

The sodium salt of dimethyl phosphite, generated with sodium methoxide in ether at ambient temperature, undergoes addition to the carbonyl carbon of benzalacetone (Eq. 22).[85] This result is in contrast to the use of other dialkyl phosphites in refluxing benzene with the same base, where Michael-type addition is reported.[86]

$$CH_3COCH{=}CHC_6H_5 + (CH_3O)_2POH \xrightarrow[NaOCH_3]{ether} CH_3C(OH)(P(O)(OCH_3)_2)CH{=}CHC_6H_5$$

(70 %) (Eq. 22)

A similar solvent-dependent variation in the site of addition is observed in the reactions of several α,β-unsaturated ketones with dimethyl phosphite and sodium methoxide. Michael-type addition is observed in benzene solution, whereas carbonyl addition occurs in ether solution.[87] In ether solution, yields of carbonyl-addition product can be increased by changing the base to diethylamine, whereas use of triethylamine gives unfavorable results (Eq. 23).[87]

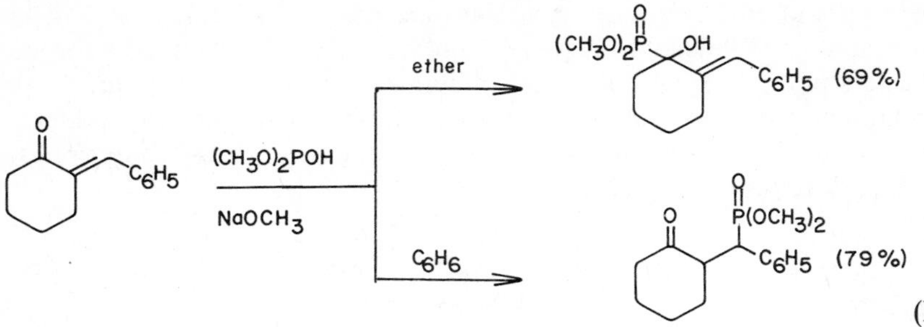

(Eq. 23)

Carbonyl Compound	Base	Yield of 1-Hydroxyalkylphosphonate (%)
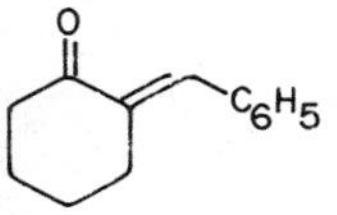 2-benzylidenecyclohexanone	$NaOCH_3$	69
	$(C_2H_5)_2NH$	100
	$(C_2H_5)_3N$	20
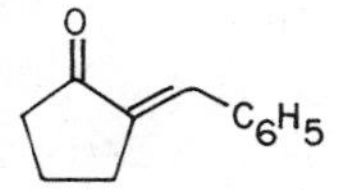 2-benzylidenecyclopentanone	$NaOCH_3$	64
	$(C_2H_5)_2NH$	100
	$(C_2H_5)_3N$	58

A further divergence in reaction course is found in the synthesis of analogs of aminoethylphosphonate with the use of dimethyl phosphite with triethylamine or sodium methoxide (Eq. 24).[88]

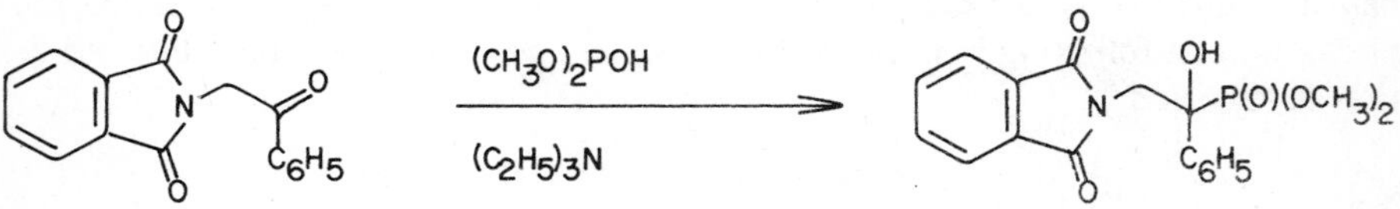

(64%) (Eq. 24)

Moderate yields of α-hydroxyalkylphosphonates are obtained when triethylamine is used. However, use of sodium methoxide leads to the formation of a phosphate triester, **3**.[88]

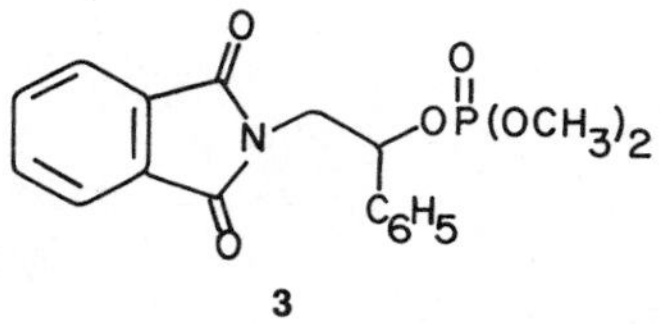

3

A variety of amines can be used to catalyze these additions. Significant levels of asymmetric induction in the product (80–85%) can be attained by using chiral amines such as quinine.[89,90]

Two approaches involving heterogeneous reaction conditions provide moderate to excellent yields of carbonyl adducts. The first involves treatment of a mixture of aldehyde or ketone and the dialkyl phosphite with a five-fold excess of a fluoride salt (potassium or cesium) at room temperature (Eq. 25).[91]

$$C_6H_5CH{=}CHCHO + (C_2H_5O)_2POH \xrightarrow[0.5\ h]{KF} C_6H_5CH{=}CHCHOHP(O)(OC_2H_5)_2 \quad (88\%) \qquad \text{(Eq. 25)}$$

The second approach uses commercial basic chromatographic alumina as the catalyst (Eq. 26).[92]

$$C_6H_5COCH_2Cl + (C_2H_5O)_2POH \xrightarrow[\text{basic } Al_2O_3]{\text{Merck 60}} C_6H_5COH(CH_2Cl)P(O)(OC_2H_5)_2 \quad (65\%) \qquad \text{(Eq. 26)}$$

The major advantage of these procedures is the relatively simple experimental conditions involved. The product is removed from the insoluble catalyst by washing with an organic solvent followed by filtration. The principal limitation to the procedures is the inefficiency of the reactions when solid or highly viscous carbonyl compounds are used.

For addition reactions to imines with dialkyl phosphites or monoalkyl phosphonites,[94–96] it appears that the imine nitrogen is sufficiently basic to promote the reaction (Eq. 27).

$$n\text{-}C_4H_9CH{=}NCH_2C_6H_5 + (C_2H_5O)_2POH \xrightarrow[0.5\ h]{100-140^\circ} n\text{-}C_4H_9CH(NHCH_2C_6H_5)P(O)(OC_2H_5)_2 \quad (65\%) \qquad \text{(Eq. 27)}$$

Moreover, in some instances bicarbonate[97] or carbonate[98] can be used to promote dialkyl phosphite addition reactions with carbonyl compounds (Eq. 28).

$$\text{tetraphenylcyclopentadienone} + (CH_3O)_2POH \xrightarrow[NaHCO_3,\ 18\ d]{\text{ether}} \text{1-hydroxy-2,3,4,5-tetraphenylcyclopenta-2,4-dienyl-}P(O)(OCH_3)_2 \quad (60\%) \qquad \text{(Eq. 28)}$$

Silylated Phosphorus Reagents. The use of silylated phosphite reagents in polar addition reactions at unsaturated carbon has been noted earlier. However, the use of such reagents represents a major advance in synthetic approach. Therefore further discussion is given here.

Several preparations of tris(trimethylsilyl) phosphite have been reported.[45–47,99,100] The most favorable method for producing tris(trimethylsilyl) phosphite of high purity and in good yield begins with the trisodium salt of phosphonoformic acid.[100] This salt, readily obtained in high purity, is decomposed to phosphorous acid of high purity, which is rendered anhydrous and silylated by treatment with chlorotrimethylsilane.

Tris(trimethylsilyl) phosphite readily undergoes polar addition reactions both at carbonyl carbon and in a conjugate manner with certain Michael-type acceptors. With simple aldehydes and ketones, addition at the carbonyl carbon proceeds in good to excellent yield and can be followed by near-quantitative conversion of the bis(trimethylsilyl) phosphonate to the free acid on water addition (Eq. 29).[49,58]

$$C_6H_5COCH_3 + [(CH_3)_3SiO]_3P \xrightarrow[2.\ H_2O]{1.\ C_6H_6} C_6H_5COH(CH_3)PO_3H_2 \quad (73\%) \qquad \text{(Eq. 29)}$$

Variable results are found with α-halocarbonyl compounds. With aliphatic α-haloketones and aldehydes, the reaction proceeds exclusively at the carbonyl carbon to generate the α-trimethylsiloxyalkylphosphonate (Eq. 30).[48]

$$(CH_3)_2CClCHO + [(CH_3)_3SiO]_3P \xrightarrow{THF} (CH_3)_2CClCH[OSi(CH_3)_3]P(O)[OSi(CH_3)_3]_2 \quad (92\%) \qquad \text{(Eq. 30)}$$

However, with α-haloaromatic ketones or carboxylates, product mixtures result from vinyl phosphate formation (Perkow reaction)[101–103] or direct halide displacement (Michaelis–Arbuzov reaction) (Eq. 31,32).[47–49,59,104]

$$BrCH_2COC_6H_5 + [(CH_3)_3SiO]_3P \xrightarrow{THF}$$

$$[(CH_3)_3SiO]_2P(O)CH_2COC_6H_5 \quad (14\%)$$

$$+ \ CH_2{=}C(C_6H_5)OP(O)[OSi(CH_3)_3]_2 \quad (61\%) \qquad \text{(Eq. 31)}$$

$$BrCH_2COCO_2C_2H_5 + [(CH_3)_3SiO]_3P \xrightarrow{THF}$$

$$CH_2{=}C(CO_2C_2H_5)OP(O)[OSi(CH_3)_3]_2 \quad (36\%)$$

$$+ \ BrCH_2C(CO_2C_2H_5)[OSi(CH_3)_3]P(O)[OSi(CH_3)_3]_2 \quad (31\%) \qquad \text{(Eq. 32)}$$

Monoesters of α-hydroxyalkylphosphonates, which otherwise can be generated only with significant difficulty, can be generated through use of this reagent.[100] The use of silyl ester linkages introduces the required selectivity into the sequence of hydrolysis, esterification, thermolysis, silylation, and addition, which would otherwise be extremely difficult. A convenient aspect of this route is the possibility of performing the entire sequence of reactions without isolation of intermediates (Eq. 33).

$$[(CH_3)_3SiO]_3P + C_2H_5O_2CCl \longrightarrow \left[C_2H_5O_2CP(O)[OSi(CH_3)_3]_2\right]$$

$$\longrightarrow \left[C_2H_5O_2CPO_3H_2\right] \longrightarrow \left[C_2H_5O_2CP(O)(OH)OR\right]$$

$$\longrightarrow \left[[(CH_3)_3SiO]_2POR\right] \longrightarrow R'CHOHP(O)(OH)OR \quad \text{(Eq. 33)}$$

The trimethylsilyl dialkyl phosphites have been more thoroughly investigated for both carbonyl–carbon and Michael-type addition reactions. These reagents are prepared conveniently either by trimethylsilyl halide reaction with the salt of the dialkyl phosphite[47,105] or with the dialkyl phosphite with a tertiary amine.[51,106]

The carbonyl adducts of these monosilyl reagents have particular utility in the synthesis of ketones. They are generated, using the adduct as an acyl anion equivalent, by alkylation of the lithiated adduct followed by cleavage of the carbon–phosphorus linkage in basic medium.[12] For example, benzyl phenyl ketone is prepared in 74% overall yield from benzaldehyde (Eq. 34).[11]

$$C_6H_5CHO + (CH_3)_3SiOP(OC_2H_5)_2 \longrightarrow C_6H_5CH[OSi(CH_3)_3]P(O)(OC_2H_5)_2 \quad (91\%)$$

$$\xrightarrow[\text{2. } C_6H_5CH_2Br]{\text{1. } LiN[C_3H_7\text{-}i]_2} C_6H_5CH_2C(C_6H_5)[OSi(CH_3)_3]P(O)(OC_2H_5)_2 \quad (85\%)$$

$$\xrightarrow[C_2H_5OH]{NaOH} C_6H_5COCH_2C_6H_5 \quad (95\%) \qquad \text{(Eq. 34)}$$

The fundamental utility of the monosilyl ester phosphites is that they can generate α-siloxyalkylphosphonates that are protected for immediate alkylation, but easily undergo selective removal of the silyl function for conversion to the α-hydroxy species and subsequent removal of the phosphorus function.

Of similar utility are the silylated phosphorodiamidites. These compounds, such as triethylsilyl *N,N,N',N'*-tetramethylphosphorodiamidite, are prepared by reaction of the silanoxide anion with the appropriate phosphorodiamido chloride (Eq. 35).[51]

$$(C_2H_5)_3SiOH \xrightarrow[\text{ether}]{NaH} [(C_2H_5)_3SiONa] \xrightarrow{ClP[N(CH_3)_2]_2} (C_2H_5)_3SiOP[N(CH_3)_2]_2 \quad (77\%) \qquad \text{(Eq. 35)}$$

After reaction with an aldehyde, either by addition at the carbonyl carbon or in a Michael-type manner, these adducts also serve as acyl–anion equivalents. They are readily alkylated and are converted to the simple carbonyl products on treatment with tetra-*n*-butylammonium fluoride (Eq. 36).[10]

$$C_6H_5CHO + (C_2H_5)_3SiOP[N(CH_3)_2]_2 \xrightarrow[0^\circ]{THF} [C_6H_5CH[OSi(C_2H_5)_3]P(O)[N(CH_3)_2]_2]$$

$$\xrightarrow[2.\ CH_3I]{1.\ n\text{-}C_4H_9Li} [C_6H_5C(CH_3)[OSi(C_2H_5)_3]P(O)[N(CH_3)_2]_2]$$

$$\xrightarrow{(n\text{-}C_4H_9)_4NF} C_6H_5COCH_3 \quad (72\%) \qquad \text{(Eq. 36)}$$

Mixed silyl aromatic phosphites have not been reported. They would be anticipated to have synthetic utility similar to the silyl alkyl phosphites and should provide an alternative route for generation of the free phosphonic acids.

Silyl esters of phosphonites would be anticipated to exhibit similar reactivity, although their utility for general synthetic purposes would be less because of the greater effort required for their preparation. Accordingly, little work has been done with such compounds. These compounds readily undergo conjugate addition to α,β-unsaturated esters,[63] nitriles,[63] ketones (Eq. 37),[63] and nitro compounds,[107] although the α,β-unsaturated aldehydes examined undergo addition only at the carbonyl carbon.[63]

$$C_6H_5(CH_2)_4P(OC_2H_5)OSi(CH_3)_3 + \text{2-cyclopentenone} \longrightarrow \text{1-}OSi(CH_3)_3\text{-3-}[P(O)(OC_2H_5)(CH_2)_4C_6H_5]\text{cyclopentene} \quad (89\%) \qquad \text{(Eq. 37)}$$

The bis(trimethylsilyl) ester of the parent phosphonous acid adds to the carbonyl carbon of chloroacetone. However, the Perkow reaction is a significant competitive pathway (Eq. 38).[108]

$$CH_3COCH_2Cl + [(CH_3)_3SiO]_2PH \longrightarrow$$

$$(CH_3)_3SiOPH(O)C(CH_3)[OSi(CH_3)_3]CH_2Cl \quad (38\%)$$

$$+ \quad (CH_3)_3SiOPH(O)OC(CH_3){=}CH_2 \quad (42\%) \qquad \text{(Eq. 38)}$$

The silyl reagent also undergoes conjugate addition to acetylenic and vinylic phosphonates in moderate yield.[109]

Nature of the Substrate

In most instances trivalent phosphorus reagents undergo addition reactions with polar unsaturated linkages, attacking as nucleophiles at electron-deficient sites. While the most common functionality undergoing such attack is the carbonyl linkage, others have been investigated and provide facile routes to organophosphorus compounds. The principal functionalities that undergo addition reactions are discussed here.

Simple Aldehydes and Ketones. Simple aldehydes and ketones undergo addition at the carbonyl carbon quite readily with the entire range of phosphorus reagents noted previously. While yields with either aldehydes or ketones are in the range of moderate to excellent, with a given phosphorus reagent aldehydes generally provide higher product yields than ketones of similar size and carbon skeleton. This is noted using fully esterified phosphorous acid reagents[49–51,110,111] as well as partially esterified acids with alumina[92] or alkali fluorides,[91] or as their anions.[35] Such observations are in keeping with the generally enhanced reactivity of aldehydes toward nucleophilic attack.

Similarly, increasing the normal polarization of the carbonyl functionality leads to improved product yields.[35] Thus the α-haloaliphatic ketones give excellent yields of carbonyl addition products,[47,91] interestingly without any significant amount of vinyl phosphate ester (Perkow reaction).[101–103] However, the two reaction pathways appear to be competitive using either α-halopyruvates with tris(trimethylsilyl) phosphite,[48] or bis(trimethylsiloxy)phosphine with simple α-haloketones.[108] Clean addition to the carbonyl–carbon of α-halopyruvates can be obtained by using dialkyl phosphites with base,[112] mixed silyl phosphite esters,[113] or diphenylphosphine oxide.[114] Further, the α-haloaromatic ketones appear to form *only* vinyl phosphate esters with tris(trimethylsilyl) phosphite[48] but exhibit only addition at the carbonyl–carbon atom with the use of dialkyl phosphites with amines.[115] Although hexafluoroacetone gives 1,3,2-dioxaphospholanes with ordinary trialkyl phosphites,[55] simple addition generating the α-siloxyalkylphosphonates in excellent yields occurs when a silyl phosphite ester is used.[56,57]

Compounds that bear a carbonyl group alpha to a carboxylate ester (pyruvate-related materials) readily undergo simple addition to the carbonyl carbon with the use of fully esterified trivalent phosphorus reagents,[58,60] although the presence of an aromatic ring favors a further rearrangement.[60] The α-bromopyruvates give mixtures of products, including those of carbonyl–carbon addition and vinyl phosphate formation (Perkow).[59] Simple carbonyl–carbon addition occurs as well with the α-ketoesters,[50,54] ketene (Eq. 39),[50] and isocyanates.[116] In the latter two instances the ultimate products correspond to those obtained by Michaelis–Arbuzov reaction of the acid halides.[117]

$$(C_2H_5O)_2POSi(CH_3)_3 + CH_2{=}C{=}O \longrightarrow (C_2H_5O)_2P(O)C[OSi(CH_3)_3]{=}CH_2$$

$$\xrightarrow{C_2H_5OH} (C_2H_5O)_2P(O)COCH_3 \quad (87\%) \qquad \text{(Eq. 39)}$$

Although variations in yields among aliphatic and aromatic aldehydes (and ketones) are found, there appears to be no fundamental difference in the product yields attainable with each, except in reaction with trialkyl phosphites.

α,β-Unsaturated Aldehydes and Ketones. When an olefinic linkage is conjugated with the carbonyl function, Michael-type addition of the phosphorus reagent, as well as addition at the carbonyl–carbon atom, can occur. Both aldehydes and ketones react readily with trialkyl phosphites in a hydroxylic solvent to give the Michael-type addition products.[32,33,118,119] Clear-cut differences in reactivity and product yield are not evident in comparison of aldehydes with ketones in this reaction. Often, the immediate product obtained is the ketal (acetal), the enol ether, or a mixture thereof rather than the free carbonyl compound. Isolation of the free carbonyl species is easily accomplished by treatment of the reaction mixture with aqueous acid. An interesting example of this is found in the reaction of trimethyl phosphite with 2,5-dicarbomethoxy-3,4-diphenylcyclopentadienone. The enol ether is formed exclusively by a 1,6-addition reaction (Eq. 40).[120]

CH₃O₂C, CO₂CH₃, C₆H₅, C₆H₅ cyclopentadienone + $(CH_3O)_3P$ $\xrightarrow{\text{room temp.}}$ enol ether product (OCH₃, CH₃O₂C, CO₂CH₃, P(O)(OCH₃)₂, C₆H₅, C₆H₅) (72%) (Eq. 40)

Other substituted cyclopentadienones produce diverse products under a variety of conditions.[97,121–124]

Pronounced differences in the chemistry of α,β-unsaturated aldehydes as compared to the ketones occur when the fully esterified phosphorous acid species contains one or more silyl functions. In reactions with dialkyl trialkylsilyl phosphites, these aldehydes undergo predominantly, if not totally, addition to the carbonyl-carbon atom.[51,61,111] However, the ketones react almost exclusively by conjugate addition.[51,61,111] Similar results are found with the use of either trienthylsilyl *N,N,N′,N′*-tetramethylphosphorodiamidite[51] or tris(trimethylsilyl) phosphite.[49] The use of trialkyl phosphites or other fully esterified phosphorus acid species with an added silyl chloride gives variable results with aldehydes, although ketones give only products of conjugate addition in good to excellent yield.[51] Silyl phosphonite esters also exhibit this diversity in reaction pathway with unsaturated carbonyl compounds. Aldehydes react exclusively at the carbonyl–carbon atom while ketones give conjugate addition products or mixtures.[63]

Unsaturated ketones give products of conjugate addition in moderate yields by reaction with dialkyl phosphite anions.[77,78,125] If reaction at more than one site is possible, phosphorus addition occurs at the less highly substituted carbon atom.[78] Unsaturated aldehydes exhibit carbonyl carbon or Michael-type addition with the anion of diphenylphosphine oxide, depending on the mode of generation of the anion.[82,84] A solvent dependence is found for the reaction of sodium salts of dimethyl phosphite with unsaturated ketones. In ether solution, addition occurs only at the carbonyl-carbon atom whereas in benzene solution conjugate addition occurs.[87]

Other α,β-Unsaturated Compounds. In addition to the α,β-unsaturated aldehydes and ketones, a variety of other types of compounds exhibit Michael-type addition of phosphorus esters. The most broadly investigated of these are the conjugated esters, although other carboxylate-related materials have been studied as well.

Maleate esters undergo conjugate addition of dialkyl phosphites in the presence of catalytic amounts of base to give the phosphonates in good to excellent yield (Eq. 41).[80,81]

$$\text{cis-}[C(=O)OCH_2CH(C_2H_5)C_4H_9\text{-}n]CH{=}CH[C(=O)OCH_2CH(C_2H_5)C_4H_9\text{-}n] + (n\text{-}C_4H_9O)_2POH \xrightarrow{Na} (n\text{-}C_4H_9O)_2P(O)CH[C(=O)OCH_2CH(C_2H_5)C_4H_9\text{-}n]CH_2C(=O)OCH_2CH(C_2H_5)C_4H_9\text{-}n \quad (96\%) \qquad \text{(Eq. 41)}$$

In the absence of base, generally sodium amide or sodium metal, only very poor yields of adduct are obtained.[80] Similar addition reactions occur with the use of secondary phosphine oxides with unsaturated monocarboxylates under basic conditions.[126] Maleimide and *N*-substituted maleimides also undergo Michael addition in the presence of phosphorous acid in the fully esterified[127] or diester–amide state.[74]

Esters of acrylic acid undergo hydrophosphinylation readily with salts of dialkyl phosphites,[128,129] trialkyl phosphites in the presence of silyl halides,[68] or silylated phosphite species.[130,131] Under the second set of conditions, acrylates substituted at either the α or the β position fail to react.[68] However, moderate yields of Michael-type adducts are obtained from such substrates using silylated phosphites (Eq. 42).[130]

$$CH_3CH_2CH{=}CHCO_2CH_3 + (C_2H_5O)_2POSi(CH_3)_3 \xrightarrow[2\ h]{120°} CH_3CH_2CH[P(O)(OC_2H_5)_2]CH_2CO_2CH_3\ (43\%)$$

(Eq. 42)

Acrylonitrile and α-substituted acrylonitriles also react under all of these conditions to give conjugate adducts in moderate to good yield.[50,76,128,132,133] Acrylamide and related 2- and 3-substituted acrylamides undergo addition readily using dialkyl phosphites with added base.[134,135] Moderate to good yields of the resultant 3-phosphonoamides are obtained. Hypophosphite monoesters similarly give products of addition to two molecules of acrylonitrile or acrylamide.[28]

When 2-bromo-3-alkoxypropionates are heated with an excess of triethyl phosphite, two competing processes occur.[136] The major route involves initial Arbuzov reaction to displace the bromide, followed immediately by elimination of alcohol. The resultant conjugated ester readily undergoes hydrophosphinylation to give 2,3-bisphosphonates in moderate yield (Eq. 43).

$$ROCH_2CHBrCO_2R' + (R''O)_3P \xrightarrow{heat} [CH_2{=}C(P(O)(OR'')_2)CO_2R'] + ROH \xrightarrow{(R''O)_3P} (R''O)_2P(O)CH_2CH[P(O)(OR'')_2]CO_2R'$$

(Eq. 43)

The minor competing route involves phosphite-induced elimination to form the acrylate, which then undergoes hydrophosphinylation to generate the β-phosphonopropionate.

Acrylic acid itself can be used in hydrophosphinylation with trialkyl phosphites, although the adduct isolated is the carboxylate ester.[137–139] Presumably the reaction proceeds by initial transesterification followed by addition of the resultant dialkyl phosphite to the acrylate ester.[137,140] This approach also can be used with α- and β-substituted acrylic acids.[137,140,141]

Nitroolefins react with fully esterified trivalent phosphorus species under protic conditions to give Michael-type adducts in moderate to good yield.[71,72,75,107,142–145] In some reactions organic nitrites are formed.[146,147] Lower yields are obtained from acyclic nitroolefins with dialkyl phosphites and base,[75] but reasonable results are obtained with cyclic carbohydrate derivatives.[148] In these latter systems, stereospecific *trans* addition is observed (Eq. 44).[148]

C_6H_5CH O O OCH_3 O_2N $\xrightarrow[(C_2H_5)_3N]{(CH_3O)_2POH}$ C_6H_5CH O O OCH_3 NO_2 $(O)P(OCH_3)_2$

(69%)

(Eq. 44)

Imines. The electron-deficient carbon atom of the imine linkage provides a site of reactivity completely analogous to the carbonyl linkage. The early reported method of Fields[94] involving dialkyl phosphite addition to either (1) preformed imines or (2) imines generated *in situ* (aldehyde + amine + phosphite) allows facile preparation of α-aminoalkylphosphonates bearing substitution on nitrogen (Eq. 45).

CHO $+ (C_2H_5)_2NH + (C_2H_5O)_2POH$ $\xrightarrow[15\ min]{<85°}$ $P(O)(OC_2H_5)_2$ $N(C_2H_5)_2$

(84%)

(Eq. 45)

Excellent yields are obtained with aldimines or ketimines of primary or secondary aliphatic amines. This includes an extensive series of diaromatic ketones having the fundamental fluorenone-type structure. These ketones are used in the preparation of α-aminophosphonic acid analogs of morphactines; these compounds have activity as plant growth regulators.[149–151] Aldimines and ketimines formed from primary aromatic amines also react under these conditions, but the yields are only moderate to good.[93,97,152–157]

The direct preparation of α-aminoalkylphosphonates without nitrogen substitution is possible by this general approach.[158–162] However, extremely stringent conditions (high temperatures, sealed tube) with the use of alcoholic ammonia are required, and only poor yields are obtained. Moderate yields (43–71%) are obtained when the aldimine hydrochloride–tin(IV) chloride complex (formed from the nitrile) is used with dialkyl phosphites.[163]

Since these target molecules have significant potential utility as analogs of α-amino acids, alternatives for their synthesis are desired. Routes to these materials are available through the use of a variety of alkyl amines that may be cleaved readily to the primary amino function after the phosphorus addition reaction is accomplished. In one approach a benzylic amine, which can be hydrogenolyzed to the free primary amino function, is used. In addition to the simple benzyl group,[95] secondary[164] and tertiary[165–168] benzylic groups can be used. Chiral α-methylbenzylamine can be used for asymmetric induction in the preparation of optically active α-aminoalkylphosphonic acids.[164–169]

Of similar utility are the diimines formally derived from 1,1-diamines. These are produced by the condensation of three molecules of aldehyde with two of ammonia. Phosphonates and phosphinates bearing an α-amino function are produced in good to excellent yields from these species by acid hydrolysis following the phosphorus addition reaction.[170–172] Benzaldazine can be used in the same manner.[96] Imine formation using *tert*-butylamine prior to phosphorus addition also allows formation of the free primary amine product in excellent yield on acidic hydrolysis (Eq. 46).[173]

$$t\text{-}C_4H_9NH_2 + H_2CO \longrightarrow t\text{-}C_4H_9N{=}CH_2 \xrightarrow{(C_2H_5O)_2POH} t\text{-}C_4H_9NHCH_2P(O)(OC_2H_5)_2$$

$$\xrightarrow{HBr} H_2NCH_2PO_3H_2 \quad (98\%) \qquad \text{(Eq. 46)}$$

Finally, an oxime can be used with a dialkyl phosphite for preparation of the α-hydroxylaminophosphonate.[174]

Fully esterified phosphites (trialkyl or triaryl) are used principally in reactions in which the imine reagent is generated *in situ* from the aldehyde and a carbonic amide. With acetic acid as catalyst, moderate to good yields of the *N*-substituted α-aminoalkylphosphonates are isolated with ureas,[175,176] thioureas,[177–180] and carbamates.[64,181] The use of an acidic catalyst (glacial acetic acid or boron trifluoride etherate) facilitates the initial condensation reaction, which is inhibited by the basic trialkyl phosphites. Triaryl phosphites are not sufficiently basic to cause this inhibition.[175] With cyclic ureas[182] and unsubstituted urea or thiourea,[175] formation of diadducts occurs readily.

Simple aldimines can be used in reaction with fully esterified phosphites. The chiral aldimines from chiral α-methylbenzylamine react with tris(trimethylsilyl) phosphite to generate chiral α-aminoalkylphosphonic acids.[183] Optical purity of 90% can be obtained with good yield by a two-step procedure.

The nature of the imine reactions described might lead one to anticipate the possibility of a Mannich-type condensation occurring with (unesterified) phosphorus acids. This does occur with formaldehyde and phosphorous acid, although mixtures of mono- and diadducts on nitrogen are produced (Eq. 47).[29]

$$RNH_2 + H_2C{=}O + H_3PO_3 \xrightarrow[H_2O]{HCl} RNHCH_2PO_3H_2 + RN(CH_2PO_3H_2)_2 \qquad \text{(Eq. 47)}$$

A report[184] that the corresponding reaction occurs with carbonyl compounds other than formaldehyde in aqueous media at low pH has been found to be in error.[185] However, direct addition of phosphorous acid to *C*-substituted imines does occur with heating in the absence of solvent[186] or with a nonprotic solvent, such as acetic anhydride.[187] The additon of hypophosphorous acid to *N*-substituted imines is of use in the preparation of 1-aminoalkylphosphonous acids as isosteric analogs of naturally occurring amino acids.[188]

The use of trivalent phosphorus acid chlorides for addition to imines is also possible for the preparation of this type of compound.[189–191] However, only moderate yields of the free acid are obtained by this method.

COMPARISON OF METHODS

This brief analysis of the carbon-phosphorus bond-forming procedures is divided into three parts. The first considers the syntheses of α-oxyphosphonates, with both free and masked hydroxyl groups. The second considers the approaches to phosphonates bearing a free α-amino group. Finally, conjugate additions generating phosphonates with an unsaturated functionality attached at the β position are reviewed.

α-Oxyphosphonates

The recent development of the use of silyl reagents in the addition of trivalent phosphorus to carbonyl carbon has provided the most efficient routes to α-oxyphosphonates and related materials.

Two distinct methods are available. The route most thoroughly investigated uses a silyl ester of the trivalent phosphorus acid, preformed by reaction of the acid with a silyl halide in the presence of base. For most systems using this route, reported product yields are greater than 80%. The use of trialkyl phosphites in the presence of silyl halides has been less thoroughly investigated. The role of the silyl halide here is not to attack the phosphite, but to intercept the initial adduct, silylating the oxide anion site (Eq. 48).

$$(RO)_3P + R'CHO \rightleftharpoons (RO)_3\overset{+}{P}-\underset{}{\overset{O^-}{\overset{|}{C}}}HR' \xrightarrow{R''_3SiX} (RO)_3\overset{+}{P}-\overset{OSiR''_3}{\overset{|}{C}}HR' + X^- \quad \text{(Eq. 48)}$$

The reported yields using this method have generally been greater than 90%. A further preferential aspect of this latter method is its "one-pot" simplicity, bypassing the separate formation of the silyl–phosphorus ester.

Both of these methods result in the formation of α-siloxy products. The silyl ether linkage can be cleaved under relatively mild conditions to generate the free hydroxyl group or can be left in place to serve as a protecting function, such as by preventing carbon–phosphorus bond fission under basic conditions while other operations are performed.

If silyl reagents are to be avoided and the α-hydroxy product is desired directly, the use of fluoride salts with the dialkyl phosphite and carbonyl compound in the absence of solvent seems to be the most efficient method.[91] Generally high yields are reported for this method, and the workup is relatively simple. Its major deficiency is seen, however, if one or both of the reactants is solid. Difficulty in obtaining and maintaining reagent contact precludes efficient reaction. In such instances use of a catalytic amount of base (alkoxide ion) with the dialkyl phosphite and carbonyl compound in a dipolar aprotic solvent can result in yields in excess of 90%, even with highly functionalized substrates.[192]

Hydroboration–oxidation of vinylphosphonates provides an alternative preparative route for α-hydroxyalkylphosphonates. The hydroboration of vinylic linkages directly attached to phosphorus proceeds regiospecifically with the attachment of boron (and subsequently oxygen) at the α position.[193,194] Use of a chiral borane reagent can also provide stereospecific introduction of the hydroxyl group.[194] A major consideration for the use of this approach as a generally competitive method for the preparation of α-hydroxyalkylphosphonates is the difficulty associated with preparation of the appropriate vinylphosphonate. An additional step is involved compared with the direct carbonyl addition methods, assuming that the appropriate carbonyl compounds are available.

α-Aminophosphonates

The use of silyl reagents for the synthesis of this category of materials has been neither as extensive nor as successful as for the α-oxyphosphonates. Rather, variable yields are reported for the relatively few syntheses performed with silyl reagents.

The most efficient method, in terms of both product yield and ease of performance, involves the addition of the parent carbonyl compound to a mixture of the amine and dialkyl phosphite in the absence of solvent.[94] In most of the examples reported, yields greater than 80% are found, with primary aromatic amines giving only slightly lower yields than aliphatic amines. Interestingly, the use of preformed imines with dialkyl phosphites exhibits a greater variation in yield, often below 70%.

Conjugate Additions

The use of silyl reagents again provides superior results for conjugate addition of trivalent phosphorus to carbonyl and related compounds bearing α,β unsaturation. Of the several techniques investigated, the use of a trivalent monobasic phosphorus reagent with a silylating agent results most consistently in yields greater than 80%. Generally, the phosphorus reagent is added to a mixture of the Michael acceptor, silyl reagent, and a tertiary amine.[63] Either the silyl–phosphite esters or trialkyl phosphites with silyl halides may also be used, but with variable results.[63]

If silyl reagents are to be avoided, the most efficient approach involves the use of a fully esterified trivalent phosphorus reagent in the presence of phenol. Alcohols can be used instead of phenol, but the yields are generally lower. The use of dialkyl phosphite with base in this type of reaction, although commonly reported, often gives low yields.

SYNTHETIC UTILITY

The polar addition of trivalent phosphorus esters and amides to unsaturated carbon sites results in the formation of functionalized organophosphorus compounds with a variety of uses. These, as noted in the introduction to this chapter, include synthetic chemical, biological, and industrial applications. Further detail regarding application in synthetic chemistry is considered here.

The utility of (α-functionalized) alkylphosphonates for further chemical syntheses has been a major driving force for the recent increased interest in their own syntheses. The phosphonyl function increases the acidity of the hydrogen at the carbon attached to phosphorus. On removal of the alpha proton with base, the resultant stabilized carbanion is capable of nucleophilic attack, including simple substitution reactions (Eq. 49).[9]

$$R'CH_2\overset{O}{\overset{\|}{P}}(OR)_2 \xrightarrow{\text{base}} R'\bar{C}H\overset{O}{\overset{\|}{P}}(OR)_2 \xrightarrow{R''X} R'R''CH\overset{O}{\overset{\|}{P}}(OR)_2 \quad \text{(Eq. 49)}$$

The presence of an oxy function at the α carbon further permits cleavage of the carbon–phosphorus bond once reaction as a carbanionic species has been completed.[12] Thus the products of phosphite addition at carbonyl carbon, the α-oxyphosphonates, can serve as synthetic equivalents of acyl anions, which are chemically equivalent to carbonyl functions of inverted polarity (umpolung). The simplicity of this characteristic is illustrated in the synthesis of geranyl phenyl ketone starting with benzaldehyde as the acyl source.[11] The α-siloxyphosphonate produced by addition of a silyl–phosphite ester to benzaldehyde is alkylated with geranyl bromide and deprotected, and the phosphorus component is removed to give the desired ketone. The overall result is as if the benzoyl anion had performed a simple displacement reaction on geranyl bromide (Eq. 50).

$$C_6H_5CHO + (C_2H_5O)_2POSi(CH_3)_3 \longrightarrow C_6H_5CH[OSi(CH_3)_3]P(O)(OC_2H_5)_2 \quad (91\,\%)$$

$$\xrightarrow[\text{2. geranyl bromide}]{\text{1. LiN}[C_3H_7\text{-}i]_2\text{, THF, }-78^\circ} C_6H_5\underset{\underset{CH_2CH=C(CH_3)CH_2CH_2CH=C(CH_3)_2}{|}}{C}[OSi(CH_3)_3]P(O)(OC_2H_5)_2 \quad (68\,\%)$$

$$\xrightarrow[\text{2. base, 30 sec}]{\text{1. acid, reflux, 1 h}} C_6H_5COCH_2CH=C(CH_3)CH_2CH_2CH=C(CH_3)_2 \quad (80\,\%)$$

equivalent to (Eq. 50)

$$\left[C_6H_5\overset{O}{\overset{\|}{C}}{:}^-\right] + (CH_3)_2C=CHCH_2CH_2C(CH_3)=CHCH_2Br \longrightarrow C_6H_5COCH_2CH=C(CH_3)CH_2CH_2CH=C(CH_3)_2$$

A wide variety of carbonyl reagents can be used in this type of reaction with alkyl and acyl halides.[11,195–198]

Stabilized anions can be generated as well from simple alkylphosphonothioates, α-chloroalkylphosphonates, or alkylphosphonamides that undergo reaction with carbonyl compounds by attacking at the carbonyl carbon. The adducts can at times be isolated or directly thermolyzed to generate alkenes, proceeding via a route akin to that involved in the Wadsworth–Emmons–Horner reactions.[5–7,199]

The stabilized anions of α-siloxyphosphonates also undergo condensation reactions with carbonyl compounds.[69] However, with these compounds α-hydroxyketones are produced in high yield in a one-pot procedure. For example, 2 equivalents of benzaldehyde are converted to benzoin in 91% yield proceeding through the α-siloxyphosphonate (Eq. 51).

$$(C_2H_5O)_3P + C_6H_5CHO + ClSi(CH_3)_3 \longrightarrow (C_2H_5O)_2P(O)CH(C_6H_5)OSi(CH_3)_3$$

$$\xrightarrow[\text{2. } C_6H_5CHO \quad \text{3. } 1N\ HCl]{\text{1. } LiN[C_3H_7\text{-}i]_2 \quad THF} C_6H_5COCHOHC_6H_5 \quad (91\%) \qquad \text{(Eq. 51)}$$

It is interesting that the intermediate adduct fragments via an acyclic route rather than through the cyclic route typical of Wadsworth–Emmons–Horner reactions.[1]

The Wadsworth–Emmons–Horner fragmentation route *is* observed in condensation reactions using the stabilized anions of substituted α-aminophosphonates.[93,200] The resultant enamines can be isolated or hydrolyzed directly to the ketones (Eq. 52).

$$(C_6H_5O)_2POH + C_6H_5CHO + H_2N\text{-}C_6H_4\text{-}NO_2\text{-}p \longrightarrow$$

$$(C_6H_5O)_2P(O)CH(C_6H_5)NH\text{-}C_6H_4\text{-}NO_2\text{-}p$$

$$\xrightarrow[\text{THF, } -78°,\ 2\text{-}NO_2\text{-}4,5\text{-}(CH_3O)_2C_6H_2CHO]{10\%\ KOH/CH_3OH} 2\text{-}NO_2\text{-}4,5\text{-}(CH_3O)_2C_6H_2CH{=}C(C_6H_5)NH\text{-}C_6H_4\text{-}NO_2\text{-}p$$

$$\xrightarrow[CH_3OH]{HCl,\ H_2O} 2\text{-}NO_2\text{-}4,5\text{-}(CH_3O)_2C_6H_2CH_2COC_6H_5 \quad (96\%) \qquad \text{(Eq. 52)}$$

The α-oxyphosphonates are also convenient precursors for the α-chlorophosphonates, which act as chloromethylene donors in the Wadsworth–Emmons–Horner reaction.[199]

Carbonyl–carbon adducts of α,β-unsaturated aldehydes provide a further synthetic utility. The α-siloxyphosphonamides, on treatment with base, react with alkyl halides or carbonyl compounds at the α-carbon atom relative to phosphorus. The resultant silylated enol phosphonamides can then be degraded efficiently to the substituted carboxylate compounds (Eq. 53).[10]

$$C_6H_5CH{=}CHCHO + (C_2H_5)_3SiOP[N(CH_3)_2]_2 \longrightarrow$$

$$C_6H_5CH{=}CHCH[OSi(C_2H_5)_3]P(O)[N(CH_3)_2]_2$$

$$\xrightarrow[\text{2. } CH_3I]{\text{1. } n\text{-}C_4H_9Li,\ \text{hexane},\ -78^\circ}$$

$$C_6H_5CH(CH_3)CH{=}C[OSi(C_2H_5)_3]P(O)[N(CH_3)_2]_2$$

$$\xrightarrow{NaOH,\ CH_3OH} \quad C_6H_5CH(CH_3)CH_2CO_2CH_3 \quad (65\%)$$

equivalent to

$$[C_6H_5\bar{C}HCH_2CO_2CH_3] + CH_3I \longrightarrow C_6H_5CH(CH_3)CH_2CO_2CH_3 \quad \text{(Eq. 53)}$$

A similar result is obtained starting with an allylic phosphate ester and base, proceeding through a phosphate–α-oxyphosphonate rearrangement.[201] Thus an unsaturated aldehyde can be functionalized to serve as a homoenolate anion equivalent. Triphenylphosphine also adds to α,β-unsaturated carbonyl compounds to generate products of use in this type of procedure.[202]

EXPERIMENTAL PROCEDURES

The following experimental procedures illustrate the variety of reactions and reagents for carbon–phosphorus bond formation by nucleophilic addition to carbon. Particular attention is given to reaction systems involving silylated phosphorus reagents and the accompanying use of silyl halides with phosphorus reagents. Examples of Michael-type additions (hydrophosphinylation) are included, as well as additions to carbonyl and imino carbons.

Diethyl 3-Trimethylsiloxy-2-cyclopenten-1-ylphosphonate (Hydrophosphinylation of an α,β-Unsaturated Ketone with a Silylated Phosphite Reagent).[62] Cyclopent-2-en-1-one (250 mg, 3.05 mmol) and trimethylsilyl diethyl phosphite (1.15 g, 5.48 mmol) (prepared from sodium diethyl phosphite and chlorotrimethylsilane) were dissolved in 10–15 mL of dry acetonitrile and placed in a sealed tube previously flushed with argon. The reaction mixture was heated at 80° for 12 hours in an oil bath. At the end of this period the tube was allowed to cool to ambient temperature, and the solvent was evaporated under reduced pressure. Removal of excess reagent was accomplished by Kugelrohr distillation at 50–70°/3–10 torr. Spectroscopically pure diethyl 3-trimethylsiloxy-2-cyclopenten-1-ylphosphonate was isolated as an oil (890 mg, 3.05 mmol, 100%). ^{1}H NMR ($CDCl_3$) δ: 0.25 (s, 9*H*), 1.20–1.35 (t, *J* = 7.5 Hz, 6*H*), 1.65–2.65 (m, 4*H*), 2.74–3.16 (m, 1*H*), 3.87–4.21 (quin, *J* = 7.5 Hz, 4*H*), 4.48–4.64 (m, 1*H*); IR ($CHCl_3$) cm^{-1}: 1620 (s), 1230 (vs), 1030 (vs).

Diethyl Hydroxymethylphosphonate (Addition of a Phosphite Diester to an Aldehyde Catalyzed by Alumina).[92] Merck-60 neutral alumina (3 g) was added to a stirred mixture of diethyl phosphite (1.38 g, 10 mmol) and paraformaldehyde (0.3 g, 10 mmol). This mixture was heated at 100° for 1.5 hours and then allowed to cool to room temperature. The reaction mixture was extracted with dichloromethane (2 × 25 mL), and the solvent was removed from the extract under reduced pressure to give pure diethyl hydroxymethylphosphonate (1.31 g, 78%) of bp 150°/0.3 torr. ^{1}H NMR ($CDCl_3$) δ: 1.35 (t, *J* = 7 Hz, 6*H*), 3.90 (d, *J* = 6 Hz, 2*H*), 4.16 (m, 4*H*), 4.37 (s, 1*H*); mass spectrum, m/z: 168.0553 (M^+) (calculated 168.0551).

Diphenyl 1-(*N*-Benzyloxycarbonyl)aminoethylphosphonate (Addition of a Phosphite Triester to a Schiff Base Formed *in situ*).[181] A mixture of triphenyl phosphite (31.0 g, 0.10 mol), freshly distilled acetaldehyde (6.6 g, 0.15 mol), benzyl carbamate (15.1 g, 0.10 mol), and glacial acetic acid (15 mL) was stirred for approximately 1 hour until the exothermic reaction subsided. The mixture was then heated at 80–85° for 1 hour followed by removal of volatile products on a rotary evaporator under reduced pressure by heating on a boiling water bath. The oily residue was dissolved in 180 mL of methanol and left for crystallization at −10°. The crystalline material was collected by filtration and recrystallized by dissolving in a minimum amount of hot chloroform followed by the addition of a four-fold volume of methanol to give 18.9 g (46%) of diphenyl 1-(*N*-benzyloxycarbonyl)aminoethylphosphonate, mp 115–117°. ^{1}H NMR ($CDCl_3$) δ: 1.90 (dd, *J* = 18 Hz, 3*H*), 4.75–5.25 (m, 1*H*), 5.5 (s, 2*H*), 6.39 (d, *J* = 10 Hz, 1*H*), 7.50–7.60 (m, 15*H*); IR (KBr) cm^{-1}: 3270, 3060, 1715, 1590, 1540, 1490, 1450, 1305, 1280, 1245, 1220, 1195, 1160, 1100, 1020, 940.

Dimethyl 2-Nitroethylphosphonate (Hydrophosphinylation of a Phosphite Triester with an α,β-Unsaturated Nitro Compound in an Alcohol Solvent).[71] A solution of nitroethylene (5.48 g, 75 mmol) (prepared from nitroethanol)[203] in dry benzene (55 mL) was added dropwise with stirring to a solution of trimethyl phosphite (7.44 g, 60 mmol) in dry *tert*-butyl alcohol (100 mL) at room temperature. The reaction was complete in 9 hours as determined by TLC, and evaporation of the solvents followed by vacuum distillation of the residue gave 10.98 g (88%) of dimethyl 2-nitroethylphosphonate, bp 124°/0.05 torr. ^1H NMR ($CDCl_3$) δ: 2.6 (m, 2*H*), 3.8 (d, $J = 12$ Hz, 6*H*), 4.66 (m, 2*H*); IR (neat) cm^{-1}: 1560, 1380, 1252, 1040.

Diethyl α-(Trimethylsiloxy)benzylphosphonate (Addition of a Phosphite Triester to an Aldehyde in the Presence of a Silyl Chloride).[69] A mixture of chlorotrimethylsilane (119.4 g, 1.1 mol), benzaldehyde (116.6 g, 1.1 mol), and triethyl phosphite (176 g, 1.0 mol) was heated at reflux for 4 hours. After removal of volatiles under reduced pressure, vacuum distillation of the residue gave 297.0 g (94%) of diethyl α-(trimethylsiloxy)benzylphosphonate, bp 95°/0.05 torr. ^1H NMR (CCl_4) δ: 0.03 (s, 9*H*), 0.9–1.4 (m, 6*H*), 3.6–4.1 (m, 4*H*), 4.85 (d, $J = 14$ Hz, 1*H*), 7.0–7.5 (m, 5*H*).

Dimethyl 1-(Trimethylsiloxy)hexylphosphonate (Addition of a Mixed Silyl Phosphite Ester to an Aliphatic Aldehyde).[51] A solution of 1-hexanal (5.00 g, 50 mmol) in benzene (20 mL) was cooled to 5°, and a solution of dimethyl trimethylsilyl phosphite (8.92 g, 49 mmol) in benzene (10 mL) was added with stirring. On completion of addition, the flask was warmed to 25° and the reaction mixture was stirred for 1 hour. Removal of the solvent under reduced pressure followed by vacuum distillation of the residue gave 11.2 g (81%) of dimethyl 1-(trimethylsiloxy)hexylphosphonate, bp 82–85°/0.03 torr. ^1H NMR (CCl_4) δ: -0.33 (s, 9*H*), 0.30–0.63 (m, 3*H*), 0.63–1.30 (m, 8*H*), 3.28 (dd, $J_{PH} = 10$ Hz, $J_{HH} = 1$ Hz, 6*H*), 3.40 (m, 1*H*); IR (neat) cm^{-1}: 1250, 1060, 1030, 850, 760.

(*Z*)-3-Triethylsiloxy-2-propenyldiphenylphosphine Oxide. (Hydrophosphinylation of a Phosphinite with an Unsaturated Aldehyde in the Presence of a Silyl Halide).[51] A solution of methyl diphenylphosphinite (5.52 g, 25.5 mmol) and triethylchlorosilane (3.85 g, 25.5 mmol) in 15 mL of benzene was cooled to 0°, after which there was added acrolein (1.43 g, 25.5 mmol). The ice bath was removed and the reaction mixture was allowed to stir at 25° for 2 hours. Removal of solvent under reduced pressure left 9.46 g (100%) of (*Z*)-3-triethylsiloxy-2-propenyldiphenylphosphine oxide as a pale-yellow viscous oil that decomposed on attempted distillation. ^1H NMR ($CDCl_3$) δ: 0.30–1.17 (m, 15*H*), 3.22 (ddt, $J_{PH} = 14.5$ Hz, $J_{HH} = 7.5$ Hz, $J'_{HH} = 1$ Hz, 2*H*), 6.27 (dtd, $J_{HH} = 6$ Hz, $J'_{HH} = 7.5$ Hz, $J_{PH} = 6$ Hz, 1*H*), 7.13–8.03 (m, 10*H*); IR (CCl_4) cm^{-1}: 3050, 1645, 1185, 1090, 685.

Diethyl 2-Carbamylethylphosphonate (Hydrophosphinylation of a Dialkyl Phosphite Anion with an α,β-Unsaturated Amide).[134] A mixture of diethyl phosphite (15.18 g, 0.11 mol) and acrylamide (7.10 g, 0.10 mol) was heated to 60–70°, and the reaction was initiated by the addition of 5 mL of 3 *M* ethanolic sodium ethoxide. The reaction was completed by heating for 1 hour at 110°. The reaction mixture was diluted with ethanol, neutralized with concentrated hydrochloric acid, and the solvent was evaporated under reduced pressure. The solid residue was recrystallized from benzene to give 15.7 g (75%) of pure diethyl 2-carbamylethylphosphonate, mp 74–76°.

Di-*n*-butyl (2-Tetrahydropyranyloxy)methylphosphonate (Addition of a Dialkyl Phosphite to an Aldehyde with an Amine Catalyst).[198] A mixture of di-*n*-butyl phosphite (155.2 g, 0.8 mol), paraformaldehyde (24 g, 0.8 mol), and triethylamine (8 g, 0.08 mol) was heated for 6 hours at 110°. Volatile materials were removed under reduced pressure. The residue was mixed with dihydropyran (33 g, 0.39 mol), diethyl ether (350 mL), and 8 drops of phosphorus oxychloride in a flask fitted with a drying tube. After 15 minutes another 33 g of dihydropyran and 8 drops of phosphorus oxychloride were added. These additions were repeated three more times at 15-minute intervals. After 2 hours an additional 1 mL of phosphorus oxychloride was added. After 4 hours the mixture was diluted with 200 mL of diethyl ether and the resultant mixture was shaken with 300 mL of 5% sodium bicarbonate solution. The organic layer was separated and washed with 100 mL of saturated sodium chloride solution. The organic layer was separated, dried over sodium sulfate, and the solvent was evaporated under reduced pressure. The residue was chromatographed through a column of silica gel (350 g, 70–230 mesh) with 80% diethyl ether in hexane. Vacuum distillation of the chromatographed material gave 155 g (50.3%) of pure di-*n*-butyl (2-tetrahydropyranyloxy)methylphosphonate, bp 138–140°/0.1 torr. ^{1}H NMR ($CDCl_3$) δ: 0.95 (t, $J = 6.5$ Hz, 6H), 1.1–2.0 (m, 14H), 3.4–4.35 (m, 8H), 4.72 (m, 1H).

Bis(trimethylsilyl) 1-Trimethylsiloxy-1-methyl-2-chloroethylphosphonate (Addition of Tris(trimethylsilyl) Phosphite to a Ketone).[47] To a solution of chloroacetone (1.85 g, 20 mmol) in dry THF was added tris(trimethylsilyl) phosphite[100] (6.26 g, 21 mmol) at room temperature over a period of 15 minutes. The reaction mixture was stirred for 3 hours. It was distilled to give 7.58 g (97%) of bis(trimethylsilyl) 1-trimethylsiloxy-1-methyl-2-chloroethylphosphonate, bp 102–103°/0.18 torr. ^{1}H NMR ($CDCl_3$) δ: 0.23 (s, 9H), 0.33 (2, 18H), 1.37 (d, $J = 16$ Hz, 3H), 3.58 (d, $J = 5$ Hz, 2H).

α-Aminobenzylphosphonic Acid (Addition of a Dialkyl Phosphite to a Schiff Base).[166] A mixture of benzaldehyde (1.06 g, 0.01 mol), 1-phenylcyclopentylamine (1.61 g, 0.01 mol), anhydrous potassium carbonate (2 g), and benzene (20 mL) was heated on a steam bath for 10 minutes. The mixture was allowed to cool, the potassium carbonate was removed by filtration, and the benzene was evaporated under reduced pressure. The residue was heated with diethyl phosphite (1.38 g, 0.01 mol) at 120–140° for 30 minutes, and the resultant mixture was refluxed with concentrated hydrochloric acid (50 mL) for 3 hours. The cooled solution was washed with benzene, and the volatile components were evaporated under reduced pressure. The residue was dissolved in ethanol and treated with propylene oxide to a pH of 6. The precipitate was recrystallized from aqueous ethanol to give 1.14 g (61%) of pure α-aminobenzylphosphonic acid, mp 280–282°.

Triethyl β-Phosphonopropanoate (Hydrophosphinylation of an α,β-Unsaturated Ester Using the Sodium Salt of a Dialkyl Phosphite).[128] Sodium (2.0 g, 0.08 g-atom) was dissolved in freshly distilled diethyl phosphite (304 g, 2.2 mol) and an equal volume of dry benzene was added. Ethyl acrylate (200 g, 2.0 mol) was then added dropwise, and the temperature was maintained at 60°. After the solution had cooled to room temperature, a slight excess of acetic acid (5 mL) was added, and the mixture was filtered. The filtrate was vacuum distilled to give 400.8 g (84%) of pure diethyl β-phosphonopropanoate, bp 109–110°/0.6 torr.

TABULAR SURVEY

The data summarized in the following tables include the reactions presented in the text and other reports of fundamentally similar reaction systems involving different substrates. The tables attempt to cover the literature published through the end of 1984, with several examples from reports in early 1985. Within each table the reactions are ordered according to increasing number of carbon atoms of the substrate. The conditions listed refer to the performance of the addition reaction alone. In certain instances the product illustrated and its yield refer to a material isolated after removal of one or more blocking groups or phosphorus ester linkages. The conditions for these deblocking procedures are not listed. Such reactions are noted by the superscript *a* following the yield.

The following abbreviations are used throughout the tables:

BDMS	*tert*-butyldimethylsilyl
DMSO	dimethyl sulfoxide
ether	diethyl ether
TES	triethylsilyl
THF	tetrahydrofuran
TMS	trimethylsilyl
TsOH	*p*-toluenesulfonic acid

TABLE I. Trivalent Phosphorus Addition Reactions at Carbonyl Carbon

No. of Carbon Atoms	Substrate	Phosphorus Reagent	Conditions	Product(s) and Yield(s)(%)	Refs.
C_1	HCHO	$(C_2H_5O)_2POH$	1. Ether, Na, room temp 2. TMSCl	$TMSOCH_2P(O)(OC_2H_5)_2$ (69)	11
			Alumina, 100°, 1.5 h	$HOCH_2P(O)(OC_2H_5)_2$ (78)	92
C_2	$CH_2{=}C{=}O$	$(C_2H_5O)_2POTMS$	20°	$(C_2H_5O)_2P(O)C(OTMS){=}CH_2$ (87)	50
	CH_3CHO	$(C_2H_5O)_2POTMS$	20°, 1 h	$(CH_3O)_2P(O)CH(OTMS)CH_3$ (54)	110
		$(C_2H_5O)_3P$	TESBr, room temp, 2 h	$(C_2H_5O)_2P(O)CH(OTES)CH_3$ (70)	67
			$(CH_3)_2SiCl_2$, 22°, 0.3 h	$(CH_3)_2Si[OCH(CH_3)P(O)(OC_2H_5)_2]_2$ (79)	67
			TMSCl, 46°, 0.5 h	$(C_2H_5O)_2P(O)CH(OTMS)CH_3$ (91)	67
		$(C_2H_5O)_2POTMS$	C_6H_6, room temp, 4 h	" (88)	11
			20°, 1 h	" (70)	50
			20°, 3 h	" (72)	110
		$(CH_3O)_2POTES$	C_6H_6, 0°, 1 h; room temp, 1 h; 50°, 15 min	$(CH_3O)_2P(O)CH(OTES)CH_3$ (78)	110
		$[(CH_3)_2N]_2POTES$	Ether, 0°, 0.5 h	$[(CH_3)_2N]_2P(O)CH(OTES)CH_3$ (83)	51
	CF_3CHO	$(C_2H_5O)_2POP(OC_2H_5)_2$		$CF_3CH[OP(OC_2H_5)_2]P(O)(OC_2H_5)_2$ (56)	204
C_3	$CH_2{=}CHCHO$	$(CH_3O)_3P$	TMSCl, 25°, 4 h	$(CH_3O)_2P(O)CH(OTMS)CH{=}CH_2$ (70)	51
		$(C_2H_5O)_2POTMS$	C_6H_6, room temp, 6 h	$(C_2H_5O)_2P(O)CH(OTMS)CH{=}CH_2$ (89)	111
		$(TMSO)_3P$	C_6H_6, room temp, 4 h	$(TMSO)_2P(O)CH(OTMS)CH{=}CH_2$ (94)	49
		$[(CH_3)_2N]_2POTES$	0°, 15 min	$[(CH_3)_2N]_2P(O)CH(OTES)CH{=}CH_2$ (90)	61
			THF, 0°, 15 min	" (90)	51
	C_2H_5CHO	$(C_2H_5O)_2POH$	KF, room temp, 1 h	$(C_2H_5O)_2P(O)CHOHC_2H_5$ (37)	91
		$(C_2H_5O)_3P$	Cl_3SiH, 45°, 3 h	$HSi[OCH(C_2H_5)P(O)(OC_2H_5)_2]_3$ (98)	67
			$Cl_2Si(CH_3)CH_2CH{=}CH_2$, 60°, 1 h	$[OCH(C_2H_5)P(O)(OC_2H_5)_2]_2$ | $Si(CH_3)CH_2CH{=}CH_2$ (100)	67
			$(CH_3)_2Si(OC_2H_5)Cl$, 100°, 1.5 h	$(CH_3)_2Si(OC_2H_5)OCH(C_2H_5)P(O)(OC_2H_5)_2$ (96)	67
			$Cl_2CHSi(CH_3)_2Cl$	$Cl_2CHSi(CH_3)_2OCH(C_2H_5)P(O)(OC_2H_5)_2$ (98)	67
		$(C_2H_5O)_2POTMS$	C_6H_6, room temp, 6 h	$(C_2H_5O)_2P(O)CH(OTMS)C_2H_5$ (72)	11
		$(TMSO)_3P$	C_6H_6, room temp, 2 h	$(TMSO)_2P(O)CH(OTMS)C_2H_5$ (85)	49

	$(CH_3)_2CO$	$(CH_3O)_2POTMS$	90°, 24 h	$(CH_3O)_2P(O)C(OTMS)(CH_3)_2$ (74)	51
			100°, 10 h	" (39)	110
		$NH_4OP(OH)OCH_2CH(C_2H_5)C_4H_9$-$n$	2-Ethylhexanol, Na, 70°, 24 h	n-$C_4H_9CH(C_2H_5)CH_2OP(O)(OH)CHOHC_4H_9$-$n$ (70)	205
		$(TMSO)_3P$	C_6H_6, room temp, 4 h	$(TMSO)_2P(O)C(OTMS)(CH_3)_2$ (90)	49
		$(C_6H_5)_2POH$	THF, CH_3MgI	$(C_6H_5)_2P(O)COH(CH_3)_2$ (92)	82
	$(CH_3)_2CS$	$(CH_3O)_3P$	Reflux, 20 h	$(CH_3O)_2P(O)C(SCH_3)(CH_3)_2$ (23), $(CH_3O)_2P(O)CSH(CH_3)_2$ (23)	206
		$(C_2H_5O)_3P$	Reflux, 20 h	$(C_2H_5O)_2P(O)C(SC_2H_5)(CH_3)_2$ (25), $(C_2H_5O)_2P(O)CSH(CH_3)_2$ (17)	206
		$(i$-$C_3H_7O)_3P$	Reflux, 20 h	$(i$-$C_3H_7O)_3P(O)CSH(CH_3)_2$ (62)	206
	CH_3COCN	$(C_2H_5O)_2POTMS$	Cl_2CH_2, room temp, 1 d	$(C_2H_5O)_2P(O)C(OTMS)(CN)CH_3$ (64)	60
	$CH_3CHClCHO$	$(TMSO)_3P$	THF, room temp, 3 h	$(TMSO)_2P(O)CH(OTMS)CHClCH_3$ (92)	47, 48
	CH_3COCH_2Cl	$(C_2H_5O)_2POH$	KF, room temp, 0.5 h	$(C_2H_5O)_2P(O)COH(CH_3)CH_2Cl$ (91)	91
		$(TMSO)_2PH$	−20°	$(TMSO)_2P(O)C(OTMS)(CH_3)CH_2Cl$ (38), $(TMSO)_2P(O)OC(CH_3){=}CH_2$ (42)	108
		$(i$-$C_3H_7O)_2POH$	KF, room temp, 6 h	$(i$-$C_3H_7O)P(O)COH(CH_3)CH_2Cl$ (60)	91
		$(C_2H_5O)_2POTMS$	Room temp, 6 h	$(C_2H_5O)_2P(O)C(OTMS)(CH_3)CH_2Cl$ (86)	47, 48
		$(TMSO)_3P$	THF, room temp, 3 h	$(TMSO)_2P(O)C(OTMS)(CH_3)CH_2Cl$ (97)	47, 48
	CH_3COCH_2Br	$(TMSO)_3P$	THF, room temp, 3 h	$(TMSO)_2P(O)C(OTMS)(CH_3)CH_2Br$ (89)	47, 48
	CH_3COCF_3	$(C_2H_5O)_2POTMS$	−25°	$(C_2H_5O)_2P(O)C(OTMS)(CF_3)CH_3$ (66)	57
	$(CF_3)_2CO$	$(CH_3O)_2POH$	25°, 3 d	$(CH_3O)_2P(O)OCH(CF_3)_2$ (94), $(CH_3O)_2P(O)COH(CF_3)_2$ (6)	207
		$(C_2H_5O)_2POH$	25°, 3 d	$(C_2H_5O)_2P(O)COH(CF_3)_2$ (12), $(C_2H_5O)_2P(O)OCH(CF_3)_2$ (88)	207
		$(C_2H_5O)_2POTMS$	−25°	$(C_2H_5O)_2P(O)C(OTMS)(CF_3)_2$ (96)	57, 207
		$(n$-$C_4H_9O)_2POH$	25°, 3 d	$(n$-$C_4H_9O)_2P(O)COH(CF_3)_2$ (95), $(n$-$C_4H_9O)_2P(O)OCH(CF_3)_2$ (5)	207
C_4	(E)-$CH_3CH{=}CHCHO$	$(C_2H_5O)_2POH$	KF, room temp, 4 h	(E)-$(C_2H_5O)_2P(O)CHOHCH{=}CHCH_3$ (78)	91
		$(CH_3O)_2POTMS$	55°, 18 h	(E)-$(CH_3O)_2P(O)CH(OTMS)CH{=}CHCH_3$ (68)	51
		$(C_2H_5O)_2POTMS$	C_6H_6, room temp, 6 h	(E)-$(C_2H_5O)_2P(O)CH(OTMS)CH{=}CHCH_3$ (73)	111
		$(TMSO)_3P$	C_6H_6, room temp, 4 h	(E)-$(TMSO)_2P(O)CH(OTMS)CH{=}CHCH_3$ (89)	49
		$C_6H_5(CH_2)_4P(OH)_2$	$CHCl_3$, TMSCl, room temp, 18 h	(E)-$C_6H_5(CH_2)_4P(O)(OH)CH(OTMS)CH{=}CHCH_3$ (88)	63
		$[(CH_3)_2N]_2POTES$	THF, 0°, 0.5 h	(E)-$[(CH_3)_2N]_2P(O)CH(OTES)CH{=}CHCH_3$ (95)	51, 61
		$C_2H_5OP(OH)(CH_2)_4C_6H_5$	$CHCl_3$, TMSCl, room temp, 18 h	(E)-$C_6H_5(CH_2)_4P(O)(OC_2H_5)CH(OTMS)CH{=}CHCH_3$ (92)	63

TABLE I. TRIVALENT PHOSPHORUS ADDITION REACTIONS AT CARBONYL CARBON (*Continued*)

No. of Carbon Atoms	Substrate	Phosphorus Reagent	Conditions	Product(s) and Yield(s)(%)	Refs.
C_4 (*Contd.*)	*i*-C_3H_7CHO	$(C_6H_5)_2POH$	THF, CH_3MgI, 0°, 1 h	(*E*)-$(C_6H_5)_2P(O)CHOHCH{=}CHCH_3$ (59)	82
		$(C_2H_5O)_2POH$	KF, room temp, 0.5 h	$(C_2H_5O)_2P(O)CHOHC_3H_7$-*i* (66)	91
		$(CH_3O)_2POTMS$	25°, 1 h	$(CH_3O)_2P(O)CH(OTMS)C_3H_7$-*i* (82)	51
		$(C_2H_5O)_2POTMS$	C_6H_6, room temp, 6 h	$(C_2H_5O)_2P(O)CH(OTMS)C_3H_7$-*i* (90), (83)	111, 11
		$(TMSO)_3P$	C_6H_6, room temp, 4 h	$(TMSO)_2P(O)CH(OTMS)C_3H_7$-*i* (82)	49
		$[(CH_3)_2N]_2POTES$	0°, 0.5 h	$[(CH_3)_2N]_2P(O)CH(OTES)C_3H_7$-*i* (90)	10
		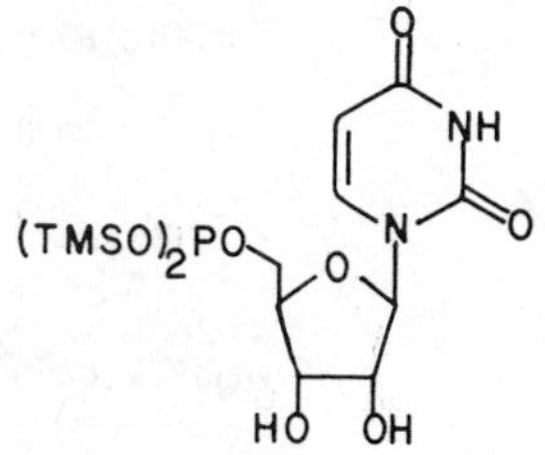 $(TMSO)_2PO$– [structure; labels: NH, N, O, HO, OH]	1. THF, room temp, 3 h 2. H_2O	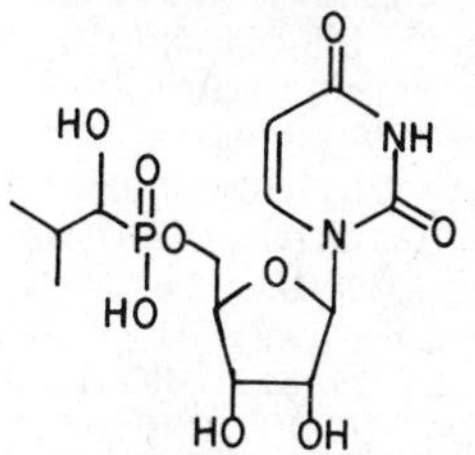 [structure; labels: HO, P, O, HO, NH, N, O, HO, OH] (84)	100
	CH_3COCH_5	$(C_2H_5O)_2POTMS$	20°, 1 h	$(C_2H_5O)_2P(O)C(OTMS)(CH_3)C_2H_5$ (55)	50
	$CH_3CSC_2H_5$	$[(CH_3O)_3P$	Reflux, 25 h	$(CH_3O)_2P(O)C(SCH_3)(CH_3)C_2H_5$ (35), $(CH_3O)_2P(O)CSH(CH_3)C_2H_5$ (18)	206
		$(C_2H_5O)_3P$	Reflux, 20 h	$(C_2H_5O)_2P(O)C(SC_2H_5)(CH_3)C_2H_5$ (17), $(C_2H_5O)_2P(O)CSH(CH_3)C_2H_5$ (24)	206
		(*i*-$C_3H_7O)_3P$	Reflux, 20 h	(*i*-$C_3H_7O)_2P(O)CSH(CH_3)C_2H_5$ (76)	206
		$(CH_3O)_2POTMS$	20°	$(CH_3O)_2P(O)C(OTMS)(CH_3)C_2H_5$ (88)	208
	$CH_3COCOCH_3$	$(C_2H_5O)_2POH$	Alumina, room temp, 24 h	$(C_2H_5O)_2P(O)COH(CH_3)COCH_3$ (24)	92
	$(CH_3)_2CClCHO$	$(TMSO)_3P$	THF, room temp, 3 h	$(TMSO)_2P(O)CH(OTMS)C(CH_3)_2Cl$ (92)	47, 48
	$CH_3COCHClCH_3$	$(C_2H_5O)_2POH$	Alumina, room temp, 72 h	$(C_2H_5O)_2P(O)COH(CH_3)CHClCH_3$ (72)	92
		$(TMSO)_3P$	THF, room temp, 3 h	$(TMSO)_2P(O)C(OTMS)(CH_3)CHClCH_3$ (92)	47, 48
	$CH_3COCO_2CH_3$	$(C_2H_5O)_2POTMS$	CH_2Cl_2, room temp, 1 d	$(C_2H_5O)_2P(O)C(OTMS)(CH_3)CO_2CH_3$ (36)	60
		$(C_6H_5)_2POH$	0°, 2 d	$(C_6H_5)_2P(O)COH(CH_3)CO_2CH_3$ (51)	114
	$CH_3COCH_2OSO_2CH_3$	$(TMSO)_3P$	THF, room temp, 3 h	$(TMSO)_2P(O)C(OTMS)(CH_3)CH_2OSO_2CH_3$ (92)	47, 48

C_5	2-furyl-CHO	$(CH_3O)_2POH$	Ether, diethylamine,	2-furyl-CH(OH)$P(O)(OCH_3)_2$ (31)	87
			Ether, CH_3OH, $NaOCH_3$	" (52)	87
	n-C_4H_9CHO	$(CH_3O)POTES$	20°, 15 h	$(CH_3O)_2P(O)CH(OTES)C_4H_9$-$n$ (58)	110
		$NH_4OP(OH)OCH_2CH(C_2H_5)C_4H_9$-$n$	2-Ethylhexanol, Na, 70°, 24 h	n-$C_4H_9CH(C_2H_5)CH_2OP(O)(OH)CHOHC_4H_9$-$n$ (72)	205
	i-$C_3H_7CSCH_3$	$(CH_3O)_3P$	Reflux, 30 h	i-$C_3H_7C(SCH_3)(CH_3)P(O)(OCH_3)_2$ (22), i-$C_3H_7CSH(CH_3)P(O)(OCH_3)_2$ (10)	206
		$(C_2H_5O)_3P$	Reflux, 35 h	i-$C_3H_7C(SC_2H_5)(CH_3)P(O)(OC_2H_5)_2$ (18), i-$C_3H_7CSH(CH_3)P(O)(OC_2H_5)_2$ (10)	206
		$(i$-$C_3H_7O)_3P$	Reflux, 25 h	i-$CH_7CSH(CH_3)P(O)(OC_3H_7$-$i)_2$ (46)	206
	$(C_2H_5)_2CS$	$(CH_3O)_3P$	Reflux, 30 h	$(C_2H_5)_2C(SCH_3)P(O)(OCH_3)_2$ (43), $(C_2H_5)_2CSHP(O)(OCH_3)_2$ (11)	206
		$(C_2H_5O)_3P$	Reflux 30 h	$(C_2H_5)_2C(SC_2H_5)P(O)(OC_2H_5)_2$ (24), $(C_2H_5)_2CSHP(O)(OC_2H_5)_2$ (29)	206
		$(i$-$C_3H_7O)_3P$	Reflux, 20 h	$(C_2H_5)_2CSHP(O)(OC_3H_7$-$i)_2$ (80)	206
	cyclopentanethione (C=S)	$(CH_3O)_3P$	Reflux, 10 h	1-CH_3S-1-$P(O)(OCH_3)_2$-cyclopentane (56)	44
		$(C_2H_5O)_3P$	Reflux, 12 h	1-C_2H_5S-1-$P(O)(OC_2H_5)_2$-cyclopentane (20), 1-HS-1-$P(O)(OC_2H_5)_2$-cyclopentane (26)	44
		$(i$-$C_3H_7O)_3P$	Reflux, 10 h	1-HS-1-$P(O)(OC_3H_7$-$i)_2$-cyclopentane (83)	44
	$(CH_3CO)_2CH_2$	$(C_2H_5O)_2POH$	KF, room temp, 0.5 h	$CH_3COCH_2COH(CH_3)P(O)(OC_2H_5)_2$ (53)	91
	$CH_3COCO_2C_2H_5$	$(C_2H_5O)_2POTMS$	CH_2Cl_2, room temp, 1 d	$(C_2H_5O)_2P(O)C(OTMS)(CH_3)CO_2C_2H_5$ (40)	60

TABLE I. Trivalent Phosphorus Addition Reactions at Carbonyl Carbon (*Continued*)

No. of Carbon Atoms	Substrate	Phosphorus Reagent	Conditions	Product(s) and Yield(s)(%)	Refs.
C_5 (*Contd.*)	$BrCH_2COCO_2C_2H_5$	$(TMSO)_2POCH_3$	C_6H_6, room temp, 5 h	$BrCH_2C(OTMS)(CO_2C_2H_5)P(O)(OTMS)OCH_3$ (8), $CH_2{=}C(CO_2C_2H_5)OP(O)(OTMS)OCH_3$ (61)	59
		$(TMSO)_2POC_2H_5$	C_6H_6, room temp, 5 h	$BrCH_2C(OTMS)(CO_2C_2H_5)P(O)(OTMS)OC_2H_5$ (20), $CH_2{=}C(CO_2C_2H_5)OP(O)(OTMS)OC_2H_5$ (57)	59
		$(TMSO)_3P$	C_6H_6, room temp, 5 h	$BrCH_2C(OTMS)(CO_2C_2H_5)P(O)(OTMS)_2$ (31), $CH_2{=}C(CO_2C_2H_5)OP(O)(OTMS)_2$ (36)	59
C_6	n-$C_5H_{11}CHO$	$(CH_3O)_2POTMS$	C_6H_6, room temp, 1 h	n-$C_5H_{11}CH(OTMS)P(O)(OCH_3)_2$ (81)	51
		$(CH_3O)_2POBDMS$	100°, 3 h	n-$C_5H_{11}CH(OBDMS)P(O)(OCH_3)_2$ (81)	51
	 O	$(CH_3O)_2POTMS$	95°, 13 h	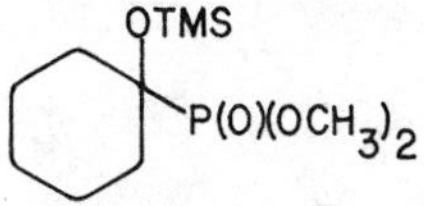 OTMS, $P(O)(OCH_3)_2$ (86)	51
		$(C_2H_5O)_2POTMS$	20°, 1 h	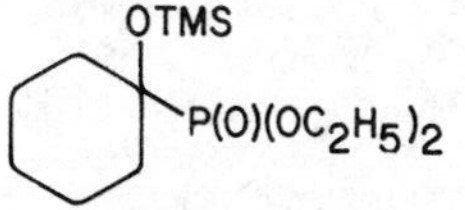 OTMS, $P(O)(OC_2H_5)_2$ (52)	50
	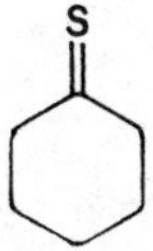 S	$(CH_3O)_3P$	C_6H_6, reflux, 20 h	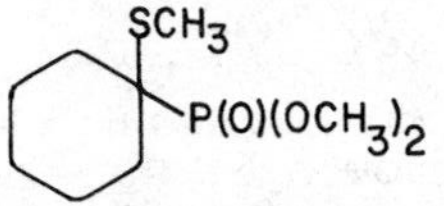 SCH_3, $P(O)(OCH_3)_2$ (59)	44
			Toluene, reflux, 20 h	" (70)	44
			Xylene, reflux, 10 h	" (68)	44
			Toluene, ethanol, reflux, 5 h	" (63)	44
		$(CH_3O)_2POTMS$	20°	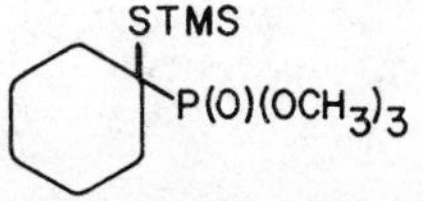 STMS, $P(O)(OCH_3)_3$ (96)	208

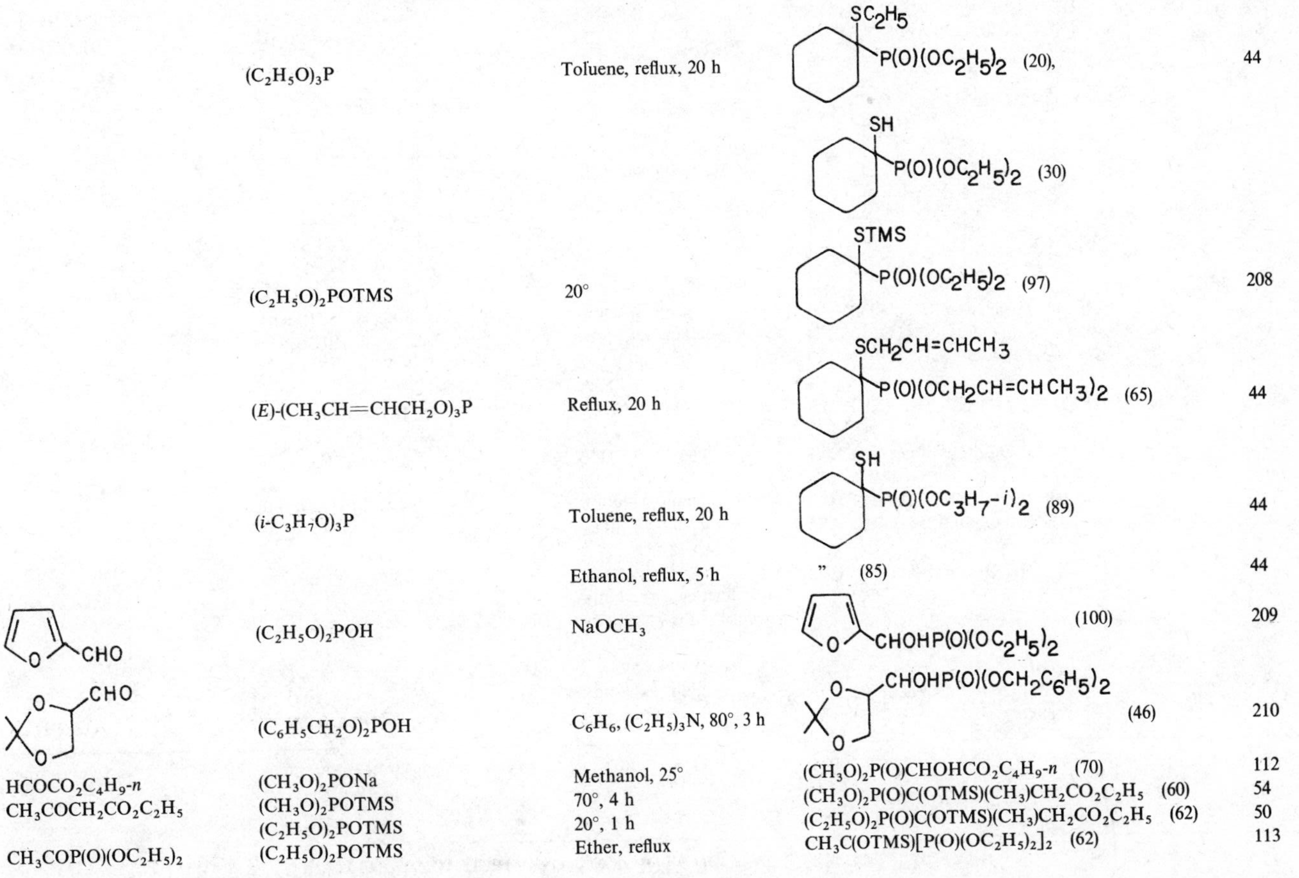

	$(C_2H_5O)_3P$	Toluene, reflux, 20 h	SC_2H_5 / $P(O)(OC_2H_5)_2$ (20),	44
			SH / $P(O)(OC_2H_5)_2$ (30)	
	$(C_2H_5O)_2POTMS$	20°	STMS / $P(O)(OC_2H_5)_2$ (97)	208
	(E)-$(CH_3CH{=}CHCH_2O)_3P$	Reflux, 20 h	$SCH_2CH{=}CHCH_3$ / $P(O)(OCH_2CH{=}CHCH_3)_2$ (65)	44
	$(i$-$C_3H_7O)_3P$	Toluene, reflux, 20 h	SH / $P(O)(OC_3H_7$-$i)_2$ (89)	44
		Ethanol, reflux, 5 h	" (85)	44
CHO	$(C_2H_5O)_2POH$	$NaOCH_3$	$CHOHP(O)(OC_2H_5)_2$ (100)	209
CHO	$(C_6H_5CH_2O)_2POH$	C_6H_6, $(C_2H_5)_3N$, 80°, 3 h	$CHOHP(O)(OCH_2C_6H_5)_2$ (46)	210
$HCOCO_2C_4H_9$-n	$(CH_3O)_2PONa$	Methanol, 25°	$(CH_3O)_2P(O)CHOHCO_2C_4H_9$-$n$ (70)	112
$CH_3COCH_2CO_2C_2H_5$	$(CH_3O)_2POTMS$	70°, 4 h	$(CH_3O)_2P(O)C(OTMS)(CH_3)CH_2CO_2C_2H_5$ (60)	54
	$(C_2H_5O)_2POTMS$	20°, 1 h	$(C_2H_5O)_2P(O)C(OTMS)(CH_3)CH_2CO_2C_2H_5$ (62)	50
$CH_3COP(O)(OC_2H_5)_2$	$(C_2H_5O)_2POTMS$	Ether, reflux	$CH_3C(OTMS)[P(O)(OC_2H_5)_2]_2$ (62)	113

TABLE I. TRIVALENT PHOSPHORUS ADDITION REACTIONS AT CARBONYL CARBON (*Continued*)

No. of Carbon Atoms	Substrate	Phosphorus Reagent	Conditions	Product(s) and Yield(s)(%)	Refs.
C_7	C_6H_5CHO	$(CH_3O)_2POH$	KF, room temp, 0.5 h	$C_6H_5CHOHP(O)(OCH_3)_2$ (88)	91
		$(C_2H_5O)_2POH$	Ether, Na, TMSCl room temp	$C_6H_5CH(OTMS)P(O)(OC_2H_5)_2$ (67)	11
			KF, room temp, 0.5 h	$C_6H_5CHOHP(O)(OC_2H_5)_2$ (98)	91
			KF/2 H_2O, room temp, 8 h	" (88)	211
			Alumina, room temp, 2 h	" (96)	92
		$(CH_3O)_2POTMS$	20°, 24 h	$C_6H_5CH(OTMS)P(O)(OCH_3)_2$ (78)	110
			C_6H_6, 25°, 24 h	" (97)	51
		$(C_2H_5O)_3P$	TMSCl, reflux, 4 h	$C_6H_5CH(OTMS)P(O)(OC_2H_5)_2$ (94)	70
		$(i\text{-}C_3H_7O)_2POH$	Alumina, room temp, 24 h	$C_6H_5CHOHP(O)(OC_3H_7\text{-}i)_2$ (90)	92
			KF, room temp, 5.5 h	" (77)	91
		$(C_2H_5O)_2POTMS$	C_6H_6, room temp, 6 h	$C_6H_5CH(OTMS)P(O)(OC_2H_5)_2$ (91)	111, 11
			50°, 1.5 h	" (66)	110
		$NH_4OP(OH)OCH_2CH(C_2H_5)C_4H_9\text{-}n$	2-Ethylhexanol, Na, 70°, 24 h	$n\text{-}C_4H_9CH(C_2H_5)CH_2OP(O)(OH)CHOHC_6H_5$ (65)	205
		$(CH_3O)_2POBDMS$	25°, 6 h	$C_6H_5CH(OBDMS)P(O)(OCH_3)_2$ (65)	51
		$(TMSO)_3P$	C_6H_6, room temp, 4 h	$C_6H_5CH(OTMS)P(O)(OTMS)_2$ (88)	49
		$[(CH_3)_2N]_2POTES$	Ether, 0°, 0.5 h	$C_6H_5CH(OTES)P(O)[(N(CH_3)_2]_2$ (92)	51
		$[(C_5H_5)Fe(C_5H_4CH_2O)]_2POH$	Ethanol, $NaOC_2H_5$, reflux, 2 h	$[(C_5H_5)Fe(C_5H_4CH_2O)]_2P(O)CHOHC_6H_5$ (41)	155
	$n\text{-}C_6H_{13}CHO$	$[(CH_3)_2N]_2POTES$	Ether, 0°, 0.5 h	$n\text{-}C_6H_{13}CH(OTES)P(O)[N(CH_3)_2]_2$ (97)	51
	$n\text{-}C_5H_{11}COCH_3$	$(CH_3O)_2POTMS$	95°, 36 h	$n\text{-}C_5H_{11}C(OTMS)(CH_3)P(O)(OCH_3)_2$ (62)	51

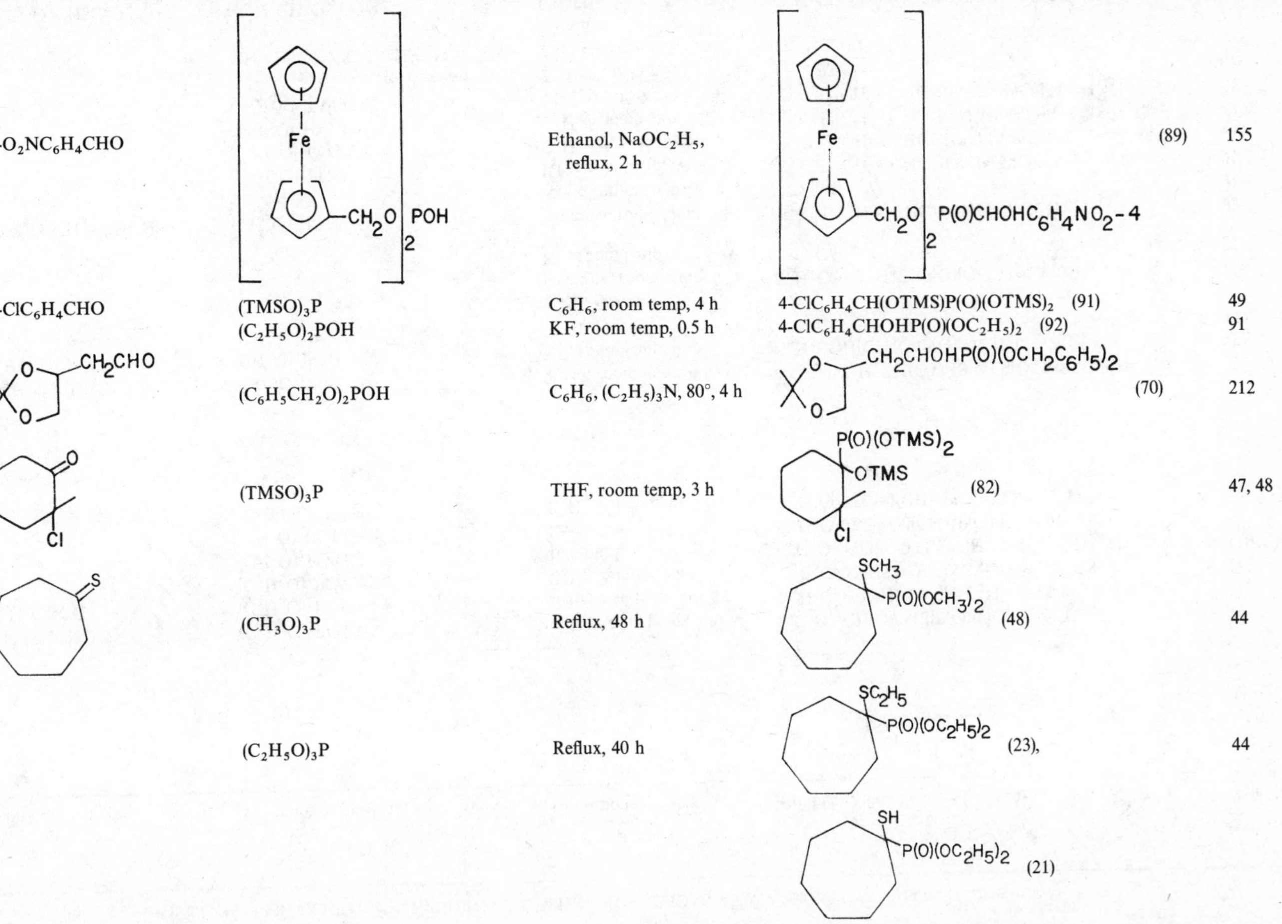

Substrate	Reagent	Conditions	Product (Yield %)	Ref.
$4\text{-}O_2NC_6H_4CHO$	$[(C_5H_5)Fe(C_5H_4CH_2O)]_2POH$	Ethanol, $NaOC_2H_5$, reflux, 2 h	$[(C_5H_5)Fe(C_5H_4CH_2O)]_2P(O)CHOHC_6H_4NO_2\text{-}4$ (89)	155
$4\text{-}ClC_6H_4CHO$	$(TMSO)_3P$	C_6H_6, room temp, 4 h	$4\text{-}ClC_6H_4CH(OTMS)P(O)(OTMS)_2$ (91)	49
	$(C_2H_5O)_2POH$	KF, room temp, 0.5 h	$4\text{-}ClC_6H_4CHOHP(O)(OC_2H_5)_2$ (92)	91
2,2-dimethyl-1,3-dioxolan-4-yl-CH_2CHO	$(C_6H_5CH_2O)_2POH$	C_6H_6, $(C_2H_5)_3N$, 80°, 4 h	2,2-dimethyl-1,3-dioxolan-4-yl-$CH_2CHOHP(O)(OCH_2C_6H_5)_2$ (70)	212
2-chloro-2-methylcyclohexanone	$(TMSO)_3P$	THF, room temp, 3 h	1-$P(O)(OTMS)_2$-1-OTMS-2-chloro-2-methylcyclohexane (82)	47, 48
cycloheptanethione	$(CH_3O)_3P$	Reflux, 48 h	1-SCH_3-1-$P(O)(OCH_3)_2$-cycloheptane (48)	44
	$(C_2H_5O)_3P$	Reflux, 40 h	1-SC_2H_5-1-$P(O)(OC_2H_5)_2$-cycloheptane (23), 1-SH-1-$P(O)(OC_2H_5)_2$-cycloheptane (21)	44

TABLE I. Trivalent Phosphorus Addition Reactions at Carbonyl Carbon (*Continued*)

No. of Carbon Atoms	Substrate	Phosphorus Reagent	Conditions	Product(s) and Yield(s)(%) (76)	Refs.
C_7 (*Contd.*)		$(i\text{-}C_3H_7O)_3P$	Reflux, 48 h	SH / $P(O)(OC_3H_7\text{-}i)$ (76)	44
C_8	$n\text{-}C_7H_{15}CHO$	$(C_2H_5O)_2POTMS$	C_6H_6, room temp, 3 h	$n\text{-}C_7H_{15}CH(OTMS)P(O)(OC_2H_5)_2$ (91)	11
	$C_6H_5COCH_3$	$(C_2H_5O)_2POH$	CsF, room temp, 0.5 h	$C_6H_5COH(CH_3)P(O)(OC_2H_5)_2$ (54)	91
		$(C_2H_5O)_2POTMS$	C_6H_6, reflux, 19 h	$C_6H_5C(OTMS)(CH_3)P(O)(OC_2H_5)_2$ (78)	111
		$(CH_3O)_2POTES$	100°, 40 h	$(CH_3O)_2P(O)C(OTES)(CH_3)C_6H_5\cdot$ (27)	110
		$(CH_3O)_2POBDMS$	120°, 48 h	$C_6H_5C(OBDMS)(CH_3)P(O)(OCH_3)_2$ (34)	51
		$(TMSO)_3P$	C_6H_6, room temp, 8 h	$C_6H_5C(OTMS)(CH_3)P(O)(OTMS)_2$ (75)	49
	O	$(C_2H_5O)_2POTMS$	180°, 18 h	OTMS / $P(O)(OC_2H_5)_2$ (87)	62
	$4\text{-}CH_3OC_6H_4CHO$	$(TMSO)_3P$	C_6H_6, room temp, 4 h	$4\text{-}CH_3OC_6H_4CH(OTMS)P(O)(OTMS)_2$ (83)	49
	$C_6H_5COCH_2Cl$	$(CH_3O)_2POH$	KF, room temp, 0.75 h	$C_6H_5COH(CH_2Cl)P(O)(OCH_3)_2$ (77)	195
		$(C_2H_5O)_2POH$	C_6H_6, $(C_2H_5)_3N$, 25°, 15 min	" (80)	115
			Alumina, room temp, 65 h	$C_6H_5COH(CH_2Cl)P(O)(OC_2H_5)_2$ (65)	92
			KF, room temp, 0.75 h	" (76)	91
C_9	$(E)\text{-}C_6H_5CH{=}CHCHO$	$(CH_3O)_2POH$	Alumina, room temp, 48 h	$(E)\text{-}C_6H_5CH{=}CHCHOHP(O)(OCH_3)_2$ (48)	92
			KF, room temp, 0.5 h	" (94)	91
		$(C_2H_5O)_2POH$	KF, room temp, 0.5 h	$(E)\text{-}C_6H_5CH{=}CHCHOHP(O)(OC_2)H_5)_2$ (88)	91
		$(C_2H_5O)_2POTMS$	C_6H_6, room temp, 6 h	$(E)\text{-}C_6H_5CH{=}CHCH(OTMS)P(O)(OC_2H_5)_2$ (85)	111
		$(TMSO)_3P$	C_6H_6, room temp, 4 h	$(E)\text{-}C_6H_5CH{=}CHCH(OTMS)P(O)(OTMS)_2$ (84)	49
		$[(CH_3)_2N]_2POTES$	THF, 0°, 0.5 h	$(E)\text{-}C_6H_5CH{=}CHCH(OTES)P(O)[N(CH_3)_2]_2$ (93)	51
			THF, 0°, 0.25 h	" (93)	61
	$(n\text{-}C_4H_9)_2CO$	$NH_4OP(OH)OCH_2CH(C_2H_5)C_4H_9\text{-}n$	2-Ethylhexanol, Na, 70°, 24 h	$n\text{-}C_4H_9CH(C_2H_5)CH_2OP(O)(OH)COH(C_4H_9\text{-}n)_2$ (76)	205
	$HCOCO_2CH_2C_6H_5$	$(C_6H_5CH_2O)_2PONa$	CH_3OH, 20°	$(C_6H_5CH_2O)_2P(O)CHOHCO_2CH_2C_6H_5$ (61)	112

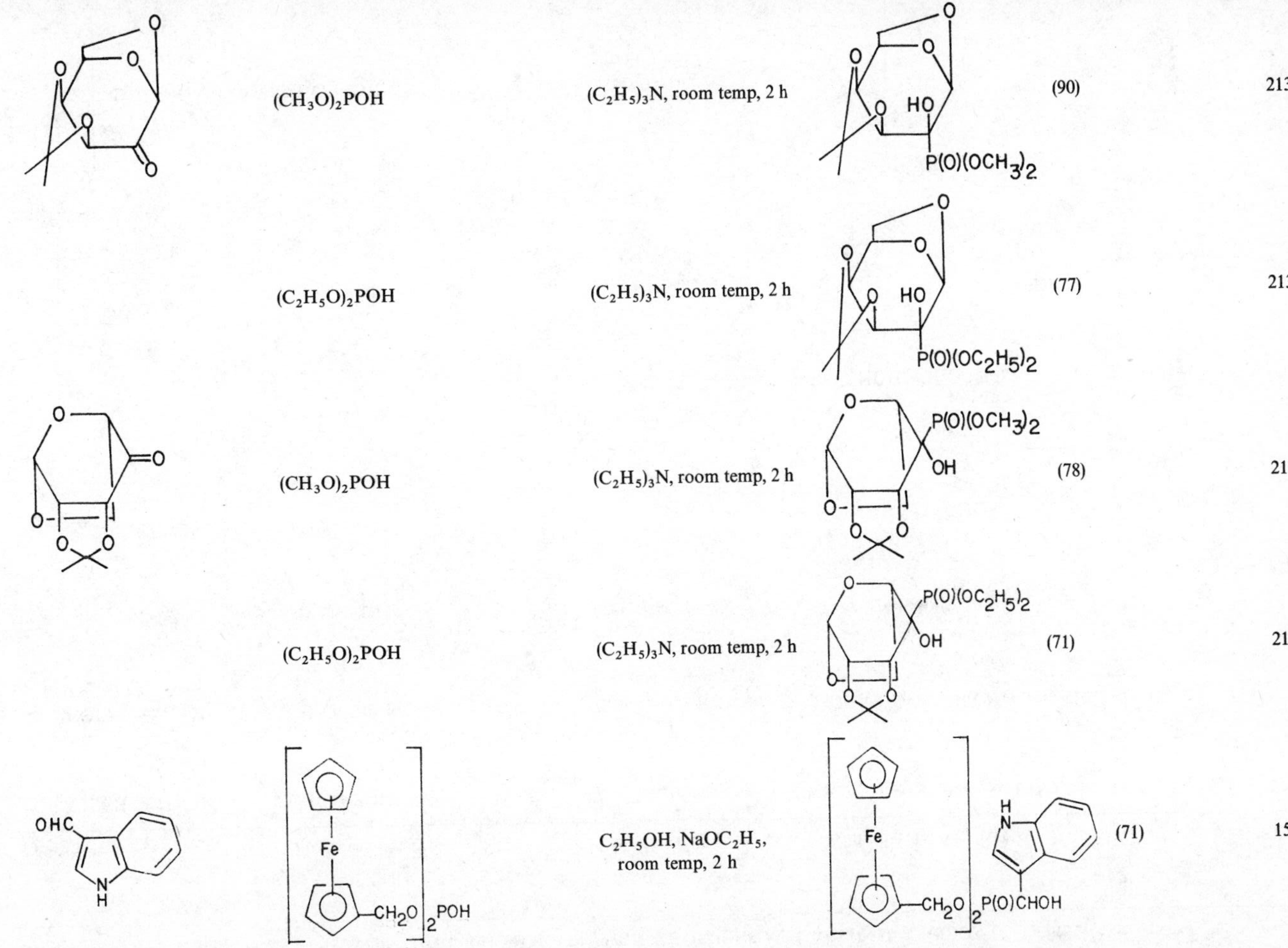

	$(CH_3O)_2POH$	$(C_2H_5)_3N$, room temp, 2 h	(90)	213
	$(C_2H_5O)_2POH$	$(C_2H_5)_3N$, room temp, 2 h	(77)	213
	$(CH_3O)_2POH$	$(C_2H_5)_3N$, room temp, 2 h	(78)	213
	$(C_2H_5O)_2POH$	$(C_2H_5)_3N$, room temp, 2 h	(71)	213
		C_2H_5OH, $NaOC_2H_5$, room temp, 2 h	(71)	155

TABLE I. TRIVALENT PHOSPHORUS ADDITION REACTIONS AT CARBONYL CARBON (*Continued*)

No. of Carbon Atoms	Substrate	Phosphorus Reagent	Conditions	Product(s) and Yield(s)(%)	Refs.
C_{10}	$CH_2CH(CH_3))CH_2CHO$ $CH_2CH{=}C(CH_3)_2$	$(CH_3O)_2POH$	Alumina, room temp, 48 h	$(CH_3O)_2P(O)CHOHCH(CH_3)CH_2CH_2CH{=}C(CH_3)_2$ (48)	92
	$CH_3COCH{=}CHC_6H_5$	$(CH_3O)_2POH$	Ether, $NaOCH_3$, room temp, 10 min	$(CH_3O)_2P(O)COH(CH_3)CH{=}CHC_6H_5$ (70)	85
	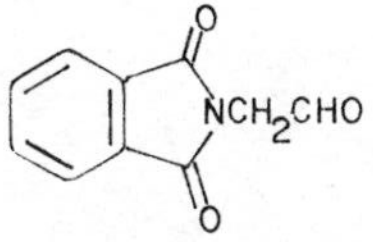 [structure: phthalimide-NCH_2CHO]	$(CH_3O)_2POH$	$(C_2H_5)_3N$, 60°, 3h	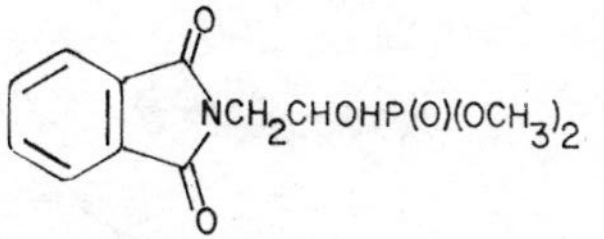 [structure: phthalimide-$NCH_2CHOHP(O)(OCH_3)_2$] (57)	88
	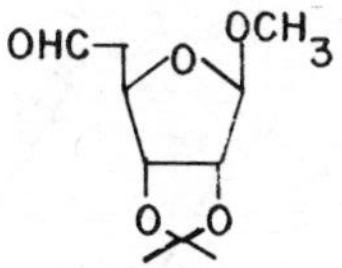 [structure: OHC–furanose, OCH_3, isopropylidene]	$(CH_3O)_2POH$	C_6H_6, CH_3OH, $NaOCH_3$, 40°, 3 h	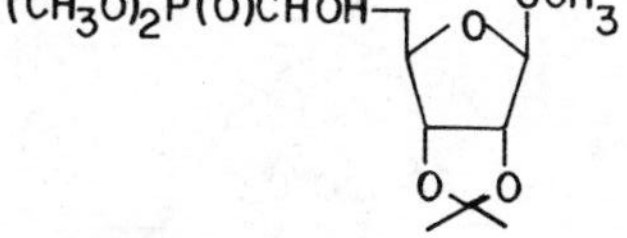 [structure: $(CH_3O)_2P(O)CHOH$–furanose, OCH_3, isopropylidene] (71)	214
	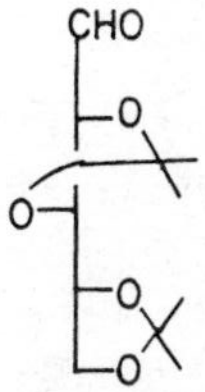 [structure: CHO, di-O-isopropylidene sugar]	$(CH_3O)_2POH$	CH_3OH, $NaOCH_3$, room temp	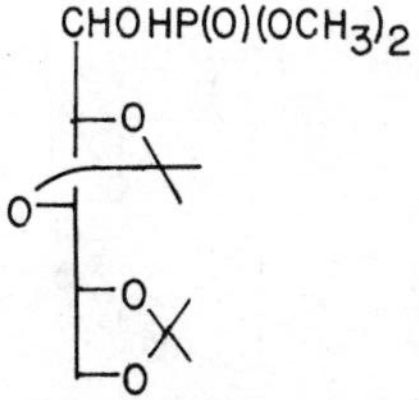 [structure: $CHOHP(O)(OCH_3)_2$, di-O-isopropylidene sugar] (86)	215
	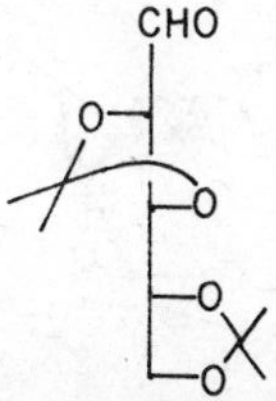 [structure: CHO, di-O-isopropylidene sugar]	$(CH_3O)_2POH$	CH_3OH, $NaOCH_3$, room temp	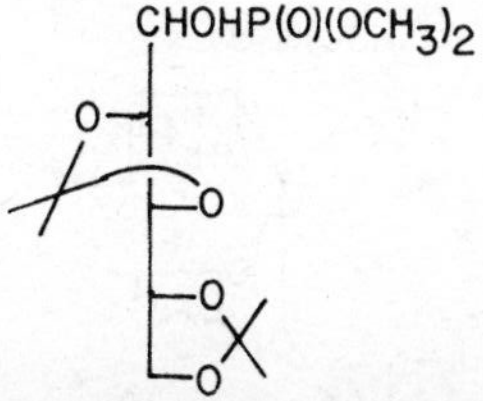 [structure: $CHOHP(O)(OCH_3)_2$, di-O-isopropylidene sugar] (67)	216

	Reactant	Phosphite	Conditions	Product (Yield, %)	Refs.
C_{11}	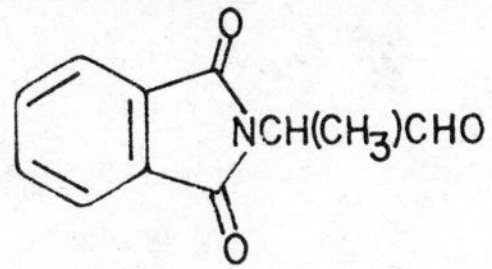 N-phthalimido-$NCH(CH_3)CHO$	$(CH_3O)_2POH$	$(C_2H_5)_3N$, 60°, 3 h	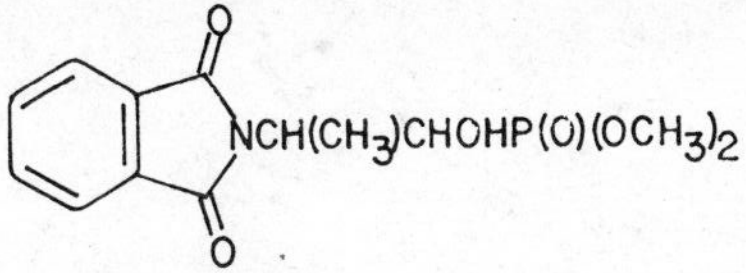 N-phthalimido-$NCH(CH_3)CHOHP(O)(OCH_3)_2$ (67)	88
	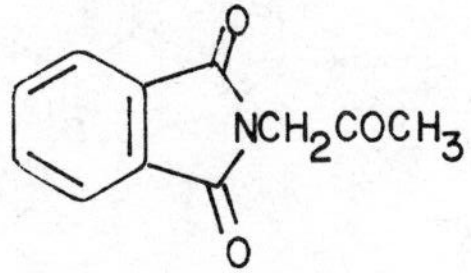 N-phthalimido-NCH_2COCH_3	$(CH_3O)_2POH$	$(C_2H_5)_3N$, 60°, 3 h	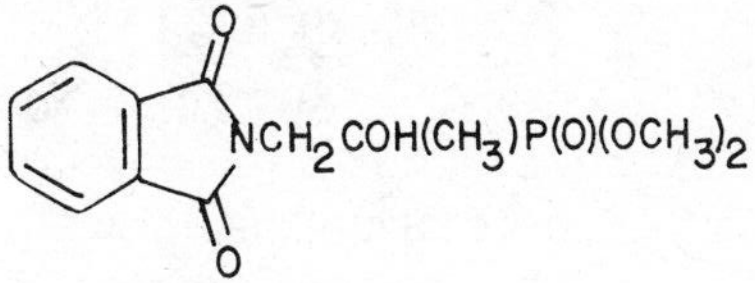 N-phthalimido-$NCH_2COH(CH_3)P(O)(OCH_3)_2$ (64)	88
	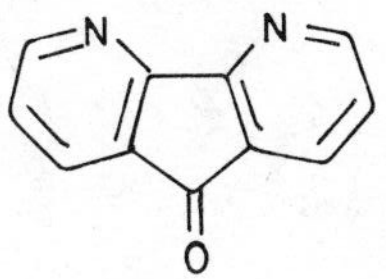 diazafluorenone	$(CH_3O)_2POH$	Toluene, $(C_2H_5)_3N$, room temp, 10 h	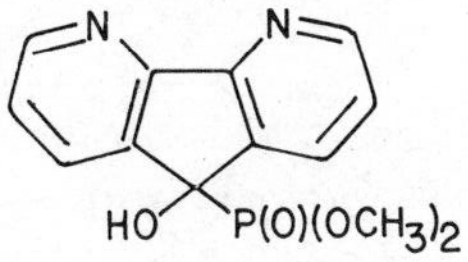 HO, $P(O)(OCH_3)_2$ adduct (85)	150
C_{12}	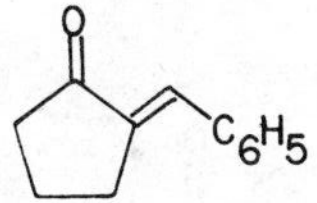 2-benzylidenecyclopentanone (C_6H_5)	$(CH_3O)_2POH$	Ether, $(C_2H_5)_2NH$, 6 h, room temp	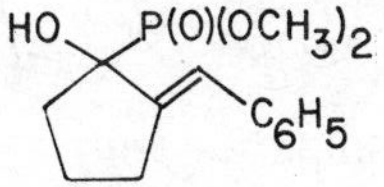 HO, $P(O)(OCH_3)_2$, C_6H_5 adduct (100)	87
			Ether, $(C_2H_5)_3N$, 6 h, room temp	" (58)	87
			Ether, CH_3OH, $NaOCH_3$, room temp, 5 min	" (64)	87
	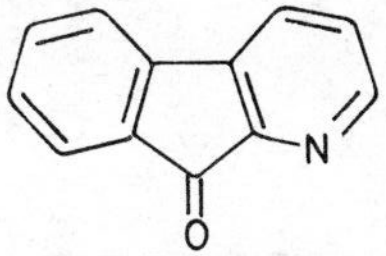 azafluorenone	$(CH_3O)_2POH$	Toluene, $(C_2H_5)_3N$, 10 h, room temp	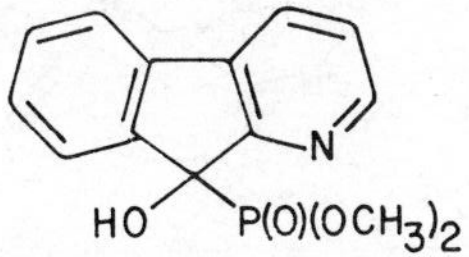 HO, $P(O)(OCH_3)_2$ adduct (90)	150
		$(C_2H_5O)_2POH$	Toluene, $(C_2H_5)_3N$, 10 h, room temp	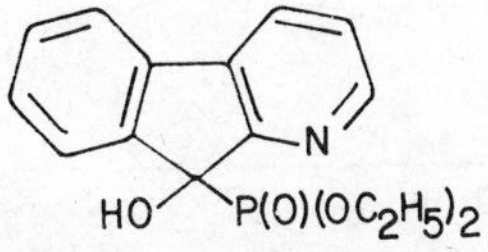 HO, $P(O)(OC_2H_5)_2$ adduct (40)	150

TABLE I. Trivalent Phosphorus Addition Reactions at Carbonyl Carbon (*Continued*)

No. of Carbon Atoms	Substrate	Phosphorus Reagent	Conditions	Product(s) and Yield(s)(%)	Refs.
C_{12} (*Contd.*)	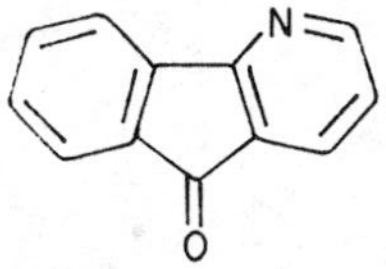	$(i\text{-}C_3H_7O)_2POH$	Toluene $(C_2H_5)_3N$, 10 h, room temp	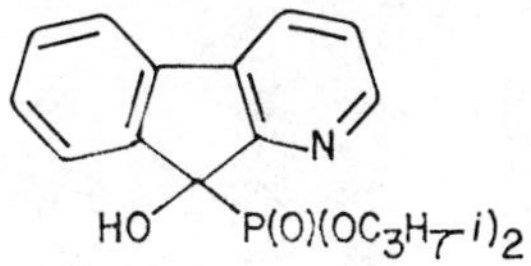 HO, $P(O)(OC_3H_7\text{-}i)_2$ (77)	150
	N, O	$(CH_3O)_2POH$	Toluene, $(C_2H_5)_3N$, 10 h, room temp	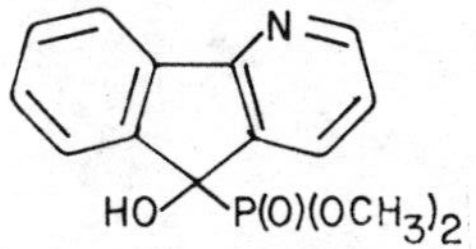 N, HO, $P(O)(OCH_3)_2$ (75)	150
		$(C_2H_5O)_2POH$	Toluene, $(C_2H_5)_3N$, 10 h, room temp	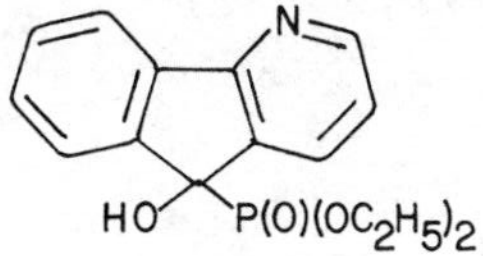 N, HO, $P(O)(OC_2H_5)_2$ (70)	150
		$(i\text{-}C_3H_7O)_2POH$	Toluene, $(C_2H_5)_3N$, 10 h, room temp	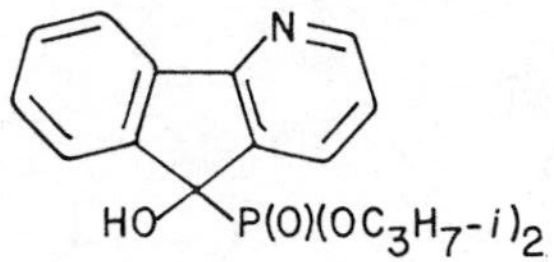 N, HO, $P(O)(OC_3H_7\text{-}i)_2$ (63)	150
C_{13}	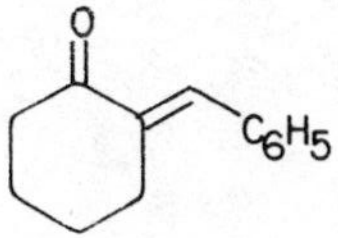 O, C_6H_5	$(CH_3O)_2POH$	Ether, $(C_2H_5)_2NH_3$ 6 h, room temp	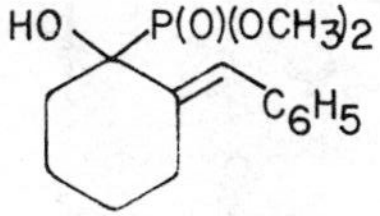 HO, $P(O)(OCH_3)_2$, C_6H_5 (100)	87
			Ether,, $(C_2H_5)_3N$, 6 h, room temp	" (20)	87
			Ether, CH_3OH, $NaOCH_3$, room temp, 5 min	" (69)	87

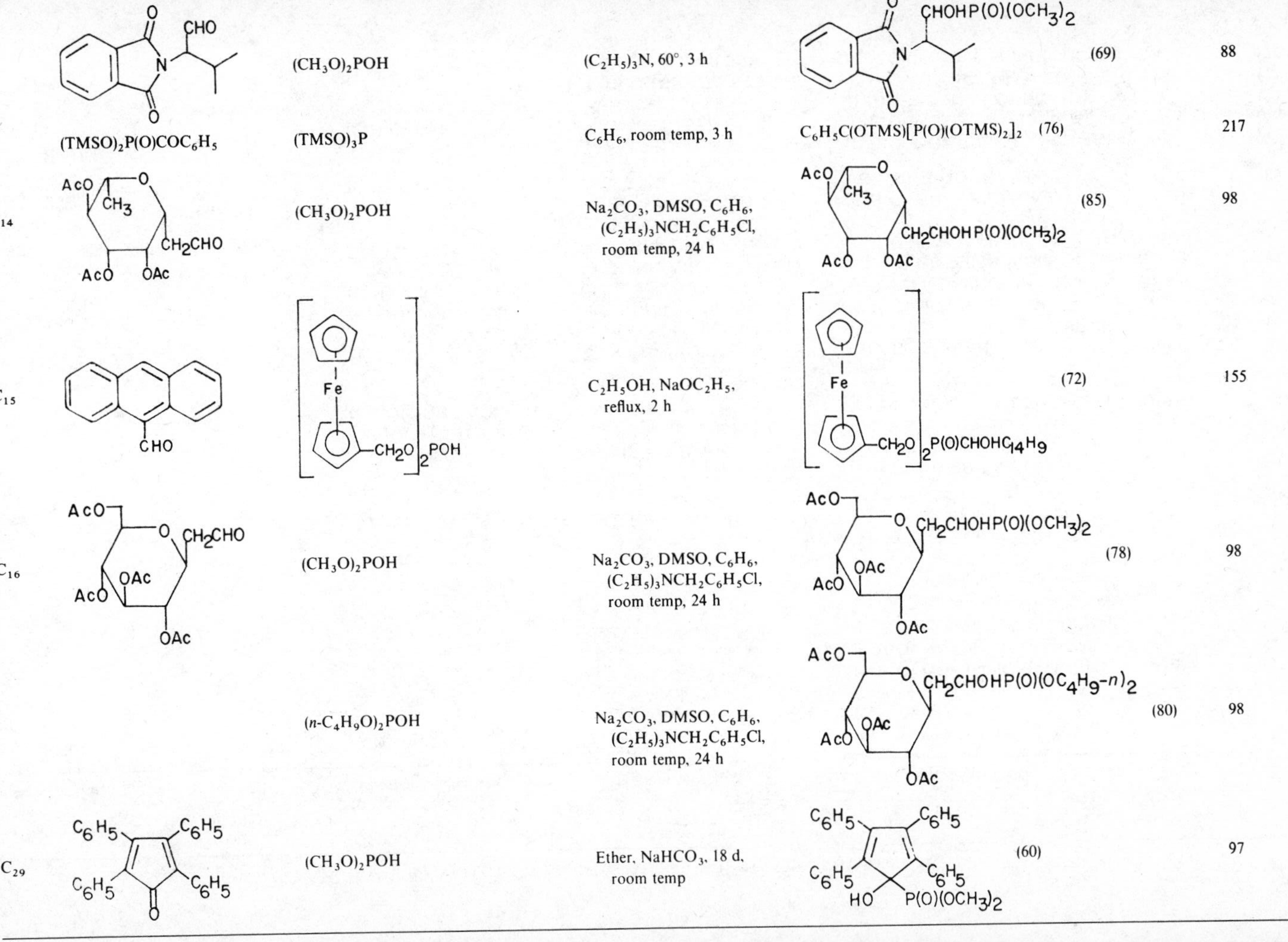

		$(CH_3O)_2POH$	$(C_2H_5)_3N$, 60°, 3 h		(69) 88
	$(TMSO)_2P(O)COC_6H_5$	$(TMSO)_3P$	C_6H_6, room temp, 3 h	$C_6H_5C(OTMS)[P(O)(OTMS)_2]_2$ (76)	217
C_{14}		$(CH_3O)_2POH$	Na_2CO_3, DMSO, C_6H_6, $(C_2H_5)_3NCH_2C_6H_5Cl$, room temp, 24 h	(85)	98
C_{15}			C_2H_5OH, $NaOC_2H_5$, reflux, 2 h	(72)	155
C_{16}		$(CH_3O)_2POH$	Na_2CO_3, DMSO, C_6H_6, $(C_2H_5)_3NCH_2C_6H_5Cl$, room temp, 24 h	(78)	98
		$(n\text{-}C_4H_9O)_2POH$	Na_2CO_3, DMSO, C_6H_6, $(C_2H_5)_3NCH_2C_6H_5Cl$, room temp, 24 h	(80)	98
C_{29}		$(CH_3O)_2POH$	Ether, $NaHCO_3$, 18 d, room temp	(60)	97

TABLE II. TRIVALENT PHOSPHORUS ADDITION REACTIONS AT IMINE CARBON

No. of Carbon Atoms	Substrate	Phosphorus Reagent	Conditions	Product(s) and Yield(s)(%)	Ref
C_1	HCHO	$(CH_3O)_2POH$	$(CH_3)_2NH$, 15 min, 85°	$(CH_3O)_2P(O)CH_2N(CH_3)_2$ (88)	94
		$(C_2H_5O)_2POH$	$(C_2H_5)_2NH$, 15 min, 85°	$(C_2H_5O)_2P(O)CH_2N(C_2H_5)_2$ (94)	94
			$(i\text{-}C_3H_7)_2NH$, 15 min, 85°	$(C_2H_5O)_2P(O)CH_2N(C_3H_7\text{-}i)_2$ (91)	94
			$(n\text{-}C_4H_9)_2NH$, 15 min, 85°	$(C_2H_5O)_2P(O)CH_2N(C_4H_9\text{-}n)_2$ (94)	94
			$(C_6H_{11})_2NH$, 15 min, 85°	$(C_2H_5O)_2P(O)CH_2N(C_6H_{11})_2$ (81)	94
			$(n\text{-}C_8H_{17})_2NH$, 15 min, 85°	$(C_2H_5O)_2P(O)CH_2N(C_8H_{17}\text{-}n)_2$ (86)	94
			Morpholine, 15 min, 85°	$(C_2H_5O)_2P(O)CH_2N$ (morpholine ring) O (92)	94
			Piperidine, 15 min, 85°	$(C_2H_5O)_2P(O)CH_2N$ (piperidine ring) (87)	94
			$C_6H_5NHC_2H_5$, 15 min, 85°	$(C_2H_5O)_2P(O)CH_2N(C_6H_5)C_2H_5$ (47)	94
		$(n\text{-}C_4H_9O)_2POH$	$(C_2H_5)_2NH$, 15 min, 85°	$(n\text{-}C_4H_9O)_2P(O)CH_2N(C_2H_5)_2$ (93)	94
		$[n\text{-}C_4H_9CH(C_2H_5)CH_2O]_2POH$	$(C_2H_5)_2NH$, 15 min, 85°	$[n\text{-}C_4H_9CH(C_2H_5)CH_2O]_2P(O)CH_2N(C_2H_5)_2$ (90)	94
		$(CH_3O)_2POH$	$H_2N(CH_2)_2NH_2$, 40°, 10 min	$[(CH_3O)_2P(O)CH_2]_2N(CH_2)_2N[CH_2P(O)(OCH_3)_2]_2$ (45)	218
C_2	CH_3CHO	H_3PO_3	$C_6H_5CH_2NHCONH_2$, 80°, 3 h	$CH_3CH(NH_2)PO_3H_2$ (10)[a]	187
			$C_6H_5CONH_2$, 80°, 3 h	" (27)[a]	187
		$(C_2H_5O)_2POH$	$(C_2H_5)_2NH$, 15 min, 85°	$(C_2H_5O)_2P(O)CH(CH_3)N(C_2H_5)_2$ (81)	94
		$(C_6H_5O)_3P$	Acetic acid, H_2NCSNH_2	$H_2O_3PCH(CH_3)CHCSNH_2$ (13)[a]	180
			$(C_6H_5)NHCSNH_2$, 80°, 30 min	$(C_6H_5O)_2P(O)CH(CH_3)NHCSNHC_6H_5$ (85)	177
			$(C_6H_5CH(CH_3)NHCSNH_2$, 80°, 30 min	$(C_6H_5O)_2P(O)CH(CH_3)NHCSNHCH(C_6H_5)CH_3$ (75)	177
			$C_6H_5CH_2OCONH_2$, 85°, 2 h	$(C_6H_5O)_2P(O)CH(CH_3)NHCO_2CH_2C_6H_5$ (46)	181
	$CH_3CH{=}NH$	$(C_2H_5O)_2POH$	100°, 3 h	$CH_3CH(NH_2)PO_3H_2$ (55)[a]	163
C_3	$CH_2{=}CHCHO$	$(C_2H_5O)_2POH$	$(C_2H_5)_2NH$, 15 min, 85°	$(C_2H_5O)_2P(O)CH[N(C_2H_5)_2]CH{=}CH_2$ (87)	94
	C_2H_5CHO	$(C_2H_5O)_2POH$	$(C_2H_5)_2NH$, 15 min, 85°	$(C_2H_5O)_2P(O)CH[N(C_2H_5)_2]C_2H_5$ (78)	94
		$(C_6H_5O)_3P$	Acetic acid, H_2NCSNH_2	$H_2O_3PCH(C_2H_5)NHCSNH_2$ (18)[a]	180
			$C_6H_5NHCSNH_2$, 80°, 30 min	$(C_6H_5O)_2P(O)CH(CH_2H_5)NHCSNHC_6H_5$ (86)	177
			$C_6H_5CH(CH_3)NHCSNH_2$, 80°, 30 min	$(C_6H_5O)_2P(O)CH(C_2H_5)NHCSNHCH(C_6H_5)CH_3$ (85)	177

	$(CH_3)_2CO$	$(C_2H_5O)_2POH$	$(C_2H_5)_2NH$, 15 min, 85°	$(C_2H_5O)_2P(O)C(CH_3)_2N(C_2H_5)_2$ (64)	94
	$CH_3CH{=}NCH_3$	$(C_2H_5O)_2POH$	85°, 15 min	$(C_2H_5O)_2P(O)CH(CH_3)NHCH_3$ (94)	94
	$C_2H_5CH{=}NH$	$(C_2H_5O)_2POH$	100°, 3 h	$C_2H_5CH(NH_2)PO_3H_2$ (47)[a]	187
	$CH_3COCH{=}NOH$	$(C_2H_5O)_2POH$	120°, 8 h	$(C_2H_5O)_2P(O)CH(NHOH)COCH_3$ (54)	206
	CH_3SCH_2CHO	$(C_6H_5O)_3POH$	30°, 1 h, $C_6H_5NHCSNH_2$	$(C_2H_5O)_2P(O)CH(CH_2SCH_3)NHCSNHC_6H_5$ (94)	219
C_4	n-C_3H_7CHO	$(C_2H_5O)_2POH$	$(C_2H_5)_2NH$, 15 min, 85°	$(C_2H_5O)_2P(O)CH[N(C_2H_5)_2]C_3H_7$-$n$ (79)	94
			$(n$-$C_4H_9)_2NH$, 15 min, 85°	$(C_2H_5O)_2P(O)CH[N(C_4H_9$-$n)_2]C_3H_7$-$n$ (91)	94
		$(C_6H_5O)_3P$	$C_6H_5NHCSNH_2$, 30 min, 80°	$(C_6H_5O)_2P(O)CH(C_3H_7$-$n)NHCSNHC_6H_5$ (89)	177
			$C_6H_5CH(CH_3)NHCSNH_2$, 80°, 30 min	$(C_6H_5O)_2P(O)CH(C_3H_7$-$n)NHCSNHCH(CH_3)C_6H_5$ (94)	177
	i-C_3H_7CHO	H_3PO_3	$C_6H_5CH_2NHCONH_2$, 80°, 3 h	i-$C_3H_7CH(NH_2)PO_3H_2$ (34)[a]	187
		$(C_6H_5O)_3P$	Acetic acid, H_2NCSNH_2	$H_2O_3PCH(C_3H_7$-$i)NHCSNH_2$ (21)[a]	180
			$C_6H_5NHCSNH_2$, 30 min, 80°	$(C_6H_5O)_2P(O)CH(C_3H_7$-$i)NHCSNHC_6H_5$ (77)	177
			$C_6H_5CH(CH_3)NHCSNH_2$, 80°, 30 min	$(C_6H_5O)_2P(O)CH(C_3H_7$-$i)NHCSNHCH(CH_3)C_6H_5$ (87)	177
			$C_6H_5CH_2OCONH_2$, 85°, 2 h	$(C_6H_5O)_2P(O)CH(C_3H_7$-$i)NHCO_2CH_2C_6H_5$ (52)	181
	n-$C_3H_7CH{=}NH$	$(C_2H_5O)_2POH$	100°, 3 h	$H_2O_3PCH(NH_2)C_3H_7$-n (55)[a]	163
	i-$C_3H_7CH{=}NH$	$(4$-$CH_3C_6H_4CH_2O)_2POH$	120°, 30 min	$H_2O_3PCH(NH_2)C_3H_7$-i (84)[a]	165
	$CH_3CH{=}NC_2H_5$	$(C_2H_5O)_2POH$	85°, 15 min	$(C_2H_5O)_2P(O)CH(CH_3)NHC_2H_5$ (91)	94
	$C_2H_5SCH_2CHO$	$(C_6H_5O)_3P$	$C_6H_5NHCSNH_2$, 30°, 1 h	$(C_6H_5O)_2P(O)CH(CH_2SCH_3)NHCSNHC_6H_5$ (92)	219
	$CH_3SCH_2CH_2CHO$	$(C_6H_5O)_3P$	$C_6H_5NHCSNH_2$, 100°, 30 min	$(C_6H_5O)_2P(O)CH(CH_2CH_2SCH_3)NHCSNHCH_6H_5$ (86)	178
			4-$O_2NC_6H_4NHCSNH_2$, 30 min, 100°	$(C_6H_5O)_2P(O)CH(CH_2CH_2SCH_3)NHCSNHC_6H_4NO_2$-4 (56)	178
			$NHCSNH_2$ (naphthyl), 100°, 30 min	$NHCSNHCH(CH_2CH_2SCH_3)P(O)(OC_6H_5)_2$ (naphthyl) (41)	178
C_5	n-C_4H_9CHO	$(C_2H_5O)_3P$	Acetic acid, H_2NCSNH_2	$H_2O_3PCH(C_4H_9$-$n)NHCSNH_2$ (22)[a]	180
			$C_6H_5NHCSNH_2$, 80°, 30 min	$(C_6H_5O)_2P(O)CH(C_4H_9$-$n)NHCSNHC_6H_5$ (81)	178
			$C_6H_5CH(CH_3)NHCSNH_2$, 80°, 30 min	$(C_6H_5O)_2P(O)CH(C_4H_9$-$n)NHCSNHCH(CH_3)C_6H_5$ (88)	177
	i-$C_3H_7CH_2CHO$	$(C_6H_5O)_3P$	$C_6H_5CH_2OCONH_2$, 85°, 2 h	$(C_6H_5O)_2P(O)CH(C_4H_9$-$i)NHCOCH_2C_6H_5$ (54)	181
	n-$C_3H_7CH{=}NCH_3$	$(C_2H_5O)_2POH$	85°, 15 min	$(C_2H_5O)_2P(O)CH(NHCH_3)C_3H_7$-$i$ (93)	94
		$(n$-$C_4H_9O)_2POH$	85°, 15 min	$(n$-$C_4H_9O)_2P(O)CH(NHCH_3)C_3H_7$-$i$ (84)	94
	$C_2H_5CH{=}NC_2H_5$	$(C_2H_5O)_2POH$	85°, 15 min	$(C_2H_5O)_2P(O)CH(NHC_2H_5)C_2H_5$ (89)	94
	$(CH_3)_2C{=}NC_2H_5$	$(C_2H_5O)_2POH$	85°, 15 min	$(C_2H_5O)_2P(O)C(CH_3)_2NHC_2H_5$ (97)	94

TABLE II. TRIVALENT PHOSPHORUS ADDITION REACTIONS AT IMINE CARBON (*Continued*)

No. of Carbon Atoms	Substrate	Phosphorus Reagent	Conditions	Product(s) and Yield(s)(%)	Ref
C_5 (*Contd.*)	2-furyl–CHO	$(C_2H_5O)_2POH$	$(C_2H_5)_2NH$, 85°, 15 min	2-furyl–$CH[N(C_2H_5)_2]P(O)(OC_2H_5)_2$ (84)	94
	2-furyl–CH=NH	$(4\text{-}CH_3C_6H_4CH_2O)_2POH$	120°, 30 min	2-furyl–$CH(NH_2)PO_3H_2$ (82)[a]	165
	2-thienyl–CHO	$(C_6H_5O)_2POH$	$4\text{-}O_2NC_6H_4NH_2$, 85°, 15 min	2-thienyl–$CH(NHC_6H_4NO_2\text{-}4)P(O)(OC_6H_5)_2$ (75)	152
	2-thienyl–CH=NH	$(4\text{-}CH_3C_6H_4CH_2O)_2POH$	120°, 30 min	2-thienyl–$CH(NH_2)PO_3H_2$ (85)[a]	165
	2-furyl–CH=NOH	$(C_2H_5O)_3P$	80°	2-furyl–$CH(NHOC_2H_5)P(O)(OC_2H_5)_2$ (39)	43
	$C_2H_5SCH_2CH_2CHO$	$(C_6H_5O)_3P$	$C_6H_5NHCSNH_2$, 100°, 30 min	$(C_6H_5O)_2P(O)CH(CH_2CH_2SC_2H_5)NHCSNHC_6H_5$ (88)	178
	$n\text{-}C_4H_9N{=}CH_2$	$(C_2H_5O)_2POH$	85°, 15 min	$(C_2H_5O)_2P(O)CH_2NHC_4H_9\text{-}n$ (89)	94
C_6	$CH_3COCH_2CH_2CO_2CH_3$	$CH_3P(OC_2H_5)OH$	C_2H_5OH, NH_3, 110°, 6 h	$CH_3P(O)(OH)C(CH_3)(NH_2)CH_2CH_2CO_2H$ (20)[a]	161
		$(C_2H_5O)_2POH$	C_2H_5OH, NH_3, 110°, 6 h	$H_2O_3PC(CH_3)(NH_2)CH_2CH_2CO_2H$ (25)[a]	161
	3-methyl-2-thienyl–CHO	$(C_6H_5O)_2POH$	$4\text{-}O_2NC_6H_4NH_2$, 85°, 15 min	3-methyl-2-thienyl–$CH(NHC_6H_4NO_2\text{-}4)P(O)(OC_6H_5)_2$ (77)	152
	5-methyl-2-thienyl–CHO	$(C_6H_5O)_2POH$	$4\text{-}O_2NC_6H_4NH_2$, 85°, 15 min	5-methyl-2-thienyl–$CH(NHC_6H_4NO_2\text{-}4)P(O)(OC_6H_5)_2$ (78)	152
C_7	C_6H_5CHO	H_3PO_3	CH_3CONH_2, 80°, 3 h	$H_2O_3PCH(NH_2)C_6H_5$ (75)[a]	187
		$(C_2H_5O)_2POH$	$C_6H_5CH(CH_3)NHCONH_2$, 4 h, reflux, BF_3–etherate	" (47)[a]	176

		$(C_6H_5O)_3P$	$CH_3NHCSNH_2$, toluene, reflux, 2 h	$H_2O_3PCH(C_6H_5)NHCSNHCH_3$ (6)[a]	179
			Acetic acid, H_2NCSNH_2	$H_2O_3PCH(C_6H_5))NHCSNH_2$ (11)[a]	180
			$C_6H_5NHCSNH_2$, 80°, 30 min	$(C_6H_5O)_2P(O)CH(C_6H_5)NHCSNHC_6H_5$ (85)	177
			$C_6H_5CH(CH_3)NHCSNH_2$, 80°, 30 min	$(C_6H_5O)_2P(O)CH(C_6H_5)NHCSNHCH(CH_3)C_6H_5$ (80)	177
			$C_6H_5OCONH_2$, 85°, 2 h	$(C_6H_5O)_2P(O)CH(C_6H_5)NHCO_2CH_2C_6H_5$ (51)	181
	$C_6H_5CH{=}NH$	$(C_2H_5O)_2POH$	100°, 3 h	$H_2O_3PCH(NH_2)C_6H_5$ (61)[a]	163
	(2-furyl)$CH{=}NC_2H_5$	$(C_2H_5O)_2POH$	85°, 15 min	(2-furyl)$CH(NHC_2H_5)P(O)(OC_2H_5)_2$ (85)	94
	(2-furyl)$CH{=}CHCHO$	$(C_6H_5O)_2POH$	$4\text{-}O_2NC_6H_4NH_2$, 85°, 15 min	(2-furyl)$CH{=}CHCH(NHC_6H_4NO_2\text{-}4)P(O)(OC_6H_5)_2$ (84)	152
	$3\text{-}O_2NC_6H_4CHO$	H_3PO_3	$C_6H_5CONH_2$, 80°, 3 h	$H_2O_3PCH(NH_2)C_6H_4NO_2\text{-}3$ (59)[a]	187
		$(C_6H_5O)_3P$	$CH_3NHCSNH_2$, toluene, reflux, 2 h	$H_2O_3PCH(C_6H_4NO_2\text{-}3)NHCSNHCH_3$ (25)[a]	179
	$4\text{-}O_2NC_6H_4CHO$	$(C_6H_5O)_3P$	$CH_3NHCSNH_2$, toluene, reflux, 2 h	$H_2O_3PCH(C_6H_4NO_2\text{-}4)NHCSNHCH_3$ (26)[a]	179
	$2\text{-}ClC_6H_4CHO$	$(C_6H_5O)_3P$	$CH_3NHCSNH_2$, toluene, reflux, 2 h	$H_2O_3PCH(C_6H_4Cl\text{-}2)NHCSNHCH_3$ (15)[a]	179
	$4\text{-}ClC_6H_4CHO$	H_3PO_3	CH_3CONH_2, 80°, 3 h	$H_2O_3PCH(NH_2)C_6H_4Cl\text{-}4$ (67)[a]	187
		$(C_6H_5O)_3P$	$CH_3NHCSNH_2$, toluene, reflux, 2 h	$H_2O_3PCH(C_6H_4Cl\text{-}4)NHCSNHCH_3$ (18)[a]	179
	$4\text{-}BrC_6H_4CHO$	$(C_6H_5O)_3P$	$CH_3NHCSNH_2$, toluene, reflux, 2 h	$H_2O_3PCH(C_6H_4Br\text{-}4)NHCSNHCH_3$ (16)[a]	179
	$C_6H_5CH{=}NH$	$(4\text{-}CH_3C_6H_4CH_2O)_2POH$	120°, 30 min	$H_2O_3PCH(NH_2)C_6H_5$ (77)[a]	165
C_8	$2\text{-}CH_3C_6H_4CHO$	$(C_6H_5O)_3P$	Acetic acid, H_2NCSNH_2	$H_2O_3PCH(C_6H_4CH_3\text{-}2)NHCSNH_2$ (11)[a]	180
	$3\text{-}CH_3C_6H_4CHO$	$(C_6H_5O)_2POH$	$4\text{-}O_2NC_6H_4NH_2$, 85°, 15 min	$(C_6H_5O)_2P(O)CH(C_6H_4CH_3\text{-}3)NHC_6H_4NO_2\text{-}4$ (78)	152
		$(C_6H_5O)_3P$	Acetic acid, H_2NCSNH_2	$H_2O_3PCH(C_6H_4CH_3\text{-}3)NHCSNH_2$ (19)[a]	180
	$4\text{-}CH_3C_6H_4CHO$	H_3PO_3	$C_6H_5CONH_2$, 80°, 3 h	$H_2O_3PCH(NH_2)C_6H_4CH_3\text{-}4$ (37)[a]	187
		$(C_6H_5O)_2POH$	$4\text{-}O_2NC_6H_4NH_2$, 85°, 15 min	$(C_6H_5O)_2P(O)CH(C_6H_4CH_3\text{-}4)NHC_6H_4NO_2\text{-}4$ (75)	152
	$C_6H_5CH_2CHO$	$(C_6H_5O)_3P$	$C_6H_5CH_2OCONH_2$, 85°, 2 h	$(C_6H_5O)_2P(O)CH(CH_2C_6H_5)NHCO_2CH_2C_6H_5$ (36)	181
			$C_6H_5NHCSNH_2$, 80°, 30 min	$(C_6H_5O)_2P(O)CH(CH_2C_6H_5)NHCSNHC_6H_5$ (70)	177
			$C_6H_5CH(CH_3)NHCSNH_2$, 80°, 30 min	$(C_6H_5O)_2P(O)CH(CH_2C_6H_5)NHCSNHCH(CH_3)C_6H_5$ (70)	177
	$C_6H_5CH{=}NCH_3$	H_3PO_3	110°, 10 min	$H_2O_3PCH(C_6H_5)NHCH_3$ (61)	186
	$3\text{-}CH_3C_6H_4CH{=}NH$	$(C_2H_5O)_2POH$	100°, 3 h	$H_2O_3PCH(NH_2)C_6H_4CH_3\text{-}3$ (51)[a]	163
	$C_6H_5CH_2CH{=}NH$	$(C_2H_5O)_2POH$	100°, 3 h	$H_2O_3PCH(NH_2)CH_2C_6H_5$ (71)[a]	163
	$2\text{-}ClC_6H_4CH{=}NCH_3$	$(C_2H_5O)_2POH$	85°, 15 min	$(C_2H_5O)_2P(O)CH(C_6H_4Cl\text{-}2)NHCH_3$ (88)	94
	(benzimidazol-2-yl)CHO	$(C_2H_5O)_2POH$	$C_6H_5NH_2$, 110°, 12 h	(benzimidazol-2-yl)$CH(NHC_2H_5)P(O)(OC_2H_5)_2$ (68)	153
	$n\text{-}C_3H_7CH{=}NC_4H_9\text{-}n$	$(C_2H_5O)_2POH$	85°, 15 min	$(C_2H_5O)_2P(O)CH(NHC_4H_9\text{-}n)C_3H_7\text{-}n$ (94)	94

TABLE II. Trivalent Phosphorus Addition Reactions at Imine Carbon (*Continued*)

No. of Carbon Atoms	Substrate	Phophorus Reagent	Conditions	Product(s) and Yield(s)(%)	Ref
C_9	$t\text{-}C_4H_9CH_2C(CH_3)_2N{=}CH_2$	$(C_2H_5O)_2POH$	85°, 15 min	$(C_2H_5O)_2P(O)CH_2NHC(CH_3)_2CH_2C_4H_9\text{-}t$ (96)	94
	$C_6H_5CH_2OCH_2CHO$	$(C_6H_5O)_3P$	$C_6H_5OCONH_2$, 85°, 2 h	$(C_6H_5O)_2P(O)CH(NHCO_2CH_2C_6H_5)CH_2OCH_2C_6H_5$ (51)	64
	$C_6H_5CH_2CH_2N{=}CH_2$	$(C_2H_5O)_2POH$	85°, 15 min	$(C_2H_5O)_2P(O)CH_2NHCH_2CH_2C_6H_5$ (91)	94
	$C_6H_5CH_2N{=}CHCH_3$	$(C_2H_5O)_2POH$	140°, 30 min	$H_2O_3PCH(CH_3)NHCH_2C_6H_5$ (58)[a]	95
		$C_6H_5P(OH)OC_2H_5$	140°, 30 min	$C_6H_5P(O)(OH)CH(CH_3)NHCH_2C_6H_5$ (43)[a]	95
	$C_6H_5CH{=}NC_2H_5$	H_3PO_3	110°, 10 min	$H_2O_3PCH(C_6H_5NHC_2H_5$ (68)	186
	$4\text{-ClC}_6H_4C(OC_2H_5){=}NH$	$(n\text{-}C_4H_9O)_2POH$	C_6H_6, Na, reflux, 5 h	$4\text{-ClC}_6H_4C(NH_2)[P(O)(OC_4H_9\text{-}n)_2]_2$ (25)	16
	$C_6H_5CH_2SCH_2CHO$	$(C_6H_5O)_3P$	$C_6H_5NHCSNH_2$, 30°, 1 h	$(C_6H_5O)_2P(O)CH(CH_2SCH_2C_6H_5)NHCSNHC_6H_5$ (97)	219
C_{10}	$(R)\text{-}C_6H_5CH(CH_3)N{=}CHCH_3$	$(TMSO)_3P$	C_6H_6, 80°, 6 h	$(TMSO)_2P(O)CH(CH_3)N(TMS)CH(CH_3)C_6H_5$ (50), 70:30 diastereomeric mixture	183
	$C_2H_5CH{=}NCH_2C_6H_5$	$(C_2H_5O)_2POH$	140°, 30 min	$H_2O_3PCH(C_2H_5)NHCH_2C_6H_5$ (55)[a]	95
		$C_6H_5P(OH)OC_2H_5$	140°, 30 min	$C_6H_5P(O)(OH)CH(C_2H_5)NHCH_2C_6H_5$ (38)[a]	95
	$3\text{-}O_2NC_6H_4CH{=}NCH_2O_2CCH_3$	$(4\text{-}CH_3C_6H_4CH_2O)_2POH$	120°, 30 min	$3\text{-}O_2NC_6H_4CH(PO_3H_2)NHCH_2CO_2CH_3$ (67)[a]	165
C_{11}	$C_6H_5CH{=}NC(CH_3)_3$	H_3PO_3	110°, 10 min	$H_2O_3PCH(C_6H_5)NHC(CH_3)_3$ (40)[a]	186
	$n\text{-}C_3H_7CH{=}NCH_2C_6H_5$	$C_2H_5P(OH)OC_2H_5$	140°, 30 min	$C_2H_5P(O)(OH)CH(C_3H_7\text{-}n)NHCH_2C_6H_5$ (41)[a]	95
		$(C_2H_5O)_2POH$	140°, 30 min	$H_2O_3PCH(C_3H_7\text{-}n)NHCH_2C_6H_5$ (65)[a]	95
		$C_6H_5P(OH)OC_2H_5$	140°, 30 min	$C_6H_5P(O)(OH)CH(C_3H_7\text{-}n)NHCH_2C_6H_5$ (39)[a]	95
	$i\text{-}C_3H_7CH{=}NCH_2C_6H_5$	H_3PO_3	110°, 10 min	$H_2O_3PCH(C_3H_7\text{-}i)NHCH_2C_6H_5$ (40)	186
	[structure: Fe, CHO]	$(C_6H_5O)_2POH$	$4\text{-}O_2NC_6H_4NH_2$, 85°, 15 min	[structure: Fe, $-CH(NHC_6H_4NO_2\text{-}4)P(O)(OC_6H_5)_2$] (86)	152
C_{12}	$n\text{-}C_4H_9CH{=}NCH_2C_6H_5$	$(C_2H_5O)_2POH$	140°, 30 min	$H_2O_3PCH(C_4H_9\text{-}n)NHCH_2C_6H_5$ (65)[a]	95
		$C_6H_5P(OH)OC_2H_5$	140°, 30 min	$C_6H_5P(O)(OH)CH(C_4H_9\text{-}n)NHCH_2C_6H_5$ (35)[a]	95
	$(R)\text{-}C_6H_5CH(CH_3)N{=}CHC_3H_7\text{-}i$	$(TMSO)_3P$	Ether, $ZnCl_2$, 20°, 10 h	$(TMSO)_2P(O)CH(C_3H_7\text{-}i)N(TMS)CH(CH_3)C_6H_5$ (70), 66: 34 diastereomeric mixture	183

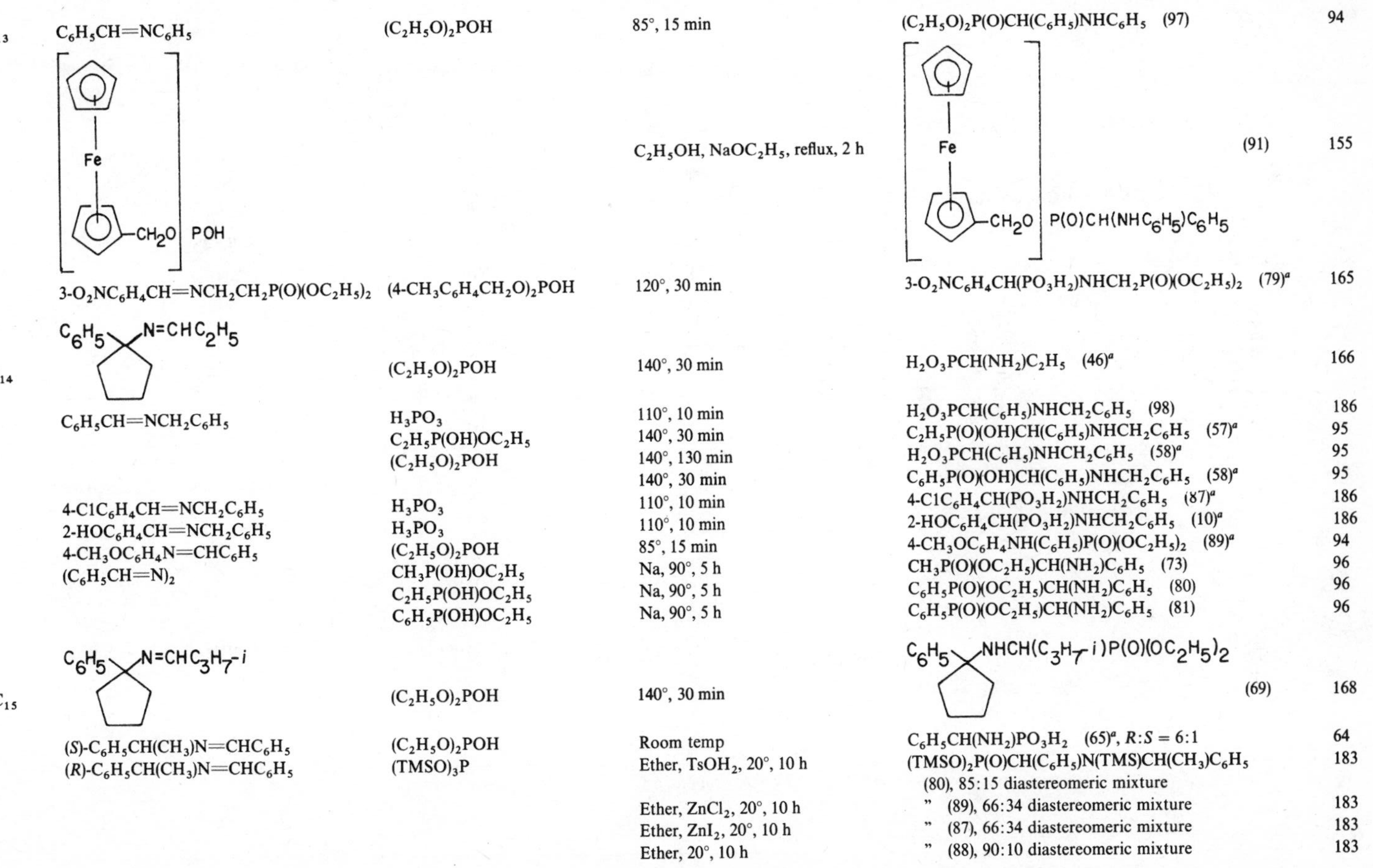

C_{13}	$C_6H_5CH{=}NC_6H_5$	$(C_2H_5O)_2POH$	85°, 15 min	$(C_2H_5O)_2P(O)CH(C_6H_5)NHC_6H_5$ (97)	94
	[Fe; CH_2O POH]		C_2H_5OH, $NaOC_2H_5$, reflux, 2 h	[Fe; CH_2O $P(O)CH(NHC_6H_5)C_6H_5$] (91)	155
	$3\text{-}O_2NC_6H_4CH{=}NCH_2CH_2P(O)(OC_2H_5)_2$	$(4\text{-}CH_3C_6H_4CH_2O)_2POH$	120°, 30 min	$3\text{-}O_2NC_6H_4CH(PO_3H_2)NHCH_2P(O)(OC_2H_5)_2$ (79)[a]	165
C_{14}	[C_6H_5; $N{=}CHC_2H_5$]	$(C_2H_5O)_2POH$	140°, 30 min	$H_2O_3PCH(NH_2)C_2H_5$ (46)[a]	166
	$C_6H_5CH{=}NCH_2C_6H_5$	H_3PO_3	110°, 10 min	$H_2O_3PCH(C_6H_5)NHCH_2C_6H_5$ (98)	186
		$C_2H_5P(OH)OC_2H_5$	140°, 30 min	$C_2H_5P(O)(OH)CH(C_6H_5)NHCH_2C_6H_5$ (57)[a]	95
		$(C_2H_5O)_2POH$	140°, 130 min	$H_2O_3PCH(C_6H_5)NHCH_2C_6H_5$ (58)[a]	95
			140°, 30 min	$C_6H_5P(O)(OH)CH(C_6H_5)NHCH_2C_6H_5$ (58)[a]	95
	$4\text{-}ClC_6H_4CH{=}NCH_2C_6H_5$	H_3PO_3	110°, 10 min	$4\text{-}ClC_6H_4CH(PO_3H_2)NHCH_2C_6H_5$ (87)[a]	186
	$2\text{-}HOC_6H_4CH{=}NCH_2C_6H_5$	H_3PO_3	110°, 10 min	$2\text{-}HOC_6H_4CH(PO_3H_2)NHCH_2C_6H_5$ (10)[a]	186
	$4\text{-}CH_3OC_6H_4N{=}CHC_6H_5$	$(C_2H_5O)_2POH$	85°, 15 min	$4\text{-}CH_3OC_6H_4NH(C_6H_5)P(O)(OC_2H_5)_2$ (89)[a]	94
	$(C_6H_5CH{=}N)_2$	$CH_3P(OH)OC_2H_5$	Na, 90°, 5 h	$CH_3P(O)(OC_2H_5)CH(NH_2)C_6H_5$ (73)	96
		$C_2H_5P(OH)OC_2H_5$	Na, 90°, 5 h	$C_6H_5P(O)(OC_2H_5)CH(NH_2)C_6H_5$ (80)	96
		$C_6H_5P(OH)OC_2H_5$	Na, 90°, 5 h	$C_6H_5P(O)(OC_2H_5)CH(NH_2)C_6H_5$ (81)	96
C_{15}	[C_6H_5; $N{=}CHC_3H_7\text{-}i$]	$(C_2H_5O)_2POH$	140°, 30 min	[C_6H_5; $NHCH(C_3H_7\text{-}i)P(O)(OC_2H_5)_2$] (69)	168
	$(S)\text{-}C_6H_5CH(CH_3)N{=}CHC_6H_5$	$(C_2H_5O)_2POH$	Room temp	$C_6H_5CH(NH_2)PO_3H_2$ (65)[a], $R{:}S = 6{:}1$	64
	$(R)\text{-}C_6H_5CH(CH_3)N{=}CHC_6H_5$	$(TMSO)_3P$	Ether, $TsOH_2$, 20°, 10 h	$(TMSO)_2P(O)CH(C_6H_5)N(TMS)CH(CH_3)C_6H_5$ (80), 85:15 diastereomeric mixture	183
			Ether, $ZnCl_2$, 20°, 10 h	" (89), 66:34 diastereomeric mixture	183
			Ether, ZnI_2, 20°, 10 h	" (87), 66:34 diastereomeric mixture	183
			Ether, 20°, 10 h	" (88), 90:10 diastereomeric mixture	183

TABLE II. Trivalent Phosphorus Addition Reactions at Imine Carbon (*Continued*)

No. of Carbon Atoms	Substrate	Phophorus Reagent	Conditions	Product(s) and Yield(s)(%)	Ref
C_{16}	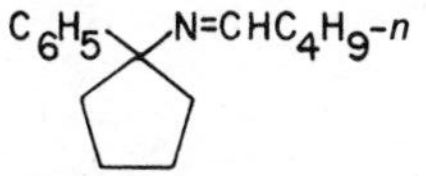 C_6H_5, $N{=}CHC_4H_9$-*n* (cyclopentane)	$(C_2H_5O)_2POH$	140°, 30 min	$H_2O_3PCH(NH_2)C_4H_9$-*n* (60)[a]	166
	$C_6H_5CH{=}NCH(C_2H_5)C_6H_5$	$(C_2H_5O)_2POH$	140°, 30 min	$H_2O_3PCH(NH_2)C_6H_5$ (61)[a]	166
	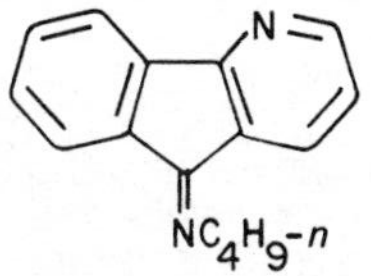 N, NC_4H_9-*n*	$(C_2H_5O)_2POH$	100°, 9 h	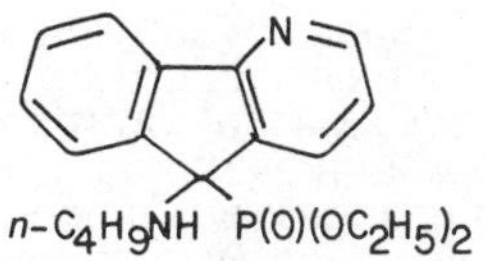 N, *n*-C_4H_9NH $P(O)(OC_2H_5)_2$ (10)	161
	3-$HOC_6H_4CH{=}NCH(C_6H_5)C_2H_5$	$(C_2H_5O)_2POH$	140°, 30 min	3-$HOC_6H_4CH(NH_2)PO_3H_2$ (69)[a]	166
	4-$O_2NC_6H_4CH{=}NCH(C_6H_5)C_2H_5$	$(C_2H_5O)_2POH$	140°, 30 min	4-$O_2NC_6H_4CH(NH_2)PO_3H_2$ (54)[a]	166
	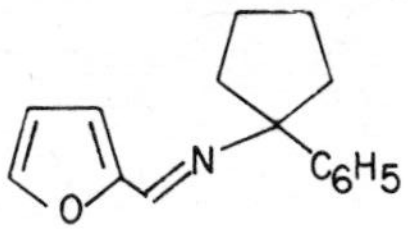 O, N, C_6H_5	$(C_2H_5O)_2POH$	140°, 30 min	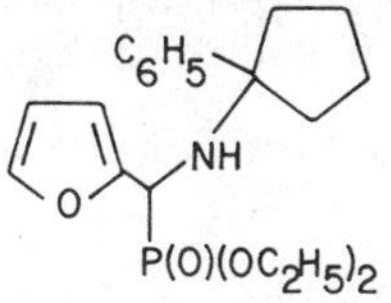 C_6H_5, NH, O, $P(O)(OC_2H_5)_2$ (79)	168
C_{17}	4-$CH_3C_6H_4CH{=}NCH(C_6H_5)C_2H_5$	$(C_2H_5O)_2POH$	140°, 30 min	4-$CH_3C_6H_4CH(NH_2)PO_3H_2$ (46)[a]	166
	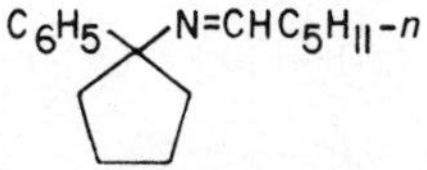 C_6H_5, $N{=}CHC_5H_{11}$-*n* (cyclopentane)	$(C_2H_5O)_2POH$	140°, 30 min	*n*-$C_5H_{11}CH(NH_2)PO_3H_2$ (62)[a]	166
	3-$O_2NC_6H_4CH{=}NC(CH_3)(C_6H_5)C_2H_5$	$(C_2H_5O)_2POH$	140°, 30 min	3-$O_2NC_6H_4CH(NH_2)PO_3H_2$ (50)[a]	166
	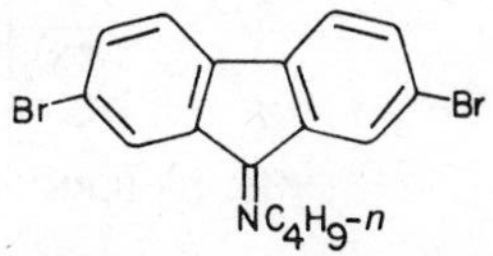 Br, Br, NC_4H_9-*n*	$(C_2H_5O)_2POH$	100°, 8 h	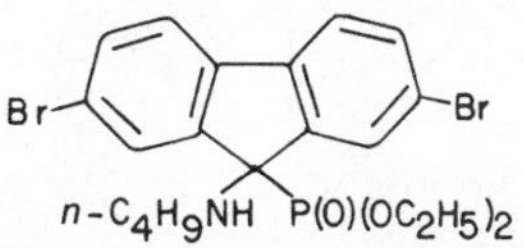 Br, Br, *n*-C_4H_9NH $P(O)(OC_2H_5)_2$ (30)	154

	2-O_2N-9-fluorenylidene=NC_4H_9-*n*	$(C_2H_5O)_2POH$	100°, 8 h	2-O_2N-9-(*n*-C_4H_9NH)-9-[$P(O)(OC_2H_5)_2$]fluorene (65)	154
C_{18}	1-C_6H_5-1-($N=CHC_6H_5$)cyclopentane	$(C_2H_5O)_2POH$	140°, 30 min	1-C_6H_5-1-[$NHCH(C_6H_5)P(O)(OC_2H_5)_2$]cyclopentane (83)	168
	3-$O_2NC_6H_4CH=N$–(1-C_6H_5-cyclopentyl)	$(C_2H_5O)_2POH$	140°, 30 min	3-$O_2NC_6H_4CH[P(O)(OC_2H_5)_2]NH$–(1-$C_6H_5$-cyclopentyl) (74)	168
	4-$BrC_6H_4CH=N$–(1-C_6H_5-cyclopentyl)	$(C_2H_5O)_2POH$	140°, 30 min	4-$BrC_6H_4CH[P(O)(OC_2H_5)_2]NH$–(1-$C_6H_5$-cyclopentyl) (63)	166
C_{19}	2-Br-9-fluorenylidene=NC_6H_5	$(C_2H_5O)_2POH$	100°, 8 h	2-Br-9-(C_6H_5NH)-9-[$P(O)(OC_2H_5)_2$]fluorene (56)	154
	2-NO_2-9-fluorenylidene=NC_6H_5	$(C_2H_5O)_2POH$	100°, 8 h	2-NO_2-9-(C_6H_5NH)-9-[$P(O)(OC_2H_5)_2$]fluorene (76)	154
C_{20}	2-$C_{10}H_7CH{=}NCH(C_6H_5)C_2H_5$	$(C_2H_5O)_2POH$	140°, 30 min	2-$C_{10}H_7CH(NH_2)PO_3H_2$ (66)[a]	166
C_{21}	$(C_6H_5CH{=}N)_2CHC_6H_5$	$(C_2H_5O)_2POH$	$C_2H_5)_3N$, H_2O, 100°, 3 h	$(C_2H_5O)_2P(O)CH(NH_2)C_6H_5$ (96)[a]	171
		$C_6H_5P(OH)OC_2H_5$	$(C_2H_5)_3N$, 55–100°	$C_6H_5P(O)(OC_2H_5)CH(NH_2)C_6H_5$ (85)[a]	170

TABLE III. HYDROPHOSPHINYLATION REACTIONS

No. of Carbon Atoms	Substrate	Phosphorus Reagent	Conditions	Product(s) and Yield(s)(%)	Ref.
C_2	$CH_2{=}CHNO_2$	$(CH_3O)_3P$	C_6H_6, $(CH_3)_3COH$, 9 h, room temp	$(CH_3O)_2P(O)CH_2CH_2NO_2$ (88)	71
		$(C_6H_5O)_3P$	C_6H_6, $(CH_3)_3COH$, 9 h, room temp	$(C_6H_5O)_2P(O)CH_2CH_2NO_2$ (65)	71
C_3	$CH_2{=}CHCN$	$CH_3OP(O)H_2$	CH_3OH, $NaOCH_3$, HCO_2CH_3, 4 h, reflux	$CH_3OP(O)(CH_2CH_2CN)_2$ (83)	220
		$CH_3P(OH)OC_2H_5$	$CH_3CON(TMS)_2$, CH_2Cl_2, 18 h, room temp	$CH_3P(O)(OC_2H_5)CH_2CH(TMS)CN$ (63)	221
		$(CH_3O)_3P$	TMSCl, 80°, 2 h	$(CH_3O)_2P(O)CH_2CH(TMS)CN$ (30)	132
		$(C_2H_5O)_2POH$	C_6H_6, Na, 60°	$(C_2H_5O)_2P(O)CH_2CH_2CN$ (78)	128
		$(CH_3O)_2POTMS$	2 h, 120°	$(CH_3O)_2P(O)CH_2CH(TMS)CN$ (46)	132
		$(C_2H_5O)_3P$	C_2H_5OH, 2 h, 0°; 19 h, room temp	$(C_2H_5O)_2P(O)CH_2CH_2CN$ (44)	33
			TMSCl, 80°, 2 h	$(C_2H_5O)_2P(O)CH_2CH(TMS)CN$ (52)	132
		$(C_2H_5O)_2POTMS$	0°	" (70)	126
			1 h, 20°	" (70)	50
			120°, 2 h	" (62)	132
		$(n\text{-}C_3H_7O)_3P$	TMSCl, 80°, 2 h	$(n\text{-}C_3H_7O)_2P(O)CH_2CH(TMS)CN$ (63)	132
		$(i\text{-}C_3H_7O)_3P$	TMSCl, 80°, 2 h	$(i\text{-}C_3H_7O)_2P(O)CH_2CH(TMS)CN$ (43)	132
		$NH_4OP(OH)OCH_2CH(C_2H_5)C_4H_9\text{-}n$	Na, 2-ethylhexanol, 70°, 8 h	$n\text{-}C_4H_9CH(C_2H_5)CH_2OP(O)(OH)CH_2CH_2CN$ (80)	205
		$(i\text{-}C_3H_7O)_2POTMS$	120°, 2 h	$(i\text{-}C_3H_7O)_2P(O)CH_2CH(TMS)CN$ (55)	132
		$(TMSO)_3P$	120°, 2 h	$(TMSO)_2P(O)CH_2CH(TMS)CN$ (52)	132
		$C_6H_5(CH_2)_4P(OH)_2$	TMSCl, $CHCl_3$, 18 h, room temp	$C_6H_5(CH_2)_4P(O)(OTMS)CH_2CH(TMS)CN$ (90)	63
			$CH_3CON(TMS)_2$, CH_2Cl_2, room temp, 18 h	" (74)	63
		$(n\text{-}C_4H_9O)_2POTMS$	120°, 2 h	$(n\text{-}C_4H_9O)_2P(O)CH_2CH(TMS)CN$ (72)	132
		$C_6H_5(CH_2)_4P(OH)OC_2H_5$	TMSCl, $CHCl_3$, 18 h, room temp	$C_6H_5(CH_2)_4P(O)(OC_2H_5)CH_2CH(TMS)CN$ (97)	63

		$(n\text{-}C_4H_9O)_3P$	TMSCl, 80°, 2 h	$(n\text{-}C_4H_9O)_2P(O)CH_2CH(TMS)CN$ (67)	132
	$CH_2{=}CHCHO$	$(CH_3O)_3P$	$(C_2H_5O)_2P(O)Cl$, 3 h, reflux	$(CH_3O)_2P(O)CH_2CH{=}CHOP(O)(OC_2H_5)_2$ (73)	73
		$(CH_3O)_2POTMS$	25°, 12 h	$(CH_3O)_2P(O)CH_2CH{=}CHOTMS$ (31)	51
		$(C_2H_5O)_3P$	$(CH_3O)_2P(O)Cl$, 3 h, reflux	$(C_2H_5O)_2P(O)CH_2CH{=}CHOP(O)(OCH_3)_2$ (72)	73
			$(C_2H_5O)_2P(O)Cl$, 3 h, reflux	$(C_2H_5O)_2P(O)CH_2CH{=}CHOP(O)(OC_2H_5)_2$ (61)	73
		$(C_6H_5)_2POCH_3$	TESCl, 100°, 30 min	$(C_6H_5)_2P(O)CH_2CH{=}CHOTES$ (100)	51
	$CH_2{=}CHCO_2H$	$(C_2H_5O)_3P$	100°, 30 min	$(C_2H_5O)_2P(O)CH_2CH_2CO_2H$ (82)[a]	138
			TMSCl, reflux, 1 h	$(C_2H_5O)_2P(O)CH_2CH_2CO_2TMS$ (96)	68
		$NH_4OP(OH)OCH_2CH(C_2H_5)C_4H_9\text{-}n$	Na, 2-ethylhexanol, 70°, 8 h	$n\text{-}C_4H_9CH(C_2H_5)CH_2OP(O)(OH)CH_2CH_2CO_2H$ (58)	205
	$CH_2{=}CHCONH_2$	$CH_3OP(O)H_2$	CH_3OH, $NaOCH_3$, HCO_2CH_3, reflux, 4 h	$CH_3OP(O)(CH_2CH_2CONH_2)_2$ (52)	220
		$(C_2H_5O)_2POH$	$NaOC_2H_5$, 110°, 1 h	$(C_2H_5O)_2P(O)CH_2CH_2CONH_2$ (75)	134
		$(C_2H_5O)_3P$	C_2H_5OH, 0°, 2 h; room temp, 19 h	" (42)	33
	$Cl_3CCH{=}CHNO_2$	$CH_3P(OCH_3)_2$	Ether, 25°, 18 h	$CH_3P(O)(OCH_3)C(CH_2NO_2){=}CCl_2$ (50)	107
		$(CH_3O)_3P$	Ether, 25°, 18 h	$(CH_3O)_2P(O)C(CH_2NO_2){=}CCl_2$ (51)	107
		$(CH_3O)_2POTMS$	Ether, 25°, 18 h	$(CH_3O)_2P(O)CH(CCl_3)CH_2NO_2$ (86)	107
		$C_6H_5P(OCH_3)OTMS$	Ether, 25°, 18 h	$C_6H_5P(O)(OCH_3)CH(CCl_3)CH_2NO_2$ (50)	107
C_4	$CH_3CH{=}CHCHO$	$(CH_3O)_2POTMS$	55°, 18 h	$(CH_3O)_2P(O)CH(CH_3)CH{=}CHOTMS$ (23)	51
		$(C_2H_5O)_3P$	C_2H_5OH, 0°, 2 h; room temp, 19 h	$(C_2H_5O)_2P(O)CH(CH_3)CH_2CHO$ (59)	33
			C_6H_5OH, 100°, 24 h	" (82)	33
			$(C_2H_5O)_2P(O)Cl$, reflux, 3 h	$(C_2H_5O)_2P(O)CH(CH_3)CH{=}CHOP(O)(OC_2H_5)_2$ (52)	73
	$CH_2{=}C(CH_3)CHO$	$(C_2H_5O)_3P$	$(CH_3O)_2P(O)Cl$, reflux, 3 h	$(C_2H_5O)_2P(O)CH_2C(CH_3){=}CHOP(O)(OCH_3)_2$ (54)	73
			$(C_2H_5O)_2P(O)Cl$, reflux, 3 h	$(C_2H_5O)_2P(O)CH_2C(CH_3){=}CHOP(O)(OC_2H_5)_2$ (68)	73

TABLE III. HYDROPHOSPHINYLATION REACTIONS (*Continued*)

No. of Carbon Atoms	Substrate	Phosphorus Reagent	Conditions	Product(s) and Yield(s)(%)	Ref.
C_4 (*Contd.*)	$CH_2{=}CHCOCH_3$	$(CH_3O)_3P$	TMSCl, 100°, 2 h	$(CH_3O)_2P(O)CH_2CH{=}C(OTMS)CH_3$ (79)	51
			TESCl, 100°, 2 h	$(CH_3O)_2P(O)CH_2CH{=}C(OTES)CH_3$ (76)	51
		$(CH_3O)_2POTMS$	50°, 6 h	" (88)	51
		$(C_2H_5O)_3P$	C_2H_5OH, 0°, 2 h; room temp, 19 h	$(C_2H_5O)_2P(O)CH_2CH_2COCH_3$ (73)	33
			C_6H_5OH, 100°, 24 h	" (76)	33
		$(C_2H_5O)_2POTMS$	C_6H_6 room temp, 22 h	$(C_2H_5O)_2P(O)CH_2CH{=}C(OTMS)CH_3$ (84)	111
			C_6H_6, reflux, 6 h	" (91)	111
		$(CH_3O)_2POTES$	100°, 3 h	$(CH_3O)_2P(O)CH_2CH{=}C(OTES)CH_3$ (36)	51
		$NH_4OP(OH)OCH_2CH(C_2H_5)C_4H_9$-*n*	Na, 2-ethylhexanol, 70°, 7 h	*n*-$C_4H_9CH(C_2H_5)CH_2OP(O)(OH)CH_2CH_2COCH_3$ (66)	205
		$(TMSO)_3P$	C_6H_6, room temp, 4 h	$(TMSO)_2P(O)CH_2CH{=}C(OTMS)CH_3$ (84)	49
		$[(CH_3)_2N]_2POTES$	THF, 0°, 30 min	$[(CH_3)_2N]_2P(O)CH_2CH{=}C(OTES)CH_3$ (82)	51
		$(C_6H_5)_2POCH_3$	TESCl, 100°, 30 min	$(C_6H_5)_2P(O)CH_2CH{=}C(OTES)CH_3$ (100)	51
	$CH_2{=}CHCO_2CH_3$	$(C_2H_5O)_3P$	TMSCl, reflux, 1 h	$(C_2H_5O)_2P(O)CH_2CH{=}C(OTMS)OCH_3$ (58)	68
		$(C_2H_5O)_2POTMS$	0°	$(C_2H_5O)_2P(O)CH_2CH(TMS)CO_2CH_3$ (75)	126
			20°, 1 h	$(C_2H_5O)_2P(O)CH_2CH_2CO_2CH_3$ (75)[a]	50
			120°, 2 h	$(C_2H_5O)_2P(O)CH_2CH{=}C(OTMS)OCH_3$ (87)	130
		$(i\text{-}C_3H_7O)_3P$	TMSCl, reflux, 1 h	$(i\text{-}C_3H_7O)_2P(O)CH_2CH{=}C(OTMS)OCH_3$ (62)	68
		$(i\text{-}C_3H_7O)_2POTMS$	120°, 2 h	$(i\text{-}C_3H_7O)_2P(O)CH_2CH{=}C(OTMS)OCH_3$ (73)	130
		$C_6H_5(CH_2)_4P(OH)_2$	TMSC1, $CHCl_3$, 18 h, room temp	$C_6H_5(CH_2)_4P(O)(OTMS)CH_2CH{=}C(OTMS)OCH_3$ (88)	63
		$C_6H_5(CH_2)_4P(OH)OC_2H_5$	TMSCl, $CHCl_3$, 18 h, room temp	$C_6H_5(CH_2)_4P(O)(OC_2H_5)CH_2CH{=}C(OTMS)OCH_3$ (87)	63
			$CH_3CON(TMS)_2$, CH_2Cl_2, room temp, 18 h	" (69)	63
	$CH_3CH{=}CHCONH_2$	$(C_2H_5O)_2POH$	$NaOC_2H_5$, 110°, 1 h	$(C_2H_5O)_2P(O)CH(CH_3)CH_2CONH_2$ (80)	142
	$CH_2{=}C(CH_3)CONH_2$	$(C_2H_5O)_2POH$	$NaOC_2H_5$, 110°, 1 h	$(C_2H_5O)_2P(O)CH_2CH(CH_3)CONH_2$ (80)	142

	Reactant	Phosphorus reagent	Conditions	Product (Yield, %)	Refs.
	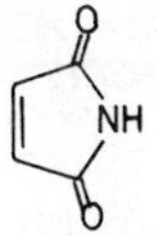 maleimide (structure)	$(C_2H_5O)_2POTMS$		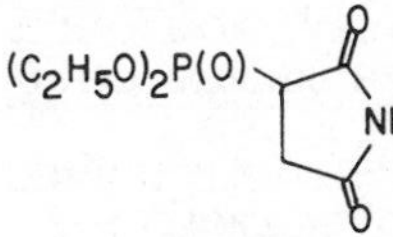 3-[$(C_2H_5O)_2P(O)$]succinimide (structure) (48)	127
	$NCCH{=}CHCN$	$(C_2H_5O)_3P$	C_2H_5OH, 0°, 2 h; room temp, 19 h	$(C_2H_5O)_2P(O)CH(CN)CH_2CN$ (55)	33
C_5	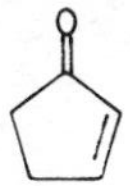 2-cyclopentenone (structure)	$(C_2H_5O)_3P$	C_6H_5OH, 100°, 24 h	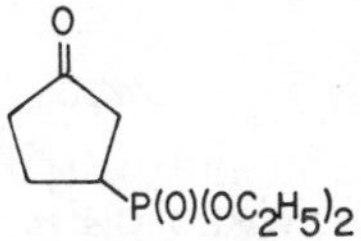 3-[$P(O)(OC_2H_5)_2$]cyclopentanone (structure) (72)	33
		$(C_2H_5O)_2POTMS$	180°, 12 h	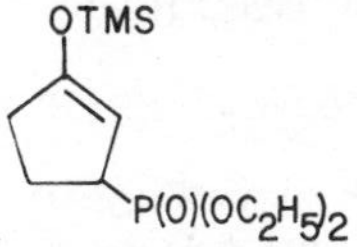 1-OTMS-3-[$P(O)(OC_2H_5)_2$]cyclopentene (structure) (100)	62
		$C_6H_5(CH_2)_4P(OH)_2$	TMSCl, $CHCl_3$, 18 h, room temp	 $C_6H_5(CH_2)_4P(O)(OTMS)$-(3-OTMS-cyclopent-2-enyl) (structure) (77)	63
		$C_6H_5(CH_2)_4P(OH)OC_2H_5$	TMSCl, $CHCl_3$, 18 h, room temp	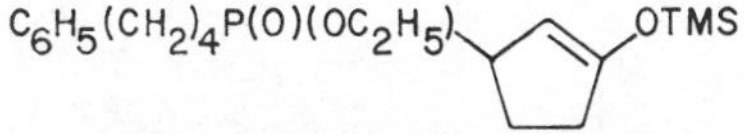 $C_6H_5(CH_2)_4P(O)(OC_2H_5)$-(3-OTMS-cyclopent-2-enyl) (structure) (89)	63
	$CH_3CH{=}CHCOCH_3$	$(CH_3O)_2POTMS$	80°, 24 h	$(CH_3O)_2P(O)CH(CH_3)CH_2CH{=}C(OTMS)CH_3$ (64)	51
		$(C_6H_5)_2POH$	THF, NaH, 0°, 1 h	$(C_6H_5)_2P(O)CH(CH_3)CH_2COCH_3$ (58)	84
	$CH_2{=}CHCOC_2H_5$	$(C_6H_5)_2POH$	THF, NaH, 0°, 1 h	$(C_6H_5)_2P(O)CH_2CH_2COC_2H_5$ (55)	84
	$CH_2{=}CHCO_2C_2H_5$	$(C_2H_5O)_2POH$	C_6H_6, Na, 60°	$(C_2H_5O)_2P(O)CH_2CH_2CO_2C_2H_5$ (84)	84
		$(C_2H_5O)_3P$	C_2H_5OH, 0°, 2 h; room temp, 19 h	" (50)	33
			C_6H_5OH, 100°, 24 h	" (90)	33
			TMSCl, reflux, 1 h	$(C_2H_5O)_2P(O)CH_2CH{=}C(OTMS)OC_2H_5$ (86)	68
		$(C_2H_5O)_2POTMS$	20°, 1 h	$(C_2H_5O)_2P(O)CH_2CH_2CO_2C_2H_5$ (65)[a]	50
			100°, 1 h	$(C_2H_5O)_2P(O)CH_2CH{=}C(OTMS)OC_2H_5$ (65)	131
			120°, 2 h	" (86)	130

TABLE III. Hydrophosphinylation Reactions (*Continued*)

No. of Carbon Atoms	Substrate	Phosphorus Reagent	Conditions	Product(s) and Yield(s)(%)	Ref.
C_5 (*Contd.*)		$C_6H_5P(OH)CH_2C_6H_5$	THF, NaH, 65°	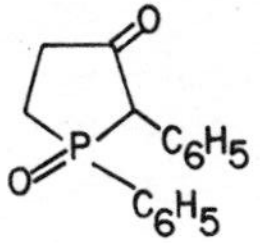 C_6H_5, C_6H_5 (38)	126
		$(C_6H_5CH_2)_2POH$	THF, NaH, 65°	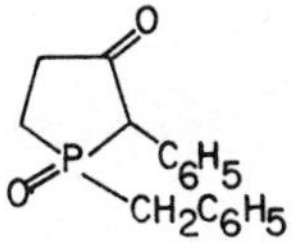 C_6H_5, $CH_2C_6H_5$ (41)	126
	$CH_3CH{=}CHCO_2CH_3$	$(C_2H_5O)_2POTMS$	120°, 2 h	$(C_2H_5O)_2P(O)CH(CH_3)CH{=}C(OTMS)OCH_3$ (56)	130
	$CH_2{=}C(CH_3)CO_2CH_3$	$(C_2H_5O)_2POTMS$	20°, 1 h	$(C_2H_5O)_2P(O)CH_2CH(CH_3)CO_2CH_3$ (51)[a]	50
		$C_6H_5(CH_2)_4P(OH)OC_2H_5$	TMSCl, $CHCl_3$, 18 h, room temp	$C_6H_5(CH_2)_4P(O)(OC_2H_5)CH_2C(CH_3){=}C(OTMS)OCH_3$ (93)	63
		$C_6H_5(CH_2)_4P(OH)_2$	TMSCl, $CHCl_3$, 18 h, room temp	$C_6H_5(CH_2)_4P(O)(OTMS)CH_2C(CH_3){=}C(OTMS)OCH_3$ (88)	63
	$CH_2{=}C(CO_2H)NHCOCH_3$	$(C_2H_5O)_3P$	130°, 1 h	$H_2O_3PCH_2CH(NH_2)CO_2H$ (75)[a]	141
		$CH_3P(OC_2H_5)_2$	70°	$CH_3P(O)(OH)CH_2CH(NH_2)CO_2H$ (65)[a]	141
		$C_2H_5P(OC_2H_5)_2$	70°	$C_2H_5P(O)(OH)CH_2CH(NH_2)CO_2H$ (60)[a]	141
C_6	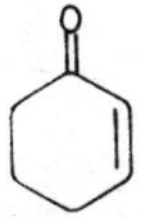	$(C_2H_5O)_2POTMS$	180°, 12 h	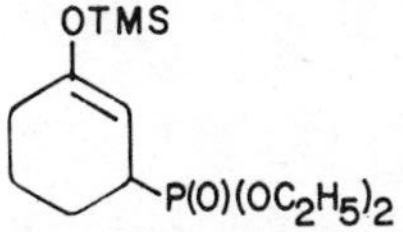 OTMS, $P(O)(OC_2H_5)_2$ (93)	62
	$(CH_3)_2C{=}CHCOCH_3$	$(C_2H_5O)_2POTMS$	140°, 4 h	$(C_2H_5O)_2P(O)C(CH_3)_2CH{=}C(OTMS)CH_3$ (50)	131
		$C_6H_5(CH_2)_4P(OH)_2$	TMSCl, $CHCl_3$, 18 h, room temp	$C_6H_5(CH_2)_4P(O)(OTMS)C(CH_3)_2CH{=}C(OTMS)CH_3$ (37)	63
	$C_2H_5CH{=}CHCO_2CH_3$	$(C_2H_5O)_2POTMS$	120°, 2 h	$(C_2H_5O)_2P(O)CH(C_2H_5)CH{=}C(OTMS)OCH_3$ (43)	130
	$CH_2{=}CHCO_2C_3H_7$-*n*	$(C_2H_5O)_3P$	TMSCl, reflux, 1 h	$(C_2H_5O)_2P(O)CH_2CH{=}C(OTMS)OC_3H_7$-*n* (79)	68

	Substrate	Reagent	Conditions	Product (Yield, %)	Refs.
	$CH_3CH{=}CHCO_2C_2H_5$	$(C_2H_5O)_2POTMS$	C_6H_6, reflux, 16 h	$(C_2H_5O)_2P(O)CH(CH_3)CH{=}C(OTMS)OC_2H_5$ (16)	111
		$(TMSO)_3P$	Dioxane, reflux, 7 h	$(TMSO)_2P(O)CH(CH_3)CH{=}C(OTMS)OC_2H_5$ (71)	49
		$(n\text{-}C_4H_9O)_2POTMS$	120°, 2 h	$(n\text{-}C_4H_9O)_2P(O)CH(CH_3)CH{=}C(OTMS)OC_2H_5$ (63)	130
	$CH_2{=}C(CH_3)CO_2C_2H_5$	$C_6H_5CH_2P(C_6H_5)OH$	THF, NaH, 65°	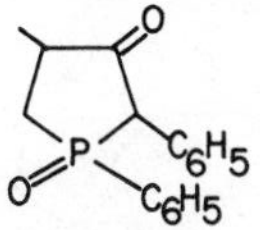 O, P, C_6H_5, C_6H_5 (45)	126
		$(C_6H_5CH_2)_2POH$		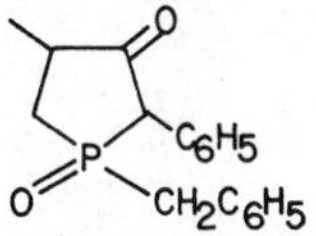 O, P, C_6H_5, $CH_2C_6H_5$ (65)	126
	$CH_2{=}CHCOCH_2O_2CCH_3$	$(C_2H_5O)_3P$	C_2H_5OH, room temp, 12 h	$(C_2H_5O)_2P(O)CH_2CH_2COCH_2O_2CCH_3$ (70)[a]	118
	$(CH_3OCOCH{=})_2$	$(C_2H_5O)_2PONa$	50°, 30 min	$(C_2H_5O)_2P(O)CH(CO_2CH_3)CH_2CO_2CH_3$ (76)	81
		$(C_2H_5O)_2POTMS$	120°, 2 h	$(C_2H_5O)_2P(O)CH(CO_2CH_3)CH{=}C(OTMS)OCH_3$ (52)	130
	$CH_3C(NH_2){=}CHCO_2C_2H_5$	$CH_3P(OH)OC_2H_5$	140°, 4 h	$CH_3P(O)(OH)C(CH_3)(NH_2)CH_2CO_2H$ (28)[a]	161
		$(C_2H_5O)_2POH$	140°, 4 h	$H_2O_3PC(CH_3)(NH_2)CH_2CO_2H$ (34)[a]	161
C_7	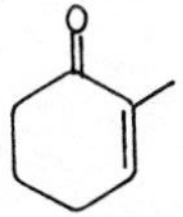 O	$(C_2H_5O)_2POTMS$	180°, 12 h	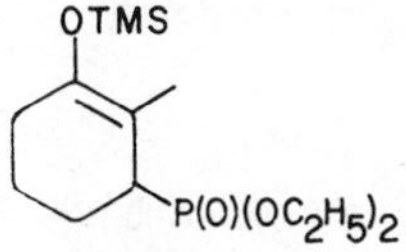 OTMS, $P(O)(OC_2H_5)_2$ (93)	62
	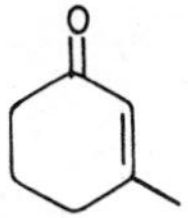 O	$(C_2H_5O)_2POTMS$	180°, 12 h	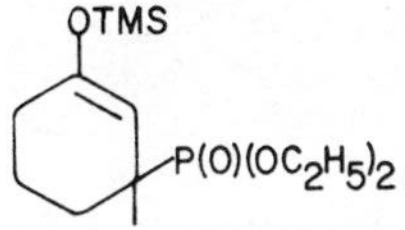 OTMS, $P(O)(OC_2H_5)_2$ (95)	62
	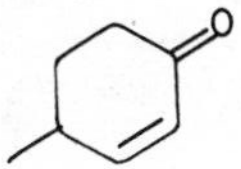 O	$(C_2H_5O)_2POTMS$	180°, 12 h	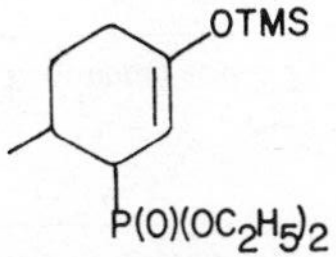 OTMS, $P(O)(OC_2H_5)_2$ (60)	62

TABLE III. HYDROPHOSPHINYLATION REACTIONS (*Continued*)

No. of Carbon Atoms	Substrate	Phosphorus Reagent	Conditions	Product(s) and Yield(s)(%)	Ref.
C_7 (*Contd.*)	$CH_2{=}CHCO_2C_4H_9$-*n*	$(C_2H_5O)_3P$	TMSCl, reflux, 1 h	$(C_2H_5O)_2P(O)CH_2CH{=}C(OTMS)OC_4H_9$-*n* (75)	68
		$(C_2H_5O)_2POTMS$	120°, 2 h	" (82)	130
	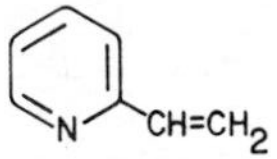 CH=CH$_2$	$NH_4OP(OH)OCH_2CH(C_2H_5)C_4H_9$-*n*	2-Ethylhexanol, Na, 100° 15 h	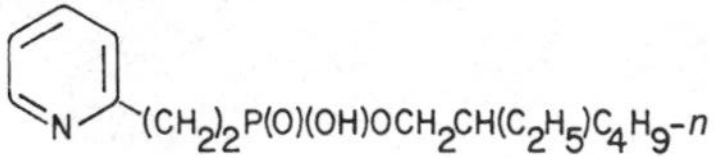 $(CH_2)_2P(O)(OH)OCH_2CH(C_2H_5)C_4H_9$-*n* (65)	205
	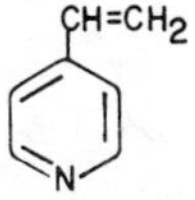 CH=CH$_2$	$NH_4OP(OH)OCH_2CH(C_2H_5)C_4H_9$-*n*	2-Ethylhexanol, Na, 100°, 15 h	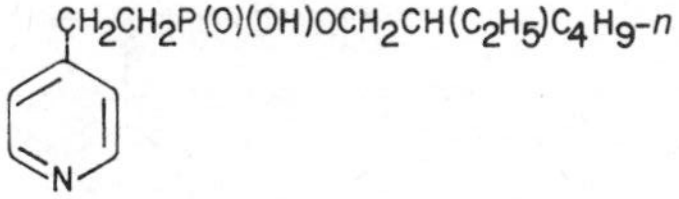 $CH_2CH_2P(O)(OH)OCH_2CH(C_2H_5)C_4H_9$-*n* (67)	205
C_8	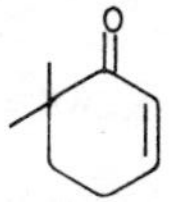	$(C_2H_5O)_2POTMS$	180°, 12 h	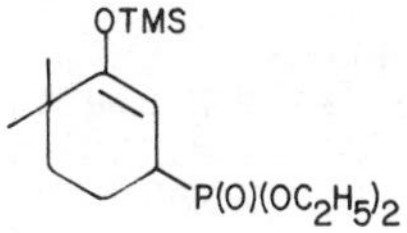 OTMS, $P(O)(OC_2H_5)_2$ (90)	62
	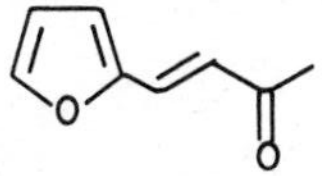	$(CH_3O)_2POH$	C_6H_6, CH_3OH, Na	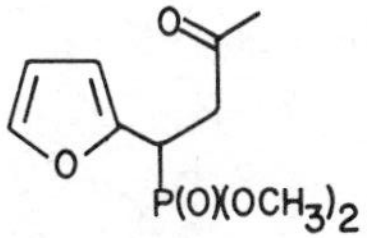 $P(O)(OCH_3)_2$ (63)	87
	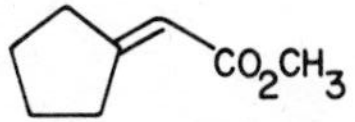 CO_2CH_3	$C_6H_5CH_2P(OH)C_6H_5$	THF, NaH, 65°	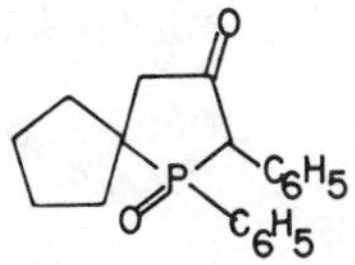 C_6H_5, C_6H_5 (58)	126
		$(C_6H_5CH_2)_2POH$	THF, NaH, 65°	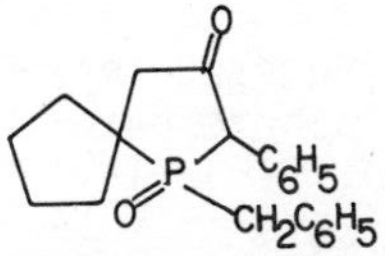 C_6H_5, $CH_2C_6H_5$ (70)	126

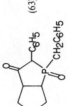

$CO_2C_2H_5$	$C_6H_5CH_2P(OH)C_6H_5$	THF, NaH, 65°	O, C_6H_5, P, O, C_6H_5 (43)	126
	$(C_6H_5CH_2)_2POH$	THF, NaH, 65°	O, C_6H_5, P, O, $CH_2C_6H_5$ (63)	126
CO_2CH_3	$C_6H_5CH_2P(OH)C_6H_5$	THF, NaH, 65°	O, C_6H_5, P, O, C_6H_5 (50)	126
	$(C_6H_5CH_2)_2POH$	THF, NaH, 65°	O, C_6H_5, P, O, $CH_2C_6H_5$ (64)	126
$(CH_3CO)_2C{=}CHOC_2H_5$	$(C_2H_5O)_2POH$	Ether, $NaOC_2H_5$, reflux, 5 h	$(CH_3CO)_2C{=}CHP(O)(OC_2H_5)_2$ (80)	222
	$(C_2H_5O)_2PONa$	Ether, reflux, 2 h	" (61)	222
	$(n\text{-}C_4H_9O)_2POH$	Ether, $NaOC_2H_5$, reflux, 5 h	$(CH_3CO)_2C{=}CHP(O)(OC_4H_9\text{-}n)$ (75)	222
	$(n\text{-}C_4H_9O)_2PONa$	Ether, reflux, 2 h	" (33)	222
O, CO_2CH_3, O	$(CH_3O)_3P$	TMSCl, CH_2Cl_2, 18 h, room temp	OH, CO_2CH_3, $P(O)(OCH_3)_2$, OH (48)	223

TABLE III. HYDROPHOSPHINYLATION REACTIONS (*Continued*)

No. of Carbon Atoms	Substrate	Phosphorus Reagent	Conditions	Product(s) and Yield(s)(%)	Ref.
C_8 (*Contd.*)	$(C_2H_5O_2CCH{=})_2$	$(C_2H_5O)_2PONa$	50°, 30 min	$(C_2H_5O)_2P(O)CH(CO_2C_2H_5)CH_2CO_2C_2H_5$ (82)	81
		$(C_2H_5O)_3P$	C_6H_5OH, 100°, 24 h	" (93)	33
	CH=CH$_2$; N	$NH_4OP(OH)OCH_2CH(C_2H_5)C_4H_9$-*n*	2-Ethylhexanol, Na, 100°, 15 h	$(CH_2)_2P(O)(OH)OCH_2CH(C_2H_5)C_4H_9$-*n*; N	205
	$C_6H_5CH{=}CHNO_2$	$(CH_3O)_3P$	CH_3CN, 65°, 20 h	$(CH_3O)_2P(O)CH(C_6H_5)CH_2NO_2$ (26)	75
		$(C_2H_5O)_3P$	CH_3CN, 50°, 30 min	$(C_2H_5O)_2P(O)CH(C_6H_5)CH_2NO_2$ (61)	75
		$(C_2H_5O)_2PC_6H_5$	CH_3CN, 60°, 6 h	$C_6H_5P(O)(OC_2H_5)CH(C_6H_5)CH_2NO_2$ (32)	75
	C_2H_5O; O; CN; C_2H_5O	$(C_2H_5O)_2POH$	Ether, $NaOC_2H_5$, reflux, 5 h	$(C_2H_5O)_2P(O)CH{=}C(CN)CO_2C_2H_5$ (79)	222
		$(C_2H_5O)_2PONa$	Ether, reflux, 2 h	" (54)	222
	$4\text{-}ClC_6H_4CH{=}CHNO_2$	$(CH_3O)_3P$	CH_3CN, 65°, 20 h	$(CH_3O)_2P(O)CH(C_6H_4Cl\text{-}4)CH_2NO_2$ (51)	75
C_9	$C_6H_5CH{=}CHCHO$	$(C_2H_5O)_3P$	C_2H_5OH, 0°, 2 h; 19 h, room temp	$(C_2H_5O)_2P(O)CH(C_6H_5)CH_2CHO$ (61)	33
		$(C_2H_5O)_3P$	$(C_2H_5O)_2P(O)Cl$ reflux, 3 h	$(C_2H_5O)_2P(O)CH(C_6H_5)CH{=}CHOP(O)(OC_2H_5)_2$ (26)	73
	O	$(C_2H_5O)_2POTMS$	180°, 12 h	OTMS; $P(O)(OC_2H_5)_2$ (82)	62
	O; O; OC_2H_5; OC_2H_5	$(C_2H_5O)_2PONa$	Ether, reflux, 2 h	$(C_2H_5O)_2P(O)CH{=}C(COCH_3)CO_2C_2H_5$ (36)	222
		$(C_2H_5O)_2POH$	Ether, $NaOC_2H_5$, reflux 5 h	" (76)	222

	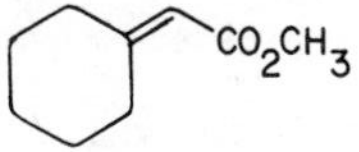	$C_6H_5CH_2P(OH)C_6H_5$	THF, NaH, 65°	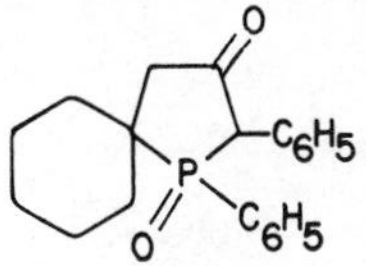 (70)	126
		$(C_6H_5CH_2)_2POH$	THF, Na, 65°	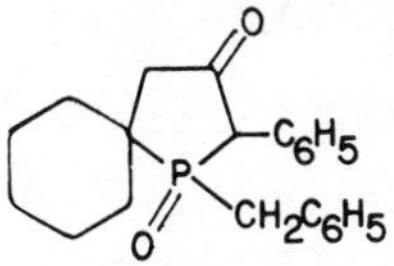 (61)	126
	$C_6H_5CH{=}CHCH{=}NOH$	$(C_2H_5O)_3P$	60°	$(C_2H_5O)_2P(O)CH(C_6H_5)CH_2CH{=}NOC_2H_5$ (44)	224
	$C_6H_5CH{=}CHCONH_2$	$(C_2H_5O)_2POH$	$NaOC_2H_5$, 110°, 1 h	$(C_2H_5O)_2P(O)CH(C_6H_5)CH_2CONH_2$ (81)	134
	4-$CH_3OC_6H_4CH{=}CHNO_2$	$(CH_3O)_3P$	CH_3CN, 65°, 20 h	$(CH_3O)_2P(O)CH(C_6H_4OCH_3\text{-}4)CH_2NO_2$ (27)	75
		$(C_2H_5O)_3P$	CH_3CN, 80°, 2 h	$(C_2H_5O)_2P(O)CH(C_6H_4OCH_3\text{-}4)CH_2NO_2$ (51)	75
	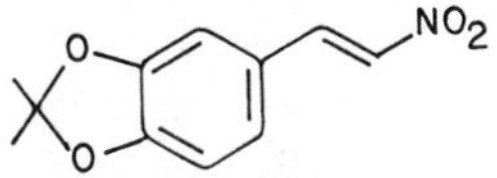	$(CH_3O)_3P$	CH_3CN, 65°, 20 h	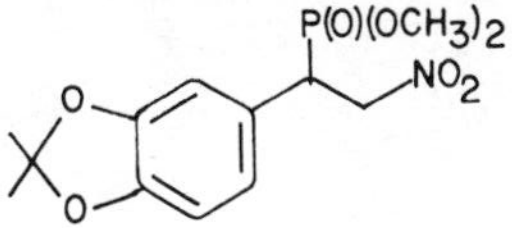 (70)	75
		$(C_2H_5O)_3P$	CH_3CN, 80°, 2 h	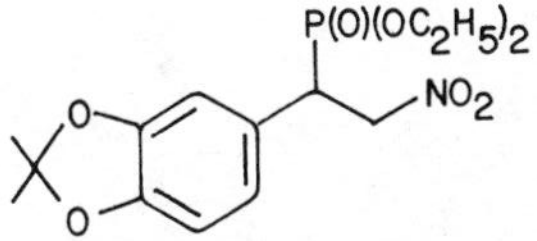 (70)	75
	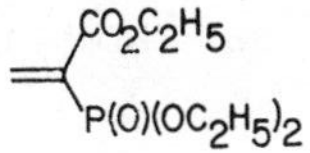	$(C_2H_5O)_3P$	C_2H_5OH, 110°, 2 h	$(C_2H_5O)_2P(O)CH_2CH[P(O)(OC_2H_5)_2]CO_2C_2H_5$ (62)	136
C_{10}	$C_6H_5CH{=}CHCOCH_3$	$(C_2H_5O)_3P$	C_6H_5OH, 100°, 24 h	$(C_2H_5O)_2P(O)CH(C_6H_5)CH_2COCH_3$ (89)	33
	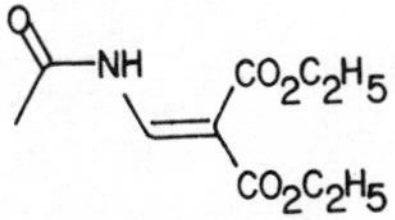	$C_2H_5OP(OH)CH_3$	C_2H_5OH, $NaOC_2H_5$, 100°, 1 h	$CH_3P(O)(OH)CH(NH_2)CH_2CO_2H$ (65)[a]	141

TABLE III. HYDROPHOSPHINYLATION REACTIONS (*Continued*)

No. of Carbon Atoms	Substrate	Phosphorus Reagent	Conditions	Product(s) and Yield(s)(%)	Ref.
C_{10} (*Contd.*)		$C_2H_5OP(OH)C_2H_5$	C_2H_5OH, $NaOC_2H_5$, 100°, 1 h	$C_2H_5P(O)(OH)CH(NH_2)CH_2CO_2H$ (70)[a]	141
		$(C_2H_5O)_2POH$	C_2H_5OH, $NaOC_2H_5$, 100°, 1 h	$H_2O_3PCH(NH_2)CH_2CO_2H$ (75)[a]	141
C_{11}	$C_6H_5CH{=}CHCO_2C_2H_5$	$(C_2H_5O)_3P$	C_6H_5OH, 100°, 24 h	$(C_2H_5O)_2P(O)CH(C_6H_5)CH_2CO_2C_2H_5$ (50)	33
		$C_6H_5CH_2P(OH)C_6H_5$	THF, Na, 65°	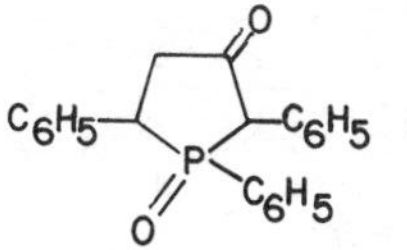 (56)	126
		$(C_6H_5CH_2)_2POH$	THF, Na, 65°	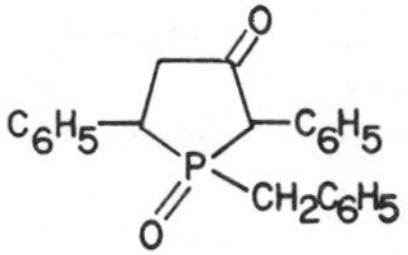 (72)	126
	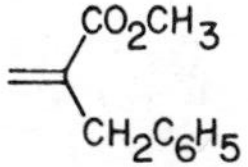	$CH_3P(OH)OC_2H_5$	$CH_3CON(TMS)_2$, CH_2Cl_2, room temp, 18 h	$CH_3P(O)(OC_2H_5)CH_2C(CH_2C_6H_5){=}C(OTMS)OCH_3$ (81)	63
	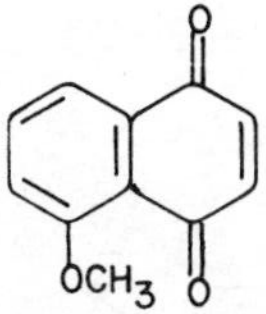	$(CH_3O)_3P$	TMSCl, CH_2Cl_2, 14 d, room temp	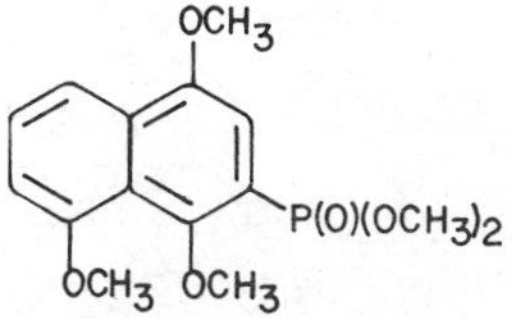 (59)	223
	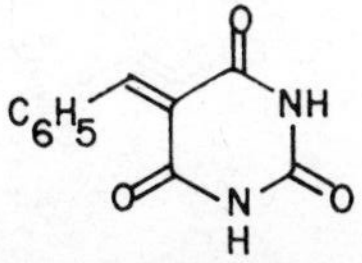	$(C_2H_5O)_2POTMS$	Ether	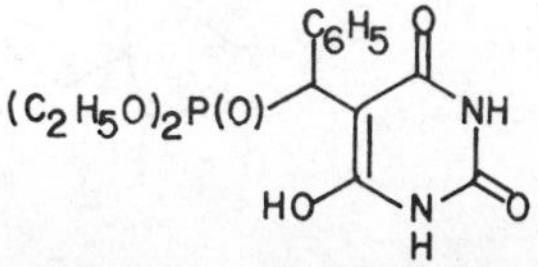 (60)	225

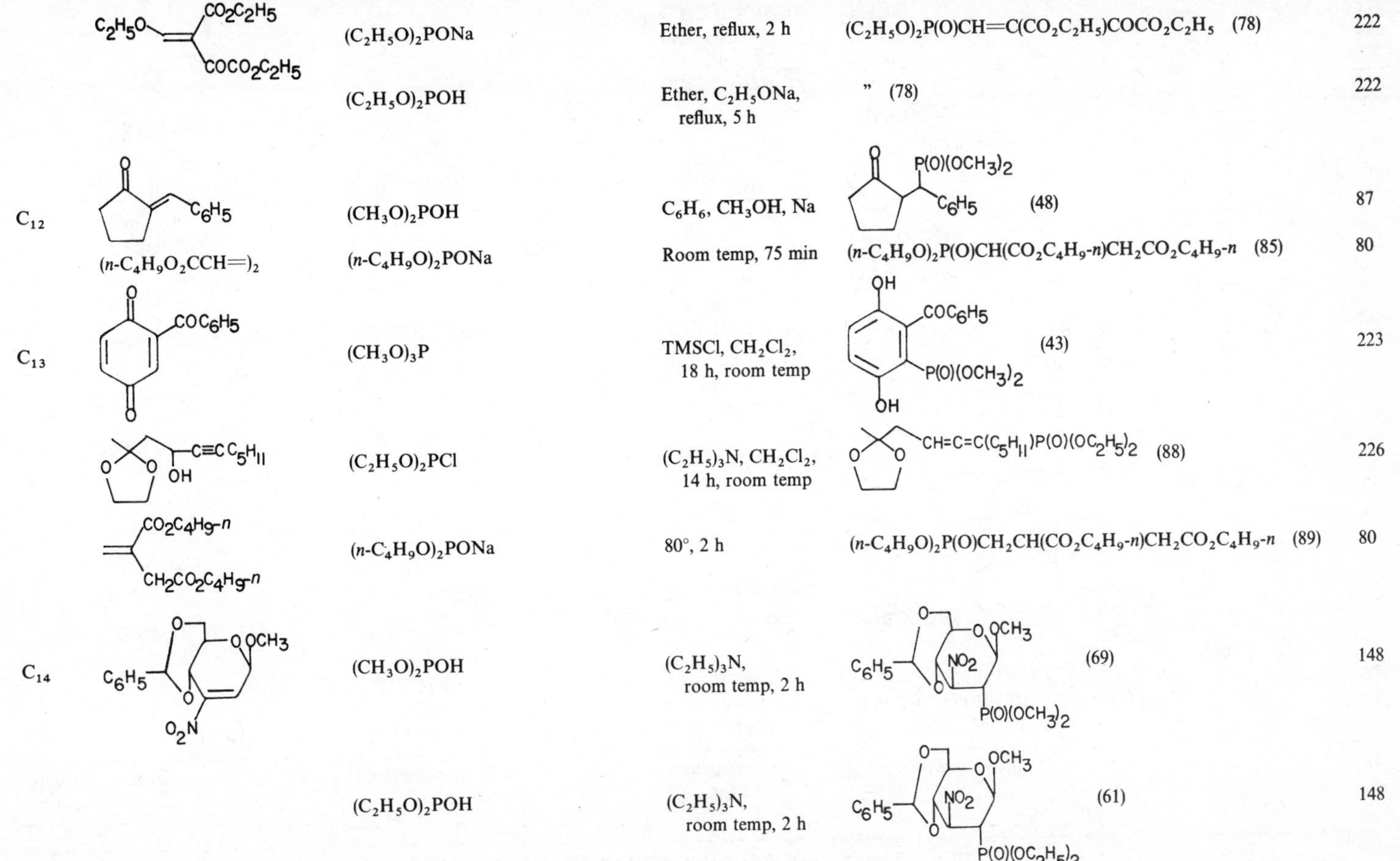

	Substrate	Reagent	Conditions	Product (Yield %)	Refs.
		$(C_2H_5O)_2PONa$	Ether, reflux, 2 h	$(C_2H_5O)_2P(O)CH{=}C(CO_2C_2H_5)COCO_2C_2H_5$ (78)	222
		$(C_2H_5O)_2POH$	Ether, C_2H_5ONa, reflux, 5 h	" (78)	222
C_{12}		$(CH_3O)_2POH$	C_6H_6, CH_3OH, Na	(48)	87
	$(n\text{-}C_4H_9O_2CCH{=})_2$	$(n\text{-}C_4H_9O)_2PONa$	Room temp, 75 min	$(n\text{-}C_4H_9O)_2P(O)CH(CO_2C_4H_9\text{-}n)CH_2CO_2C_4H_9\text{-}n$ (85)	80
C_{13}		$(CH_3O)_3P$	TMSCl, CH_2Cl_2, 18 h, room temp	(43)	223
		$(C_2H_5O)_2PCl$	$(C_2H_5)_3N$, CH_2Cl_2, 14 h, room temp	(88)	226
		$(n\text{-}C_4H_9O)_2PONa$	80°, 2 h	$(n\text{-}C_4H_9O)_2P(O)CH_2CH(CO_2C_4H_9\text{-}n)CH_2CO_2C_4H_9\text{-}n$ (89)	80
C_{14}		$(CH_3O)_2POH$	$(C_2H_5)_3N$, room temp, 2 h	(69)	148
		$(C_2H_5O)_2POH$	$(C_2H_5)_3N$, room temp, 2 h	(61)	148

TABLE III. HYDROPHOSPHINYLATION REACTIONS (*Continued*)

No. of Carbon Atoms	Substrate	Phosphorus Reagent	Conditions	Product(s) and Yield(s)(%)	Ref.
C_{15}	$C_6H_5COCH{=}CHC_6H_5$	$(CH_3O)_2POH$	C_6H_6, CH_3OH, Na	$(CH_3O)_2P(O)CH(C_6H_5)CH_2COC_6H_5$ (13)	87
		$(TMSO)_3P$	C_6H_6, reflux, 4 h	$(TMSO)_2P(O)CH(C_6H_5)CH{=}C(OTMS)C_6H_5$ (75)	49
		$(C_6H_5)_2POH$	Ether, room temp, 24 h	$(C_6H_5)_2P(O)CH(C_6H_5)CH_2COC_6H_5$ (62)	227
C_{16}	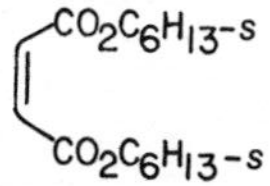 $CO_2C_6H_{13}$-s / $CO_2C_6H_{13}$-s	$(n\text{-}C_4H_9O)_2POH$	Room temp, 75 min	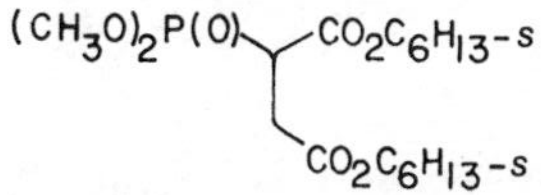 $(CH_3O)_2P(O)$, $CO_2C_6H_{13}$-s, $CO_2C_6H_{13}$-s (96)	80
C_{17}	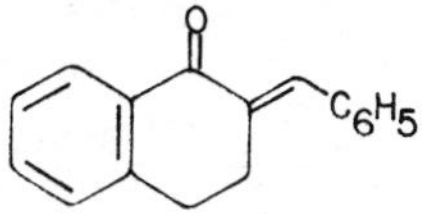 C_6H_5	$(CH_3O)_2POH$	C_6H_6, CH_3OH, Na	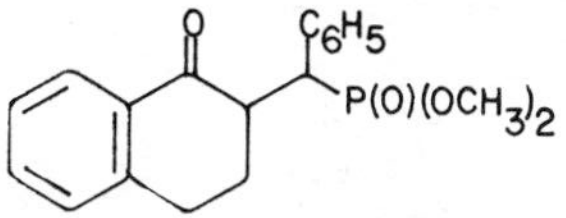 C_6H_5, $P(O)(OCH_3)_2$ (8)	87
	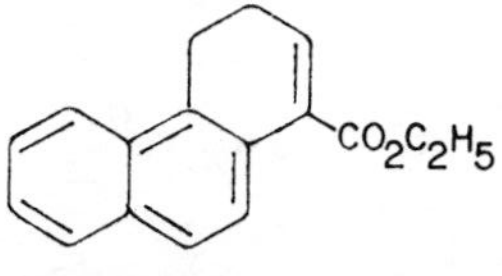 $CO_2C_2H_5$	$(C_6H_5CH_2)_2POH$	THF, Na, 65°	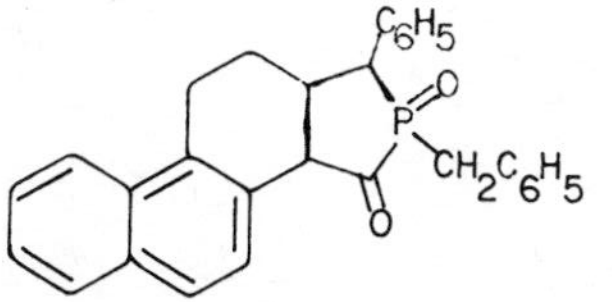 C_6H_5, $CH_2C_6H_5$ (20)	126
C_{18}	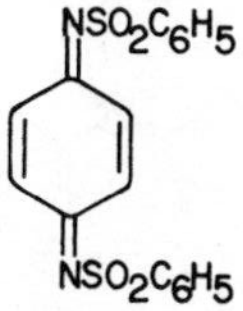 $NSO_2C_6H_5$ / $NSO_2C_6H_5$	$(C_2H_5O)_2POH$	25°, 48 h	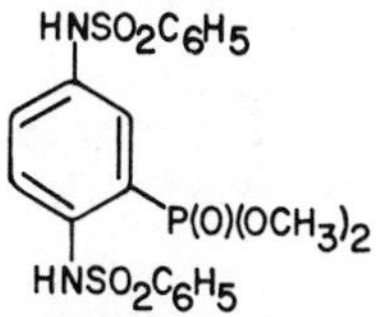 $HNSO_2C_6H_5$, $P(O)(OCH_3)_2$, $HNSO_2C_6H_5$ (40)	228
			C_6H_6, 25°, 72 h	" (45)	228
			110°, 2 h	" (20)	228
			140°, 5 h	" (15)	228
			C_6H_6, 10 h, reflux	" (30)	228

	Reactant	Reagent	Conditions	Product (Yield, %)	Refs.
		$(CH_3O)_2POH$	25°, 48 h	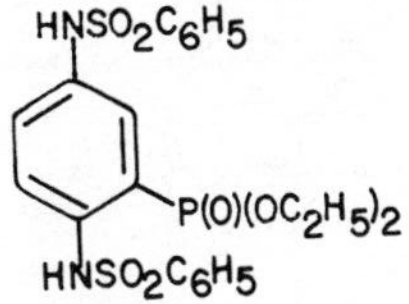 $HNSO_2C_6H_5$, $P(O)(OC_2H_5)_2$, $HNSO_2C_6H_5$ (45)	228
			C_6H_6, 25°, 72 h	" (45)	228
			110°, 2 h	" (25)	228
			140°, 2 h	" (20)	228
			C_6H_6, reflux, 10 h	" (45)	228
C_{19}	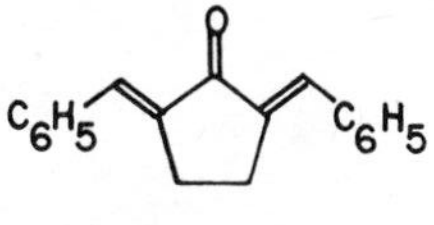 O, C_6H_5, C_6H_5	$(CH_3O)_3P$	Acetic acid, 74–120°	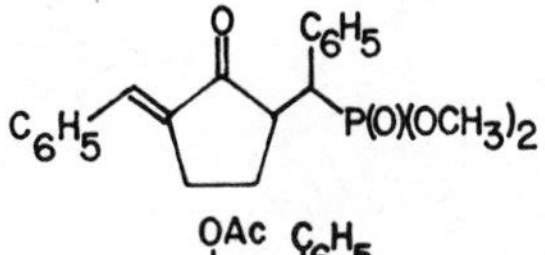 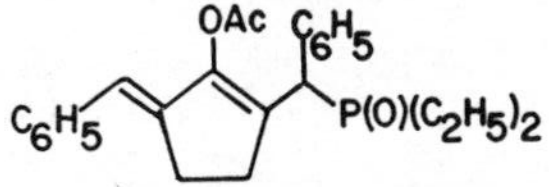 O, C_6H_5, C_6H_5, $P(O)(OCH_3)_2$ (42), OAc, C_6H_5, C_6H_5, $P(O)(C_2H_5)_2$ (9)	119
C_{20}	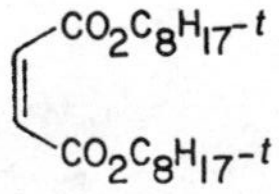 $CO_2C_8H_{17}$-*t*, $CO_2C_8H_{17}$-*t*	$(n\text{-}C_4H_9O)_2PONa$	70°, 2 h	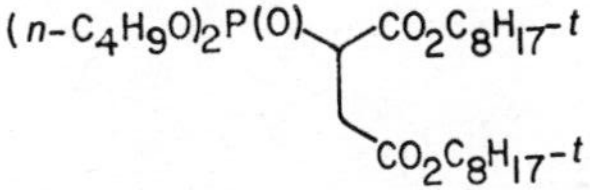 $(n\text{-}C_4H_9O)_2P(O)$, $CO_2C_8H_{17}$-*t*, $CO_2C_8H_{17}$-*t* (95)	148
C_{22}	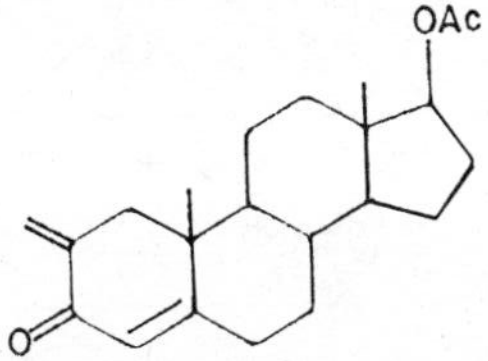 OAc	$(CH_3O)_3P$	C_6H_5OH	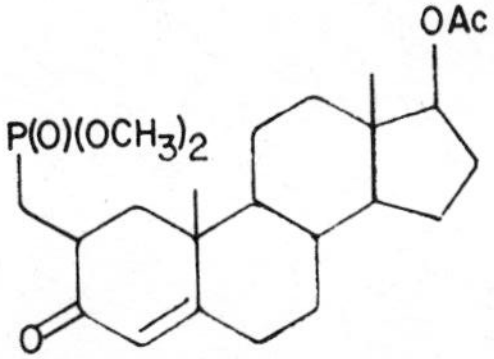 OAc, $P(O)(OCH_3)_2$ (63)	229

TABLE III. HYDROPHOSPHINYLATION REACTIONS (*Continued*)

No. of Carbon Atoms	Substrate	Phosphorus Reagent	Conditions	Product(s) and Yield(s)(%)	Ref.
C_{25}	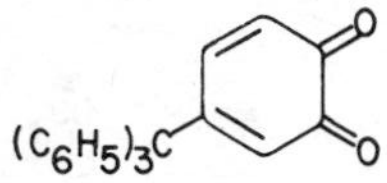 $(C_6H_5)_3C$, O, O	$(CH_3O)_2POH$	C_6H_6, reflux, 2 h	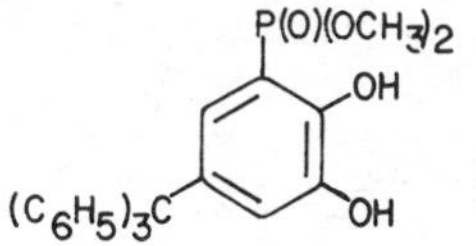 $P(O)(OCH_3)_2$, OH, OH, $(C_6H_5)_3C$ (90)	230
		$(C_2H_5O)_2POH$	C_6H_6, reflux, 12 h	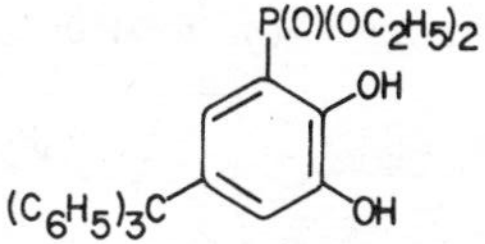 $P(O)(OC_2H_5)_2$, OH, OH, $(C_6H_5)_3C$ (92)	230
		$(i\text{-}C_3H_7O)_2POH$	C_6H_6, reflux, 12 h	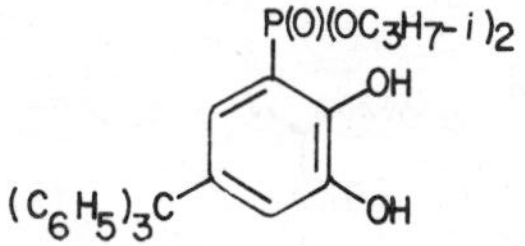 $P(O)(OC_3H_7\text{-}i)_2$, OH, OH, $(C_6H_5)_3C$ (85)	230
C_{28}	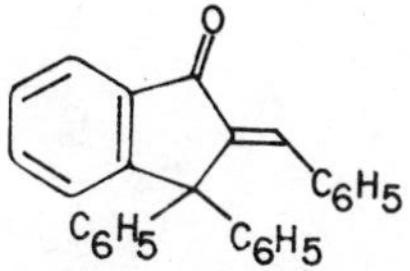 O, C_6H_5, C_6H_5, C_6H_5	$(CH_3O)_2POH$	C_6H_6, CH_3OH, Na	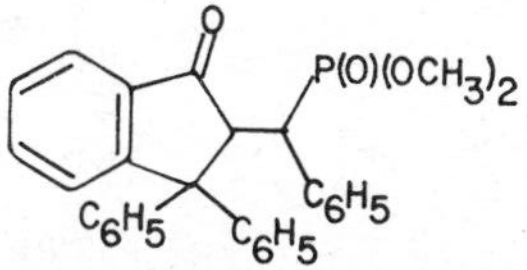 O, $P(O)(OCH_3)_2$, C_6H_5, C_6H_5, C_6H_5 (15)	87
C_{29}	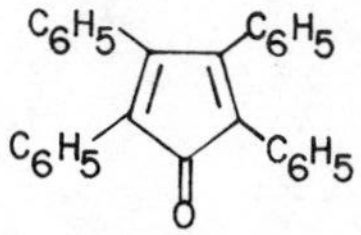 C_6H_5, C_6H_5, C_6H_5, C_6H_5, O	$(CH_3O)_2POH$	170°, 20 min	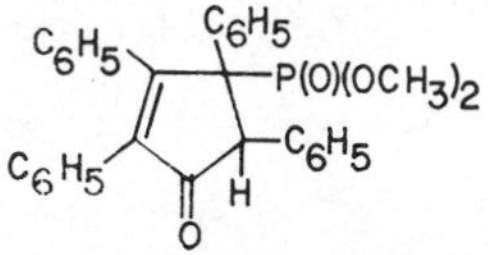 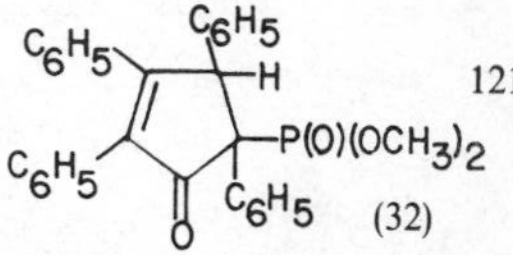 C_6H_5, C_6H_5, $P(O)(OCH_3)_2$, C_6H_5, C_6H_5, H, O (25), C_6H_5, C_6H_5, H, $P(O)(OCH_3)_2$, C_6H_5, C_6H_5, O (32)	121

REFERENCES

[1] W. S. Wadsworth, Jr., *Org. React.*, **25**, 73 (1977).
[2] W. S. Wadsworth and W. D. Emmons, . *J. Am. Chem. Soc.*, **83**, 1733 (1961).
[3] L. Horner, H. Hoffmann, and H. G. Wippel, *Chem. Ber.*, **91**, 61 (1958).
[4] L. Horner, H. Hoffmann, W. Klinke, H. Ertel, and V. G. Toscano, *Chem. Ber.*, **95**, 581 (1962).
[5] E. J. Corey and G. T. Kwiatkowski, *J. Am. Chem. Soc.*, **88**, 5652 (1966).
[6] E. J. Corey and G. T. Kwiatkowski, *J. Am. Chem. Soc.*, **88**, 5653 (1966).
[7] E. J. Corey and G. T. Kwiatkowski, *J. Am. Chem. Soc.*, **88**, 5654 (1966).
[8] H. Takahashi, K. Fujiwara, and M. Ohta, *Bull. Chem. Soc. Jpn.*, **35**, 1498 (1962).
[9] E. J. Corey and R. P. Volante, *J. Am. Chem. Soc.*, **98**, 1291 (1976).
[10] D. A. Evans, J. M. Takacs, and K. M. Hurst, *J. Am. Chem. Soc.*, **101**, 371 (1979).
[11] M. Sekine, M. Nakajima, A. Kume, A. Hashizume, and T. Hata, *Bull. Chem. Soc. Jpn.*, **55**, 224 (1982).
[12] L. H. Horner and H. Roder, *Chem. Ber.*, **103**, 2984 (1970).
[13] D. Hendlin, E. O. Stapley, M. Jackson, H. Wallick, A. K. Miller, F. J. Wolf, T. W. Miller, L. Chaiet, F. M. Kahan, E. L. Foltz, H. B. Woodruff, J. M. Mata, S. Hernandez, and S. Mochales, *Science*, **166**, 122 (1969).
[14] M. Horiguchi and H. Kandatsu, *Nature* (*London*), **184**, 901 (1959).
[15] R. Engel, in *The Role of Phosphonates in Living Systems*, R. L. Hilderbrand, Ed., CRC Press, Boca Raton, FL, 1983, Chapter 5.
[16] C. L. Bentzen, L. N. Mong, and E. Niesor, U. S. Patent 4,416,877 (1983) [*C.A.*, **100**, 114996t (1984)].
[17] N. L. Shipkowitz, R. R. Bowen, R. N. Appell, C. W. Nordeen, L. R. Overby, W. R. Roderick, J. B. Sleicher, and A. M. vonEsch, *Appl. Microbiol.*, **26**, 264 (1973).
[18] F. Ramirez, A. V. Patwardhan, and S. R. Heller, *J. Am. Chem. Soc.*, **86**, 514 (1964).
[19] V. A. Ginsberg and A. Y. Jakubovich, *Zh. Obshch. Khim.*, **30**, 3979 (1960) [*C.A.*, **55**, 22099 (1961)].
[20] J. E. Baldwin, *J. Chem. Soc., Chem. Commun.*, **1976**, 734.
[21] F. Ramirez, S. B. Bhatia, and C. P. Smith, *Tetrahedron*, **23**, 2067 (1967).
[22] A. J. Muller, K. Nishiyama, G. W. Griffin, K. Ishikawa, and D. M. Gilson, *J. Org. Chem.*, **47**, 2342 (1982).
[23] F. Ramirez, S. B. Bhatia, A. V. Patwardhan, and C. P. Smith, *J. Org. Chem.*, **32**, 3547 (1967).
[24] F. Ramirez, S. B. Bhatia, A. V. Patwardhan, and C. P. Smith, *J. Org. Chem.*, **32**, 2194 (1967).
[25] F. Ramirez, S. B. Bhatia, and C. P. Smith, *J. Am. Chem. Soc.*, **89**, 3030 (1967).
[26] F. Ramirez, S. B. Bhatia, and C. P. Smith, *J. Am. Chem. Soc.*, **89**, 3026 (1967).
[27] F. Ramirez, *Acc. Chem. Res.*, **1**, 168 (1967).
[28] T. K. Gazizov, A. M. Pundyk, Y. I. Sudarev, V. B. Podobedov, V. I. Kovalenko, and A. N. Pudovik, *Bull. Acad. Sci. USSR Div. Chem. Sci.* (*Engl. Transl.*), **27**, 2319 (1978).
[29] N. A. Vorontsova, O. N. Vlasov, and N. N. Melnikov, *J. Gen. Chem. USSR* (*Engl. Transl.*), **52**, 392 (1982).
[30] V. S. Abramov, *Dokl. Akad. Nauk SSSR*, **95**, 991 (1954) [*C.A.*, **49**, 6084d (1955)].
[31] L. H. Sommer, *Stereochemistry, Mechanism and Silicon*, McGraw-Hill, New York, 1965, pp. 176–178.
[32] R. G. Harvey and E. V. Jensen, *Tetrahedron Lett.*, **1963**, 1801.
[33] R. G. Harvey, *Tetrahedron*, **22**, 2561 (1966).
[34] Y. A. Strepikheev, L. V. Kovalenko, A. V. Batalina, and A. I. Livshits, *J. Gen. Chem. USSR* (*Engl. Transl.*), **46**, 2364 (1976).
[35] V. S. Abramov, *Zh. Obshch. Khim.*, **22**, 647 (1952) [*C.A.*, **47**, 5351c (1953)].
[36] F. Ramirez and S. Dershowitz, *J. Org. Chem.*, **22**, 956 (1957).
[37] F. Ramirez and S. Dershowitz, *J. Org. Chem.*, **22**, 1282 (1957).
[38] F. Ramirez and S. Dershowitz, *J. Org. Chem.*, **23**, 778 (1958).
[39] F. Ramirez and S. Dershowitz, *J. Am. Chem. Soc.*, **81**, 587 (1959).
[40] F. Ramirez, E. H. Chen, and S. Dershowitz, *J. Am. Chem. Soc.*, **81**, 4338 (1959),

[41] E. A. C. Lucken, F. Ramirez, V. P. Catto, D. Ruhm, and S. Dershowitz, *Tetrahedron*, **22**, 637 (1966).

[42] F. Ramirez, H. J. Kugler, and C. P. Smith, *Tetrahedron Lett.*, **1965**, 261.

[43] M. P. Osipova, L. V. Kuzmina, and V. A. Kukhtin, *J. Gen. Chem. USSR (Engl. Transl.)*, **52**, 392 (1982).

[44] S. Yoneda, T. Kawase, and Z. -i. Yoshida, *J. Org. Chem.*, **43**, 1980 (1978).

[45] N. F. Orlov, B. L. Kaufman, L. Sukhi, L. N. Selsarand, and E. V. Sudakova, *Khim. Prim. Soedin.*, **1966**, 111 [*C.A.*, **72**, 21738y (1970)].

[46] M. A. Belokrinitskii and N. F. Orlov, *Kremniiorg. Mater.*, **1971**, 145 [*C.A.*, **78**, 29929f (1973)].

[47] M. Sekine, H. Yamagata, and T. Hata, *Tetrahedron Lett.*, **1979**, 3013.

[48] M. Sekine, K. Okimoto, and T. Hata, *J. Am. Chem. Soc.*, **100**, 1001 (1978).

[49] M. Sekine, I. Yamamoto, A. Hashizume, and T. Hata, *Chem. Lett.*, **1977**, 485.

[50] Z. S. Novikova, S. N. Mososhina, T. S. Sapozhnikova, and I. F. Lutsenko, *J. Gen. Chem. USSR (Engl. Transl.)*, **41**, 2655 (1971).

[51] D. A. Evans, K. M. Hurst, and J. M. Takacs, *J. Am. Chem. Soc.*, **100**, 3467 (1978).

[52] R. Rabinowitz, *J. Org. Chem.*, **28**, 2975 (1963).

[53] T. Morita, Y. Okamoto, and H. Sakurai, *Bull. Chem. Soc. Jpn.*, **51**, 2169 (1978).

[54] E. K. Ofitserova, O. E. Ivanova, E. N. Ofitserova, I. V. Konovalova, and A. N. Pudovik, *J. Gen. Chem. USSR (Engl. Transl.)*, **51**, 390 (1981).

[55] F. Ramirez, A. S. Gulati, and C. P. Smith, *J. Am. Chem. Soc.*, **89**, 6283 (1967).

[56] A. N. Pudovik, T. K. Gazizov, and A. M. Kibardin, *J. Gen. Chem. USSR (Engl. Transl.)*, **44**, 1170 (1974).

[57] A. N. Kibardin, T. K. Gazizov, and A. N. Pudovik, *J. Gen. Chem. USSR (Engl. Transl.)*, **45**, 1947 (1975).

[58] M. Sekine, T. Futatsugi, and T. Hata, *J. Org. Chem.*, **47**, 3453 (1982).

[59] M. Sekine, T. Tetsuaki, K. Yamada, and T. Hata, *J. Chem. Soc., Perkin Trans. 1*, **1982**, 2509.

[60] I. V. Konovalova, L. A. Burnaeva, N. S. Saifullina, and A. N. Pudovik, *J. Gen. Chem. USSR (Engl. Transl.)*, **46**, 17 (1976).

[61] D. A. Evans, K. M. Hurst, L. K. Truesdale, and J. M. Takacs, *Tetrahedron Lett.*, **1977**, 2495.

[62] D. Liotta, U. Sunay, and S. Ginsberg, *J. Org. Chem.*, **47**, 2227 (1982).

[63] J. K. Thottathil, D. E. Ryono, C. A. Przybyla, J. L. Moniot, and R. Neubeck, *Tetrahedron Lett.*, **25**, 4741 (1984).

[64] B. Lejczak, P. Kafarski, M. Soroka, and P. Mastalerz, *Synthesis*, **1984**, 577.

[65] A. K. Jung and R. Engel, *J. Org. Chem.*, **40**, 244 (1975).

[66] A. K. Jung and R. Engel, *J. Org. Chem.*, **40**, 3652 (1975).

[67] G. H. Birum and G. A. Richardson, U.S. Patent 3,113,139 (1963) [*C.A.*, **60**, 5551d (1964)].

[68] Y. Okamoto, T. Azuhata, and H. Sakurai, *Chem. Lett.*, **1981**, 1265.

[69] R. E. Koenigkramer and H. Zimmer, *J. Org. Chem.*, **45**, 3994 (1980).

[70] R. E. Koenigkramer and H. Zimmer, *Tetrahedron Lett.*, **21**, 1017 (1980).

[71] D. Ranganathan, C. B. Rao, and S. Ranganathan, *J. Chem. Soc., Chem. Commun.*, **1979**, 975.

[72] D. Ranganathan, C. B. Rao, S. Ranganathan, A. K. Mehrota, and R. Iyengar, *J. Org. Chem.*, **45**, 1185 (1980).

[73] Y. Okamoto, *Chem. Lett.*, **1984**, 87.

[74] A. N. Pudovik, E. S. Batyeva, and Y. N. Girfanova, *J. Gen. Chem. USSR (Engl. Transl.)*, **43**, 1681 (1974).

[75] H. Teichmann, W. Thierfelder, and A. Weight, *J. Prakt. Chem.*, **319**, 207 (1977).

[76] Y. K. Gusev, V. N. Chistokletov, and A. A. Petrov, *J. Gen. Chem. USSR (Engl. Transl.)*, **47**, 39 (1977).

[77] A. N. Pudovik and B. A. Arbuzov, *Dokl. Akad. Nauk SSSR*, **73**, 327 (1950) [*C.A.*, **45**, 7517i (1951)].

[78] A. N. Pudovik and B. A. Arbuzov, *Zh. Obshch. Khim.*, **21**, 382 (1951) [*C.A.*, **45**, 2853b (1951)].

[79] L. Evelyn, L. D. Hall, L. Lynn, P. R. Steiner, and D. H. Stokes, *Carbohydr. Res.*, **27**, 21 (1973).

[80] F. Johnston, U.S. Patent. 2,754,319 (1956) [*C.A.*, **51**, 466b (1957)].

[81] S. Linke, J. Kurz, D. Lipinski, and W. Gau, *Justus Liebigs Ann. Chem.*, **1980**, 542.

[82] P. F. Cann, S. Warren, and M. R. Williams, *J. Chem. Soc., Perkin Trans. 1*, **1972**, 2377.

[83] P. F. Cann, D. Howells, and S. Warren, *J. Chem. Soc., Perkin Trans. 2*, **1972**, 304.

[84] A. Bell, A. H. Davidson, C. Earnshaw, H. K. Norrish, R. S. Torr, and S. Warren, *J. Chem. Soc., Chem. Commun.*, **1978**, 988.

[85] B. A. Arbuzov, V. M. Zoroastrova, G. A. Tudril, and A. V. Fuzhenkova, *Bull. Acad. Sci. USSR, Div. Chem. Sci.*, (*Engl. Transl.*), **1974**, 2541.

[86] R. S. Tewari and R. Shukla, *Indian J. Chem.*, **10**, 823 (1972).

[87] B. A. Arbuzov, A. V. Fuzhenkova, G. A. Tudril, and V. M. Zoroastrova, *Bull. Acad. Sci. USSR, Div. Chem. Sci.*, **1975**, 1285.

[88] T. Tone, Y. Okamoto, and H. Sakurai, *Chem. Lett.*, **1978**, 1349.

[89] H. Wynberg and A. A. Smaardijk, *Tetrahedron Lett.*, **24**, 5899 (1983).

[90] A. A. Smaardijk, S. Noorda, F. vanBolhuis, and H. Wynberg. *Tetrahedron Lett.*, **26**, 493 (1985).

[91] F. Texier-Boullet and A. Foucaud, *Synthesis*, **1982**, 165.

[92] F. Texier-Boullet and A. Foucaud, *Synthesis*, **1982**, 916.

[93] P. D. Seemuth and H. Zimmer, *J. Org. Chem.*, **43**, 3063 (1978).

[94] E. K. Fields, *J. Am. Chem. Soc.*, **74**, 1528 (1952).

[95] R. Tyka, *Tetrahedron Lett.*, **1970**, 677.

[96] C. Wasielewski, K. Antczak, and J. Rachon, *Z. Chem.*, **19**, 253 (1979).

[97] J. A. Miller, G. M. Stevenson, and B. C. Williams, *J. Chem. Soc.* (*C*), **1971**, 2714.

[98] Y. A. Zhdanov, Z. I. Glebova, and L. A. Uzlova, *J. Gen. Chem. USSR* (*Engl. Transl.*), **51**, 1428 (1981).

[99] M. S. Voronkov and Y. I. Skorik, *Zh. Obshch. Khim.*, **35**, 106 (1965) [*C.A.*, **62**, 13173 (1965)].

[100] M. Sekine, H. Mori, and T. Hata, *Bull. Chem. Soc. Jpn.*, **55**, 239 (1982).

[101] I. J. Borowitz, S. Firstenberg, E. W. R. Casper, and R. K. Croucj, *J. Org. Chem.*, **36**, 3282(1971).

[102] I. J. Borowitz, S. Firstenberg, G. B. Borowitz, and D. Schuessler, *J. Am. Chem. Soc.*, **94**, 1623 (1972).

[103] E. M. Gaydov and J. P. Bianchini, *Can. J. Chem.*, **54**, 3626 (1976).

[104] A. K. Bhattacharya and G. Thyagarajan, *Chem. Rev.*, **81**, 415 (1981).

[105] E. F. Burgerenko, E. A. Chernyshev, and E. M. Popov, *Izv. Akad. Nauk SSSR, Ser. Khim.*, **1966**, 1391 [*C.A.*, **66**, 76078q (1967)].

[106] L. V. Nesterov, N. E. Krepysheva, R. A. Sabirova, and G. N. Romanova, *J. Gen. Chem. USSR* (*Engl. Transl.*), **41**, 2449 (1971).

[107] E. E. Borisova, R. D. Gareev, and I. M. Shermergorn, *J. Gen. Chem. USSR* (*Engl. Transl.*), **50**, 1786 (1980).

[108] A. N. Pudovik, G. V. Romanov, and R. Y. Nazmutdinov, *J. Gen. Chem. USSR* (*Engl. Transl.*), **44**, 212 (1974).

[109] N. G. Kushainova, G. V. Romanov, R. Y. Nazmutdinov, and A. N. Pudovik, *J. Gen. Chem. USSR* (*Engl. Transl.*), **51**, 1893 (1981).

[110] L. V. Nesterov, N. E. Krepysheva, R. A. Sabirova, and G. N. Romanova, *J. Gen. Chem. USSR* (*Engl. Transl.*), **41**, 2474 (1971).

[111] T. Hata, A. Hashizume, M. Nakajima, and M. Sekine, *Tetrahedron Lett.*, **1978**, 363.

[112] F. Lieb, H. Oediger, and G. Streissle, U.S. Patent 4,340,599 (1982) [*C.A.*, **95**, 81215k (1981)].

[113] A. N. Pudovik, E. S. Batyeva, and G. U. Zameletdinova, *J. Gen. Chem. USSR* (*Engl. Transl.*), **43**, 676 (1973).

[114] A. N. Pudovik, I. V. Guryanov, G. V. Romanov, and A. A. Lapin, *J. Gen. Chem. USSR* (*Engl. Transl.*), **41**, 710 (1971).

[115] M. A. Ruveda and S. A. deLicastro, *Tetrahedron*, **28**, 6013 (1972).

[116] M. Sekine, H. Yamagata, and T. Hata, *Tetrahedron Lett.*, **1979**, 3013.

[117] N. M. Mirsalikhova, L. A. Baranova, V. L. Tunitskaya, and N. N. Gulyaev, *Biokhimiya*, **46**, 258 (1981) [*C.A.*, **94**, 134965t (1981)].

[118] S. L. Goldstein, D. Braksmayer, B. E. Tropp, and R. Engel, *J. Med. Chem.*, **17**, 363 (1974).

[119] B. A. Arbuzov, G. A. Tudril, and A. V. Fuzhenkova, *Bull. Acad. Sci. USSR. Div. Chem. Sci.* (*Engl. Transl.*), **1980**, 294.

[120] B. A. Arbuzov, A. V. Fuzhenkova, and N. F. Rozhkova, *J. Gen. Chem. USSR* (*Engl. Transl.*), **52**, 10 (1982).

[121] B. A. Arbuzov, A. V. Fuzhenkova, and N. I. Galyantdinov, *Bull. Acad. Sci. USSR, Div. Chem. Sci.* (*Engl. Transl.*), **1978**, 381.

[122] B. A. Arbuzov, A. V. Fuzhenkova, N. I. Galyantdinov, and R. F. Shaikullina, *Bull. Acad. Sci. USSR, Div. Chem. Sci.* (*Engl. Transl.*), **1980**, 826.

[123] B. A. Arbuzov, A. V. Fuzhenkova, and N. I. Galyantdinov, *Bull. Acad. Sci. USSR, Div. Chem. Sci.* (*Engl. Transl.*), **1980**, 651.

[124] M. J. Gallagher and I. D. Jenkins, *J. Chem. Soc.* (*C*), **1971**, 210.

[125] B. A. Arbuzov, G. A. Tudril, and A. V. Fuzhenkova, *Bull. Acad. Sci. USSR, Div. Chem. Sci.* (*Engl. Transl.*), **1979**, 1466.

[126] R. Bodalski and K. Pietrusiewicz, *Tetrahedron Lett.*, **1972**, 4209.

[127] A. N. Pudovik, E. S. Batyeva, and G. U. Zameletdinova, *J. Gen. Chem. USSR* (*Engl. Transl.*), **45**, 922 (1975).

[128] A. F. Isbell, J. P. Berry, and L. W. Tansey, *J. Org. Chem.*, **37**, 4399 (1972).

[129] J. D. Smith and J. H. Law, *Biochemistry*, **9**, 2152 (1970).

[130] Y. Okamoto and H. Sakurai, *Synthesis*, **1982**, 497.

[131] Z. S. Novikova and I. F. Lutsenko, *J. Gen. Chem. USSR* (*Engl. Transl.*), **40**, 2110 (1970).

[132] M. Nakano, Y. Okamoto, and S. Sakurai, *Synthesis*, **1982**, 915.

[133] J. M. J. Trochet, J. -R. Neser, L. Gonzalez, and E. J. Charollais, *Helv. Chim. Acta*, **62**, 2022 (1979).

[134] J. Barycki, P. Mastalerz, and M. Soroka, *Tetrahedron Lett.*, **1970**, 3147.

[135] B. S. Sharawat, I. Handa, and H. L. Bhatnagar, *Indian J. Chem., Sect. A*, **16A**, 306 (1978).

[136] Y. Okamoto, T. Tone, and H. Saurai, *Bull. Chem. Soc. Jpn.*, **54**, 303 (1981).

[137] J. R. Chambers and A. F. Isbell, *J. Org. Chem.*, **29**, 832 (1964).

[138] H. B. F. Dixon and M. J. Sparkes, *Biochem. J.*, **141**, 715 (1974).

[139] G. Kamai and V. A. Kukhtin, *Dokl. Akad. Nauk SSSR*, **109**, 91 (1956) [*C.A.*, **51**, 1827g (1957)].

[140] M. Horiguchi and H. Rosenberg, *Biochim. Biophys. Acta*, **404**, 333 (1975).

[141] M. Soroka and P. Mastalerz, *Rocz. Chem.*, **50**, 661 (1976).

[142] R. D. Gareev and A. N. Pudovik, *J. Gen. Chem. USSR* (*Engl. Transl.*), **48**, 450 (1978).

[143] H. Paulsen, W. Greve, and H. Kuhne, *Tetrahedron Lett.*, **1971**, 2109.

[144] H. Teichmann, W. Thierfelder, A. Weigt, and E. Schafer, *DE* Patent 130,354 (1978) [*C.A.*, **91**, 20716v (1979)].

[145] R. D. Gareev. A. G. Abulkhanov, and A. N. Pudovik, *J. Gen. Chem. USSR* (*Engl. Transl.*), **48**, 245 (1978).

[146] R. D. Gareev, G. M. Loginova, A. G. Abulkhanov, and A. N. Pudovik, *J. Gen. Chem. USSR* (*Engl. Transl.*), **46**, 2284 (1976).

[147] W. E. Kruger and R. J. Maloney, *J. Org. Chem.*, **38**, 4208 (1973).

[148] H. Paulsen and W. Greve, *Chem. Ber.*, **106**, 2114 (1973).

[149] R. Gancarz, P. Kafarski, B. Lejczak, P. Mastalerz, J. S. Wieczorek, E. Przybyla, and W. Czerwinski, *Phosphorus Sulfur*, **18**, 373 (1983).

[150] J. S. Wieczorek, R. Boduszek, and R. Gancarz, *J. Prakt. Chem.*, **326**, 349 (1984).

[151] J. S. Wieczorek and R. Gancarz, *Rocz. Chem.*, **50**, 2171 (1976).

[152] H. Zimmer, R. E. Koenigkramer, R. M. Cepulis, and D. M. Nene, *J. Org. Chem.*, **45**, 2018 (1980).

[153] G. L. Mastevosyan, R. M. Matyushicheva, S. N. Vodovatova, and P. M. Zavlin, *J. Gen. Chem. USSR* (*Engl. Transl.*), **51**, 636 (1981).

[154] R. Gancarz and J. S. Wieczorek, *J. Prakt. Chem.*, **322**, 213 (1980).

[155] I. V. Boev and A. V. Dombrovskii, *J. Gen. Chem. USSR* (*Engl. Transl.*), **49**, 1246 (1979).

[156] V. R. Gaertner, European Patent 0,007,684 (1983) [*C.A.*, **93**, 114699k (1980)].

[157] N. S. Kozlov, V. D. Pak, and E. S. Elin, *Zh. Obshch. Khim. SSSR*, **39**, 2407 (1969) [*C.A.*, **72**, 79156 (1970)].

[158] M. I. Kabachnik and T. Y. Medved, *Dokl. Akad. Nauk SSSR*, **83**, 689 (1952) [*C.A.*, **47**, 2724h (1953)].

[159] T. Y. Medved and M. I. Kabachnik, *Dokl. Akad. Nauk SSSR*, **84**, 717 (1952) [*C.A.*, **47**, 3226g (1953)].

[160] M. I. Kabachnik and T. Y. Medved, *Izv. Akad. Nauk SSSR*, **1953**, 868 [*C.A.*, **49**, 840c (1955)].

[161] E. Gruszecka, M. Soroka, and P. Mastalerz, *Pol. J. Chem.*, **53**, 2327 (1979).

[162] M. Soroka and P. Mastalerz, *Rocz. Chem.*, **48**, 1119 (1974).

[163] R. Gancarz and J. S. Wieczorek, *Synthesis*, **1977**, 625.

[164] T. Glowiak, W. Sawka-Dobrowolska, J. Kowalik, P. Mastalerz, M. Soroka, and J. Zon, *Tetrahedron Lett.*, **1977**, 3965.

[165] J. Lukszo, J. Kowalik, and P. Mastalerz, *Chem. Lett.*, **1978**, 1103.

[166] J. Lukszo and R. Tyka, *Synthesis*, **1977**, 239.

[167] J. Lukszo and R. Tyka, *Pol. J. Chem.*, **52**, 321 (1978).

[168] J. Lukszo and R. Tyka, *Pol. J. Chem.*, **52**, 959 (1978).

[169] W. F. Gilmore and H. A. McBride, *J. Am. Chem. Soc.*, **94**, 4361 (1972).

[170] Y. P. Belov, G. B. Rakhnovich, V. A. Davankov, N. N. Godovikov, G. G. Aleksandrov, and Y. T. Struchkov, *Bull. Acad. Sci. USSR, Div. Chem. Sci.* (*Engl. Transl.*), **1980**, 832.

[171] S. V. Rogozhin, V. A. Davankov, and Y. P. Belov, *Bull. Acad. Sci. USSR, Div. Chem. Sci.* (*Engl. Transl.*), **1973**, 926.

[172] K. Issleib, K. -P. Dopfer, and A. Balszuweit, *Z. Anorg. Allg. Chem.*, **444**, 249 (1978).

[173] K. Moedritzer, *Synth. React. Inorg. Metal-Org. Chem.*, **2**, 317 (1972).

[174] M. P. Osipova, P. M. Lukin, and V. A. Kukhtin, *J. Gen. Chem. USSR* (*Engl. Transl.*), **52**, 393 (1982).

[175] G. A. Birum, *J. Org. Chem.*, **39**, 209 (1974).

[176] J. W. Huber and W. F. Gilmore, *Tetrahedron Lett.*, **1979**, 3049.

[177] Z. H. Kudzin and W. J. Stec, *Synthesis*, **1978**, 469.

[178] C. C. Tam, K. L. Mattocks, and M. Tischler, *Synthesis*, **1982**, 188.

[179] J. Oleksyszyn and R. Tyka, *Pol. J. Chem.*, **52**, 1949 (1978).

[180] J. Oleksyszyn, R. Tyka, and P. Mastalerz, *Synthesis*, **1977**, 571.

[181] J. Oleksyszyn, L. Subotkowska, and P. Mastalerz, *Synthesis*, **1979**, 985

[182] J. Mikroyannidis, *Phosphorus Sulfur*, **12**, 249 (1982).

[183] J. Zon *Pol. J. Chem.*, **55**, 643 (1981).

[184] W. Szczepaniak and J. Siepak, *Rocz. Chem.*, **47**, 929 (1973).

[185] D. Redmore, *J. Org. Chem.*, **43**, 996 (1978).

[186] D. Redmore, *J. Org. Chem.*, **43**, 992 (1978).

[187] J. Oleksyszyn and E. Gruszecka, *Tetrahedron Lett.*, **22**, 3537 (1981).

[188] E. K. Baylis, C. D. Campbell, and J. G. Dingwall, *J. Chem. Soc., Perkin Trans. 1*, **1984**, 2845.

[189] J. Oleksyszyn, *Synthesis*, **1980**, 722.

[190] W. Subotkowski, R. Tyka, and P. Mastalerz, *Pol. J. Chem.*, **54**, 503 (1980).

[191] J. Oleksyszyn, R. Tyka, and P. Mastalerz, *Synthesis*, **1978**, 479.

[192] R. Engel, unpublished results.

[193] A. Hampton, F. Perini, and P. J. Harper, *Biochemistry*, **12**, 1730 (1973).

[194] N. Lalinde, B. E. Tropp, and R. Engel, *Tetrahedron*, **39**, 2369 (1983).

[195] M. Sekine, M. Nakajima, and T. Hata, *J. Org. Chem.*, **46**, 4030 (1981).

[196] M. Sekine, M. Nakajima, and T. Hata, *Bull. Chem. Soc. Jpn.*, **54**, 218 (1981).

[197] T. Hata, M. Nakajima, and M. Sekine, *Tetrahedron Lett.*, **1979**, 2047.

[198] A. F. Kluge and I. S. Cloudsdale, *J. Org. Chem.*, **44**, 4847 (1979).

[199] H. Zimmer, K. R. Hickey, and R. J. Schumacher, *Chimia*, **28**, 656 (1974).

[200] H. Zimmer and D. N. Nene, *Chimia*, **31**, 330 (1977).

[201] G. Sturtz, B. Corbel, and J. -P. Paugam, *Tetrahedron Lett.*, **1976**, 47.

[202] H. G. Cristau, J. P. Vors, C. Niangoran, and H. Cristol, in *Phosphorus Chemistry. Proceedings of the* 1981 *International Conference*, L. D. Quin and J. G. Verkade, Eds. American Chemical Society, Washington, DC, 1981, pp. 59–62.

[203] S. Ranganathan, D. Ranganathan, and A. K. Mehrotra, *J. Am. Chem. Soc.*, **96**, 5261 (1974).

[204] V. L. Foss, N. V. Lukashev, Y. E. Tsvetkov, and I. F. Lutsenko, *J. Gen. Chem. USSR* (*Engl. Transl.*), **52**, 1942 (1982).

[205] B. N. Laskorin, V. V. Yakshin, and V. B. Bulgakova, *J. Gen. Chem. USSR (Engl. Transl.)*, **46**, 2372 (1972).
[206] T. Kawase, S. Yoneda, and Z. -i. Yoshida, *Bull. Chem. Soc. Jpn.*, **52**, 3342 (1979).
[207] A. F. Jensen and R. Pollitt, *Can. J. Chem.*, **48**, 1987 (1970).
[208] M. G. Zimin, A. R. Burilov, and A. N. Pudovik, *J. Gen. Chem. USSR (Engl. Transl.)*, **51**, 1841 (1981).
[209] E. Castagnino, S. Corsano, and G. P. Strappavecchia, *Tetrahedron Lett.*, **26**, 93 (1985).
[210] P. R. Adams, R. Harrison, and T. D, Inch, *Biochem. J.*, **141**, 729 (1974).
[211] F. Texier-Boullet and A. Foucaud, *Tetrahedron Lett.*, **21**, 2161 (1980).
[212] K. -C. Tang, B. E. Tropp, and R. Engel, *Tetrahedron*, **34**, 2873 (1978).
[213] H. Paulsen and W. Greve, *Chem. Ber.*, **106**, 2124 (1973).
[214] B. Mlotkowska, B. E. Tropp, and R. Engel, *Carbohydr. Res.*, **117**, 95 (1983).
[215] H. Paulsen and H. Kuhne, *Chem. Ber.*, **107**, 2635 (1974).
[216] H. Paulsen and H. Kuhne, *Chem. Ber.*, **108**, 1239 (1975).
[217] M. Sekine and T. Hata, *J. Chem. Soc., Chem. Commun.*, **1978**, 285.
[218] F. Kruger and D. Palleduhn, *DE Patent 274, 1504* (1981) [*C.A.*, **91**, 20708u (1979)].
[219] Z. H. Kudzin, *Synthesis*, **1981**, 643.
[220] L. Maier, *Helv. Chim. Acta*, **56**, 489 (1973).
[221] E. J. Glamkowski, G. Gal, R. Purick, A. J. Davidson, and M. Sletzinger, *J. Org. Chem.*, **35**, 3510 (1970).
[222] N. Kreutzkamp and H. Schindler, *Chem. Ber.*, **92**, 1695 (1959).
[223] R. O. Duthaler, P. A. Lyle, and C. Heuberger, *Helv. Chim. Acta*, **67**, 1406 (1984).
[224] M. P. Osipova, L. A. Mikhailova, and V. A. Kukhtin, *J. Gen. Chem. USSR (Engl. Transl.)*, **52**, 394 (1982).
[225] A. N. Pudovik, E. S. Batyeva, and G. V. Zameletdinova, *J. Gen. Chem. USSR (Engl. Transl.)*, **43**, 944 (1973).
[226] H.-J. Altenbach and R. Korff, *Angew. Chem., Int. Ed. Engl.*, **21**, 371 (1982).
[227] J. A. Miller and D. Stewart, *J. Chem. Soc., Perkin Trans. 1*, 1977, 2416.
[228] M. M. Sidky, M. R. Mahran, and M. F. Zayed, *Phosphorus Sulfur*, **9**, 337 (1981).
[229] J.-L. Bravet, C. Benezra, and J.-P. Weniger, *Steroids*, **19**, 101 (1972).
[230] M. M. Sidky and F. H. Osman, *Tetrahedron*, **29**, 1725 (1973).

CHAPTER 3

REDUCTIONS BY METAL ALKOXYALUMINUM HYDRIDES. PART II. CARBOXYLIC ACIDS AND DERIVATIVES, NITROGEN COMPOUNDS, AND SULFUR COMPOUNDS

JAROSLAV MÁLEK*

Institute of Chemical Process Fundamentals, Czechoslovak Academy of Sciences, Prague, Czechoslovakia

CONTENTS

PAGE

* Deceased, November 20, 1986.

INTRODUCTION

In continuation of the previous review on reductions by metal alkoxyaluminum hydrides,[1] this chapter is devoted to reductions of carboxylic acids and their derivatives, open-chain and heterocyclic nitrogen compounds, and open-chain and heterocyclic sulfur compounds by alkoxyaluminum hydrides, metal alkoxyaluminum hydrides, and chiral metal alkoxyaluminohydride complexes. The review presented in this chapter covers the literature through December 1985. Emphasis is placed on the scope, limitations, and the synthetic utility of metal alkoxyaluminum hydrides. Currently accepted views on reaction mechanisms are mentioned briefly. Discussion of alternative methods of reduction, particularly those using other metal hydrides or complex metal hydrides,[2,3] is limited to examples of the most important transformations of functional groups; reducing properties of these hydrides (see p. 3 in Ref. 1) and hydride reagents such as borane–dimethyl sulfide,[4–7] amine boranes,[8] haloboranes,[7] organoboranes,[6,7,9] sodium borohydride–pyridine,[10] sodium borohydride–triphenyl borate,[11] sodium borohydride–transition-metal complexes,[10,12,13] potassium triisopropoxyborohydride,[14a] and sodium triacyloxyborohydrides[14b] have been reported in several papers and reviews. Recent monographs and reviews[14c,*d*,15–25] summarize reductions with metal alkoxyaluminum hydrides[14c,d] and asymmetric reductions with chiral lithium alkoxyaluminum hydrides[1,26–36] and chiral boron reagents.[37–45]

MECHANISM

Partial Reductions of Functional Groups

The striking feature of partial reductions of unsymmetrical cyclic anhydrides by $LiAlH_4$,[46–55] $LiBH_4$,[56] $NaBH_4$,[47,48,51,53–57] and

$LiAlH(OC_4H_9\text{-}t)_3$[47–50,52,58] to form γ-lactones is the predominant hydride attack at the more sterically hindered anhydride carbonyl group. The mechanism suggested for the reduction of 2-methylmaleic anhydride involves initial complexing of the solvated cation with the less hindered C-4 carbonyl function, predominant hydride attack at the more hindered C-1 carbonyl and formation of the aldehyde intermediate **1**, which is protonated to give the hydroxylactone **2** or undergoes reduction to form the lactone **3**.[46,49,52,59–61] The regioselectivity of reductions by $LiAlH_4$, $NaBH_4$, and $LiAlH(OC_4H_9\text{-}t)_3$ often varies with the anhydride structure, and the mechanism is still a subject of discussion.[49,51,53,60–62] Moreover, bulky complex metal hydrides such as lithium tri-*sec*-butylborohydride[54–57] or lithium trisiamylborohydride[54] lead in some instances to a complete reversal of the regioselectivity and produce γ-lactones resulting from preferential hydride attack at the less hindered anhydride carbonyl group.

88% $\xrightarrow{LiAlHX_3}$ Li^+ H AlX_3^- $\longrightarrow$

1 $\xrightarrow{H_3O^+}$ **2** (OH)

1 $\xrightarrow{1.\ LiAlHX_3,\ 2.\ H_3O^+}$ **3**

$X = H, OC_4H_9\text{-}t$

The partial reduction of phenyl esters of carboxylic acids by $LiAlH(OC_4H_9\text{-}t)_3$ has been rationalized in terms of an intermediate complex **4** that is relatively stable toward further reduction.

$$RCO_2C_6H_5 \xrightarrow[THF,\ 0^\circ]{LiAlH(OC_4H_9\text{-}t)_3} \left[\begin{array}{l} RCHOC_6H_5 \\ \quad | \\ OAl(OC_4H_9\text{-}t)_3 \end{array}\right]^- Li^+ \xrightarrow{H_3O^+} RCHO + C_6H_5OH$$

4

Therefore, it undergoes protonation instead to yield the corresponding aldehyde.[7,59,63–67] A similar mechanism presumably applies also in the partial reduction of alkyl esters by $NaAlH_2(OCH_2CH_2OCH_3)_2$[67–72] and its complexes with morpholine or 1-methylhexahydropyrazine.[73] The reductive transformation of diethyl perfluoroglutarate by $NaAlH_2(OCH_2CH_2OCH_3)_2$ follows another reaction path involving intramolecular cyclization of the initial reduction product **5** to form the ethoxy lactone **6**, which by further reduction yields the cyclic hemiacetal **7**.[74,75]

$$\text{RO}_2\text{C}-\text{CF}_2\text{CF}_2\text{CF}_2-\text{CO}_2\text{R} \xrightarrow[\text{THF } C_6H_6,\ -70^\circ]{\text{H}^-} \left[\textbf{5}\right] \longrightarrow \left[\textbf{6}\right] \xrightarrow[\text{THF } C_6H_6,\ -70^\circ]{\text{H}^-} \textbf{7}\ (87\%)$$

R = C_2H_5

The partial reduction of acid chlorides by $LiAlH(OC_4H_9\text{-}t)_3$[7,76–79] or $NaAlH(OC_4H_9\text{-}t)_3$[80] presumably proceeds according to Eq. 1.

$$\text{RCOCl} \xrightarrow[(CH_3OCH_2CH_2)_2O,\ -78^\circ]{MAlH(OC_4H_9\text{-}t)_3} \left[\begin{array}{l}\text{RCHCl}\\ \quad|\\ \text{OAl}(OC_4H_9\text{-}t)_3\end{array}\right]^- M^+ \xrightarrow{H_2O} \qquad \text{(Eq. 1)}$$

$$\text{RCHO} + \text{MCl} + \text{Al}(OC_4H_9\text{-}t)_3$$

M = Li, Na

Solvation of the lithium cation and its coordination to the imine nitrogen atom in the complex **8** have been suggested to explain a relatively high complex stability and the solvent effect[81,82] observed particularly in partial reductions of aromatic and α-branched aliphatic nitriles by $LiAlH(OC_2H_5)_3$.[81–85]

$$\text{RCN} \xrightarrow[(C_2H_5)_2O,\ 0^\circ]{LiAlH(OC_2H_5)_3} \left\{\begin{array}{c}\text{Li}^+(\text{S})\\ \downarrow\\ \text{RCH}{=}\text{N}[\text{Al}(OC_2H_5)_3]^-\end{array}\right\} \xrightarrow{H_3O^+} \text{RCHO}$$

8

Reductions of some alicyclic nitriles such as the sterically unhindered equatorial nitrile **9a** or the hindered axial nitrile **9b** are accompanied by a base-induced partial inversion of configuration, and the product composition is determined by the rate ratio of reduction to isomerization.[86]

$t\text{-}C_4H_9$–[cyclohexane]–H, CN (**9a**) ⟶

$t\text{-}C_4H_9$–[cyclohexane]–H, CHO (**10**) + $t\text{-}C_4H_9$–[cyclohexane]–CHO, H (**11**)

(**10**:**11** = 95:5)

$t\text{-}C_4H_9$–[cyclohexane]–CN, H (**9b**) ⟶ **11** + **10** + **9b** + **9a**

(**11**:**10** = 75:25)
(**9b**:**9a** = 75:25)

The mechanism suggested for hydride reductions of *N*,*N*-disubstituted carboxamides involves initial formation of an amino alcohol complex **12** that, depending on the amide structure, nature of the hydride, and reaction conditions, can react further along three paths: (1) hydrolysis to give the aldehyde and the secondary amine, (2) nucleophilic hydride attack on the carbon–oxygen bond to yield the tertiary amine **12c**, and (3) nucleophilic hydride attack on the carbon–nitrogen bond followed by hydrolysis to produce the alcohol and the secondary amine. The prerequisite for the preferential formation of aldehydes is that the rate of hydrolysis of complex **12** be faster than its conversion to **12a**, **12b**, and **12c**.[87,88]

$$RCHO + R^1R^2NH$$

$$\uparrow H_2O$$

$$RCONR^1R^2 \xrightarrow{MAlHX_3} \left[\begin{matrix} OAlX_3 \\ | \\ RCHNR^1R^2 \end{matrix}\right]^- M^+ \xrightarrow{MAlHX_3} RCH_2NR^1R^2$$

12 **12c**

$$\downarrow MAlHX_3$$

$$[RCH_2OAlX_3]^-M^+ + [R^1R^2NAlX_3]^-M^+$$

12a **12b**

$$\downarrow H_2O$$

$$RCH_2OH + R^1R^2NH$$

$M = Li, Na$
$X = H, C_2H_5O, t\text{-}C_4H_9O$

Experimental data on the partial reduction of *N*,*N*-dimethylcarboxamides,[88–93] *N*-methylanilides,[88,94,95] *N*-acylaziridines, -piperidines,

-pyrrolidines,[88] -imidazolines,[96,97] -saccharins,[98] *N*-methyl-*N*-(2-pyridyl)-carboxamides,[99] 3-acylthiazolidine-2-thiones,[100–102] and *N*-acylproline esters[103,104] by various metal alkoxyaluminum hydrides can all be accommodated by this mechanism.

Hydride Addition to α,β-Unsaturated Esters, Nitriles, and Sulfides

Enhanced 1,2 addition of $LiAlH_3(OC_2H_5)$* to aliphatic[105] and alicyclic[106] α,β-unsaturated esters to form the corresponding allylic alcohols (Eq. 2) has been interpreted in terms of greater hardness of the hydride and the C-2 position compared with C-4 in a conjugated carbonyl system[107–109] (see p. 11 in Ref. 1).

$$RCH{=}CHCO_2R' \longrightarrow RCH{=}CHCH_2OH \qquad \text{(Eq. 2)}$$

On the other hand, 1,4 addition should predominate in reductions of α,β-unsaturated esters by soft hydrides.[108,109] Nevertheless, despite the higher softness of Na^+ compared with Li^+,[108] $NaAlH_2(OCH_2CH_2OCH_3)_2$ behaves as a hard hydride and favors 1,2 addition in these reactions.[110]

Reduction of a carbon–carbon double bond in the α,β-unsaturated cyanoacetate **13** (only one isomer is shown) with $LiAlH(OC_4H_9\text{-}t)_3$ proceeds with a marked preference for equatorial attack and yields the axial isomer **14** as the major product; addition of $NaBH_4$ follows the opposite stereochemistry.

CN H $CO_2C_2H_5$ **13** ⟶ H H $CH(CN)CO_2C_2H_5$ **14** + H $CH(CN)CO_2C_2H_5$ H **15**

Reducing Agent	Isomer Ratio **14**[a]:**15**[a]	Yield (%)
$LiAlH(OC_4H_9\text{-}t)_3$ in THF[b]	67:33	60
$NaBH_4$ in ethanol[b]	50:50	74
$NaBH_4$ in THF[c]	27:73	70

[a] The product is a mixture of diastereomers.
[b] The reactions were carried out at 0° for 1 hour.
[c] The reaction was carried out at 0° for 2 hours.

* The formula is used for simplicity; recent results show that this hydride is unstable and that the actual reducing agent is instead a mixture of $LiAlH(OC_2H_5)_3$ and $LiAlH_4$ (see pp. 4–7 in Ref. 1); as observed experimentally, even $LiAlH(OC_2H_5)_3$ is not a uniform compound.[81,82,85a,85b,88]

This result has been interpreted in terms of a more reactantlike transition state in $LiAlH(OC_4H_9\text{-}t)_3$ reductions.[111] The behavior of $NaBH_4$ in conjugate additions to cyclohexylidenecyanoacetates has been rationalized on the basis of a cyclic six-center transition state.[112]

The 1,4 addition of $LiAlH(OCH_3)_3$–CuBr–2-butanol or $NaAlH_2 \cdot (OCH_2CH_2OCH_3)_2$–CuBr–2-butanol complexes (tentatively formulated as $MCuH_2$) to acyclic α,β-unsaturated esters **16** presumably proceeds by way of electron transfer to form a radical anion **17** (R^1 = alkyl); this can afford the enolate **18** either directly or indirectly by coupling with the hydride-complex radical cation to give the covalently bonded copper complex **19** followed by reductive elimination of copper hydride. As shown by deuterium labeling experiments, the hydrogen (Hβ) that adds to the β-carbon atom in **17** originates from the hydride-complex species.[113,114]

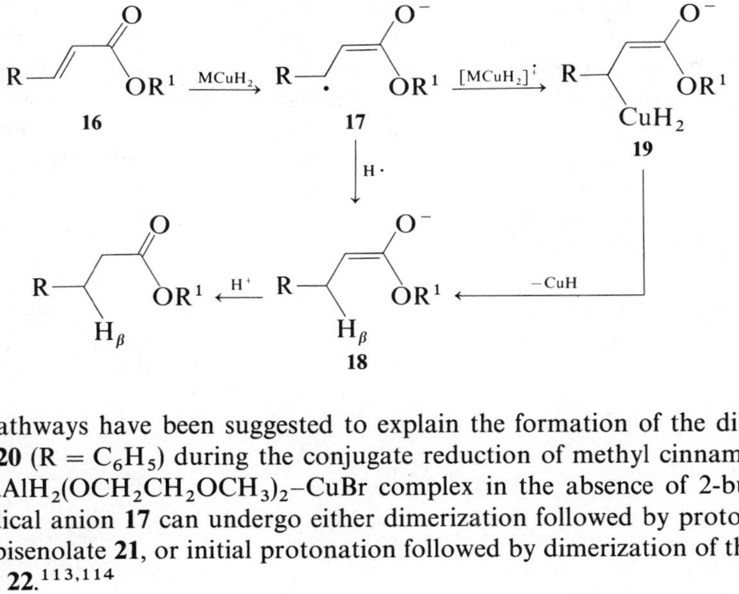

16 17 19 18

Two pathways have been suggested to explain the formation of the dihydrodimer **20** ($R = C_6H_5$) during the conjugate reduction of methyl cinnamate by the $NaAlH_2(OCH_2CH_2OCH_3)_2$–CuBr complex in the absence of 2-butanol; the radical anion **17** can undergo either dimerization followed by protonation of the bisenolate **21**, or initial protonation followed by dimerization of the enol radical **22**.[113,114]

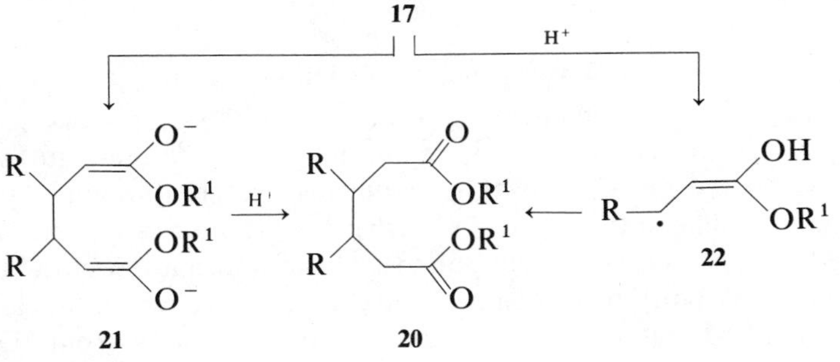

17 21 20 22

A mechanism similar to that indicated by formulas **16** → **19**[115] presumably applies in 1,4 additions of the $NaAlH_2(OCH_2CH_2OCH_3)_2$–CuBr–2-butanol complex to α,β-unsaturated nitriles **23** (Eq. 3).[115–117]

$$\text{R–CH=CH–C≡N}\ (\mathbf{23}) \xrightarrow{MCuH_2} \text{R–}\dot{\text{C}}\text{H–CH=C=N}^- \xrightarrow{[MCuH_2]^{+}} \text{R–CH(CuH}_2\text{)–CH=C=N}^- \xrightarrow{-CuH} \text{R–CH(H}_\beta\text{)–CH=C=N}^- \longrightarrow \text{R–CH(H}_\beta\text{)–CH}_2\text{–C≡N} \qquad \text{(Eq. 3)}$$

The mechanism of the reduction of 1-alkynyl sulfides by the $LiAlH(OCH_3)_3$–CuBr complex (equivalent amount of cuprous bromide; path A) involves attack of the hydride ion on the β-carbon atom to form the adduct **24**, which on decomposition by water gives *cis*-1-alkenyl sulfides **25** (X = H) with greater than 95% stereoselectivity. α-Deuterated *cis*-1-alkenyl sulfides **25** (X = D) are the products of decomposition by deuterium oxide. Reductions with catalytic amounts of cuprous bromide (path B) presumably proceed by way of a rapid transfer of cuprous bromide from the adduct **24** to $LiAlH(OCH_3)_3$, formation of the adduct **26**, and regeneration of the starting hydride complex. The presence of cuprous bromide has a profound impact on stereochemistry, since reductions by $LiAlH(OCH_3)_3$ alone or $LiAlH_4$ give almost exclusively the *trans*-1-alkenyl sulfides.[118]

$$RC{\equiv}CSR^1 \xrightarrow[\text{(path A or B)}]{LiAlH(OCH_3)_3-CuBr} \left[\text{(R)(H)}\overset{\beta}{C}{=}\overset{\alpha}{C}\text{(SR}^1\text{)(CuBr}\cdot Al(OCH_3)_3)\right]^- Li^+\ (\mathbf{24}) \xrightarrow[\text{(path A)}]{H_2O} \text{(R)(H)C=C(SR}^1\text{)(X)}\ (\mathbf{25})$$

$$\mathbf{24} \xrightarrow[\text{(path B)}]{LiAlH(OCH_3)_3} LiAlH(OCH_3)_3\text{–CuBr} + \left[\text{(R)(H)C=C(SR}^1\text{)(Al(OCH}_3)_3)\right]^- Li^+\ (\mathbf{26}) \xrightarrow[\text{(path B)}]{H_2O} \mathbf{25}$$

$$LiAlH(OCH_3)_3\text{–CuBr} \xrightarrow{\text{(path B)}} \text{(returns to first step)}$$

Hydrogenolytic Reactions

Aromatic carboxylic acids bearing electron-withdrawing groups in *ortho* or *para* positions, their alkyl esters,[119–122] and metal salts[123] undergo hydrogenolysis by $NaAlH_2(OCH_2CH_2OCH_3)_2$ at elevated temperatures. The proposed mechanism is similar to that suggested earlier for the hydrogenolysis of amino-substituted aromatic carbonyl compounds:[124] rapid formation of the carbinol, cleavage of the carbon–oxygen bond to give a resonance-stabilized carbonium ion, and rapid hydride attack on the latter to form a methyl group (Eq. 4).

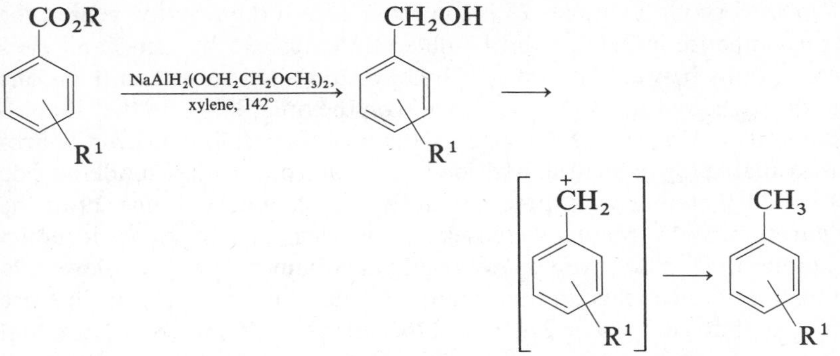

(Eq. 4)

R = H, alkyl, Na, BrMg—; R^1 = NH_2, $N(CH_3)_2$, OCH_3, OH

Hydrogenolysis of esters such as 3-carbethoxy-4-methylpyrrole by $NaAlH_2$-$(OCH_2CH_2OCH_3)_2$ yielding 3,4-dimethylpyrrole[125] seems to be explained rather on the basis of the vinylogy principle formulated by Gaylord.[126]

Hydride Addition to Carbon–Nitrogen Double Bonds

Reaction of the $LiAlH_4$–3-*O*-benzyl-1,2-cyclohexylidene-α-D-glucofuranose (1:1) complex **27a** (X = H_a) with imines **28** proceed apparently by way of a preferential transfer of the more active, less sterically hindered hydrogen H_a to the carbon atom of the imino group (path A); this approach gives optically active secondary amines **29** of the (*S*) configuration.

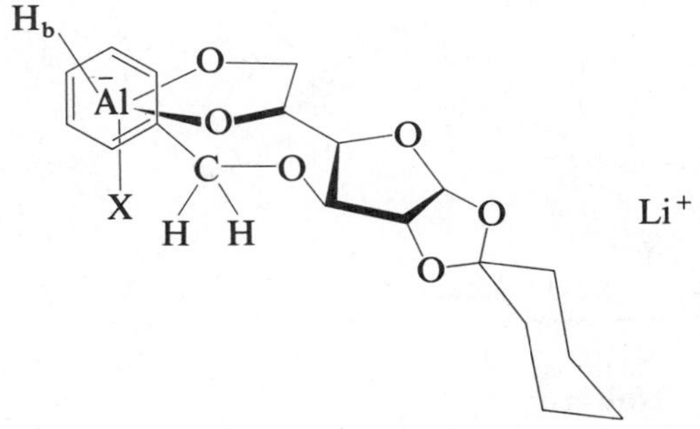

27a,b

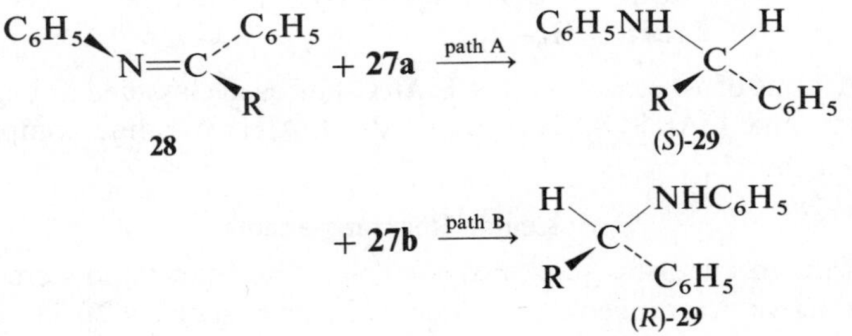

R = alkyl

As the *N*-phenyl group in imines **28** is too remote from the reaction center, the extent of asymmetric induction will be determined instead by steric and electronic interactions between the imine *C*-phenyl and alkyl substituents on one hand and the carbohydrate 3-*O*-benzyl group on the other hand. In the ethanol-modified complex **27b** ($X = OC_2H_5$), only the more sterically hindered hydrogen H_b is available for reduction; hydride approach from the less hindered side (path B) would therefore yield predominantly the secondary amine **29** of the (*R*) configuration.[127,128] Similar stereoselectivity observed in asymmetric reductions of ketones,[129–131] ketoximes, and *O*-alkyl ketoximes[132] by complexes **27a** and **27b** justifies the conclusion that analogous mechanisms apply in all these reactions and that complexes **27a** and **27b** can be used for configurational correlations.

Stereoselectivity of hydride additions to the carbon–nitrogen double bond in ketimines or their iminium salts derived from alkyl-substituted cyclohexanones and cyclopentanones is affected significantly by steric demands of substrates as well as complex metal hydrides. Whereas 3- and 4-alkylcyclohexanone imines and iminium salts are attacked by the more bulky reagents predominantly from the equatorial side to yield the axial amines, the 2-alkyl derivatives give *cis* amines as major products, regardless of hydride bulkiness.[133,134]

t-C_4H_9 (cyclohexylidene) $\overset{+}{N}(CH_3)_2$ ClO_4^- (Ax, Eq) $\longrightarrow$

t-C_4H_9 (cyclohexyl, axial $N(CH_3)_2$) **A** + t-C_4H_9 (cyclohexyl, equatorial $N(CH_3)_2$) **B**

Reagent	**A** : **B**
$NaBH_4$	16 : 84
$NaBH_3(CN)$	42 : 58
$LiAlH_4$	37 : 63
$NaAlH_2(OCH_2CH_2OCH_3)_2$	72 : 28
$LiBH(C_4H_9\text{-}s)_3$	94 : 6

In reductions of α-keto ketimines, $LiAlH_4$ has been classified as hard, $NaBH_4$ as medium, and $LiAlH(OC_4H_9-t)_3$ or the $LiAlH_4$-pyridine complex as soft hydride reagents.[135]

Molecular Rearrangements

One of the mechanisms suggested to explain the course and stereochemistry of the reductive rearrangement of δ-enol lactones such as **30** involves intramolecular transfer of the trialkoxyaluminum fragment to the enolate oxygen followed by cyclic rearrangement which, on the basis of models, should produce exclusively the axial keto aluminate **32**. Transformation of the latter depends apparently on the pH of the medium;[136,137] whereas decomposition with a strong acid yields the thermodynamically less stable axial bridged keto alcohol **33a** as the single product,[136] workup with water gives the axial (**33a**) and equatorial (**33b**) isomers in a 95:5 ratio (overall yield 90%).[136–138]

$LiAlH(OC_4H_9\text{-}t)_3$, THF, −70°

30 **31** **32**

$X = t\text{-}C_4H_9O$

33a: R = OH, R^1 = H (86%)
33b: R = H, R^1 = OH (4%)

The course of reduction of δ-enol lactones having an exocyclic carbon–carbon double bond can change significantly with the lactone structure[136,137,139–143a] and apparently depends on the relative rates of reduction and rearrangement of the intermediate complex **31**.[137] Interestingly, δ-thioenol lactones bearing an exocyclic carbon–carbon double bond undergo only partial reduction to form hydroxythiacycloalkenes.[143b]

Formation of aziridines and secondary amines can compete with simple reduction of ketoximes and their *O*-alkyl derivatives by $LiAlH_4$ or $NaAlH_2(OCH_2CH_2OCH_3)_2$ in tetrahydrofuran, yielding primary amines.[59,138,144–149] Reduction of the *anti*-oxime **34a** and the *syn*-oxime **34b** illustrates some of the proposed reaction pathways leading to these products. Intermolecular hydride attack from the least hindered side gives the nitrene **35a**; its reduction by $LiAlH_4$ (X = H) via insertion into an aluminium–hydrogen bond gives the primary amine **37** as the major product (82%).

Nitrene insertion into a carbon–hydrogen bond is preferred in reductions by the more bulky $NaAlH_2(OCH_2CH_2OCH_3)_2$ ($X = OCH_2CH_2OCH_3$). Thus the *anti*-oxime **34a** gives the *cis*-aziridine **36a** by concerted insertion of the nitrene into a benzylic carbon–hydrogen bond and 2-benzylaziridine **36c** by insertion into a methyl carbon–hydrogen bond in a 9:1 ratio, along with lesser amounts of the primary amine **37**. In reductions of the sterically more congested *syn*-oxime **34b**, some of the nitrene rotamer **35b** is proposed to be generated as well, resulting in the formation of the *trans*-aziridine **36b** in addition to the *cis* isomer **36a**.[150]

An alternative mechanism involves proton elimination from the methylene group of the *syn*-oxime **34b** by intramolecular hydride transfer or 1,4-sigmatropic rearrangement to form the unsaturated nitrene **38**, cyclization of the latter to give the azirine **39**,[150] and azirine reduction[151] producing the *cis*-aziridine **36a**.[150]

34b → **38** → **39** $\xrightarrow{\bar{A}lH_2X_2}$ **36a**

Similar proton elimination from the α-methyl group in the *anti*-oxime **34a** followed by formation of an unsaturated nitrene, its cyclization, and azirine reduction would lead to 2-benzylaziridine **36c**. In contrast to tetrahydrofuran or 1,2-dimethoxyethane, diethyl ether and bis(2-methoxyethyl) ether suppress the aziridine formation completely; this strong solvent effect has been attributed to efficient solvation by both former solvents, thus facilitating elimination of the $OAlHX_2$ group with formation of the nitrene.[150]

Reduction of aryl alkyl ketoximes to secondary amines is considered to proceed by way of the nitrene **40**, the Schiff base **41** formed by a 1,2 shift of the aryl group, and the complex **42** resulting from reduction of the Schiff base.[145,150]

40 → **41** $\xrightarrow{\bar{A}lH_2X_2}$ **42** $\xrightarrow{H_2O}$ ArNHCHR(H)

The reaction of *anti*-oximes of 2-cyclohexen-1-ones with $NaAlH_2(OCH_2CH_2OCH_3)_2$ appears to follow a mechanism suggested for the reduction of aryl alkyl ketoximes[147] and gives predominantly α,β-unsaturated aziridines by way of cyclization involving the methylene group adjacent to the oxime carbon atom. The reduction of *syn*-oximes of 2-cyclohexen-1-ones by the same hydride evidently follows another reaction path as indicated in Eqs. 5 and 6. Specific formation of aziridines **43** and **44** has been interpreted as due to (1) high steric requirements of $NaAlH_2(OCH_2CH_2OCH_3)_2$ that hinder the addition of hydride ion to the conjugated carbon–carbon double bond and (2) basicity or hardness of the hydride, that is, its capability of eliminating mobile protons from the methylene (Eq. 5) or methyl group (Eq. 6) attached to the conjugated carbon–carbon double bond. Noteworthy here is the weak solvent effect.[152–154]

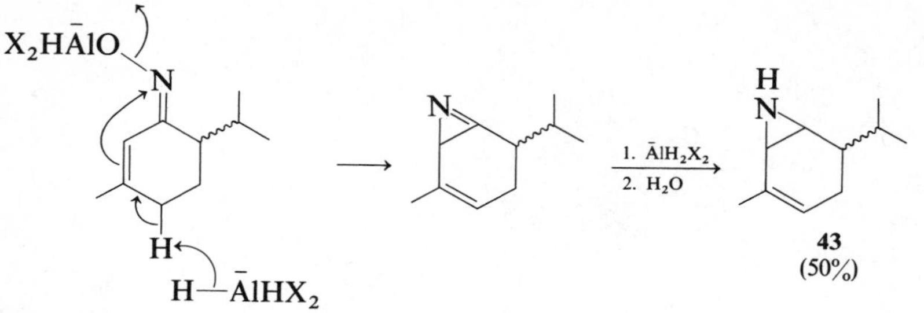

(Eq. 5)

43 (50%)

(Eq. 6)

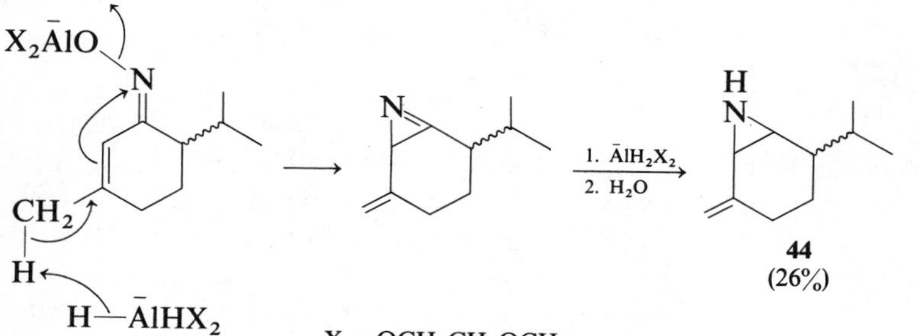

44 (26%)

X = $OCH_2CH_2OCH_3$

Berbine methiodides such as the *trans*-methiodide **45** easily undergo the Stevens rearrangement on reaction with $NaAlH_2(OCH_2CH_2OCH_3)_2$;[155–158] the reaction follows the dissociation–recombination pathway involving either ion-pair or radical-pair intermediates[159] and gives the spirobenzylisoquinoline **46** along with the 8β-methylberbine derivative **47**. The hydride acts as a strong

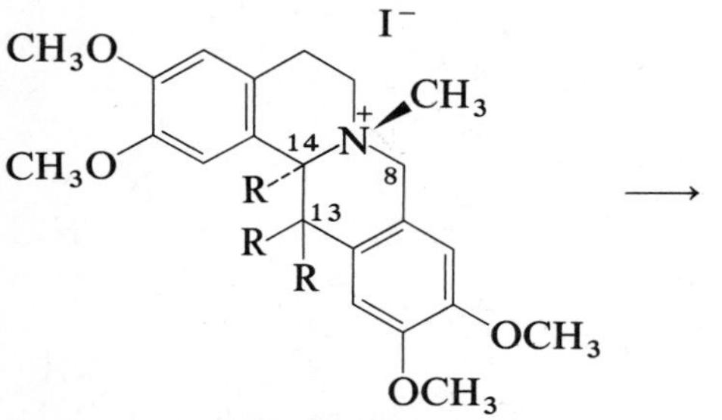

dl-**45**: R = H
dl-**45a**: R = D

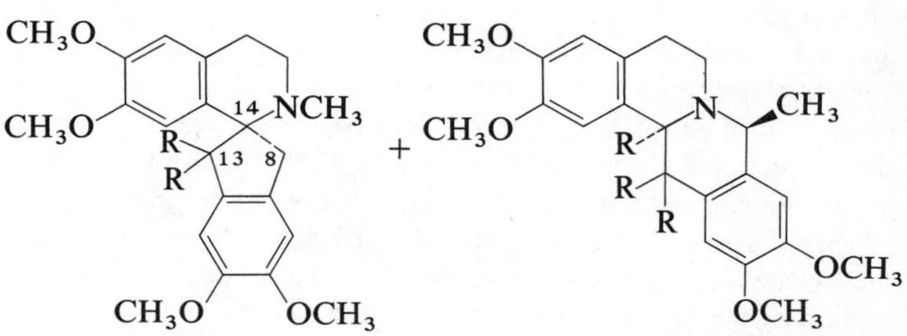

46: R = H (77%)
46a: R = D

47: R = H (6%)
47a: R = D

base, abstracting the quasi-axial hydrogen atoms at C-8 and C-14 positions of berbine methiodides; the results of the rearrangement of optically active quaternary salts and of the trideuteriomethiodide **45a** to form deuterium-labeled compounds **46a** and **47a** support this conclusion.[157]

The sulfone bis(methanesulfonate) **48** undergoes rearrangement on reaction with $LiAlH_4$ or $NaAlH_2(OCH_2CH_2OCH_3)_2$. A mechanism that could account for the formation of (1*S*,6*R*)-*S*,*S*-dioxide **49** and its (4*R*)-4-hydroxy derivative **50** involves proton abstraction at C-2, elimination of the mesyloxy group at C-7, and cleavage of the ether linkage at C-1 to give the bicyclic intermediate **48a**; reductive removal of the mesyloxy group by an intramolecular S_N2 reaction would produce the intermediate **48b**. At this stage, elimination from **48b** followed by reduction to form the *S*,*S*-dioxide **49** can compete with decomposition of **48b** to yield the alcohol **50**. The effect of the sulfone group in **48** is essential, since no rearrangement occurs when the sulfide corresponding to the *S*,*S*-dioxide **48** is reduced under similar reaction conditions.[160]

MsO 7 6 OMs 1 2 SO_2 O 5 4

48

$X_2\bar{A}l$—H OMs O H S O_2 H

48a

X_2AlO H S O_2 H

48b

H_3O^+

HO H S O_2 H

50

H^-

49

Ms = CH_3SO_2—; X = H, $CH_3OCH_2CH_2O$—

$LiAlH_4$: **49**:**50** = 89:11

$NaAlH_2(OCH_2CH_2OCH_3)_2$: **49**:**50** = 40:60

SCOPE AND LIMITATIONS

Carboxylic Acids and Derivatives

Reduction of Carboxylic Acids. Simple carboxylic acids are reduced to primary alcohols by $LiAlH(OCH_3)_3$,[65,161] calcium alkoxyaluminohydrides,[162–165] or $NaAlH_2(OCH_2CH_2OCH_3)_2$[67–70,166–169] (Eq. 7)[170–172] as readily as with $LiAlH_4$;[65,173,174] $NaAlH_2(OCH_2CH_2OCH_3)_2$, $LiAlH_4$,[123] and diborane[175] show comparable reactivity in transforming alkali metal and halomagnesium salts of carboxylic acids into the corresponding alcohols (cf., however, p. 478 in Ref. 173).

$$i\text{-}C_3H_7(CH_2)_3\overset{*}{C}H(CH_3)CH_2CO_2H \xrightarrow[C_6H_6,\ \text{reflux}]{NaAlH_2(OCH_2CH_2OCH_3)_2} i\text{-}C_3H_7(CH_2)_3\overset{*}{C}H(CH_3)(CH_2)_2OH \quad \text{(Eq. 7)}$$

(3*R*)-(+) → (3*R*)-(+) (97%)

In reductions of poorly soluble carboxylic acids such as 3-hydroxybenzoic acid[119,121] or 9,9′-spirobifluorene-2,2′-dicarboxylic acid,[176] $NaAlH_2(OCH_2CH_2OCH_3)_2$ is a more useful reagent than $LiAlH_4$.

$$3\text{-}HOC_6H_4CO_2H \xrightarrow[\text{xylene, reflux}]{NaAlH_2(OCH_2CH_2OCH_3)_2} 3\text{-}HOC_6H_4CH_2OH\ (72\%)$$

$$\text{9,9'-spirobifluorene-2,2'-}(CO_2H)_2 \xrightarrow[C_6H_6,\ \text{reflux}]{NaAlH_2(OCH_2CH_2OCH_3)_2} \text{9,9'-spirobifluorene-2,2'-}(CH_2OH)_2\ (84\%)$$

γ-Keto carboxylic acids are reduced by $NaAlH_2(OCH_2CH_2\ OCH_3)_2$ either to diols or γ-lactones (Eq. 8)[167]

$$CH_3CO(CH_2)_2CO_2H \xrightarrow[C_6H_6,\ 80^\circ]{NaAlH_2(OCH_2CH_2OCH_3)_2} CH_3CHOH(CH_2)_3OH\ (67\%)$$

$$CH_3CO(CH_2)_2CO_2H \xrightarrow[C_6H_6,\ -20^\circ]{NaAlH_2(OCH_2CH_2OCH_3)_2} \gamma\text{-valerolactone}\ (57\%) \quad \text{(Eq. 8)}$$

In the reduction of the *cis* γ-keto acid **51** to the isomeric γ-lactones **52a** and **52b**, $LiBH(C_2H_5)_3$ and $LiAlH(OC_4H_9\text{-}t)_3$ show the greatest selectivity; catalytic reduction on platinum oxide gives the isomeric lactones in a reversed ratio (26:74).[177]

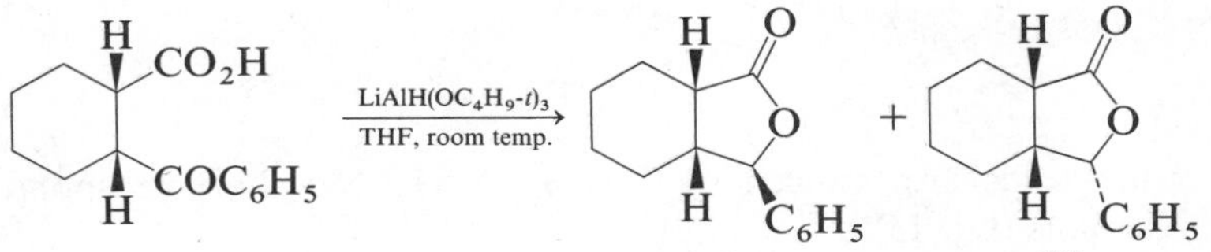

$LiAlH(OC_4H_9\text{-}t)_3$, THF, room temp.

51 → 52a + 52b

52a:**52b** = 94:6

The carboxyl group is not attacked by $NaAlH_2(OCH_2CH_2OCH_3)_2$ at low temperatures (Eq. 9).[178–180]

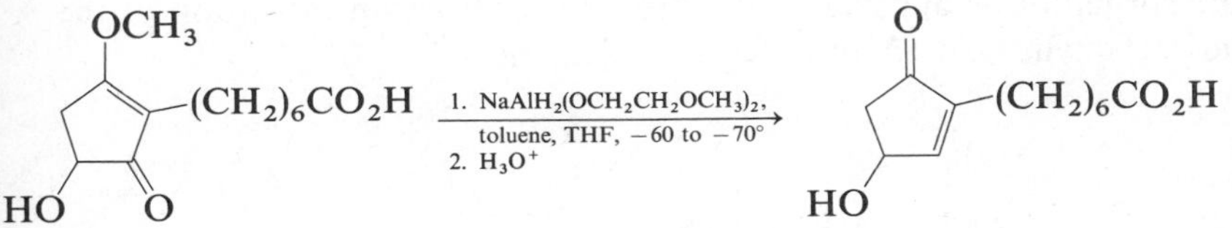

1. $NaAlH_2(OCH_2CH_2OCH_3)_2$, toluene, THF, −60 to −70°
2. H_3O^+

(Eq. 9)

Alternatively, $LiAlH(OC_4H_9\text{-}t)_3$,[63,65,181,182] $NaBH_4$,[183] or $LiBH(C_2H_5)_3$[184] can be used for selective reductions of aldehyde or ketone functions in the presence of a carboxyl. On the other hand, selective transformation of a carboxyl into a hydroxymethyl group in ester–carboxylic acids (Eq. 10)[185] and lactone–carboxylic acids (Eq. 11)[186] is possible via intermediate acyl chlorides.

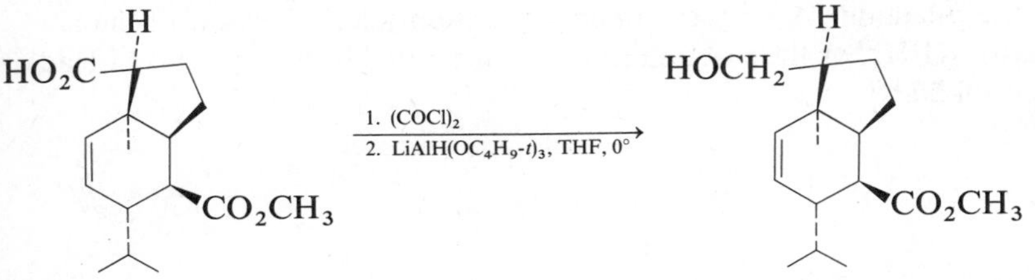

1. $(COCl)_2$
2. $LiAlH(OC_4H_9\text{-}t)_3$, THF, 0°

(90%)

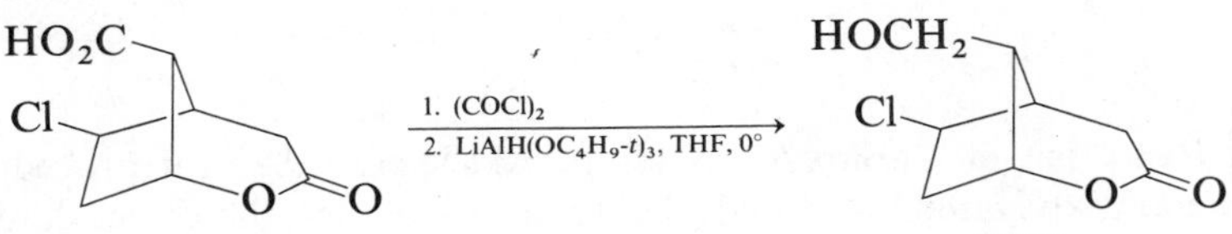

(Eq. 10)

1. $(COCl)_2$
2. $LiAlH(OC_4H_9\text{-}t)_3$, THF, 0°

(52%)

(Eq. 11)

Aromatic carboxylic acids substituted in the *ortho* or *para* positions by a hydroxyl or amino group undergo hydrogenolysis on reaction with $NaAlH_2$-$(OCH_2CH_2OCH_3)_2$ at elevated temperatures,[119–122] as shown in Eq. 12.[122,187]

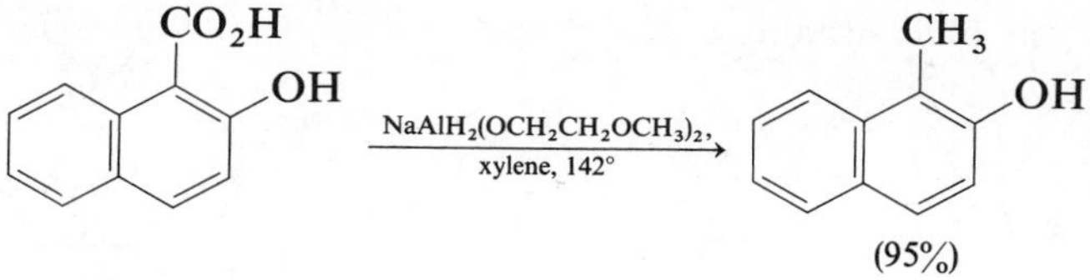

$$\text{2-hydroxy-1-naphthoic acid (CO}_2\text{H, OH)} \xrightarrow[\text{xylene, 142°}]{\text{NaAlH}_2(\text{OCH}_2\text{CH}_2\text{OCH}_3)_2,} \text{1-methyl-2-naphthol (CH}_3\text{, OH) (95\%)} \quad \text{(Eq. 12)}$$

α-Amino acids are reduced by $LiAlH_2(OCH_3)_2$ to the corresponding α-amino alcohols (Eq. 13).[188]

$$\underset{(R)\text{-}(-)}{C_6H_5CH(NH_2)CO_2H} \xrightarrow{\text{LiAlH}_2(\text{OCH}_3)_2} \underset{\substack{(R)\text{-}(-)\\(90\%)\\(\text{optical purity } 99.8\%)}}{C_6H_5CH(NH_2)CH_2OH} \quad \text{(Eq. 13)}$$

Lactam formation is apparently the first step in the transformation of the *cis* amino carboxylic acid **53** into the bicyclic amine.[189]

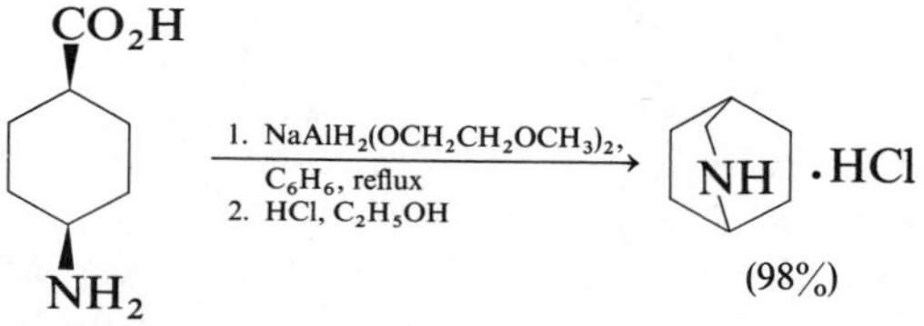

$$\underset{\mathbf{53}}{\text{cis-4-aminocyclohexanecarboxylic acid (CO}_2\text{H, NH}_2)} \xrightarrow[\text{2. HCl, C}_2\text{H}_5\text{OH}]{\text{1. NaAlH}_2(\text{OCH}_2\text{CH}_2\text{OCH}_3)_2,\ \text{C}_6\text{H}_6\text{, reflux}} \text{NH} \cdot \text{HCl (98\%)}$$

Like $LiAlH_4$, $NaAlH_2(OCH_2CH_2OCH_3)_2$ fails to reduce the carboxylic group in the β-lactam **54**; only treatment with diborane, reduction of the acyl azide **55** with KBH_4, or the reaction sequence indicated in Eqs. 10 and 11 leads to the alcohol **56**.[190]

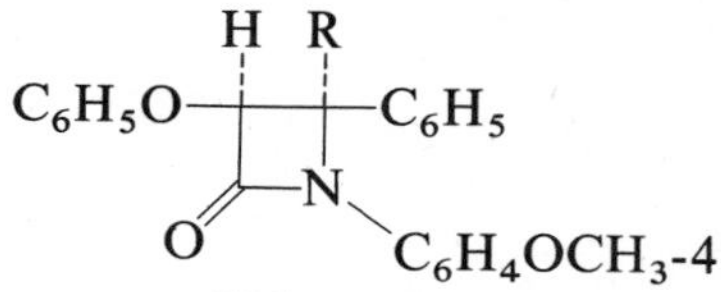

β-Lactam: C_6H_5O and H on C-3, R and C_6H_5 on C-4, C=O, N–$C_6H_4OCH_3$-4

54: R = CO_2H
55: R = CON_3
56: R = CH_2OH

Partial Reduction of Carboxylic Acids to Aldehydes. Several procedures utilizing metal alkoxyaluminum hydrides are available for the direct conversion of aliphatic, alicyclic, aromatic, and α,β-unsaturated carboxylic acids to aldehydes.

Reaction of carboxylic acids with 1,1′-carbonyldiimidazole produces *N*-acylimidazoles that easily undergo partial reduction by $LiAlH(OC_4H_9\text{-}t)_3$[96] to give aldehydes in high yields (Eq. 14).[97]

$$\text{OAc-cyclopentane (CO}_2\text{H, (CH}_2)_6\text{CO}_2\text{CH}_3\text{, dioxolane)} \xrightarrow[\text{2. LiAlH(OC}_4\text{H}_9\text{-}t)_3]{\text{1. 1,1'-carbonyldiimidazole}} \text{OAc-cyclopentane (CHO, (CH}_2)_6\text{CO}_2\text{CH}_3\text{, dioxolane)} \quad (81\%) \qquad \text{(Eq. 14)}$$

Treating carboxylic acids with γ-saccharin chloride rapidly gives *N*-acylsaccharins, which are reduced *in situ* (~2 hours) to aldehydes in 63–80% yields (Eq. 15). Chloro and nitro groups are not attacked under the conditions of the partial reduction.[98]

$$RCO_2H + \text{(γ-saccharin chloride, Cl, N, S, O}_2) \xrightarrow{CH_2Cl_2,\ 0^\circ} \text{(saccharin, O, NCOR, S, O}_2) \xrightarrow[\text{2. }H_3O^+]{\text{1. }NaAlH_2(OCH_2CH_2OCH_3)_2,\ C_6H_6,\ 0\text{-}5^\circ \text{ or } -70^\circ}$$

$$RCHO + \text{(saccharin, O, NH, S, O}_2) \qquad \text{(Eq. 15)}$$

γ-Saccharin chloride can be prepared by chlorination of saccharin with phosphorus pentachloride.[191]

Aliphatic and aromatic carboxylic acids react with *N*,*N*-dimethylchloromethyleniminium chloride (**57**) to form carboxymethyleniminium chlorides **58**; $LiAlH(OC_4H_9\text{-}t)_3$ reduction of these salts *in situ* (10 minutes) and in the presence of copper(I) iodide as catalyst followed by decomposition of the betaine **59** affords aldehydes in 55–90% yields.

$$HCON(CH_3)_2 + (COCl)_2 \xrightarrow{CH_2Cl_2,\ 0^\circ} \underset{\mathbf{57}}{ClCH{=}\overset{+}{N}(CH_3)_2Cl^-} \xrightarrow[\text{THF-CH}_3\text{CN, } -30^\circ]{RCO_2H,\ \text{pyridine,}}$$

$$\underset{\mathbf{58}}{R{-}C(=O){-}O{-}CH{=}\overset{+}{N}(CH_3)_2\ Cl^-} \xrightarrow[\text{THF, } -78^\circ]{LiAlH(OC_4H_9\text{-}t)_3\text{-CuI}} \underset{\mathbf{59}}{H(R)C(O^-){-}O{-}CH{=}\overset{+}{N}(CH_3)_2} \xrightarrow{H_3O^+}$$

$$RCHO + HCON(CH_3)_2$$

Chloro, bromo, cyano, nitro, and ester groups, as well as conjugated carbon–carbon double bonds, are not reduced by the hydride under these conditions. The high reactivity of carboxymethyleniminium salts **58** makes it possible to reduce keto acids to keto aldehydes without protecting the ketone function (Eq. 16).[192]

$$C_2H_5CO(CH_2)_4CO_2H \longrightarrow C_2H_5CO(CH_2)_4CHO \quad (71\%) \qquad \text{(Eq. 16)}$$

Alternatively, bis(*N*-methylpiperazinyl)aluminum hydride reduces carboxylic acids to give the corresponding aldehydes in 62–95% yields.[193]

Reduction of Carboxylic Acid Anhydrides. The reduction of simple aliphatic and aromatic acid anhydrides by $LiAlH(OCH_3)_3$ in tetrahydrofuran,[65,161] $NaAlH_2(OCH_2CH_2OCH_3)_2$ in benzene,[168] or calcium alkoxyaluminum hydrides in toluene[165] at elevated temperatures gives primary alcohols in high yields. The complete reduction of cyclic anhydrides, as shown in Eq. 17,[194] by $LiAlH(OCH_3)_3$,[65,161] AlH_3,[65] or $NaAlH_2(OCH_2CH_2OCH_3)_2$[167,168] often proceeds more rapidly than with $LiAlH_4$.[65,174]

$NaAlH_2(OCH_2CH_2OCH_3)_2$, C_6H_6, reflux → CH_2OH / CH_2OH (97%) (Eq. 17)

Partial Reduction of Cyclic Anhydrides to Lactones and Hydroxylactones. The low-temperature reduction of cyclic anhydrides by complex metal hydrides[63,65,77,174] leads to lactones, hydroxylactones, or mixtures of both. Whereas formation of hydroxylactones predominates in reductions by $LiAlH(OC_4H_9\text{-}t)_3$[47–50,52,58,195] (Eq. 18)[196] and sometimes also by $LiAlH_4$,[49,50,52] lactones are the major products of reductions with $NaBH_4$,[47,48,51–57,196,197] $LiAlH_4$,[46–48,53–55] $LiBH_4$,[56,198] $LiBH(C_2H_5)_3$,[54,56,184] $LiBH(C_4H_9\text{-}s)_3$,[54,55–57] $KBH(C_4H_9\text{-}s)_3$,[57] and lithium trisiamylborohydride.[54]

OAc, O, O, R, O → ($LiAlH(OC_4H_9\text{-}t)_3$, THF, −20°) → OAc, O, O, R, OH (83%) (Eq. 18)

R = CH_2Cl

The effect of the hydride and of the reaction conditions on the product distribution in the reduction of an unsymmetrical cyclic anhydride such as **60** is illustrated in Eq. 19.

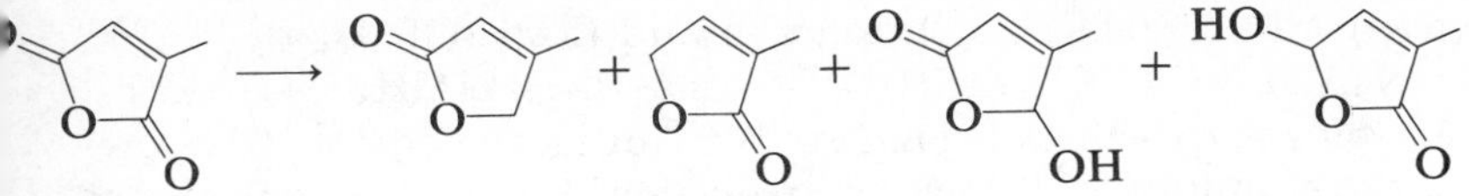

60 → A + B + C + D (Eq. 19)

	Product ratio				Refs.
	A	B	C	D	
$NaBH_4$	85	15	0	0	51, 61
$LiAlH_4$	90	10	0	0	51, 61
$KBH(C_4H_9\text{-}s)_3$	0	~100	0	0	57
$LiAlH(OC_4H_9\text{-}t)_3$, $(C_2H_5)_2O$[a]	0	0	89	11	58
$LiAlH(OC_4H_9\text{-}t)_3$, DME[b,c]	35	3	53	9	49, 52

[a] The reaction was carried at −55° to room temperature for 3 hours.
[b] The abbreviation DME denotes 1,2-dimethoxyethane.
[c] The reaction was carried out at −30° for 1 hour and then at 25° for 12 hours.

Reduction of Lactones and Lactols. Simple aliphatic and aromatic lactones are reduced by $LiAlH(OCH_3)_3$[65,161] to diols as rapidly as with $LiAlH_4$,[65,174] AlH_3,[65] or $LiBH(C_2H_5)_3$;[184] calcium alkoxyaluminum hydrides such as $Ca[AlH_2(OC_4H_9\text{-}i)_2]_2$[165] and $NaAlH_2(OCH_2CH_2OCH_3)_2$[167,168] require longer reaction times (Eq. 20)[199] or elevated temperatures[199b–d] to obtain diols in high yields.

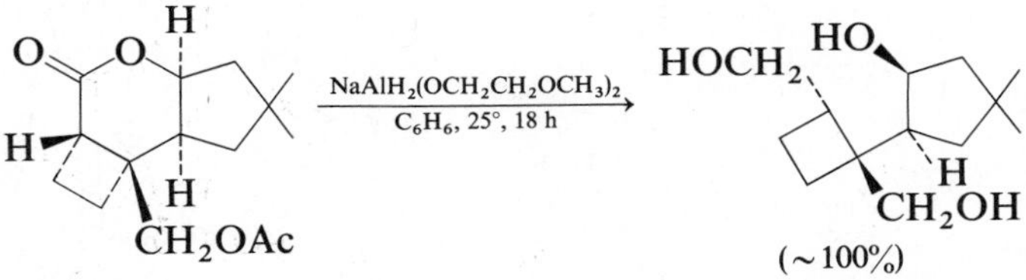

$\xrightarrow[C_6H_6,\ 25°,\ 18\ h]{NaAlH_2(OCH_2CH_2OCH_3)_2}$ (~100%) (Eq. 20)

Aromatic lactones such as **61**, which bear a vicinal electron-donating amino group, undergo hydrogenolysis rather than reduction under forcing conditions. This conversion of **61** can be effected with neither $LiAlH_4$ nor diborane.[200–204]

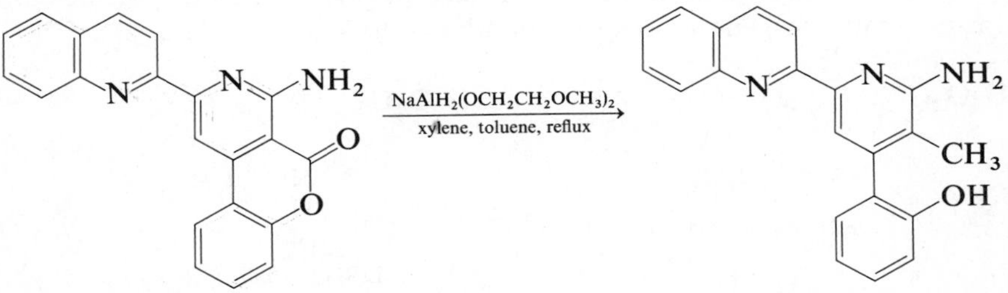

61 $\xrightarrow[\text{xylene, toluene, reflux}]{NaAlH_2(OCH_2CH_2OCH_3)_2}$ (78%)

The relatively greater stability of lactones toward $LiAlH(OC_4H_9\text{-}t)_3$,[1,63,65,186,205–208] $NaAlH_2(OCH_2CH_2OCH_3)_2$,[209] and the $LiAlH_4$-(*S*)-(−)-2,2′-dihydroxy-1,1′-binaphthyl-ethanol complex[210–216] or its analogs[217,218] at low temperatures can be utilized for the selective reduction of the ketone or aldehyde function in keto or formyl lactones. Translactonization is sterically favored following the reduction of the keto δ-lactone **62**.[219]

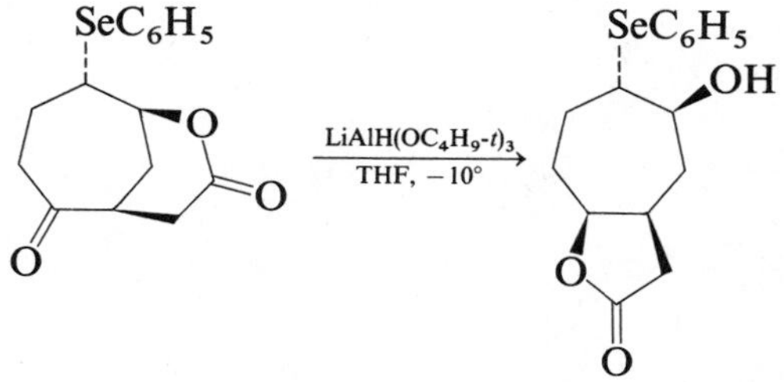

SeC_6H_5 … SeC_6H_5, OH

$\xrightarrow[\text{THF, } -10°]{LiAlH(OC_4H_9\text{-}t)_3}$

62 (83%)

At elevated temperatures, however, lactones such as **63** are reduced to diols by $LiAlH(OC_4H_9\text{-}t)_3$; $LiAlH_4$ affords a mixture of isomeric hemiketals **64**.[220]

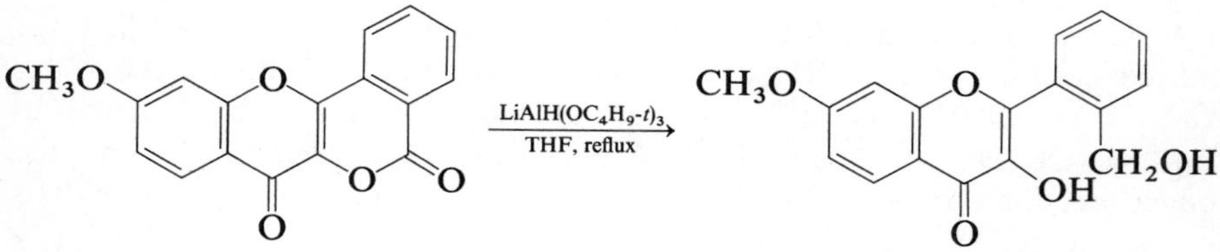

CH_3O … $\xrightarrow[\text{THF, reflux}]{LiAlH(OC_4H_9\text{-}t)_3}$ CH_3O … CH_2OH, OH

63 (51%)

63 $\xrightarrow[\text{THF, reflux}]{LiAlH_4}$ CH_3O … OH

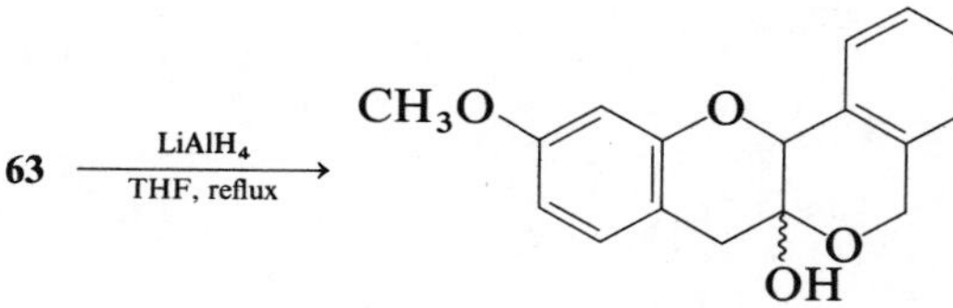

64
cis:*trans* = 83:17

Lactols are usually stable toward $LiAlH(OC_4H_9\text{-}t)_3$, and $LiBH_4$ is therefore a more suitable reagent for cleavage of a lactol to an alcohol (Eq. 21).[221a]

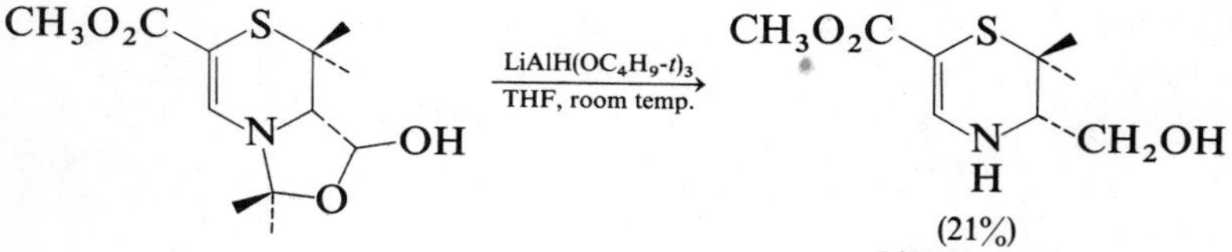

CH_3O_2C … OH $\xrightarrow[\text{THF, room temp.}]{LiAlH(OC_4H_9\text{-}t)_3}$ CH_3O_2C … CH_2OH (Eq. 21)

(21%)
$LiBH_4$: ~99%

Partial Reduction of Lactones to Lactols. The conversion of lactones to lactols by means of metal alkoxyaluminum hydrides[221b–e] often provides a valuable alternative to the known method using $AlH(OC_4H_9\text{-}i)_2$.[222] With $LiAlH(OC_4H_9\text{-}t)_3$[223] only the γ-lactone is reduced in the bicyclic dilactone **65**.[224]

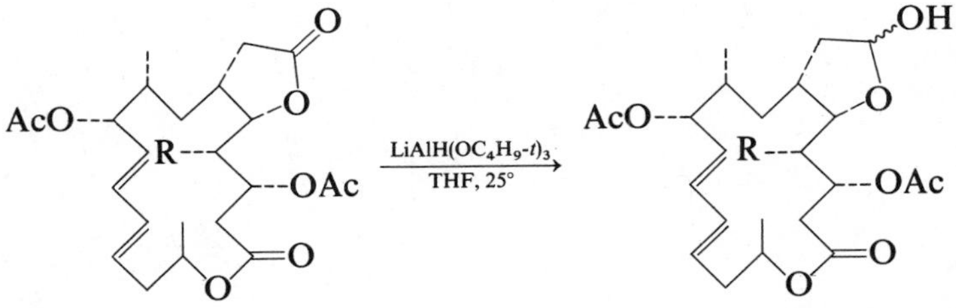

AcO, R, OAc, O; **65** —$LiAlH(OC_4H_9\text{-}t)_3$, THF, 25°→ AcO, R, OH, OAc, O (~100%)

R = CH_3O

The conversion of some terpene lactones to lactols is effected by $LiAlH(OC_2H_5)_3$ (Eq. 22).[255]

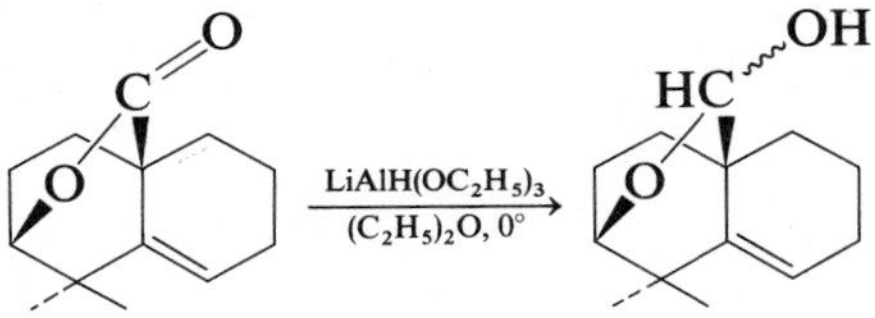

$$\xrightarrow[(C_2H_5)_2O,\ 0^\circ]{LiAlH(OC_2H_5)_3} \quad (89\%) \qquad \text{(Eq. 22)}$$

High stereospecificity in the lactone reduction can be achieved by using $NaAlH_2(OCH_2CH_2OCH_3)_2$ modified with ethanol;[73] noteworthy is the stability of the ester group in the ester lactone **66** under these reaction conditions.[226,227]

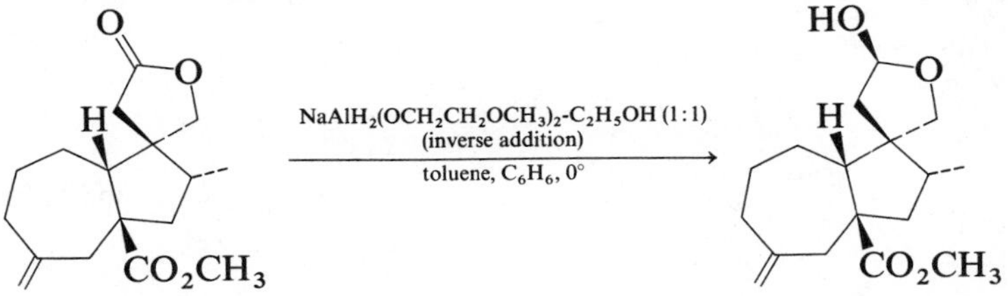

66 (CO_2CH_3) —$NaAlH_2(OCH_2CH_2OCH_3)_2\text{-}C_2H_5OH$ (1:1) (inverse addition), toluene, C_6H_6, 0°→ HO, H, CO_2CH_3 (83%)

$NaAlH_2(OCH_2CH_2OCH_3)_2$ modified by 2-propanol (Eq. 23)[228] or the same hydride in pyridine solution (Eq. 24)[229] can be used for selective conversions of lactones to lactols in the presence of an amide group.

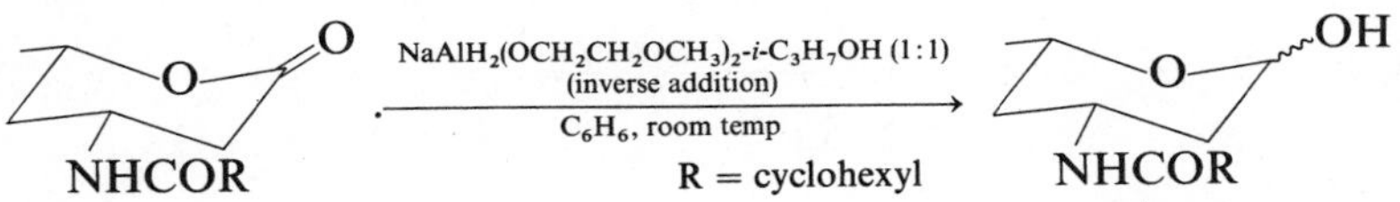

NHCOR —$NaAlH_2(OCH_2CH_2OCH_3)_2\text{-}i\text{-}C_3H_7OH$ (1:1) (inverse addition), C_6H_6, room temp→ NHCOR, OH (67%) (Eq. 23)

R = cyclohexyl

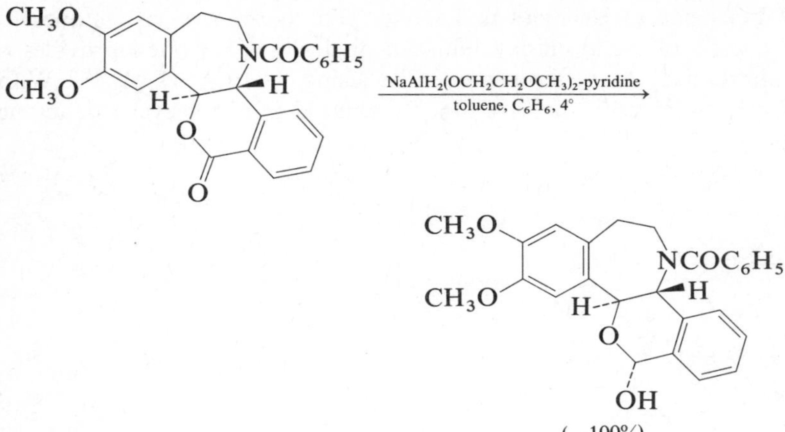

$$\xrightarrow[\text{toluene, } C_6H_6,\ 4^\circ]{NaAlH_2(OCH_2CH_2OCH_3)_2\text{-pyridine}}$$

(~100%) (Eq. 24)

The low-temperature partial reduction of the cyanolactone **67** by $NaAlH_2(OCH_2CH_2OCH_3)_2$ at prolonged reaction time gives the corresponding cyanolactol; ester and amido groups in lactones of similar structure are not reduced under these conditions. Interestingly, $AlH(C_4H_9\text{-}i)_2$ is less selective and attacks ester, cyano, and amido functions even at low temperature.[230]

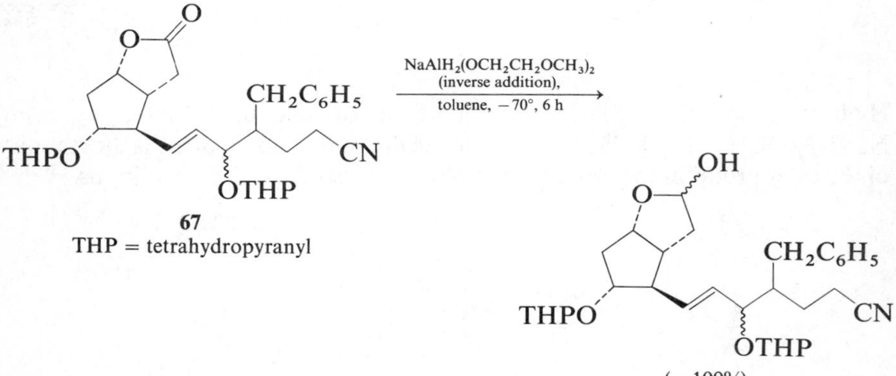

67
THP = tetrahydropyranyl

$$\xrightarrow[\text{toluene, } -70^\circ,\ 6\ h]{NaAlH_2(OCH_2CH_2OCH_3)_2 \text{ (inverse addition),}}$$

(~100%)

Reduction of Unsaturated Lactones. Fused furan derivatives can be prepared by reduction of α,β-unsaturated γ-lactones with $NaAlH_2(OCH_2CH_2OCH_3)_2$ (Eq. 25).[231]

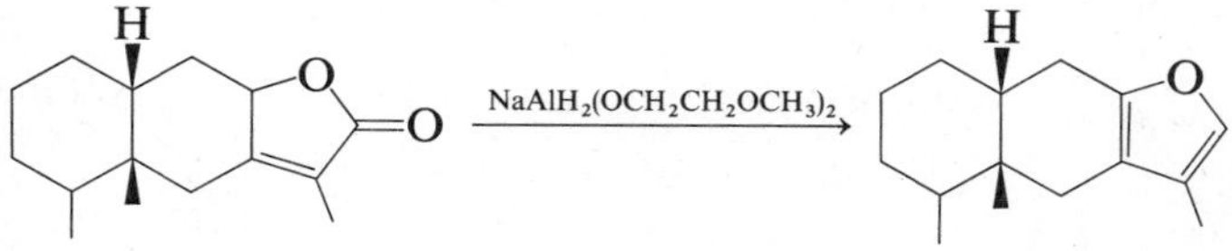

$$\xrightarrow{NaAlH_2(OCH_2CH_2OCH_3)_2}$$

(~100%) (Eq. 25)

dl-Menthofuran (**68**) can be obtained similarly by using $LiAlH_4$–2-propanol in excess;[232] $AlH(C_4H_9\text{-}i)_2$ is also useful in this reaction.[222]

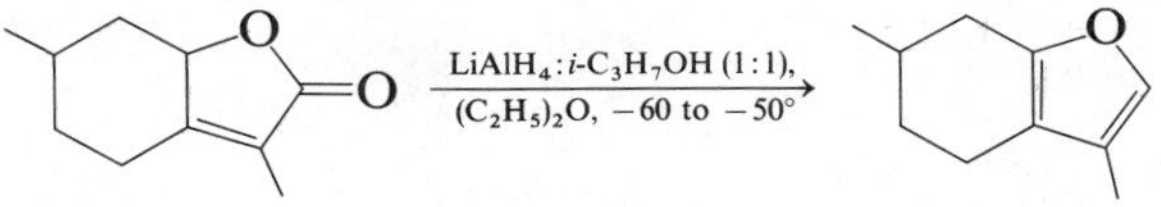
O
O
LiAlH4 : i-C3H7OH (1 : 1),
(C2H5)2O, −60 to −50°
O

68 (75%)

The $NaAlH_2(OCH_2CH_2OCH_3)_2$–CuBr–2-butanol complex[113,114] is an effective reagent in the conjugate reduction of α,β-unsaturated γ-lactones **69**. The reaction represents one step in the preparation of optically active 4-alkenyl-γ-lactones **70** from optically active alkyl 4-hydroxy-2-alkynoates.[233]

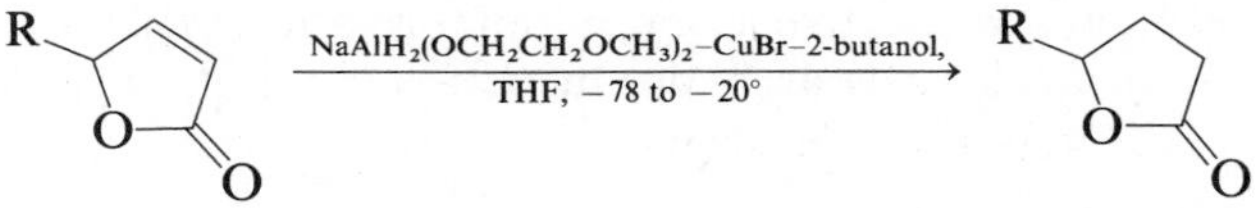
R
O
O
NaAlH2(OCH2CH2OCH3)2–CuBr–2-butanol,
THF, −78 to −20°
R
O
O

69 **70** (~100%)

R = *cis*-*n*-$C_5H_{11}CH{=}CHCH_2$—,
cis-*n*-$C_8H_{17}CH{=}CH$—

α,β-Unsaturated δ-lactones are converted by metal alkoxyaluminum hydrides into either unsaturated diols (Eq. 26)[234] or δ-lactols (Eq. 27).[235]

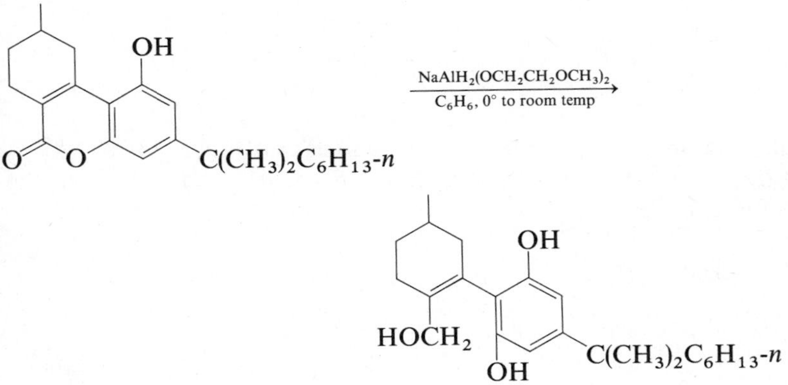
OH
O
O
C(CH3)2C6H13-n
NaAlH2(OCH2CH2OCH3)2
C6H6, 0° to room temp
OH
HOCH2
OH
C(CH3)2C6H13-n

(Eq. 26)

(98%)

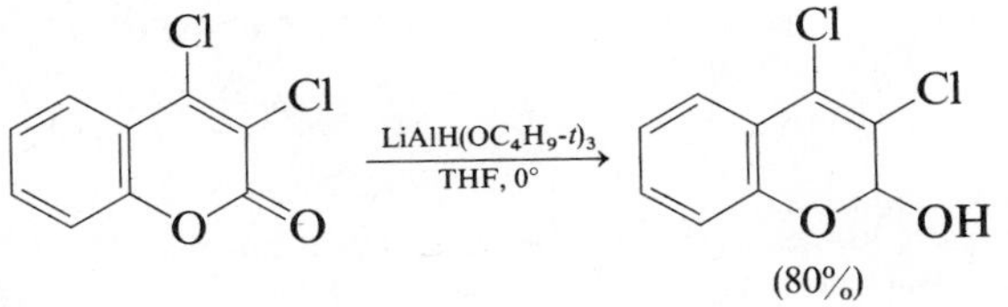
Cl
Cl
O
O
LiAlH(OC4H9-t)3
THF, 0°
Cl
Cl
O
OH

(Eq. 27)

(80%)

In the presence of $LiAlH(OC_4H_9\text{-}t)_3$ and at low temperature, δ-enol lactones with an exocyclic carbon–carbon double bond undergo reductive rearrangement to give bridged ketols[143a,236a–c] (Eq. 28);[236c] $LiAlH(OC_4H_9\text{-}t)_3$ at elevated temperature[143a,236d] and $LiAlH_4$[138] afford the corresponding diols.

CH_3O_2C … OCH_3 $\xrightarrow[\text{THF, } -70^\circ]{LiAlH(OC_4H_9\text{-}t)_3}$ CH_3O_2C … OCH_3, HO (91%) (Eq. 28)

The reaction is often less selective as a result of formation of saturated lactones, keto aldehydes, or keto acids as byproducts;[137,139,141,142] in these reactions $AlH(C_4H_9\text{-}i)_2$ is a more suitable reagent.[222]

Reduction of Carboxylic Acid Esters. Alkyl esters of carboxylic acids react with $LiAlH(OCH_3)_3$[65,81,161] at 0° as rapidly as with $LiAlH_4$[65,81,173,174] to form primary alcohols in high yields,* and the former hydride thus finds use more as a selective reducing agent (Eq. 29).[237]

CH_3O_2C, H, OH, O $\xrightarrow[\text{THF, } 23^\circ]{LiAlH(OCH_3)_3}$ $HOCH_2$, H, OH, O (93%) (Eq. 29)

Calcium alkoxyaluminum hydrides[165] and $NaAlH_2(OCH_2CH_2OCH_3)_2$[67,68,167,168,238] (Eq. 30)[239] exhibit somewhat lower reactivity than $LiAlH_4$ in the reduction of esters.

$$C_2H_5O_2C\underset{\substack{|\\OCH(OC_2H_5)CH_3}}{C}HCH_2CO_2C_2H_5 \xrightarrow[\text{ether, reflux, 2 h}]{NaAlH_2(OCH_2CH_2OCH_3)_2} HOCH_2\underset{\substack{|\\OCH(OC_2H_5)CH_3}}{C}H(CH_2)_2OH \qquad \text{(Eq. 30)}$$

(2S) → (2S) (95%)

* Differences in reactivity can occur at low temperatures; in contrast to $LiAlH_4$, $LiAlH(OCH_3)_3$ does not reduce ethyl benzoate at −80°.[81]

Ortho and *para* amino- and hydroxy-substituted benzoic esters undergo either reduction or hydrogenolysis by $NaAlH_2(OCH_2CH_2OCH_3)_2$, depending on reaction conditions[119–121] (Eq. 31).[240]

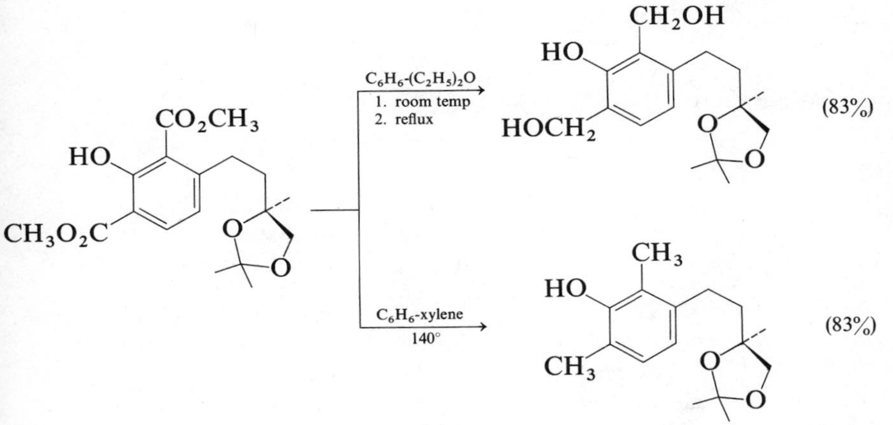

C_6H_6-$(C_2H_5)_2O$: 1. room temp; 2. reflux → (83%)

C_6H_6-xylene, 140° → (83%)

(Eq. 31)

The aromatic vicinal diol diacetate **71** gives *o*-quinone **72** on reaction with $NaAlH_2(OCH_2CH_2OCH_3)_2$ and air.[241]

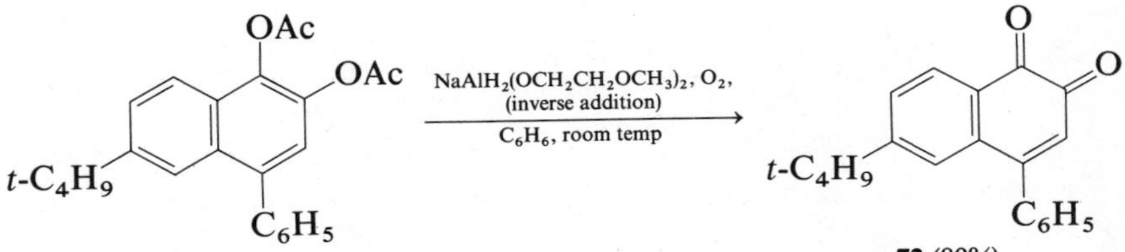

$NaAlH_2(OCH_2CH_2OCH_3)_2$, O_2, (inverse addition), C_6H_6, room temp

71 → **72** (80%)

Replacement of $LiAlH_4$ by $NaAlH_2(OCH_2CH_2OCH_3)_2$ minimizes formation of the undesired reduced product **73**[242a,b] in the malonic enolate reduction–elimination reaction.[243–245]

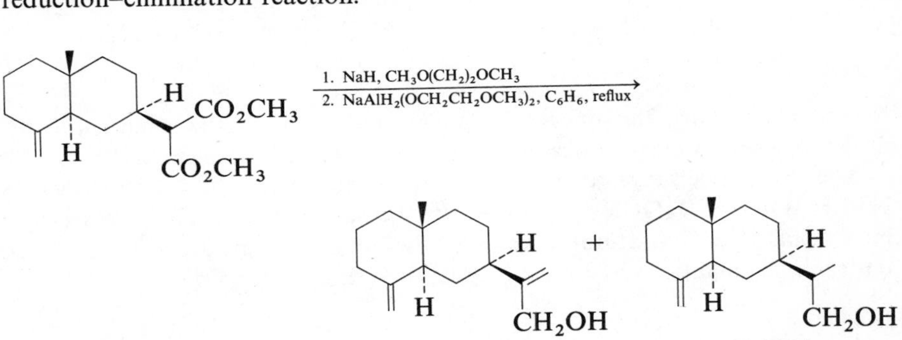

1. NaH, $CH_3O(CH_2)_2OCH_3$
2. $NaAlH_2(OCH_2CH_2OCH_3)_2$, C_6H_6, reflux

(74%) + **73** (13%)

Ester groups in amido esters can be reduced selectively by $NaAlH_2(OCH_2CH_2OCH_3)_2$ at lower temperatures and shorter reaction time (Eq. 32).[246]

$$\underset{dl\text{-}erythro}{C_6H_5CHOH\underset{|}{C}HCO_2C_2H_5 \atop \quad NHCOC_6H_5} \xrightarrow[\text{toluene, } 0^\circ,\ 1\text{ h}]{NaAlH_2(OCH_2CH_2OCH_3)_2} \underset{dl\text{-}erythro\ (76\%)}{C_6H_5CHOH\underset{|}{C}HCH_2OH \atop \quad NHCOC_6H_5} \quad \text{(Eq. 32)}$$

Generally, $LiAlH(OC_4H_9\text{-}t)_3$ reacts extremely slowly with alkyl and cycloalkyl esters of carboxylic acids under mild conditions (0° to room temperature);[63,65,247–255] see, however, Ref. 256. This low reactivity can be utilized with advantage for selective reductions of other reducible groups in functionalized esters (Eq. 33).[257]

$$\text{2-oxo-cyclopentane (with } CO_2CH_3 \text{ and cis-pentenyl side chain)} \xrightarrow[\text{THF, } 0\text{–}5^\circ]{LiAlH(OC_4H_9\text{-}t)_3} \text{2-hydroxy-cyclopentane (with } CO_2CH_3\text{)}\ (99\%) \quad \text{(Eq. 33)}$$

Reductions of keto esters with chiral lithium alkoxyaluminum hydrides[258,259] such as the Mosher–Yamaguchi complex (Eq. 34)[1,260–262] also proceed with preservation of the ester group and give optically active hydroxycarboxylic acid esters (Eq. 34).[263]

$$n\text{-}C_7H_{15}COCO_2CH_3 \xrightarrow[\text{1,2-diphenyl-3-methyl-2-butanol,* } (C_2H_5)_2O]{LiAlH_4\text{-}(2S,3R)\text{-}(+)\text{-4-dimethylamino-}} (+)\text{-}n\text{-}C_7H_{15}CHOHCO_2CH_3\ (62\%\ \text{e.e.}) \quad \text{(Eq. 34)}$$

However, ketol acetates undergo complete reduction by the same chiral hydride complex to yield optically active diols (see, e.g., p. 63 in Ref. 1).[264–266]

The low-temperature reduction of enol esters of alicyclic 1,3-diketones by $NaAlH_2(OCH_2CH_2OCH_3)_2$ followed by an acid-catalyzed allylic rearrangement leads to α,β-unsaturated ketones; the alkyl ester group is not attacked under these conditions (see p. 84 in Ref. 1)[267] (Eq. 35).[268,269]

* CHIRALD™, Aldrich Chemical Company.

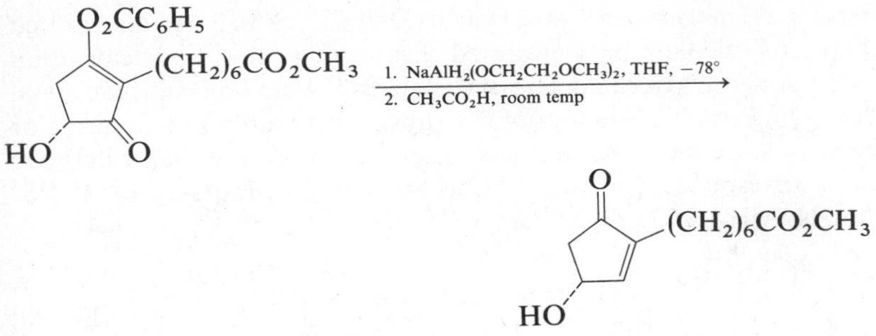

$O_2CC_6H_5$, $(CH_2)_6CO_2CH_3$, HO, O

1. $NaAlH_2(OCH_2CH_2OCH_3)_2$, THF, −78°
2. CH_3CO_2H, room temp

O, $(CH_2)_6CO_2CH_3$, HO

(60%)

(Eq. 35)

Reduction with $LiAlH(OC_4H_9\text{-}t)_3$ at elevated temperatures represents a synthetically useful procedure.[270–275] For instance an alkyl ester can thus be reduced selectively in the presence of a peroxide group (Eq. 36).[272]

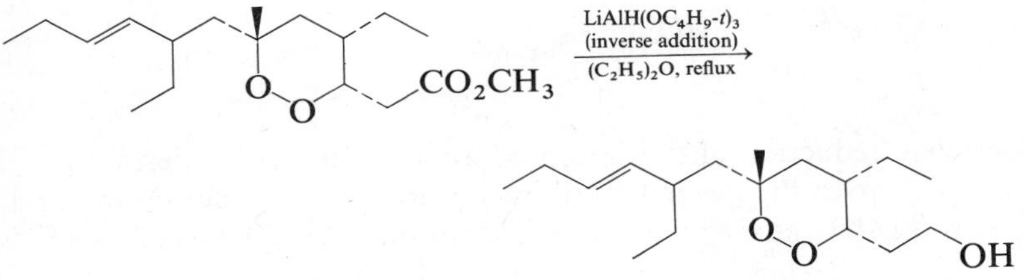

O, O, CO_2CH_3

$LiAlH(OC_4H_9\text{-}t)_3$
(inverse addition)
$(C_2H_5)_2O$, reflux

O, O, OH

(85%)

(Eq. 36)

Ester reductions by this hydride under forcing conditions are sometimes accompanied by cleavage of a carbon–carbon bond and an elimination reaction (Eq. 37).[271]

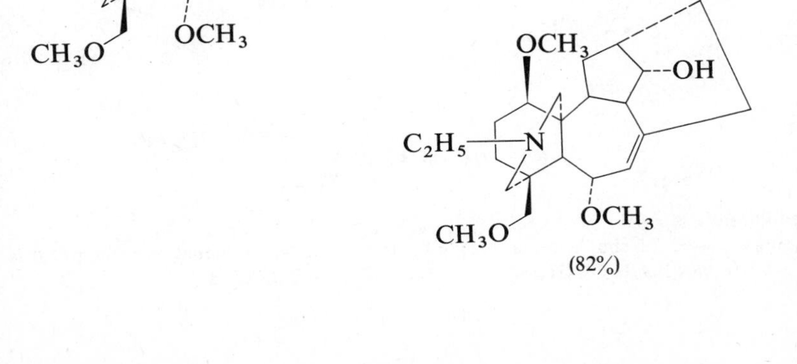

OCH_3, OCH_3, OAc, C_2H_5, N, OAc, OCH_3, CH_3O

$LiAlH(OC_4H_9\text{-}t)_3$
$(CH_3OCH_2CH_2)_2O$, 140°

OCH_3, OCH_3, OH, C_2H_5, N, OCH_3, CH_3O

(82%)

(Eq. 37)

Reduction of α,β-Unsaturated Carboxylic Acids and Esters. An unsaturated alcohol, a saturated alcohol, or a saturated carboxylic acid or ester can result from reduction of α,β-unsaturated carboxylic acids or esters by complex metal hydrides, depending on the structure of the substrate, nature of the hydride or solvent, and reaction conditions. A wide variety of α,β-unsaturated esters are reduced to the corresponding allylic alcohols by using $LiAlH_3(OC_2H_5)$*[105,106] (Eq. 38)[276] or $NaAlH_2(OCH_2CH_2OCH_3)_2$[277a–f] (Eq. 39).[110]

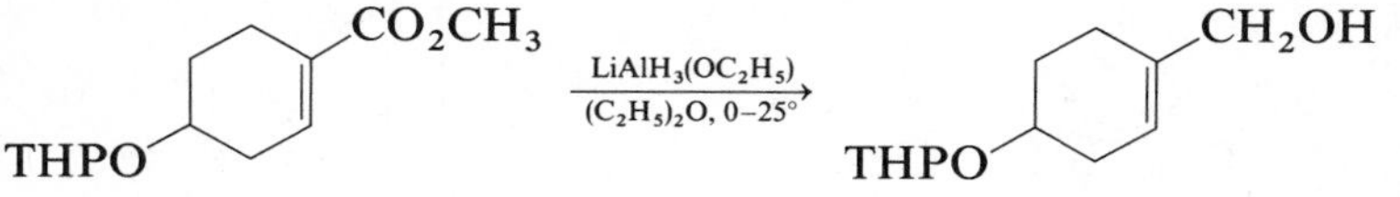

THPO–C₆H₈–CO₂CH₃ $\xrightarrow[(C_2H_5)_2O,\ 0\text{–}25^\circ]{LiAlH_3(OC_2H_5)}$ THPO–C₆H₈–CH₂OH (97%) (Eq. 38)

CH_3O_2C ... CO_2CH_3 $\xrightarrow[C_6H_6,\ 25^\circ]{NaAlH_2(OCH_2CH_2OCH_3)_2}$ HO ... OH (96%) (Eq. 39)

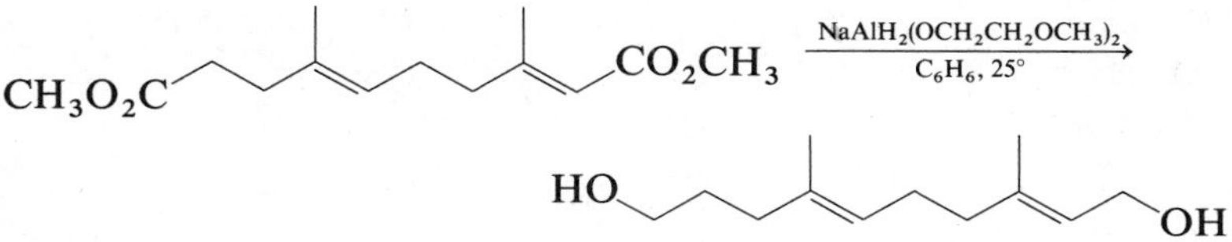

Although selective reduction of the saturated ester group in the diester **74** by $LiAlH_2(OC_2H_5)_2$† gives the desired hydroxy α,β-unsaturated ester **75** in good yield, the use of $LiAlH_4$ at -78° is preferred, affording **75** in 81% yield.[277g,278]

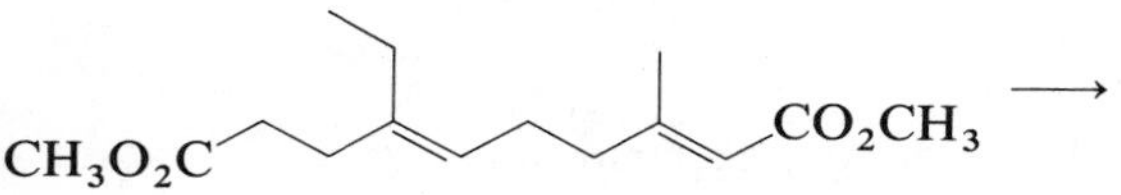

CH_3O_2C ... CO_2CH_3 (**74**) $\longrightarrow$ HO ... CO_2CH_3 (**75**)

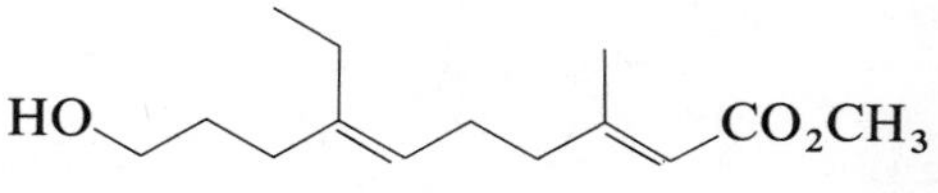

Alternatively, the lower reactivity of the *tert*-butoxycarbonyl group can be utilized for the selective reduction of a conjugated methyl ester group in diesters (Eq. 40).[279]

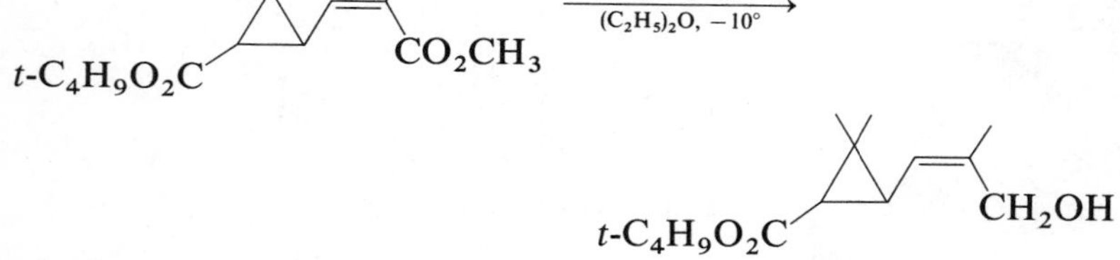

$t\text{-}C_4H_9O_2C$ ... CO_2CH_3 $\xrightarrow[(C_2H_5)_2O,\ -10^\circ]{LiAlH_3(OC_2H_5)(1\ eq)}$ $t\text{-}C_4H_9O_2C$ ... CH_2OH (89%) (Eq. 40)

* See the footnote on p. 256.

† The formula is used for simplicity; according to recent results, the actual reducing agent is presumably a mixture of $LiAlH(OC_2H_5)_3$ and $LiAlH_4$ (see pp. 4–7 in Ref. 1).

Selective 1,2 reduction of an α,β-unsaturated ester has been performed by using $LiAlH_2(OCH_3)_2$ or $LiAlD_2(OCH_3)_2$ (Eq. 41).[280]

$$\xrightarrow[\text{2. } (CH_3CO)_2O\text{, pyridine}]{\text{1. } LiAlD_2(OCH_3)_2\text{, THF, } -20 \text{ to } 0°}$$

(Eq. 41)

(91%)

The conjugate reduction of acyclic α,β-unsaturated esters by the $NaAlH_2(OCH_2CH_2OCH_3)_2$–CuBr–2-butanol complex (Eq. 42) gives the saturated esters in 82–92% yields.[113,114]

$$(CH_3)_2C{=}CHCO_2CH_3 \xrightarrow[\text{THF, } -78 \text{ to } -20°]{NaAlH_2(OCH_2CH_2OCH_3)_2\text{–CuBr (1:1), 2-butanol,}} i\text{-}C_4H_9CO_2CH_3 \ (92\%) \qquad \text{(Eq. 42)}$$

Reduction of methyl α-alkynoates (Eq. 43) is less selective and produces a mixture of *cis* and *trans* α,β-unsaturated esters along with the saturated ester.[113,114]

$$C_6H_5C{\equiv}CCO_2CH_3 \xrightarrow[C_6H_6\text{, THF, } -78 \text{ to } -20°]{NaAlH_2(OCH_2CH_2OCH_3)_2\text{–CuBr(1:1), 2-butanol}}$$

$$\underset{(47\%)}{C_6H_5CH{=}CHCO_2CH_3 \ (cis)} + \underset{(21\%)}{C_6H_5CH{=}CHCO_2CH_3 \ (trans)} + \underset{(6\%)}{C_6H_5(CH_2)_2CO_2CH_3} \qquad \text{(Eq. 43)}$$

Nitrile and ester groups are stable under conditions of the conjugate reduction; ketones, aldehydes, and bromides are attacked by the hydride complex and thus limit its use.[114] Alternatively, iron pentacarbonyl–sodium hydroxide[281,282] and $LiBH(C_4H_9\text{-}s)_3$[283] are recommended for 1,4 reductions of α,β-unsaturated esters.

Partial Reduction of Carboxylic Acid Esters to Aldehydes. Reduction of carboxylic acid esters by metal alkoxyaluminum hydrides provides a valuable alternative to the preparation of aldehydes from esters by low-temperature reductions with $AlH(C_4H_9\text{-}i)_2$,[59,67,222,284–286] $NaAlH_4$,[67,287,288] and $NaAlH_2(C_4H_9\text{-}i)_2$.[289] Methyl and 2-methoxyethyl esters of aliphatic carboxylic acids undergo partial reduction with $NaAlH_2(OCH_2CH_2OCH_3)_2$ to form aldehydes in 70–90% yields (Eq. 44).[67–72]

$$n\text{-}C_5H_{11}CO_2CH_3 \xrightarrow[C_6H_6,\ (C_2H_5)_2O,\ -70^\circ]{NaAlH_2(OCH_2CH_2OCH_3)_2\ \text{(inverse addition)}} n\text{-}C_5H_{11}CHO\ (86\%) \qquad \text{(Eq. 44)}$$

Reduction of the diester **76** gives the dialdehyde in the form of a dihydrate **77**.[74,75]

$$(CF_2)_4(CO_2CH_3)_2\ (\mathbf{76}) \xrightarrow[THF,\ C_6H_6,\ -70^\circ]{NaAlH_2(OCH_2CH_2OCH_3)_2} (CF_2)_4[CH(OH)_2]_2\ (\mathbf{77}\ (96\%))$$

Since the hydride does not react with phenyl[68] or *tert*-butyl carboxylates at low temperature,[242b] selective reductions of diesters can be performed by using this procedure (Eq. 45).[290]

$$\text{1-}[(CH_2)_6CO_2C_4H_9\text{-}t]\text{-5-}(CO_2CH_3)\text{-2-pyrrolidinone} \xrightarrow[THF,\ -78^\circ]{NaAlH_2(OCH_2CH_2OCH_3)_2} \text{1-}[(CH_2)_6CO_2C_4H_9\text{-}t]\text{-5-}(CHO)\text{-2-pyrrolidinone}\ (90\%) \qquad \text{(Eq. 45)}$$

Under the conditions of partial reduction, the hydride tolerates not only amide[246] but also nitrile groups; the acetoxy nitrile **78** is transformed into the desired hydroxy nitrile via the stable hemiacetal **79**.[291]

$$\mathbf{78}\ (CH_2CN,\ OAc,\ C_4H_9\text{-}t\ \text{cyclohexane}) \xrightarrow[C_6H_6,\ (C_2H_5)_2O,\ -70\ \text{to}\ -40^\circ]{NaAlH_2(OCH_2CH_2OCH_3)_2} \mathbf{79}\ (CH_2CN,\ OCHOHCH_3,\ C_4H_9\text{-}t) \xrightarrow[50-60^\circ]{H_3O^+} (CH_2CN,\ OH,\ C_4H_9\text{-}t)$$

The yields of aldehydes in partial reductions of alkyl arene- and aralkanecarboxylates can be significantly improved by modifying the parent hydride with heterocyclic bases[73] (Eqs. 46[292] and 47[73]).

CO_2CH_3 (3,5-dimethoxy-4-fluoro, CH_3O, OCH_3, F) $\xrightarrow[\text{toluene, 0 to } -5^\circ]{NaAlH_2(OCH_2CH_2OCH_3)_2\text{-}N\text{-methylhexahydropyrazine (inverse addition)}}$ CHO (CH_3O, OCH_3, F) (91%) (Eq. 46)

$(CH_2)_2CO_2CH_3$ (OCH_3) $\xrightarrow[\text{toluene, } -55 \text{ to } -40^\circ]{NaAlH_2(OCH_2CH_2OCH_3)_2\text{–}N\text{-methylhexahydropyrazine (inverse addition)}}$ $(CH_2)_2CHO$ (OCH_3) (88%) (Eq. 47)

The morpholine-modified hydride can be used for the partial reduction of aromatic (Eq. 48)[73] as well as heterocyclic (Eq. 49)[293] α-β-unsaturated esters; replacement of the morpholine-modified hydride in the latter reaction (Eq. 49) by $LiAlH_4$ leads to decomposition.[293]

$CH{=}CHCO_2CH_3$ (OCH_3) $\xrightarrow[\text{toluene, } C_6H_6, 0^\circ]{NaAlH_2(OCH_2CH_2OCH_3)_2\text{–morpholine (inverse addition)}}$ $CH{=}CHCHO$ (OCH_3) (84%) (Eq. 48)

CO_2CH_3 (O, N, CH_3) $\xrightarrow[\text{toluene, } -40^\circ]{NaAlH_2(OCH_2CH_2OCH_3)_2}$ CHO (O, N, CH_3) (95%) (Eq. 49)

Alternatively, $LiAlH(OC_4H_9\text{-}t)_3$ reduces phenyl esters of alicyclic, arylaliphatic, and particularly aliphatic carboxylic acids to give the corresponding aldehydes in yields of up to 77% (Eq. 50).[7,59,63–67,161,174] Alcohols are formed as byproducts.[64]

$$t\text{-}C_4H_9CO_2C_6H_5 \xrightarrow[\text{THF, 0°}]{\substack{LiAlH(OC_4H_9\text{-}t)_3 \\ \text{(inverse addition)}}} t\text{-}C_4H_9CHO \quad (67\%) \qquad \text{(Eq. 50)}$$

Reduction of Acyl Chlorides. Simple acyl chlorides are reduced rapidly by $LiAlH(OCH_3)_3$[65,161] and $LiAlH(OC_4H_9\text{-}t)_3$[63,65,77,174] at 0° to give the primary alcohols in yields comparable to those obtained by using $LiAlH_4$;[65,174] $Ca[AlH_2(OC_4H_9\text{-}i)_2]_2$[165] and $NaAlH_2(OCH_2CH_2OCH_3)_2$[67–70,167,168] are somewhat less reactive, and higher temperatures (25–105°) are needed to obtain alcohols in high yields. Consequently, $LiAlH_4$ is the reagent of choice unless the presence of other easily reducible groups or carbon–carbon double bonds requires the use of a more selective reagent (see, e.g., Refs. 185, 186, and 294a) (Eqs. 51[165] and 52[294b]).

$$C_6H_5CH{=}CHCOCl \xrightarrow[\text{toluene, 25°}]{Ca[AlH_2(OC_3H_7\text{-}i)_2]_2\cdot THF} C_6H_5CH{=}CHCH_2OH \quad (\sim 100\%) \qquad \text{(Eq. 51)}$$

$$\text{1,2-}(C_6H_5)_2\text{-cyclopropene-3-}CO_2H \xrightarrow[\substack{2.\ LiAlH(OC_4H_9\text{-}t)_3, \\ (CH_3OCH_2CH_2)_2O,\ 0°}]{1.\ (COCl)_2} \text{1,2-}(C_6H_5)_2\text{-cyclopropene-3-}CH_2OH \quad (78\%) \qquad \text{(Eq. 52)}$$

For the same purpose, $NaBH_4$,[183,295,296] $NaBH(OCH_3)_3$,[295,296] or tetra-*n*-butylammonium octahydrotriborate[297] can be used; the last reduces acyl chlorides to primary alcohols in the presence of keto and formyl groups.

Partial Reduction of Acyl Chlorides to Aldehydes. Aromatic acyl chlorides undergo reduction by $LiAlH(OC_4H_9\text{-}t)_3$ at low temperature to give aldehydes in 60–80° yields;[59,63,65,67,76–78,296] *ortho* substituents decrease the yield. Nitro,[76–78,298a] cyano, or ester groups[78] and urethanes (Eq. 53),[299] as well as conjugated carbon–carbon double bonds,[78,298a] are not affected by the hydride (Eq. 54).[298a].

$$\text{3-}C_6H_5CH_2O_2C\text{-4-}(CH_2)_2COCl\text{-oxazolidin-5-one} \xrightarrow[-78°,\ 1\ \text{h, then to room temp, 1 h}]{LiAlH(OC_4H_9\text{-}t)_3,\ THF} \text{3-}C_6H_5CH_2O_2C\text{-4-}(CH_2)_2CHO\text{-oxazolidin-5-one} \quad (85\%) \qquad \text{(Eq. 53)}$$

$$p\text{-}O_2NC_6H_4CH{=}CHCOCl \xrightarrow[(CH_3OCH_2CH_2)_2O,\ -50^\circ \text{ to room temp}]{LiAlH(OC_4H_9\text{-}t)_3 \text{ (inverse addition)}} p\text{-}O_2NC_6H_4CH{=}CHCHO \quad (84\%) \qquad \text{(Eq. 54)}$$

The lower aldehyde yield in reductions of aliphatic and alicyclic acid chlorides (30–60%)[77,78] can be improved by performing the reaction on a small scale and by strictly maintaining low temperature during the hydride addition[298b–g] (Eqs. 55[300] and 56[301]) to prevent overreduction and formation of esters as byproducts.[302]

$$CH_3O_2C(CH_2)_6COCl \xrightarrow[(CH_3OCH_2CH_2)_2O,\ -78^\circ]{LiAlH(OC_4H_9\text{-}t)_3 \text{ (inverse addition)}} CH_3O_2C(CH_2)_6CHO \quad (96\%) \qquad \text{(Eq. 55)}$$

$$\text{1-COCl-2,3-}(C_6H_5)_2\text{cyclopropene} \xrightarrow[(CH_3OCH_2CH_2)_2O,\ -75^\circ]{LiAlH(OC_4H_9\text{-}t)_3 \text{ (inverse addition)}} \text{1-CHO-2,3-}(C_6H_5)_2\text{cyclopropene} \quad (96\%) \qquad \text{(Eq. 56)}$$

α-Fluoroaldehydes are prepared by reduction of α-fluoroacyl chlorides with $LiAlH(OC_4H_9\text{-}t)_3$ (Eq. 57), which, unlike some other hydrides, does not cause hydrogenolysis of the aliphatic carbon–fluorine bond.[303]

$$RCHFCOCl \xrightarrow[THF,\ -70^\circ]{LiAlH(OC_4H_9\text{-}t)_3} RCHFCHO \quad (50\text{–}65\%) \qquad \text{(Eq. 57)}$$

$$R = H, CH_3, C_2H_5$$

Reduction of acyl chlorides by $LiAlD(OC_4H_9\text{-}t)_3$ provides access to 1-deuterium-labeled aldehydes[304–308] (Eq. 58).[306]

$$o\text{-}O_2NC_6H_4COCl \xrightarrow[(CH_3OCH_2CH_2)_2O,\ -80^\circ]{LiAlD(OC_4H_9\text{-}t)_3 \text{ (inverse addition)}} o\text{-}O_2NC_6H_4CDO \quad (78\%) \qquad \text{(Eq. 58)}$$

Alternatively, acyl chlorides can be converted into aldehydes with $NaAlH_4$,[67,80] bis(triphenylphosphine)copper(I)borohydride,[309] bis(cyanotrihydridoborato)-tetrakis(triphenylphosphine)dicopper,[310] $(n\text{-}C_4H_9)_3SnH$–palladium complexes,[311] or $NaBH_4$-pyridine in dimethylformamide solutions,[10] or by catalytic reduction.[76,78,312]

Reduction of Nitriles. Nitrile groups attached directly to an aromatic ring are reduced by $LiAlH(OCH_3)_3$,[65,81,161,174,296] $NaAlH_2(OCH_2CH_2OCH_3)_2$[67–70,94,167] (Eq. 59),[94] and $Al_2H_3(OCH_2CH_2OCH_3)_3$[313] as readily as by $LiAlH_4$,[59,65,81,173,174,296] AlH_3,[65,296] and $AlH(C_4H_9\text{-}i)_2$[222] to give primary amines in high yields.

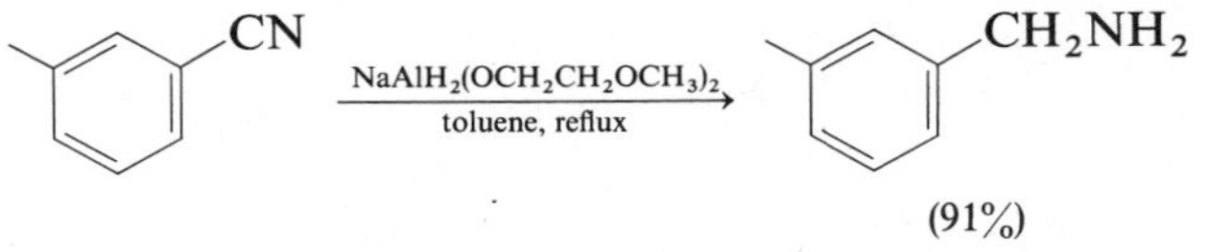
(Eq. 59)

However, aliphatic nitriles carrying relatively acidic α-hydrogen atoms react with $NaAlH_2(OCH_2CH_2OCH_3)_2$ with evolution of hydrogen and afford only traces of primary amines;[94,167] similarly, reduction of arylaliphatic nitriles by this hydride gives unsatisfactory results.[94] Since the hydrogen evolution and accompanying side reactions also complicate the $LiAlH_4$ reductions of these nitriles,[161,174,296] AlH_3,[65,296] $LiAlH_4$–$AlCl_3$,[296] BH_3-$(CH_3)_2S$,[5] $LiAlH(OCH_3)_3$,[161,174,296] or $Al_2H_3(OCH_2CH_2OCH_3)_3$[313] are more suitable reagents. The relative stability of the cyano function toward $NaAlH_2(OCH_2CH_2OCH_3)_2$[230,291] and $LiAlH(OC_4H_9\text{-}t)_3$[63,65,77,81,174,314] can be utilized for selective reductions of the carbonyl group in cyano aldehydes, cyanoketones, cyanoesters, and cyanolactones; a cyano group attached to the pyridine ring is not attacked by the former hydride.[315,316] However, the stability of the cyano group can be influenced by reaction conditions; rapid reduction with $LiAlH(OC_4H_9\text{-}t)_3$ of the 6-keto-12-cyanohydrin ether **80** gives the expected 12β-cyano-6β-hydroxy derivative (see p. 234 in Ref. 1), but prolonged treatment with the same hydride yields the imino lactone **81**.[317]

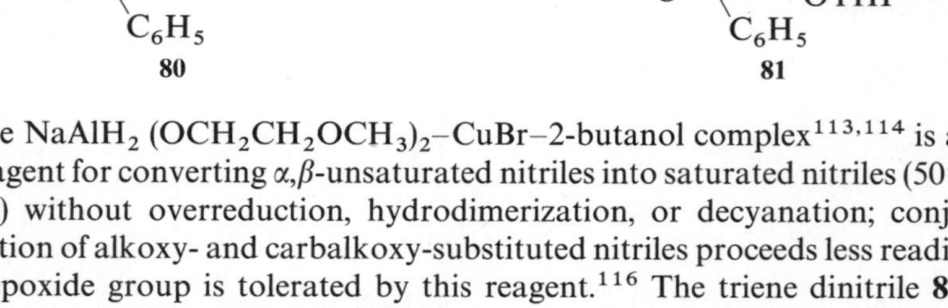
80 **81**

The $NaAlH_2(OCH_2CH_2OCH_3)_2$–CuBr–2-butanol complex[113,114] is a useful reagent for converting α,β-unsaturated nitriles into saturated nitriles (50–99% yields) without overreduction, hydrodimerization, or decyanation; conjugate reduction of alkoxy- and carbalkoxy-substituted nitriles proceeds less readily.[115] The epoxide group is tolerated by this reagent.[116] The triene dinitrile **82** can be reduced to the dinitrile **83**;[115,116] in contrast, reduction by magnesium in methanol, which is capable of transforming α,β-unsaturated nitriles to saturated nitriles,[318] gives a transannular saturated dinitrile.[115,116]

$$\xrightarrow[\text{THF, C}_6\text{H}_6\text{, } -78^\circ \text{ to room temp}]{\text{NaAlH}_2(\text{OCH}_2\text{CH}_2\text{OCH}_3)_2\text{–CuBr (1:1)–2-butanol}}$$

CHCN CHCN
82

CH_2CN CH_2CN
83
(70%)

Partial Reduction of Nitriles to Aldehydes. Transformation of nitriles into aldehydes[82,296,319] by $NaAlH(OC_2H_5)_3$,[319,320] $LiAlH(OC_2H_5)_3$,[7,59,79,81–84,296] $LiAlH_2(OC_2H_5)_2$,[82] or $NaAlH_2(OCH_2CH_2OCH_3)_2$[321] provides an alternative to the well-known methods using $AlH(C_4H_9\text{-}i)_2$[222,322] or the Stephen reduction.[323] Aromatic aldehydes and dialdehydes and certain heterocyclic aldehydes can be prepared by reduction of nitriles or dinitriles with $NaAlH(OC_2H_5)_3$ in 50–95% yields (Eq. 60).[319]

$$\text{1-Naphthyl-CN} \xrightarrow[\text{THF, } 20^\circ]{\text{NaAlH(OC}_2\text{H}_5)_3 \text{ (inverse addition)}} \text{1-Naphthyl-CHO} \ (87\%) \quad \text{(Eq. 60)}$$

Aliphatic nitriles give generally low aldehyde yields, and certain nitriles such as *o*-phthalonitrile and 9-cyanofluorene do not react at all with this hydride.[319] Partial reduction of aliphatic, alicyclic, and aromatic nitriles by $LiAlH(OC_2H_5)_3$ gives aldehydes in 55–95% yields[81–84] (Eq. 61).[82]

$$\text{C}_6\text{H}_{11}\text{CN} \xrightarrow[(\text{C}_2\text{H}_5)_2\text{O, } 0^\circ]{\text{LiAlH(OC}_2\text{H}_5)_3} \text{C}_6\text{H}_{11}\text{CHO} \ (71\%) \quad \text{(Eq. 61)}$$

Deuterium-labeled aldehydes can be prepared by reacting a nitrile with $LiAlD(OC_2H_5)_3$ (Eq. 62).[324]

$$i\text{-C}_3\text{H}_7\text{CN} \xrightarrow[(\text{C}_2\text{H}_5)_2\text{O, room temp}]{\text{LiAlD(OC}_2\text{H}_5)_3} i\text{-C}_3\text{H}_7\text{CDO} \ (61\%) \quad \text{(Eq. 62)}$$

However, the use of $LiAlH(OC_2H_5)_3$ has several disadvantages. In some instances, aldehyde yields as low as 10–20% are encountered. Compounds bearing a relatively acidic α hydrogen, such as phenylacetonitrile, react with evolution of hydrogen and fail to give an aldehyde.[81,82] Some α,β-unsaturated nitriles[324a] or α-substituted nitriles[324b,c] do not react at all with this reagent. The partial nitrile reduction can be accompanied by formation of the corresponding primary amine.[325]

In some reactions such as in reduction of the nitrile **84a**, $LiAlH_2(OC_2H_5)_2$ is a superior reagent.

$$\mathbf{84a} \xrightarrow[(C_2H_5)_2O,\ 0^\circ]{LiAlH_2(OC_2H_5)_2} \text{CHO product } (76\%) \qquad \mathbf{84b}$$

Interestingly, this reagent is unreactive toward the epimeric nitrile **84b** (see also Ref. 326), as are both $LiAlH(OC_4H_9\text{-}n)_3$[82] and $AlH(C_4H_9\text{-}i)_2$ toward **84a**.[327]

Alternatively, $NaAlH_2(OCH_2CH_2OCH_3)_2$ can be used for transforming nitriles into aldehydes (Eq. 63).[328]

$$\xrightarrow[\text{THF, toluene, } -15^\circ \text{ to room temp}]{NaAlH_2(OCH_2CH_2OCH_3)_2} \quad (80\%) \qquad \text{(Eq. 63)}$$

Schiff bases, if sufficiently stable, are the products of the partial reduction of nitriles (Eq. 64); in the presence of bases, these Schiff bases undergo cyclization to form indoloquinolines.[329]

$$\xrightarrow[\text{toluene, reflux}]{NaAlH_2(OCH_2CH_2OCH_3)_2} \qquad \text{(Eq. 64)}$$

A similar reaction leading directly to spiroindolines presumably proceeds also via formation of a Schiff base; in this case, $LiAlH_4$ gives a mixture of the spiroindoline and primary amine (Eq. 65).[330]

$$\xrightarrow[CH_3OCH_2CH_2OCH_3,\ \text{reflux}]{LiAlH(OC_2H_5)_3} \quad (92\%) \qquad \text{(Eq. 65)}$$

Cyanohydrin ethers, which are prepared from the cyanohydrin and ethyl vinyl ether in essentially quantitative yield, can be partially reduced to form the corresponding aldehyde ethers[331,332] (Eq. 66).[331]

$$i\text{-}C_3H_7CH(CN)OCH(CH_3)OC_2H_5 \xrightarrow[(C_2H_5)_2O,\ \text{toluene},\ 0^\circ]{NaAlH_2(OCH_2CH_2OCH_3)_2} i\text{-}C_3H_7CH(CHO)OCH(CH_3)OC_2H_5 \quad (70\%) \qquad \text{(Eq. 66)}$$

Reduction of Carboxylic Acid Amides. Simple secondary and tertiary amides are generally more reactive than primary amides in reductions to amines by $LiAlH_4$,[65,174] AlH_3,[65] $LiAlH(OCH_3)_3$,[65,161] or $Ca[AlH_2(OC_4H_9\text{-}i)_2]_2$.[165] Amides are stable toward $LiAlH(OC_4H_9\text{-}t)_3$,[63,65,333] higher alkoxyaluminum hydrides,[334] and $NaAlH_2(OCH_2CH_2OCH_3)_2$ at low temperatures (about -70°);[230,335] elevated temperature or longer reaction times are needed to obtain amines in high yields with the last hydride.[67–70,94,167] The reaction course and product yields are markedly influenced by the amide structure and the nature of the hydride used. Whereas $LiBH(C_4H_9\text{-}s)_3$ is unreactive toward the bicyclic amide **85**, the alkoxy hydride gives selectively the amine **86**.[336]

CON(CH₃)₂ pyridine crown ether [(CH₂)₂O]₄(CH₂)₂ **85** $\xrightarrow[C_6H_6,\ \text{reflux}]{NaAlH_2(OCH_2CH_2OCH_3)_2}$ $CH_2N(CH_3)_2$ analogue [(CH₂)₂O]₄(CH₂)₂ **86** (96%)

In contrast, reduction of the monocyclic amide **87** under the same conditions is accompanied by formation of an aldehyde as the minor product.[336]

87 (3-$CON(CH_3)_2$-2,6-di-OC_2H_5-pyridine) $\xrightarrow[C_6H_6,\ \text{reflux}]{NaAlH_2(OCH_2CH_2OCH_3)_2}$ **A** (3-$CH_2N(CH_3)_2$-2,6-di-OC_2H_5-pyridine) + **B** (3-CHO-2,6-di-OC_2H_5-pyridine)

A:**B** = 80:20

The propionamide **88** reacts with $NaAlH_2(OCH_2CH_2OCH_3)_2$ to give the *trans*-spiroiminopyrrolidine **89**, whereas AlH_3, formed *in situ* from $LiAlH_4$ and chloroform, gives the expected amine **90**.[337]

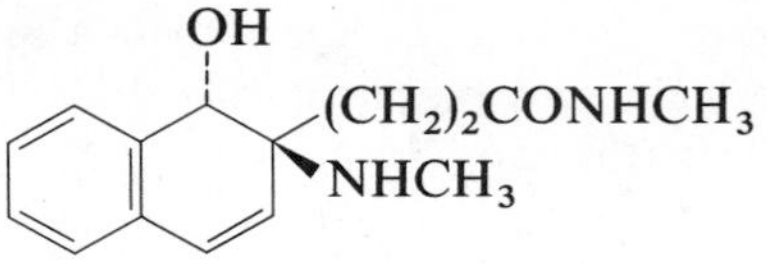

(1*RS*, 2*RS*)
88

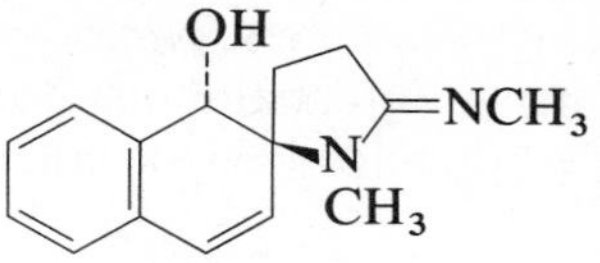

89 (45%)

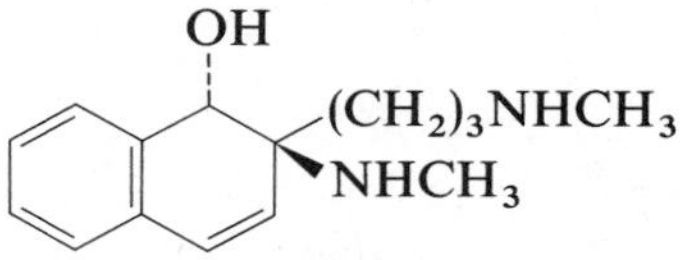

(1*RS*, 2*RS*)
90 (78%)

The amide **91** reacts with $LiAlH_3(OC_4H_9\text{-}n)$ and $LiAlH_4$ to afford the amine **92** and alcohol **93** in opposite ratios; borane also gives the cleavage product **94**.[338]

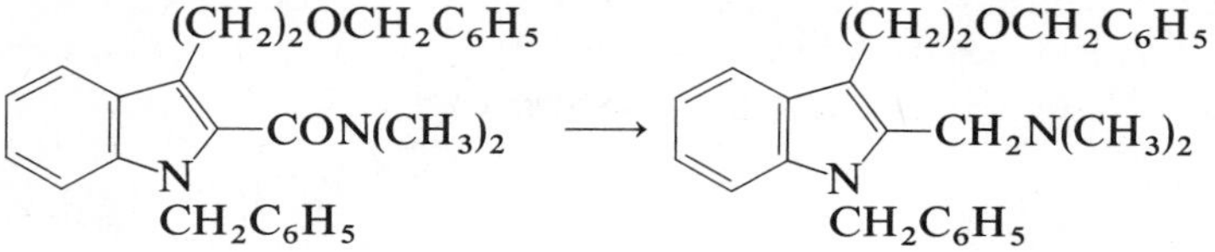

91 → **92** 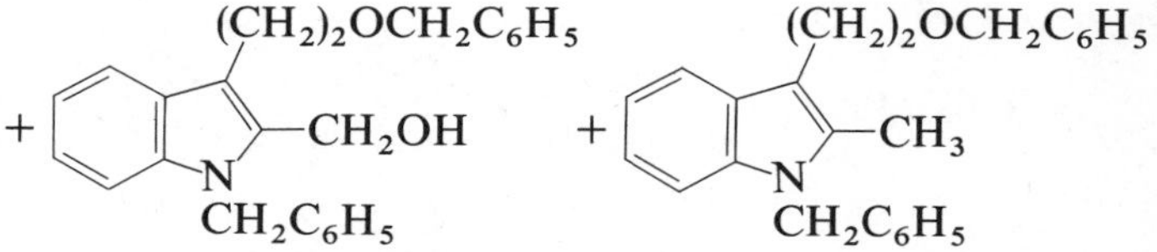 + **93** + **94**

Hydride	**92**	**93**	**94**
$LiAlH_3(OC_4H_9\text{-}n)$	(79%)	(13%)	(0%)
$LiAlH_4$	(10%)	(67%)	(0%)
BH_3	(58%)	(0%)	(—)

Reduction of *N*-formyl derivatives (Eq. 67) with $NaAlH_2(OCH_2CH_2OCH_3)_2$ is an effective method for reductive *N*-methylation of heterocyclic nitrogen bases.[339]

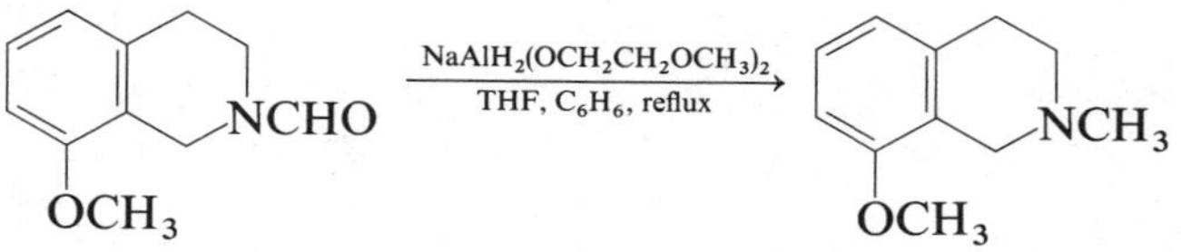

$$\xrightarrow[\text{THF, } C_6H_6\text{, reflux}]{NaAlH_2(OCH_2CH_2OCH_3)_2} \quad (73\%) \qquad \text{(Eq. 67)}$$

N-Acylated heterocyclic bases can be converted selectively into either *N*-alkyl compounds or deacylated bases by action of $NaAlH_2(OCH_2CH_2OCH_3)_2$ at proper reaction temperatures; $LiAlH_4$ is less selective, always giving a mixture of both products (Eq. 68).[340]

$NaAlH_2(OCH_2CH_2OCH_3)_2$, toluene → I ($C_2H_5$ on N) + II (H on N) (Eq. 68)

	I	II
20°	(~100%)	(0)
−40°	(0%)	(~100%)

Starting material: N-COCH₃ benzoxazine; $LiAlH_4$, $(C_2H_5)_2O$ → I + II

	I	II
30°	(64%)	(31%)
−40°	(25%)	(70%)

Low-temperature debenzoylation produces the free base and benzaldehyde in high yields (Eq. 69).[340] The principle of this procedure has been utilized with success for the synthesis of optically active α,β-epoxyaldehydes from methyl *N*-(α,β-epoxyacyl)prolinates.[103,104]

COC_6H_5 (on N) —$NaAlH_2(OCH_2CH_2OCH_3)_2$, toluene, −40°→ H (on N) (~100%) + C_6H_5CHO (~100%) (Eq. 69)

However, a modification of this method using the morpholine-complexed hydride makes it possible to reduce other functions without cleavage of the *N*-acyl group (Eq. 70).[340]

CO_2CH_3, $COCH_3$ (on N) — 1. $NaAlH_2(OCH_2CH_2OCH_3)_2$-morpholine, toluene, −40°; 2. H_3O^+ → CHO, $COCH_3$ **A** + $CH(OH)_2$, $COCH_3$ **B** (Eq. 70)

(**A** + **B** = 91%)

Partial Reduction of Carboxylic Acid Amides to Aldehydes. Controlled reduction of tertiary amides with metal alkoxyaluminum hydrides to form aldehydes provides an alternative[88] to the partial reduction of *N,N*-disubstituted amides such as *N*-acylaziridines, *N*-acylimidazoles, and *N*-acylcarbazoles, or *N*-methylanilides with $LiAlH_4$,[59,88,296,341,342] $AlH(C_4H_9\text{-}i)_2$,[222,341,343] or $NaAlH_4$.[91] Reduction of *N,N*-dimethylamides of aliphatic, alicyclic, aromatic, arylaliphatic, and some heterocyclic carboxylic acids with $LiAlH_2(OC_2H_5)_2$ or

$LiAlH(OC_2H_5)_3$ produces aldehydes in 60–90% yields (Eqs. 71,[88] 72,[344a] and 73[88,89]). The effectiveness of both these hydrides varies widely with the amide structure.[7,59,79,88,89,296,344b]

$$CH_3CHClCON(CH_3)_2 \xrightarrow[(C_2H_5)_2O,\ -30\ \text{to}\ -20^\circ]{LiAlH_2(OC_2H_5)_2\ \text{(inverse addition)}} CH_3CHClCHO \quad \text{(Eq. 71)}$$

(87% by analysis)

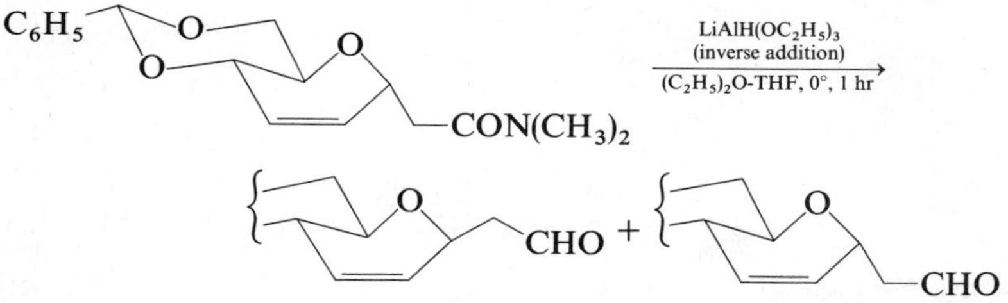

(90%) (~10%)

(Eq. 72)

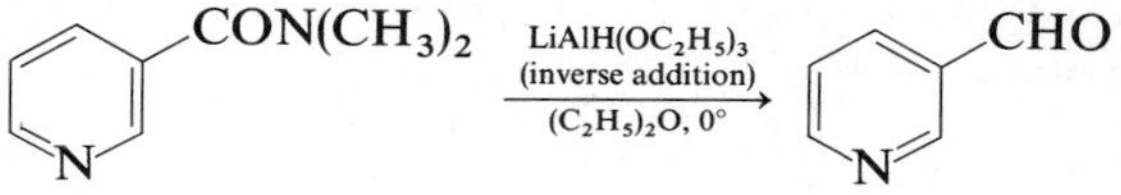

(89% by analysis)

(Eq. 73)

Combination of the partial reduction with the Huang–Minlon modification of the Wolff–Kishner reaction represents a method for transforming a carboxamide into a methyl group (Eq. 74).[345]

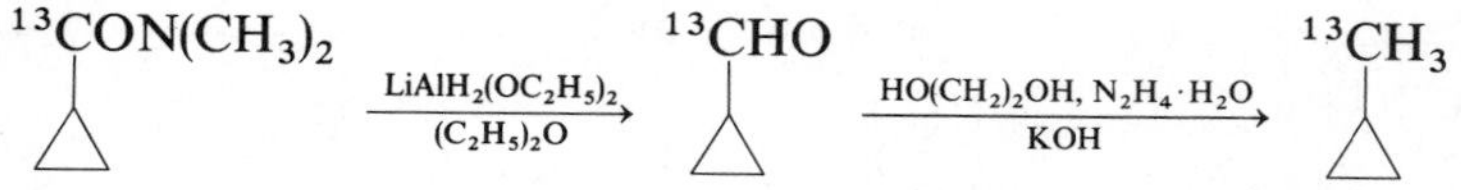

(Eq. 74)

Reduction with $LiAlD_2(OC_2H_5)_2$ provides access to 1-deuterium-labeled aldehydes; the hydroxy amide **95** thus gives the hydroxy-17-deuterioaldehyde isolated as the hemiacetal **96**.[346]

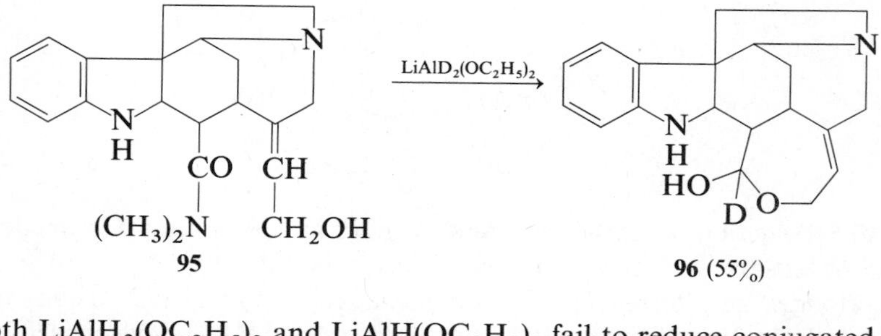

95 **96** (55%)

Both $LiAlH_2(OC_2H_5)_2$ and $LiAlH(OC_2H_5)_3$ fail to reduce conjugated *N,N*-dimethylamides to conjugated aldehydes;[88] in this case, $AlH(C_4H_9\text{-}i)_2$ is the superior reagent.[222]

In some instances, the use of $NaAlH_2(OCH_2CH_2OCH_3)_2$ can be advantageous.[94] Partial reduction of the amide **97** followed by reduction of the aldehyde (without isolation) gives the alcohol **98**; in contrast, $LiAlH_4$ and $LiAlH_2(OC_2H_5)_2$ also produce fairly large amounts of the tertiary amine.[92]

C_6H_5 $CON(CH_3)_2$ OH (S)-(+) **97** $\xrightarrow[C_6H_6,\ (C_2H_5)_2O,\ -20°]{NaAlH_2(OCH_2CH_2OCH_3)_2}$ C_6H_5 OH OH (S)-(+) **98** (73%)

Debenzoylation of the amide **99** and simultaneous formation of benzaldehyde proceeds more easily with $NaAlH_2(OCH_2CH_2OCH_3)_2$ than with $LiAlH_4$, which produces the cyclic amine **100** in lower yield (80%).[229]

CH_3O CH_3O $NCOC_6H_5$ H H CH_3O **99** $\xrightarrow[THF,\ C_6H_6,\ -15°]{NaAlH_2(OCH_2CH_2OCH_3)_2}$

CH_3O CH_3O NH H H CH_3O **100** (95%) + C_6H_5CHO

Aliphatic, alicyclic, and aromatic aldehydes are obtained in 60–80% yields by $NaAlH_2(OCH_2CH_2OCH_3)_2$ reduction of *N*-acylsaccharins (Eq. 75), which are available by reaction of sodium saccharinate with an acyl chloride; an α,β-unsaturated amide, *N*-cinnamoylsaccharin, gives cinnamaldehyde in 77% yield.

O NCOR S O_2 $\xrightarrow[C_6H_6,\ -70\ to\ 5°]{NaAlH_2(OCH_2CH_2OCH_3)_2\ (inverse\ addition)}$ RCHO + O NH S O_2 (Eq. 75)

The method thus represents an extension of the partial reduction of *N,N*-dimethylamides with $LiAlH_2(OC_2H_5)_2$ or $LiAlH(OC_2H_5)_3$.[98]

Unlike $LiAlH_4$, which converts *N*-methyl-*N*-(2-pyridyl)benzamide (**101**) into benzyl alcohol (98%), the $NaAlH_2(OCH_2CH_2OCH_3)_2$-*N*-methylhexahydropyrazine complex forms benzaldehyde. *N*-Methyl-*N*-(2-pyridyl)carboxamides appear to be generally suitable intermediates for the transformation into aldehydes with the use of the latter hydride complex.[99]

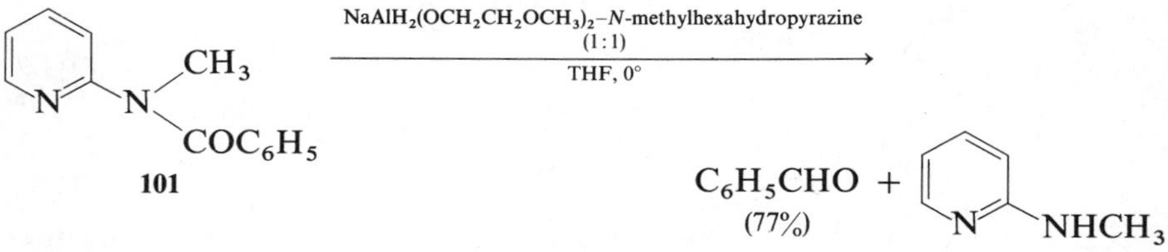

101 $\xrightarrow[\text{THF, }0^\circ]{NaAlH_2(OCH_2CH_2OCH_3)_2\text{–}N\text{-methylhexahydropyrazine (1:1)}}$ C_6H_5CHO (77%) + 2-(methylamino)pyridine ($NHCH_3$)

3-Acylthiazolidinethiones react with $LiAlH(OC_4H_9\text{-}t)_3$ (Eq. 76)[101] or $AlH(C_4H_9\text{-}i)_2$ to form aldehydes in 70–90% yields.[100–102]

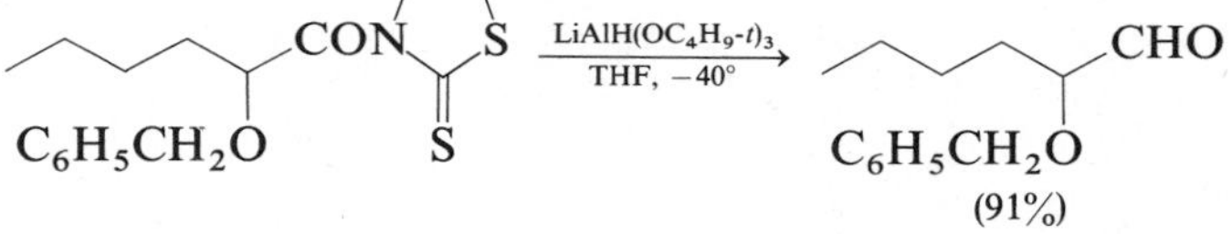

$\xrightarrow[\text{THF, }-40^\circ]{LiAlH(OC_4H_9\text{-}t)_3}$ (91%) (Eq. 76)

Reaction of *N*-acylimidazoles with $LiAlH(OC_4H_9\text{-}t)_3$[96] provides a convenient synthetic route to aldehydes containing other reducible groups (Eq. 77).[97]

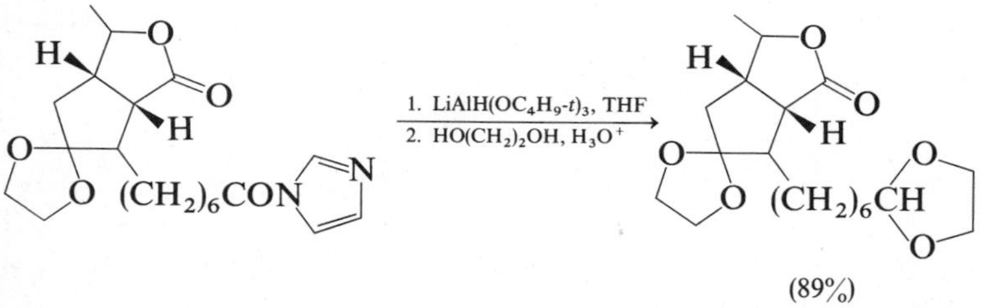

1. $LiAlH(OC_4H_9\text{-}t)_3$, THF; 2. $HO(CH_2)_2OH$, H_3O^+ (89%) (Eq. 77)

Controlled reduction of methyl *N*-(α,β-epoxyacyl)prolinates such as **102** possessing established absolute configurations gives rise to α,β-epoxy aldehydes of high enantiomeric purity.[103,104]

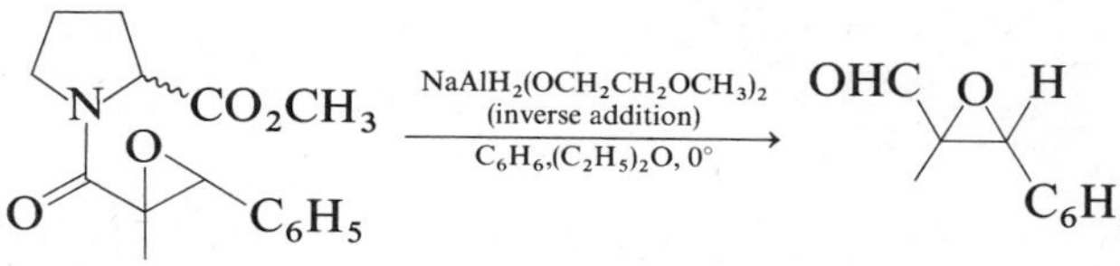

102 (2*R*, 3*S*) $\xrightarrow[C_6H_6, (C_2H_5)_2O, 0^\circ]{NaAlH_2(OCH_2CH_2OCH_3)_2 \text{ (inverse addition)}}$ (2*R*, 3*S*) (72%) (98% e.e.)

Reduction of Lactams. Simple lactams such as 1-methyl-2-piperidinone[94] or ε-caprolactam[94,167] react with $NaAlH_2(OCH_2CH_2OCH_3)_2$ to form cyclic amines in high yields (85–95%) as easily as with $LiAlH_4$,[169,341,347] $NaAlH_4$,[348] or $AlH(C_4H_9\text{-}i)_2$.[169,341,343] The nature of the products, however, can be significantly altered by the reaction conditions. The 2-pyrrolidinone carbonyl, which is stable toward $NaAlH_2(OCH_2CH_2OCH_3)_2$ at low temperatures,[290] is reduced rapidly at ambient temperature to yield a pyrrolidine.[94] On the other hand, an extended reaction at higher temperatures leads to dimerization and trimerization, presumably via the intermediate Δ^2-pyrroline (Eq. 78).[349] In high dilution, ε-caprolactam does not undergo reduction by $NaAlH_2(OCH_2CH_2OCH_3)_2$ but yields a complex with the hydride capable of initiating anionic polymerization of the lactam.[350–355]

N-CH3 2-pyrrolidinone → ($NaAlH_2(OCH_2CH_2OCH_3)_2$; C_6H_6, room temp) → N-CH3 pyrrolidine (97%)

N-CH3 2-pyrrolidinone → ($NaAlH_2(OCH_2CH_2OCH_3)_2$ (inverse addition); C_6H_6; 0°, then reflux) → dimer (22%) + trimer (26%) (Eq. 78)

Intermediates formed in reduction of lactams can be utilized in condensation reactions (Eq. 79)[356a] and for the preparation of 2-cyano-substituted cyclic amines (Eq. 80);[327] in both examples, $LiAlH_4$ is less effective.

CH_3O, CH_3O substituted tetrahydroisoquinoline lactam → ($LiAlH[OCH(CH_3)C_4H_9\text{-}t]_3$; THF, 50°) → CH_3O, CH_3O cyclized product (40%) (Eq. 79)

$C_6H_4OCH_3$-3 substituted N-CH_3 piperidinone → (1. $NaAlH_2(OCH_2CH_2OCH_3)_2$; 2. H_2SO_4, KCN) → $C_6H_4OCH_3$-3 substituted N-CH_3 2-CN piperidine (~100%) (Eq. 80)

Reduction of the diastereomeric lactams **103a,b** followed by cleavage of the intermediate products gives the keto aldehydes **104a,b**, which can be transformed into the cyclopentenones **105a,b** with high enantiomeric excess.[356b]

1. $NaAlH_2(OCH_2CH_2OCH_3)_2$ (0.67 eq), 0–25°, THF
2. $(n\text{-}C_4H_9)_4N^+ H_2PO_4^-$

KOH, C_2H_5OH

103a,b

104a (70%)
104b (73%)

105a (*S*)- (92%; ~99% e.e.)
105b (*R*)- (95%; ~99% e.e.)

a, $R = CH_2CH{=}CH_2$; $R^1 = CH_2C_6H_5$
b, $R = CH_2C_6H_5$; $R^1 = CH_2CH{=}CH_2$

In contrast to $LiAlH_4$, the less reactive $LiAlH_2(OCH_3)_2$ does not attack the ether group of the lactam **106**.[357,358]

106

$LiAlH_2(OCH_3)_2$, THF, reflux → (76%)

$LiAlH_4$, THF, reflux → (70%) + C_6H_5OH

In some cases, such as in the reduction of 3-alkyl-3-phenyl-2-indolinones, the reaction can be stopped at the carbinolamine stage (Eq. 81).[359,360]

$NaAlH_2(OCH_2CH_2OCH_3)_2$ (1 eq), C_6H_6, reflux → (85%)

$NaAlH_2(OCH_2CH_2OCH_3)_2$ (excess), C_6H_6, reflux → (84%)

(Eq. 81)

A rapid reduction of lactams with $LiAlH_2(OC_2H_5)_2$ leads to enamines;[327] in the reduction of the lactam **107**, the procedure using $AlH(C_4H_9\text{-}i)_2$[222,361] is less successful, giving a 57:43 mixture of the desired enamine **107a** and the piperidine base **107b**.[327]

$C_6H_4OCH_3$-3 | $\xrightarrow[(C_2H_5)_2O,\ 0°]{LiAlH_2(OC_2H_5)_2}$ | $C_6H_4OCH_3$-3 | $C_6H_4OCH_3$-3

N, O, CH_3 — **107**; N, CH_3 — **107a** (99%); N, CH_3 — **107b**

Conversion of α,β-unsaturated lactams such as **108** with $NaAlH_2(OCH_2CH_2OCH_3)_2$ into unsaturated bases in 60–90% yields[362,363] provides an alternative procedure to similar reductions with $AlH(C_4H_9\text{-}i)_2$.[222]

O, $NCH_2CH{=}CH_2$, N, H — **108**

1. $NaAlH_2(OCH_2CH_2OCH_3)_2$, C_6H_6, room temp
2. *cis*-$HO_2CCH{=}CHCO_2H$

$\overset{+}{N}HCH_2CH{=}CH_2$ $HO_2CCH{=}CHCO_2^-$, N, H

(80%)

The complex $LiAlH_4 \cdot 2AlCl_3$ is a more effective reagent than $LiAlH_4$ or lithium ethoxyaluminum hydrides in transforming α,β-unsaturated lactams into saturated bases (Eq. 82).[364]

$$\text{N}(C_6H_5)\text{, O} \xrightarrow[(C_2H_5)_2O]{LiAlH_3(OC_2H_5)} \text{N}(C_6H_5) \qquad \text{(Eq. 82)}$$

(84%)
$LiAlH_4 \cdot 2AlCl_3$: 94%

Lactams are generally not reduced by $LiAlH(OC_4H_9\text{-}t)_3$;[190,365–368] lumiflavins may be an exception.[369]

Whereas the hydantoin carbonyl in 5-phenyl-substituted hydantoins such as **109** is attacked by $NaAlH_2(OCH_2CH_2OCH_3)_2$ to form the 2-imidazolone **109a**,[370] the same group in 5,5-diphenyl derivatives (e.g., **110**) undergoes complete reduction that affords 2-imidazolidinone **111** as the final product.[371,372]

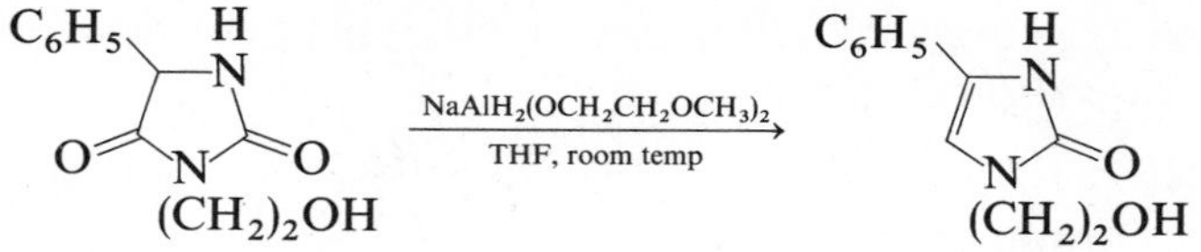

$\xrightarrow[\text{THF, room temp}]{\text{NaAlH}_2(\text{OCH}_2\text{CH}_2\text{OCH}_3)_2}$

109 → **109a** (45%)

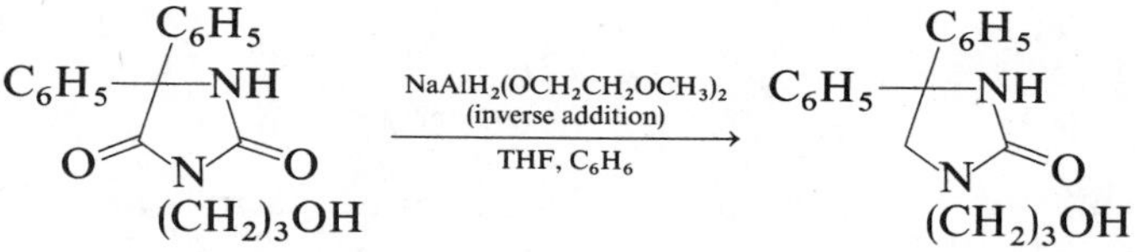

$\xrightarrow[\text{THF, C}_6\text{H}_6]{\text{NaAlH}_2(\text{OCH}_2\text{CH}_2\text{OCH}_3)_2 \text{ (inverse addition)}}$

110 → **111** (56%)

Partial reductions of the hydantoin carbonyl can be achieved by using a limited amount of the hydride (Eq. 83).[372]

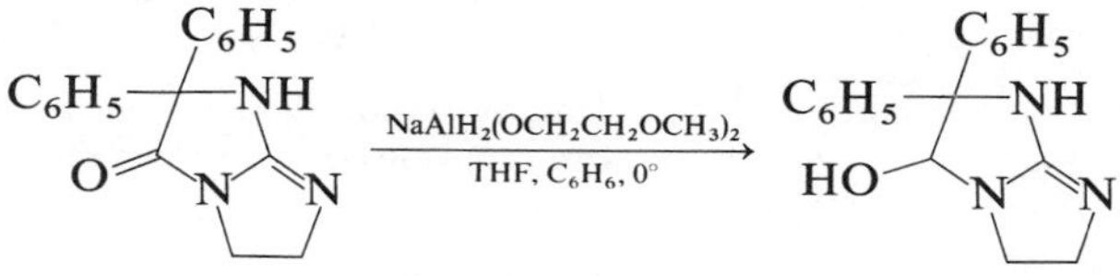

$\xrightarrow[\text{THF, C}_6\text{H}_6\text{, 0°}]{\text{NaAlH}_2(\text{OCH}_2\text{CH}_2\text{OCH}_3)_2}$ (93%) (Eq. 83)

The photochemical cyclization of enamides[373,374] in the presence of a chiral hydride complex provides access, although in modest yield, to isoquinoline alkaloids such as (−)-xylopinine (**112**), presumably via [1,5]-sigmatropic rearrangement of the intermediate **113** and reduction of the (−)-lactam **114**[375,376] (see, however, Refs. 377a and 377b).

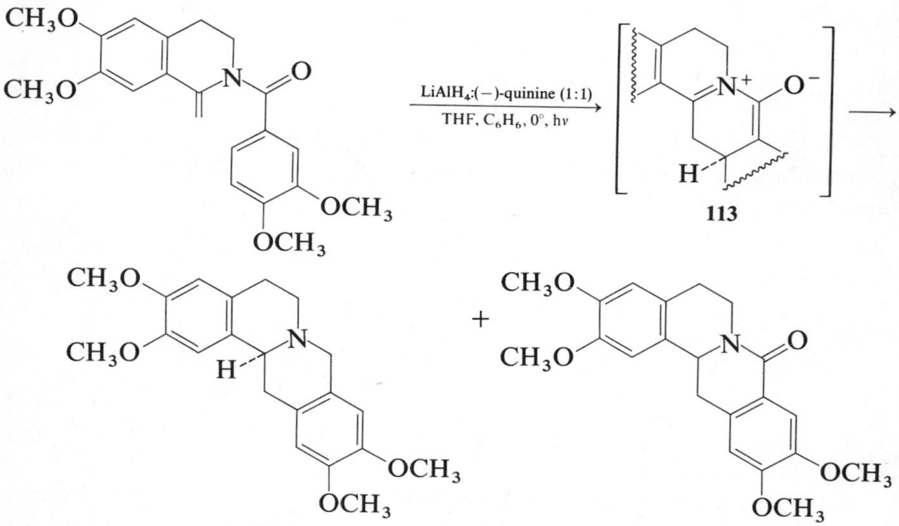

$\xrightarrow[\text{THF, C}_6\text{H}_6\text{, 0°, h}\nu]{\text{LiAlH}_4\text{:(−)-quinine (1:1)}}$ [**113**] →

112 (38%, 37% e.e.) + **114** (10%)

Reduction of Cyclic Imides. Substituted *N*-methylsuccinimides[94,167] and *N*-methylglutarimides[378] undergo reduction by $NaAlH_2(OCH_2CH_2OCH_3)_2$ to form *N*-methylpyrrolidines and *N*-methylpiperidines, respectively, in high yields. A sterically congested imide such as **115** undergoes regioselective hydride attack at the more hindered imide carbonyl group (see Ref. 60), producing the lactam **116** as the single product.[379]

$NaAlH_2(OCH_2CH_2OCH_3)_2$ / C_6H_6, toluene, reflux

115 → **116** (92%)

Successful selective reduction (85–95%) of one of the diastereotopic carbonyl groups of succinimides by $NaAlH_2(OCH_2CH_2OCH_3)_2$ or $NaAlH_2(OC_2H_5)_2$ (Eq. 84) represents the first step in their transformation into optically active lactones.[380,381]

$CH(C_6H_5)CH_2OH$ (+)-isomer → ($NaAlH_2(OCH_2CH_2OCH_3)_2$, THF, −78°) → $CH(C_6H_5)CH_2OH$ (89%) (Eq. 84)

N-Benzylphthalimide reacts with $NaAlH_2(OCH_2CH_2OCH_3)_2$ to form *N*-benzylisoindole along with lesser amounts of *N*-benzylisoindoline (**117**); $LiAlH_4$ affords isoindoline **117** as the sole product.[382]

$NCH_2C_6H_5$ → ($NaAlH_2(OCH_2CH_2OCH_3)_2$, C_6H_6, 15–20°) → $NCH_2C_6H_5$ (49%) + $NCH_2C_6H_5$ **117** (17%)

The stability of imides toward $LiAlH(OC_4H_9\text{-}t)_3$[383,384] is utilized in the diastereoselective synthesis of D-*threo*-sphingamine (Eq. 85); $LiAlH_4$ and

$NaBH_4$ are less suitable reagents because they attack one of the phthalimide carbonyls.[385]

$$\text{PhthN}\text{CH}(\text{CH}_2\text{OAc})(\text{COC}_{15}\text{H}_{31}\text{-}n) \xrightarrow[\text{THF, } 0^\circ]{\text{LiAlH(OC}_4\text{H}_9\text{-}t)_3}$$

(D)-

$$\text{PhthN}\text{CH}(\text{CH}_2\text{OAc})(\text{CHOHC}_{15}\text{H}_{31}\text{-}n) \longrightarrow \underset{\text{D-}threo}{\text{HOCH}_2\text{CH(NH}_2\text{)CHOHC}_{15}\text{H}_{31}\text{-}n} \quad \text{(Eq. 85)}$$

Miscellaneous Open-Chain Nitrogen Compounds

Reduction of Imines. Reactions of aldimines and ketimines with complex metal hydrides and metal alkoxyaluminum hydrides proceed with saturation of the carbon-nitrogen double bond[59,127,133–135,169,295,296,341] (however, see pp. 55 and 56 in Ref. 1). In some cases such as in the reduction of the ketimine **118**, where $LiAlH(OC_4H_9\text{-}t)_3$, $NaBH_4$, or 9-borabicyclo[3.3.1]nonane give no reaction, $NaAlH_2(OCH_2CH_2OCH_3)_2$ affords the diastereomeric amines in higher yield and with higher stereoselectivity than $LiAlH_4$, BH_3, or $NaBH_3(CN)$. The method has been used with success for the preparation of optically active fluorinated amines.[386]

$$C_6H_5COCF_3 + \underset{(S)\text{-}(-)}{C_6H_5CH(CH_3)NH_2} \longrightarrow$$

$$\underset{\mathbf{118}}{C_6H_5C(CF_3){=}NCH(CH_3)C_6H_5} \xrightarrow[\text{THF, } -78^\circ]{NaAlH_2(OCH_2CH_2OCH_3)_2}$$

$$\underset{\substack{(95\%) \\ \text{(diastereomer ratio} = 92:8)}}{C_6H_5CH(CF_3)NHCH(CH_3)C_6H_5} \xrightarrow[\text{Pd/C}]{H_2} \underset{\substack{(\sim 100\%) \\ (S):(R) = 92:8}}{C_6H_5CH(CF_3)NH_2}$$

Reduction of the anil **119** by $NaAlH_2(OCH_2CH_2OCH_3)_2$ to the *endo* and *exo* amines **120a** and **120b**, respectively, shows higher stereoselectivity than catalytic hydrogenation (**120a**:**120b** = 90:10) or reduction by sodium metal in ethanol (**120a**:**120b** = 55:45).[387]

CH_3 N NC_6H_5 $\xrightarrow[C_6H_6,\ \text{reflux}]{NaAlH_2(OCH_2CH_2OCH_3)_2}$

119

CH_3 N NHC_6H_5 + CH_3 N NHC_6H_5

120a **120b**

(**120a**:**120b** = 94:6)

Asymmetric reductions of ketimines by chiral $LiAlH_4$–(−)-menthol, $LiAlH_4$–(+)-borneol, and $LiAlH_4$–(−)-quinine complexes give amines in only low optical purity (1–10% e.e.).[388] Somewhat higher, albeit still unsatisfactory, optical yields of amines (9–24% e.e.) are achieved by using the $LiAlH_4$–3-*O*-benzyl-1,2-*O*-cyclohexylidene-α-D-glucofuranose complex (**27a**) or its ethanol-modified derivative (**27b**).[127,128]

Reduction of Imidoyl Chlorides. An imidoyl chloride such as **121** reacts with $LiAlH(OC_4H_9\text{-}t)_3$ to form the imine **122**, which can be transformed via acylation and hydrolysis into the amide **123**.

$N{=}CClC_2H_5$ $\xrightarrow[THF,\ -78^\circ\ \text{to room temp}]{LiAlH(OC_4H_9\text{-}t)_3}$ $N{=}CHC_2H_5$ $\xrightarrow{n\text{-}C_7H_{15}COCl}$

121 **122** (65%)

$NCOC_7H_{15}\text{-}n$ / $CH{=}CHCH_3$ $\xrightarrow[HCl]{CH_3OH}$ H $NCOC_7H_{15}\text{-}n$

(68%) **123** (90%)

Since imidoyl chlorides can be prepared from amides, the reaction sequence represents a general method for acyl exchange in secondary amides.[389]

Reduction of Oximes. The course of hydride reductions of oximes depends markedly on the oxime structure and the nature of the hydride used. Whereas $LiAlH(OCH_3)_3$[65,161] and $LiAlH(OC_4H_9\text{-}t)_3$[63,65] react with oximes only with hydrogen evolution, diborane yields hydroxylamines,[296,390,391] and $LiAlH_4$,[65,174] AlH_3,[65] or $NaAlH_2(OCH_2CH_2OCH_3)_2$[94,167] reduce aldoximes and aliphatic or simple alicyclic ketoximes to primary amines. Some oximes such as **124** undergo rearrangement[59,145,150] prior to reduction and form secondary amines, in this case the cyclic amine **125**; $LiAlH_4$ gives the same product in lower yield (~32%).[392]

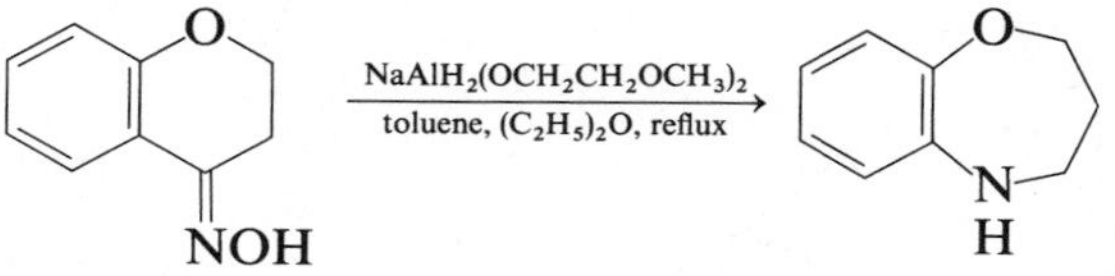

$NaAlH_2(OCH_2CH_2OCH_3)_2$
toluene, $(C_2H_5)_2O$, reflux

NOH

124

N
H

125 (65%)

In contrast to $LiAlH_4$, which reduces substituted alicyclic ketoximes to primary amines, $NaAlH_2(OCH_2CH_2OCH_3)_2$ yields aziridines (Eq. 86)[393] or primary amines (Eq. 87).[189]

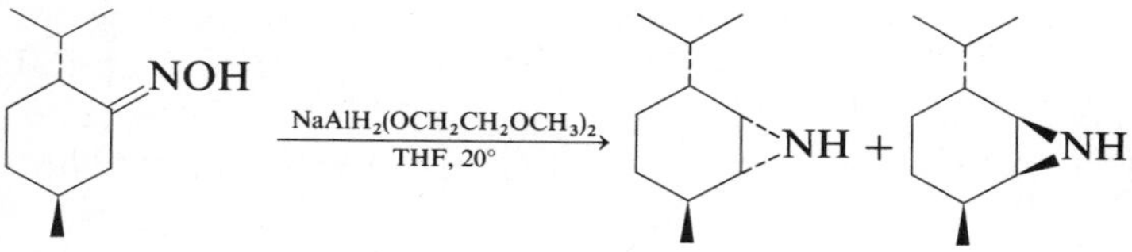

NOH

$NaAlH_2(OCH_2CH_2OCH_3)_2$
THF, 20°

NH + NH (Eq. 86)

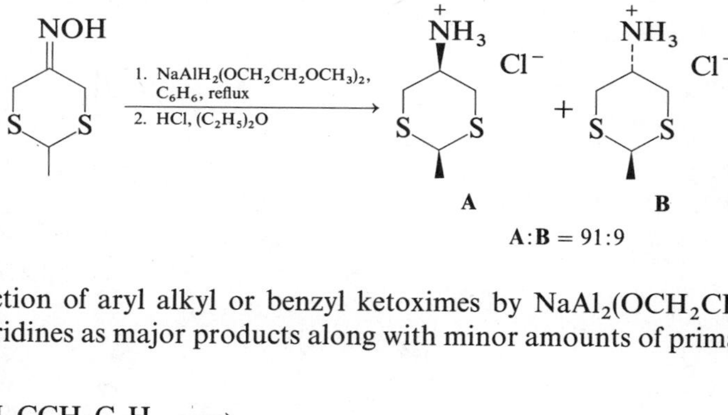

(55%) (8%)

NOH

1. $NaAlH_2(OCH_2CH_2OCH_3)_2$, C_6H_6, reflux
2. HCl, $(C_2H_5)_2O$

$\overset{+}{N}H_3$ Cl^- + $\overset{+}{N}H_3$ Cl^- (Eq. 87)

A **B**

A:**B** = 91:9

Reduction of aryl alkyl or benzyl ketoximes by $NaAl_2(OCH_2CH_2OCH_3)_2$ gives aziridines as major products along with minor amounts of primary amines (Eq. 88).

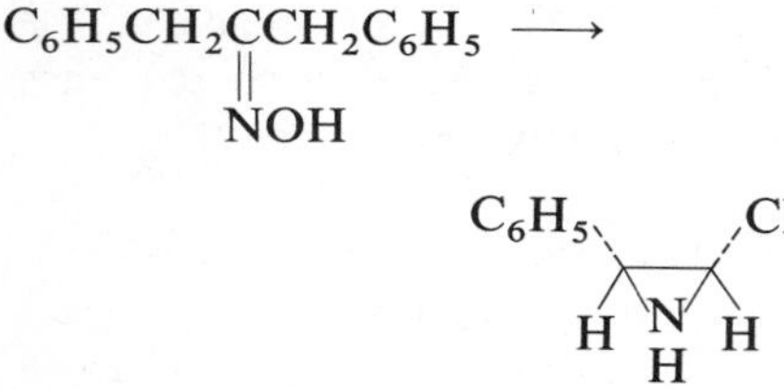

$C_6H_5CH_2C(=NOH)CH_2C_6H_5$ ⟶

C_6H_5 $CH_2C_6H_5$

H N H
H

+ $(C_6H_5CH_2)_2CHNH_2$ (Eq. 88)

A **B**

	(**A**:**B**)
$NaAlH_2(OCH_2CH_2OCH_3)_2$, THF, C_6H_6, reflux	(91:8)
$LiAlH_4$	(80:20)

In some reactions secondary amines can occur as byproducts. *O*-Alkyl oximes yield more primary amines at the expense of aziridines. When compared with $NaAlH_2(OCH_2CH_2OCH_3)_2$ in this series, $LiAlH_4$ generally produces more primary amine and less aziridine.[150]

The effect of the hydride nature on the product distribution in reductions of α,β-unsaturated alicyclic ketoximes is illustrated by the reduction of the *anti*-oxime **126**, which by action of $NaAlH_2(OCH_2CH_2OCH_3)_2$ affords the aziridines **127**, **128**, and **129** free of primary and secondary amines; under similar conditions, $LiAlH_4$ forms the aziridine **127** (56%) along with the saturated aziridine (mixture of *cis* and *trans* isomers; 6%), an unsaturated primary amine (28%), and 3,5,5-trimethyl-1-azacycloheptane (10%).[152,153]

126 $\xrightarrow[\text{THF, reflux}]{NaAlH_2(OCH_2CH_2OCH_3)_2}$ **127** (88%) + **128** (4%) + **129** (8%)

In the asymmetric reduction of ketoximes and their *O*-methyl or *O*-tetrahydropyranyl derivatives by the chiral $LiAlH_4$–3-*O*-benzyl-1,2-*O*-cyclohexylidene-α-D-glucofuranose (1:1) complex (**27a**), the (*S*)-amine (10–56% e.e.) is the predominating enantiomer (Eq. 89).

$$(CH_3)(C_6H_{11})C{=}NOH \xrightarrow[(C_2H_5)_2O,\ \text{reflux}]{\textbf{27a}} H_2N(H)C(CH_3)(C_6H_{11}) \qquad \text{(Eq. 89)}$$

(*S*)-(+)
(~70%) (56% e.e.)

The ethanol-modified $LiAlH_4$–monosaccharide complex (1:1:1) **27b** yields predominantly the (*R*)-amine (13–19% e.e.) (Eq. 90).[132,394]

$$(C_6H_5)(CH_3)C{=}NOCH_3 \xrightarrow[(C_2H_5)_2O,\ \text{reflux}]{\textbf{27b}} H(NH_2)C(CH_3)(C_6H_5) \qquad \text{(Eq. 90)}$$

(*R*)-(+)
(~55%) (19% e.e.)

Reduction of *N*-Alkylidenephosphinic Amides. Various (*R*)-*N*-alkyldiphenylphosphinic amides are prepared in 21–75% yields and 8–34% enantiomeric excess by reduction of prochiral *N*-alkylidenediphenylphosphinic amides with a chiral $LiAlH_4$ complex (Eq. 91). Dephosphinylation with hydrochloric acid leads to (*R*)-amine hydrochlorides.[395]

$$CH_3(C_6H_5CH_2)C{=}NP(O)(C_6H_5)_2 \xrightarrow[\text{THF-}(C_2H_5)_2O\text{, room temp}]{LiAlH_4\text{-}(-)\text{-quinine (1:1)}} CH_3(C_6H_5CH_2)CHNHP(O)(C_6H_5)_2 \quad \text{(Eq. 91)}$$

(*R*)-(−)
(21%) (34% e.e.)

Reduction of *N*-Alkylidenesulfinamides. In reductions of *dl*-*N*-alkylidenesulfinamide **130** by chiral lithium alkoxyaluminohydride complexes, $LiAlH_4$–(−)-menthol (1:2) gives the *N*-alkylsulfinamide **131** in highest yield, with greatest stereoselectivity, but with only very low asymmetric induction.[396]

$$C_6H_5(1\text{-}C_{10}H_7)C{=}NS(O)C_6H_4CH_3\text{-}p \xrightarrow[(C_2H_5)_2O\text{, room temp}]{LiAlH_2(OR^*)_2} C_6H_5(1\text{-}C_{10}H_7)CHNHS(O)C_6H_4CH_3\text{-}p$$

130 R* = (−)-menthyl

131 (90%) (diastereomer ratio: 96:4)

Reduction of Isocyanates and Isothiocyanates. Whereas $LiAlH_4$,[65,169,174] AlH_3,[65] $LiAlH(OCH_3)_3$,[65,161] and $NaAlH_2(OCH_2CH_2OCH_3)_2$[397] reduce isocyanates to form substituted *N*-methylamines, $LiAlH(OC_4H_9\text{-}t)_3$ gives formamides in high yields.[63,398] Since isocyanates easily undergo dimerization or trimerization even by mildly basic reagents,[399–402] the product formation can depend on reaction conditions (Eqs. 92 and 93).

2,2-Diphenyl-1-methylcyclopropyl-NCO, (*R*)-(−) $\xrightarrow[\text{THF}]{LiAlH(OC_4H_9\text{-}t)_3}$

−15°: 2,2-diphenyl-1-methylcyclopropyl-NHCHO, (*R*)-(−) (80%) (Eq. 92)

25°: (2,2-diphenyl-1-methylcyclopropyl)N(CHO)C(O)NH(2,2-diphenyl-1-methylcyclopropyl) (83%) (Eq. 93)

The product in Eq. 93 seems to have arisen by the addition of the formamide to the isocyanate. Dehydration of formamides provides easy access to aliphatic, vinylic, and aromatic isonitriles.[398] A modification of the procedure given by Eq. 92, using a short reaction time, is useful for the preparation of various 4-substituted *N*-(*β*-styryl)formamides (Eq. 94).[403,404]

$$4\text{-}CH_3O C_6H_4 CH{=}CHNCO \xrightarrow[\text{THF, 15 min}]{LiAlH(OC_4H_9\text{-}t)_3} 4\text{-}CH_3O C_6H_4 CH{=}CHNHCHO \ (80\%) \qquad \text{(Eq. 94)}$$

Thiocyanates are stable toward $LiAlH(OC_4H_9\text{-}t)_3$.[405a–c] Thioformamides are produced by rapid reduction of isothiocyanates with $NaAlH_2(OCH_2CH_2OCH_3)_2$ (Eq. 95).[405b]

$$\text{2-morpholino-4-quinazolinyl-NCS} \xrightarrow[(C_2H_5)_2O,\ C_6H_6]{NaAlH_2(OCH_2CH_2OCH_3)_2} \text{2-morpholino-4-quinazolinyl-NHCHS} \ (60\%) \qquad \text{(Eq. 95)}$$

Reduction of Urethanes and Cyanamides. Urethanes readily undergo reduction by $LiAlH_4$,[59,169,341,406] $AlH(C_4H_9\text{-}i)_2$,[406] or $NaAlH_2(OCH_2CH_2OCH_3)_2$[406,407a] to form substituted *N*-methylamines (Eq. 96).[407b]

$$3,5\text{-}(CH_3O)_2C_6H_3CH_2CH_2NHCO_2C_2H_5 \xrightarrow[(C_2H_5)_2O,\ C_6H_6]{NaAlH_2(OCH_2CH_2OCH_3)_2} 3,5\text{-}(CH_3O)_2C_6H_3CH_2CH_2NHCH_3 \ (83\%) \qquad \text{(Eq. 96)}$$

Occasionally, reduction of unsaturated urethanes by $LiAlH_4$ or $NaAlH_2(OCH_2CH_2OCH_3)_2$ proceeds with concomitant, presumably intramolecular, hydride addition to the carbon–carbon double bond (see also p. 90 in Ref. 59) (Eq. 97).[408]

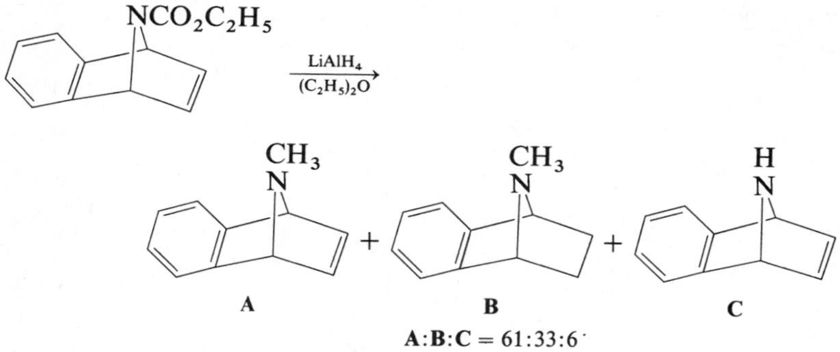

$NCO_2C_2H_5$ —$\xrightarrow[(C_2H_5)_2O]{LiAlH_4}$ A + B + C (Eq. 97)

A (N–CH_3), B (N–CH_3), C (N–H)

A:B:C = 61:33:6

The stability of urethanes toward $LiAlH_2(OC_4H_9\text{-}t)_2$,[406] $LiAlH(OC_4H_9\text{-}t)_3$,[406,409] or $NaBH_4$[409] can be utilized for selective reductions of a keto function in keto urethanes.

N-Cyano derivatives undergo decyanation on treatment with $NaAlH_2(OCH_2CH_2OCH_3)_2$ at high temperatures (Eq. 98).[410]

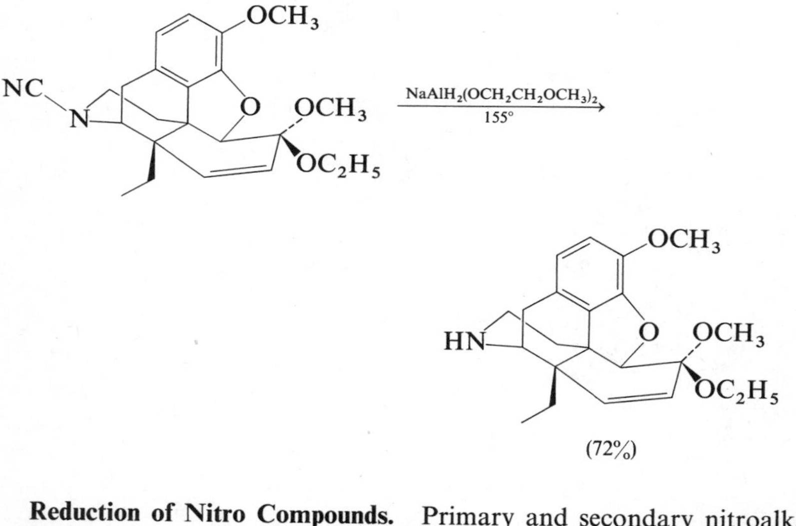

NC–N ... OCH_3, OCH_3, OC_2H_5 $\xrightarrow[155°]{NaAlH_2(OCH_2CH_2OCH_3)_2}$ HN ... OCH_3, OCH_3, OC_2H_5 (72%) (Eq. 98)

Reduction of Nitro Compounds. Primary and secondary nitroalkanes are reduced by $LiAlH_4$,[59,65,169,174,296] $LiAlH(OCH_3)_3$,[65,161] $NaAlH(OC_2H_5)_3$,[319] or $NaAlH_2(OCH_2CH_2OCH_3)_2$[167,411] with evolution of hydrogen to form primary amines. Reaction of 1-nitropropane with $LiAlH(OC_4H_9\text{-}t)_3$ involves a relatively rapid hydrogen evolution but no reduction.[63,65] Nitroarenes, which are stable toward $LiAlH(OC_4H_9\text{-}t)_3$,[63,65] AlH_3,[65] or $LiAlH_4$ on silica gel,[412] react with $NaAlH_2(OCH_2CH_2OCH_3)_2$ to yield either azoxyarenes,[411,413] azoarenes,[238,413] or hydrazoarenes,[167,411] depending on the reaction conditions (Eq. 99).[413]

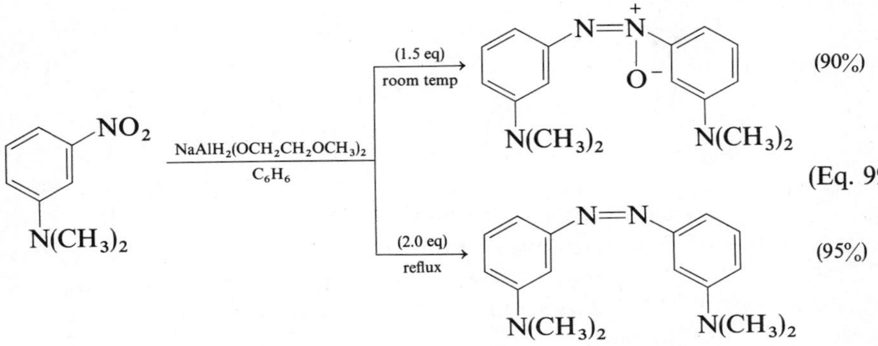

(1.5 eq), room temp → (90%)

$NaAlH_2(OCH_2CH_2OCH_3)_2$, C_6H_6

(Eq. 99)

(2.0 eq), reflux → (95%)

Nitrobenzene is reduced by $LiAlH(OCH_3)_3$ either to azoxybenzene (0°, 3 hours) or azobenzene (25°, 24 hours).[65,161] An unusual reaction course is seen in the reduction of the highly hindered nitro derivative **132**; by comparison, $LiAlH_4$ gives the oxime (10%) along with 2,4,6-tri-*tert*-butylaniline (6%) and unreacted starting material.[414]

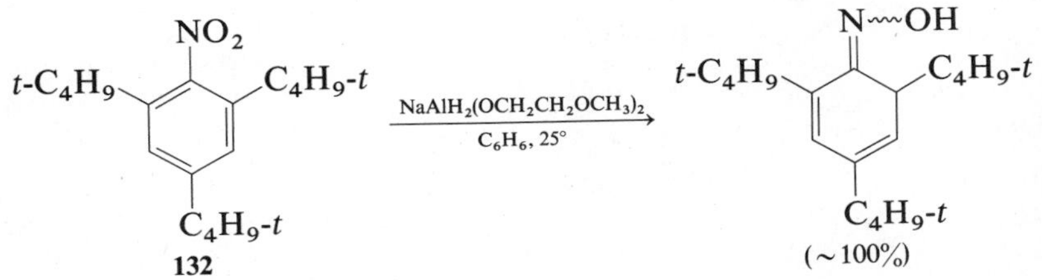

$NaAlH_2(OCH_2CH_2OCH_3)_2$, C_6H_6, 25°

132 (~100%)

β-Arylethylamines are prepared in good yields by reduction of β-nitrostyrenes (Eq. 100).[415]

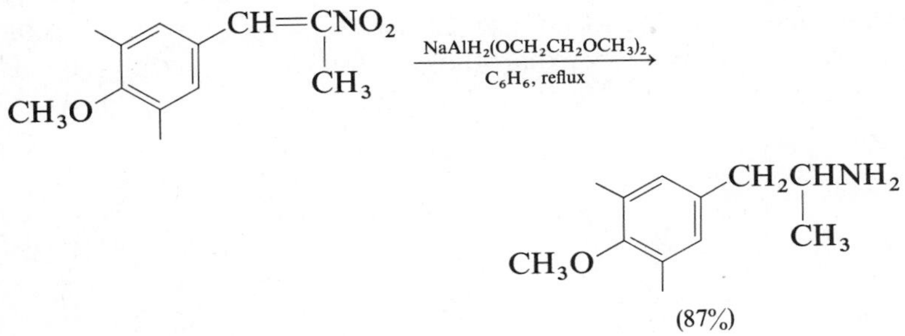

$NaAlH_2(OCH_2CH_2OCH_3)_2$, C_6H_6, reflux

(Eq. 100)

(87%)

Reduction of *N*-Nitroso, Azoxy, Azo, Azido, and Nitrimino Compounds and Nitrate Esters. *N*-Nitroso-1,3-oxazines are reduced to the *N*-amino derivatives by $NaAlH_2(OCH_2CH_2OCH_3)_2$ (Eq. 101); $LiAlH_4$ cleaves the carbon–oxygen bond in the heterocyclic ring and produces *N*-(3-hydroxypropyl)-*N*-benzylhydrazine.[416]

$$\text{2-}C_6H_5\text{-3-nitroso-tetrahydro-1,3-oxazine} \xrightarrow[C_6H_6,\ 10°]{NaAlH_2(OCH_2CH_2OCH_3)_2} \text{3-amino-2-}C_6H_5\text{-tetrahydro-1,3-oxazine}\ (65\%) \qquad \text{(Eq. 101)}$$

Azoxybenzene and azobenzene, which are stable toward $LiAlH(OC_4H_9\text{-}t)_3$[63,65] and AlH_3,[65] react with $LiAlH_4$[65,174] or $LiAlH(OCH_3)_3$[65,161,174] to form hydrazobenzene. The nitrate ester group, $—ONO_2$, is inert to $LiAlH(OC_4H_9\text{-}t)_3$.[417] Azido compounds, which are completely unreactive toward the latter hydride,[418–420] undergo reduction by $NaAlH_2(OCH_2CH_2OCH_3)_2$ to produce primary amines.[421] In the nitrimine group, $>C{=}NNO_2$, only the carbon–nitrogen double bond is attacked by $LiAlH(OC_4H_9\text{-}t)_3$, yielding a nitramine $>CHNHNO_2$.[422,423]

Reduction of Open-Chain Ammonium, Hydrazonium, and Dithiocarbamidium Salts. Hydroxy-substituted acetylenic quaternary iodides such as **133** are converted by $NaAlH_2(OCH_2CH_2OCH_3)_2$ or $LiAlH_4$ into α-allenic alcohols.[424]

$$\text{6-}CH_3O\text{-2-naphthyl-}COH(CH_3)C{\equiv}CCH_2\overset{+}{N}(CH_3)_3I^- \ (\mathbf{133}) \xrightarrow[THF,\ 20–30°]{NaAlH_2(OCH_2CH_2OCH_3)_2} \text{6-}CH_3O\text{-2-naphthyl-}COH(CH_3)CH{=}C{=}CH_2\ (84\%)$$

Hydrazonium methiodides derived from alkyl-substituted alicyclic ketones undergo a slow reduction by $NaAlH_2(OCH_2CH_2OCH_3)_2$ to give *cis* and *trans* aziridines in yields that are higher than those obtained by reacting the corresponding oxime with the same hydride (Eq. 102).[393]

$$\text{cyclohexylidene}{=}N\overset{+}{N}(CH_3)_3I^- \xrightarrow[THF,\ 70°]{NaAlH_2(OCH_2CH_2OCH_3)_2} \text{aziridine NH}\ (70\%) + \text{aziridine NH}\ (16\%) \qquad \text{(Eq. 102)}$$

Formation of α,β-unsaturated aziridines by reduction of *anti*-hydrazonium methiodides of α,β-unsaturated cyclohexanones is enhanced by using partially hydrolyzed $NaAlH_2(OCH_2CH_2OCH_3)_2$ (Eq. 103).[154]

$$\text{1-methyl-cyclohexenylidene hydrazonium } N^{+}(CH_3)_3\ I^- \xrightarrow{NaAlH_2(OCH_2CH_2OCH_3)_2-H_2O\ (7:3)} \text{unsaturated aziridine (NH)} \quad (97\%) \qquad \text{(Eq. 103)}$$

syn-Hydrazonium methiodides give either saturated or α,β-unsaturatedd aziridines as major products, depending on the type of the hydride used (Eq. 104).[154]

$$(CH_3)_3\overset{+}{N}-N{=}\text{(3,5,5-trimethylcyclohex-2-enylidene)}\ I^- \xrightarrow{NaAlH_2(OCH_2CH_2OCH_3)_2} \mathbf{A} + \mathbf{B} \quad (\mathbf{A}:\mathbf{B} = 95:5)$$

$$\xrightarrow{NaAlH_2(OCH_2CH_2OCH_3)_2-H_2O\ (7:3)} \mathbf{B} + \mathbf{A} + \mathbf{C} \quad (\mathbf{B}:\mathbf{A}:\mathbf{C} = 83:8:8) \qquad \text{(Eq. 104)}$$

Reduction of dithiocarbamidium salts by $NaAlH_2(OCH_2CH_2OCH_3)_2$ affords formamide thioacetals in moderate yields (36–71%) (Eq. 105); $LiAlH_4$ and $NaBH_4$ are less effective reagents.[425]

$$(CH_3S)_2C{=}\overset{+}{N}(CH_3)_2\ ClO_4^- \xrightarrow[THF,\ C_6H_6,\ -5^\circ]{NaAlH_2(OCH_2CH_2OCH_3)_2\ \text{(inverse addition)}} (CH_3S)_2CH{-}N(CH_3)_2 \quad (38\%) \qquad \text{(Eq. 105)}$$

Heterocyclic Nitrogen Compounds

Reduction of Heterocyclic Bases and Their Quaternary Salts. Azirines react readily with $NaAlH_2(OCH_2CH_2OCH_3)_2$ to form aziridines in high to almost quantitative yields (Eq. 106).[151,426]

$$\text{2-}C_6H_5\text{-3-}CONH_2\text{-3-}C_6H_5\text{-azirine} \xrightarrow[C_6H_6,\ 0^\circ]{NaAlH_2(OCH_2CH_2OCH_3)_2} \text{2-}C_6H_5\text{-3-}CONH_2\text{-3-}C_6H_5\text{-aziridine (N–H)} \quad (71\%) \qquad \text{(Eq. 106)}$$

2-Alkyl– or 2-aryl–substituted 1-methyl-Δ^1-dihydropyrrolinium perchlorates (Eq. 107)[427,428] and 1-methyl-Δ^1-tetrahydropyridinium perchlorates (Eq. 108)[427,429] are reduced by chiral lithium alkoxyaluminum hydrides to form optically active 2-alkyl– or 2-aryl–substituted 1-methylpyrrolidines and 1-methylpiperidines, respectively, in 75–95% chemical yields; only modest asymmetric induction is noted in these reactions.

$CH_2C_6H_5$, CH_3, ClO_4^- $\xrightarrow[(C_2H_5)_2O,\ reflux]{LiAlH_3(OR^*)}$ $CH_2C_6H_5$, H, CH_3 (Eq. 107)

R* = (−)-menthyl (*R*)-(+) (89%) (6% e.e.)

C_3H_7-*n*, CH_3, ClO_4^- $\xrightarrow[(C_2H_5)_2O,\ reflux]{LiAlH_3\ (OR^*)}$ C_3H_7-*n*, H, CH_3 (Eq. 108)

R* = (−)-menthyl (*S*)-(+) (87%)(12% e.e.)

High stereoselectivity can be achieved in reductions of some iminium salts by $LiAlH(OC_4H_9\text{-}t)_3$. Reaction of the quaternary salt **134** with this hydride leads to the *cis* base **135** as the single product; $NaBH_4$ gives the *cis* and *trans* isomers in a 80:20 ratio.[430,431]

ClO_4^-, N, H, N^+, CH_3O_2C, SCH_3 $\xrightarrow[THF,\ 0°]{LiAlH(OC_4H_9\text{-}t)_3\ (inverse\ addition)}$ H, N, N, H, CH_3O_2C, SCH_3

134 **135** (98%)

Pyridine is stable toward $LiAlH(OCH_3)_3$[65,161] and $LiAlH(OC_4H_9\text{-}t)_3$.[63,65] Reductions of pyridinium and alkylpyridinium methiodides by $LiAlH_4$,[432,433] AlH_3,[434,435] $NaBH_4$,[436,437] $NaAlH_4$, or $NaAlH_2(OCH_2CH_2OCH_3)_2$[438] give mixtures of 1-methyl-3-piperideines and 1-methylpiperidines (Eq. 109); $NaBH_4$ affords highest product yields.[437,438] In $LiAlH_4$,[433] AlH_3,[435] and particularly in $NaAlH_4$[438] reductions, the formation of piperideines and piperidines is accompanied by fission of the pyridine ring to yield dienylamines. However, reduction of *N*-alkoxycarbonyl, *N*-aryloxycarbonyl, or *N*-acylpyridinium salts by the $LiAlH(OC_4H_9\text{-}t)_3$–CuBr complex leads regioselectively, although in only moderate yields, to N-alkoxycarbonyl, N-aryloxycarbonyl, or N-acyl-1,4-dihydropyridines.[439]

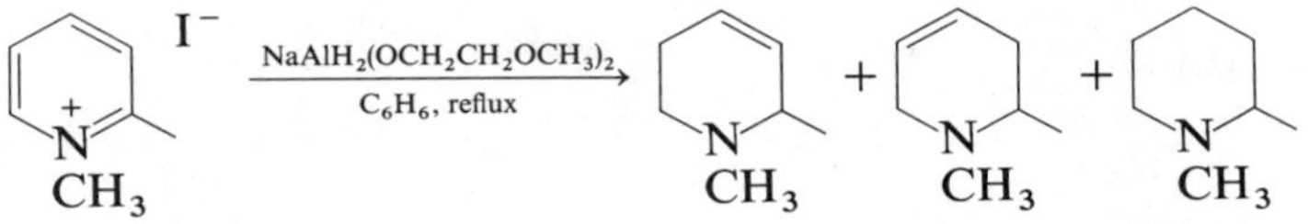

(Eq. 109)

$NaAlH_2(OCH_2CH_2OCH_3)_2$, C_6H_6, reflux

(29%) (13%) (8%)

Cl^- $\xrightarrow{H^-}$ $CO_2C_6H_5$ **A** + **B**

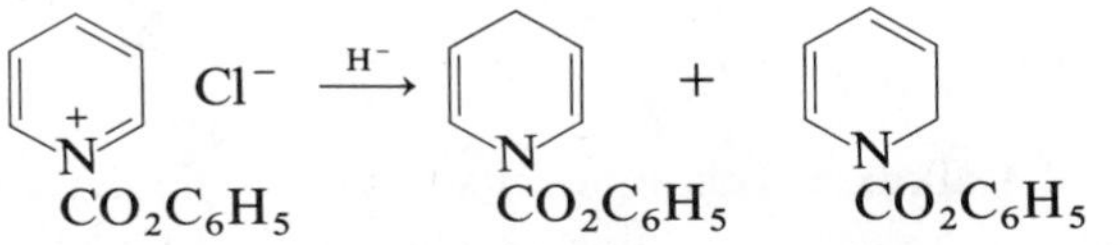

	Molar Ratio **A**:**B**
$NaBH_4$	60:40
$NaBH_3(CN)$	60:40
$LiAlH(OCH_3)_3$–CuBr	79:21
$LiAlH(OC_4H_9\text{-}t)_3$–CuBr	~100:0 (**A** = 65% yield)

Transformation of the quaternary salt **136** proceeds uniformly to give the enamine **137**;[440] in a similar case, the tetrahydro derivative is the product of reduction by $NaBH_4$.[157]

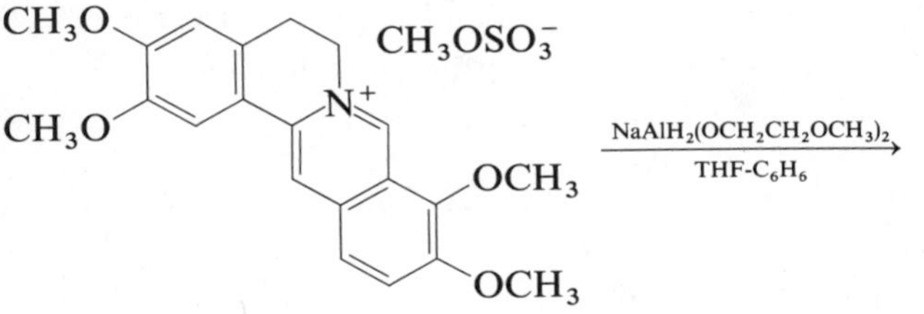

$NaAlH_2(OCH_2CH_2OCH_3)_2$, THF-$C_6H_6$

136

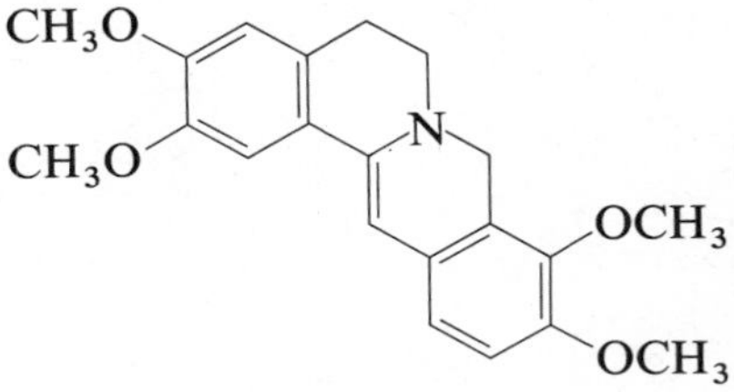

137 (92%)

Pyridine N-oxide and its alkyl derivatives react with $LiAlH_4$,[65,174,441] $LiAlH(OCH_3)_3$,[65,161] AlH_3, $NaBH_4$, and $NaAlH_2(OCH_2CH_2OCH_3)_2$[441] to form 3-piperideines, piperidines, and pyridines; AlH_3 is the superior reagent, giving the highest yield of product mixtures (Eq. 110).[441]

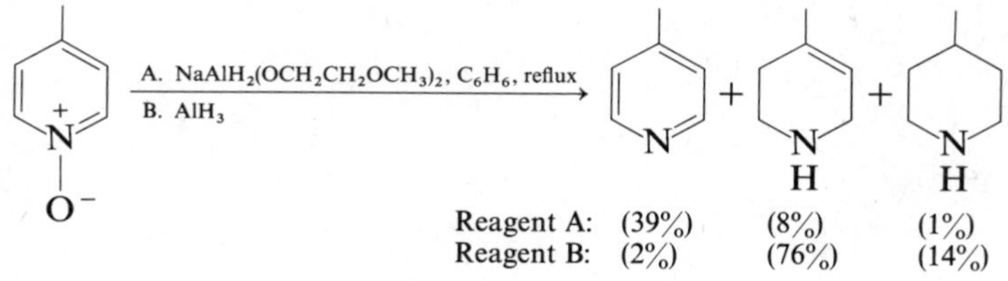

A. $NaAlH_2(OCH_2CH_2OCH_3)_2$, C_6H_6, reflux
B. AlH_3

(Eq. 110)

Reagent A: (39%) (8%) (1%)
Reagent B: (2%) (76%) (14%)

The reduction of 1-alkoxypyridinium salts follows a similar course.[441]

Heterocyclic *N*-oxides are stable toward $LiAlH(OC_2H_5)_3$;[442] $NaAlH_2(OCH_2CH_2OCH_3)_2$ behaves similarly at shorter reaction times or at low temperature (Eq. 111).[443,444]

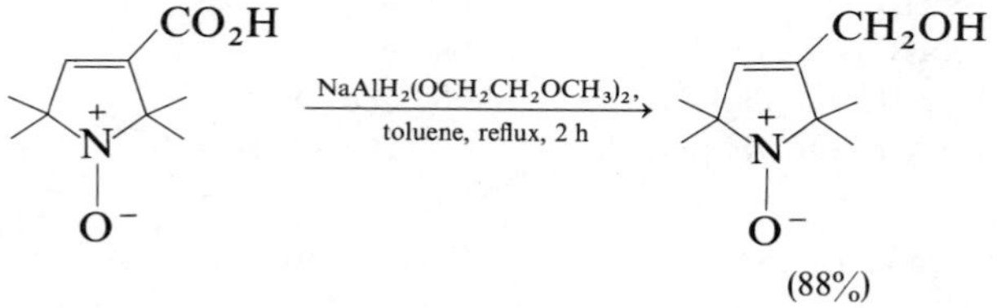

CO_2H → $NaAlH_2(OCH_2CH_2OCH_3)_2$, toluene, reflux, 2 h → CH_2OH (88%)

(Eq. 111)

Various *N*-(carbalkoxy)-1,2-dihydroquinolines are prepared in 85–95% yields from quinolines via quinoline-*N*-boranes and sequential treatment with $NaAlH_2(OCH_2CH_2OCH_3)_2$ and methyl chloroformate (Eq. 112).[445,446]

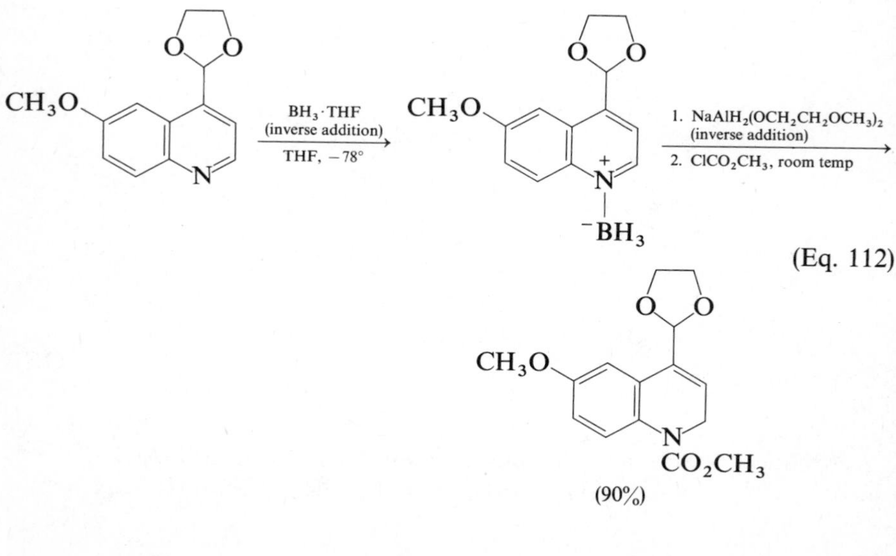

CH_3O — $BH_3 \cdot THF$ (inverse addition), THF, −78° → CH_3O, $^{-}BH_3$ — 1. $NaAlH_2(OCH_2CH_2OCH_3)_2$ (inverse addition) 2. $ClCO_2CH_3$, room temp →

(Eq. 112)

CH_3O, CO_2CH_3 (90%)

Partial reduction of 4(1*H*)-pyridones to the 2,3-dihydro derivatives (60–90% yield) is achieved with $LiAlH(OC_2H_5)_3$ (Eq. 113)[447] or $NaAlH_2(OCH_2CH_2OCH_3)_2$.[448]

$$\text{1-}C_6H_5\text{-4(1}H\text{)-pyridone} \xrightarrow[\text{THF, } 15^\circ]{LiAlH(OC_2H_5)_3} \text{1-}C_6H_5\text{-2,3-dihydro-4-pyridone } (88\%) \quad \text{(Eq. 113)}$$

Reduction of di- and trisubstituted *N*-hydroxyimidazolinium salts by $NaAlH_2(OCH_2CH_2OCH_3)_2$ is utilized in the preparation of imidazoles (overall yield 50–80%) from alkenes (Eq. 114).[449]

$$CH_3CH{=}CH_2 \xrightarrow[CH_3CN,\ -15^\circ \text{ to } 0^\circ]{NOBF_4} \text{N-hydroxyimidazolinium } BF_4^- \xrightarrow{NaAlH_2(OCH_2CH_2OCH_3)_2} \text{imidazole } (80\%) \quad \text{(Eq. 114)}$$

2,3,5-Trialkylisoxazolinium salts with electron-withdrawing substituents at C-4 are reduced by $LiAlH_4$, $NaBH_4$, or $LiAlH(OC_4H_9\text{-}t)_3$ to afford 4-substituted isoxazolidines (Eq. 115) or mixtures of 3- and 4-isoxazolines (Eq. 116).[450]

$$O_2N\text{-isoxazolinium } (N\text{-}C_2H_5)\ BF_4^- \xrightarrow[\text{THF, } 0^\circ]{LiAlH(OC_4H_9\text{-}t)_3} O_2N\text{-isoxazolidine } (N\text{-}C_2H_5)\ (60\%) \quad \text{(Eq. 115)}$$

$$NC\text{-isoxazolinium } (N\text{-}C_2H_5)\ BF_4^- \xrightarrow[\text{THF, } 0^\circ]{LiAlH(OC_4H_9\text{-}t)_3} NC\text{-4-isoxazoline } (N\text{-}C_2H_5)\ (43\%) + NC\text{-3-isoxazoline } (N\text{-}C_2H_5)\ (24\%) \quad \text{(Eq. 116)}$$

Hydride reduction of the isoxazoline **138** followed by benzoylation gives rise to diastereomeric α-amino-γ-hydroxyamides **139a** and **139b**. The diastereoselectivity is markedly increased when $LiAlH_4$ is replaced by $NaAlH_2(OCH_2CH_2OCH_3)_2$ or $LiAlH_2(OC_4H_9\text{-}t)_2$.[451]

R = 4-$CH_3OC_6H_4$

138

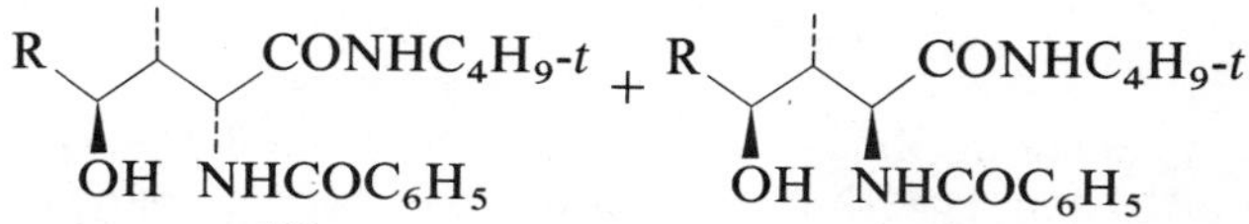

1. Hydride
2. C_6H_5COCl

139a + **139b**

Hydride	Yield (%)	Ratio **139a**:**139b**
$LiAlH_4$	(96)	69:31
$NaAlH_2(OCH_2CH_2OCH_3)_2$	(87)	94:6
$LiAlH_2(OC_4H_9\text{-}t)_2$	(85)	95:5

Sulfur Compounds

Reactions of Thiols and Sulfides. In the presence of $LiAlH(OC_4H_9\text{-}t)_3$, benzenethiol liberates hydrogen substantially more easily than 1-hexanethiol;[63,65] $LiAlH(OCH_3)_3$[65,161] resembles $LiAlH_4$[65,174] and AlH_3,[65] all yielding rapidly 1 equivalent of hydrogen in the presence of aliphatic as well as aromatic thiols.[452]

p-Tolyl methyl sulfide[65] and phenyl *n*-propyl sulfide[63] are inert to $LiAlH(OC_4H_9\text{-}t)_3$, $LiAlH_4$,[174] $LiAlH(OCH_3)_3$,[65,161] AlH_3,[65] $NaBH_4$,[295] and $LiBH(C_2H_5)_3$.[184] An aryl cycloalkyl sulfide shows similar stability[453a] toward the complex $LiAlH(OCH_3)_3$–CuI.[113,114] Alkyl 1-alkynyl sulfides are converted by the $LiAlH(OCH_3)_3$–CuBr complex into alkyl *cis*-1-alkenyl sulfides (~90%) and by $LiAlH_4$ into alkyl *trans*-1-alkenyl sulfides (~100%).[118]

Reduction of Disulfides. Di-*n*-butyl disulfide is reduced by $LiAlH(OCH_3)_3$[65,161] under hydrogen evolution to *n*-butanethiol as rapidly as with $LiAlH_4$[65,174] and much more rapidly than by AlH_3;[65] essentially no reduction occurs with $LiAlH(OC_4H_9\text{-}t)_3$.[63,65,453b] The more reactive diphenyl disulfide is reduced rapidly to benzenethiol by $LiAlH_4$,[65,174] $LiAlH(OCH_3)_3$,[65,161] AlH_3[65] and very rapidly by $LiBH(C_2H_5)_3$;[184] $LiAlH(OC_4H_9\text{-}t)_3$ reacts with the aromatic disulfide only with difficulty.[63,65] Since $NaBH_4$ also reduces disulfides to thiols,[295,452] $LiAlH(OC_4H_9\text{-}t)_3$ offers the possibility of selective reductions in the presence of a disulfide linkage.[63,453b]

Reduction of Sulfoxides. Two equivalents of hydride are required for the reduction of dimethyl sulfoxide to dimethyl sulfide with AlH_3,[65] $LiAlH(OCH_3)_3$,[65,161] and $LiAlH_4$ (see footnote in Ref. 454);[65,174,455,456] the sulfoxide is not attacked by $LiAlH(OC_4H_9\text{-}t)_3$ or $LiBH(C_2H_5)_3$.[184] Rapid reduction of aliphatic, aromatic, and heterocyclic sulfoxides by $NaAlH_2(OCH_2CH_2OCH_3)_2$ gives sulfides in 70–98% yields (Eq. 117).[457]

$$C_6H_5S(O)CH_3 \xrightarrow[C_6H_6,\ \text{reflux}]{NaAlH_2(OCH_2CH_2OCH_3)_2,} \underset{(98\%)}{C_6H_5SCH_3} \qquad \text{(Eq. 117)}$$

Alternatively, $NaBH_4$–$CoCl_2$,[12] $LiAlH_4$–$TiCl_4$,[458] and low-valence species of group IV, V, and VI metals[459] can be used as reducing agents. However, exceptions can occur; $NaAlH_2(OCH_2CH_2OCH_3)_2$ as well as $LiAlH_4$ fail to reduce 2-methyl-2-vinylthiacyclohexane 1-oxide to the corresponding sulfide.[460]

Attempted kinetic resolution of racemic dialkyl and aryl alkyl sulfoxides by incomplete reduction with chiral lithium alkoxyaluminum hydrides (Eq. 118) gives (*R*) or (*S*) sulfoxides in only low optical yields.[461]

$$2\,n\text{-}C_4H_9S(O)CH_3 \xrightarrow[(C_2H_5)_2O,\ \text{reflux}]{LiAlH_4-(+)\text{-quinidine}\ (1:1)}$$

$$\underset{\substack{(R)\text{-}(-)\\(50\%,\ 2.7\%\ \text{e.e.})}}{n\text{-}C_4H_9\overset{*}{S}(O)CH_3} + n\text{-}C_4H_9SCH_3 \qquad \text{(Eq. 118)}$$

Reduction of Sulfones. In contrast to AlH_3[65] and $LiAlH(OC_4H_9\text{-}t)_3$,[63,65] which do not react with diphenyl sulfone at all, $LiBH(C_2H_5)_3$ causes alkylation to form ethylbenzene (75%);[184] $NaAlH_2(OCH_2CH_2OCH_3)_2$ is either inert[160,462] or, like $LiAlH(OCH_3)_3$,[65,161] gives sulfides only in low yields.[457] The ease of reduction of sulfones to sulfides by $LiAlH_4$[65,174] appears to decrease on going from four- or five-membered cyclic to aromatic, to six-membered cyclic, to aliphatic sulfones.[454,463,464] Like $LiAlH_4$, $NaAlH_2(OCH_2CH_2OCH_3)_2$ affords thia-2-cyclohexene as the major product of reduction of thiacyclohexane 1,1-dioxide (Eq. 119), presumably via a sulfone dianion and sulfide anion.[463] 2-Methyl-2-vinylthiacyclopentane-1,1-dioxide does not react with these hydrides.[460]

$$\text{thiacyclohexane 1,1-dioxide } (SO_2) \xrightarrow[C_6H_6,\ \text{reflux}]{NaAlH_2(OCH_2CH_2OCH_3)_2} \text{thia-2-cyclohexene (S)} + \text{thiacyclohexane (S)} \qquad \text{(Eq. 119)}$$

The 2- and 4-hydroxybenzyl phenyl sulfones undergo hydrogenolysis on reaction with $NaAlH_2(OCH_2CH_2OCH_3)_2$ (Eq. 120) or $LiBH_4$.[465a]

$$\text{3,5-di-}i\text{-}C_3H_7\text{-4-HO-}C_6H_2CH(C_6H_5)SO_2C_6H_5 \xrightarrow[C_6H_6,\ 25^\circ]{NaAlH_2(OCH_2CH_2OCH_3)_2} \text{3,5-di-}i\text{-}C_3H_7\text{-4-HO-}C_6H_2CH_2C_6H_5 \quad \text{(Eq. 120)}$$

Reduction of Alkane- and Arenesulfonic Acids and Esters. Methanesulfonic acid and *p*-toluenesulfonic acid interact with $LiAlH_4$,[65,174] $LiBH(C_2H_5)_3$,[184] AlH_3,[65] $LiAlH(OCH_3)_3$,[65,161] and $LiAlH(OC_4H_9\text{-}t)_3$[63,65] only with evolution of hydrogen. Alkyl and cycloalkyl methanesulfonates (mesylates) and *p*-toluenesulfonates (tosylates) are essentially inert toward $LiAlH(OCH_3)_3$[65,161] or

$$c\text{-}C_6H_{11}OSO_2C_6H_4CH_3\text{-}p \longrightarrow \underset{\mathbf{A}}{\text{cyclohexane}} + \underset{\mathbf{B}}{\text{cyclohexene}} + \underset{\mathbf{C}}{\text{cyclohexanol}} \quad \text{(Eq. 121)}$$

	Analytical Normalized Yields (%)			Refs.
	A	B	C	
$LiAlH_4$, THF, 25°	54	26	20	174, 470
LM-9BBN,[a] THF, 0°	88	12	0	469
$LiBH(C_2H_5)_3$, THF, 0°	84	16	0	184, 469, 470
$LiBH(C_4H_9\text{-}s)_3$, THF, 25°	52	48	0	470
$NaBH(C_2H_5)_3$, THF, 25°	83	17	0	470
$NaBH(C_2H_5)_3$, C_6H_6, 25°	18	82	0	469
$LiBHSia_3$,[b] THF, 25°	20	80	0	470
$LiCuH(C_4H_9\text{-}n)$	80[c]	0	0	471
$LiAlH(OCH_3)_3$–CuI, THF	98[c]	0	0	472a

[a] LM-9-BBN = lithium B-methyl-9-borabicyclo[3.3.1]nonyl hydride.
[b] $LiBHSia_3$ = lithium trisiamylborohydride.
[c] The analytical yield is not normalized.

$LiAlH(OC_4H_9\text{-}t)_3$[63,65,466,467] (however, see Ref. 465b) and many other hydride reagents such as AlH_3,[65] borane–tetrahydrofuran,[468] 9-borabicyclo[3.3.1]nonane,[469] thexylborane, disiamylborane, and $LiBH_4$.[470] The effect of various hydrides on the reductive displacement of a secondary tosyloxy group in the less reactive cyclohexyl tosylate, and formation of the alkene and alcohol as byproducts, is summarized in the table accompanying Eq. 121.

The high effectiveness of $LiBH(C_2H_5)_3$[469,470] or the $LiAlH(OCH_3)_3$–CuI complex[472a] is also seen in reductions of aliphatic primary (Eq. 122) and secondary sulfonates.

$$n\text{-}C_7H_{15}CH_2OSO_2CH_3 \xrightarrow[\text{THF, room temp}]{LiAlH(OCH_3)_3\text{–CuI (2:1)}} \underset{(95\%)}{n\text{-}C_8H_{18}} \qquad \text{(Eq. 122)}$$

$NaAlH_2(OCH_2CH_2OCH_3)_2$ is less effective in these reactions.[473a] As shown in reductions with $NaBH(C_2H_5)_3$, the product distribution can be influenced by solvents;[469,470] replacement of diethyl ether by bis(2-methoxyethyl) ether leads to resistance of the tosyloxy group toward the action of $LiAlH_4$.[473b] However, the stability of this group toward metal tri-*tert*-butoxyaluminum hydrides can be weakened by neighboring functional groups.[474,475]

The ability of $NaAlH_2(OCH_2CH_2OCH_3)_2$ to coordinate with oxygen functions[476–481] can be utilized for regioselective reduction of the epoxytosylate **140** to form the desired diol **141a**, presumably by way of an alkoxyaluminum hydride intermediate involving the C-6 hydroxyl group.[482,483]

HOCH₂–O–CH₂OTs (**140**) ⟶ **141a** + **141b**

$NaAlH_2(OCH_2CH_2OCH_3)_2$, toluene, reflux: **141a**:**141b** = 92:8 (~100%)
$LiAlH_4$: **141a**:**141b** = 75:25

Both $LiCuH(C_4H_9\text{-}n)$[471] and $NaBH_3(CN)$[484,485] tolerate a number of other reducible groups and are therefore suitable reducing agents for selective displacement of sulfonyloxy groups in functionalized tosylates and mesylates.

Lithium trimethoxyaluminum hydride is a superior reagent for reduction of the chiral acetylenic mesylate **142** to the chiral allene **143** by a preferred *anti* substitution mechanism.[486] This result is in accordance with the predominant *anti* substitution in reductions of chiral tosylates and camphorsulfonate esters of 1,3-di-*tert*-butyl-2-propyn-1-o1 by $LiAlH_2(OCH_3)_2$.[467]

$$CH_3\text{---}\underset{\underset{H}{|}}{\overset{\overset{OMs}{|}}{C}}C\equiv CC_6H_{13}\text{-}n \xrightarrow[\text{THF, }-20^\circ]{LiAlH(OCH_3)_3}$$

(*S*)-(−)
142

$$\underset{CH_3}{\overset{H}{}}C{=}C{=}C\underset{C_6H_{13}\text{-}n}{\overset{H}{}} + C_2H_5C\equiv CC_6H_{13}\text{-}n$$

(*S*)-(+)
143
(79%, 73% e.e.)

144 (15%)

Reduction of the racemic mesylate **142** by $LiAlH_4$ or AlH_3 yields only 3-decyne (**144**)(83–89%).[486] Transformations of acetylenic mesylates into allenes can be carried out also with $NaAlH_2(OCH_2CH_2OCH_3)_2$.[487]

Reduction of Sulfonamides and Thioimides. Alkane- and arenesulfonamides derived from aliphatic, alicyclic, and aromatic primary and secondary amines or cyclic amines undergo reductive cleavage by $NaAlH_2(OCH_2CH_2OCH_3)_2$ to form the corresponding amines in 55–98% yields[488–491] (Eq. 123);[490] the reductions proceed more easily than with $LiAlH_4$.[488]

$$\text{(cyclohexane, } OC_2H_5\text{, } NHSO_2CH_2C_6H_5) \xrightarrow[C_6H_6\text{, reflux}]{NaAlH_2(OCH_2CH_2OCH_3)_2} \text{(cyclohexane, } OC_2H_5\text{, } NH_2) \quad (98\%) \qquad \text{(Eq. 123)}$$

Lower yields are observed in reactions where alkene formation by elimination of the *p*-toluenesulfonamide group competes with the amine regeneration. The aziridine ring, which is stable toward $NaAlH_2(OCH_2CH_2OCH_3)_2$ at low temperatures,[151,426] is cleaved under the conditions of the sulfonamide reduction[488,490] (Eq. 124).[488]

$$\text{N-Ts aziridine} \xrightarrow{NaAlH_2(OCH_2CH_2OCH_3)_2} C_2H_5NHTs + C_2H_5NH_2 \qquad \text{(Eq. 124)}$$

	C_2H_5NHTs	$C_2H_5NH_2$
C_6H_6, reflux:	~100%	0
Toluene, reflux:	0	57%

N-Tosylazetidines are stable toward $NaAlH_2(OCH_2CH_2OCH_3)_2$ in boiling benzene.[488]

Primary triflamides (trifluoromethanesulfonamides), which afford stable salts with $LiAlH_4$ and do not undergo reduction by this hydride, are reduced by $NaAlH_2(OCH_2CH_2OCH_3)_2$ in minutes to give the parent primary amine in higher than 90% yield. On the other hand, secondary triflamides, which are stable toward $NaAlH_2(OCH_2CH_2OCH_3)_2$, are reduced rapidly by $LiAlH_4$ to form secondary amines in almost quantitative yields.[492–494]

N-(α-Chlorocycloalkylthio)imides such as **145** yield cycloalkene sulfides (e.g., **146**).[495]

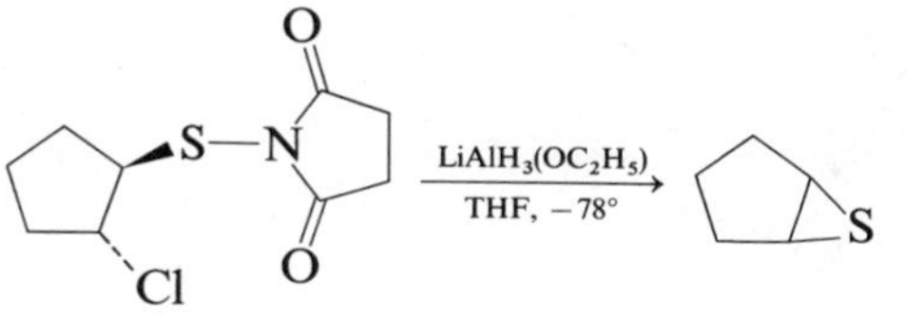

$LiAlH_3(OC_2H_5)$, THF, −78°

145 **146** (58%)

Reduction of Sulfur Heterocyclic Compounds. α-Hydroxyepisulfides such as **147** are reduced by $NaAlH_2(OCH_2CH_2OCH_3)_2$ to give β-hydroxythiols with high regioselectivity (see Refs. 476–481 and 496–499a,b).[500]

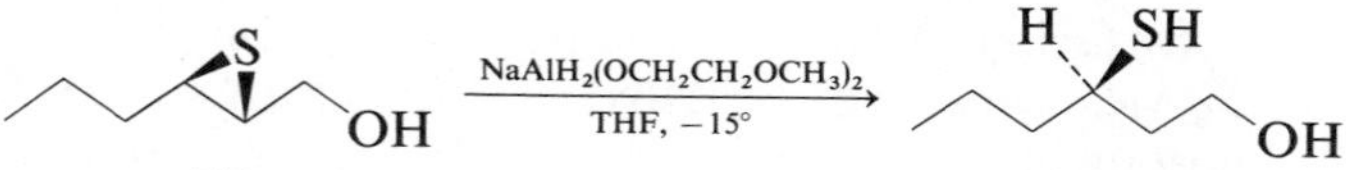

$NaAlH_2(OCH_2CH_2OCH_3)_2$, THF, −15°

147 (*R*)

2-Dimethylamino-1,3-dithiolan-2-ylium perchlorate reacts with complex metal hydrides to form the formamide thioacetal (Eq. 125); $NaBH_4$ produces this compound in 60% and $LiAlH_4$ in 20% yields.[425]

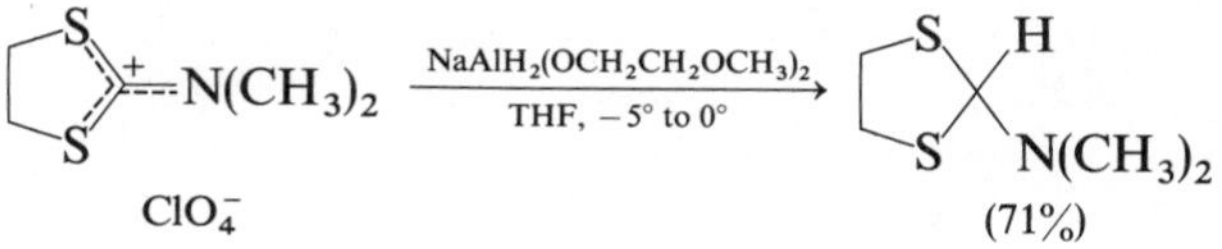

$NaAlH_2(OCH_2CH_2OCH_3)_2$, THF, −5° to 0° (Eq. 125)

ClO_4^- (71%)

Reduction of 2-methylthio-3-methyl-4,5-dihydrothiazolium iodide by $NaAlH_2(OCH_2CH_2OCH_3)_2$ is temperature-dependent and gives either the expected 2-methylthio-3-methylthiazolidine or a mixture of 3-methylthiazolidine-2-thione (**148**) and 3-methylthiazolidine (**149**). The thione **148** is stable toward the hydride[425] (see also Refs. 100–102).

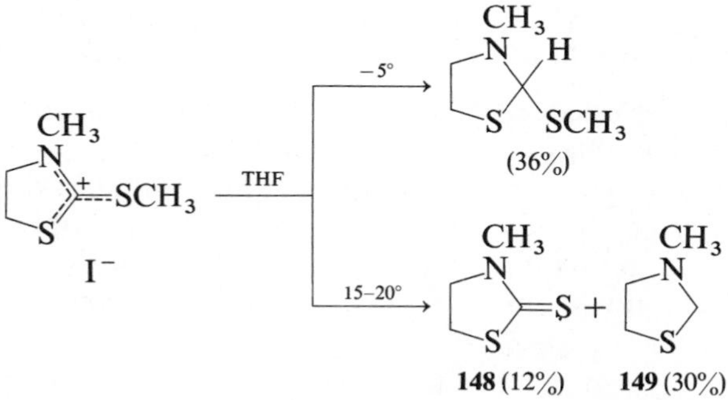

THF; −5° → (36%); 15–20° →

I^-

148 (12%) **149** (30%)

The penam derivative **150** undergoes cleavage on reaction with LiAlH-(OC_4H_9-*t*)$_3$ to yield the isothiazolone **151** as the major product.[501]

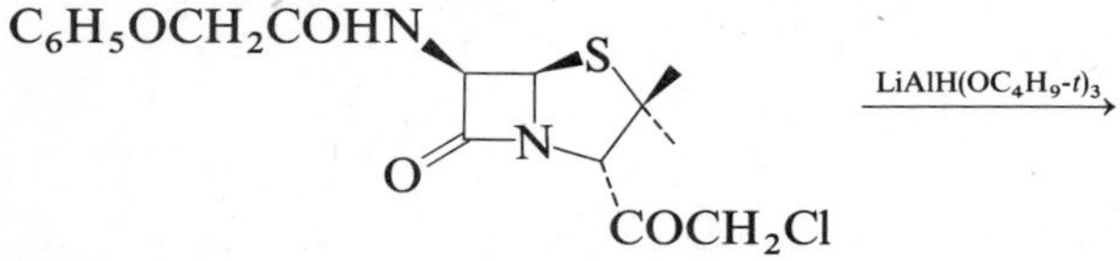

$C_6H_5OCH_2COHN$ ... $\xrightarrow{LiAlH(OC_4H_9\text{-}t)_3}$

$COCH_2Cl$

150

$C_6H_5OCH_2COHN$— ... $COCH_2Cl$

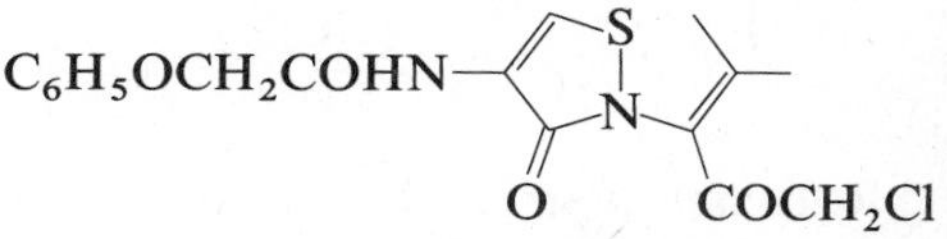

151

Treatment of the penam sulfone **152** with a large excess of LiAlH(OC_4H_9-*t*)$_3$ or KBH_4 leads to fission of the thiazolidine ring to form the 2-azetidinone **153** in 4 and 12% yields, respectively; KBH_4 in slight excess reduces the penam sulfone **152** almost quantitatively to the alcohol **154**, which is considered to be an intermediate in the transformation of **152** into the 2-azetidinone **153.**[502]

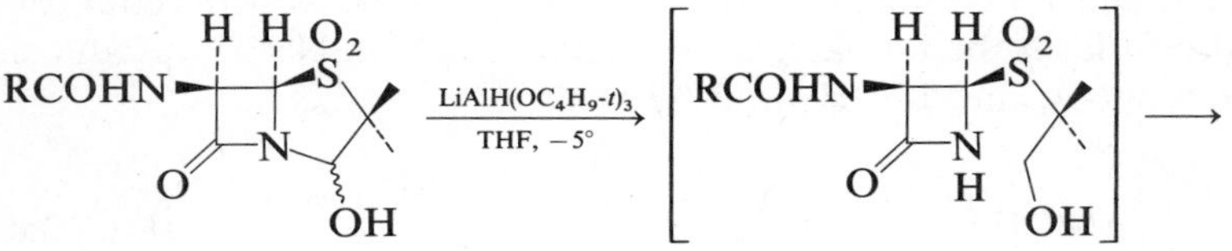

RCOHN ... $\xrightarrow[\text{THF, }-5^\circ]{LiAlH(OC_4H_9\text{-}t)_3}$ [RCOHN ...] $\longrightarrow$

152 **154**

RCOHN ... NH

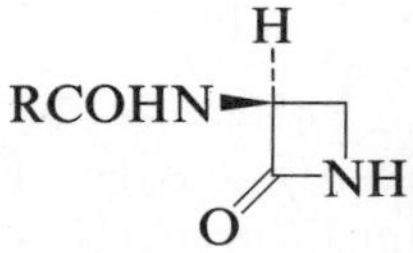

153

REDUCTION OF VARIOUS FUNCTIONAL GROUPS WITH SELECTED METAL ALKOXYALUMINUM HYDRIDES AND COMPLEX METAL HYDRIDES

The reactions of selected metal alkoxyaluminum hydrides and complex metal hydrides with various functional organic groups are summarized in the following table. The symbol (+) indicates positive reaction and the symbol (−), very slow or insignificant reaction; the symbol (±) denotes positive or negative reaction, depending on the substrate structure and reaction conditions. When the literature lacks any information about the hydride reactivity the symbol 0 is used.

	SMEAH[a]	LTBA[a]	LTMA[a]	CALH[b]	$NaBH_4$	$LiAlH_4$
Haloalkanes or Haloarenes → alkanes or arenes	±	−	±	±	−	±
Aldehydes → alcohols	+	+	+	+	+	+
Ketones → alcohols	+	+	+	+	+	+
Epoxides → alcohols	±[c]	−	±	+	−	+
Carboxylic acids → alcohols	+	−	+	+	−	+
Carboxylic acids → aldehydes	+[d]	+[e,f]	0	0	−	+[e]
Esters → alcohols	+[g]	−[h]	+	+	−	+
Esters → aldehydes	+[i]	+[j]	0	0	−	−
Cyclic anhydrides → diols	+	−	+	+	−	+
Cyclic anhydrides → lactones and/or hydroxylactones	−	+	0	0	+	+
Lactones → diols	+	±	+	+	−	+
Lactones → lactols	+[k]	±	−	0	−	−
Acyl chlorides → alcohols	+	+	+	+	+	+
Acyl chlorides → aldehydes	−	+[l]	−	0	−	−
Nitriles → amines	±	−	+	+	−	+
Nitriles → aldehydes	+[m]	−	−	0	−	−
Amides → amines	±	−	+	+	−	+
Amides → aldehydes	+[n]	−	−	0	−	+[n]
Lactams → cyclic amines	±	−	0	+	−	+
Imides → cyclic amines or other products	+[o]	−	0	0	−	+
Imines → amines	+	+	0	0	+	+
Azido compounds → amines	+	−	0	0	−	+
Oximes → amines or other products	±	−	0	+	−	+
Nitro compounds → amines or other products	±	−	±	0	−	+

(*Continued*)

(*Continued*)	SMEAH[a]	LTBA[a]	LTMA[a]	CALH[b]	$NaBH_4$	$LiAlH_4$
Urethanes → substituted methylamines	+	−	0	+	−	+
Disulfides → thiols	+	−	+	0	−	+
Sulfoxides → sulfides	+	−	+	0	−	+
Sulfones → sulfides	±	−	−	0	−	+
Alkanesulfonates → alkanes or alcohols	+	−	+[p]	0	−	+
Sulfonamides → amines	+	−	0	0	−	+

[a] Abbreviations used for the metal hydrides are listed in the "Tabular Survey" section on p. 334.

[b] The abbreviation CALH denotes calcium alkoxyaluminum hydrides.

[c] No reduction takes place at 0°.

[d] The acids are treated with γ-saccharin chloride and the resulting *N*-acylsaccharins are reduced at 0–5°.

[e] The acids are treated with 1,1′-carbonyldiimidazole, and the resulting *N*-acylimidazoles are reduced with the hydride.

[f] The acids are treated with the reaction product from dimethylformamide and oxalyl chloride in the presence of pyridine, and the intermediates are reduced *in situ* at −70°.

[g] No reduction takes place at about −78°; *tert*-butyl and phenyl esters are generally stable toward the hydride at low temperatures.

[h] The reaction proceeds readily at 37–140°.

[i] The reaction proceeds at −40 to −70° or at 0 to −55° by using the hydride modified by *N*-methylhexahydropyrazine or morpholine.

[j] Phenyl esters are used as the reactants.

[k] Either the unmodified hydride at −70° or the hydride modified with ethanol, or pyridine, or 2-propanol is used at 0° to room temperature.

[l] The reaction is carried out at −78° to room temperature.

[m] The reaction is carried out at −15° to room temperature.

[n] The reduction of dimethylamides, *N*-methylanilides, or 1-acylaziridines is carried out at −15°, 0°, or room temperature.

[o] Reduction of *N*-substituted cyclic imides at −78° gives hydroxylactams.

[p] The reaction is carried out with the LTMA-CuI complex and gives predominantly alkanes.

EXPERIMENTAL CONSIDERATIONS

References to the preparation and properties of metal alkoxyaluminum hydrides, purification of solvents and chemicals, apparatus used for reductions, and the selection of reaction conditions are given in Ref. 1. Dry methanol, ethanol, and *tert*-butyl alcohol for the preparation of the corresponding metal alkoxyaluminum hydrides are obtained best by drying the alcohols with molecular sieves.[503]

An alternative method for preparing a standardized solution of $LiAlH(OC_4H_9\text{-}t)_3$ in bis(2-methoxyethyl) ether, useful for partial reductions of acyl chlorides, involves extraction of the commercial $LiAlH_4$ with diethyl ether in a Soxhlet apparatus, reaction of the ethereal solution of $LiAlH_4$ with 3 equivalents of dry *tert*-butyl alcohol, removal of diethyl ether in a stream of dry nitrogen, and dissolution of the alkoxyhydride in bis(2-methoxyethyl) ether.[504] Preparations of hydride reagents such as $NaAlH_2(OCH_2CH_2OCH_3)_2$*–copper(I) bromide—2-butanol,[113–115] $NaAlH_2(OCH_2CH_2OCH_3)_2$–morpholine,[73,293] $NaAlH_2(OCH_2CH_2OCH_3)_2$–*N*-methylhexahydropyrazine,[73,292] and $NaAlH_2(OCH_2CH_2OCH_3)_2$–ethanol[73,227] are described in detail in the "Experimental Procedures" section. The $LiAlH(OCH_3)_3$–copper(I) iodide (2:1) reagent is prepared as follows. Concentrated (0.9–1.1 *M*) $LiAlH(OCH_3)_3$ (54.5 mmol) solution in tetrahydrofuran is added dropwise in an atmosphere of argon and under stirring to an ice-cooled suspension of copper(I) iodide (5.18 g, 27.2 mmol) in dry tetrahydrofuran (10 mL). The stirred suspension becomes gelatinous, and tetrahydrofuran is added in 5-mL portions to facilitate stirring. A total of 150 mL is usually required. After the addition is complete, the resulting dark-brown mixture is stirred at 0° for 30 minutes. The $LiAlD(OCH_3)_3$–copper(I) iodide reagent is prepared from $LiAlD(OCH_3)_3$ and copper(I) iodide in a similar way.[472a,b]

Calcium alkoxyaluminum hydrides are available either from commercial sources[505 †] or can be prepared by partial alcoholysis of calcium aluminum hydride[162–164] and by reaction of sodium aluminum hydride with calcium chloride and alcohols.[163,164] The preparation of $Ca[AlH_2(OC_3H_7\text{-}i)_2]_2 \cdot 2\,THF$ by an alternative route from sodium aluminum hydride, calcium chloride, sodium isopropoxide, and aluminum chloride is as follows. A solution of aluminum chloride (1.59 g, 11.9 mmol) in tetrahydrofuran (30 mL) is added slowly under nitrogen to a stirred suspension of sodium aluminum hydride (0.64 g, 11.9 mmol), powdered calcium chloride (1.98 g, 17.8 mmol), and sodium isopropoxide (3.90 g, 47.5 mmol) in dry tetrahydrofuran (65 mL). The mixture is stirred at room temperature for 1 hour and then under reflux for about 0.5 hour until the Ca:Al molar ratio in the solution is 0.48. The mixture is filtered under nitrogen and the filtrate is evaporated *in vacuo*. The resulting white solid is dried *in vacuo* (0.01 mm) at room temperature for 2 hours to give 4.9 g (85% yield) of product. For $C_{20}H_{48}Al_2CaO_6$ (478.6) calculated: Al 11.3 and Ca 8.4%; molar ratio H_{active}:Al = 2.00; found: Al 12.4 and Ca 8.8%; molar ratio H_{active}:Al = 1.83.[164]

EXPERIMENTAL PROCEDURES

***dl*-9,9′-Spirobifluorene-2,2′-dimethanol [Reduction of a Dicarboxylic Acid to a Diol by $NaAlH_2(OCH_2CH_2OCH_3)_2$].**[176] A 70% solution of $NaAlH_2(OCH_2CH_2OCH_3)_2$ (10 g, 35 mmol) in benzene was added under nitrogen slowly and dropwise to a suspension of 2.0 g (5 mmol) of 9,9′-spirobifluorene-2,2′-dicarboxylic acid in 20 mL of benzene. After 2 hours of heating under reflux,

* In addition to references given in Ref. 1, $NaAlH_2(OCH_2CH_2OCH_3)_2$ can also be purchased from Hexcel Corporation, Zeeland, Michigan (USA).[2]

† Manufactured by ANIC, Italy.

the dicarboxylic acid dissolved in the reaction mixture. The mixture was cooled to 10°, decomposed by addition of water, acidified with concentrated hydrochloric acid, and extracted with chloroform. The organic phase was separated, washed with water, and dried over anhydrous magnesium sulfate, and the solvents were evaporated. Crystallization of the residue from benzene gave 1.57 g (84%) of product, mp 254–255°; IR (KBr): 3400, 1612, 1465, 1451, and 1031 cm^{-1}; NMR ($CDCl_3$) δ: 1.99 (s, 2OH), 4.30 (s, $2CH_2O$), 6.57–7.82 (14 ArH); mass spectrum, m/z 376 (M^+).

3-Phenylpropanal [Conversion of a Carboxylic Acid to an Aldehyde via Reduction of the Carboxymethyleniminium Chloride *in situ* by $LiAlH(OC_4H_9\text{-}t)_3$].[192] Oxalyl chloride (0.5 mL) was added to a solution of *N*,*N*-dimethylformamide (1.5 g, 18 mmol) in dichloromethane (3 mL) at 0°. After this solution was stirred for 1 hour, the solvent was removed under reduced pressure. To the *N*,*N*-dimethylchloromethyleniminium chloride thus obtained as white powder was added acetonitrile (3 mL), dry tetrahydrofuran (5 mL), and then a solution of 3-phenylpropionic acid (0.30 g, 2 mmol) and pyridine (1.6 g, 19 mmol) in dry tetrahydrofuran (3 mL) at −30°, and the reaction mixture was stirred at this temperature for 1 hour. To the carboxymethyleniminium chloride thus formed was added under nitrogen a suspension of 10 mol% (0.038 g, 0.2 mmol) of copper(I) iodide in dry tetrahydrofuran and a tetrahydrofuran solution of $LiAlH(OC_4H_9\text{-}t)_3$ (2.6 mL of a 1.54 *M* solution, 4 mmol) at −78°. The mixture was stirred for 10 minutes, and the reaction was then quenched by addition of aqueous 2 *N* hydrochloric acid. The aqueous layer was extracted with diethyl ether, and the extract was washed with aqueous sodium bicarbonate and dried over anhydrous magnesium sulfate. The solvent was removed by evaporation; TLC of the residue on silica gel plates (eluted with hexane–diethyl ether, 3:2) gave 0.217 g (81% yield) of the product.

3′,4,5,5′,6′,7′,8′,8a′-Octahydro-2′-methylspiro[furan-3(2*H*),1′(2′*H*)-naphthalen]-5-ol [Partial Reduction of a Lactone to a Lactol (Hemiacetal) by the $NaAlH_2(OCH_2CH_2OCH_3)_2$–Ethanol Reagent].[73,227] To a solution of $NaAlH_2(OCH_2CH_2OCH_3)_2$ (50% solution in benzene) (2 mL, 5.1 mmol) in 2 mL of dry toluene stirred under nitrogen at 0° was added dropwise by means of a syringe a mixture of dry ethanol (0.18 mL, 3 mmol) and dry toluene (1 mL). Part of this reagent solution (0.5 mL) was added to an ice-cooled solution of 3′,5′,6′,7′,8′,8a′-hexahydro-2′-methyl-(1α′,2α′,8aα′)spiro[furan-(3(2*H*),1′(2′*H*)-naphthalen]-5(4*H*)-one (25 mg, 0.114 mmol) in dry toluene (1 mL). After 1 hour a further amount of the reagent solution (1.5 mL) was added, the mixture was allowed to stand for 30 minutes, and the reaction was quenched by addition of water. Benzene extraction and column chromatography on silica gel gave 24 mg (96% yield) of the product, mp 141–143°, and 1 mg (~4% yield) of (1α,2α,8aα)-1,2,3,5,6,7,8,8a-octahydro-1-hydroxymethyl-2-methyl-1-naphthaleneethanol as byproduct.

***tert*-Butyl 8-{3-Hydroxy-7-(2-tetrahydropyranyloxy)-2-oxa-6-bicyclo[3.3.0]-octyl}-6-(2-tetrahydropyranyloxy)-7-octenoate [Reduction of a Lactone Ester to a Lactol (Hemiacetal) Ester by $NaAlH_2(OCH_2CH_2OCH_3)_2$].**[230] To a solution of 47.8 g (91.0 mmol) of *tert*-butyl 6-(2-tetrahydropyranyloxy)-8-{3-oxo-7-(2-tetrahydropyranyloxy)-2-oxa-6-bicyclo-[3.3.0]octyl}-7-octenoate in 1 L of dry toluene was added dropwise under argon 116 mL (415 mmol) of a 70% solution of $NaAlH_2(OCH_2CH_2OCH_3)_2$ in toluene at −70°. The mixture was stirred at this temperature for 6 hours and then decomposed with a mixture of 400 mL of methanol and 400 mL of water at −70°. The resulting mixture was diluted at 20° with 400 mL of toluene and 1.2 L of a saturated sodium chloride solution and extracted with ethyl acetate. The extract was washed with a saturated sodium chloride solution and dried over anhydrous sodium sulfate. Evaporation of the solvents under reduced pressure gave 48.19 g (~100% yield) of the product; NMR ($CDCl_3$) δ: 5.27–5.70 (m, 2*H*, CH=CH), 4.50–4.74 (m, 2*H*, OCHO), 1.43 [s, 9*H*, $C(CH_3)_3$]; IR (film): 3370 (OH), 1730 ($CO_2C_4H_9$-*t*), and 980 (CH=CH) cm^{-1}.

(*S*)-(+)-2,3,6-Trimethyl-5-(2,2,4-trimethyl-1,3-dioxolane-4-ethyl)phenol [Hydrogenolysis of an Aromatic Ester Group to a Methyl Group by $NaAlH_2(OCH_2CH_2OCH_3)_2$].[240] A solution of 1.02 g (2.79 mmol) of dimethyl (*S*)-(+) 2-hydroxy-6-methyl-4-(2,2,4-trimethyl-1,3-dioxolane-4-ethyl)-1,3-benzenedicarboxylate in 5 mL of xylene was added under argon dropwise over a 5-minute period to a stirred solution of 6 mL (2.17 mmol) of $NaAlH_2(OCH_2CH_2OCH_3)_2$ (70% solution in benzene) in 5 mL of xylene. The resulting solution was stirred and refluxed for 3.75 hours and then cooled to 10°, at which point a solution of 1.16 mL of concentrated sulfuric acid in 5 mL of water was cautiously added dropwise. The resulting slurry was diluted with 23 mL of methanol and refluxed with stirring for 10 minutes. After cooling, the slurry was filtered and the granular solid was washed successively with methanol and ether. The filtrates and washings were combined and concentrated *in vacuo*. The residue was taken up in ether, and the solution was washed with saturated brine, dried over anhydrous magnesium sulfate, and filtered. Evaporation of the solvent gave 769 mg of a yellow oil that was chromatographed on EM silica gel 60 (0.063–0.2 mm; 30 g). Elution with 4:1, 2:1, and 1:1 hexane–diethyl ether afforded 640 mg (83% yield) of the product as a colorless oil that crystallized. A portion of this material was recrystallized from hexane, giving a colorless solid, mp 60–62°, $[\alpha]_D^{25}$ + 5.00° (c 2.0, $CHCl_3$); IR $(CHCl)_3$:3610 cm^{-1} (OH); NMR ($CDCl_3$) δ: 6.55 (s, 1*H*, ArH), 4.80 (br s, 1*H*, OH), 3.78 (m, 2*H*, CH_2O), 2.19, 2.15, 2.10 (3 s, 9*H*, $ArCH_3$), 1.41 [s, 6*H*, $C(CH_3)_2$], and 1.35 (s, 3*H*, CH_3); UV (95% C_2H_5OH) nm max (ε): 274 (955), 281 (985), and sh 223 (9500).

4-Fluoro-3,5-dimethoxybenzaldehyde [Partial Reduction of an Aromatic Ester to an Aldehyde by the $NaAlH_2(OCH_2CH_2OCH_3)_2$–*N*-Methylhexahydropyrazine Reagent].[292] *N*-Methylhexahydropyrazine (10.9 mL, 98 mmol) dissolved in 48 mL of toluene was added under nitrogen at 0° to a solution of

$NaAlH_2(OCH_2CH_2OCH_3)_2$ (25.6 g, 89 mmol) in toluene (ca. 70% solution). This reagent was added dropwise with stirring to a solution of 4.1 g (19 mmol) of methyl 4-fluoro-3,5-dimethoxybenzoate in 90 mL of toluene at 0 to $-5°$. The mixture was stirred for an additional 30 minutes, cooled to $-10°$, and 20 mL of water was added. The precipitate was removed by filtration and washed with toluene. The combined filtrate and washings were washed with 1 *N* hydrochloric acid followed by water, dried, and evaporated. Crystallization of the residue from heptane gave 3.3 g (91% yield) of the pure product, mp 97–98°; IR (KBr): 2860, 1700, 1616, 1241, and 1133 cm^{-1}; NMR ($CDCl_3$) δ: 9.95 (s, 1*H*), 7.18 (d, $J_{HF} = 12.2$ Hz), and 3.97 (s, 6*H*).

***dl*-4-Tetrahydropyranyloxy-1-cyclohexenemethanol [Reduction of an α,β-Unsaturated Ester to an Allylic Alcohol by $LiAlH_3(OC_2H_5)$].**[276] To a stirred suspension of 4.56 g (0.120 mol) of $LiAlH_4$ in 250 mL of dry diethyl ether was added 5.52 g (0.120 mol) of dry ethanol, the mixture was cooled to 0°, 20.1 g (83.6 mmol) of methyl *dl*-4-tetrahydropyranyloxy-1-cyclohexene-1-carboxylate was added, and the mixture was stirred at this temperature for 1 hour. After additional stirring at 25° for 2 hours, the mixture was decomposed by addition of 23 mL of 1 *N* sodium hydroxide and worked up to give 17.2 g (97% yield) of an oily product showing no C=O absorption near 1700 cm^{-1} in the IR spectrum.

3-Ethyl-7,11-dimethyl-2-*trans*,6-*trans*,10-dodecatrien-1-ol [Reduction of an α,β-Unsaturated Ester to an Allylic Alcohol by $NaAlH_2(OCH_2CH_2OCH_3)_2$].[506] A solution of $NaAlH_2(OCH_2CH_2OCH_3)_2$ in benzene (3.58 *M*, 0.691 mL, 2.47 mmol) was added under nitrogen to a stirred solution of ethyl 3-ethyl-7,11-dimethyl-2-*trans*,6-*trans*,10-dodecatrienoate (627 mg, 2.25 mmol) in 5 mL of dry diethyl ether at 0°. The mixture was allowed to warm to room temperature and was stirred for 2.5 hours. Water was added to the mixture at 0° until precipitation occurred, and the precipitate was removed by filtration. The solvent was evaporated; bulb-to-bulb distillation of the residue gave 501 mg (94% yield) of the pure product; NMR ($CDCl_3$) δ: 0.99 (t, J = 7.5 Hz, 3*H*, CH_2CH_3), 1.58 and 1.69 (2 s, 9*H*, allyl CH_3), 1.83–2.35 (m, 10*H*, allyl CH_2), 2.39 (s, 1*H*, exchanged by D_2O, OH), 4.15 (d, *J* = 7 Hz, 2*H*, CH_2O), 5.15 (br s, 2*H*, vinyl H), and 5.40 (t, J = 7 Hz, 1*H*, C-2 vinyl H); IR (film): 3310 and 1675 cm^{-1}; mass spectrum, m/z 237 (MH^+).

***trans*-7-Methyl-2-phenyl-3,5,6-octatrien-2-ol [Reduction of a 4,5-Alkadien-2-yn-l-yl Acetate to a *trans*-2,4,5-Alkatrien-l-ol by $LiAlH_3(OCH_3)$].**[507] To a 0.67 *M* solution of $LiAlH_4$ (20 mL, 13.4 mmol) in dry tetrahydrofuran stirred under argon was added slowly via syringe 0.43 g (13.4 mmol) of dry methanol at 0°. After additional stirring for 15 minutes, a solution of 2.54 g (10 mmol) of 7-methyl-2-phenyl-5,6-octadien-3-yn-2-yl acetate in 2.5 mL of dry tetrahydrofuran was added to the reagent, and the mixture was heated to 70° for 2 hours.

The resulting mixture was cooled to 0°, decomposed by addition of a solution of tetrahydrofuran in water (1:1), and the precipitate was removed by filtration. The product (1.71 g, 80%) was obtained by column chromatography on silica gel by using petroleum ether–diethyl ether (90:10) as the eluent; IR: 1950 (C=C=C), 3040–3020 (=CH), 1640 (C=C), 990–970 (*trans* CH=CH), and 850 (=C—H) cm^{-1}; NMR δ: 1.55 (s, 3*H*), 1.66 (d, J = 3 Hz, 6*H*), 5.4–6.2 (m, 3*H*), and 7.25 (m, 5*H*).

Methyl 3-(3,4,5-Trimethoxyphenyl)propionate [Reduction of an α,β-Unsaturated Ester to a Saturated Ester by the $NaAlH_2(OCH_2CH_2OCH_3)_2$–Copper(I) Bromide–2-Butanol Reagent].[114] A solution of $NaAlH_2(OCH_2CH_2OCH_3)_2$ in benzene (11.2 mL, 40 mmol) was added dropwise to a suspension of copper(I) bromide (5.70 g, 40 mmol) in tetrahydrofuran (75 mL) maintained at −5 to 0° (bath temperature) under argon. After 30 minutes at 0°, the mixture was cooled to −78° and 2-butanol (8.0 mL, 90 mmol) was added rapidly. Within 10 minutes, a solution of methyl 3,4,5-trimethoxycinnamate (1.26 g, 5.0 mmol) in tetrahydrofuran (10 mL) was added rapidly. The reaction mixture was stirred at −78° for 15 minutes and at −20° for 2 hours. Then 10 mL of water was added and the mixture was poured into a saturated aqueous ammonium chloride solution. The organic layer was washed with water and concentrated by rotary evaporation. Distillation of the residue in a shortpath still gave 1.23 g (97% yield) of the pure product as a colorless liquid; IR (neat) 1740 cm^{-1}; NMR ($CDCl_3$) δ: 6.42 (s, 2*H*, ArH), 3.80 (s, 9*H*, OCH_3), and 2.4–3.1 (m, 4*H*, CH_2CH_2); mass spectrum, m/z: M^+ calculated for $C_{13}H_{18}O_5$, 254.1167; found, 254.1154.

4-Methyl-4*H*-1,4-benzoxazine-2-carboxaldehyde [Partial Reduction of an α,β-Unsaturated Ester to an α,β-Unsaturated Aldehyde by the $NaAlH_2(OCH_2CH_2OCH_3)_2$–Morpholine Reagent].[293] A 70% toluene solution (3.2 mL) of $NaAlH_2(OCH_2CH_2OCH_3)_2$ (11 mmol) was diluted with 4.8 mL of dry toluene, and to this solution was added dropwise under nitrogen 1.07 g (12 mmol) of morpholine in 6 mL of dry toluene at 0°; 4.6 mL of this reagent was added dropwise in an atmosphere of nitrogen to a solution of 103 mg (0.5 mmol) of methyl 4-methyl-4*H*-1,4-benzoxazine-2-carboxylate in 6 mL of dry toluene at −40°, and the mixture was stirred at this temperature for an additional 20 minutes. The reaction mixture was decomposed by addition of water and extracted with toluene. The organic layer was separated, the solvents were evaporated under reduced pressure, and the oily residue was subjected to preparative TLC on silica gel 60 F_{254} (Merck, 20 cm × 20 cm × 2 mm). Elution with benzene–ethyl acetate (1:9) gave 83 mg (95% yield) of product ($R_f = 0.20$) as yellow crystals, mp 142–144° (ethyl acetate); IR (KBr): 1670 (C=O), 1640 (C=C), 1590, and 1500 cm^{-1}; NMR ($CDCl_3$) δ:2.98 (s, 3*H*, NCH_3), 6.11 (s, 1*H*, CH), 6.28–6.50 (m, 1*H*, ArH), 6.58–6.85 (m, 3*H*, ArH), and 8.46 (s, 1*H*, HC=O); mass spectrum, m/z (relative intensity): 175 (M^+, 100), 160 (24), and 146 (72).

Methyl 8-Oxooctanoate [Partial Reduction of an Acyl Chloride Ester to an Aldehyde Ester by $LiAlH(OC_4H_9\text{-}t)_3$].[300] To a stirred suspension of 0.86 g (22.7 mmol) of $LiAlH_4$ in 20 mL of bis(2-methoxyethyl) ether cooled to 5–10° was added under nitrogen dropwise 5 mL of *tert*-butyl alcohol. When the hydrogen evolution ceased, the solution of $LiAlH(OC_4H_9\text{-}t)_3$ was stirred for an additional 30–60 minutes and transferred under nitrogen into a dropping funnel. It was then added dropwise under nitrogen during 1 hour to a stirred solution of 5.0 g (24.2 mmol) of methyl 7-chloroformylheptanoate in 20–30 mL of bis(2-methoxyethyl) ether maintained at −78° (acetone–dry ice bath). The reaction mixture was stirred at the same temperature for an additional 60 minutes and allowed to warm to 20°. About 50 g of ice and 1 mL of concentrated hydrochloric acid were added and the mixture was extracted with diethyl ether. The ether solution was washed several times with water and the organic phase was dried over anhydrous sodium sulfate. Evaporation of solvents (12 mm) at 30–40° (water bath temperature) left 4 g (96% yield)* of the crude product as a yellowish liquid. Kugelrohr distillation (0.2 mm) at 80° (air bath temperature) afforded the analytically pure product (single peak in GLC on SE-30, 140°); IR (film): 2710 (CHO), 1720–1740 (CO_2CH_3 + HCO), and 1200–1150 (CO_2CH_3) cm^{-1}; mass spectrum, m/z (relative intensity): 172 (M^+), 141 (21), 129 (45), 97 (68), 87 (95), and 74 (100); 2,4-dinitrophenylhydrazone, mp 81–83°.

Benzaldehyde-d_1[Conversion of an Acyl Chloride to an Aldehyde-d_1 by Reduction with $LiAlD(OC_4H_9\text{-}t)_3$].[308] To a well-stirred suspension of $LiAlD_4$ (2.52 g, 60 mmol) in dry diethyl ether (150 mL) a solution of freshly distilled dry *tert*-butyl alcohol (13.83 g, 189 mmol) in dry diethyl ether (60 mL) was added at such a rate as to maintain a gentle reflux. The mixture was stirred for an additional 15 minutes at room temperature and the solvent and excess *tert*-butyl alcohol were then removed, affording 13.1 g (86% yield) of $LiAlD(OC_4H_9\text{-}t)_3$. A slurry of the hydride (13.1 g, 51.4 mmol) in tetrahydrofuran (50 mL) was added to a well-stirred solution of benzoyl chloride (6.85 g, 48.7 mmol) in dry tetrahydrofuran (21 mL) maintained at −78°. The mixture was stirred at this temperature for 1 hour and brought to room temperature by removing the acetone–dry ice bath. The mixture was then poured onto crushed ice (100 g), and the product was extracted several times with diethyl ether. Diethyl ether and tetrahydrofuran were removed by evaporation *in vacuo*, and the residue was redissolved in diethyl ether (100 mL) and washed twice with water (60 mL) to remove any trace of tetrahydrofuran. The ethereal solution was dried over anhydrous magnesium sulfate, filtered, and evaporated to dryness, yielding 5.1 g of the crude product. This was purified by conversion

* The high yields of aldehyde obtained by reacting 5–10 g of the acyl chloride with the hydride decreased markedly when the reaction was performed on a larger scale; the acyl chloride was reduced even at −40° to the alcohol, which reacted with the acyl chloride to form an ester.[300] Presumably, more efficient mixing is necessary[508] to achieve immediate contact of the acyl chloride with the hydride and to prevent side reactions (see also Refs. 509 and 510).

into the bisulfite addition compound with 40% sodium bisulfite solution. The solid was filtered, washed with diethyl ether, and reconverted to the product with 10% sodium bicarbonate solution. Extraction with diethyl ether, drying over anhydrous magnesium sulfate, filtration, and removal of the solvent afforded 3.9 g (74% yield) of the pure product as a colorless liquid; IR (neat): 2100, 2500 (both C–D stretching), and 1680 (C=O) cm^{-1}; NMR δ 7.42–7.92 (m, 5*H*, ArH).

2-Methyl-3-*n*-propylhexanenitrile [Selective Reduction of an α,β-Unsaturated Nitrile to a Saturated Nitrile by the $NaAlH_2(OCH_2CH_2OCH_3)_2$–Copper(I) Bromide–2-Butanol Reagent].[115] To 1.86 g (13.0 mmol) of copper(I) bromide in dry tetrahydrofuran (10 mL) cooled to 0° was added under nitrogen 7.4 mL (26.0 mmol) of $NaAlH_2(OCH_2CH_2OCH_3)_2$ (3.5 *M* solution in benzene). The resulting dark solution was stirred at 0° for 30 minutes and cooled to −78°. 2-Butanol (2.3 mL, 26.0 mmol) was cautiously introduced via syringe, followed by a solution of 2-methyl-3-*n*-propyl-2-hexenenitrile (0.197 g, 1.3 mmol) in dry tetrahydrofuran (5 mL). After 2 hours at −78°, the reaction mixture was maintained at room temperature for 4 hours, and 6 mL of a saturated solution of ammonium chloride was added. The mixture was extracted with dichloromethane, the organic phase was separated, dried, and filtered; and the solvents were evaporated. The residue was purified by column chromatography on silica gel (elution with diethyl ether–hexane, 1:1), which afforded 0.166 g (83% yield) of the pure product; IR (neat): 2960, 2870, 2220, 1455, and 1380 cm^{-1}; NMR ($CDCl_3$) δ: 1.50–1.10 (m, 13*H*) and 0.90 (m, 6*H*); mass spectrum, m/z 153.1517 (calculated), 153.1522 (observed).

1-Methyl-1-isochromanecarboxaldehyde [Partial Reduction of a Nitrile to an Aldehyde by $NaAlH_2(OCH_2CH_2OCH_3)_2$].[328] To a solution of 3.4 g (20 mmol) of 1-methyl-1-isochromanecarbonitrile in 50 mL of dry tetrahydrofuran was added at −15° under nitrogen 4.04 g (20 mmol) of $NaAlH_2(OCH_2CH_2OCH_3)_2$ as a 70% solution in toluene. The mixture was stirred at this temperature for 1 hour and then at room temperature for 1 hour. The resulting mixture was cooled to 0° and added to a dilute sulfuric acid solution. The organic layer was separated and the aqueous layer was extracted with diethyl ether. The combined organic phase was washed with sodium bicarbonate solution and brine, dried over anhydrous magnesium sulfate, and concentrated. Distillation of the oily residue *in vacuo* gave 2.8 g (80% yield) of the product, bp 56–58° (air-bath temperature) (0.01 mm); NMR ($CDCl_3$) δ: 1.52 (s, CH_3), 2.6–2.9 (m, CH_2), 3.8–4.2 (m, OCH_2), 7.1 (m, 4*H*, ArH), and 9.56 (s, OCH).

2-(2-Methoxyphenylthio)benzaldehyde [Partial Reduction of an Aromatic Dimethylamide to an Aldehyde by $LiAlH(OC_2H_5)_3$].[511] To a stirred suspension of 6.0 g (0.158 mol) of $LiAlH_4$ in 100 mL of dry diethyl ether was added dropwise during 30 minutes 3.9 g (0.044 mol) of ethyl acetate at 0–7°. The mixture was stirred for an additional 30 minutes, and over 5 minutes a solution of 30.2 g (0.105 mol) of *N*,*N*-dimethyl-2-(2-methoxyphenylthio)benzamide in a mixture of 150 mL of dry tetrahydrofuran and 100 mL of dry diethyl ether was added at −10 to 0°. The reaction mixture was stirred for 6 hours, allowed to stand overnight at room temperature, and decomposed with a 30% solution of sulfuric acid (100 mL). The mixture was filtered and the organic layer of the filtrate was separated, dried over anhydrous magnesium sulfate, and evaporated to give 22.8 g (89% yield) of the crude product; crystallization from ethanol afforded pure product, mp 107.5–109.5°; UV nm (log ε): 233 (4.23), 340 (3.49), 271 (inflection, 3.85), and 284 (3.74); IR (Nujol): 762, 1028, 1253, 1280, 1469, 1480, 1561, 1590, 1681, 1700, 2765, 3020, and 3065 cm^{-1}; NMR ($CHCl_3$, $CFCl_3$) δ: 10.35 (s, 1*H*, CHO), 7.80 (m, 1*H*, H_6 of benzaldehyde), 6.80 − 7.50 (m, 7*H*, remaining ArH), and 3.76 (s, 3*H*, OCH_3).

(+)-2-Methyl-3-phenyl-(2*R*,3*S*)-epoxypropanal [Partial Reduction of an Epoxyamide to an Epoxyaldehyde by $NaAlH_2(OCH_2CH_2OCH_3)_2$].[103] A solution of $NaAlH_2(OCH_2CH_2OCH_3)_2$ (70% in benzene) (9.16 mL, 32.7 mmol) in dry diethyl ether (9.2 mL) was added under argon dropwise over 5 minutes to a solution of methyl *N*-[2-methyl-3-phenyl-(2*R*,3*S*)-epoxypropionyl]prolinate (4.30 g, 14.9 mmol) in dry diethyl ether (130 mL) at 0°. After stirring at this temperature for 1 hour, the reaction was quenched by addition of a saturated solution of ammonium chloride (34 mL). The resulting mixture was stirred at 0° for 30 minutes, filtered through a pad of Celite, and diluted with diethyl ether. The ether solution was washed successively with 20% ammonium chloride solution and saturated sodium chloride solution. Filtration and evaporation of solvents gave the crude product as a brown oil, which was purified by column chromatography on silica gel, using diethyl ether as the eluent, to give 1.73 g (72% yield) of the product as a pale brown oil, $[\alpha]_D^{20}$ + 167° (c = 1.01, $CHCl_3$). Distillation *in vacuo* afforded the pure product as a colorless oil, bp 75–82° (0.9 mm), $[\alpha]_D^{20}$ + 182° (c = 2.00, $CHCl_3$), with an optical purity of 98%; IR ($CHCl_3$): 1730 (CHO), 890 and 850 (epoxide group) cm^{-1}; NMR ($CDCl_3$) δ: 1.22 (s, 3*H*, CH_3), 4.27 (s, 1*H*, CH), 7.30 (s, 5*H*, C_6H_5), and 9.00 (s, 1*H*, CHO); mass spectrum, m/z 162 (M^+), 145, and 133.

dl-*N*-(2,2-Diphenyl-1-methylcyclopropyl)formamide [Reduction of an Isocyanate to a Formamide by $LiAlH(OC_4H_9\text{-}t)_3$].[398] *dl*-2,2-Diphenyl-1-methylcyclopropyl-1-isocyanate (16.6 g, 0.066 mol) dissolved in 100 mL of dry tetrahydrofuran was added during 3 hours to a solution of 25 g (0.1 mol) of $LiAlH(OC_4H_9\text{-}t)_3$ in 150 mL of dry tetrahydrofuran at −15°. After 2 hours of additional stirring, 50 mL of 50% formic acid was added with fast mechanical stirring at the same temperature. The mixture was taken up in diethyl ether, washed with dilute hydrochloric acid and saturated sodium carbonate solution, and dried over anhydrous magnesium sulfate. Evaporation of the solvents gave 17 g of the crude product, which was crystallized from chloroform–hexane to give 14.2 g (85% yield) of the product, mp 114.0–114.5°. Recrystallization from chloroform–hexane gave the pure product, mp 115.5–116.5°; IR (CCl_4): 3415 (w, d), 2750 (w), 1704 (s), and 1215 cm^{-1}; NMR δ: 1.41 (s, 3*H*, CH_3), 1.3–1.9 (m, 2*H*, CH_2), 5.95 (br s, 1*H*, NH), 7.1–7.7 (m, 10*H*, ArH), and 7.82 (1*H*, CHO).

8-Chloro-10-(4-methylhexahydropyrazino)-10,11-dihydrodibenzo[*b,f*]thiepin [Reduction of a Heterocyclic *N*-Carbalkoxy Derivative (Urethane) to a Heterocyclic *N*-Methyl Base by $NaAlH_2(OCH_2CH_2OCH_3)_2$].[512] A solution of $NaAlH_2(OCH_2CH_2OCH_3)_2$ (47 mmol) in benzene (15 mL) was added dropwise to a solution of 9 g (14.9 mmol) of 8-chloro-10-(4-ethoxycarbonylhexahydropyrazino)-10,11-dihydrodibenzo[*b,f*]thiepin in 60 mL of dry benzene and the mixture was heated to 60° for 5 hours. After cooling, 18 mL of water was added dropwise and the mixture was allowed to stand overnight. The organic phase was washed with 15% sodium hydroxide solution and water, dried over potassium carbonate, and evaporated. Crystallization of the residue from ethanol gave 4.80 g (92% yield) of the product, mp 99–100°.

4-Hydroxy-4-(6-methoxy-2-naphthalenyl)-1,2-pentadiene [Reduction of an Acetylenic Quaternary Amino Alcohol to an α-Allenic Alcohol by $NaAlH_2(OCH_2CH_2OCH_3)_2$].[424] A solution of $NaAlH_2(OCH_2CH_2OCH_3)_2$ (265 g, 1.32 mol) in dry tetrahydrofuran (2900 mL) was added to a stirred suspension of 4-hydroxy-4-(6-methoxy-2-naphthalenyl)-1-trimethylamino-2-pentyne iodide (281 g, 0.66 mol) in dry tetrahydrofuran (1900 mL) at 20–30° over a period of 30 minutes. Stirring was continued for 2 hours at 25°, and 2 *N* aqueous sodium hydroxide (1100 mL) was added with external cooling, whereupon two phases separated. From the upper layer, most of the tetrahydrofuran was removed by evaporation under reduced pressure. The lower layer was extracted with three 700-mL portions of benzene, and the benzene extracts were combined with the residue from the upper layer. This mixture was washed with three 250-mL portions of water, dried over anhydrous sodium sulfate, and evaporated *in vacuo* to give 134 g (84% yield) of the pure product as a slowly crystallizing oil, mp 52–54°.

Methyl 1-Ethyl-1,2,3,4,6,7,12,12b-octahydro-α-methylthioindolo [2,3-*a*]-quinolizine-1-propanoate [Stereoselective Reduction of an Ester of a Heterocyclic Iminium Salt to an Ester of a Saturated Base by $LiAlH(OC_4H_9\text{-}t)_3$].[430,431] To a solution of 1-ethyl-2,3,4,6,7,12-hexahydro-1-[3-methoxy-2-(methylthio)-3-oxopropyl]-1*H*-indolo [2,3-*a*] quinolizin-5-ium perchlorate (0.328 g, 0.68 mmol) in 1.5 mL of dry tetrahydrofuran was added under nitrogen a solution of 0.535 g (2.11 mmol) of $LiAlH(OC_4H_9\text{-}t)_3$ in 2.5 mL of dry tetrahydrofuran at 4°. The mixture was stirred at 22° for 3 hours, and the reaction was quenched by adding a mixture of a saturated sodium sulfate solution and dichloromethane. The organic layer was separated, dried by filtration through anhydrous magnesium sulfate, and then evaporated to dryness under reduced pressure. Column chromatography on alumina (Brinkmann, activity grade II, 2 g, eluted with diethyl ether–chloroform 1:1) gave 0.259 g (98.5% yield) of the crystalline product, mp

101–103°; IR ($CHCl_3$): 3500 (m), 2940 (s), 1725 (s), 1460 (m), and 1155 (s) cm^{-1}; NMR ($CDCl_3$ δ: 1.08 (dt, 3*H*), 1.66 (m, 6*H*), 1.95 (s, 3*H*), 2.80 (m, 9*H*), 3.58 (s, 3*H*), 3.60 (m, 1*H*), 7.20 (m, 4*H*), and 7.71 (br s, 1*H*); mass spectrum, m/z 386.

Methyl 1,2-Dihydro-4-chloro-6-methoxyquinoline-1-carboxylate [Conversion of a Quinoline to an *N*-Carbalkoxy-1,2-dihydroquinoline via Reduction of a Quinoline *N*-Borane by $NaAlH_2(OCH_2CH_2OCH_3)_2$].[445] To 0.194 g (1 mmol) of 4-chloro-6-methoxyquinoline in 20 mL of dry tetrahydrofuran was added 1 mL of 1.0 *M* borane-tetrahydrofuran complex (1 mmol) in tetrahydrofuran at −78°. After 30 minutes, 0.57 mL (2 mmol) of $NaAlH_2(OCH_2CH_2OCH_3)_2$ (3.5 *M* in benzene) diluted with 2 mL of dry tetrahydrofuran was added. The resulting deep yellow reaction mixture was stirred at −78° for 30 minutes. Methyl chloroformate (1.14 g, 12 mmol) was then added all at once, and the dry-ice bath was removed. After 18 hours at room temperature, the pale-yellow reaction mixture was cooled to 0° and 2 mL of water was added. The solvent was decanted from the aluminum salts into a separatory funnel containing 100 mL of water, and the product was extracted with three 20-mL portions of dichloromethane. The combined extracts were dried over anhydrous sodium sulfate and evaporated. The crude product was dissolved in 5 mL of dichloromethane and filtered through a small plug of silica gel to remove aluminum salts carried into the organic layer during extraction. The dichloromethane was removed by evaporation, and residual solvent as well as volatile carbonates were removed under reduced pressure (0.1 mm, room temperature, 12 hours). Bulb-to-bulb distillation of the crude material at 145–150° (0.1 mm) gave 0.225 g (88% yield) of the pure product as a pale-yellow oil; IR (neat) (major bands): 761, 873, 1035, 1056, 1136, 1164, 1192, 1230, 1249, 1266, 1285, 1344, 1381, 1449, 1493, 1571, 1607, and 1719 cm^{-1}; NMR ($CDCl_3$) δ: 3.77 (s, 3*H*, C-6 OCH_3), 3.81 (s, 3*H*, CO_2CH_3), 4.41 (d, $J = 4.8$ Hz, 2*H*, C-2 H, 6.15 (t, $J = 4.8$ Hz, C-3 H), 6.83 (dd, $J = 2.8$ and 9.0 Hz, 1*H*, C-7 H), 7.12 (d, $J = 2.8$ Hz, 1*H*, C-5 H), and 7.47 (d, $J = 9.0$ Hz, C-8 H).

1,3α-Dimethyl-2β-cyano-4α-(3-methoxyphenyl)piperidine [Conversion of a Lactam to a Cyano-Substituted Base via Reduction by $LiAlH_2(OC_2H_5)_2$ and Cyanation].[327] A solution of 3.0 g (12.8 mmol) of *cis*-1,3-dimethyl-4-(3-methoxyphenyl)-2-piperidone in dry diethyl ether (34 mL) was cooled in an ice bath and treated with 1.33 *M* $LiAlH_2(OC_2H_5)_2$ in diethyl ether (10.0 mL, 13.3 mmol) over 3 minutes. The resulting suspension was pipetted during 10 minutes into a vigorously stirred, cooled (ice bath) solution of 8.30 g (128 mmol) of potassium cyanide in 1.5 *M* sulfuric acid (112 mL). The mixture was allowed to warm to room temperature over 14 hours, degassed for 30 minutes with nitrogen, poured into 3.75 *M* aqueous sodium hydroxide (100 mL), and extracted with three 75-mL portions of chloroform. The combined organic extracts were dried and evaporated to afford 3.12 g (~ 100% yield) of the product, mp 77–78° after recrystallization from dichloromethane–hexane; NMR ($CDCl_3$)

δ: 7.16 (m, 1*H*), 6.71 (m, 3*H*), 3.80 (s, 3*H*), 3.72 (m, 1*H*), 1.4–3.3 (m, 6*H*), 2.40 (s, 3*H*), 0.93 (d, J = 7 Hz), 3*H*).

(*S*)-(+)-2,3-Decadiene [Reduction of a Chiral Propargyl Mesylate to a Chiral Allene by $LiAlH(OCH_3)_3$].[486] A solution of 2.32 g (10 mmol) of (*S*)-(−)-3-decyn-2-yl methanesulfonate in 10 mL of dry tetrahydrofuran was added under nitrogen to a stirred solution of 3.84 g (30 mmol) of $LiAlH(OCH_3)_3$ in 50 mL of dry tetrahydrofuran at −20°. The reaction was followed by GLC, and after stirring at this temperature for 48 hours, the mixture was decomposed by addition of aqueous ammonium chloride solution. The resulting mixture was extracted with light petroleum and the extract was subjected to primary purification on a silica gel column, using light petroleum as the eluent. The solution was concentrated, and preparative GLC of the residue on a 6-m 20% Carbowax 20M column gave 1.09 g (79 % yield) of the product, $[\alpha]_D^{22}$ + 52.9° (c 2.0, CH_3OH)(73% e.e.); 3-decyne (0.21 g, 15% yield) was obtained as byproduct The ratio of retention times for the alkadiene and alkyne at 170° was 1:1.16.

Di-*n*-butyl Sulfide [Reduction of a Sulfoxide to a Sulfide by $NaAlH_2(OCH_2CH_2OCH_3)_2$].[457] A solution of $NaAlH_2(OCH_2CH_2OCH_3)_2$ (4.3 mL) in benzene (3.46 *M*) was added slowly under stirring and in an atmosphere of nitrogen to 1.62 g (10 mmol) of di-*n*-butyl sulfoxide in 10 mL of dry benzene. The resulting mixture was stirred under reflux for 30 minutes, cooled, and poured into 1 *N* hydrochloric acid. The organic layer was separated and the aqueous phase was extracted with 25 mL of chloroform. The combined organic layer and extracts were dried over anhydrous magnesium sulfate and evaporated to dryness. Distillation of the residue (single spot in TLC) under normal pressure gave 1.43 g (98% yield) of the product, bp 183–186° (760 mm).

TABULAR SURVEY

Hydride reductions of carboxylic acids and their derivatives, nitrogen compounds, and sulfur compounds are grouped in Tables I–XXXI and follow the discussion in the "Scope and Limitations" section. Polyfunctional compounds in which more than one group is reduced by a hydride can be found in tables pertaining to reduction of each specific functional group. Steroids, carbohydrates, and organometallic compounds containing the functionalities covered in this chapter are not included in the tabular survey. Because of the lack of experimental material, reductions of imidoyl chlorides; *N*-alkylidenesulfinamides; isothiocyanates; azo, azoxy, and nitrilimino compounds; and nitrate esters as well as open-chain dithiocarbamidium salts, sulfides, and sulfones are not included in the tables.

Within each table the compounds are listed according to increasing carbon number and complexity of the molecular formula using the *Chemical Abstracts* convention. If a reference contains more than one set of conditions for the reduction of one reactant, only that providing the best product yield is presented.

Yields and percent enantiomeric excess (if available) are given in parentheses; numbers not in parentheses are the product ratios. A dash indicates that no yield is given in the reference. When there is more than one reference for a given reactant–hydride reaction, the first reference is the one that gives the best recorded yield. In some cases yields have been calculated by the author from the literature data.

The literature has been reviewed through December 1985. Examples of reductions published during the last quarter of 1985 are not given in the tables but cited in an addendum at the end of the tables. A list of standard abbreviations used throughout the tables follows:

Ac	acetyl
BCGF	3-*O*-benzyl-1,2-*O*-cyclohexylidene-α-D-glucofuranose
BME	bis(2-methoxyethyl) ether
(+)-BN	(+)-borneol
(−)-BN	(−)-borneol
Bz	benzoyl
(−)-CND	(−)-cinchonidine
(*R*)-(+)-DBN	(*R*)-(+)-2,2′-dihydroxy-1,1′-binaphthyl
(*S*)-(−)-DBN	(*S*)-(−)-2,2′-dihydroxy-1,1′-binaphthyl
(+)-DMDB	(2*S*,3*R*)-(+)-4-dimethylamino-3-methyl-1,2-diphenyl-2-butanol (CHIRALD™)
DME	1,2-dimethoxyethane
d.e.	percent diastereomeric excess
e.e.	percent enantiomeric excess
ether	diethyl ether
EtOH	ethanol
IA	inverse addition
LAH	lithium aluminum hydride
LTBA	lithium tri-*tert*-butoxyaluminum hydride
LTMA	lithium trimethoxyaluminum hydride
MHP	*N*-methylhexahydropyrazine
MPL	morpholine
Ms	methanesulfonyl
(−)-MTH	(−)-menthol
(−)-NME	(−)-*N*-methylephedrine
o.p.	percent optical purity
opt. y.	percent optical yield
Py	pyridine
(+)-QND	(+)-quinidine
(−)-QN	(−)-quinine
SMEAH	sodium bis(2-methoxyethoxy)aluminum hydride
THF	tetrahydrofuran
THP	tetrahydropyranyl
Ts	*p*-toluenesulfonyl

TABLE I. Reduction of Carboxylic Acids

No. of Carbon Atoms	Reactant	Conditions	Product(s) and Yield(s)(%)	Refs.
C_4	$HO_2CCH_2CH(NH_2)CO_2H$	SMEAH (3 eq), C_6H_6	$HO(CH_2)_2CH(NH_2)CH_2OH$[a] (—)	513
	$H_2NCOCH_2CH(NH_2)CO_2H$	1. SMEAH (1.5 eq), C_6H_6, reflux, 1 h 2. H_3O^+	$HO_2CCH_2CH(NH_2)CH_2OH$[a] (—)	513
		SMEAH (3 eq), C_6H_6	$HO(CH_2)_2CH(NH_2)CH_2OH$[a] (—)	513
	n-$C_3H_7CO_2H$	$Ca(AlH_2[OCH_2CH(C_2H_5)C_4H_9]_2)_2 \cdot THF$, hexane, reflux, 2.5 h	n-C_4H_9OH (92)	165
C_5	$CH_3CO(CH_2)_2CO_2H$	SMEAH, C_6H_6, 80°, 12 min	$CH_3CHOH(CH_2)_3OH$ (67)	168
		SMEAH, C_6H_6, −20°, 1.5 h	O, O (57)	168
	$HO_2C(CH_2)_2CH(NH_2)CO_2H$	SMEAH (3 eq), C_6H_6	$HO(CH_2)_3CH(NH_2)CH_2OH$ (—)	513
	$H_2NCO(CH_2)_2CH(NH_2)CO_2H$	1. SMEAH (1.5 eq), C_6H_6, reflux, 1 h 2. H_3O^+	$HO_2C(CH_2)_2CH(NH_2)CH_2OH$[a] (—)	513
		SMEAH (3 eq), C_6H_6	$HO(CH_2)_3CH(NH_2)CH_2OH$ (—)	513
C_6	OH, N, CO_2H	SMEAH, xylene, reflux, 7 h	OH, N, CH_3 (28)	514
	n-$C_5H_{11}CO_2H$	$Ca[AlH_2(OC_3H_7\text{-}i)_2]_2 \cdot THF$, toluene, 85°, 2 h	n-$C_6H_{13}OH$ (~99)	165
C_7	$C_6H_5CO_2MgBr$	SMEAH, C_6H_6, reflux, 1 h	$C_6H_5CH_2OH$ (79)	123
	$C_6H_5CO_2Na$	SMEAH, C_6H_6, reflux, 2 h	" (89)	123
	4-$HOC_6H_4CO_2Na$	SMEAH, xylene, reflux, 8 h	4-$HOC_6H_4CH_3$ (44)	123
	4-$H_2NC_6H_4CO_2Na$	SMEAH, xylene, reflux, 3 h	4-$H_2NC_6H_4CH_3$ (38)	123

TABLE I. REDUCTION OF CARBOXYLIC ACIDS (*Continued*)

No. of Carbon Atoms	Reactant	Conditions	Product(s) and Yield(s)(%)	Refs.
C_7 (*Contd.*)	$C_6H_5CO_2H$	SMEAH, C_6H_6, 80°, 1.5 h	$C_6H_5CH_2OH$ (97)	167, 168
		$Ca[AlH_2OC_4H_9\text{-}i)_2]_2 \cdot$THF, toluene, 80°, 2 h	" (68)	165
	$2\text{-}HOC_6H_4CO_2H$	SMEAH, xylene, reflux, 7 h	$2\text{-}HOC_6H_4CH_3$ (88)	119, 121
	$3\text{-}HOC_6H_4CO_2H$	SMEAH, xylene, reflux, 2 h	$3\text{-}HOC_6H_4CH_2OH$ (74)	121, 119
	$4\text{-}HOC_6H_4CO_2H$	SMEAH, xylene, reflux, 6 h	$4\text{-}HOC_6H_4CH_3$ (91)	121, 119
	$2\text{-}H_2NC_6H_4CO_2H$	SMEAH, C_6H_6	$2\text{-}H_2NC_6H_4CH_2OH$ (—)	515
		SMEAH, xylene, reflux, 3 h	$2\text{-}H_2NC_6H_4CH_3$ (71)	120
	$4\text{-}H_2NC_6H_4CO_2H$	", ", ", "	$4\text{-}H_2NC_6H_4CH_3$ (94)	120
	CO_2H O	SMEAH, ether, room temp, 4 h	CH_2OH O (56)	516
	CO_2H O	", ", ", "	CH_2OH O (67)	516
	$C_6H_{11}CO_2H$	SMEAH, THF, overnight	$C_6H_{11}CH_2OH$ (82)	517
	$HO_2CC(CH_3)_2(CH_2)_2CO_2H$	SMEAH, C_6H_6–THF, reflux, 2 h	$HOCH_2C(CH_3)_2(CH_2)_3OH$ (65	518
	CO_2H H_2N	1. SMEAH, toluene, 80°, 2 h 2. HCl, EtOH	CH_2OH $H_3\overset{+}{N}$ Cl^- (67)	189
C_8	$C_6H_4(CO_2Na)_2\text{-}1,2$	SMEAH, C_6H_6, reflux, 7 h	$C_6H_4(CH_2OH)_2\text{-}1,2$ (82)	123
	$4\text{-}ClC_6H_4CH_2CO_2H$	SMEAH, C_6H_6, room temp, 3.5 h	$4\text{-}ClC_6H_4(CH_2)_2OH$ (65)	519

	Substrate	Conditions	Product(s) (Yield, %)	Refs.
	2-$CH_3SC_6H_4CO_2H$	SMEAH, C_6H_6, 50–55°, 1 h, then room temp, 3 h	2-$CH_3SC_6H_4CH_2OH$ (87)	520
	3-$CH_3SC_6H_4CO_2H$	", ", ", ", "	3-$CH_3SC_6H_4CH_2OH$ (80)	520
	4-$CH_3OC_6H_4CO_2H$	SMEAH, xylene, reflux, 2 h	4-$CH_3OC_6H_4CH_2OH$ (55) + 4-$CH_3OC_6H_4CH_3$ (18) + 4-$HOC_6H_4CH_3$ (20)	121
	n-$C_7H_{15}CO_2MgBr$	SMEAH, C_6H_6, reflux, 1 h	n-$C_8H_{17}OH$ (85)	123
C_9	4-$CH_3C_6H_4CH_2CO_2H$	SMEAH, C_6H_6, room temp, 3.5 h	4-$CH_3C_6H_4(CH_2)_2OH$ (85)	519
	4-$CH_3OC_6H_4CH_2CO_2H$	SMEAH, C_6H_6, room temp, 3.5 h	4-$CH_3OC_6H_4(CH_2)_2OH$ (89)	519
	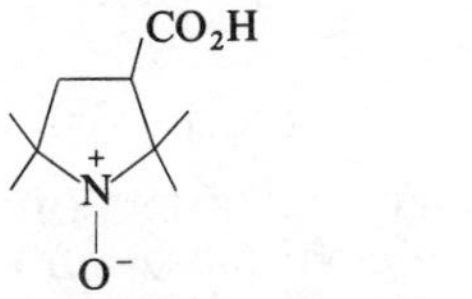 3-CO_2H-2,2,5,5-tetramethylpyrrolidine N^+–O^- (structure)	SMEAH, toluene, reflux, 2 h	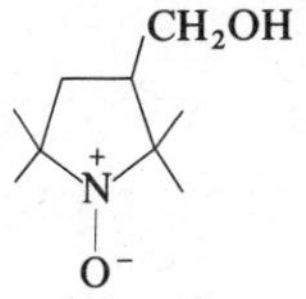 3-CH_2OH-2,2,5,5-tetramethylpyrrolidine N^+–O^- (structure) (85)	444
	n-$C_8H_{17}CO_2H$	SMEAH, C_6H_6, 80–85°, 1 h	n-$C_9H_{19}OH$ (92)	168
C_{10}	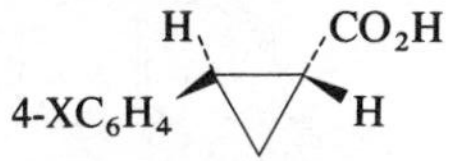 2-(4-XC_6H_4)cyclopropane-1-CO_2H (structure) (1R, 2R)-(−) X = Br, Cl, H	SMEAH, toluene, reflux, 2 h	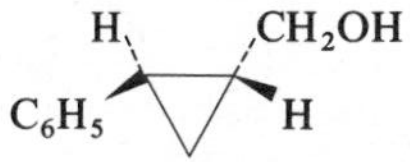 2-C_6H_5-cyclopropane-1-CH_2OH (structure) (1R, 2R)-(−) (54–82)	522, 523
	2-$NC(CH_2)_2SC_6H_4CO_2H$	SMEAH, C_6H_6, room temp, 3 h	$(2\text{-}HOCH_2C_6H_4S)_2$ (40)	524
	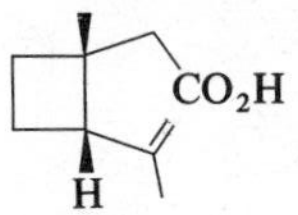 bicyclic cyclobutane CO_2H (structure, H)	SMEAH, C_6H_6–ether	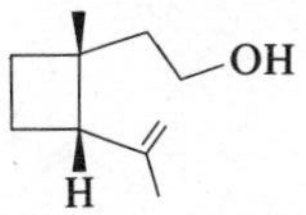 bicyclic cyclobutane OH (structure, H) (—)	525
	$(CH_3)_2C{=}CH(CH_2)_2CH(CH_3)CH_2CO_2H$ (R)-(+)	SMEAH, C_6H_6–ether, 22°, 18 h	$(CH_3)_2C{=}CH(CH_2)_2CH(CH_3)(CH_2)_2OH$ (R)-(+) (97)	170

TABLE I. REDUCTION OF CARBOXYLIC ACIDS (*Continued*)

No. of Carbon Atoms	Reactant	Conditions	Product(s) and Yield(s)(%)	Refs.
C_{11}	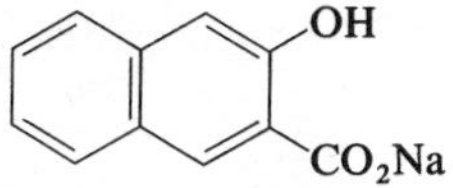 OH, CO_2Na	SMEAH, C_6H_6, reflux, 6 h	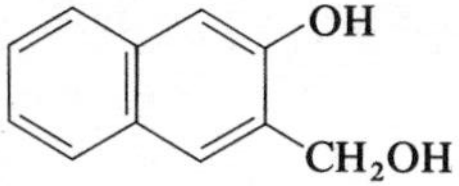 OH, CH_2OH (81)	123
	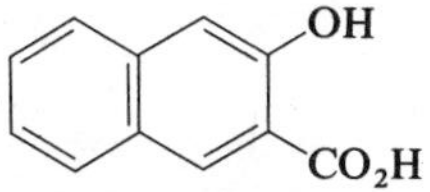 OH, CO_2H	SMEAH, C_6H_6–ether, 35°, 1 h	" (86)	122
	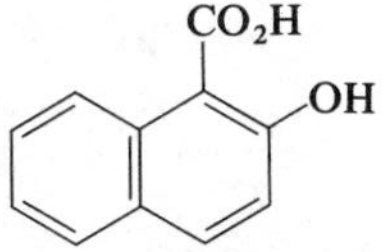 CO_2H, OH	SMEAH, xylene, 142°, 2.25 h	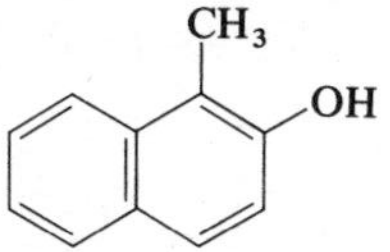 CH_3, OH (95)	122
	$CH_2{=}CH(CH_2)_8CO_2Na$	SMEAH, C_6H_6, reflux, 15 min	$CH_2{=}CH(CH_2)_9OH$ (98)	123
	$CH_2{=}CH(CH_2)_8CO_2H$	SMEAH, C_6H_6, 80°, 1 h	" (94)	168, 526
	$HO_2C(CH_2)_2C(C_2H_5)_2(CH_2)_2CO_2H$	SMEAH, C_6H_6, reflux, 1 h	$HO(CH_2)_3C(C_2H_5)_2(CH_2)_3OH$ (91)	527
C_{12}	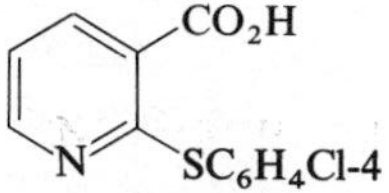 CO_2H, N, SC_6H_4Cl-4	SMEAH, C_6H_6, room temp, 30 min	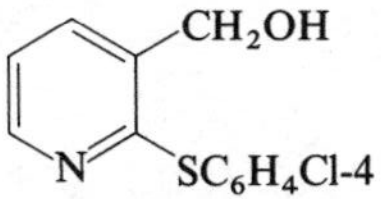 CH_2OH, N, SC_6H_4Cl-4 (98)	528, 529
	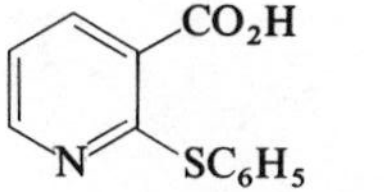 CO_2H, N, SC_6H_5	SMEAH, C_6H_6, room temp, 2 h	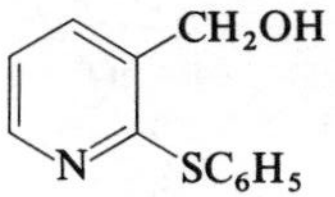 CH_2OH, N, SC_6H_5 (81)	530, 531
	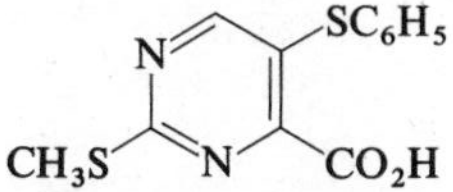 N, SC_6H_5, CH_3S, N, CO_2H	SMEAH, C_6H_6, room temp, 7 h	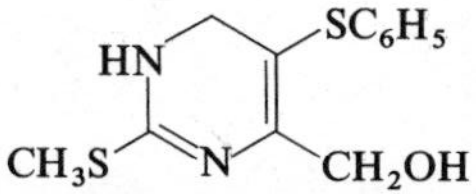 HN, SC_6H_5, CH_3S, N, CH_2OH (25)	524

	Reactant	Conditions	Product(s) and Yield(s) (%)	Refs.
	1-Naphthyl-CH_2CO_2H	SMEAH, C_6H_6, room temp, 3.5 h	1-Naphthyl-$(CH_2)_2OH$ (71)	519
	HO-benzocycloheptene-CO_2H	SMEAH, C_6H_6, reflux, 3 h	HO-benzocycloheptene-CH_2OH (54)	532
	Spiro-dithiolane cyclohexene (S, S), CO_2H side chain	SMEAH, THF–C_6H_6, 0°, 4.25 h	Spiro-dithiolane cyclohexene (S, S), OH side chain	
	d-(+)-		*d*-(+)- (~100)	533, 534
	l-(−)-		*l*-(−)- (~100)	533, 534
	$CH_3(CH{=}CH)_2(CH_2)_6CO_2H$ *trans*-8,*trans*-10:*cis*-8,*trans*-10 = 75:25	SMEAH	$CH_3(CH{=}CH)_2(CH_2)_7OH$ (—) *trans*-8,*trans*-10:*cis*-8,*trans*-10 = 75:25	535
	$C_2H_5(CH{=}CH)_2(CH_2)_5CO_2H$ *trans*-7,*cis*,*trans*-9	SMEAH, C_6H_6, reflux, 1 h	$C_2H_5(CH{=}CH)_2(CH_2)_6OH$ (—) *trans*-7,*cis*,*trans*-9	536, 537
	n-$C_3H_7C{\equiv}C(CH_2)_6CO_2H$	SMEAH, C_6H_6	*n*-$C_3H_7C{\equiv}C(CH_2)_7OH$ (96)	538
C_{13}	F, F, SC_6H_4Cl-4, CO_2H (benzene)	SMEAH, C_6H_6, 15°, 3 h	F, F, SC_6H_4Cl-4, CH_2OH (benzene) (77)	539
	Cl, SC_6H_4F-3, CO_2H (benzene)	SMEAH, C_6H_6, 40°, then room temp, 2.5 h	Cl, SC_6H_4F-3, CH_2OH (benzene) (80)	540
	C_6H_5S, CO_2H, Cl (benzene)	SMEAH, C_6H_6, room temp, 3 h	C_6H_5S, CH_2OH, Cl (benzene) (84)	541

TABLE I. REDUCTION OF CARBOXYLIC ACIDS (*Continued*)

No. of Carbon Atoms	Reactant	Conditions	Product(s) and Yield(s)(%)	Refs.
C_{13} (*Contd.*)	2-X-$C_6H_4CO_2H$	SMEAH, C_6H_6, room temp, 3 h	2-X-$C_6H_4CH_2OH$	
	X = $SC_6H_3Cl_2$-2,3	"	(44)	542
	X = $SC_6H_3Cl_2$-2,4	"	(71)	543
	X = $SC_6H_3Cl_2$-2,5	"	(90)	543
	X = $SC_6H_3Cl_2$-3,4	"	(81)	544
	X = SC_6H_4Br-4	"	(—)	545
	X = SC_6H_4Cl-4	"	(88)	545, 512
	X = SeC_6H_4Cl-4	"	(79)	546
	X = SC_6H_4F-4	"	(—)	545
	X = OC_6H_4Cl-4	SMEAH, C_6H_6, room temp, overnight	(97)	547, 545
	X = OC_6H_4F-4	"	(98)	547
	X = SC_6H_5	SMEAH, C_6H_6, room temp, 3 h	(—)	545
	X = OC_6H_5	"	(—)	545, 547
	3-$C_6H_5SC_6H_4CO_2H$	SMEAH, C_6H_6, room temp, 3 h	3-$C_6H_5SC_6H_4CH_2OH$ (79)	541
C_{14}	2-(SC_6H_4Cl-4)-4-CH_3O-5-F-$C_6H_2CO_2H$	SMEAH, C_6H_6, room temp	2-(SC_6H_4Cl-4)-4-CH_3O-5-F-$C_6H_2CH_2OH$ (92)	540
	2-(4-Cl-3-F-C_6H_3S)-5-CH_3O-$C_6H_3CO_2H$	SMEAH, C_6H_6, room temp, 3 h	2-(4-Cl-3-F-C_6H_3S)-5-CH_3O-$C_6H_3CH_2OH$ (—)	548

$(2\text{-}HO_2CC_6H_4S)_2$		SMEAH, C_6H_6, reflux, 30 min	$(2\text{-}HOCH_2C_6H_4S)_2$ (67)	524
$2\text{-}HO_2CC_6H_4CH_2C_6H_4Cl\text{-}4$		SMEAH, C_6H_6, room temp, 3 h	$2\text{-}HOCH_2C_6H_4CH_2C_6H_4Cl\text{-}4$ (—)	545
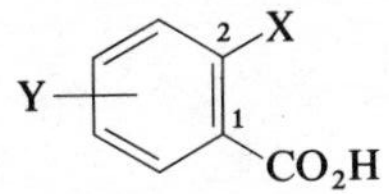 Y, X (2), CO_2H (1)		SMEAH, C_6H_6, room temp,	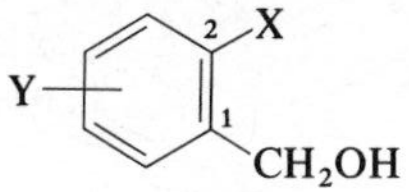 Y, X (2), CH_2OH (1)	
$X = SC_6H_4Cl\text{-}4$	$Y = OCH_3\text{-}4$	2 h	(89)	549
X = "	$Y = OCH_3\text{-}5$	2 h	(83)	549
X 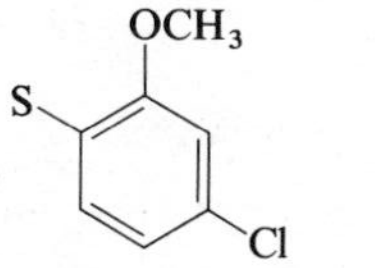 = S, OCH_3, Cl	Y = H	3 h	(89)	550
X 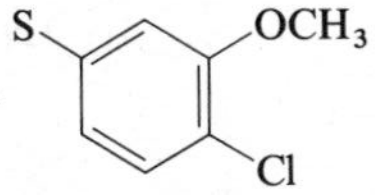 = S, OCH_3, Cl	Y = H	3 h	(91)	551
$X = SC_6H_4F\text{-}4$	$Y = CH_3\text{-}5$	3 h	(—)	552
$X = SC_6H_4SCH_3\text{-}4$	Y = F-4	5 h	(65)	553
X = "	Y = F-5	5 h	(89)	553
$X = CH_2C_6H_5$	Y = H	3 h	(76)	554, 545
$X = SC_6H_4CH_3\text{-}4$	Y = H	3 h	(—)	545
$X = SC_6H_4SeCH_3\text{-}4$	Y = H	3 h	(90)	555
$X = SC_6H_5$	$Y = SCH_3\text{-}5$	13.5 h	(97)	556
$X = SC_6H_4SCH_3\text{-}4$	Y = H	3 h	(—)	545
$X = OC_6H_4CH_3\text{-}4$	Y = H	3 h	(—)	545
$X = OC_6H_4SCH_3\text{-}4$	Y = H	3 h	(—)	545
$X = SC_6H_4OCH_3\text{-}2$	Y = H	2.5 h	(98)	511
$X = SC_6H_4OCH_3\text{-}3$	Y = H	3 h	(71)	557, 558
$X = SC_6H_4OCH_3\text{-}4$	Y = H	4 h	(87)	559, 545
$X = SC_6H_5$	$Y = OCH_3\text{-}4$	3.5 h	(91)	560, 561
$X = OC_6H_4OCH_3\text{-}4$	Y = H	3 h	(—)	545
$X = SC_6H_4(SO_2CH_3)\text{-}4$	Y = H	3 h	(—)	545

TABLE I. REDUCTION OF CARBOXYLIC ACIDS (*Continued*)

No. of Carbon Atoms	Reactant	Conditions	Product(s) and Yield(s)(%)	Refs.
C_{14} (*Contd.*)	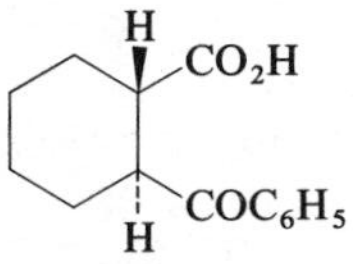 H, CO_2H, H, COC_6H_5	LTBA, THF, room temp, 4 h	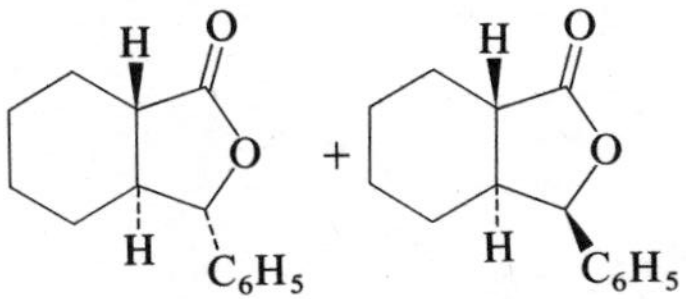 H, O, O, H, C_6H_5 **I** + H, O, O, H, C_6H_5 **II** (—) (I:II = 52:48)	177
	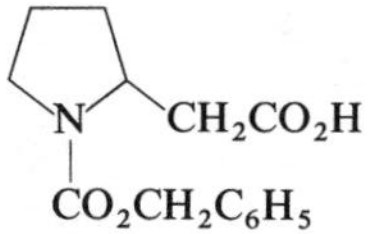 N, CH_2CO_2H, $CO_2CH_2C_6H_5$	SMEAH, C_6H_6, reflux, 3 h	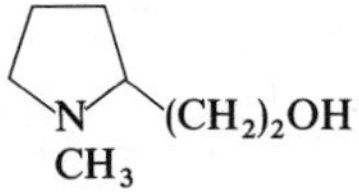 N, $(CH_2)_2OH$, CH_3	
	(+)-		(+)- (77)	562
	(−)-		(−)- " (—)	562
	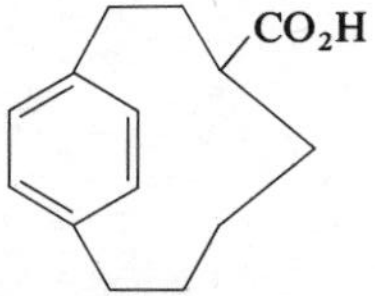 CO_2H	SMEAH, C_6H_6, reflux, 1 h	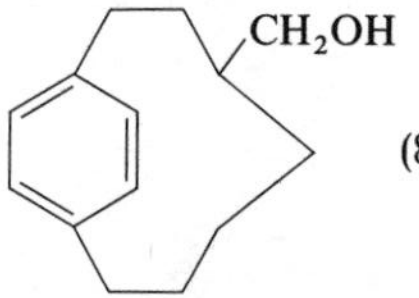 CH_2OH (84)	563, 564
	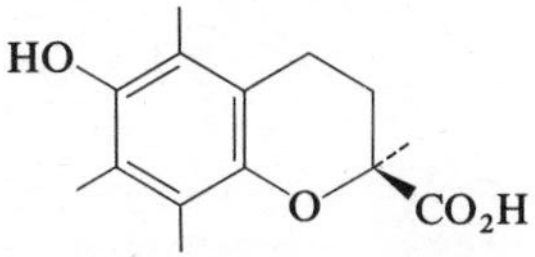 HO, O, CO_2H	SMEAH, THF–C_6H_6, room temp, 1 h	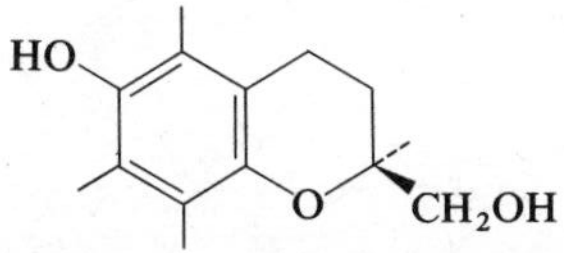 HO, O, CH_2OH (86)	240

	Substrate	Conditions	Product(s) (Yield, %)	Refs.
	OH, CO_2H, OH, O	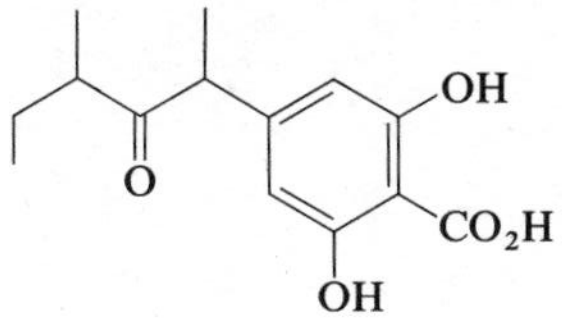 SMEAH, xylene, 140°, 6 h	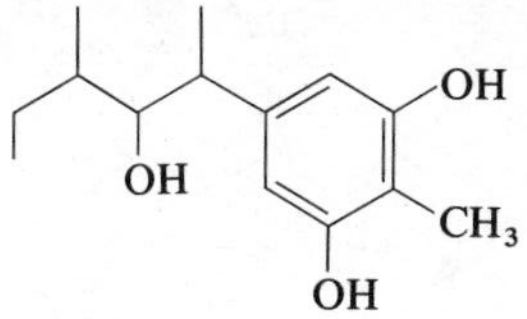 OH, OH, CH_3, OH (55)	565
C_{15}	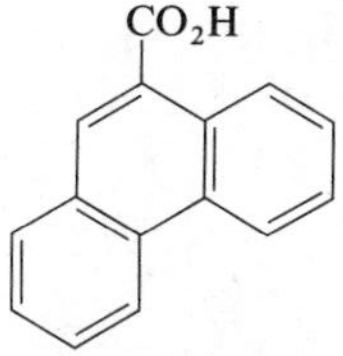 CO_2H	SMEAH, THF–C_6H_6, reflux, 2 h	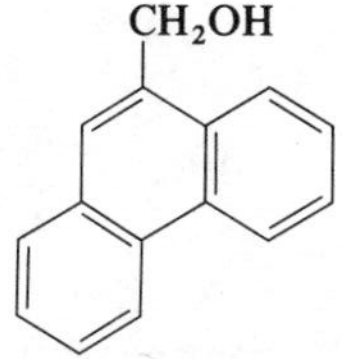 CH_2OH (63)	201, 566
	Y, 2, X, 1, CO_2H	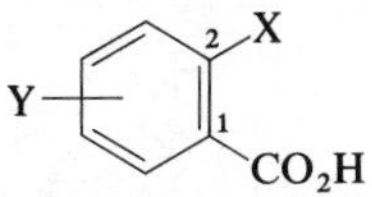 SMEAH, C_6H_6	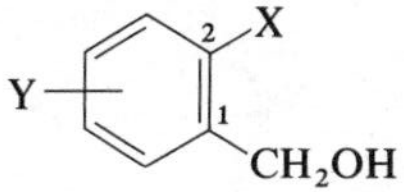 Y, 2, X, 1, CH_2OH	
	X = $CH_2C_6H_4SCH_3$-4 Y = H	room temp, 3 h	(—)	545
	X = $SC_6H_4C_2H_5$-4 Y = H	room temp, 3 h	(89)	567
	X = $SC_6H_4SCH_3$-4 Y = OCH_3-4	room temp, 4 h	(92)	568
	X = $SC_6H_4OCH_3$-3 Y = OCH_3-4	20°, 16 h	(88)	569
	X = $SC_6H_4OCH_3$-4 Y = OCH_3-4	35–45°, 4.75 h	(79)	570
	X = $SC_6H_4[SO_2N(CH_3)_2]$-4 Y = H	room temp, 3 h	(—)	545
	CO_2H, N, $SC_6H_4(C_3H_7\text{-}i)$-4	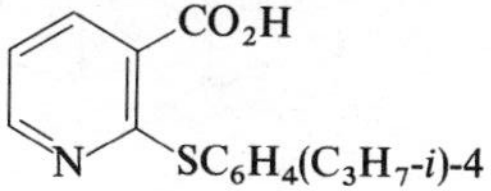 SMEAH, C_6H_6, room temp, 30 min	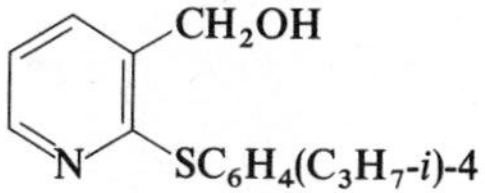 CH_2OH, N, $SC_6H_4(C_3H_7\text{-}i)$-4 (85)	528

TABLE I. REDUCTION OF CARBOXYLIC ACIDS (*Continued*)

No. of Carbon Atoms	Reactant	Conditions	Product(s) and Yield(s)(%)	Refs.
C_{16}	9-CO_2H, 3-OCH_3 phenanthrene	SMEAH, THF–C_6H_6, reflux, 2 h	9-CH_2OH, 3-OCH_3 phenanthrene (68)	201, 566
	$(4\text{-}FC_6H_4)_2CH(CH_2)_2CO_2H$	SMEAH, C_6H_6, 2 h	$(4\text{-}FC_6H_4)_2CH(CH_2)_3OH$ (90)	571, 572, 573
	2-(indan-5-yl-S)-$C_6H_4CO_2H$	SMEAH, C_6H_6	2-(indan-5-yl-S)-$C_6H_4CH_2OH$ (84)	574
	$(C_6H_5)_2CH(CH_2)_2CO_2H$	SMEAH, C_6H_6	$(C_6H_5)_2CH(CH_2)_3OH$ (97)	571
	2-X-$C_6H_4CO_2H$	SMEAH, C_6H_6, room temp, 3 h	2-X-$C_6H_4CH_2OH$	
	X = $SC_6H_4(C_3H_7\text{-}i)\text{-}4$		(94)	567
	X = $SC_6H_4(SC_3H_7\text{-}n)\text{-}4$		(84)	575
C_{17}	2-(1-naphthyl-S)-$C_6H_4CO_2H$	SMEAH, C_6H_6, room temp, 4 h	2-(1-naphthyl-S)-$C_6H_4CH_2OH$ (81)	576

Reactant	Conditions	Product(s) and Yield(s) (%)	Refs.
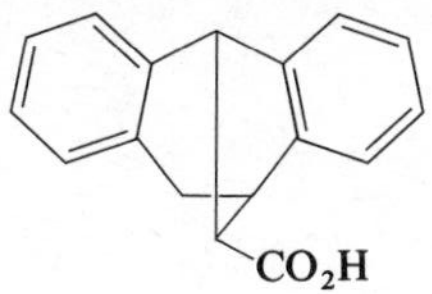 CO_2H	SMEAH	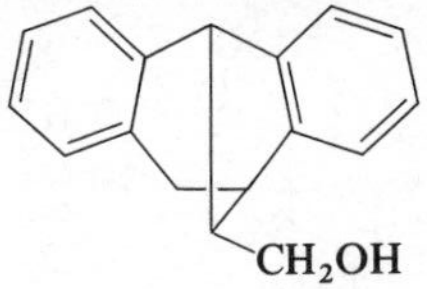 CH_2OH (—)	577
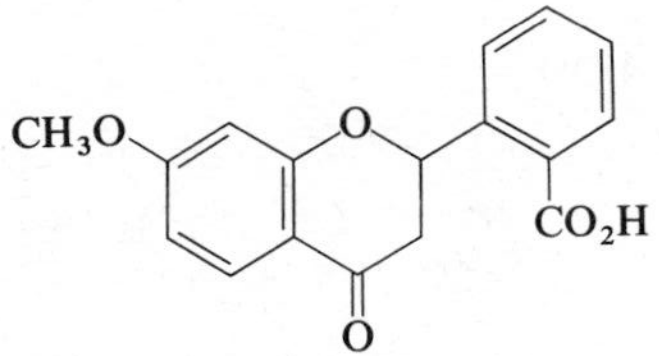 CH_3O, O, CO_2H, O	1. SMEAH, C_6H_6–ether, reflux, 3 h 2. SiO_2	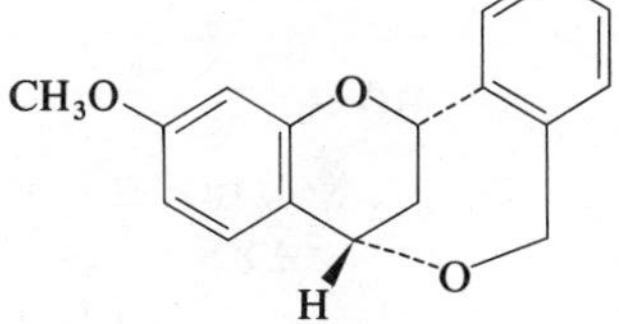 CH_3O, O, O, H (54)	578
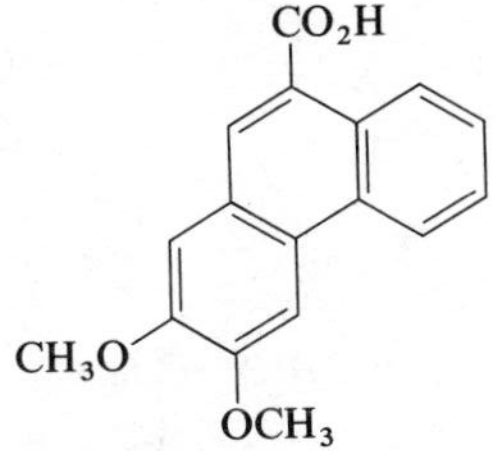 CO_2H, CH_3O, OCH_3	SMEAH, THF–C_6H_6, reflux, 2 h	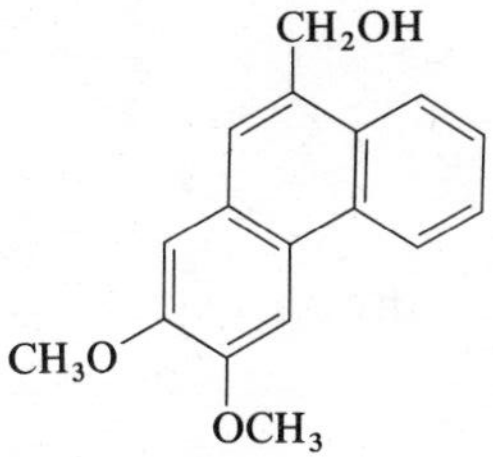 CH_2OH, CH_3O, OCH_3 (60)	566
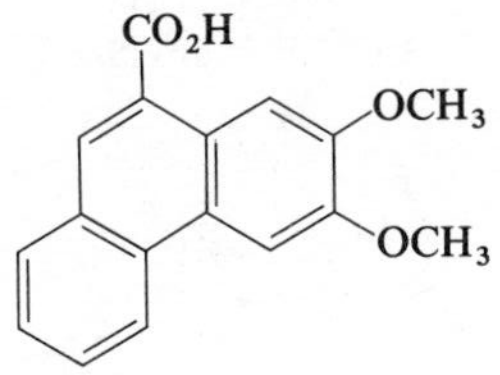 CO_2H, OCH_3, OCH_3	SMEAH, THF–C_6H_6, reflux, 2 h	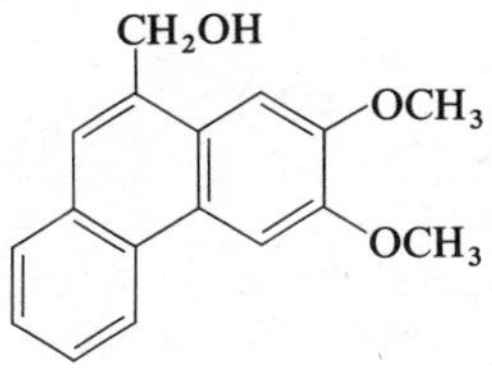 CH_2OH, OCH_3, OCH_3 (63)	566

TABLE I. REDUCTION OF CARBOXYLIC ACIDS (*Continued*)

No. of Carbon Atoms	Reactant	Conditions	Product(s) and Yield(s)(%)	Refs.
C_{17} (*Contd.*)	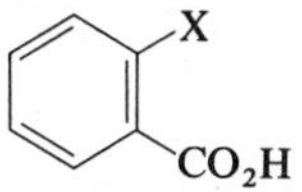 2-X-$C_6H_4CO_2H$	SMEAH, C_6H_6, room temp, 3 h	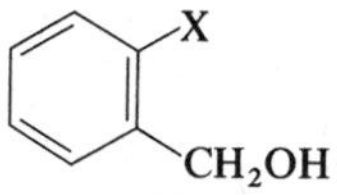 2-X-$C_6H_4CH_2OH$	
	X = $SC_6H_4(C_4H_9$-*n*)-4		(92)	567
	X = $SC_6H_4(C_4H_9$-*t*)-4		(—)	545
C_{18}	X = $SC_6H_4(C_5H_9)$-4		(82)	567
	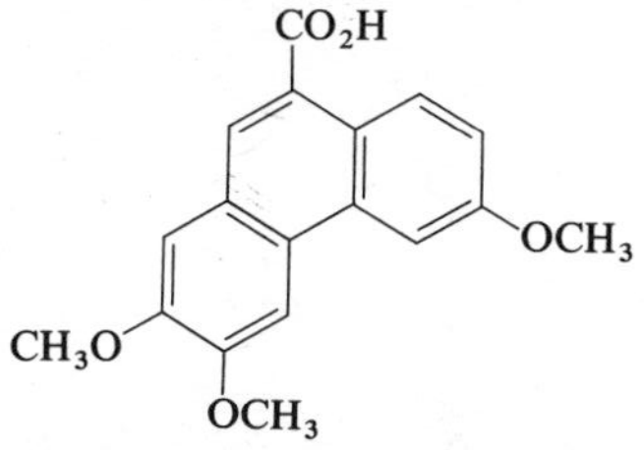 3,6,7-Trimethoxyphenanthrene-9-CO_2H (CO_2H; OCH_3; CH_3O; OCH_3)	SMEAH, THF–C_6H_6, reflux, 2 h	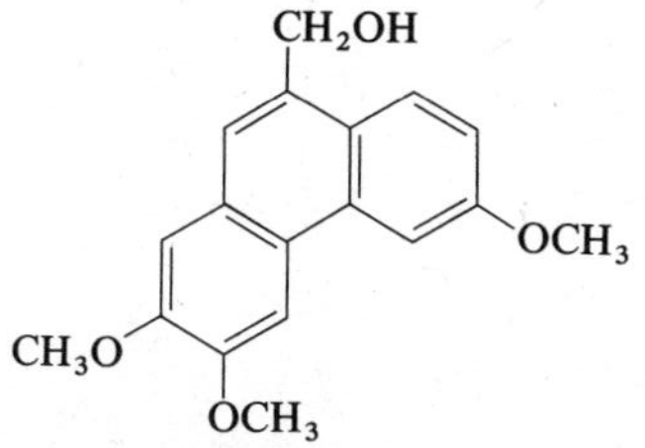 3,6,7-Trimethoxyphenanthrene-9-CH_2OH (CH_2OH; OCH_3; CH_3O; OCH_3) (60)	566
	n-$C_{17}H_{35}CO_2Na$	SMEAH, C_6H_6, reflux, 1 h	*n*-$C_{18}H_{37}OH$ (96)	123
C_{19}	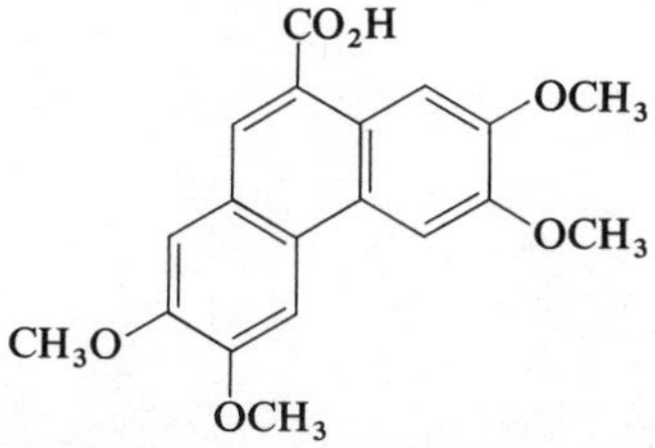 2,3,6,7-Tetramethoxyphenanthrene-9-CO_2H (CO_2H; OCH_3; OCH_3; CH_3O; OCH_3)	SMEAH, THF–C_6H_6, reflux, 2 h	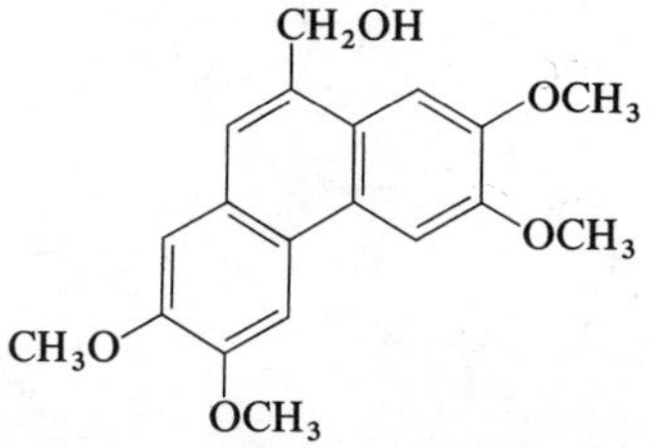 2,3,6,7-Tetramethoxyphenanthrene-9-CH_2OH (CH_2OH; OCH_3; OCH_3; CH_3O; OCH_3) (60)	566

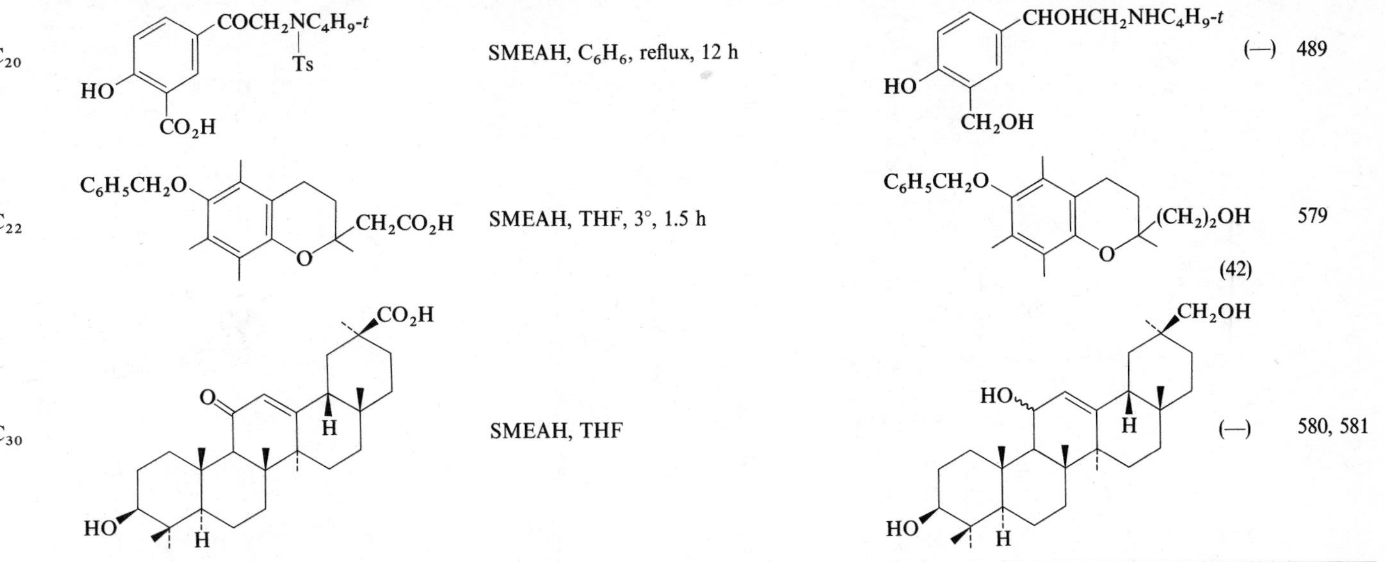

	Reactant	Conditions	Product(s) and Yield(s) (%)	Refs.
C_{20}	HO, CO_2H, $COCH_2N(Ts)C_4H_9$-*t*	SMEAH, C_6H_6, reflux, 12 h	HO, CH_2OH, $CHOHCH_2NHC_4H_9$-*t* (—)	489
C_{22}	$C_6H_5CH_2O$, CH_2CO_2H	SMEAH, THF, 3°, 1.5 h	$C_6H_5CH_2O$, $(CH_2)_2OH$ (42)	579
C_{30}	CO_2H, HO, H	SMEAH, THF	CH_2OH, HO, H (—)	580, 581

[a] This result was not confirmed in an independent experiment.[582]

TABLE II. Partial Reduction of Carboxylic Acids to Aldehydes

No. of Carbon Atoms	Reactant	Conditions[a]	Product(s) and Yield(s) (%)	Ref.
C_2	CH_3CO_2H	A	CH_3CHO (76)	98
C_3	$C_2H_5CO_2H$	A	C_2H_5CHO (78)	98
C_4	n-$C_3H_7CO_2H$	A	n-C_3H_7CHO (78)	98
C_5	Furan-2-carboxylic acid	B	Furan-2-carboxaldehyde (70)	192
C_6	Pyridine-3-carboxylic acid	B	Pyridine-3-carboxaldehyde (55)	192
	$Br(CH_2)_5CO_2H$	B	$Br(CH_2)_5CHO$ (73)	192
C_7	2-$ClC_6H_4CO_2H$	A	2-ClC_6H_4CHO (65)	98
	4-$ClC_6H_4CO_2H$	B	4-ClC_6H_4CHO (79)	192
		A	" (72)	98
	4-$O_2NC_6H_4CO_2H$	B	4-$O_2NC_6H_4CHO$ (81)	192
		A	" (75)	98
	$C_6H_5CO_2H$	B	C_6H_5CHO (78)	192
		A	" (80)	98
	$NC(CH_2)_5CO_2H$	B	$NC(CH_2)_5CHO$ (80)	192
	$C_6H_{11}CO_2H$	B	$C_6H_{11}CHO$ (84)	192
		A	" (75)	98
	$CH_3O_2C(CH_2)_4CO_2H$	B	$CH_3O_2C(CH_2)_4CHO$ (64)	192
	n-$C_6H_{13}CO_2H$	B	n-$C_6H_{13}CHO$ (90)	192
C_8	2-$CH_3C_6H_4CO_2H$	A	2-$CH_3C_6H_4CHO$ (63)	98
	4-$CH_3C_6H_4CO_2H$	A	4-$CH_3C_6H_4CHO$ (70)	98
	4-$CH_3OC_6H_4CO_2H$	B	4-$CH_3OC_6H_4CHO$ (70)	192
	CO_2H	B	CHO (82)	192

C_9	$C_6H_5CH{=}CHCO_2H$	B	$C_6H_5CH{=}CHCHO$ (83)	192
		A	" (77)	98
	$4\text{-}HOC_6H_4CH{=}CHCO_2H$	B	$4\text{-}HOC_6H_4CH{=}CHCHO$ (80)	583
C_{10}	$3\text{-}CH_3O\text{-}4\text{-}HOC_6H_3CH{=}CHCO_2H$	B	$3\text{-}CH_3O\text{-}4\text{-}HOC_6H_3CH{=}CHCHO$ (71)	583
	$(CH_3)_2C{=}CH(CH_2)_2C(CH_3){=}CHCO_2H$ (*trans*:*cis* = 90:10)	B	$(CH_3)_2C{=}CH(CH_2)_2C(CH_3){=}CHCHO$ (82) (*trans*:*cis* = 90:10)	192
	$(CH_3)_2C{=}CH(CH_2)_2CH(CH_3)CH_2CO_2H$	B	$(CH_3)_2C{=}CH(CH_2)_2CH(CH_3)CH_2CHO$ (80)	192
C_{11}	$3,5\text{-}(CH_3O)_2\text{-}4\text{-}HOC_6H_2CH{=}CHCO_2H$	B	$3,5\text{-}(CH_3O)_2\text{-}4\text{-}HOC_6H_2CH{=}CHCHO$ (74)	583
	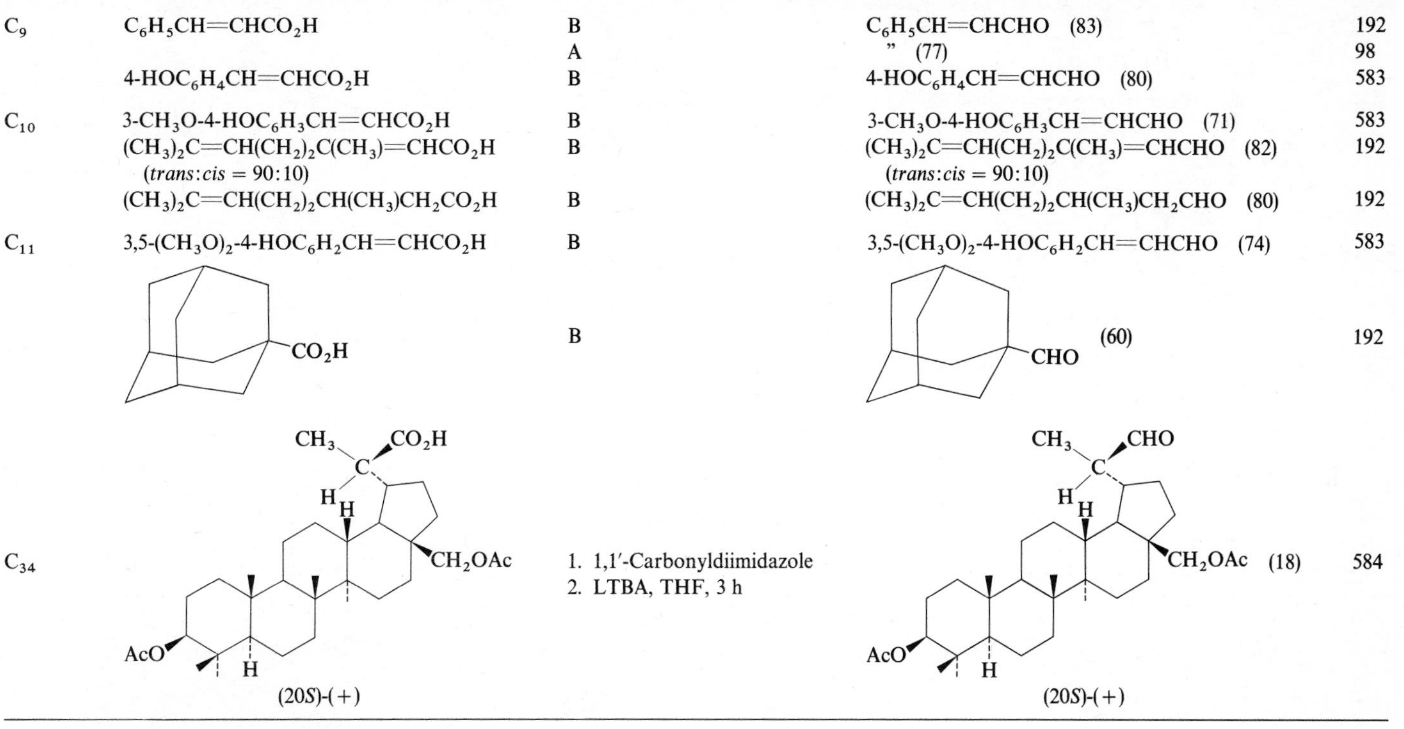	B	(60)	192
C_{34}	(20*S*)-(+)	1. 1,1′-Carbonyldiimidazole 2. LTBA, THF, 3 h	(18) (20*S*)-(+)	584

[a] Reaction conditions were as follows:

A. 1. γ-Saccharin chloride, CH_2Cl_2, 0°
   2. SMEAH, C_6H_6, 0–5° to −70°
   3. H_3O^+

B. 1. $ClCH{=}\overset{+}{N}(CH_3)_2\ Cl^-$, Py, THF, CH_3CN, −30°
   2. LTBA–CuI, THF, −78°, 10 min
   3. H_3O^+

TABLE III. REDUCTION OF CARBOXYLIC ACID ANHYDRIDES

No. of Carbon Atoms	Reactant	Conditions	Product(s) and Yield(s) (%)	Refs.
C_4	$(CH_3CO)_2O$	SMEAH, C_6H_6, reflux, 2 h	C_2H_5OH (91)	238
C_6	$(C_2H_5CO)_2O$	$Ca[AlH_2(OC_6H_{11})_2]_2 \cdot THF$, C_6H_6, reflux, 2.5 h	n-C_3H_7OH (~99)	165
C_8	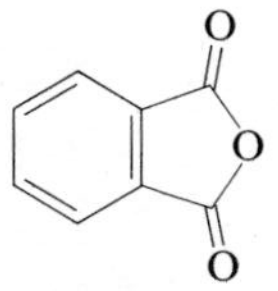 [phthalic anhydride structure]	SMEAH, C_6H_6, 80°, 1.5 h	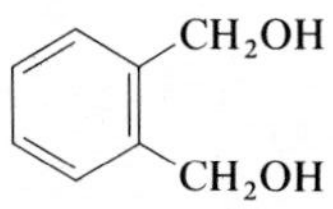 [1,2-benzenedimethanol structure] CH_2OH, CH_2OH (89)	168, 167
C_{10}	$(t$-$C_4H_9CO)_2O$	$Ca[AlH_2(OC_3H_7$-$i)_2]_2 \cdot THF$, toluene, 80°, 2 h	t-$C_4H_9CH_2OH$ (91)	165
C_{11}	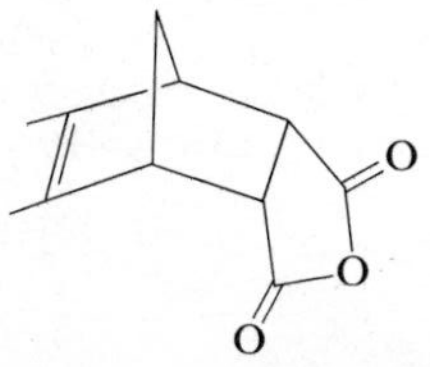 [bicyclic anhydride structure]	SMEAH, C_6H_6–ether, reflux, 1 h	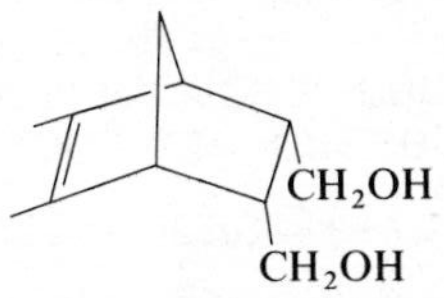 [bicyclic diol structure] CH_2OH, CH_2OH (69)	585

C_{12}	$(n\text{-}C_5H_{11}CO)_2O$	SMEAH, C_6H_6, 80°	$n\text{-}C_6H_{13}OH$ (85)	68
C_{14}	$(C_6H_5CO)_2O$	SMEAH, C_6H_6, 80°, 1.25 h	$C_6H_5CH_2OH$ (89)	168
		$Ca[AlH_2(OC_3H_7\text{-}i)_2]_2 \cdot THF$, toluene, 100°, 3 h	" (92)	165
		$NaAlH(OC_2H_5)_3$, THF, 65°, 4 h	" (79)	319
C_{15}	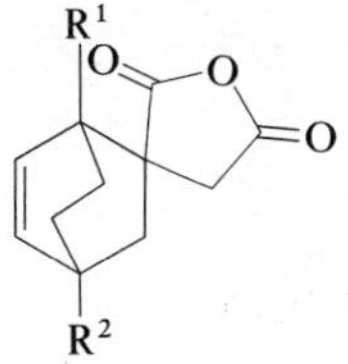 R[1], O, O, O, R[2]	SMEAH, toluene	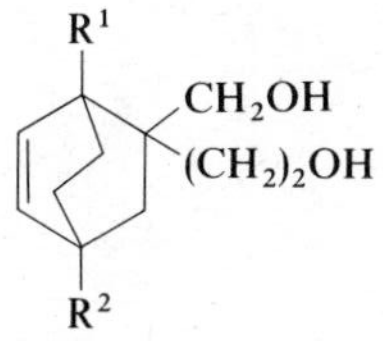 R[1], CH_2OH, $(CH_2)_2OH$, R[2]	
	$R^1 = CH_3$; $R^2 = i\text{-}C_3H_7$		(76)	586
	$R^1 = i\text{-}C_3H_7$; $R^2 = CH_3$		(—)	586
C_{20}	$(n\text{-}C_9H_{19}CO)_2O$	SMEAH, C_6H_6, 80°	$n\text{-}C_{10}H_{21}OH$ (86)	68

TABLE IV. PARTIAL REDUCTION OF CARBOXYLIC ACID ANHYDRIDES TO LACTONES AND HYDROXYLACTONES

No. of Carbon Atoms	Reactant	Conditions	Product(s) and Yield(s) (%)	Refs.
C_5	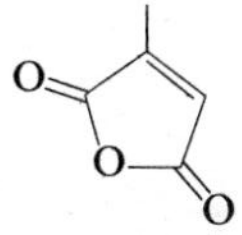 O, O, O	LTMA (IA), ether, 0°, 2.5 h	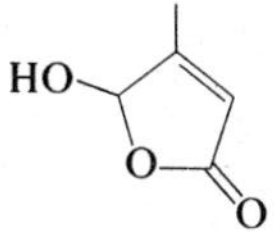 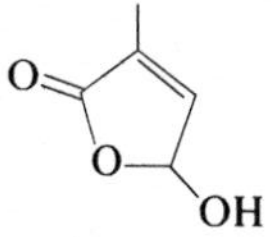 HO, O, O (I) + O, O, OH (II) (—) (I:II = 75:25)	58
C_6	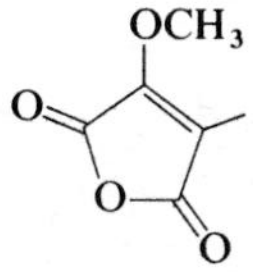 OCH_3, O, O, O	LTBA (IA), DME, −30°, 1 h, then 25°, 12 h	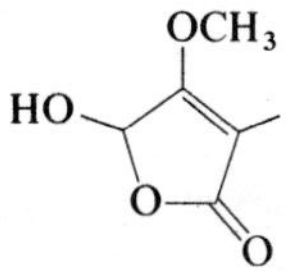 HO, OCH_3, O, O (45) 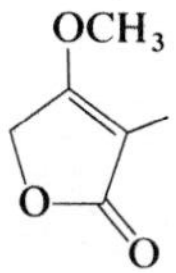 + OCH_3, O, O (24)	52, 49
C_8	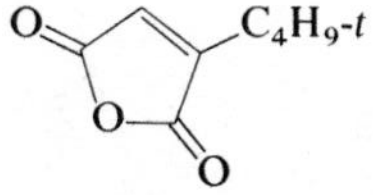 O, O, O, C_4H_9-*t*	LTBA, THF, reflux, 1 h	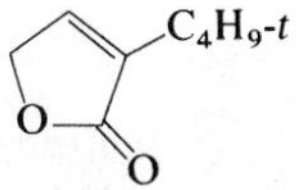 C_4H_9-*t*, O, O (40) 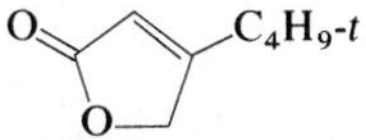 + O, O, C_4H_9-*t* (20) + C_4H_9-*t*, O, O (—) + O, O, C_4H_9-*t* (—) 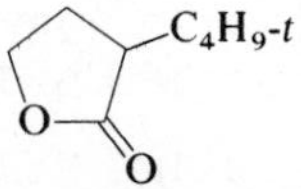 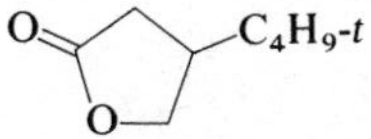	587a

	Reactant	Conditions	Product(s) and Yield(s) (%)	Refs.
	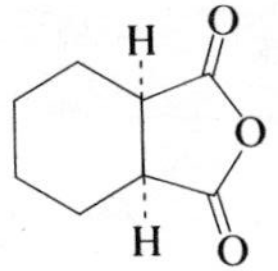	LTBA (IA), THF, -20 to $-30°$, 6 h	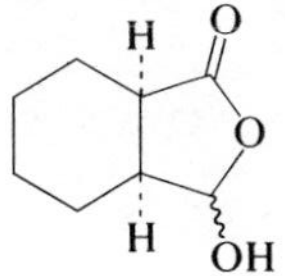 (63) + 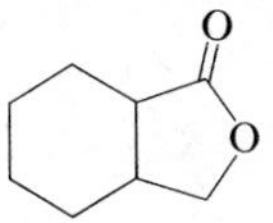 (23)	587b
C_{10}	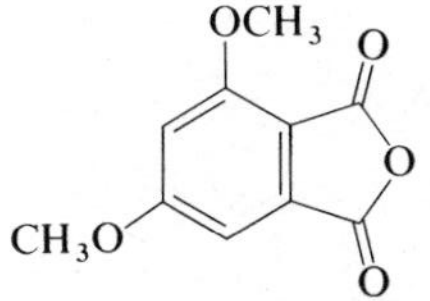	LTBA, THF, 20°, 18 h	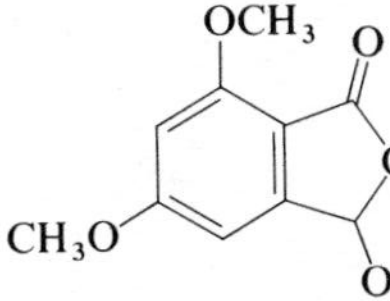 (33) + 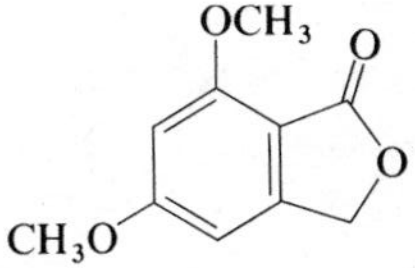 (—)	195
	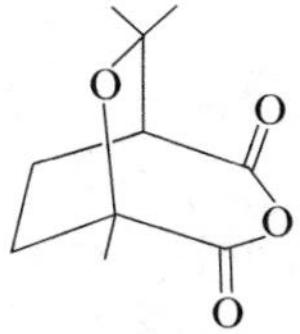	SMEAH, C_6H_6, 2 h	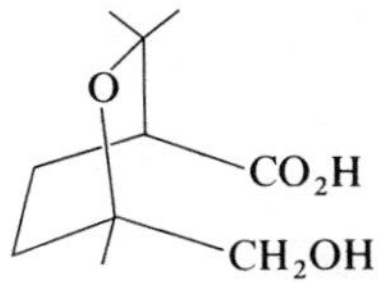 (25)	588
C_{11}	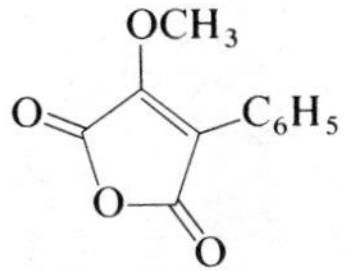	LTBA (IA), DME, $-30°$, 1 h, then 25°, 12 h	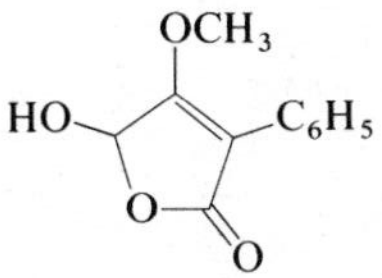 (43) + 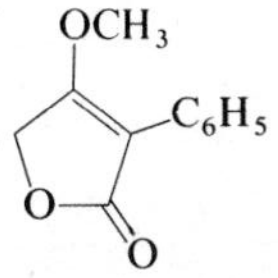 (21)	52, 49

TABLE IV. PARTIAL REDUCTION OF CARBOXYLIC ACID ANHYDRIDES TO LACTONES AND HYDROXYLACTONES (*Continued*)

No. of Carbon Atoms	Reactant	Conditions	Product(s) and Yield(s) (%)	Refs.
C_{12}	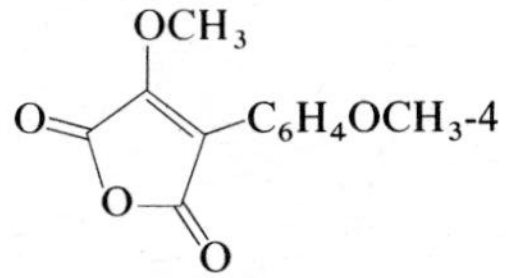 OCH_3, $C_6H_4OCH_3$-4, O, O, O	LTBA (IA), DME, −30°, 1 h, then 25°, 12 h	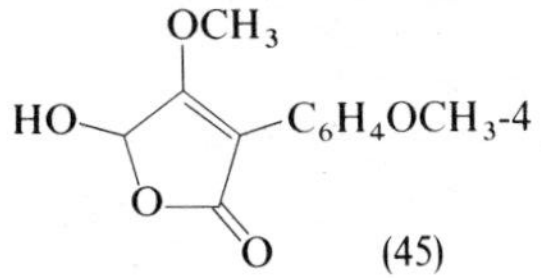 OCH_3, HO, $C_6H_4OCH_3$-4, O, O (45) 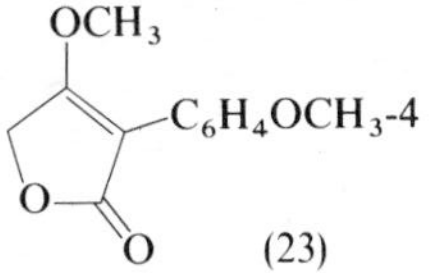 + OCH_3, $C_6H_4OCH_3$-4, O, O (23)	52, 50
C_{13}	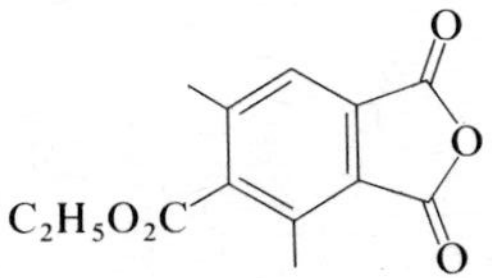 $C_2H_5O_2C$, O, O, O	LTBA (IA), THF, room temp, 3 h	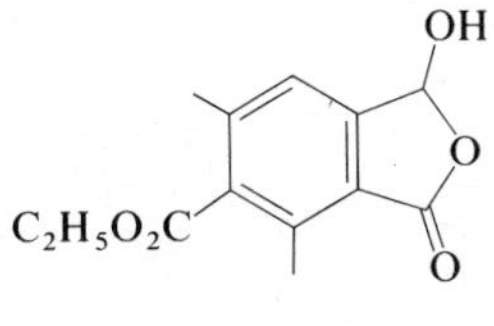 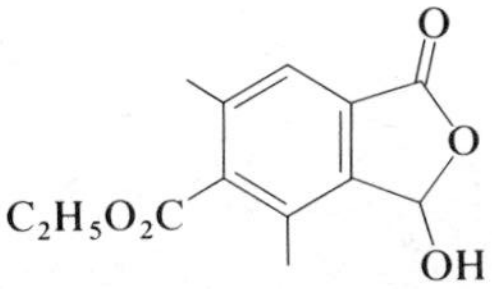 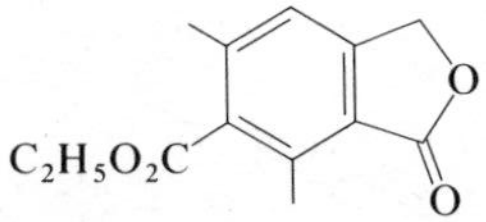 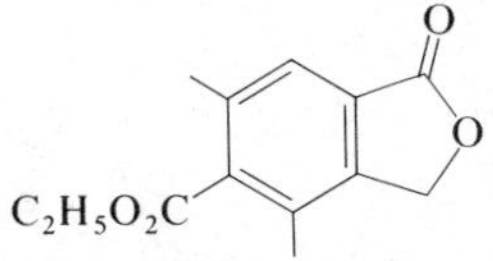 OH, $C_2H_5O_2C$, O, O **I** + O, $C_2H_5O_2C$, O, OH **II** + $C_2H_5O_2C$, O, O **III** + O, $C_2H_5O_2C$, O **IV** (**I** + **II** = 49) (**III** + **IV** = ~41)	589

C_{14}	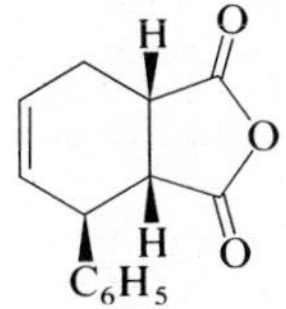 H, O, O, O, H, C_6H_5	SMEAH, toluene, room temp, 40 min	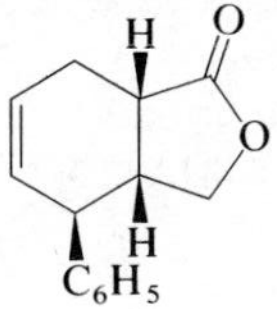 H, O, O, H, C_6H_5 (22) 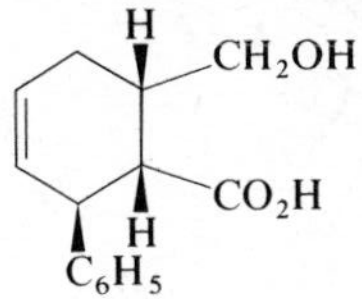 + H, CH_2OH, CO_2H, H, C_6H_5 (21)	590
C_{15}	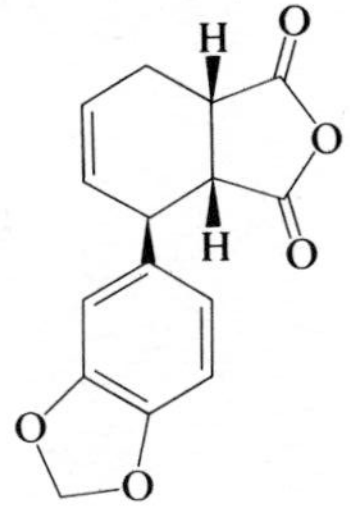 H, O, O, O, H, O, O	SMEAH, toluene, room temp, 40 min	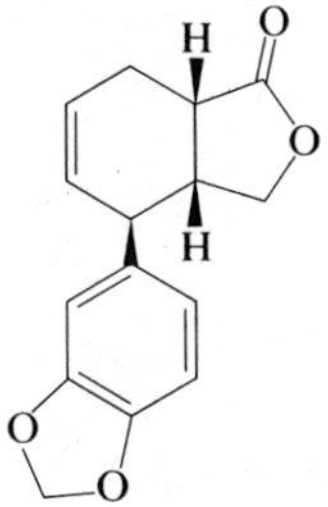 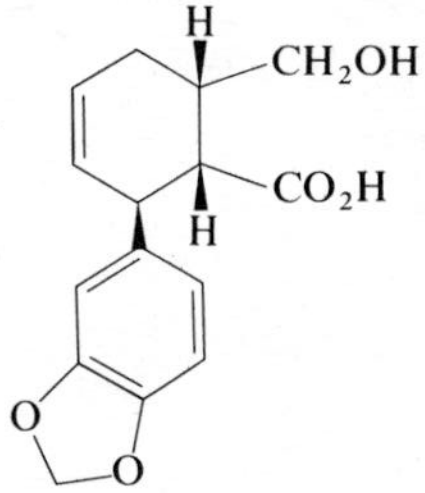 H, O, O, H, O, O (—) + H, CH_2OH, CO_2H, H, O, O (—)	590
C_{17}	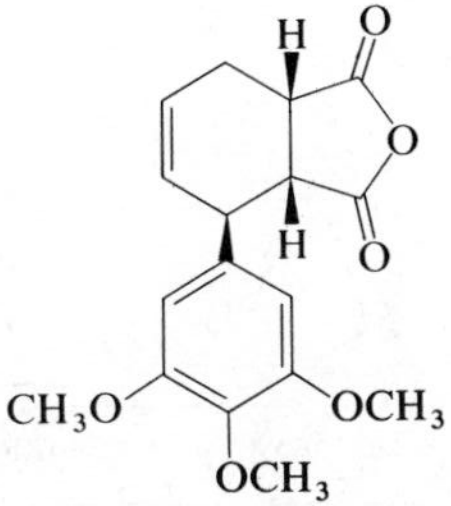 H, O, O, O, H, CH_3O, OCH_3, OCH_3	SMEAH, toluene, room temp, 40 min	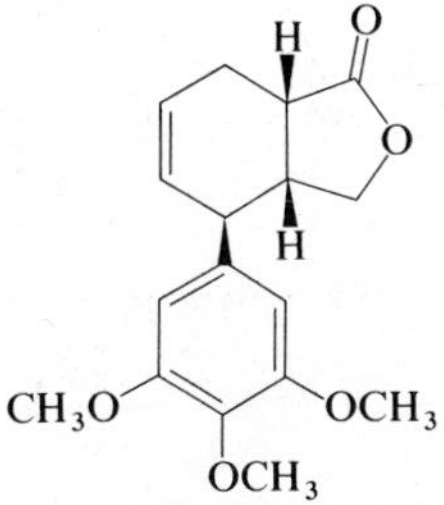 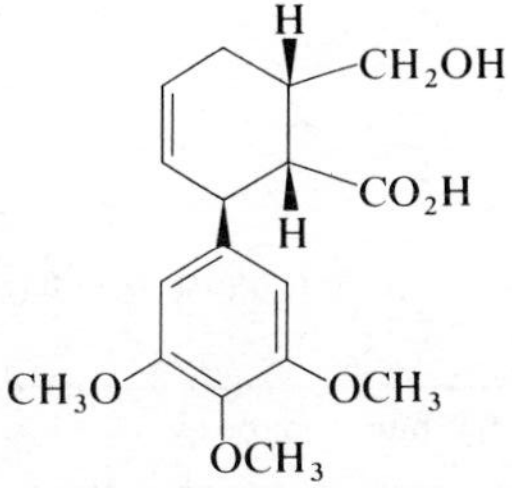 H, O, O, H, CH_3O, OCH_3, OCH_3 + H, CH_2OH, CO_2H, H, CH_3O, OCH_3, OCH_3 (—)	590

TABLE V. REDUCTION OF LACTONES AND LACTOLS (HEMIACETALS) TO DIOLS

No. of Carbon Atoms	Reactant	Conditions	Product(s) and Yield(s) (%)	Ref.
C_4	O, O	SMEAH, C_6H_6, 80°, 20 min	$HO(CH_2)_4OH$ (78)	168
		$Ca[AlH_2(OC_4H_9\text{-}i)_2]_2 \cdot THF$, toluene, 20°, 1 h	" (~99)	165
C_7	C_2H_5, O, O	SMEAH, C_6H_6, 80°, 1 h	$C_2H_5CHOH(CH_2)_4OH$ (60)	168
C_{10}	$CH_2CO_2CH_3$, O, O	SMEAH	$CH_2CO_2CH_3$, $HOCH_2$, CH_2OH (—)	591
	O, O	$LiAlH(OC_2H_5)_3$, ether, room temp	OH, OH (—)	592

	Reactant	Conditions	Product(s) and Yield(s) (%)		Refs.
C_{12}	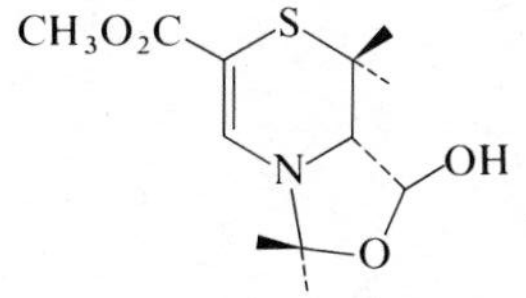 CH_3O_2C, S, N, OH, O	$LiAlD(OC_4H_9\text{-}t)_3$, THF, room temp, 1 h	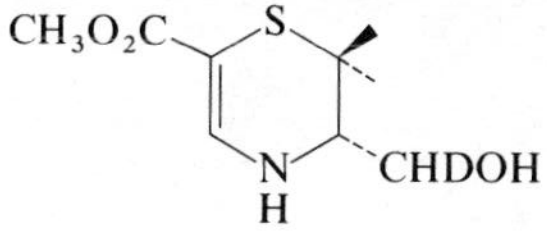 CH_3O_2C, S, N, H, CHDOH	(—)	221a
C_{15}	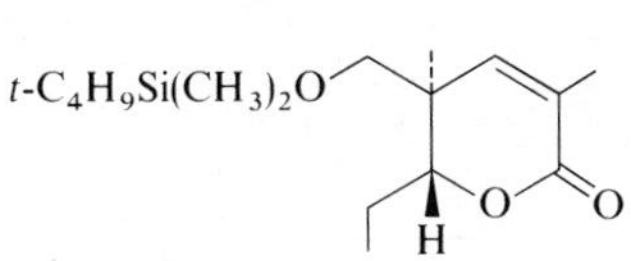 $t\text{-}C_4H_9Si(CH_3)_2O$, O, O, H	1. LTMA, THF, 0° 2. CH_3COCH_3, H_3O^+, $CuSO_4$	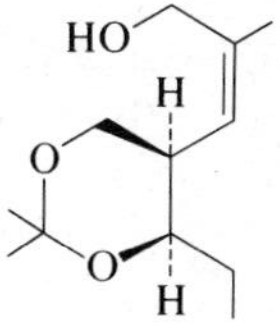 HO, H, O, O, H	(81)	593
C_{20}	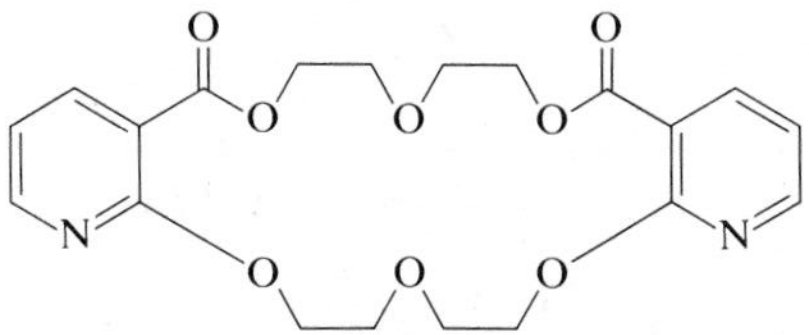 O, O, O, O, O, N, N, O, O, O	SMEAH, C_6H_6, reflux, 1 h	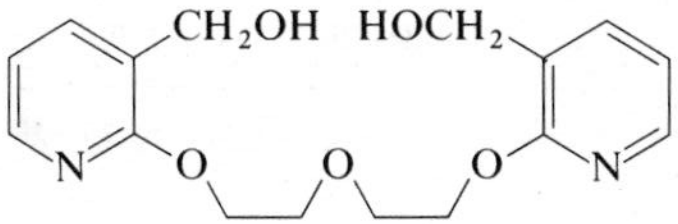 CH_2OH $HOCH_2$, N, O, O, O, N	(52)	594
	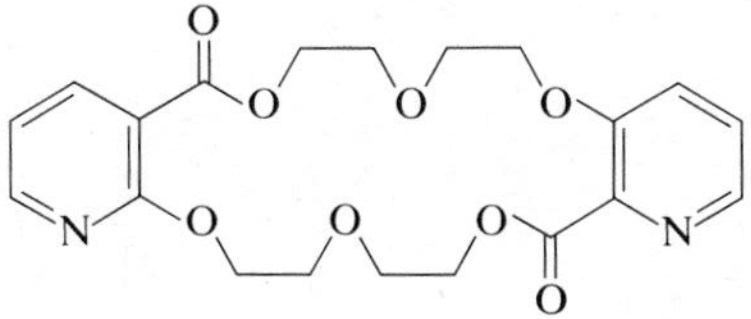 O, O, O, O, N, O, O, O, N, O	SMEAH, C_6H_6, reflux, 1 h	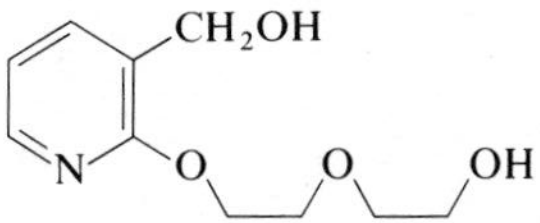 CH_2OH, N, O, O, OH	(65)	594

TABLE V. REDUCTION OF LACTONES AND LACTOLS (HEMIACETALS) TO DIOLS (*Continued*)

No. of Carbon Atoms	Reactant	Conditions	Product(s) and Yield(s) (%)	Ref.
C_{21}	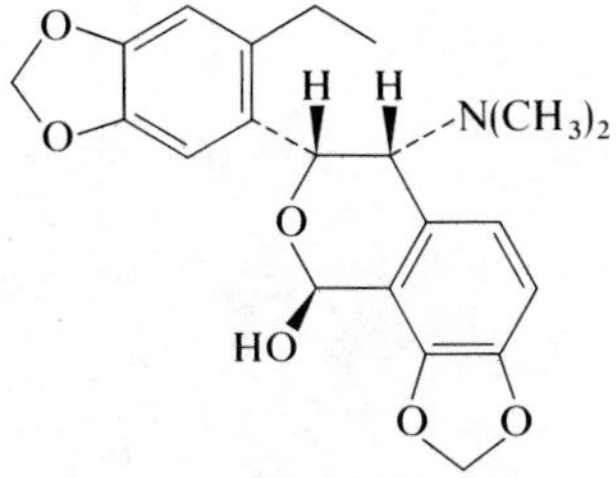 H H N(CH$_3$)$_2$ HO	SMEAH, C_6H_6, reflux, 2 h	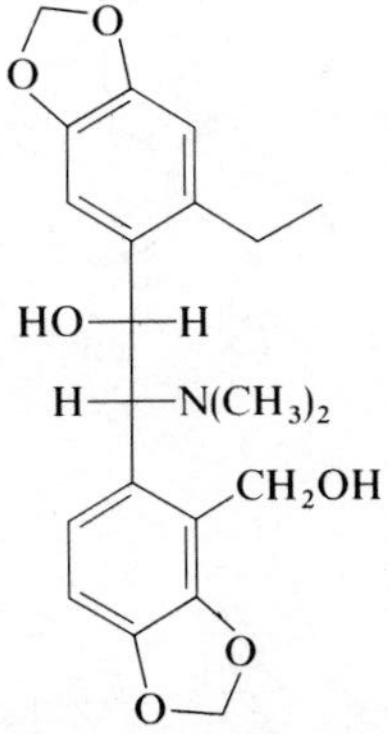 HO–H H–N(CH$_3$)$_2$ CH$_2$OH (95)	595

	Reactant	Conditions	Product(s) and Yield(s) (%)	Refs.
C_{22}	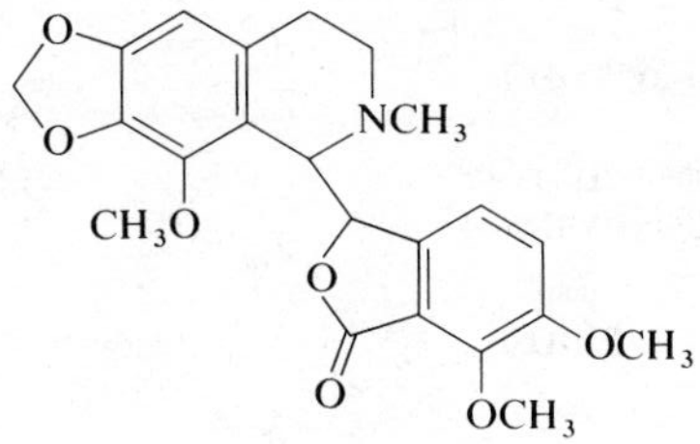 O, O, NCH_3, CH_3O, O, O, OCH_3, OCH_3	SMEAH, C_6H_6	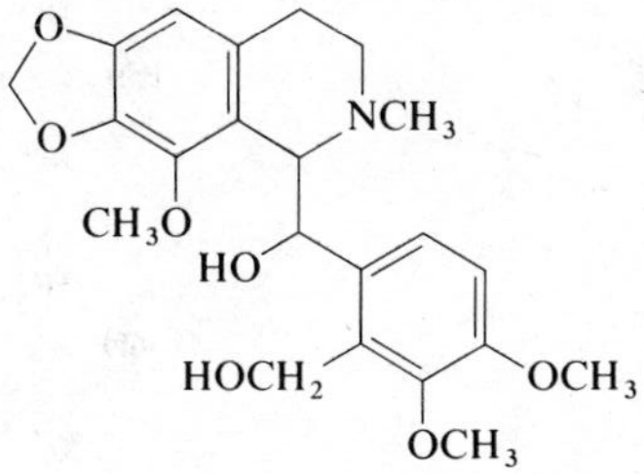 O, O, NCH_3, CH_3O, HO, $HOCH_2$, OCH_3, OCH_3 (—)	596
C_{29}	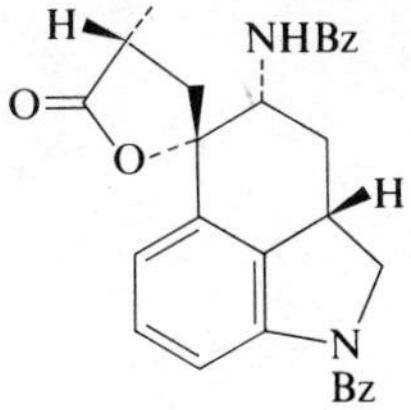 H, NHBz, O, O, H, N, Bz	1. SMEAH (IA), THF, 0°, 2 h 2. BzCl, Py	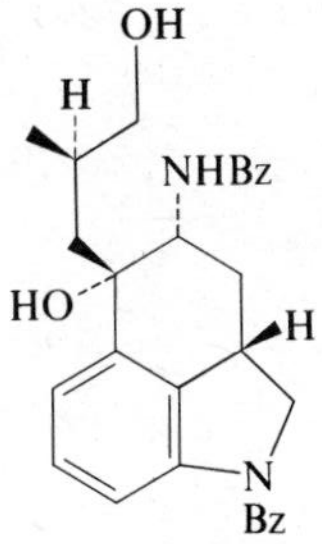 OH, H, NHBz, HO, H, N, Bz (75)	597, 598

TABLE VI. PARTIAL REDUCTION OF LACTONES TO LACTOLS (HEMIACETALS)

Reactant	Conditions	Product(s) and Yield(s) (%)	Refs.
C_6			
	LTBA, THF, −70°	(30)	599
	", ", "	(16)	599
C_7			
	", ", "	(36)	599
C_8			
n-C_3H_7	", ", "	n-C_3H_7 (44)	599
C_9			
	LTBA (IA), THF, −60°, 75 min, then −60° to room temp, 1.5 h	(49)	223
	SMEAH–EtOH (1:1)(IA), C_6H_6–toluene, −60°, 15 min	" (64)	73
	LTBA (IA), THF, −50°	(64)	600

Reactant	Conditions	Product (Yield, %)	Refs.
C_{10}			
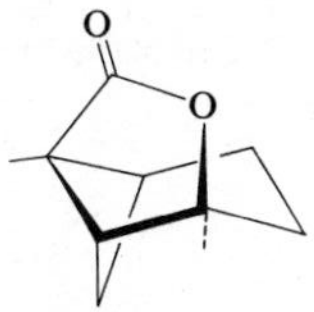 O	$LiAlH(OC_2H_5)_3$, ether, $-22°$	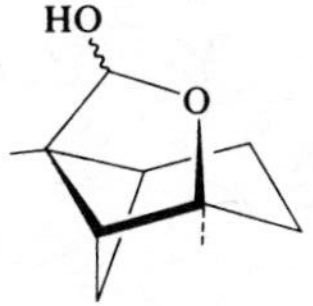 HO (97)	592, 601
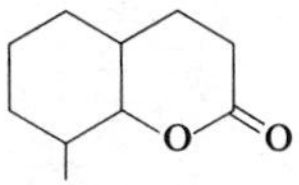 O	LTBA (IA), THF, $-50°$	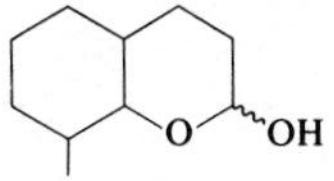 OH (42)	600
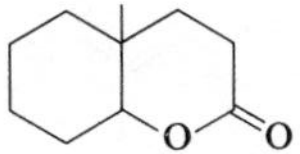 O	", ", "	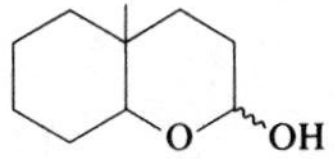 OH (44)	600
C_{11}			
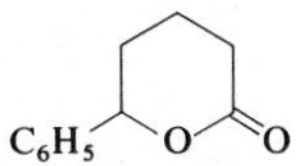 C_6H_5 O	LTBA, THF, $-70°$	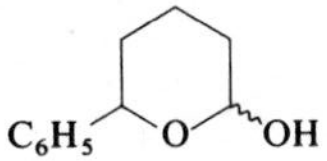 C_6H_5 OH (51)	599
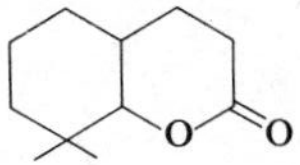 O	LTBA (IA), THF, $-50°$	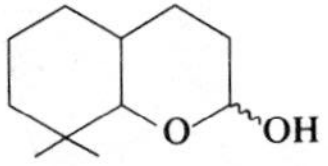 OH (33)	600
C_{12}			
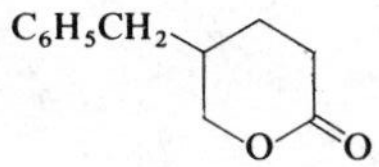 $C_6H_5CH_2$ O	LTBA, THF, $-70°$	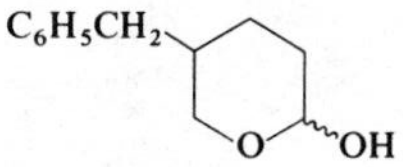 $C_6H_5CH_2$ OH (62)	599
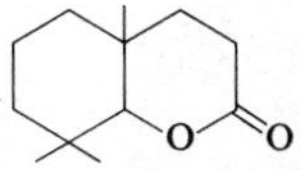 O	LTBA (IA), THF, $-50°$	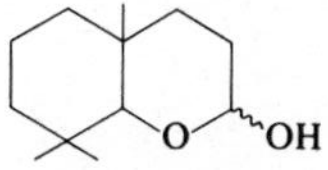 OH (52)	600

TABLE VI. PARTIAL REDUCTION OF LACTONES TO LACTOL (HEMIACETALS) (*Continued*)

Reactant	Conditions	Product(s) and Yield(s) (%)	Refs.
C_{13}			
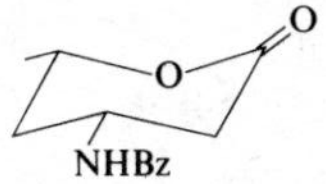 NHBz	SMEAH-i-C_3H_7OH (1:1) (IA), room temp, 1.75 h	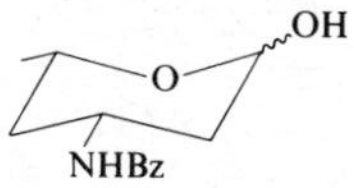 NHBz, OH (61)	228
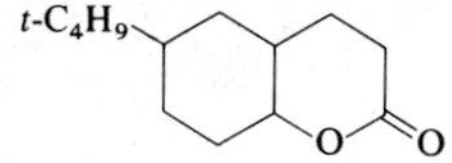 t-C_4H_9	LTBA (IA), THF, −50°	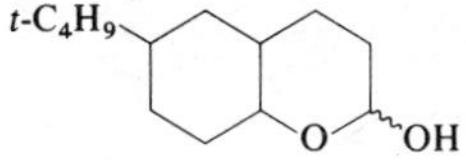 t-C_4H_9, OH (12)	600
C_{14}			
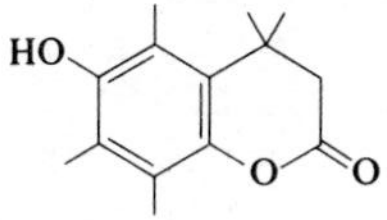 HO	SMEAH	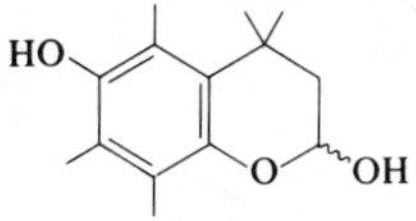 HO, OH (—)	602
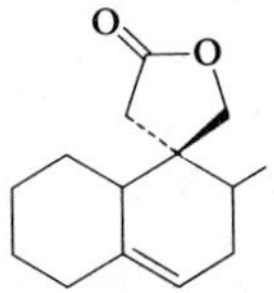	SMEAH (IA), C_6H_6–THF, −65°, 1 h	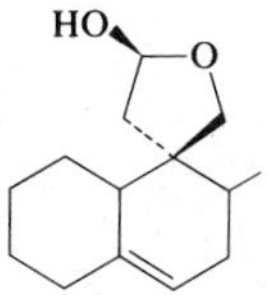 HO (20) 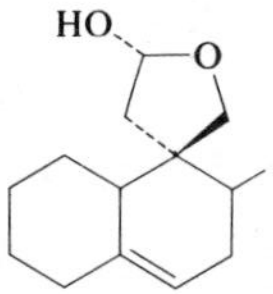 + HO (44) 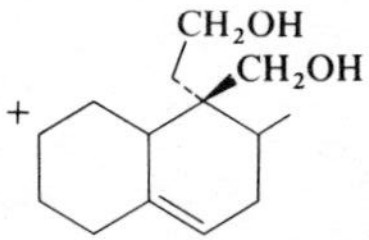 + CH_2OH, CH_2OH (27)	227
C_{15}			
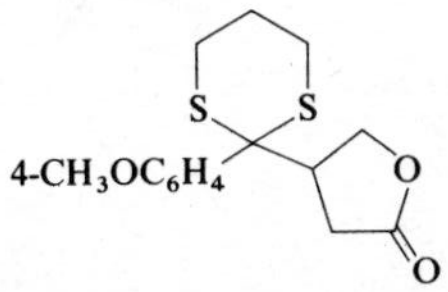 4-$CH_3OC_6H_4$	SMEAH–C_2H_5OH (1:1) (IA), C_6H_6-toluene, 0°, 25 min	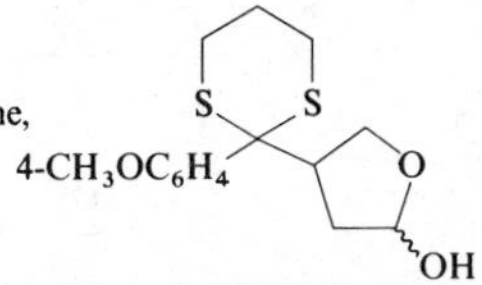 4-$CH_3OC_6H_4$, OH (84)	73

Substrate	Reagent and conditions	Product	Yield (%)	Ref.
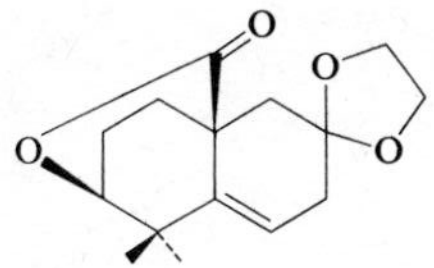	LTBA, ether, reflux, 9 h	OH	(79)	603
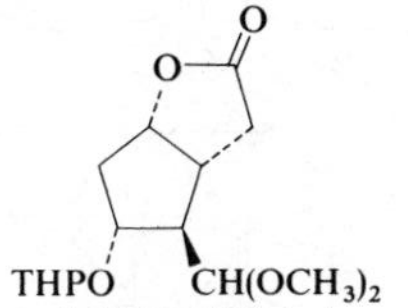 THPO CH$(OCH_3)_2$ (1*S*,5*R*,6*R*,7*R*)-(—)	SMEAH (IA), toluene, $-28°$, 2 h	OH, THPO CH$(OCH_3)_2$	(75)	604
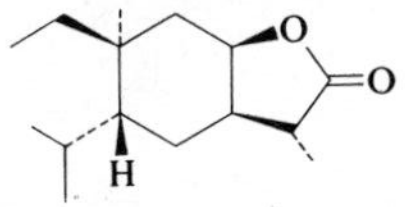 H	1. SMEAH, C_6H_6–THF, 0–5° 2. $N_2H_4 \cdot H_2O$, KOH	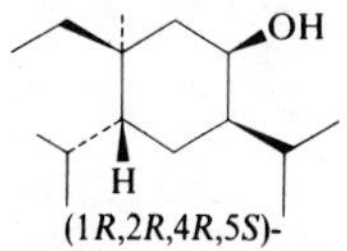 OH, H (1*R*,2*R*,4*R*,5*S*)-	(—)	605
C_{17}				
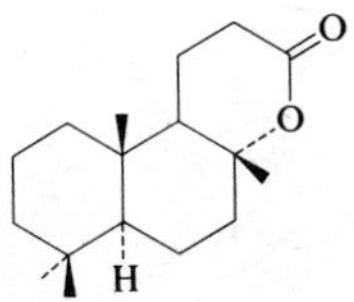 H	$LiAlH_2(OC_2H_5)_2$, THF, 0°, 30 min, then reflux, 1 h	OH, H	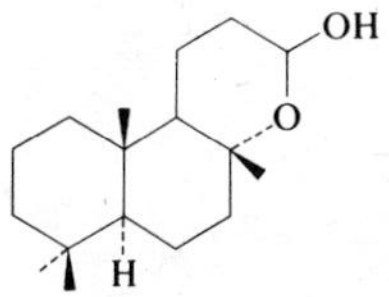 (22)	606
C_{20}				
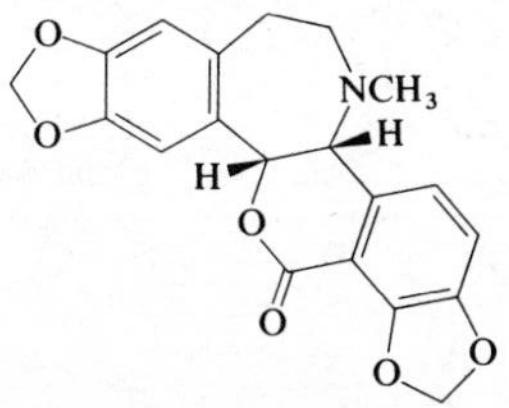 NCH_3, H, H (4b*R*,10b*R*)-(+)-	1. SMEAH-Py, THF, $-70°$ 2. $HC(OCH_3)_3$, CH_3OH, H_3O^+	NCH_3, H, H, CH_3O (+)-	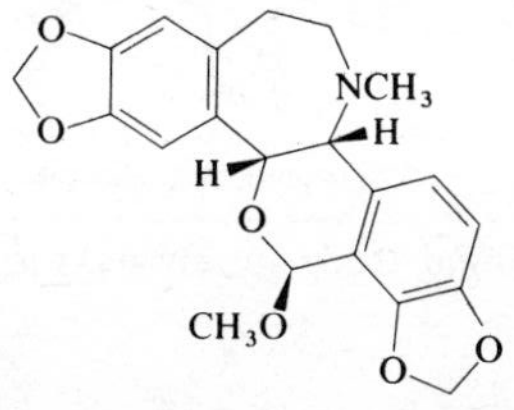 (60)	607a, 607b

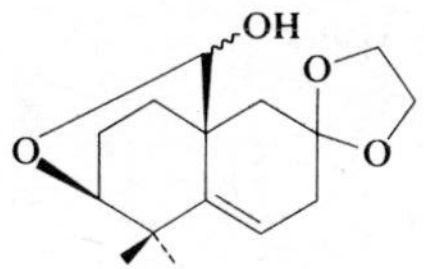

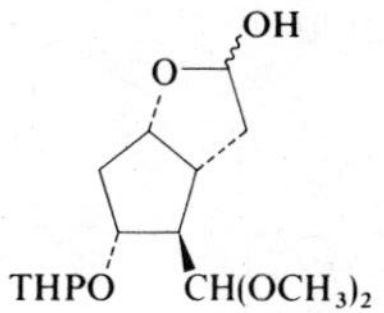

TABLE VI. PARTIAL REDUCTION OF LACTONES TO LACTOLS (HEMIACETALS) (*Continued*)

Reactant	Conditions	Product(s) and Yield(s) (%)	Refs.
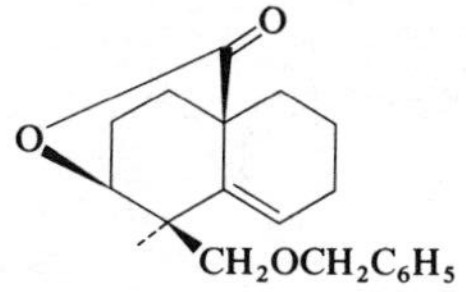 O; $CH_2OCH_2C_6H_5$	$LiAlH(OC_2H_5)_3$, ether, 0°, 5.5 h	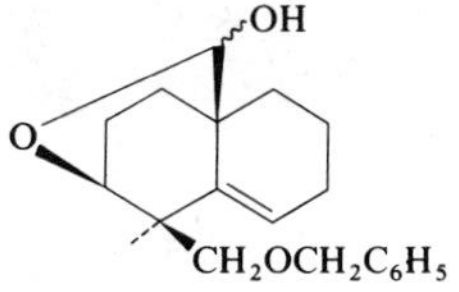 OH; O; $CH_2OCH_2C_6H_5$ (88)	225, 607c
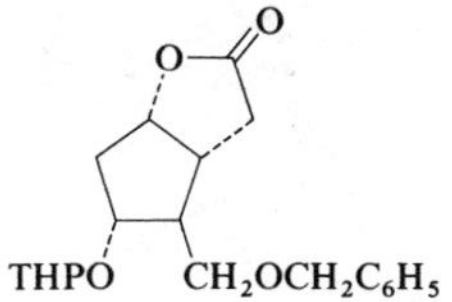 O; THPO $CH_2OCH_2C_6H_5$	SMEAH, toluene, −65 to −70°, 5 h	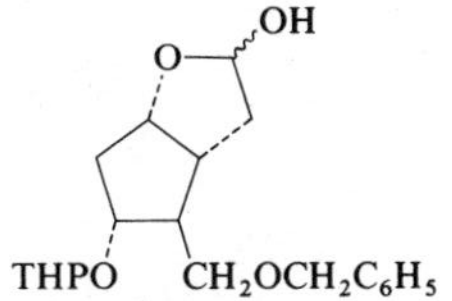 OH; THPO $CH_2OCH_2C_6H_5$ (—)	608
C_{21}			
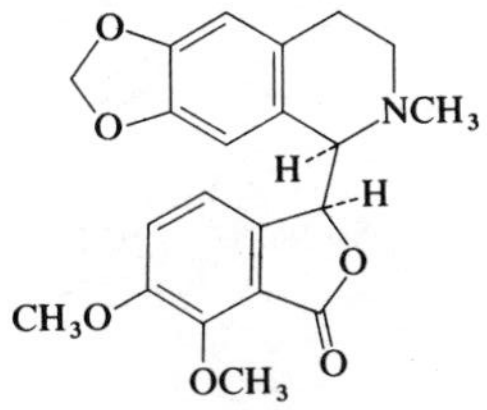 NCH_3; H; H; CH_3O; OCH_3; O	SMEAH	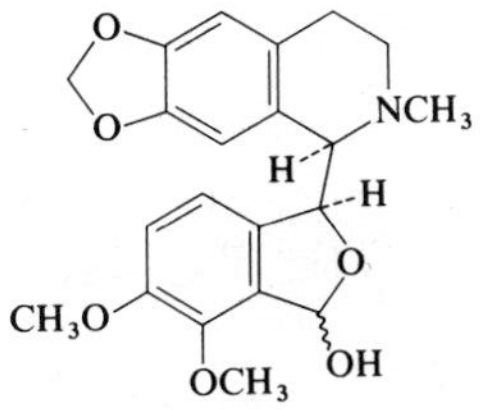 NCH_3; H; H; CH_3O; OCH_3; OH (—)	609
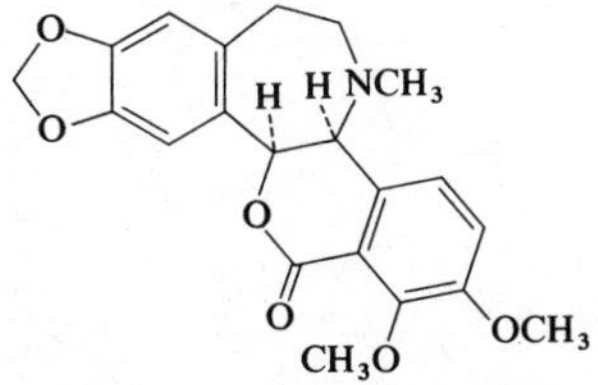 H H NCH_3; O; CH_3O; OCH_3	SMEAH, THF–C_6H_6, 4°	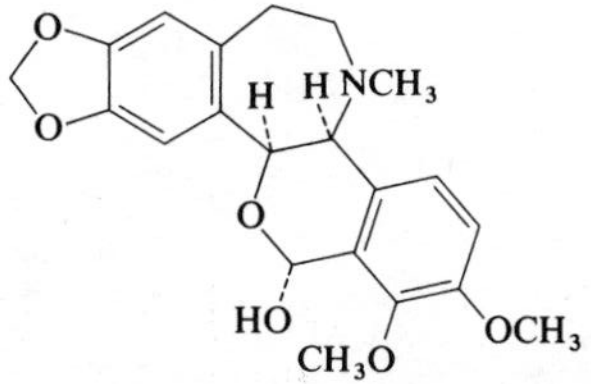 H H NCH_3; HO; CH_3O; OCH_3 (62)	610

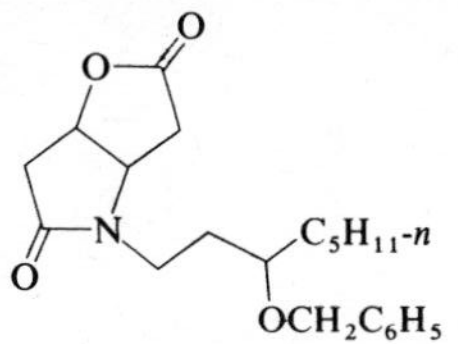	LTBA, THF, 0–20°, 3 h	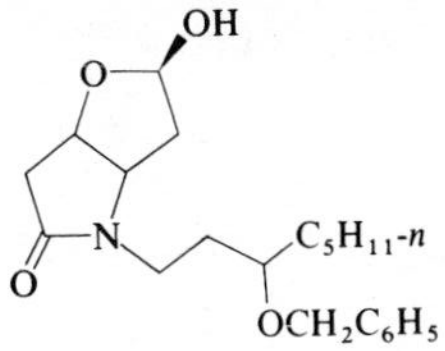	(63)	611
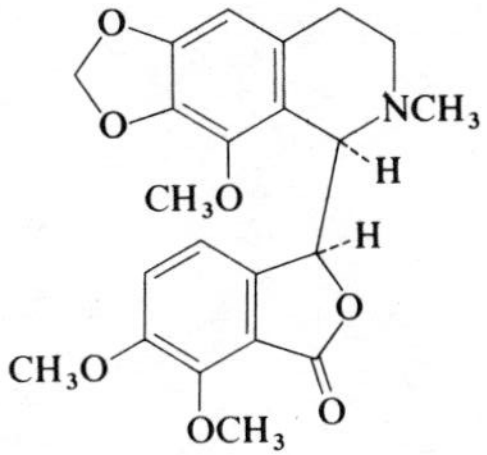	SMEAH	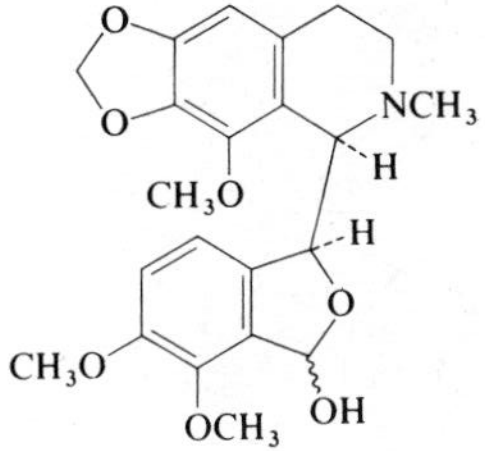	(—)	609
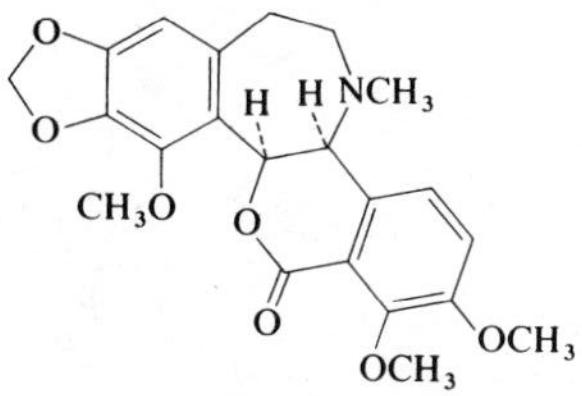	SMEAH, Py, C_6H_6–THF, 4°, 3.5 h	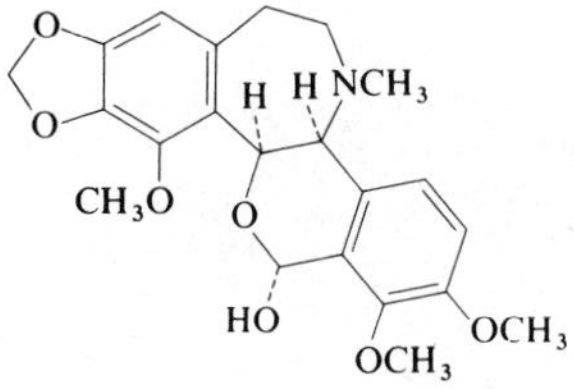	(64)	612, 613
		+ 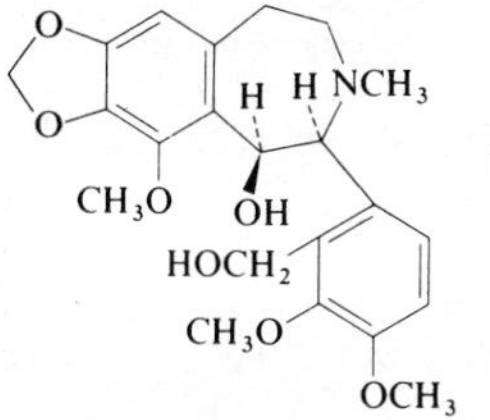	(15)	

TABLE VI. PARTIAL REDUCTION OF LACTONES TO LACTOLS (HEMIACETALS) (*Continued*)

Reactant	Conditions	Product(s) and Yield(s) (%)	Refs.
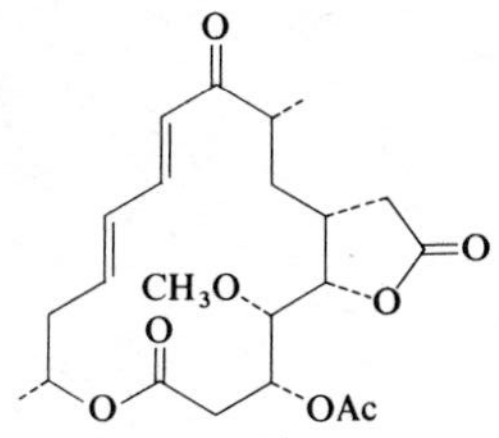 [structure: O, CH_3O, O, OAc]	LTBA, THF, 25°	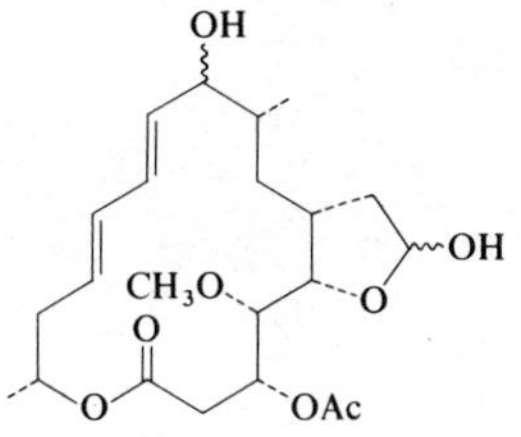 [structure: OH, CH_3O, O, OH, OAc] (50)	614
C_{24}			
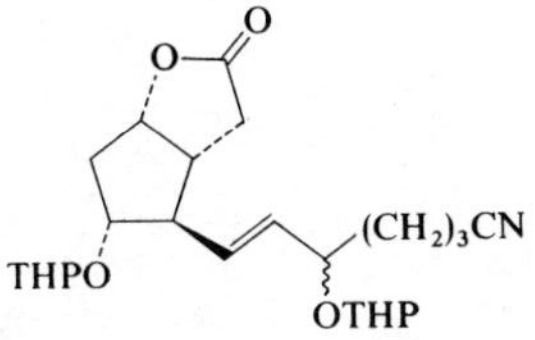 [structure: O, THPO, $(CH_2)_3CN$, OTHP]	SMEAH (IA), toluene, −70°, 6 h	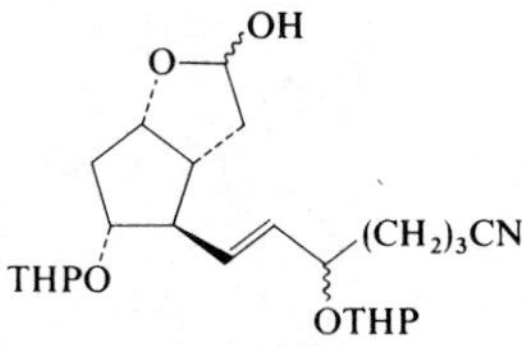 [structure: OH, THPO, $(CH_2)_3CN$, OTHP] (~100)	230
C_{25}			
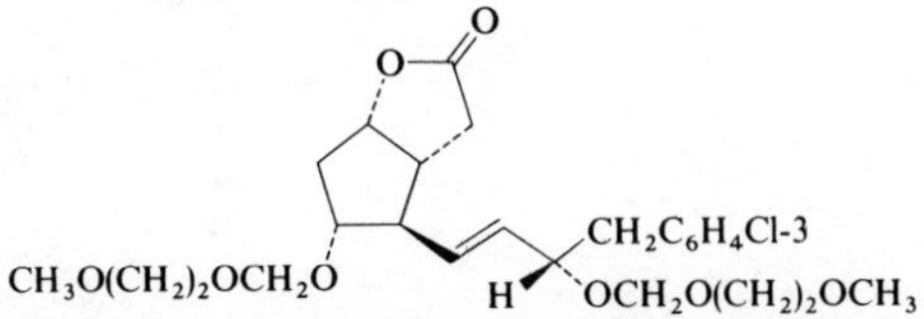 [structure: O, $CH_3O(CH_2)_2OCH_2O$, $CH_2C_6H_4Cl$-3, H, $OCH_2O(CH_2)_2OCH_3$]	SMEAH, toluene	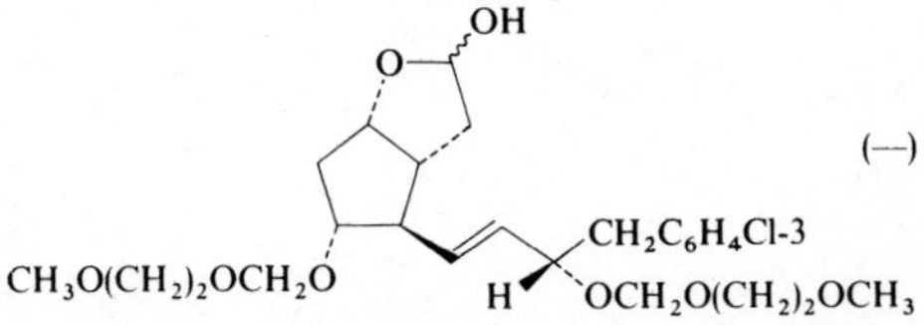 [structure: OH, $CH_3O(CH_2)_2OCH_2O$, $CH_2C_6H_4Cl$-3, H, $OCH_2O(CH_2)_2OCH_3$] (—)	615

Substrate	Conditions	Product	(Yield, %)	Ref.
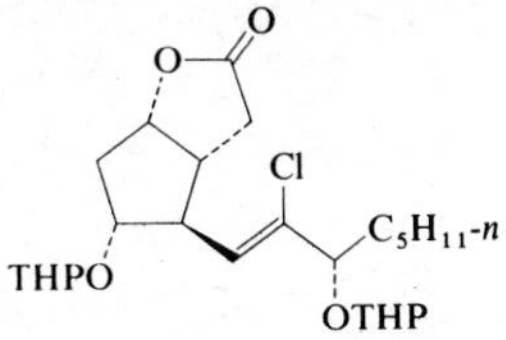 THPO, Cl, C_5H_{11}-*n*, OTHP	SMEAH, toluene, −60 to −70°, 4 h	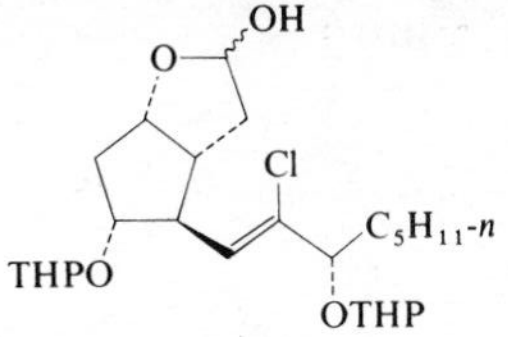 OH, THPO, Cl, C_5H_{11}-*n*, OTHP	(—)	616
C_{27} 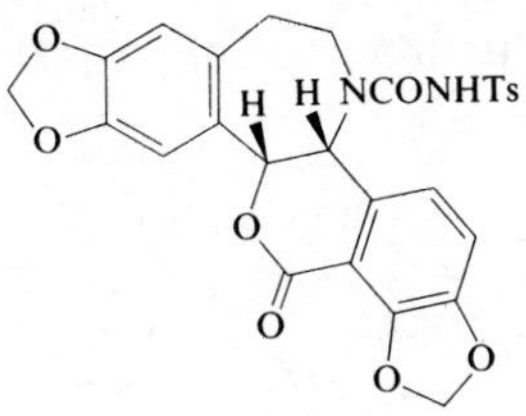 H H NCONHTs, (6*R*,15*R*)-(+)	1. SMEAH, C_6H_6–THF, 0°, 30 min 2. $HC(OCH_3)_3$, H_3O^+	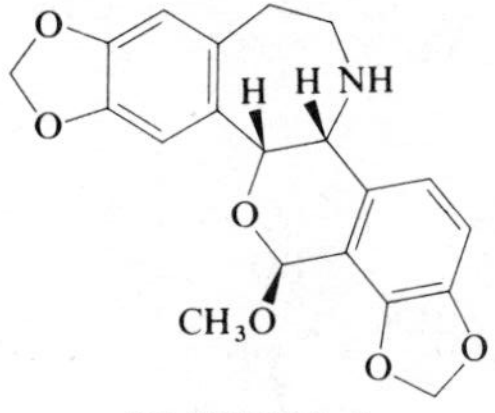 H H NH, CH_3O, (6*R*,8*S*,15*R*)-(+)	(38)	617
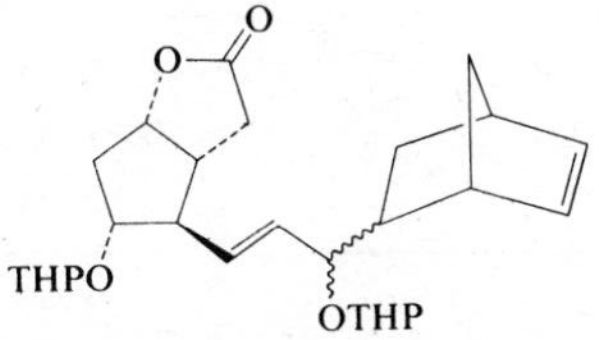 THPO, OTHP	SMEAH (IA), toluene, −70°, 6 h	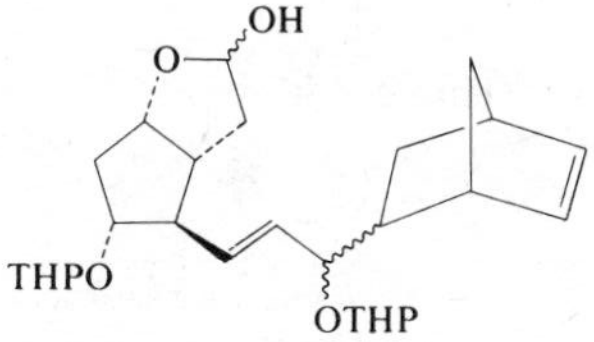 OH, THPO, OTHP	(∼100)	230

TABLE VI. PARTIAL REDUCTION OF LACTONES TO LACTOLS (HEMIACETALS) (*Continued*)

Reactant	Conditions	Product(s) and Yield(s) (%)	Refs.
C_{28}			
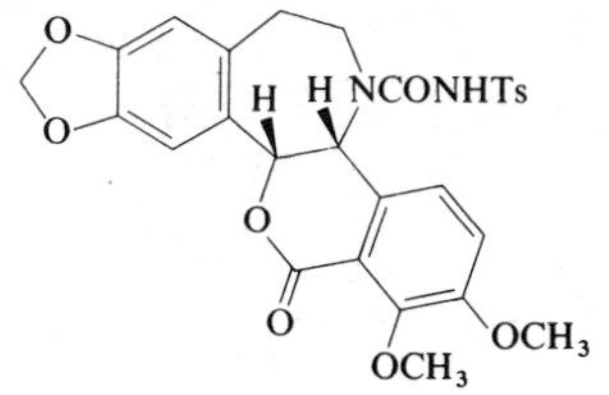 H H NCONHTs; O; O; OCH_3; OCH_3	SMEAH, THF–C_6H_6, 0°	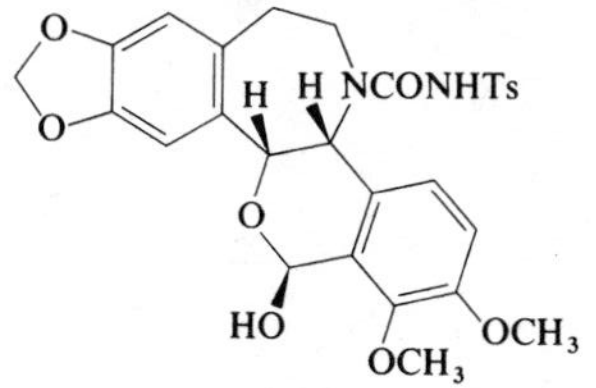 H H NCONHTs; HO; OCH_3; OCH_3 (83)	618
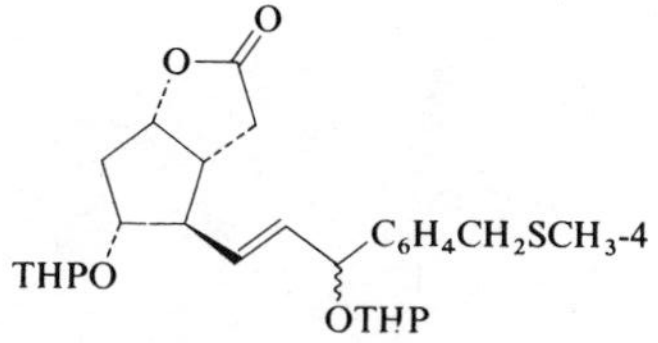 THPO; $C_6H_4CH_2SCH_3$-4; OTHP	SMEAH (IA), toluene, −70°, 6 h	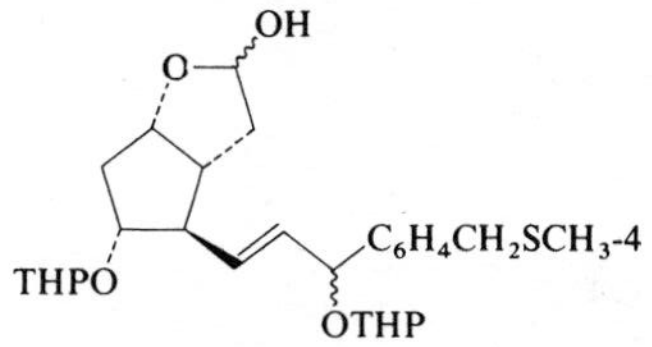 OH; THPO; $C_6H_4CH_2SCH_3$-4; OTHP (~100)	230
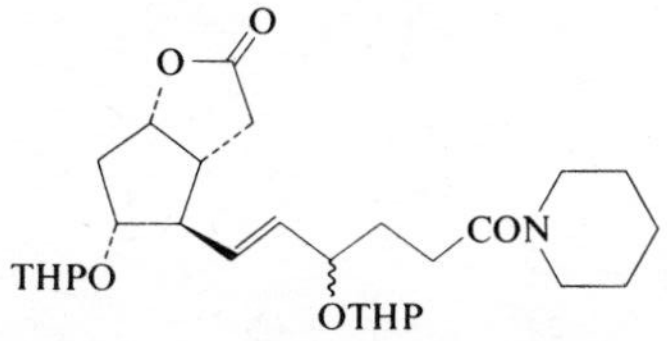 THPO; CON; OTHP	SMEAH (IA), toluene, −70°, 6 h	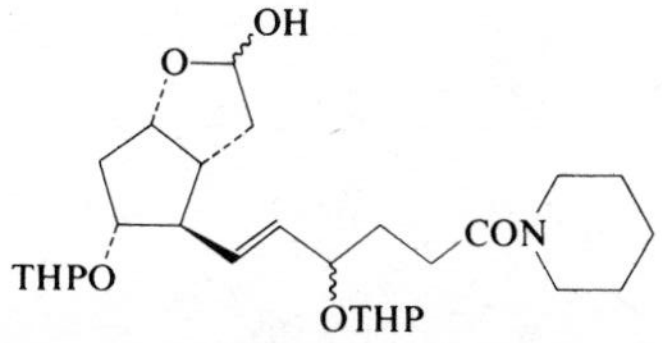 OH; THPO; CON; OTHP (~100)	230

C_{30} 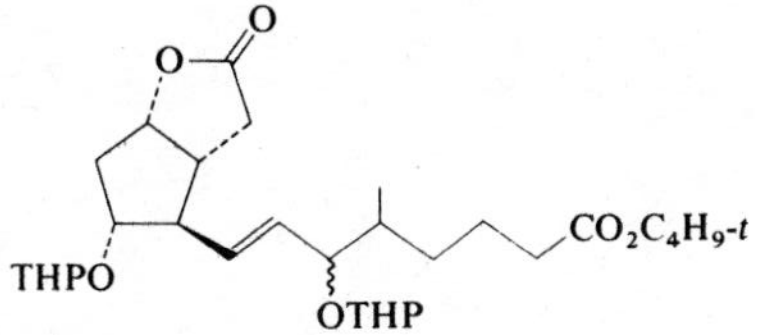	THPO, O, O, $CO_2C_4H_9\text{-}t$, OTHP	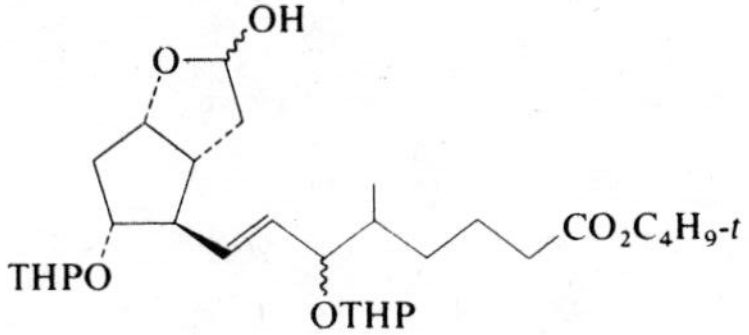 ", ", ", "	OH, O, THPO, $CO_2C_4H_9\text{-}t$, OTHP (~100)	230
C_{32} 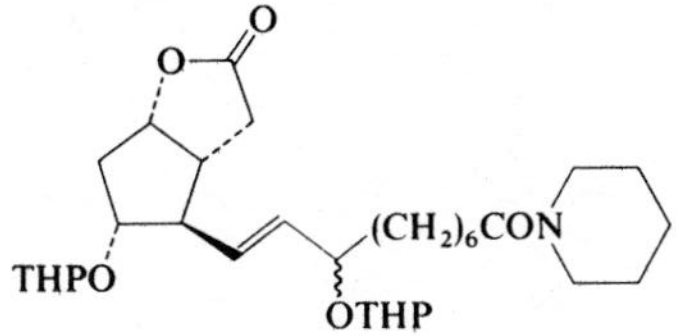	THPO, O, O, $(CH_2)_6CON$, OTHP	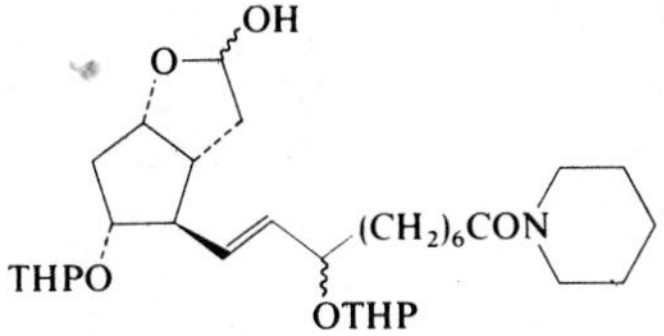 ", ", ", "	OH, O, THPO, $(CH_2)_6CON$, OTHP (~100)	230
C_{34} 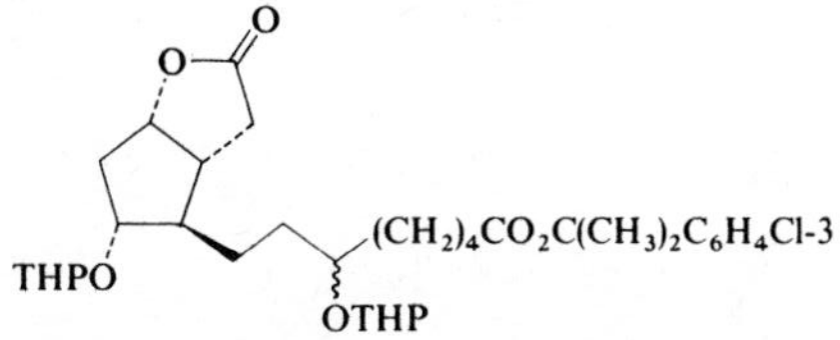	THPO, O, O, $(CH_2)_4CO_2C(CH_3)_2C_6H_4Cl\text{-}3$, OTHP	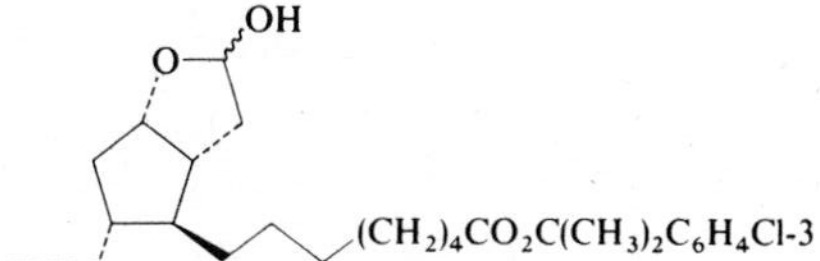 ", ", ", "	OH, O, THPO, $(CH_2)_4CO_2C(CH_3)_2C_6H_4Cl\text{-}3$, OTHP (~100)	230

TABLE VII. Reduction of Unsaturated Lactones

No. of Carbon Atoms	Reactant	Conditions	Product(s) and Yield(s) (%)	Refs.
C_9	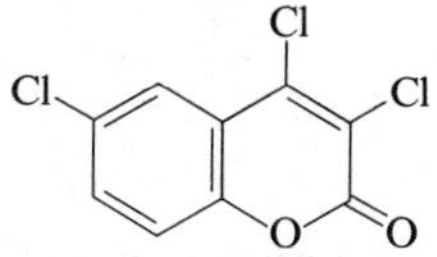 Cl, Cl, Cl	LTBA, THF, 0°, 6 h	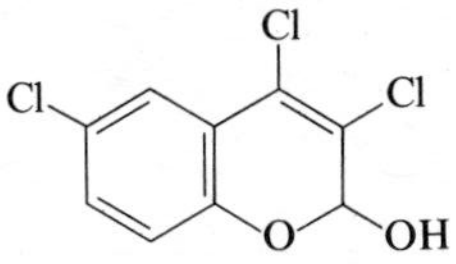 Cl, Cl, Cl, OH (62)	235,619
	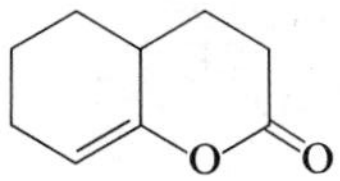 O, O	LTBA (IA), THF, −70°, then room temp, 15 h	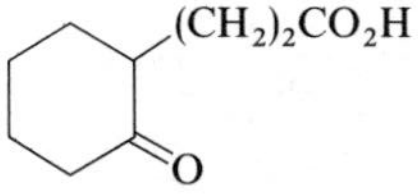 $(CH_2)_2CO_2H$, O (47) 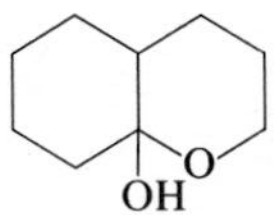 + OH, O (21) 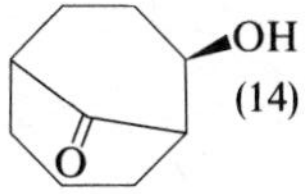 + OH (14) 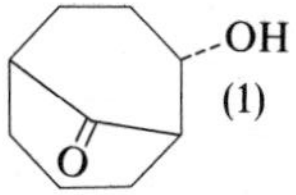 + OH (1)	137
	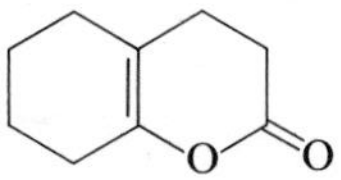 O, O	", ", ", ", ", "	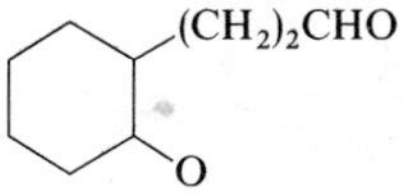 $(CH_2)_2CHO$, O (67)	137
	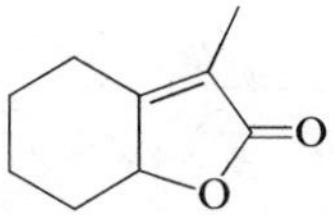 O, O	LAH + 2-propanol (1:1), ether, −60 to −50°, 3 h	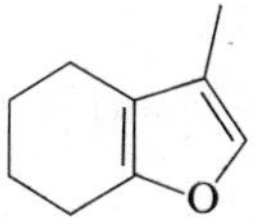 O (57)	232
C_{10}	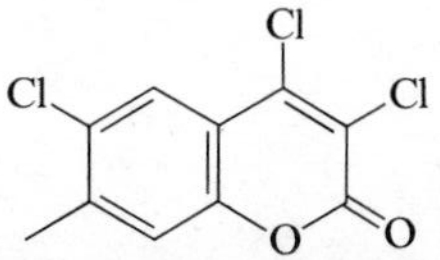 Cl, Cl, Cl	LTBA, THF, 0°, 6 h	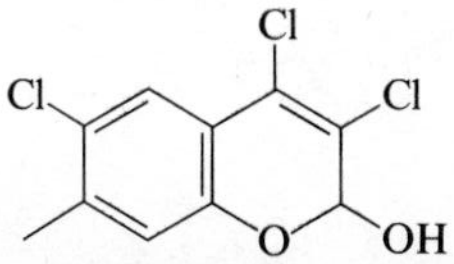 Cl, Cl, Cl, OH (75)	235

	Reactant	Conditions	Product(s) and Yield(s) (%)	Refs.
	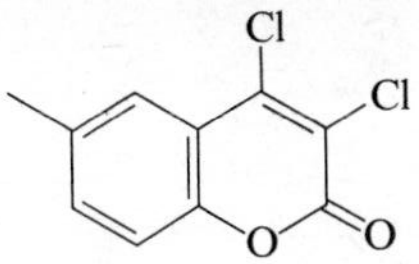	LTBA, THF. 0°, 6 h	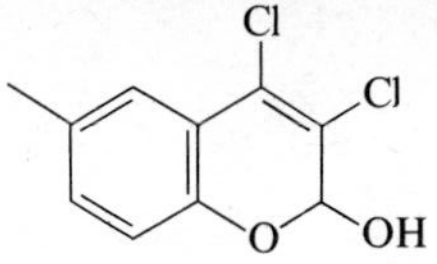 (70)	
C_{12}	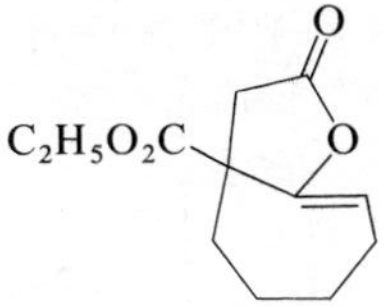 $C_2H_5O_2C$	LTBA (IA). THF, −80°, then room temp, 3 h	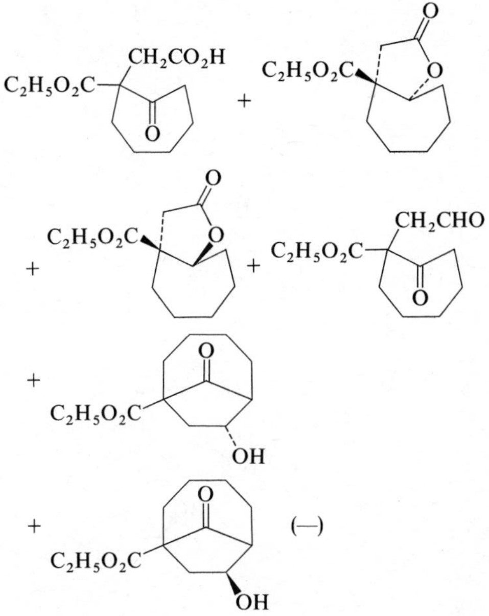 $C_2H_5O_2C$, CH_2CO_2H + $C_2H_5O_2C$ + $C_2H_5O_2C$ + $C_2H_5O_2C$, CH_2CHO + $C_2H_5O_2C$, OH + $C_2H_5O_2C$, OH (—)	139, 620
	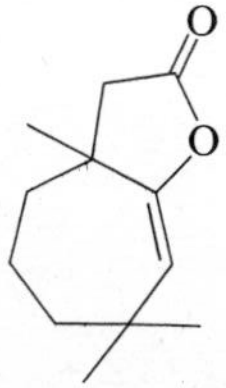	LTBA	No reaction	621

TABLE VII. REDUCTION OF UNSATURATED LACTONES (*Continued*)

No. of Carbon Atoms	Reactant	Conditions	Product(s) and Yield(s) (%)	Refs.
C_{14}	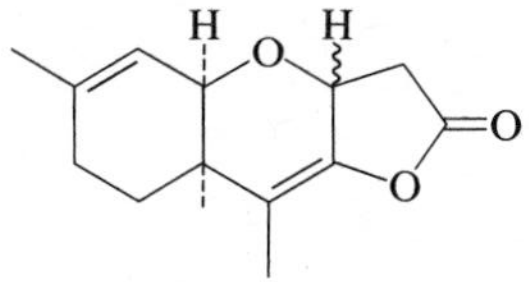	LTBA, ether, −78°, 45 min	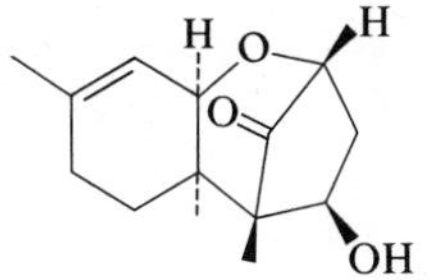 (7) 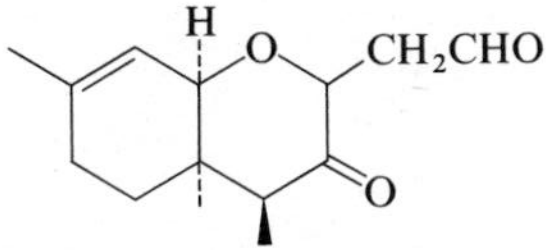 + CH_2CHO (—)	141, 143, 140
C_{15}	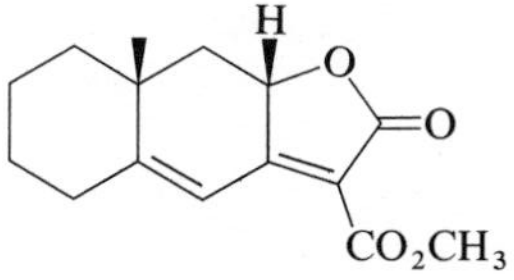 CO_2CH_3	SMEAH, THF, −40 to −20°, 45 min	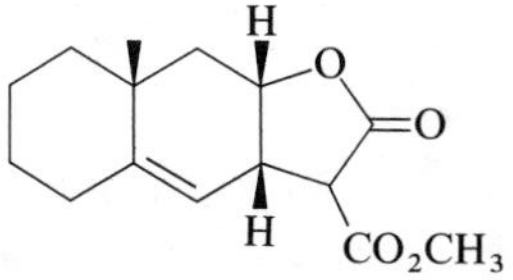 CO_2CH_3 (—)	622
C_{16}	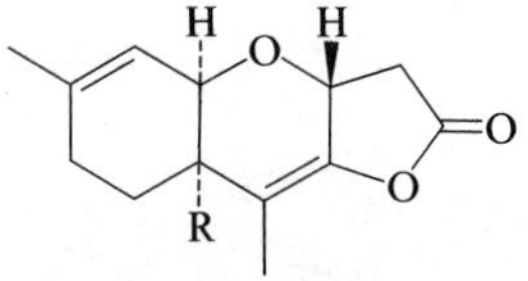 R = $CH_3OCH_2OCH_2$—	LTBA, ether, −78°, then room temp, 45 min	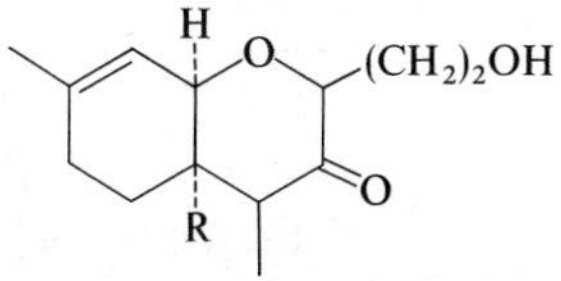 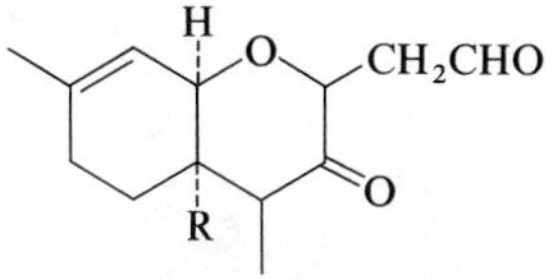 $(CH_2)_2OH$ + CH_2CHO	142

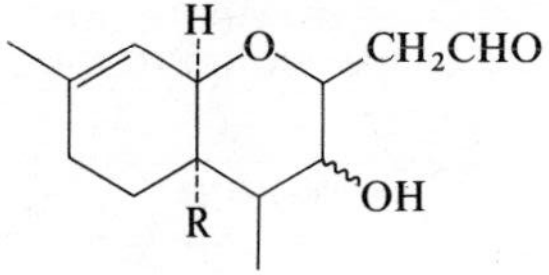

	Reactant	Conditions	Product(s) (Yield, %)	Ref.
C_{20}	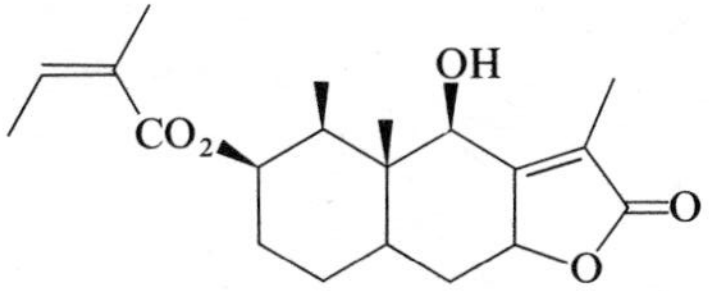	SMEAH, toluene	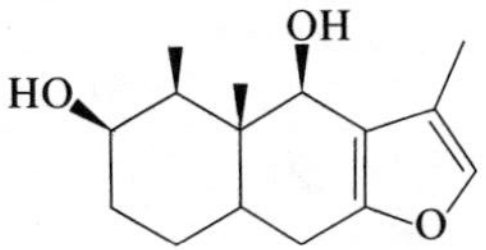 (26) + (—)	623
C_{27}	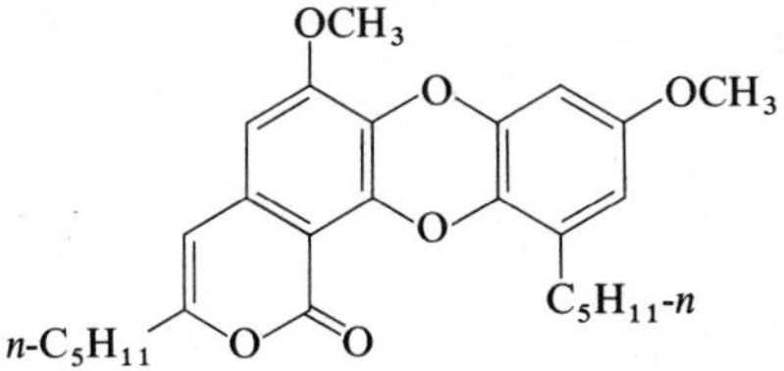	SMEAH, C_6H_6, reflux, 1.5 h	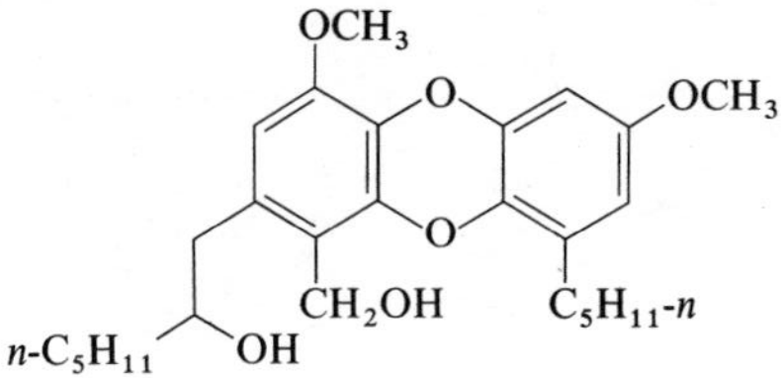 (22)	624
C_{28}	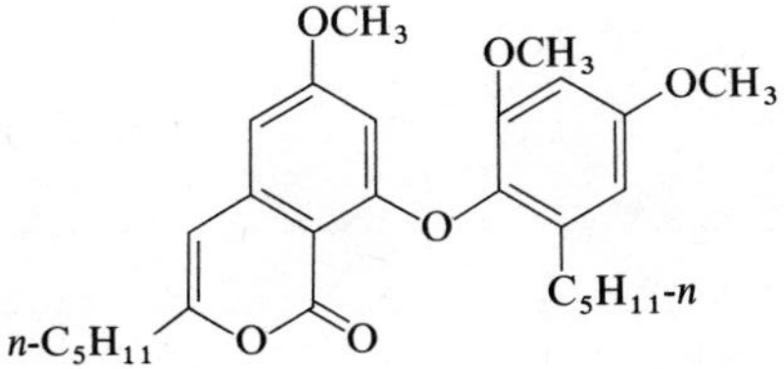	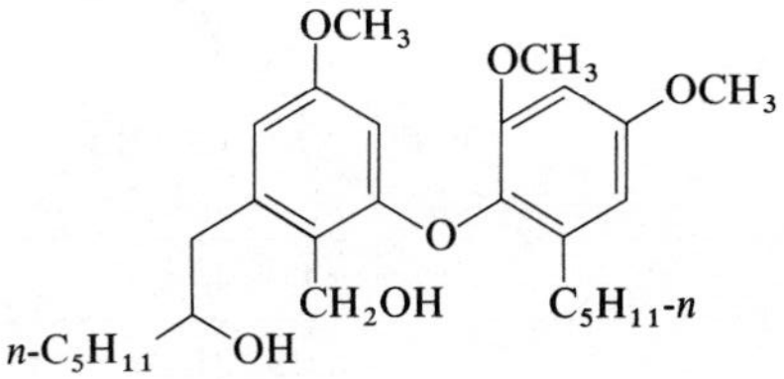 ", ", ", "	(48)	624

TABLE VIII. Reduction of Carboxylic Acid Esters

Reactant	Conditions	Product(s) and Yield(s) (%)	Refs.
C_4			
$CClF_2CF_2CO_2CH_3$	SMEAH, C_6H_6	$CClF_2CF_2CH_2OH$ (—)	625
C_6			
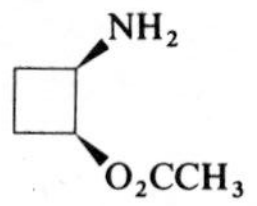 NH_2, O_2CCH_3 (cyclobutane structure)	SMEAH, C_6H_6, reflux, 4 h	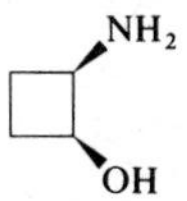 NH_2, OH (cyclobutane structure) (46)	626
n-$C_3H_7CO_2C_2H_5$	SMEAH, C_6H_6, reflux, 2 h	n-C_4H_9OH (89)	238, 68
	$Ca[AlH_2(OC_6H_{11})_2]_2 \cdot THF$, C_6H_6, 20°, 1 h	" (~99)	165
	$NaAlH(OC_2H_5)_3$, THF, 30°, 4 h	" (80)	319
C_7			
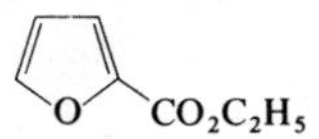 O, $CO_2C_2H_5$ (furan structure)	SMEAH, C_6H_6, 30°	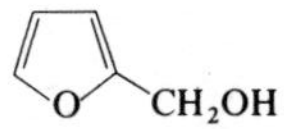 O, CH_2OH (furan structure) (85)	68
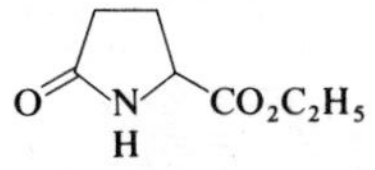 O, N, H, $CO_2C_2H_5$ (pyrrolidinone structure)	SMEAH, C_6H_6	No reaction	627
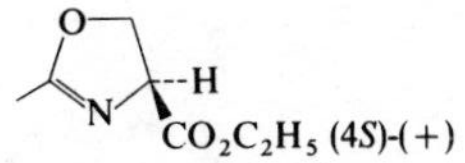 O, N, H, $CO_2C_2H_5$ (4*S*)-(+) (oxazoline structure)	SMEAH, C_6H_6–ether, 0°	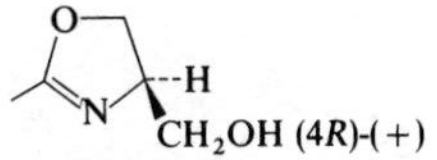 O, N, H, CH_2OH (4*R*)-(+) (oxazoline structure) (73)	628
C_8			
2-$HOC_6H_4CO_2CH_3$	SMEAH, C_6H_6, 80°, 1 h	2-$HOC_6H_4CH_2OH$ (87)	119, 121
	SMEAH, xylene, reflux, 2 h	2-$HOC_6H_4CH_3$ (96)	119, 121

Substrate	Reagent and Conditions	Product(s) and Yield(s) (%)	Refs.
3,5-$(HO)_2C_6H_3CO_2CH_3$	1. $(CH_3)_3SiNHSi(CH_3)_3$, $(CH_3)_3SiCl$ 2. SMEAH 3. H_3O^+	3,5-$(HO)_2C_6H_3CH_2OH$ (—)	629
$CO_2C_2H_5$ (pyridine ring, N)	SMEAH (IA), C_6H_6, 80°, 15 min	CH_2OH (pyridine ring, N) (82)	167, 168
	$Ca[AlH_2(OC_4H_9\text{-}i)_2]_2 \cdot$THF, toluene, 80°, 30 min	" (73)	165
Br, Br; $CO_2C_2H_5$ (cyclopropane)	SMEAH, C_6H_6	Br; $CO_2C_2H_5$ (cyclopropane) + Br; CH_2OH (cyclopropane) (60)	630
CO_2CH_3, CO_2CH_3 (cyclobutane)	SMEAH	CH_2OH, CH_2OH (cyclobutane) (—)	631
$CH_3O_2C(CH_2)_4CO_2CH_3$	SMEAH, C_6H_6, 80°, 30 min	$HO(CH_2)_6OH$ (92)	168
n-$C_5H_{11}CO_2C_2H_5$	SMEAH, C_6H_6, 80°	n-$C_6H_{13}OH$ (96)	68
	$Ca[AlH_2(OC_3H_7\text{-}i)_2]_2 \cdot$THF, toluene, 25°, 1 h	" (92)	165
	$Al_2H_3(OCH_2CH_2OCH_3)_3$, C_6H_6, 80°, 2 h	" (98)	313
	$NaAlH_2(OCH_2CH_2OC_3H_7\text{-}i)_2$, C_6H_6, 25°, 2 h	" (~99)	632
	$NaAlH[OCH_2CH_2N(CH_3)_2]_3$, C_6H_6, 80°, 2 h	n-$C_6H_{13}OH$ (95)	633
	$NaAl_2H_4[OCH_2CH_2N(CH_3)_2]_3$, C_6H_6, 80°, 1 h	" (99)	634
C_9			
2,6-F_2-4-$CH_3OC_6H_2CO_2CH_3$	SMEAH, C_6H_6	2,6-F_2-4-$CH_3OC_6H_2CH_2OH$ (—)	635a
CO_2CH_3, CO_2CH_3 (pyridine ring, N)	SMEAH, C_6H_6, reflux, 1 h	CH_2OH, CH_3 (pyridine ring, N) (6)	635b

TABLE VIII. Reduction of Carboxylic Acid Esters (*Continued*)

Reactant	Conditions	Product(s) and Yield(s) (%)	Refs.
$C_6H_5CO_2C_2H_5$	SMEAH, C_6H_6, 80°, 30 min	$C_6H_5CH_2OH$ (91)	168
	$Ca[AlH_2(OC_4H_9\text{-}i)_2]_2 \cdot THF$, toluene, 60°, 15 min	" (~99)	165
	$NaAlH(OC_2H_5)_3$, THF, 65°, 3 h	" (76)	319
	$NaAlH[OCH_2CH_2N(CH_3)_2]_3$, C_6H_6, 25°, 1.5 h	" (23)	633
4-$HOC_6H_4CO_2C_2H_5$	SMEAH, C_6H_6, 80°, 15 min	4-$HOC_6H_4CH_2OH$ (93)	119, 121
	SMEAH, xylene, 141°, 2 h	4-$HOC_6H_4CH_3$ (92)	119, 121
4-$CH_3OC_6H_4CO_2CH_3$	SMEAH, C_6H_6	4-$CH_3OC_6H_4CH_2OH$ (75)	636a
2-HO-4-$CH_3OC_6H_3CO_2CH_3$	SMEAH, ether, reflux, 30 min	2-HO-4-$CH_3OC_6H_3CH_2OH$ (97)	636b
2-$H_2NC_6H_4CO_2C_2H_5$	SMEAH, C_6H_6, 40°, 1 h	2-$H_2NC_6H_4CH_2OH$ (14)	120
	SMEAH, xylene, 141°, 1 h	2-$H_2NC_6H_4CH_3$ (88)	120
4-$H_2NC_6H_4CO_2C_2H_5$	", ", ", "	4-$H_2NC_6H_4CH_3$ (93)	120
3,5-$(H_2N)_2C_6H_3CO_2C_2H_5$	SMEAH, C_6H_6–THF, 40°, 4 h, then room temp, 12 h	3,5-$(H_2N)_2C_6H_3CH_2OH$ (82)	637a
$C_2H_5O_2C$, N, O, OC_2H_5	$Ca[AlH_2(OC_3H_7\text{-}i)_2]_2 \cdot THF$, toluene, room temp, 4 h	$HOCH_2$, N, O, OC_2H_5 (99)	505
$C_6H_{11}CO_2C_2H_5$	$Ca[AlH_2(OCH_2CH(C_2H_5)C_4H_9)_2]_2 \cdot THF$, toluene, 20°, 1 h	$C_6H_{11}CH_2OH$ (~99)	165
C_{10}			
CO_2CH_3, N, H	SMEAH, C_6H_6–toluene, room temp, 12 h	CH_2OH, N, H (85)	638
$C_6H_4(CO_2CH_3)_2$-1,2	SMEAH, toluene, 110°	$C_6H_4(CH_2OH)_2$-1,2 (89)	68
$C_6H_4(CO_2CH_3)_2$-1,4	SMEAH, C_6H_6, 80°, 1 h	$C_6H_4(CH_2OH)_2$-1,4 (87)	168

3-CH_3O-4-Cl-5-$NHCH_3$-$C_6H_2CO_2CH_3$	SMEAH	3-CH_3O-4-Cl-5-$NHCH_3$-$C_6H_2CH_2OH$ (—)	639a
$(CH_3)_2C{=}C{=}CHC{\equiv}CCH_2O_2CCH_3$	$LiAlH_3(OCH_3)$, THF, 70–80°, 1–3 h	$(CH_3)_2C{=}C{=}CHCH{=}CHCH_2OH$ (90)	507
2-CH_3O-3-$CH_3C_6H_3CO_2CH_3$	SMEAH, C_6H_6, reflux, 2 h	2-CH_3O-3-$CH_3C_6H_3CH_2OH$ (98)	639b
3-cyclopentenyl-$(CH_2)_2CO_2C_2H_5$	LAH + EtOH	3-cyclopentenyl-$(CH_2)_3OH$ (—)	640
$HC(CO_2C_2H_5)_3$	$LiAlH_2(OCH_2CH_2OCH_3)_2$	$HC(CH_2OH)_3$ (7)	641
$Cl(CH_2)_3CH{=}C(CH_3)CH_2CH_2CO_2CH_3$	SMEAH, C_6H_6–THF	$Cl(CH_2)_3CH{=}C(CH_3)CH_2CH_2CH_2OH$ (98)	642
n-$C_7H_{15}CO_2C_2H_5$	SMEAH, C_6H_6, 80°	n-$C_8H_{17}OH$ (96)	68
C_{11}			
3-CH_3O-4-Cl-5-$NHCH_3$-$C_6H_2CO_2C_2H_5$	1. SMEAH, C_6H_6 2. HCl	3-CH_3O-4-Cl-5-$\overset{+}{N}H_2CH_3$-$C_6H_2CH_2OH$ Cl^- (51)	643a
4-H_2N-2-CO_2CH_3-$C_6H_3CH_2CO_2CH_3$	SMEAH, THF, C_6H_6, reflux, 7.5 h	4-H_2N-2-CH_2OH-$C_6H_3(CH_2)_2OH$ (93)	643b
3,4,5-$(CH_3O)_3C_6H_2CO_2CH_3$	SMEAH, C_6H_6, 0°, then room temp, 1 h	3,4,5-$(CH_3O)_3C_6H_2CH_2OH$ (~100)	644

TABLE VIII. REDUCTION OF CARBOXYLIC ACID ESTERS (*Continued*)

Reactant	Conditions	Product(s) and Yield(s) (%)	Refs.
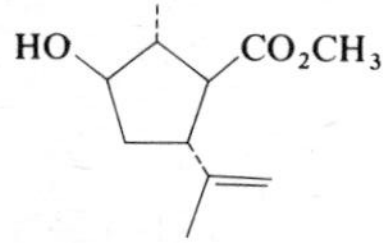 [structure: HO, CO_2CH_3 cyclopentane with isopropenyl]	SMEAH	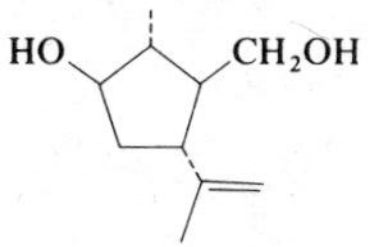 [structure: HO, CH_2OH cyclopentane with isopropenyl] (—)	645
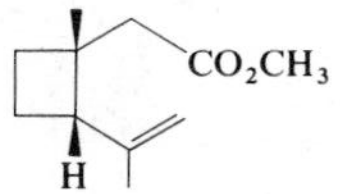 [structure: cyclobutane with CO_2CH_3, H, isopropenyl]	SMEAH, C_6H_6–ether	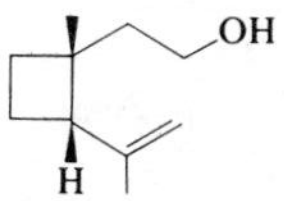 [structure: cyclobutane with OH, H, isopropenyl] (—)	525
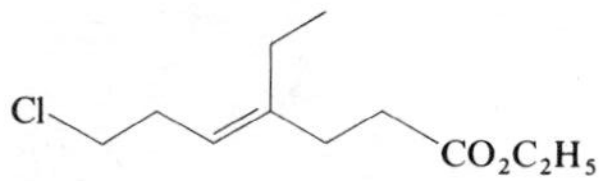 Cl-[chain]=C(ethyl)-[chain]-$CO_2C_2H_5$	SMEAH, THF–ether, −15 to −10°, 5 h	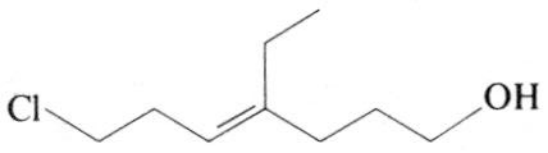 Cl-[chain]=C(ethyl)-[chain]-OH (87)	277g, 278
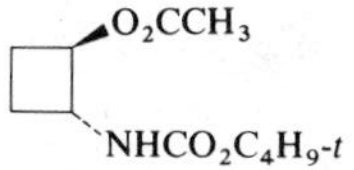 [structure: cyclobutane with O_2CCH_3, $NHCO_2C_4H_9$-*t*]	SMEAH, C_6H_6, reflux, 24 h	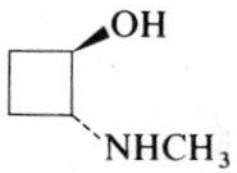 [structure: cyclobutane with OH, $NHCH_3$] (67)	646
C_{12}			
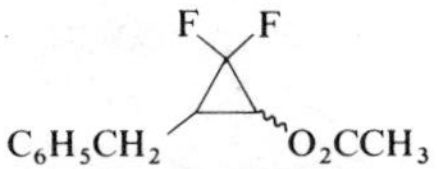 [structure: F, F cyclopropane with $C_6H_5CH_2$, O_2CCH_3] (*cis*: *trans* = 3:97)	SMEAH, C_6H_6–toluene	$C_6H_5CH_2CH{=}CFCH_2OH$ (99) (*cis*:*trans* = 70:30)	647
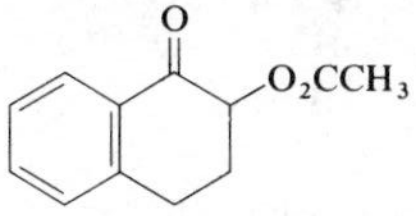 [structure: 1-tetralone with 2-O_2CCH_3]	LAH + (+)-DMDB (1:2), ether, 0°, 16 h	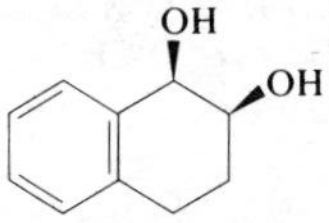 [structure: tetralin-1,2-diol, OH, OH] (1*R*,2*S*)-(−) (30)(20% opt. y.) 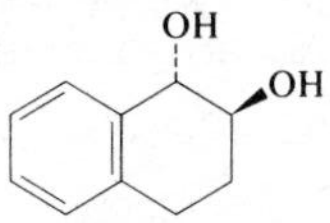 + [structure: tetralin-1,2-diol, OH, OH] (1*R*,2*R*)-(+) (5)(62% opt. y.)	264, 266

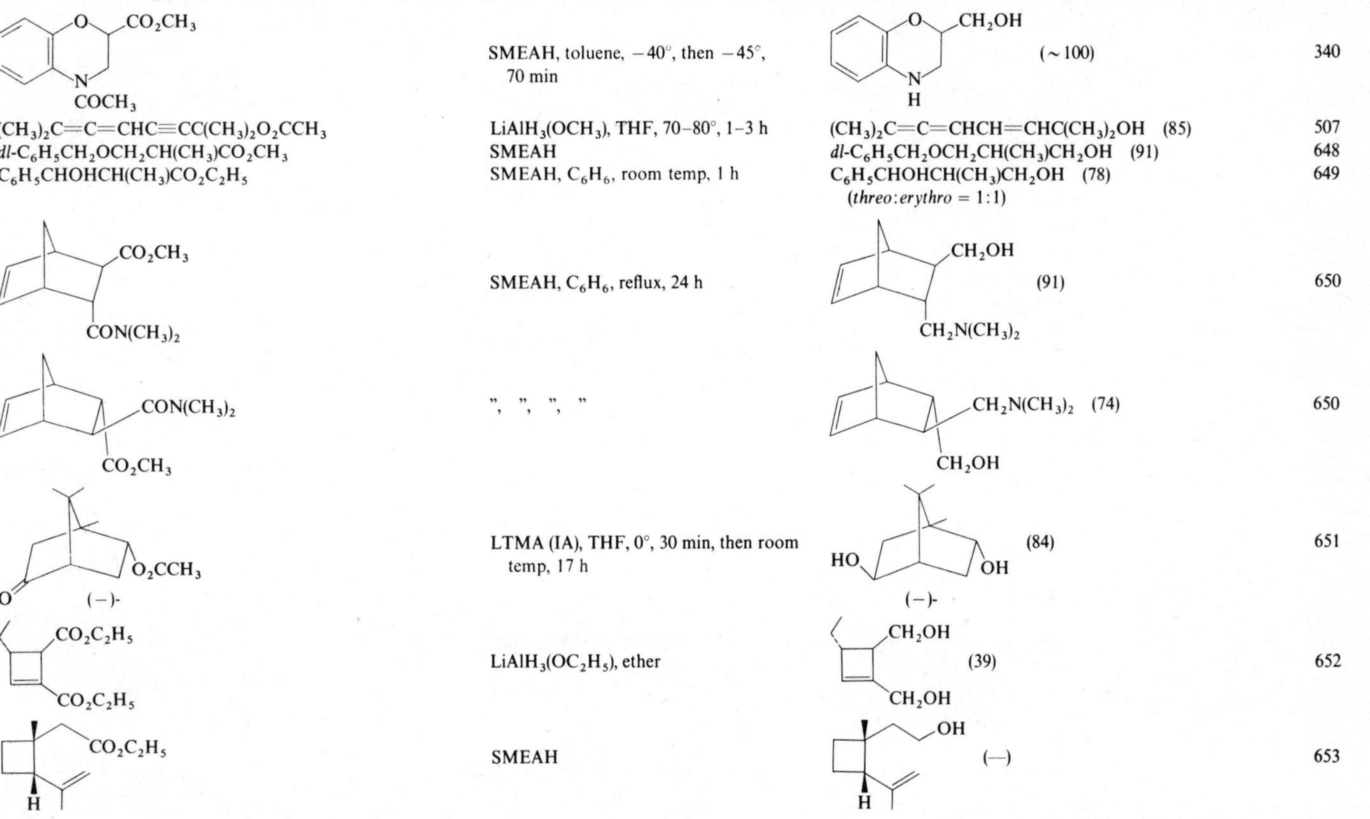

Reactant	Conditions	Product(s) and Yield(s) (%)	Refs.
	SMEAH, toluene, −40°, then −45°, 70 min	(~100)	340
$(CH_3)_2C{=}C{=}CHC{\equiv}CC(CH_3)_2O_2CCH_3$	$LiAlH_3(OCH_3)$, THF, 70–80°, 1–3 h	$(CH_3)_2C{=}C{=}CHCH{=}CHC(CH_3)_2OH$ (85)	507
dl-$C_6H_5CH_2OCH_2CH(CH_3)CO_2CH_3$	SMEAH	*dl*-$C_6H_5CH_2OCH_2CH(CH_3)CH_2OH$ (91)	648
$C_6H_5CHOHCH(CH_3)CO_2C_2H_5$	SMEAH, C_6H_6, room temp, 1 h	$C_6H_5CHOHCH(CH_3)CH_2OH$ (78) (*threo*:*erythro* = 1:1)	649
	SMEAH, C_6H_6, reflux, 24 h	(91)	650
	", ", ", "	(74)	650
(−)-	LTMA (IA), THF, 0°, 30 min, then room temp, 17 h	(−)- (84)	651
	$LiAlH_3(OC_2H_5)$, ether	(39)	652
	SMEAH	(—)	653

TABLE VIII. REDUCTION OF CARBOXYLIC ACID ESTERS (*Continued*)

Reactant	Conditions	Product(s) and Yield(s) (%)	Refs.
$-CO_2C_2H_5$	SMEAH, C_6H_6–ether, reflux, 2 h, then room temp, 12 h	$\sim CH_2OH$ (48) (*syn*:*anti* = 66:34)	654
$(CH_2)_2CO_2CH_3$, CH_3O	SMEAH, C_6H_6	$(CH_2)_3OH$, CH_3O (99)	655
$CO_2C_2H_5$, OH, HO, $CO_2C_2H_5$	1. LTBA, THF, reflux, 22 h 2. NaF, MeOH, HCl	OH, OH, HO, HO (50)	270
n-$C_9H_{19}CO_2C_2H_5$	SMEAH, C_6H_6, 80°	n-$C_{10}H_{21}OH$ (89)	68
$CO_2C_2H_5$, OH, OH	SMEAH, C_6H_6, 0°, 1 h	OH, OH, OH (—)	656a
C_{13}			
N, CHO, CO_2CH_3	SMEAH, toluene, room temp, 1.5 h	N, CH_2OH, CH_2OH (47)	656b

Substrate	Conditions	Product (Yield, %)	Refs.
	SMEAH (IA), toluene, room temp. 3 h	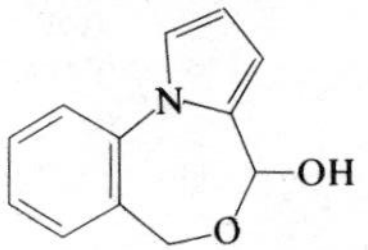 OH (26)	656b
$(CH_3)_2C{=}C{=}CHC{\equiv}CCH{=}CHCH(CH_3)O_2CCH_3$	$LiAlH_3(OCH_3)$, THF, reflux, 2 h	$(CH_3)_2C{=}C{=}CHCH{=}C{=}CHCH_2CH(CH_3)OH$ (70)	657, 658
$(CH_3)_2C{=}C{=}CHC{\equiv}CC(CH_3){=}CHCH_2O_2CCH_3$	", ", ", "	$(CH_3)_2C{=}C{=}CHCH{=}C{=}C(CH_3)(CH_2)_2OH$ (80)	657, 658
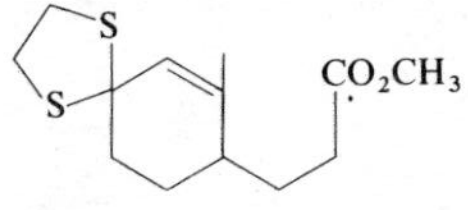 S, S, CO_2CH_3	SMEAH, THF, 0°, 4 h	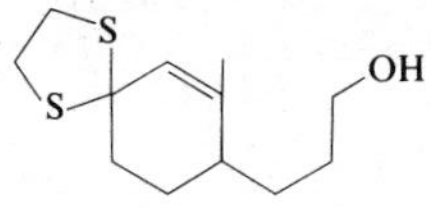 S, S, OH (~100)	533, 534
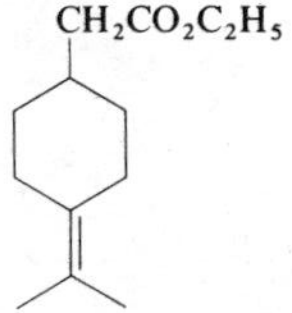 $CH_2CO_2C_2H_5$	SMEAH, C_6H_6, 20°, 10 h	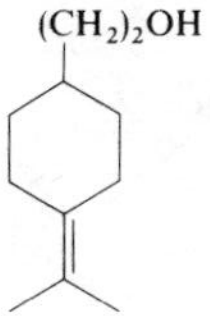 $(CH_2)_2OH$ (~100)	659
C_{14}			
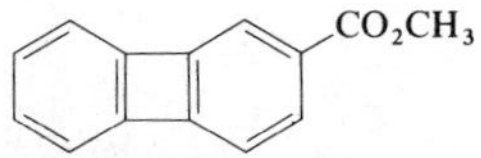 CO_2CH_3	SMEAH, C_6H_6, 40°	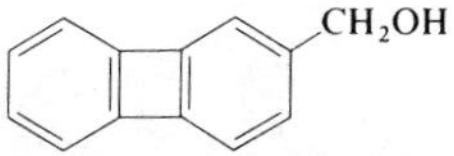 CH_2OH (~100)	660
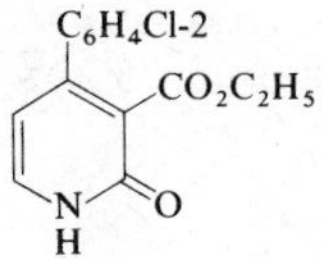 C_6H_4Cl-2, $CO_2C_2H_5$, N, H, O	SMEAH (IA), THF, reflux	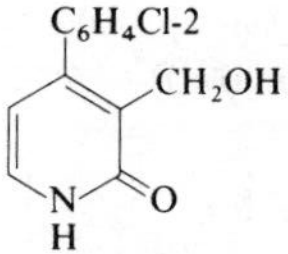 C_6H_4Cl-2, CH_2OH, N, H, O (94)	661
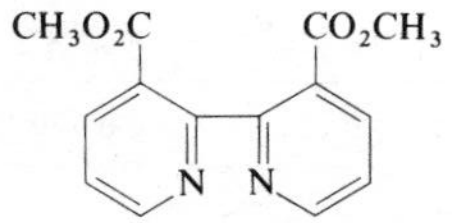 CH_3O_2C, CO_2CH_3, N, N	SMEAH, C_6H_6, 0°, 1.5 h	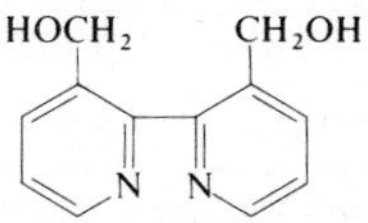 $HOCH_2$, CH_2OH, N, N (84)	662

TABLE VIII. Reduction of Carboxylic Acid Esters (*Continued*)

Reactant	Conditions	Product(s) and Yield(s) (%)	Refs.
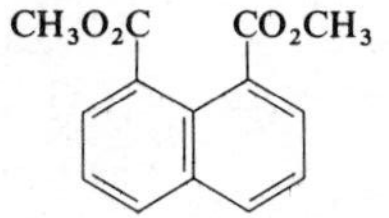 CH_3O_2C CO_2CH_3 (1,8-naphthalene diester)	SMEAH, C_6H_6, room temp, 40 min, then reflux, 1.5 h	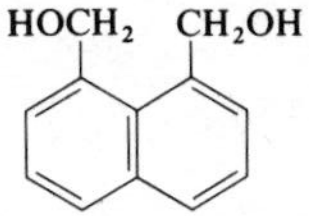 $HOCH_2$ CH_2OH (1,8-naphthalene diol) (69)	663a
2,6-$F_2C_6H_3COCH(CH_2CH{=}CH_2)CO_2C_2H_5$	1. LAH, THF, $-70°$, 30 min 2. SMEAH, toluene	2,6-$F_2C_6H_3CHOHCH(CH_2CH{=}CH_2)CH_2OH$ (90) *threo*- (~ 100)	663b
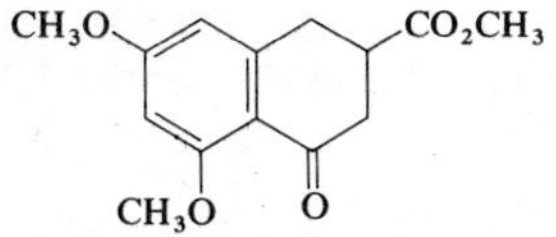 CH_3O, CO_2CH_3, CH_3O, O (tetralone ester)	1. 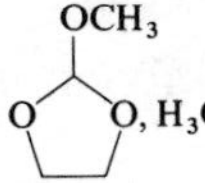 (OCH_3 dioxolane: O, O), H_3O^+ 2. SMEAH, C_6H_6, 30°, 1 h	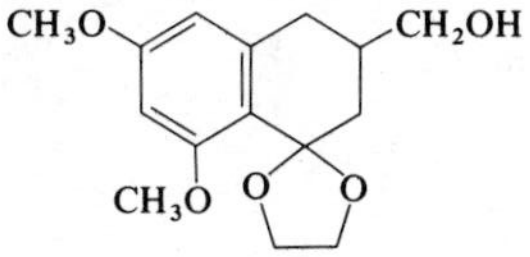 CH_3O, CH_2OH, CH_3O, O, O (ketal)	664
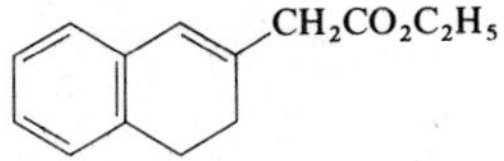 $CH_2CO_2C_2H_5$ (dihydronaphthalene)	$LiAlH_3(OC_2H_5)$(IA), ether, 2.5 h	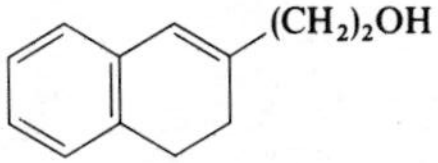 $(CH_2)_2OH$ (dihydronaphthalene) (69)	665
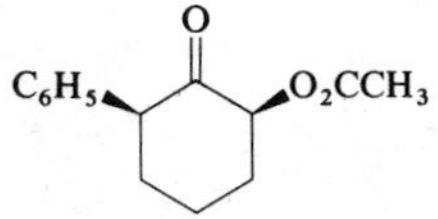 C_6H_5, O, O_2CCH_3 (cyclohexanone)	LAH + (−)-DMDB, ether, 0°, 16 h	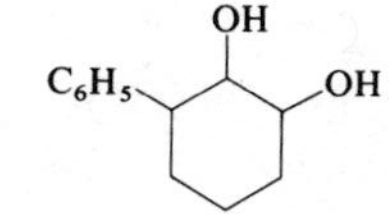 C_6H_5, OH, OH (cyclohexanediol) **I** (1*S*,2*R*,3*S*)-(+) (38; 46% opt.y.) + (1*R*,2*R*,3*R*)-(+) (21; 72% opt.y.)	265, 264, 266
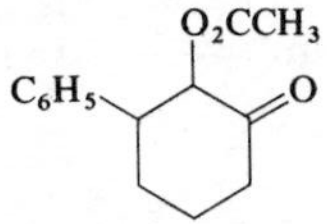 O_2CCH_3, C_6H_5, O (cyclohexanone)	LAH + (−)-DMDB, ether, 0°, 16 h	**I** (28; 5% opt.y.) (1*S*,2*S*,3*S*)-(−)	265, 264
$(CH_3)_2C{=}C{=}CHC{\equiv}CC(CH_3){=}C(CH_3)CH_2O_2CCH_3$	$LiAlH_3(OCH_3)$, THF, reflux, 2 h	$(CH_3)_2C{=}C{=}CHCH{=}C{=}C(CH_3)CH(CH_3)CH_2OH$ (65)	657, 658

Reactant	Conditions	Product (Yield, %)	Refs.
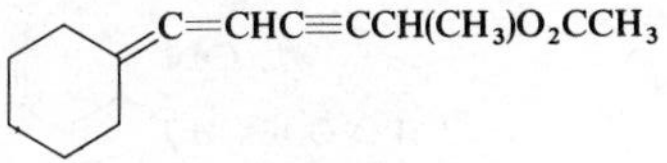 $C{=}CHC{\equiv}CCH(CH_3)O_2CCH_3$ (cyclohexylidene)	$LiAlH_3(OCH_3)$, THF, 70–80°, 1–3 h	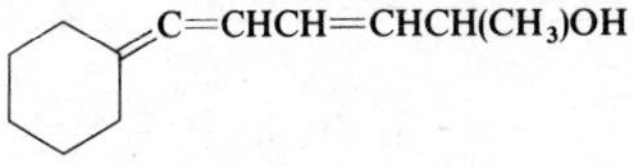 $C{=}CHCH{=}CHCH(CH_3)OH$ (cyclohexylidene) (85)	507
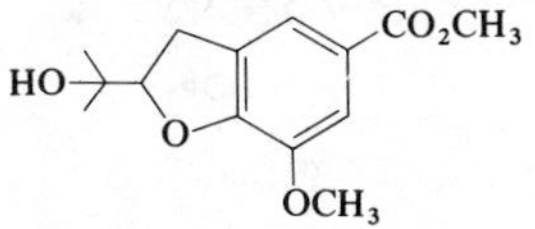 HO, O, CO_2CH_3, OCH_3	SMEAH	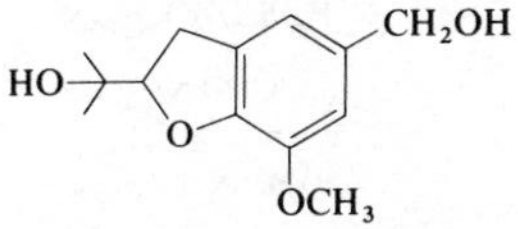 HO, O, CH_2OH, OCH_3 (—)	666
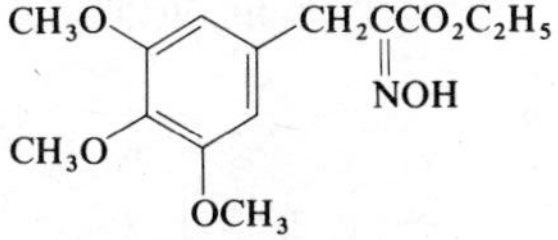 CH_3O, CH_3O, OCH_3, $CH_2CCO_2C_2H_5$ (=NOH)	1. SMEAH, C_6H_6, 45–50°, 40 min 2. HCl	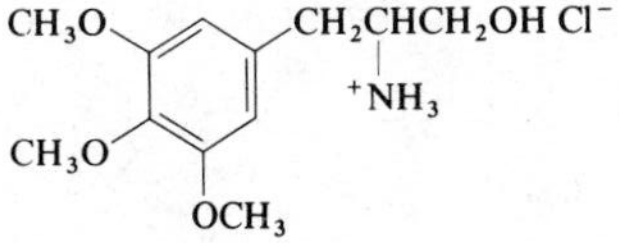 CH_3O, CH_3O, OCH_3, $CH_2CHCH_2OH\ Cl^-$ ($^+NH_3$) (44)	667a
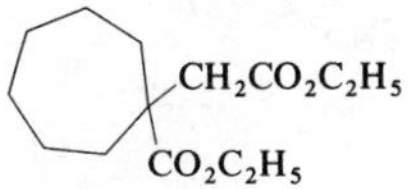 $CH_2CO_2C_2H_5$, $CO_2C_2H_5$ (cycloheptane)	SMEAH (IA), toluene, reflux, 2 h	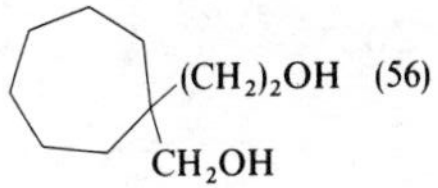 $(CH_2)_2OH$, CH_2OH (cycloheptane) (56)	667b
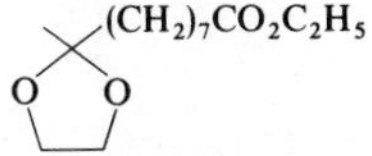 $(CH_2)_7CO_2C_2H_5$ (O, O; dioxolane)	1. SMEAH, ether 2. H_3O^+	$CH_3CO(CH_2)_8OH$ (~100)	668
n-$C_{11}H_{23}CO_2C_2H_5$	SMEAH, C_6H_6, 30–90°, 30 min	n-$C_{12}H_{25}OH$ (98)	167, 168
C_{15}			
2-CH_3O-3-$C_6H_5C_6H_3CO_2CH_3$	SMEAH, THF, reflux, 15 h	2-CH_3O-3-$C_6H_5C_6H_3CH_2OH$ (87)	669
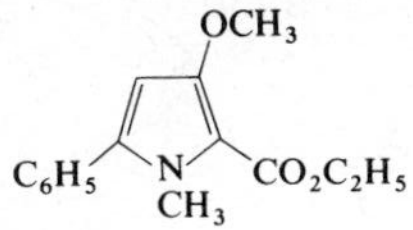 OCH_3, C_6H_5, N, CH_3, $CO_2C_2H_5$ (pyrrole)	LTBA	No reaction	670
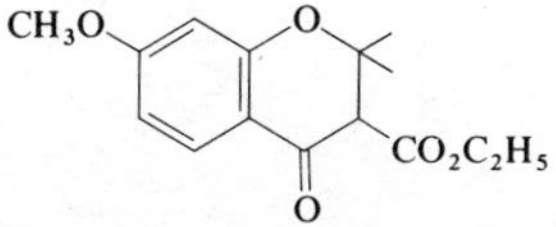 CH_3O, O, $CO_2C_2H_5$, O (chromanone)	LTBA, THF, 48 h	No reaction	671

TABLE VIII. REDUCTION OF CARBOXYLIC ACID ESTERS (*Continued*)

Reactant	Conditions	Product(s) and Yield(s) (%)	Refs.
$(CH_3)_2C{=}C{=}CHC{\equiv}CC(CH_3){=}C(CH_3)CH(CH_3)O_2CCH_3$	$LiAlH_3(OCH_3)$, THF, reflux, 2 h	$(CH_3)_2C{=}C{=}CHCH{=}C{=}C(CH_3)CH(CH_3)CH(CH_3)OH$ (60)	657, 658
1-O_2CCH_3-1-[$C{\equiv}CCH{=}C{=}C(CH_3)_2$]cyclohexane	$LiAlH_3(OCH_3)$, THF, 70–80°, 1.3 h	1-OH-1-[$CH{=}CHCH{=}C{=}C(CH_3)_2$]cyclohexane (75)	507
cyclohexylidene$=C{=}CHC{\equiv}CC(CH_3)_2O_2CCH_3$	$LiAlH_3(OCH_3)$, THF, 70–80°, 1.3 h	cyclohexylidene$=C{=}CHCH{=}CHC(CH_3)_2OH$ (80)	507
HO–C(CH_3)_2–dihydrobenzofuran–$(CH_2)_2CO_2CH_3$	SMEAH, C_6H_6	HO–C(CH_3)_2–dihydrobenzofuran–$(CH_2)_3OH$ (—)	666
4-(piperidin-1-yl)$C_6H_4CH_2CO_2C_2H_5$	SMEAH	4-(piperidin-1-yl)$C_6H_4(CH_2)_2OH$ (38)	672
$4\text{-}CH_3C_6H_4CH(N(CH_3)_2)CH(CO_2CH_3)_2$	SMEAH, C_6H_6, reflux, 2 h	$4\text{-}CH_3C_6H_4CH(N(CH_3)_2)CH(CH_2OH)_2$ (16)	673, 674
2-pyridyl–$CH(N(CH_3)_2)CH(CO_2C_2H_5)_2$	", ", ", "	2-pyridyl–$CH(N(CH_3)_2)CH(CH_2OH)_2$ (94)	674
3-pyridyl–$CH(N(CH_3)_2)CH(CO_2C_2H_5)_2$	", ", ", "	3-pyridyl–$CH(N(CH_3)_2)CH(CH_2OH)_2$ (60)	674

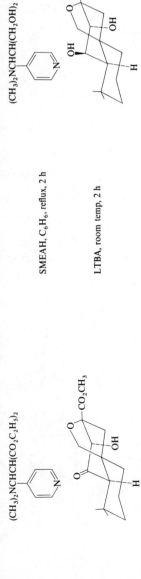

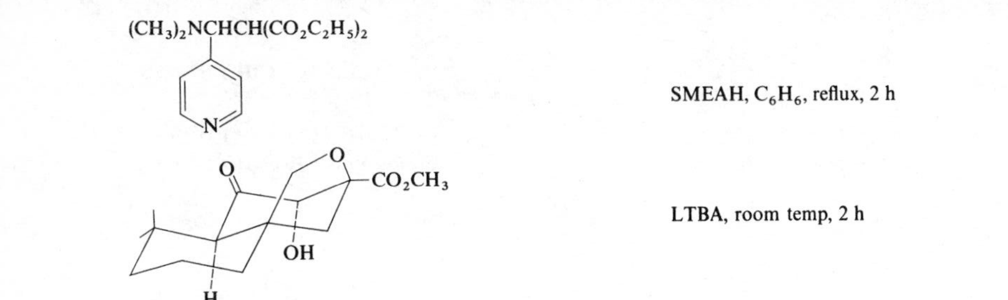

Reactant	Conditions	Product(s) and Yield(s) (%)		Refs.
$(CH_3)_2NCHCH(CO_2C_2H_5)_2$ (4-pyridyl)	SMEAH, C_6H_6, reflux, 2 h	$(CH_3)_2NCHCH(CH_2OH)_2$ (4-pyridyl)	(17)	674
(keto lactone, CO_2CH_3, OH, H)	LTBA, room temp, 2 h	(OH, CO_2CH_3, OH, H) + (OH, CH_2OH, OH, H)	(—)	675
C_{16}				
3-Br-9-CO_2CH_3-phenanthrene	SMEAH	3-Br-9-CH_2OH-phenanthrene	(—)	676
$2-CH_3OC_6H_4COCH(CH_2CH{=}CH_2)CO_2C_2H_5$	1. $Zn(BH_4)_2$, THF, 25°, 30 min 2. SMEAH, toluene	$2-CH_3OC_6H_4CHOHCH(CH_2CH{=}CH_2)CH_2OH$ (*erythro*:*threo* = 94:6)	(90)	663b
$C_6H_{10}{=}C{=}CHC{\equiv}CCH{=}CHCH(CH_3)O_2CCH_3$	$LiAlH_3(OCH_3)$, THF, reflux, 2 h	$C_6H_{10}{=}C{=}CHCH{=}C{=}CHCH_2CH(CH_3)OH$	(55)	657, 658
	$LiAlH_3(OCH_3)$, THF, reflux, 10 h	$C_6H_{10}{=}CHCH{=}CHCH{=}CHCH_2CH(CH_3)OH$	(64)	658

TABLE VIII. Reduction of Carboxylic Acid Esters (*Continued*)

Reactant	Conditions	Product(s) and Yield(s) (%)	Refs.
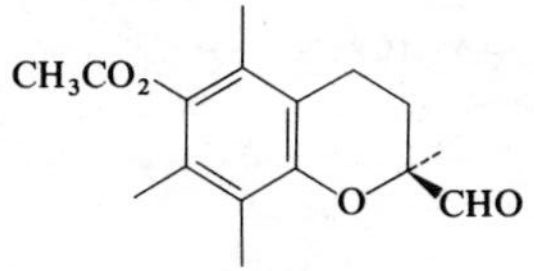 (S)-chroman structure: CH_3CO_2-substituted ring with CHO	SMEAH, THF–C_6H_6, room temp, 1 h	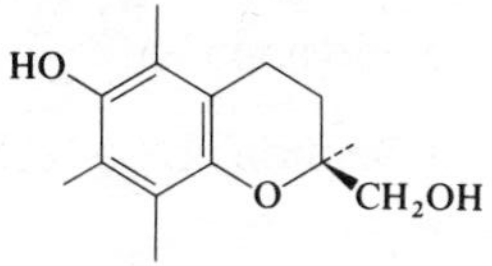 HO-substituted chroman with CH_2OH (92)	240
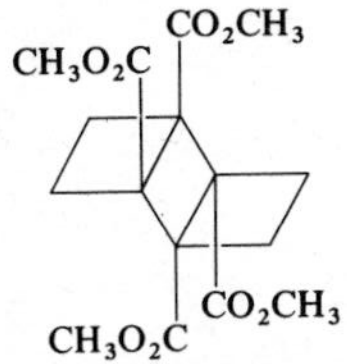 Cage structure with two CO_2CH_3 and two CH_3O_2C groups	LTMA, THF, 0°, 2 h	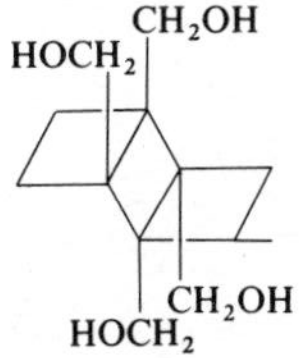 Cage structure with two CH_2OH and two $HOCH_2$ groups (72)	677
2,4-$Cl_2C_6H_3CHN(CH_3)_2CH(CO_2C_2H_5)_2$	SMEAH, C_6H_6, room temp, 12 h	2,4-$Cl_2C_6H_3CHN(CH_3)_2CH(CH_2OH)_2$ (87)	673, 674
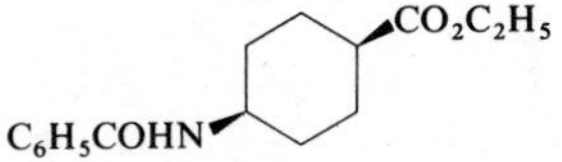 C_6H_5COHN-cyclohexane-$CO_2C_2H_5$	SMEAH, C_6H_6, reflux, 2 h	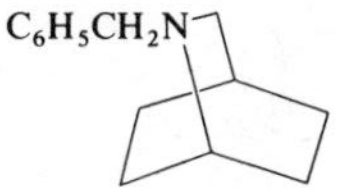 $C_6H_5CH_2N$ bicyclic amine (18) 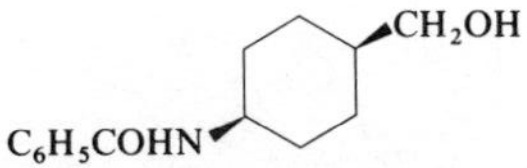 + C_6H_5COHN-cyclohexane-CH_2OH (9)	189
2-$ClC_6H_4CHN(CH_3)_2CH(CO_2C_2H_5)_2$	SMEAH, C_6H_6, reflux, 12 h	2-$ClC_6H_4CHN(CH_3)_2CH(CH_2OH)_2$ (53)	673, 674
3-$ClC_6H_4CH(N(CH_3)_2)CH(CO_2C_2H_5)_2$	SMEAH, C_6H_6, room temp, 12 h	3-$ClC_6H_4CH(N(CH_3)_2)CH(CH_2OH)_2$ (53)	673, 674
4-$ClC_6H_4CH(N(CH_3)_2)CH(CO_2C_2H_5)_2$	", ", " "	4-$ClC_6H_4CH(N(CH_3)_2)CH(CH_2OH)_2$ (59)	673, 674

Reactant	Conditions	Product(s) and Yield(s) (%)	Refs.
$4\text{-}FC_6H_4CH(N(CH_3)_2)CH(CO_2C_2H_5)_2$	SMEAH, C_6H_6, room temp, 12 h	$4\text{-}FC_6H_4CH(N(CH_3)_2)CH(CH_2OH)_2$ (40)	673, 674
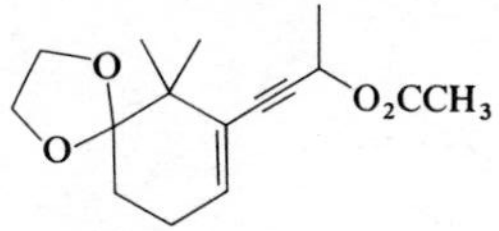 (structure) O_2CCH_3	SMEAH	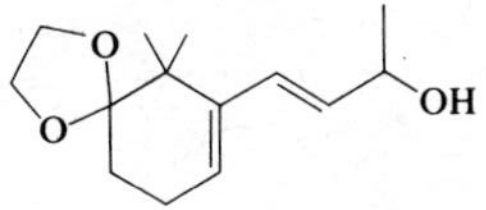 (structure) OH (—)	678
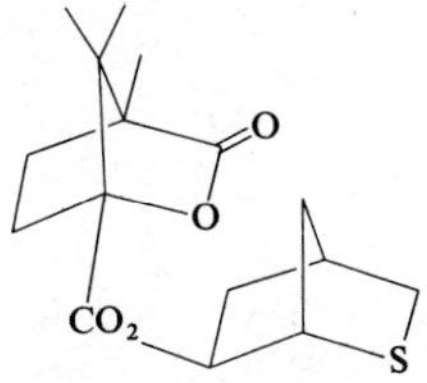 (structure) CO_2, S, (−)-	SMEAH (IA), C_6H_6, 5–10°, overnight	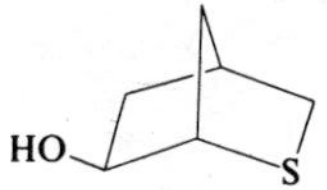 HO (structure) S (∼100), (−)-	679
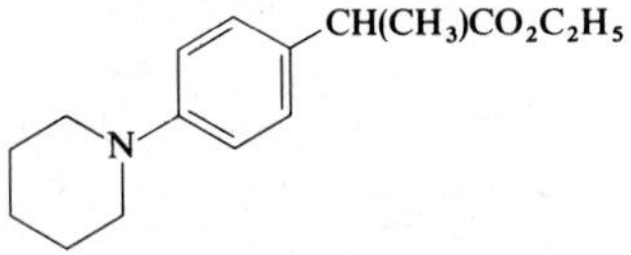 (structure) $CH(CH_3)CO_2C_2H_5$	SMEAH, C_6H_6	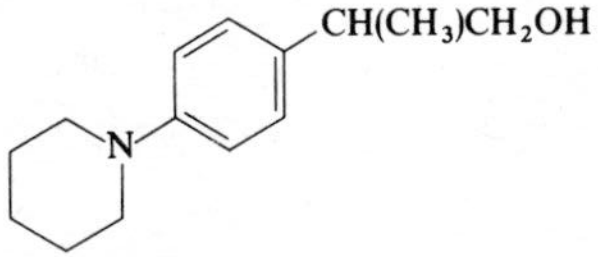 (structure) $CH(CH_3)CH_2OH$ (67)	672
$C_6H_5CH(N(CH_3)_2)CH(CO_2C_2H_5)_2$	SMEAH, C_6H_6, room temp, 12 h	$C_6H_5CH(N(CH_3)_2)CH(CH_2OH)_2$ (49)	673, 674
C_{17}			
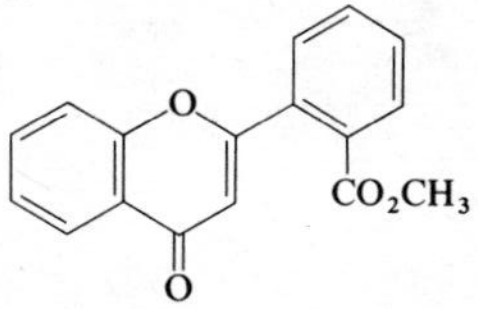 (structure) CO_2CH_3	SMEAH, C_6H_6–ether, room temp, 1 h	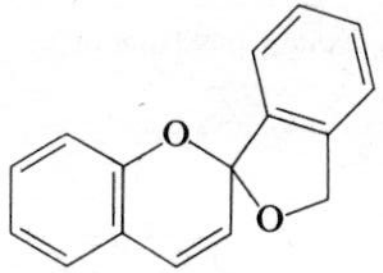 (structure) (86)	680

TABLE VIII. Reduction of Carboxylic Acid Esters (*Continued*)

Reactant	Conditions	Product(s) and Yield(s) (%)	Refs.
$(CH_3)_2C{=}C{=}CHC{\equiv}CCH(C_6H_5)O_2CCH_3$	$LiAlH_3(OCH_3)$, THF, 70–80°, 1–3 h	$(CH_3)_2C{=}C{=}CHCH{=}CHCH(C_6H_5)OH$ (60)	507
2-$CH_3C_6H_4CH(N(CH_3)_2)CH(CO_2C_2H_5)_2$	SMEAH, C_6H_6, reflux, 2 h	2-$CH_3C_6H_4CH(N(CH_3)_2)CH(CH_2OH)_2$ (74)	674, 673
$ArCH(N(CH_3)_2)CH(CO_2C_2H_5)_2$	SMEAH, C_6H_6, room temp, 16 h	$ArCH(N(CH_3)_2)CH(CH_2OH)_2$	
Ar = 3-$CH_3C_6H_4$		(25)	674, 673
Ar = 4-$CH_3C_6H_4$		(16)	674
Ar = 3-$CH_3OC_6H_4$		(27)	674, 673
Ar = 4-$CH_3OC_6H_4$		(27)	674, 673
$CH_2{=}CHCH((CH_2)_6CN)CH(S\text{-}t\text{-}C_4H_9)CO_2CH_3$	SMEAH, toluene, reflux	$CH_2{=}CHCH((CH_2)_6CN)CH(S\text{-}t\text{-}C_4H_9)CH_2OH$ (—)	681
C_{18}			
3,6-Bis(trifluoromethyl)-9-(CO_2CH_3)phenanthrene	SMEAH	3,6-Bis(trifluoromethyl)-9-(CH_2OH)phenanthrene (—)	676

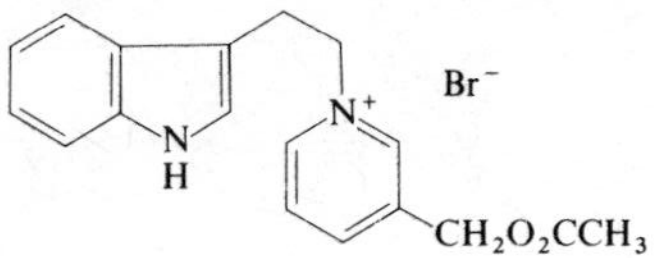 [indole–CH$_2$CH$_2$–N$^+$-pyridinium] 3-$CH_2O_2CCH_3$, Br^-	LTBA, THF, reflux, 2 h	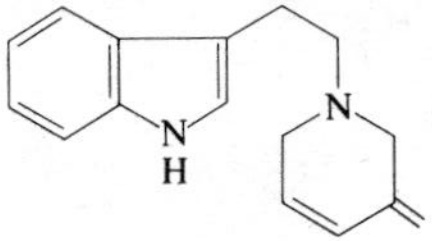 [indole–CH$_2$CH$_2$–N-tetrahydropyridine, methylene] (31)	682
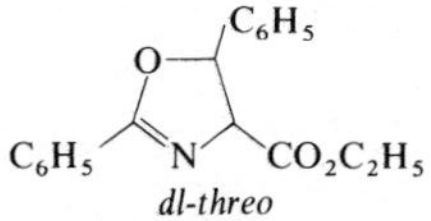 [oxazoline] 2-C_6H_5, 5-C_6H_5, 4-$CO_2C_2H_5$ *dl-threo*	SMEAH, C_6H_6–toluene–ether, 0°, 1 h	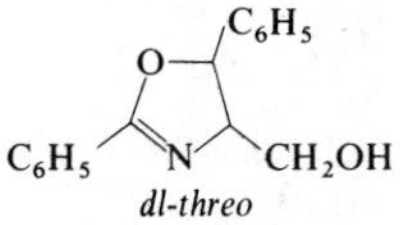 [oxazoline] 2-C_6H_5, 5-C_6H_5, 4-CH_2OH *dl-threo* (84)	246
$C_6H_5CHOHCH(NHBz)CO_2C_2H_5$ *dl-threo*	SMEAH, toluene, 0°, 1 h	$C_6H_5CHOHCH(NHBz)CH_2OH$ *dl-threo* (66)	246
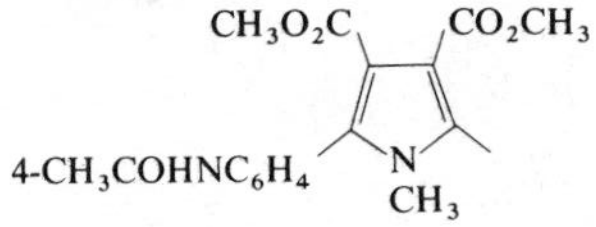 [pyrrole] 2-(4-$CH_3COHNC_6H_4$), 3-CH_3O_2C, 4-CO_2CH_3, 1-CH_3, 5-methyl	$LiAlH_2(OC_4H_9\text{-}t)_2$	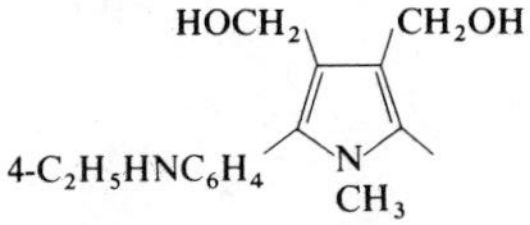 [pyrrole] 2-(4-$C_2H_5HNC_6H_4$), 3-$HOCH_2$, 4-CH_2OH, 1-CH_3, 5-methyl (—)	683
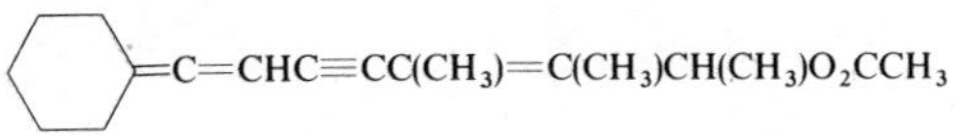 [cyclohexylidene]$=C=CHC\equiv CC(CH_3)=C(CH_3)CH(CH_3)O_2CCH_3$	$LiAlH_3(OCH_3)$, THF, reflux, 2 h	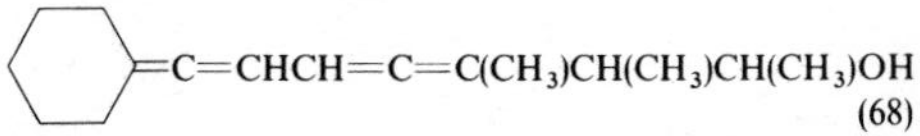 [cyclohexylidene]$=C=CHCH=C=C(CH_3)CH(CH_3)CH(CH_3)OH$ (68)	658, 657
	$LiAlH_3(OCH_3)$, THF, reflux, 10 h	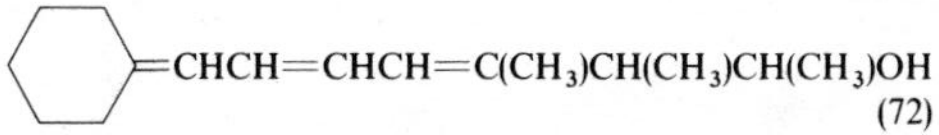 [cyclohexylidene]$=CHCH=CHCH=C(CH_3)CH(CH_3)CH(CH_3)OH$ (72)	658
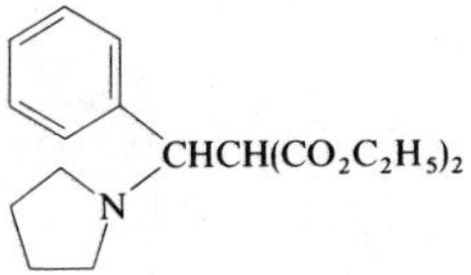 C_6H_5(N-pyrrolidinyl)$CHCH(CO_2C_2H_5)_2$	SMEAH, C_6H_6, room temp, 16 h	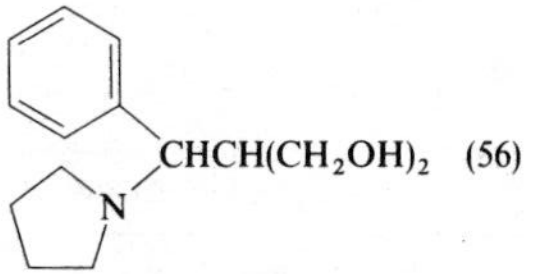 C_6H_5(N-pyrrolidinyl)$CHCH(CH_2OH)_2$ (56)	674

TABLE VIII. REDUCTION OF CARBOXYLIC ACID ESTERS (*Continued*)

Reactant	Conditions	Product(s) and Yield(s) (%)	Refs.
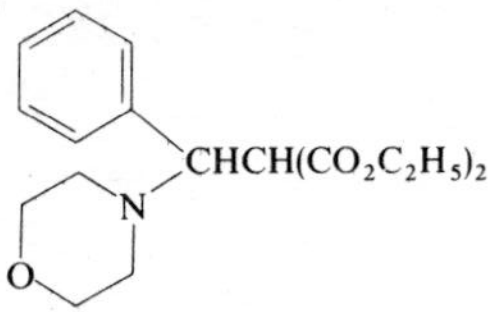 $C_6H_5CH(NC_4H_8O)CH(CO_2C_2H_5)_2$ (morpholino structure)	SMEAH, C_6H_6, room temp, 16 h	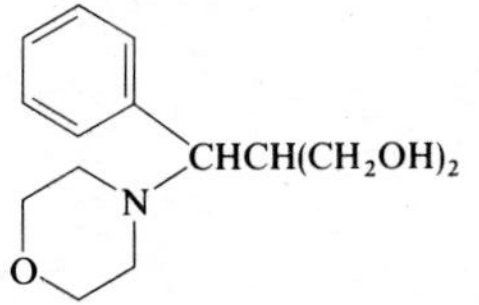 $C_6H_5CH(NC_4H_8O)CH(CH_2OH)_2$ (85)	674
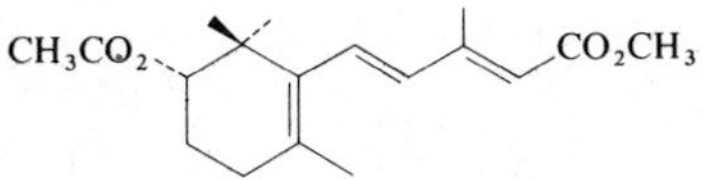 CH_3CO_2- (cyclohexenyl dienoate, CO_2CH_3) (structure)	SMEAH	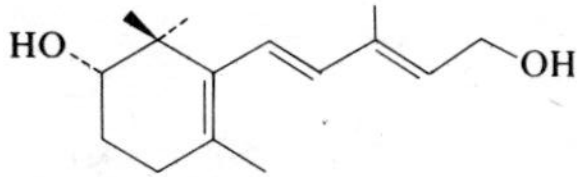 HO- (cyclohexenyl dienol, OH) (structure) (—)	678
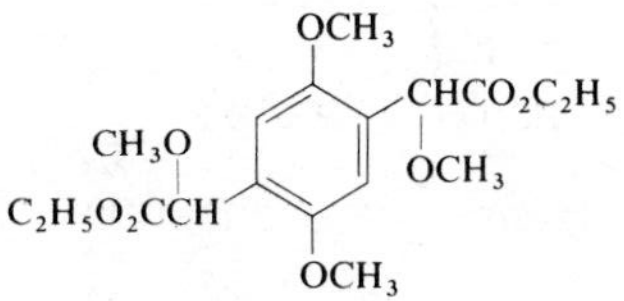 $C_2H_5O_2CCH(OCH_3)$-, $CHCO_2C_2H_5(OCH_3)$, OCH_3, OCH_3 (structure)	SMEAH (IA), THF–C_6H_6, reflux, 2.5 h	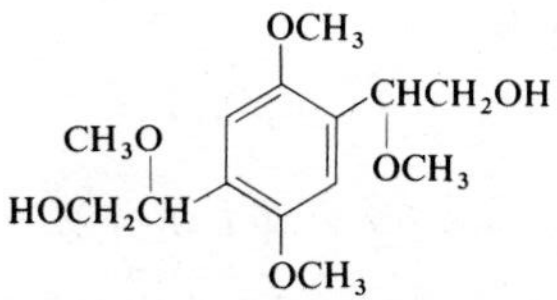 $HOCH_2CH(OCH_3)$-, $CHCH_2OH(OCH_3)$, OCH_3, OCH_3 (structure) (93)	684
$C_6H_5CHCH(CO_2C_2H_5)_2$ with $N(C_2H_5)_2$	SMEAH, C_6H_6, room temp, 16 h	$C_6H_5CHCH(CH_2OH)_2$ with $N(C_2H_5)_2$ (56)	674
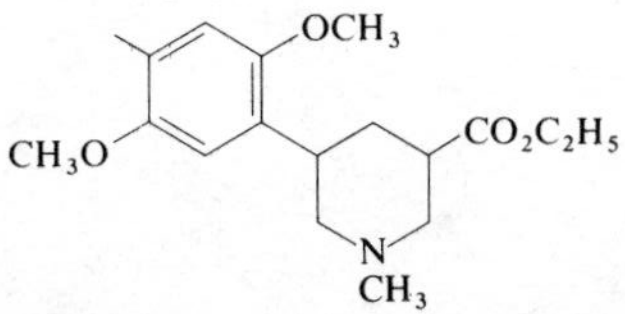 CH_3O, OCH_3 aryl-piperidine, $CO_2C_2H_5$, N-CH_3 (structure)	SMEAH, C_6H_6, reflux, 1 h	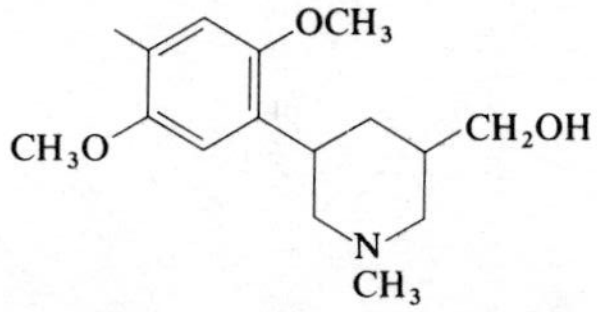 CH_3O, OCH_3 aryl-piperidine, CH_2OH, N-CH_3 (structure) (91)	685

Reactant	Conditions	Product(s) and Yield(s) (%)	Refs.
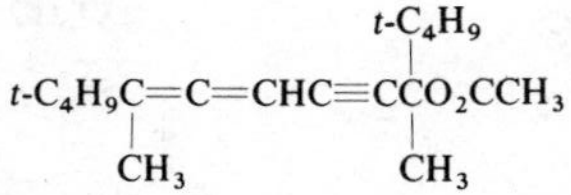 $t\text{-}C_4H_9C(CH_3){=}C{=}CHC{\equiv}CCO_2C(CH_3)(t\text{-}C_4H_9)$	$LiAlH_3(OCH_3)$, THF, 70–80°, 1–3 h	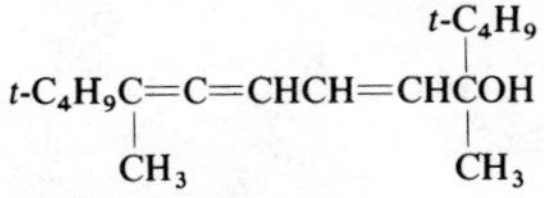 $t\text{-}C_4H_9C(CH_3){=}C{=}CHCH{=}CHC(OH)(CH_3)(t\text{-}C_4H_9)$ (83)	507
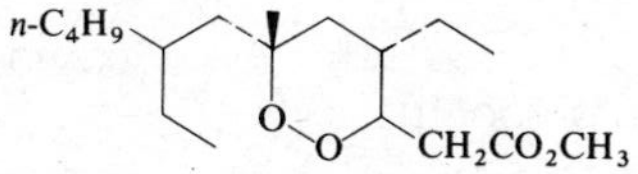 $n\text{-}C_4H_9$ … O–O … $CH_2CO_2CH_3$	LTBA, ether, reflux, 1.5 h	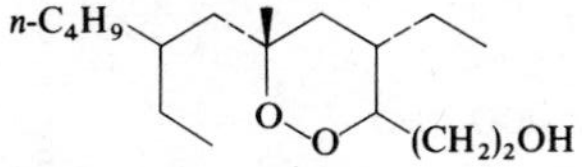 $n\text{-}C_4H_9$ … O–O … $(CH_2)_2OH$ (—)	273
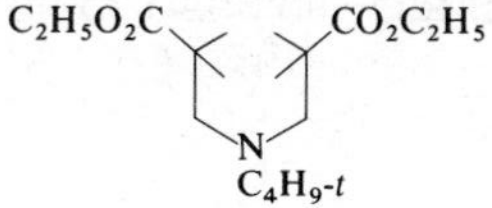 $C_2H_5O_2C$ … $CO_2C_2H_5$ … N … $C_4H_9\text{-}t$	LTBA	No reaction	686
C_{19}			
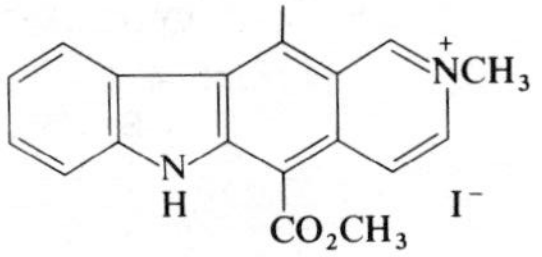 $\overset{+}{N}CH_3$ … N H … CO_2CH_3 I^-	1. SMEAH, xylene, 130° 2. 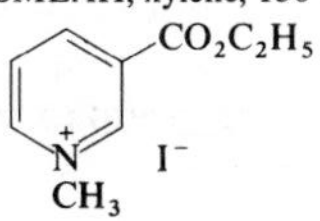 $CO_2C_2H_5$ … $\overset{+}{N}$ CH_3 I^-	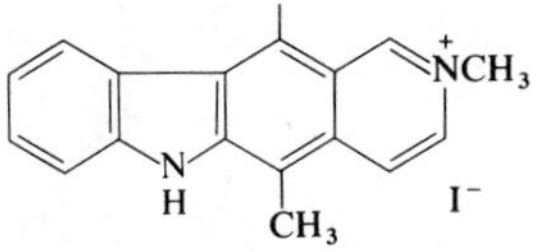 $\overset{+}{N}CH_3$ … N H … CH_3 I^- (85)	687
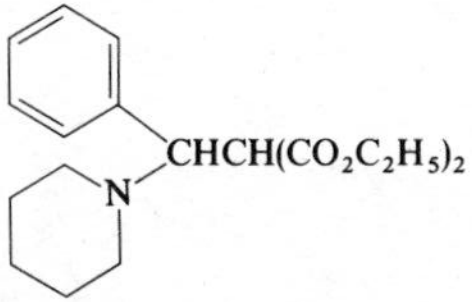 $CHCH(CO_2C_2H_5)_2$ … N	SMEAH, C_6H_6, room temp, 16 h	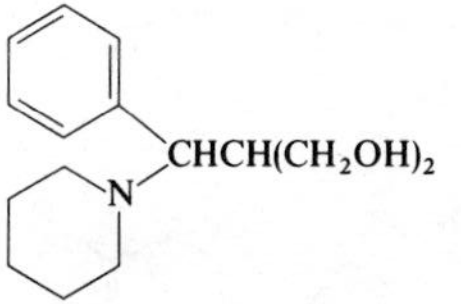 $CHCH(CH_2OH)_2$ … N (52)	674
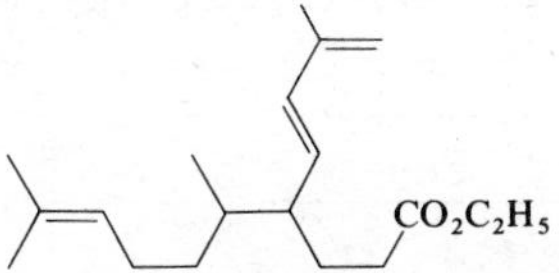 … $CO_2C_2H_5$	LAH + EtOH (IA), ether, 0°, 1 h	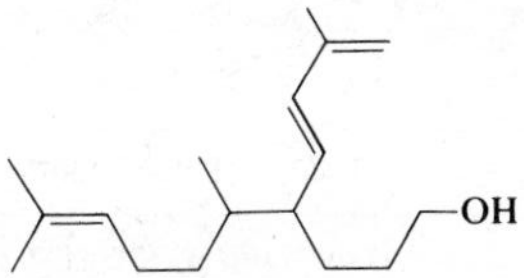 … OH (80)	688

TABLE VIII. REDUCTION OF CARBOXYLIC ACID ESTERS (*Continued*)

Reactant	Conditions	Product(s) and Yield(s) (%)	Refs.
C_{20}			
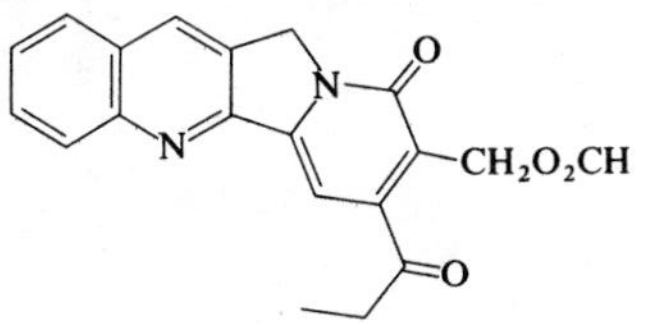 [structure] $-CH_2O_2CH$	LAH + BCGF, ether, reflux, 2 h	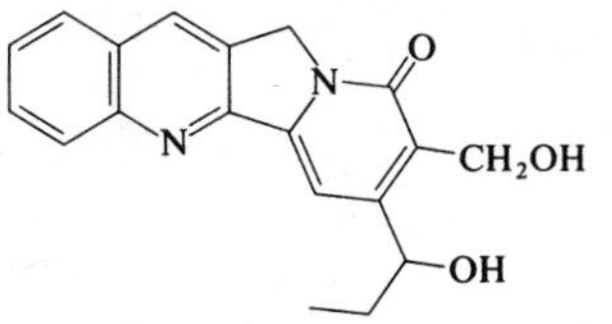 [structure] $-CH_2OH$ (—)	689
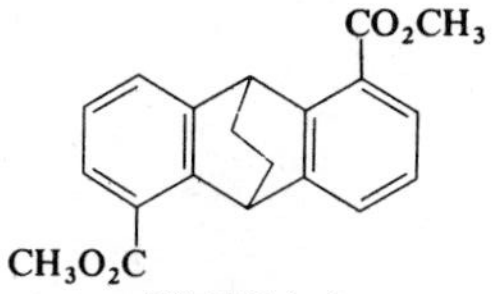 [structure, CO_2CH_3, CH_3O_2C] (9*R* 10*R*)-(+)	SMEAH (IA), C_6H_6, room temp, 2.5 h	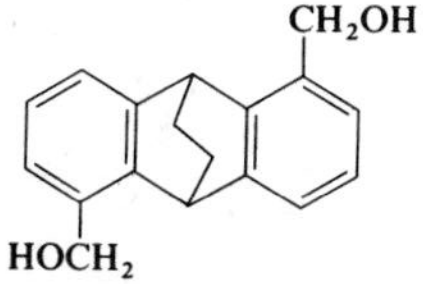 [structure, CH_2OH, $HOCH_2$] (~100)	690, 691
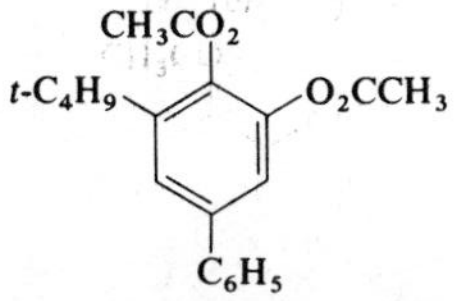 [structure, CH_3CO_2, t-C_4H_9, O_2CCH_3, C_6H_5]	SMEAH, C_6H_6, room temp, 24 h	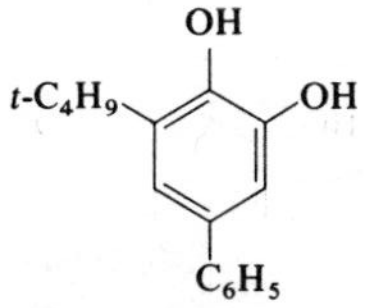 [structure, OH, t-C_4H_9, OH, C_6H_5] (69)	241
$CH_3C{\equiv}C(CH_2)_2CH{=}C(CH_3)CH_2CH(OCH_2C_6H_5)CO_2C_2H_5$	SMEAH	$CH_3C{\equiv}C(CH_2)_2CH{=}C(CH_3)CH_2CH(OCH_2C_6H_5)CH_2OH$ (—)	692
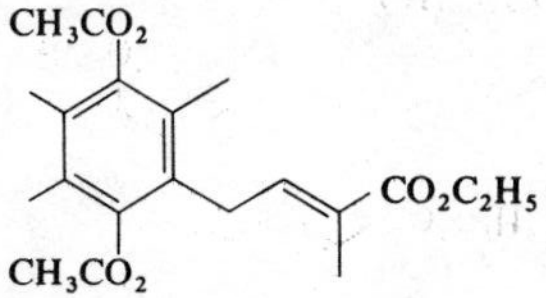 [structure, CH_3CO_2, $CO_2C_2H_5$, CH_3CO_2]	1. SMEAH (IA), C_6H_6, room temp, 2 h 2. $FeCl_3$, CH_3OH	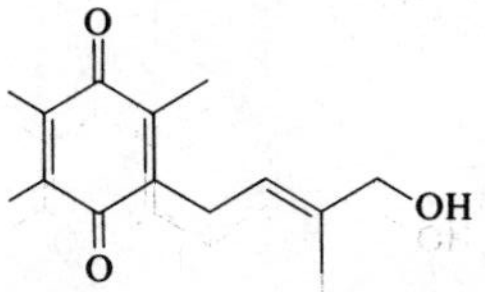 [structure, O, OH, O] (—)	693

Reactant	Conditions	Product (% yield)	Refs.
$n\text{-}C_8H_{17}CH{=}CH(CH_2)_7CO_2C_2H_5$	SMEAH, C_6H_6, 80°	$n\text{-}C_8H_{17}CH{=}CH(CH_2)_8OH$ (87)	68
$n\text{-}C_{17}H_{35}CO_2C_2H_5$	$NaAlH(OC_2H_5)_3$, THF, 30°, 4 h	$n\text{-}C_{18}H_{37}OH$ (91)	319
	SMEAH, C_6H_6, 80°	" (92)	68
(structure: CO_2, OH, O)	SMEAH, toluene	(structure: HO, OH, O) (26)	623
C_{21}			
(structure: O, N, N, CH_3, $CO_2C_2H_5$)	SMEAH, THF, room temp	(structure: N, N, CH_3) (57)	694, 695
(structure: $(CH_2)_4O_2CCH_3$, O, N, C_6H_5, C_6H_5, N, H, O)	SMEAH (IA), THF–C_6H_6	(structure: $(CH_2)_4OH$, N, C_6H_5, C_6H_5, N, H, O) (56)	372
(structure: N, H, N, H, CO_2CH_3, $HOCH_2$)	SMEAH	(structure: N, H, N, H, CH_2OH, $HOCH_2$) (—)	696
(structure: N, H, N, H, CH_2OH, CH_3O_2C)	"	" (—)	696
$C_6H_5CHOHCHCO_2C_2H_5$ (with $NHCOC_9H_{19}\text{-}n$) *dl-threo*	SMEAH, toluene, 0°, 1 h	$C_6H_5CHOHCHCH_2OH$ (with $NHCOC_9H_{19}\text{-}n$) *dl-threo* (—)	246

TABLE VIII. REDUCTION OF CARBOXYLIC ACID ESTERS (*Continued*)

Reactant	Conditions	Product(s) and Yield(s) (%)	Refs.
C_{22}			
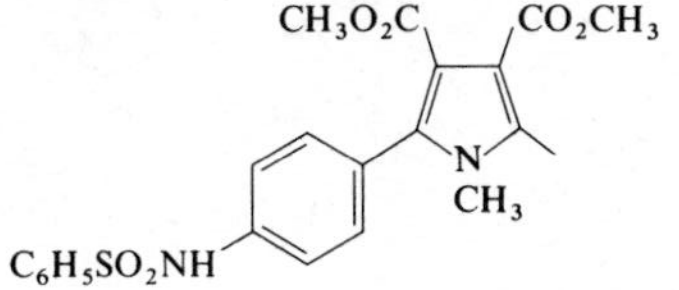 CH_3O_2C, CO_2CH_3, N, CH_3, $C_6H_5SO_2NH$	$LiAlH_2(OC_4H_9\text{-}t)_2$	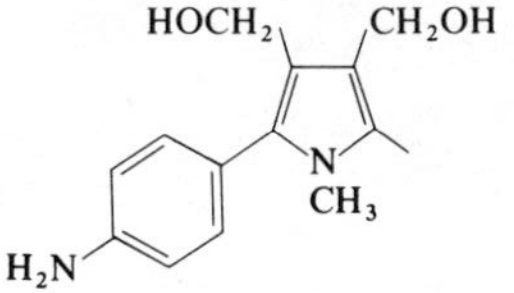 $HOCH_2$, CH_2OH, N, CH_3, H_2N (—)	683
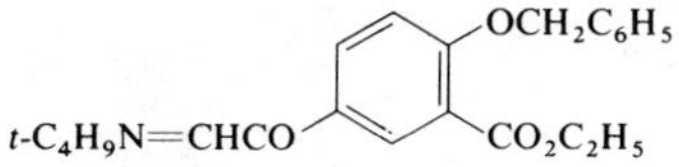 $OCH_2C_6H_5$, $t\text{-}C_4H_9N{=}CHCO$, $CO_2C_2H_5$	SMEAH, C_6H_6, reflux, 1 h	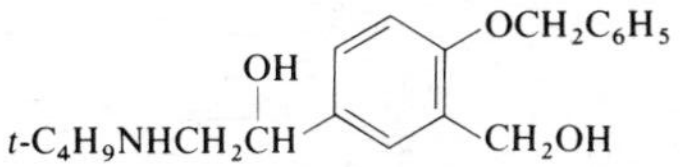 $OCH_2C_6H_5$, OH, $t\text{-}C_4H_9NHCH_2CH$, CH_2OH (—)	697
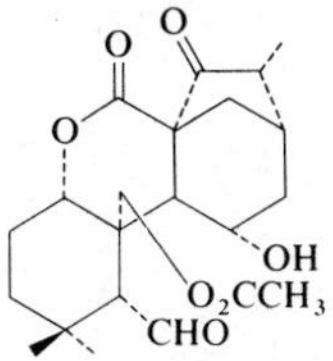 O, O, O, OH, O_2CCH_3, CHO	1. LAH + MeOH, 0°, 4 h 2. H_3O^+	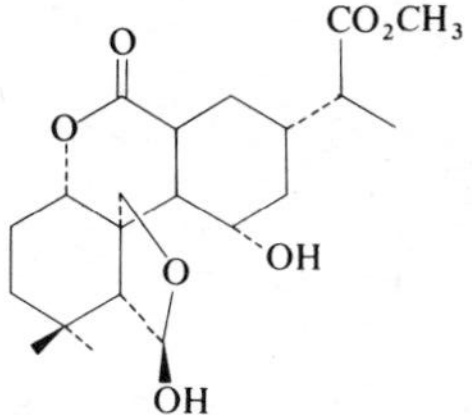 CO_2CH_3, O, O, OH, O, OH (28)	698
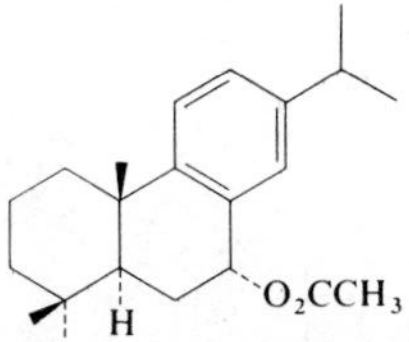 H, O_2CCH_3	SMEAH, C_6H_6–ether, 4 h	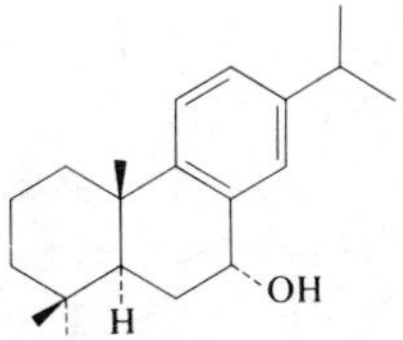 H, OH (91)	699
$n\text{-}C_5H_{11}(CH{=}CHCH_2)_3(CH_2)_4CHCO_2CH_3$ (all-*cis*), CO_2H	LTBA, BME–ether, 0–40°, 7.5 h	$n\text{-}C_5H_{11}(CH{=}CHCH_2)_3(CH_2)_4CHCH_2OH$, CO_2H (35)	275

C_{23}

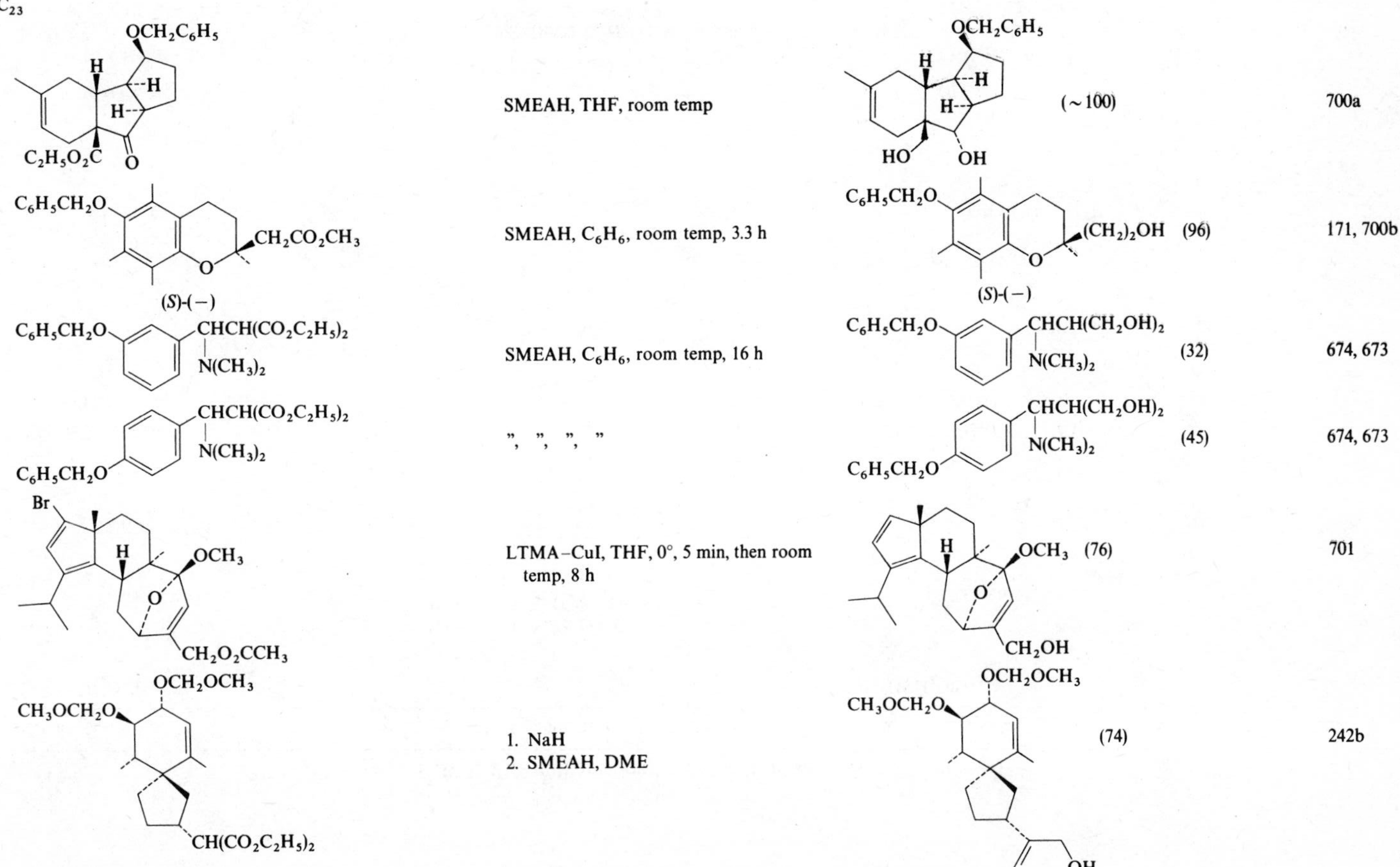

Conditions	Product (Yield, %)	Refs.
SMEAH, THF, room temp	(~100)	700a
SMEAH, C_6H_6, room temp, 3.3 h	(S)-(−) $(CH_2)_2OH$ (96)	171, 700b
SMEAH, C_6H_6, room temp, 16 h	(32)	674, 673
" " " "	(45)	674, 673
LTMA–CuI, THF, 0°, 5 min, then room temp, 8 h	(76)	701
1. NaH 2. SMEAH, DME	(74)	242b

TABLE VIII. REDUCTION OF CARBOXYLIC ACID ESTERS (*Continued*)

Reactant	Conditions	Product(s) and Yield(s) (%)	Refs.
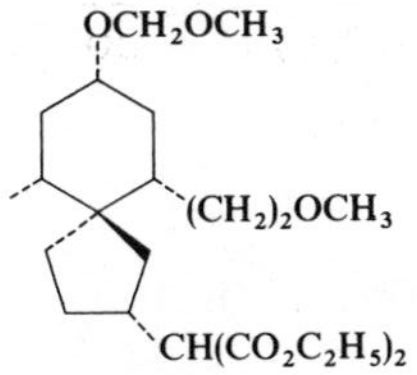 Spiro compound: OCH_2OCH_3, $(CH_2)_2OCH_3$, $CH(CO_2C_2H_5)_2$	1. NaH, DME, reflux 2. SMEAH, DME, reflux	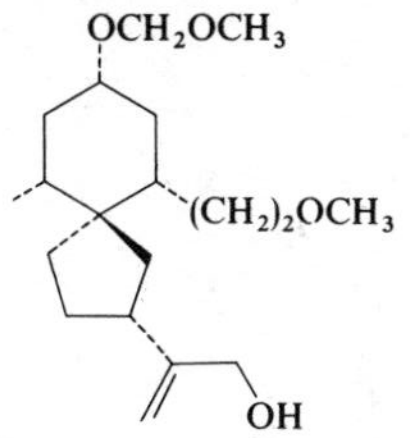 Spiro compound: OCH_2OCH_3, $(CH_2)_2OCH_3$, $C(=CH_2)CH_2OH$ (50)	242c
C_{24}			
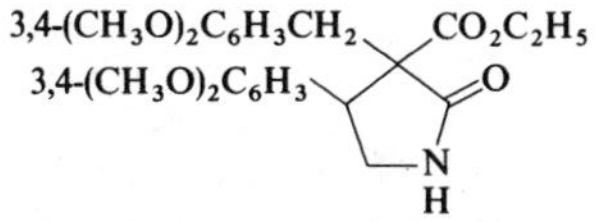 3,4-$(CH_3O)_2C_6H_3CH_2$, $CO_2C_2H_5$, 3,4-$(CH_3O)_2C_6H_3$ pyrrolidinone (N–H, C=O)	SMEAH, C_6H_6, ~5°	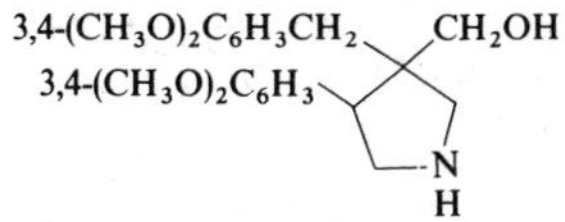 3,4-$(CH_3O)_2C_6H_3CH_2$, CH_2OH, 3,4-$(CH_3O)_2C_6H_3$ pyrrolidine (N–H) (—)	702
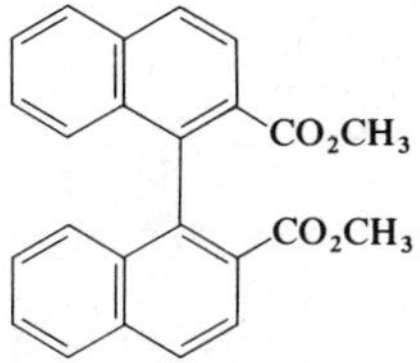 1,1'-Binaphthyl-2,2'-di(CO_2CH_3)	LTBA, C_6H_6–ether	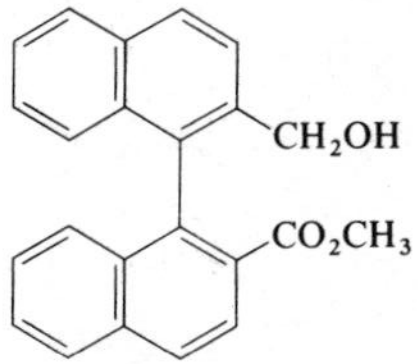 1,1'-Binaphthyl-2-CH_2OH-2'-CO_2CH_3 (—)	703a, 703b
C_{26}			
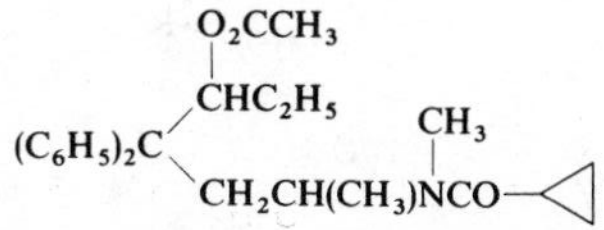 $(C_6H_5)_2C(CH(O_2CCH_3)C_2H_5)CH_2CH(CH_3)N(CH_3)CO$-cyclopropyl	SMEAH, C_6H_6, reflux, 30 min	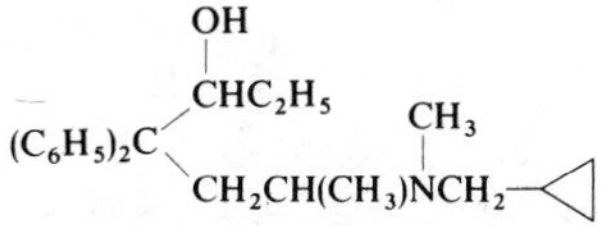 $(C_6H_5)_2C(CH(OH)C_2H_5)CH_2CH(CH_3)N(CH_3)CH_2$-cyclopropyl (78)	704a

Substrate	Conditions	Product(s) (Yield, %)	Refs.
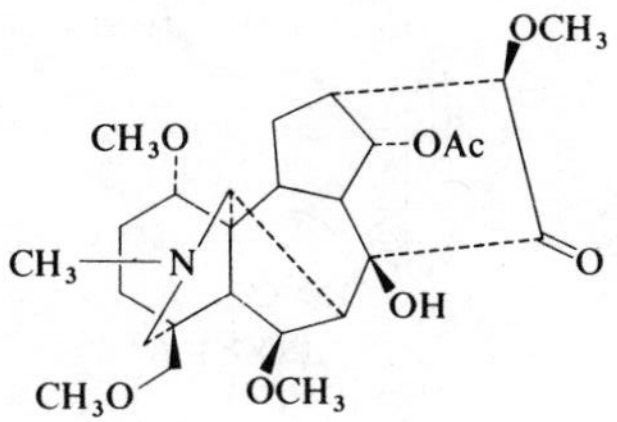 CH_3O, CH_3–N, CH_3O, OCH_3, OAc, OH, OCH_3, =O	LTMA, −70 to −50°, 1 h, then room temp, 1 h	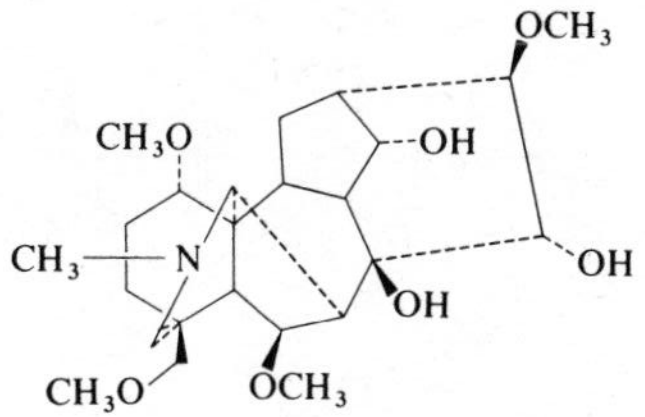 CH_3O, CH_3–N, CH_3O, OCH_3, OH, OH, OCH_3, OH (79)	704b
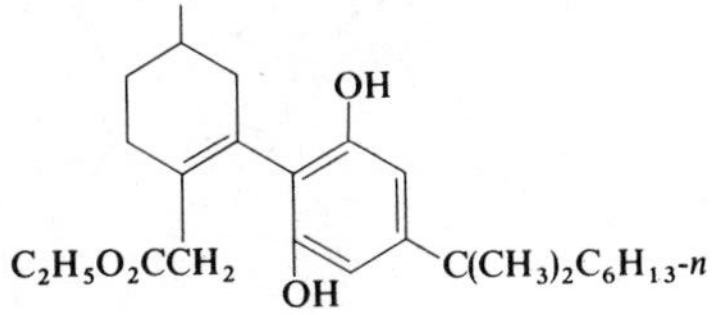 $C_2H_5O_2CCH_2$, OH, OH, $C(CH_3)_2C_6H_{13}$-*n*	SMEAH, C_6H_6, 0–10°, 15 min, then room temp, 1 h	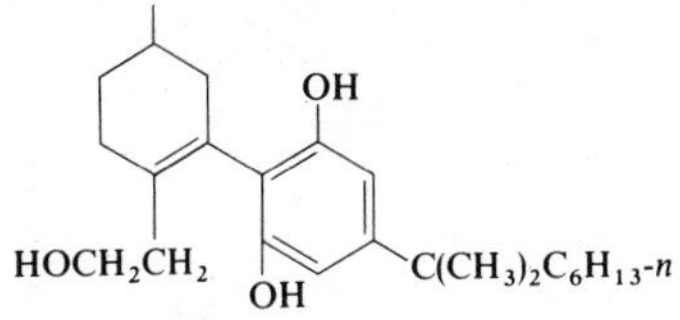 $HOCH_2CH_2$, OH, OH, $C(CH_3)_2C_6H_{13}$-*n* (~100)	234
C_{27}			
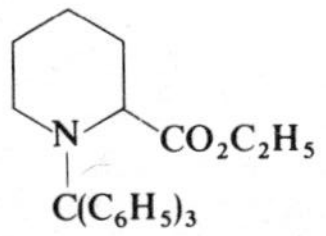 N–$C(C_6H_5)_3$, $CO_2C_2H_5$	SMEAH, C_6H_6, 40°, 3 h, then reflux, 4 h	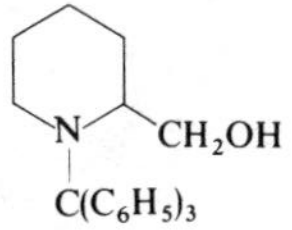 N–$C(C_6H_5)_3$, CH_2OH (81)	705
C_{28}			
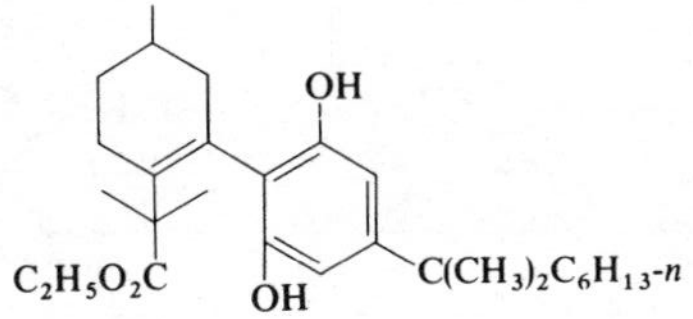 $C_2H_5O_2C$, OH, OH, $C(CH_3)_2C_6H_{13}$-*n*	SMEAH, C_6H_6, room temp, 3 h	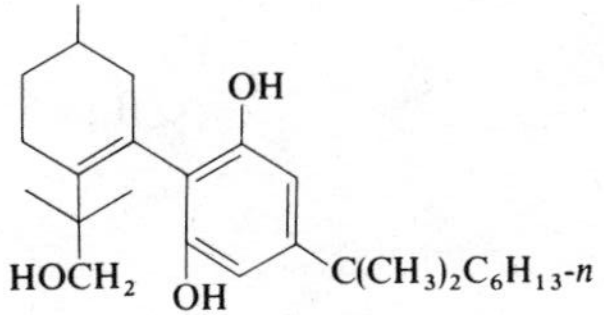 $HOCH_2$, OH, OH, $C(CH_3)_2C_6H_{13}$-*n* (85)	234

TABLE VIII. REDUCTION OF CARBOXYLIC ACID ESTERS (*Continued*)

Reactant	Conditions	Product(s) and Yield(s) (%)	Refs.
C_{29}			
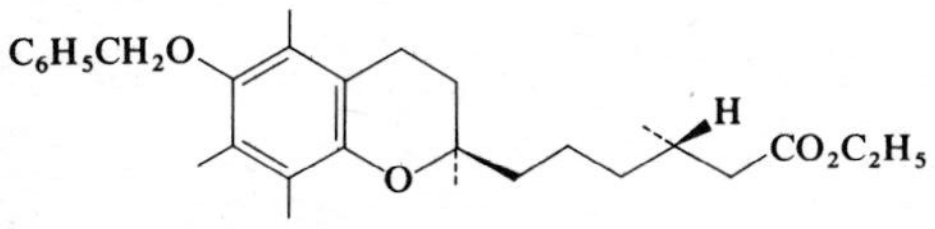 $C_6H_5CH_2O$ … $CO_2C_2H_5$ (2*R*,3*S*)-	SMEAH, C_6H_6–ether, reflux, 3 h	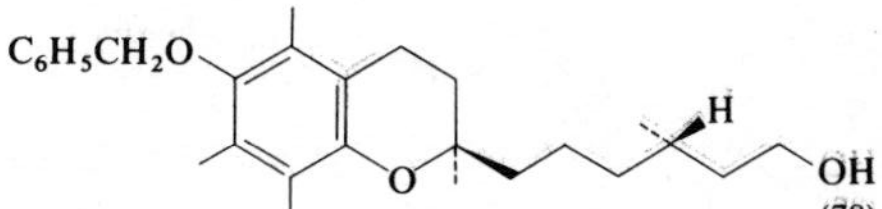 $C_6H_5CH_2O$ … OH (78)	706, 707
C_{31}			
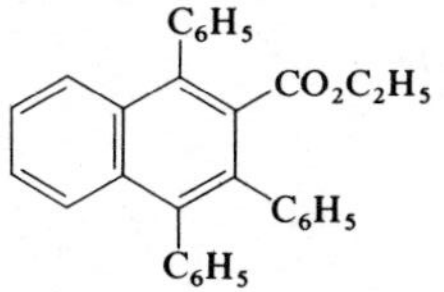 C_6H_5, $CO_2C_2H_5$, C_6H_5, C_6H_5	SMEAH, C_6H_6, reflux, 5 h	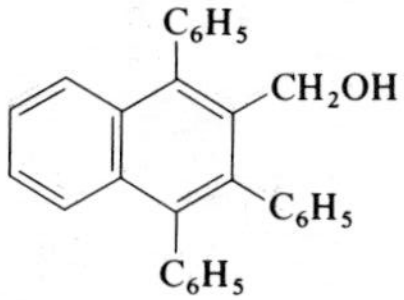 C_6H_5, CH_2OH, C_6H_5, C_6H_5 (92)	708
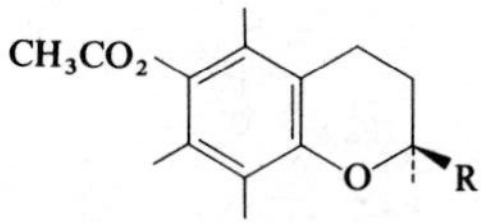 CH_3CO_2 … R; R = $[(CH_2)_3CH(CH_3)]_3CH_3$ (2*R*,4′*R*,8′*R*)	SMEAH, ether, 75 min	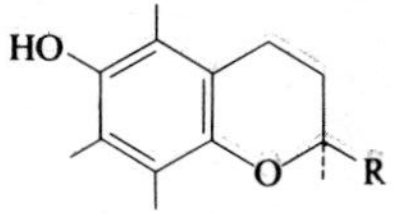 HO … R (~100)	709, 710
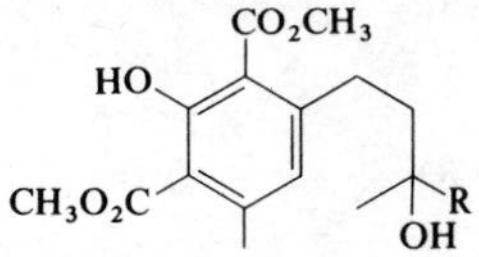 CO_2CH_3, HO, CH_3O_2C, R, OH; R = $[(CH_2)_3CH(CH_3)]_3CH_3$	SMEAH, xylene, 10°, 10 min, then reflux, 90 min	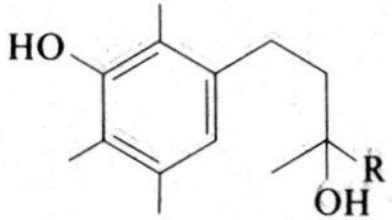 HO … R, OH (74)	711, 712

Substrate	Conditions	Product(s) (Yield, %)	Ref.
C_{32}			
(structure: CH_3O, OH, OBz, CH_3–N, O_2CCH_3, CH_3O, OCH_3)	LTBA, BME, 140°, 20 min	(structure: CH_3O, OH, OH, CH_3–N, CH_3O, OCH_3) (44) + (structure: CH_3O, OH, OH, CH_3–N, CH_3O, OCH_3) (8)	274
C_n			
$\left[-CH_2C(CH_3)(CO_2CH_3)-\right]_n$	SMEAH-*N*-methylmorpholine–toluene, 100°, 12 h	$\left[-CH_2C(CH_3)(CH_2OH)-\right]_n$ (62)	713
	$LiAlH(OC_2H_5)_3$, THF, reflux, 4 h	" (—)	714a
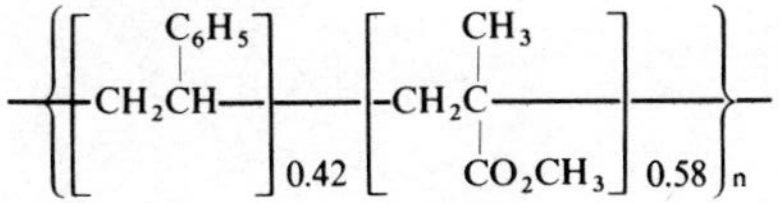 $\left\{\left[-CH_2CH(C_6H_5)-\right]_{0.42}\left[-CH_2C(CH_3)(CO_2CH_3)-\right]_{0.58}\right\}_n$	SMEAH, toluene, reflux, 4 h	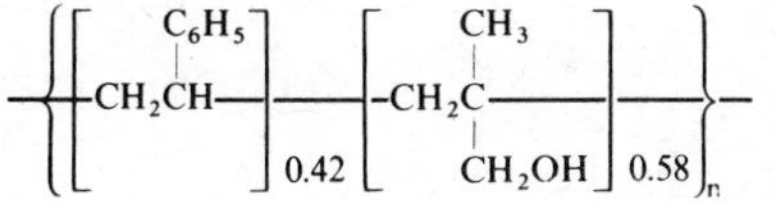 $\left\{\left[-CH_2CH(C_6H_5)-\right]_{0.42}\left[-CH_2C(CH_3)(CH_2OH)-\right]_{0.58}\right\}_n$ (~100)	714b

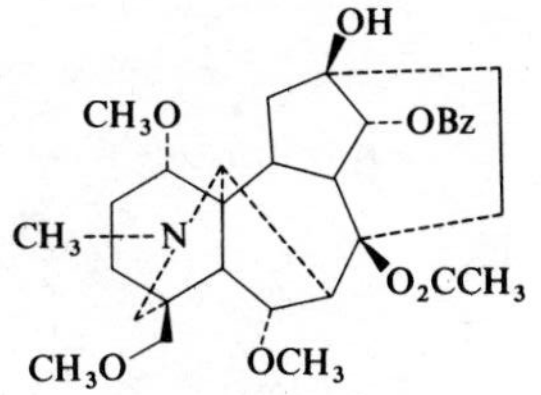

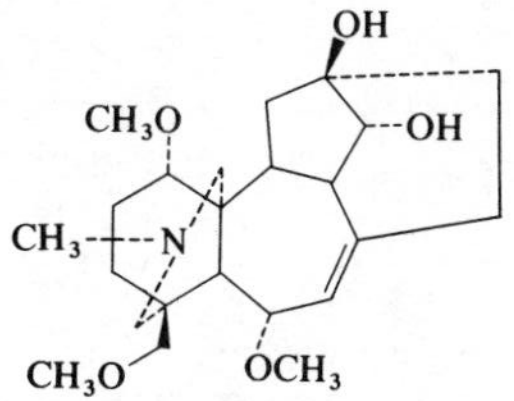

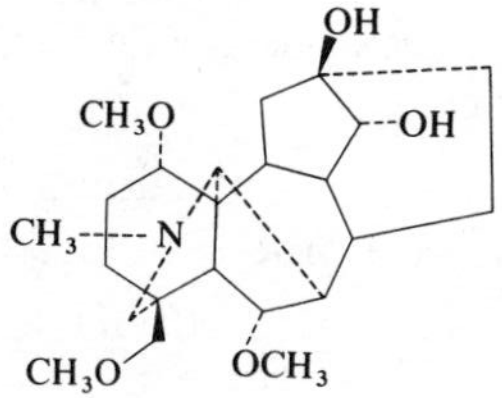

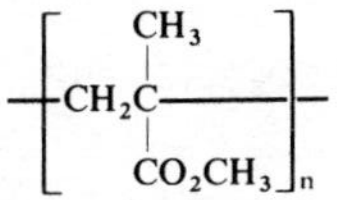

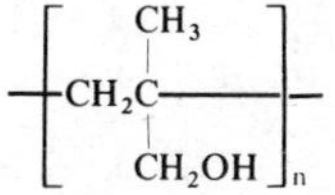

TABLE IX. REDUCTION OF α,β-UNSATURATED CARBOXYLIC ACIDS AND ESTERS

Reactant	Conditions	Product(s) and Yield(s) (%)	Refs.
C_3			
$HC{\equiv}CCO_2H$	1. $LiAlD_3(OC_2H_5)$, ether, 0–5°, 12 h 2. Aqueous HBr, 15 min	$CD_2{=}CDCD_2Br$(67) + n-C_3D_7Br (—)	715a
C_5			
$(CH_3)_2C{=}CHCO_2H$	SMEAH, C_6H_6, reflux, 1.5 h, then RT, 12 h	$(CH_3)_2C{=}CHCH_2OH$ (38)	715b
$CH_3CH{=}CHCO_2CH_3$ *cis-*	SMEAH, C_6H_6–ether, 15–20°, 1 h	$CH_3CH{=}CHCH_2OH$ (—) *cis-*	716
	SMEAH–CuBr–2-butanol, THF, −78°, 10 min, then −20°, 1 h	n-$C_3H_7CO_2CH_3$ (84)	113, 114
$CH_2{=}C(CH_3)CO_2CH_3$	SMEAH, C_6H_6–ether	$CH_2{=}C(CH_3)CH_2OH$ (—)	717
C_8			
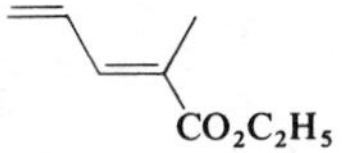 $CO_2C_2H_5$	$LiAlH_3(OC_2H_5)$(IA), ether, −5°	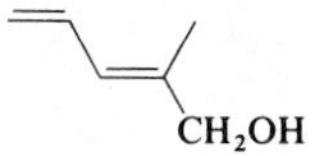 CH_2OH (—)	718
 CO_2CH_3	$LiAlH_3(OC_2H_5)$	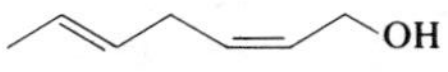 OH (—)	719
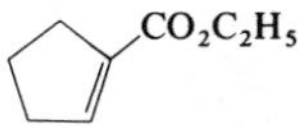 $CO_2C_2H_5$	$LiAlH_3(OC_2H_5)$, ether-THF, reflux, 3 h	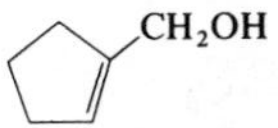 CH_2OH (93)	720
$(CH_3)_2C{=}C(CH_3)CO_2C_2H_5$	SMEAH (IA), C_6H_6–THF, 0°, 5.5 h	$(CH_3)_2C{=}C(CH_3)CH_2OH$ (87)	721
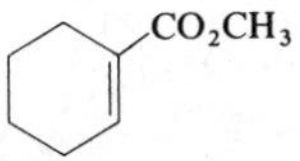 CO_2CH_3	SMEAH, C_6H_6, 80°	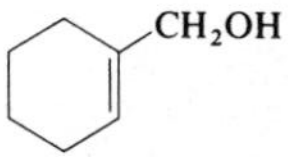 CH_2OH (95)	68
C_9			
$C_6H_5CH{=}CHCO_2H$	SMEAH, C_6H_6, 80°, 30 min	$C_6H_5(CH_2)_3OH$ (22)	168
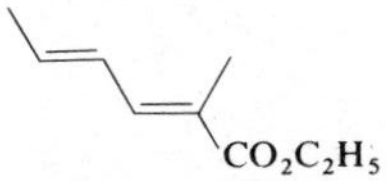 $CO_2C_2H_5$	$LiAlH_3(OC_2H_5)$ (IA), ether, −5°	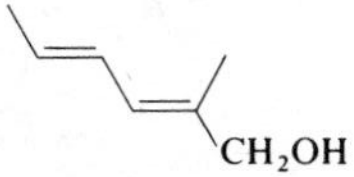 CH_2OH (—)	718

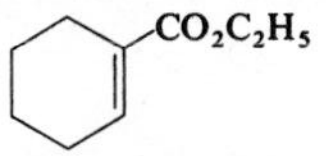 $CO_2C_2H_5$ (cyclohexenyl)	$LiAlH_2(OC_2H_5)_2$	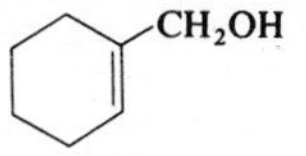 CH_2OH (cyclohexenyl) (—)	720
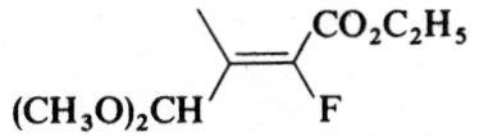 $(CH_3O)_2CH$, $CO_2C_2H_5$, F	SMEAH, C_6H_6–ether, $-70°$, then $-20°$, 1 h	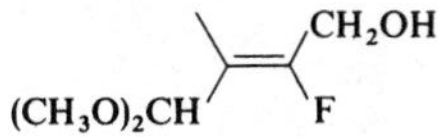 $(CH_3O)_2CH$, CH_2OH, F (—)	722
C_{10}			
$C_6H_5CH{=}CHCO_2CH_3$	SMEAH–CuBr–2-butanol, THF, $-78°$, 10 min, then $-20°$, 1 h	$C_6H_5(CH_2)_2CO_2CH_3$ (82) + $[CH_3O_2CCH_2CH(C_6H_5){-}]_2$ *meso-* (—)	114, 113
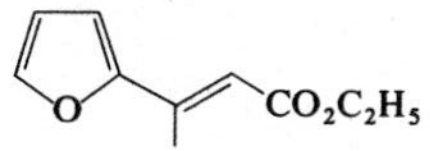 O (furyl), $CO_2C_2H_5$	SMEAH, C_6H_6, room temp, 2 h	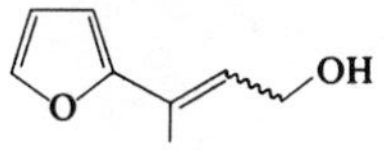 O (furyl), OH (—) (*trans*:*cis* = 67:33)	723
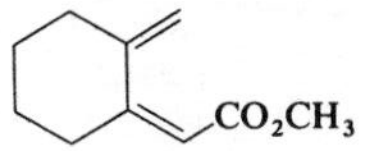 CO_2CH_3	SMEAH, C_6H_6, 10–20°, 1.5 h	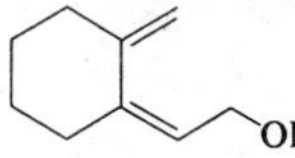 OH (95)	724a
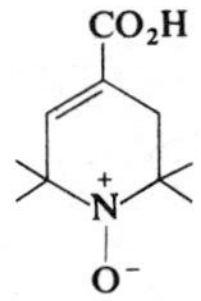 CO_2H, $\overset{+}{N}$, O^-	SMEAH (IA), toluene, reflux, 2 h	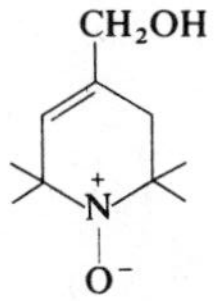 CH_2OH, $\overset{+}{N}$, O^- (59)	724b
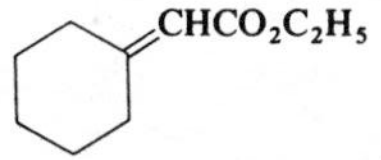 $CHCO_2C_2H_5$ (cyclohexylidene)	SMEAH (IA), C_6H_6, 0°, then room temp, 6 h	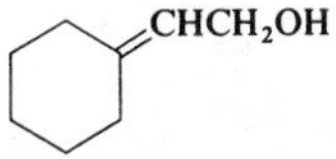 $CHCH_2OH$ (cyclohexylidene) (70)	724c
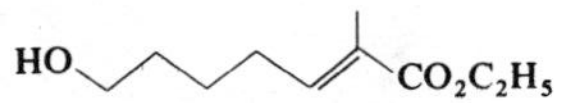 HO, $CO_2C_2H_5$	SMEAH, C_6H_6	 HO, OH (68)	725
C_{11}			
$C_6H_5CH{=}CHCO_2C_2H_5$	SMEAH (1A), C_6H_6, 15–20°, 45 min	$C_6H_5CH{=}CHCH_2OH$ (76) + $C_6H_5CH{=}CHCHO$ (2)	168, 167
	SMEAH, C_6H_6, 15–20°, 2 h	$C_6H_5(CH_2)_3OH$ (45)	168, 167
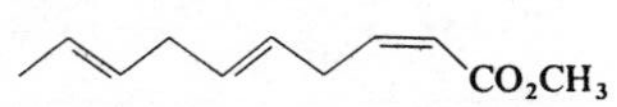 CO_2CH_3	$LiAlH_3(OC_2H_5)$	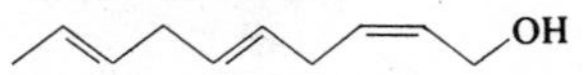 OH (—)	719

TABLE IX. REDUCTION OF α,β-UNSATURATED CARBOXYLIC ACIDS AND ESTERS (*Continued*)

Reactant	Conditions	Product(s) and Yield(s) (%)	Refs.
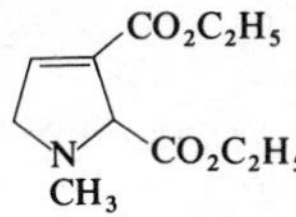 1-methyl-2,3-dihydropyrrole with $CO_2C_2H_5$ groups at 2 and 3 (N–CH_3)	SMEAH, C_6H_6–ether, room temp, 15 min	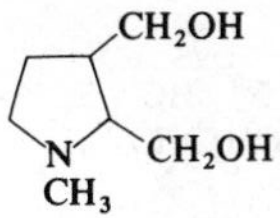 1-methyl-2,3-bis(hydroxymethyl)pyrrolidine, CH_2OH, CH_2OH, N–CH_3 (56)	726
$(CH_3)_2C{=}CH(CH_2)_2C(CH_3){=}CHCO_2CH_3$ (*trans*:*cis* = 91:9)	$LiAlH_3(OC_2H_5)$ (IA)	$(CH_3)_2C{=}CH(CH_2)_2C(CH_3){=}CHCH_2OH$ (*trans*:*cis* = 91:9) (90)	727
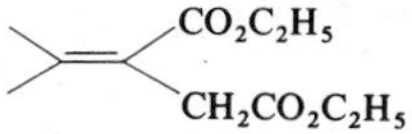 $(CH_3)_2C{=}C(CO_2C_2H_5)CH_2CO_2C_2H_5$	$LiAlH(OC_2H_5)_3$, ether, room temp	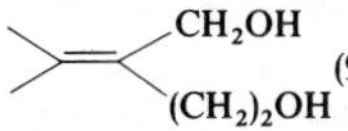 $(CH_3)_2C{=}C(CH_2OH)(CH_2)_2OH$ (90)	728, 729
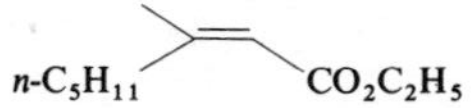 n-C_5H_{11}, $CO_2C_2H_5$ (alkene isomer)	$LiAlH_3(OC_2H_5)$ (IA), ether, reflux, 24 h	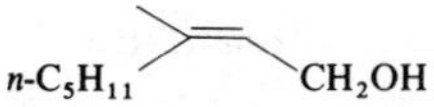 n-C_5H_{11}, CH_2OH (97)	730
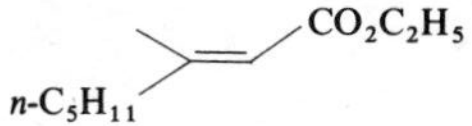 n-C_5H_{11}, $CO_2C_2H_5$ (alkene isomer)	" " " "	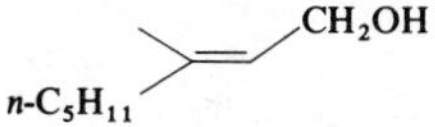 n-C_5H_{11}, CH_2OH (95)	730
C_{12}			
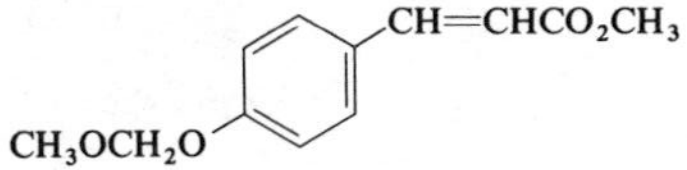 p-CH_3OCH_2O–C_6H_4–$CH{=}CHCO_2CH_3$	$LiAlH_3(OC_2H_5)$ (IA), ether, room temp, 4 h 40 min	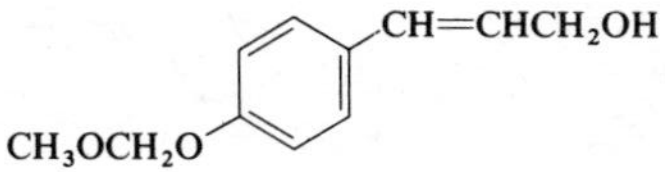 p-CH_3OCH_2O–C_6H_4–$CH{=}CHCH_2OH$ (68)	731
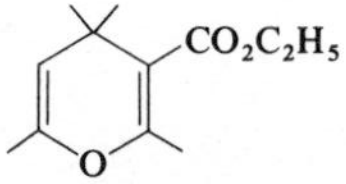 2,4,4,6-tetramethyl-4H-pyran-3-$CO_2C_2H_5$	$LiAlH_3(OC_2H_5)$, ether, reflux, 30 min	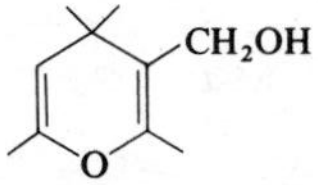 2,4,4,6-tetramethyl-4H-pyran-3-CH_2OH (82)	732
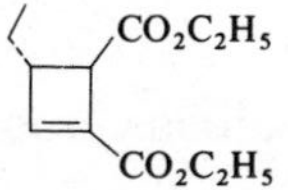 vinyl-cyclobutene with $CO_2C_2H_5$, $CO_2C_2H_5$	$LiAlH_3(OC_2H_5)$, ether	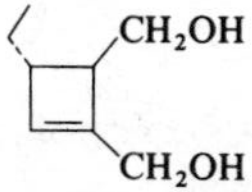 vinyl-cyclobutene with CH_2OH, CH_2OH (39)	652
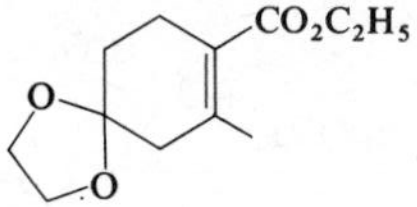 1,4-dioxaspiro[4.5]dec-7-ene, 7-methyl-8-$CO_2C_2H_5$	1. SMEAH (IA), C_6H_6–THF, room temp, overnight 2. H_3O^+	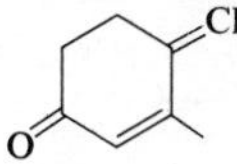 3-methyl-4-methylenecyclohex-2-enone, CH_2, O (89)	534
	SMEAH, C_6H_6, room temp, 31 h	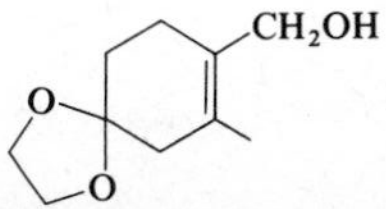 1,4-dioxaspiro[4.5]dec-7-ene, 7-methyl-8-CH_2OH (74)	733, 734

Reactant	Conditions	Product(s) (Yield, %)	Refs.
C_{13}			
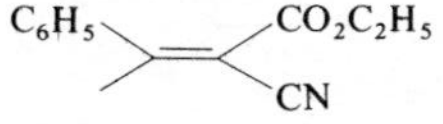 C_6H_5, $CO_2C_2H_5$, CN (substituted alkene)	1. LAH + (−)-QN (1:1), ether 2. Aqueous KOH 3. Heat	$C_6H_5CH(CH_3)CH_2CO_2H$ (*R*)-(−) (72) (11% e.e.)	735
	1. LAH + (+)-QND (1:1), ether 2. Aqueous KOH 3. Heat	" (*S*)-(+) (60) (1% e.e.)	735
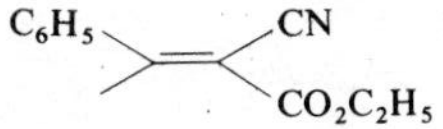 C_6H_5, CN, $CO_2C_2H_5$ (substituted alkene)	1. LAH + (−)-CND (1:1), ether 2. Aqueous KOH 3. Heat	" (*S*)-(+) (74) (3% e. e.)	735
	1. LAH + (+)-QND (1:1), ether 2. Aqueous KOH 3. Heat	" (*R*)-(−) (67) (12% e.e.)	735
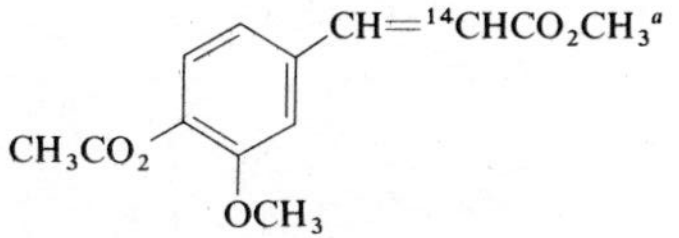 CH_3CO_2, OCH_3, $CH{=}^{14}CHCO_2CH_3$ [a] (aryl)	SMEAH, toluene, 80°, 20 min	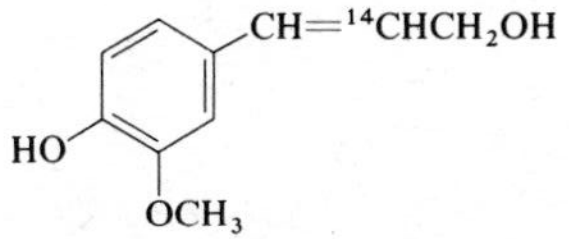 HO, OCH_3, $CH{=}^{14}CHCH_2OH$ (aryl) (—)	736
3,4,5-$(CH_3O)_3C_6H_2CH{=}CHCO_2CH_3$	SMEAH–CuBr–2-butanol, THF, −78°, 15 min, then −20°, 2 h	3,4,5-$(CH_3O)_3C_6H_2(CH_2)_2CO_2CH_3$ (97)	114, 113
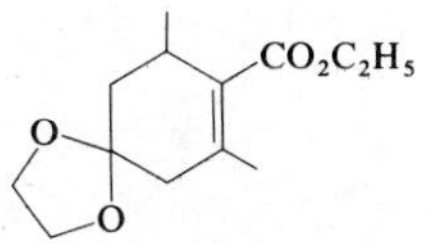 $CO_2C_2H_5$ (dioxolane ketal, cyclohexene)	$LiAlH_3(OC_2H_5)$ (IA)	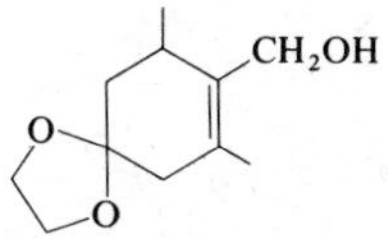 CH_2OH (dioxolane ketal, cyclohexene) (50)	737
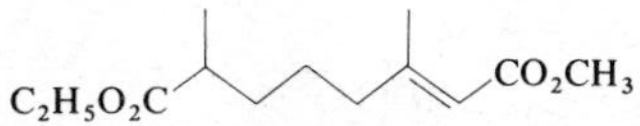 $C_2H_5O_2C$ … CO_2CH_3	$LiAlH_3(OC_2H_5)$	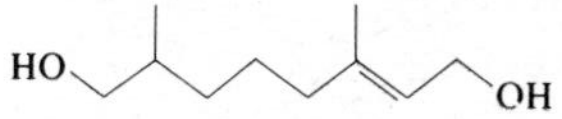 HO … OH (95)	738
C_{14}			
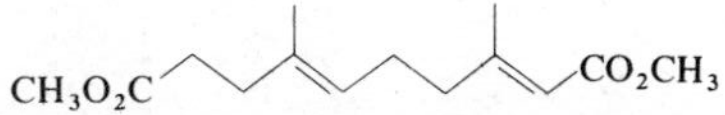 CH_3O_2C … CO_2CH_3	SMEAH, C_6H_6, 25°, 10 min	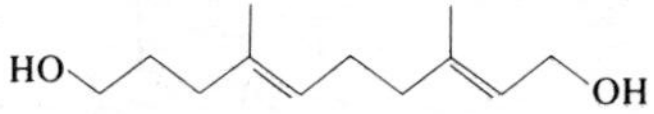 HO … OH (96)	110
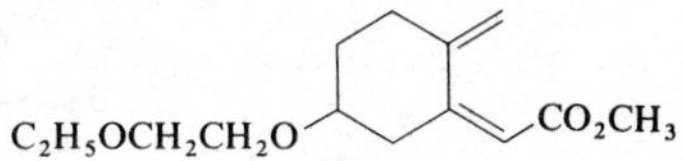 $C_2H_5OCH_2CH_2O$ … CO_2CH_3 *dl*-	1. SMEAH, ether, 22°, 1.5 h 2. H_3O^+ 3. 4-$O_2NC_6H_4COCl$, Py 4. KOH, EtOH	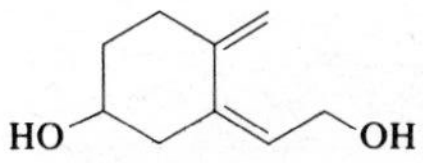 HO … OH *dl-cis* (39)	276

TABLE IX. REDUCTION OF α,β-UNSATURATED CARBOXYLIC ACIDS AND ESTERS (*Continued*)

Reactant	Conditions	Product(s) and Yield(s) (%)	Refs.
C_{15}			
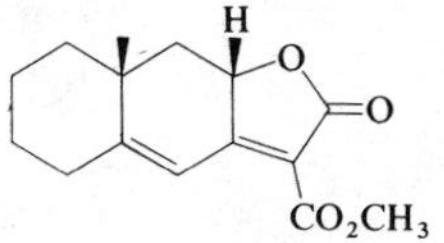 CO_2CH_3	SMEAH, THF, −40 to −20°, 45 min	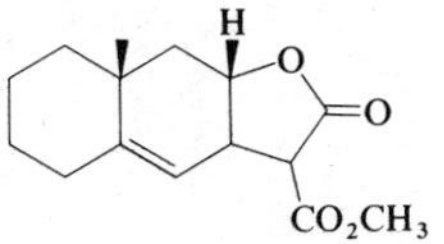 CO_2CH_3 (—)	622
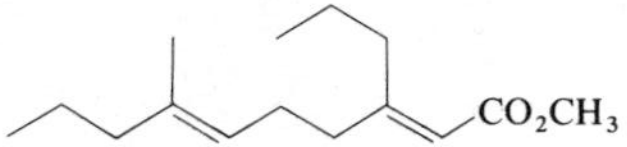 CO_2CH_3	$LiAlH_3(OC_2H_5)$, ether	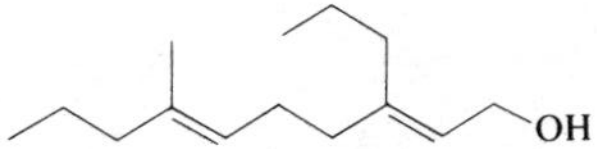 OH (90)	105
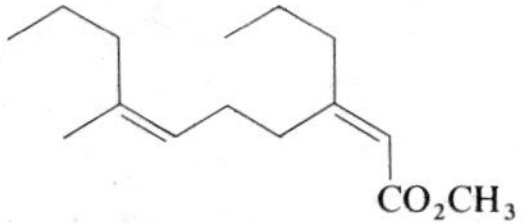 CO_2CH_3	", "	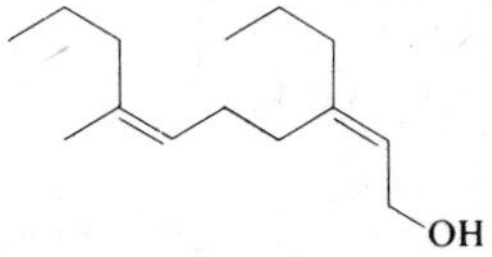 OH (90)	105
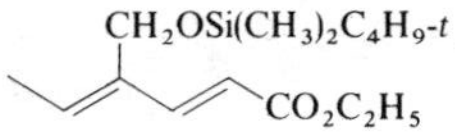 $CH_2OSi(CH_3)_2C_4H_9$-*t* $CO_2C_2H_5$	SMEAH, toluene, 0°, 45 min	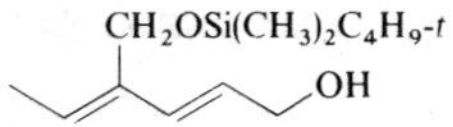 $CH_2OSi(CH_3)_2C_4H_9$-*t* OH (89)	739
C_{16}			
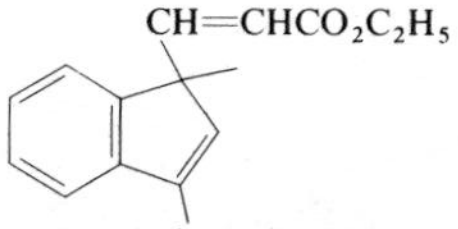 $CH{=}CHCO_2C_2H_5$ (+)-*trans*	$LiAlH_3(OC_2H_5)$, ether, 0°, 30 min	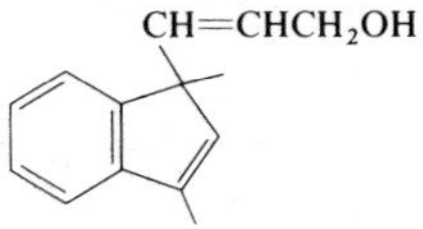 $CH{=}CHCH_2OH$ (+)-*trans* (37)	740
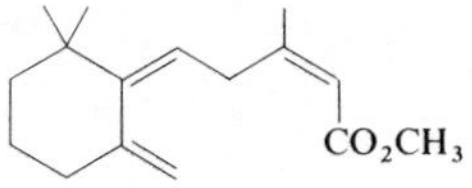 CO_2CH_3	1. SMEAH (IA), C_6H_6–ether, room temp, 30 min 2. $P(C_6H_5)_3Br_2$ 3. $OHCC(CH_3){=}CHCO_2C_4H_9$-*n*, KOH 4. SMEAH (IA), C_6H_6–ether, room temp, 1 h 5. MnO_2	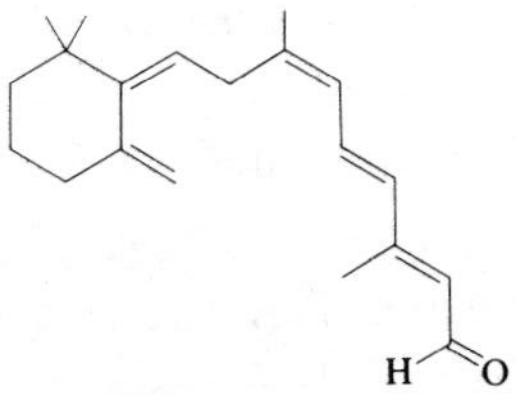 H O (9)	741

Substrate	Reagent and conditions	Product(s) (Yield, %)	Ref.
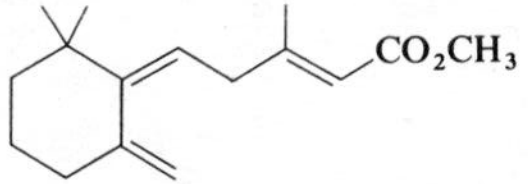 CO_2CH_3	1. SMEAH (IA), C_6H_6–ether, room temp, 30 min 2. $P(C_6H_5)_3Br_2$ 3. $OHCC(CH_3){=}CHCO_2C_4H_9$-*n*, KOH	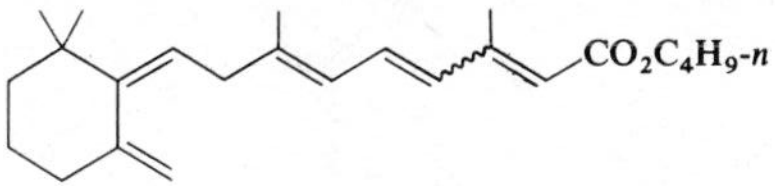 $CO_2C_4H_9$-*n* (21)	741
C_{17}			
$4\text{-}(C_6H_5CH_2O)C_6H_4CH{=}CHCO_2CH_3$	$LiAlH_3(OC_2H_5)$(IA), ether, room temp, 5.5 h	$4\text{-}(C_6H_5CH_2O)C_6H_4CH{=}CHCH_2OH$ (65)	731
 $CO_2C_2H_5$		 CR_2OH	
2-*trans*,6-*trans*	$LiAlD_4 + EtOH(1{:}0.2)$ (IA), THF, room temp, 1 h	2-*trans*,6-*trans* R = D (—)	742
	$LiAlH_3(OC_2H_5)$, ether, 20°, 2 h	" " " R = H (71)	743a
2-*cis*,6-*trans*	" " " "	2-*cis*,6-*trans* " R = H (—)	743a
	$LiAlD_4 + EtOH(1{:}0.2)$ (IA). THF, room temp, 1 h	" " " R = D (—)	742
2-*cis*(24%),*trans*(76%),6-*trans*	$LiAlH_3(OC_4H_9\text{-}t$, ether, 0°, 2.5 h	2-*cis*(30%)*trans*(70%), 6-*trans* " R = H (79)	743b
2-*cis*,6-*cis*	$LiAlH_3(OC_2H_5)$, ether, 20°, 2 h	2-*cis*,6-*cis* " R = H (—)	743a
2-*trans*,6-*cis*	" " " "	2-*trans*,6-*cis* " R = H (—)	743a
cis-*i*-$C_3H_7(CH_2)_3CH(CH_3)(CH_2)_3C(CH_3){=}CHCO_2C_2H_5$	LAH + EtOH (1:0.2) (IA), THF, room temp, 1 h	*i*-$C_3H_7(CH_2)_3CH(CH_3)(CH_2)_3C(CH_3){=}CHCH_2OH$ (95) *cis*- **I** + *i*-$C_3H_7(CH_2)_3CH(CH_3)(CH_2)_3CH(CH_3)(CH_2)_2OH$ (3) **II**	742
trans- "	", ", ", ", ", "	*trans*-**I** (89) + **II** (11)	742
C_{18}			
cis-$C_6H_5COC(C_6H_5){=}CHCO_2C_2H_5$	$LiAlH_3(OC_2H_5)$	*cis*-$C_6H_5CHOHC(C_6H_5){=}CHCH_2OH$ + $C_6H_5COCH(C_6H_5)(CH_2)_2OH$ (~100)	744

TABLE IX. Reduction of α,β-Unsaturated Carboxylic Acids and Esters (*Continued*)

Reactant	Conditions	Product(s) and Yield(s) (%)	Refs.
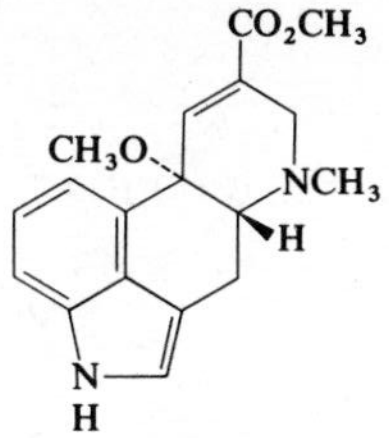 CO_2CH_3, CH_3O, NCH_3, H, N, H	SMEAH (IA), THF–C_6H_6, room temp, 30 min	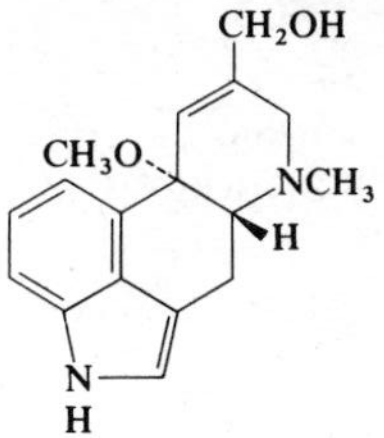 CH_2OH, CH_3O, NCH_3, H, N, H (—)	745, 746
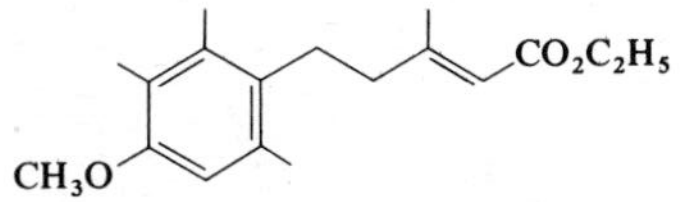 CH_3O, $CO_2C_2H_5$	SMEAH, ether, 35°, 4 h	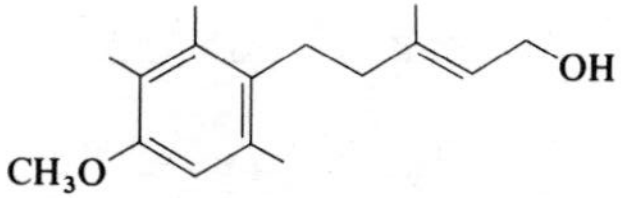 CH_3O, OH (74)	747
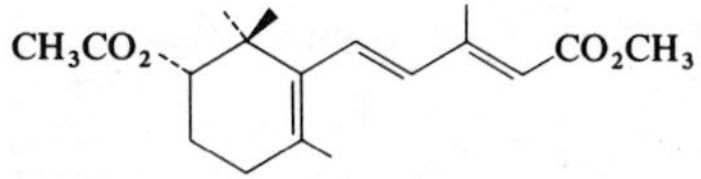 CH_3CO_2, CO_2CH_3	SMEAH	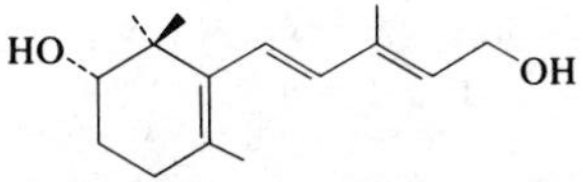 HO, OH (—)	678
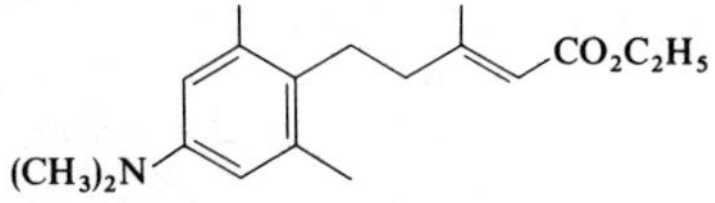 $(CH_3)_2N$, $CO_2C_2H_5$	SMEAH, ether, 35°, 4 h	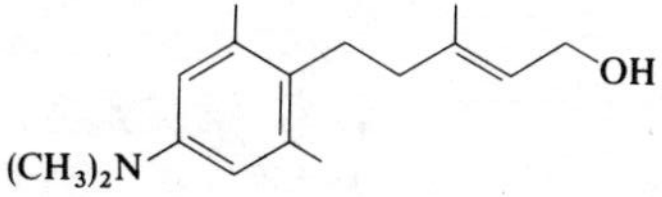 $(CH_3)_2N$, OH (65)	747
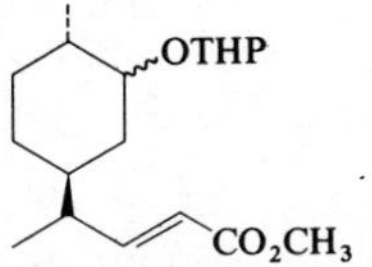 OTHP, CO_2CH_3	1. SMEAH, toluene, room temp to 50°, 30 min 2. H_3O^+ 3. MnO_2, room temp, 4 h 4. H_3O^+	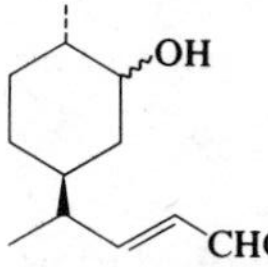 OH, CHO (78)	748
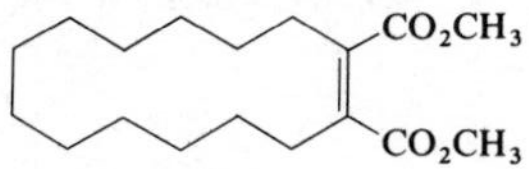 CO_2CH_3, CO_2CH_3	$LiAlH_3(OC_2H_5)$, ether, −20°	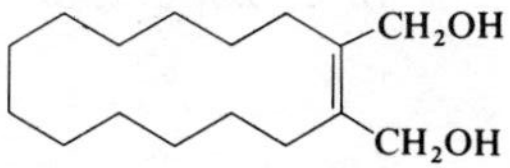 CH_2OH, CH_2OH (—)	749
	$LiAlH_3(OC_2H_5)$ (IA), −20°	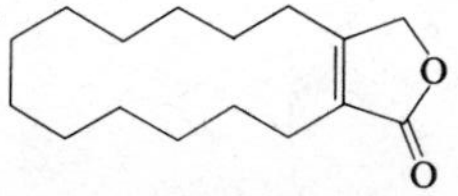 O, O (60)	749

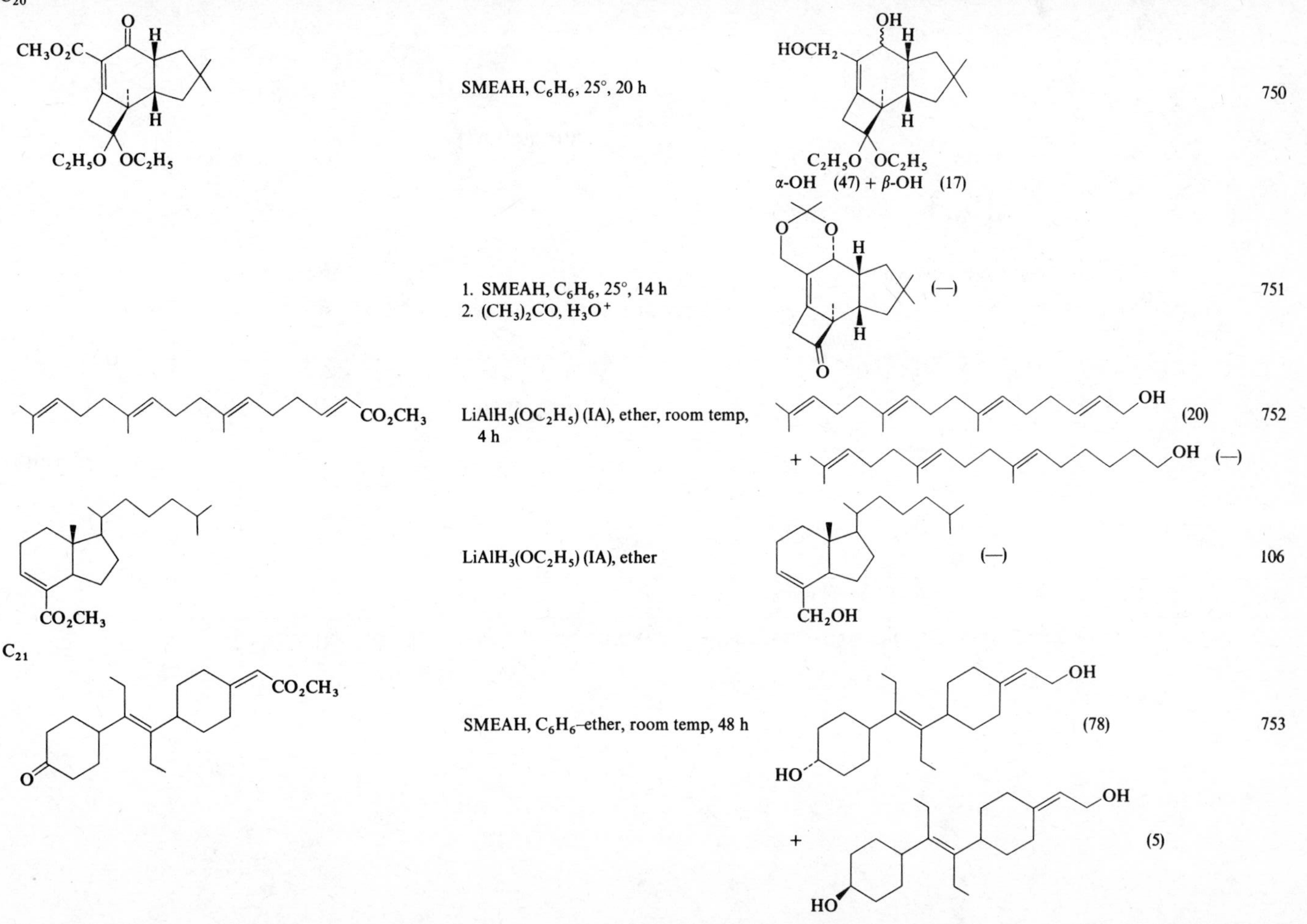

C20
CH3O2C
C2H5O OC2H5
SMEAH, C6H6, 25°, 20 h
HOCH2
OH
C2H5O OC2H5
α-OH (47) + β-OH (17)
750
1. SMEAH, C6H6, 25°, 14 h
2. (CH3)2CO, H3O+
(—)
751
CO2CH3
LiAlH3(OC2H5) (IA), ether, room temp, 4 h
OH (20)
+ OH (—)
752
CO2CH3
LiAlH3(OC2H5) (IA), ether
CH2OH
(—)
106
C21
CO2CH3
SMEAH, C6H6–ether, room temp, 48 h
HO
OH
(78)
+ HO
OH
(5)
753

TABLE IX. REDUCTION OF α,β-UNSATURATED CARBOXYLIC ACIDS AND ESTERS (*Continued*)

Reactant	Conditions	Product(s) and Yield(s) (%)	Refs.
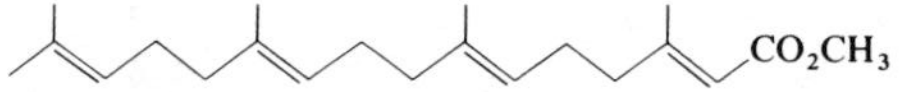 CO_2CH_3	$LiAlH_3(OC_2H_5)$ (IA), ether, room temp, 4 h	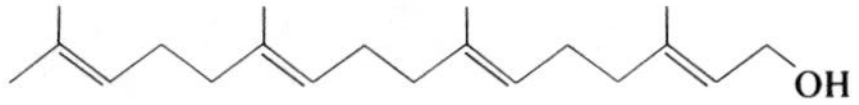 OH (50)	752
C_{22} 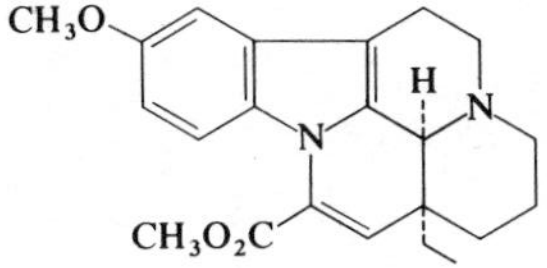		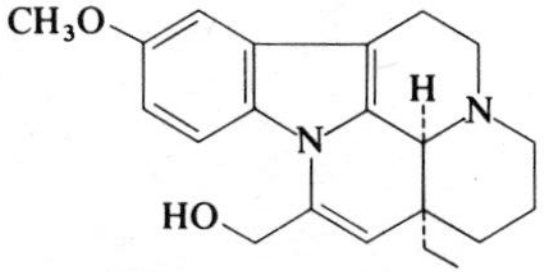	
CH_3O, H, N, N, CH_3O_2C	SMEAH	CH_3O, H, N, N, HO (—)	754
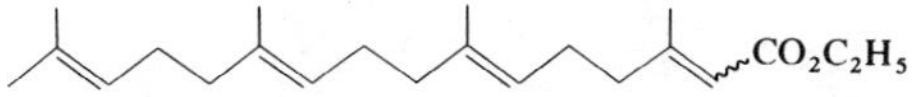 $CO_2C_2H_5$ 2-*cis*(24%),*trans*(76%),10-*trans*-	$LiAlH_3(OC_4H_9\text{-}t)$, ether, 0°, 2.5 h	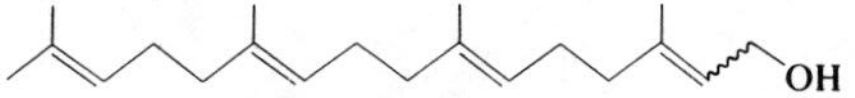 OH (88) 2-*cis*(30%),*trans*(70%),10-*trans*-	743b
C_{23}			
CH_3O, CH_3O, $CO_2C_2H_5$, CO_2H, O, O	$LiAlH_2(OC_2H_5)_2$	CH_3O, CH_3O, CH_2OH, CO_2H, O, O (44)	755a

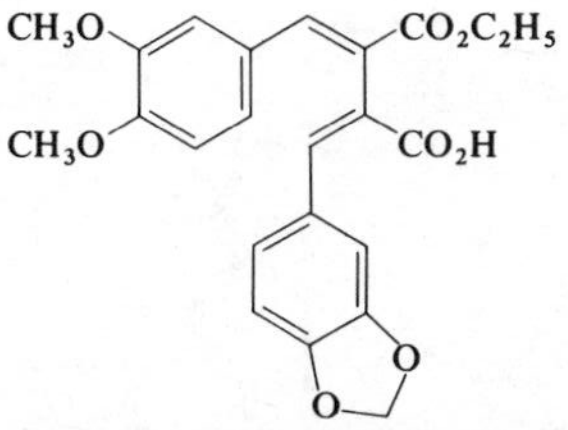

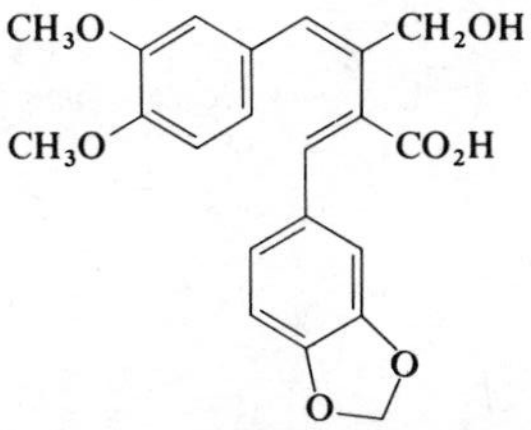

Reactant	Conditions	Product(s) (Yield, %)	Refs.
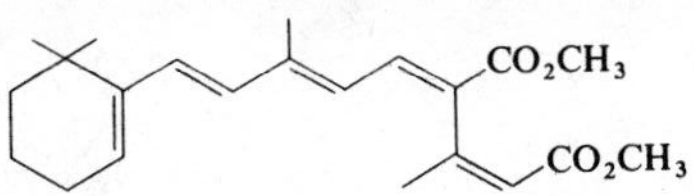 CO_2CH_3, CO_2CH_3	SMEAH (IA), toluene–C_6H_{14}, $-5°$, 30 min	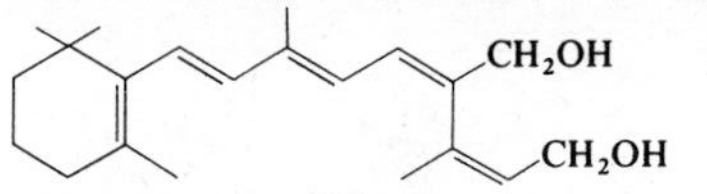 CH_2OH, CH_2OH (58) 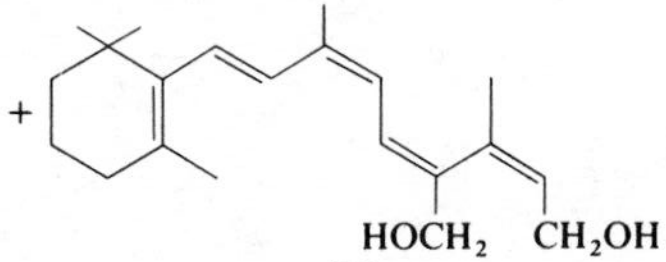 + $HOCH_2$, CH_2OH (12)	755b
C_{24}			
$CO_2C_4H_9$-*n*	1. SMEAH (IA), C_6H_6–ether, room temp, 1 h 2. MnO_2	CHO (13) + CHO (4)	741
C_{25}			
CO_2CH_3, H, H, X, OR	1. $LiAlH_2(OCH_3)_2$, THF, $-20°$, 30 min, then -20 to $0°$, 2.5 h 2. Ac_2O, Py	CH_2OAc, H, H, X, OR	
$X = CD_3$; R 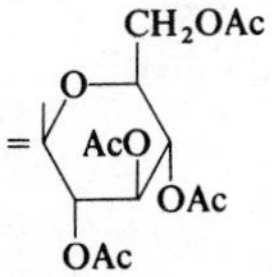 = (tetra-O-acetyl sugar: CH_2OAc, AcO, OAc, OAc)		(78)	280
$X = CH_2T$; "		(—)	280
$X = CH_3$; "		(49)	280

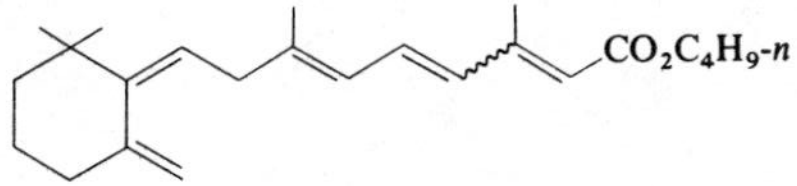

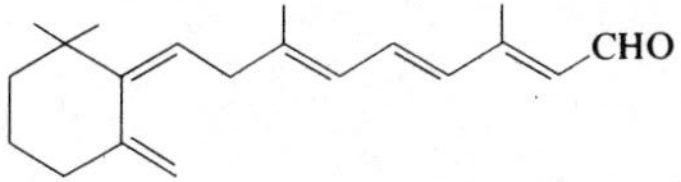

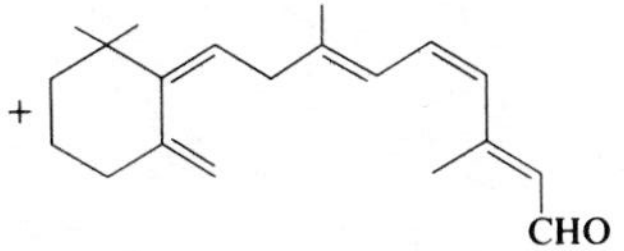

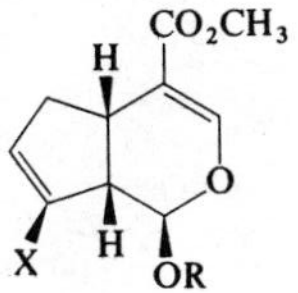

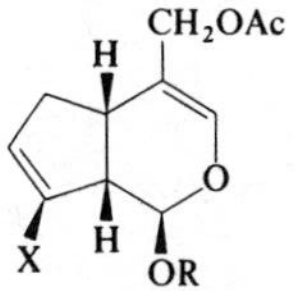

TABLE IX. REDUCTION OF α,β-UNSATURATED CARBOXYLIC ACIDS AND ESTERS (*Continued*)

Reactant	Conditions	Product(s) and Yield(s) (%)	Refs.
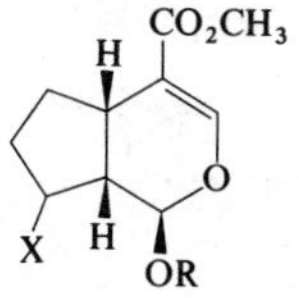 CO_2CH_3, H, H, X, OR	1. $LiAlH_2(OCH_3)_2$, THF, −20°, 30 min, −20° to 0°, 2.5 h 2. Ac_2O, Py	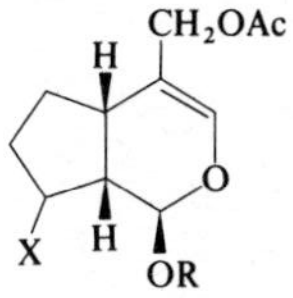 CH_2OAc, H, H, X, OR	
X = CH_2T, R 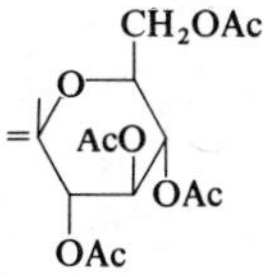 = CH_2OAc, AcO, OAc, OAc		(—)	280
X = CH_3 "		(85)	280
X = HO, CH_3 "		(~100)	280
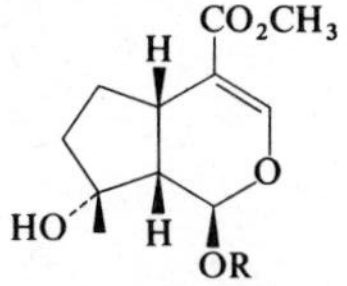 CO_2CH_3, H, H, HO, OR R = CH_2OAc, AcO, OAc, OAc 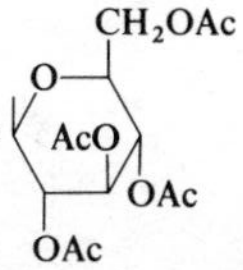	1. $LiAlH_2(OCH_3)_2$, THF, −15°, 1.5 h 2. AcOH, 55–60°, 45 min 3. Ac_2O, Py	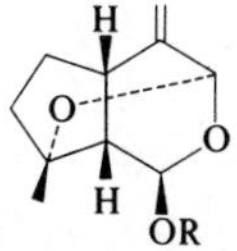 H, H, O, OR (17)	756a

Substrate	Conditions	Product(s) and Yield(s) (%)	Refs.
C_{26} H, O, OH, H, CH_3O_2C, OCH_3	SMEAH, THF, 1 h	H, O, OH, H, HO, OCH_3 (80)	756b
C_{32} CO_2CH_3, $t\text{-}C_4H_9(C_6H_5)_2SiO$, O, O, OCH_3	SMEAH	OH, $t\text{-}C_4H_9(C_6H_5)_2SiO$, O, O, OCH_3 (—)	639a

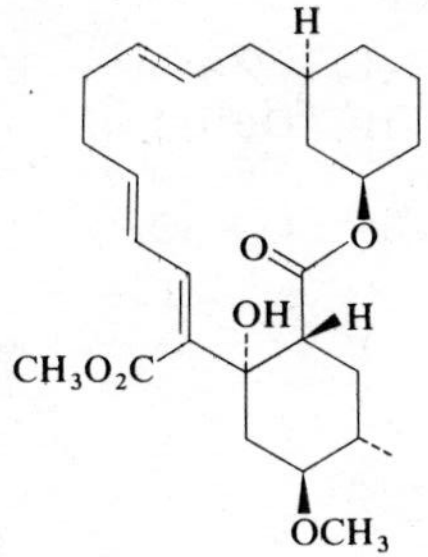

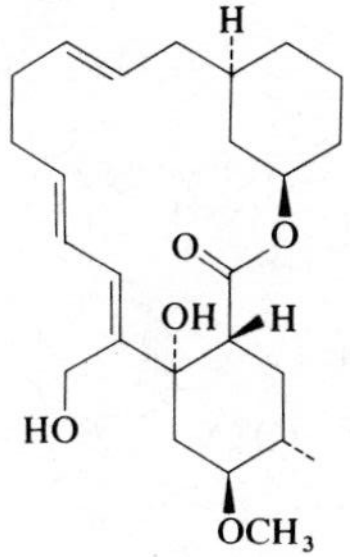

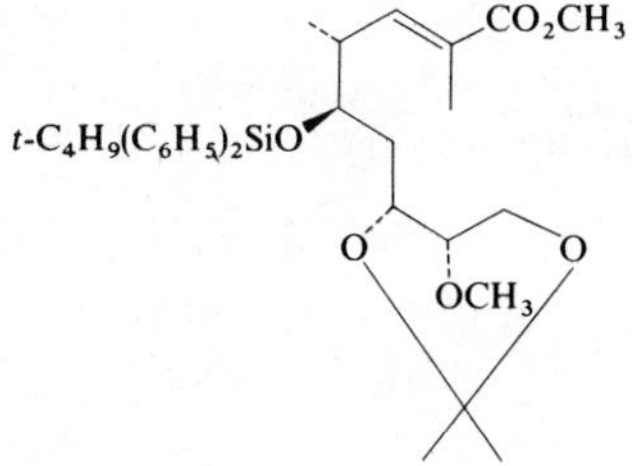

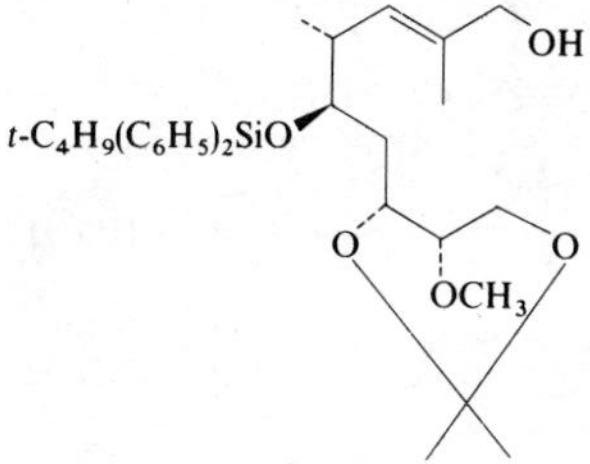

[a] This reaction was also carried out with ^{14}C in the methyl ester carboxyl carbon.

TABLE X. PARTIAL REDUCTION OF CARBOXYLIC ACID ESTERS TO ALDEHYDES

No. of Carbon Atoms	Reactant	Conditions	Product(s) and Yield(s) (%)	Refs.
C_2	HCO_2CH_3	SMEAH (IA), ether, −70°, 3 h	HCHO (86)	68
C_3	$CH_3CO_2CH_3$	SMEAH (IA), ether, −70°, 4 h	CH_3CHO (92)	68
		$NaAlH_2[OCH_2CH_2N(CH_3)_2]_2$	" (92)	68
C_5	$i\text{-}C_3H_7CO_2CH_3$	SMEAH (IA), ether, −70°, 5 h	$i\text{-}C_3H_7CHO$ (77)	72, 68
	$CH_3CO_2(CH_2)_2OCH_3$	SMEAH (IA), ether, −70°, 2 h	CH_3CHO (88)	68
C_6	$t\text{-}C_4H_9CO_2CH_3$	SMEAH (IA), ether, −70°, 5 h	$t\text{-}C_4H_9CHO$ (81)	68
C_7	Methyl pyridine-3-carboxylate	SMEAH (IA), ether, −70°, 5 h	Pyridine-3-carboxaldehyde (—)	72
	$CH_3C{\equiv}C(CH_2)_2CO_2CH_3$	SMEAH (IA), THF–C_6H_6, −70°, 6 h	$CH_3C{\equiv}C(CH_2)_2CHO$ (**I**) + $CH_3C{\equiv}C(CH_2)_3OH$ (**II**) (—) (**I**:**II** = 85:15)	757
	$n\text{-}C_5H_{11}CO_2CH_3$	SMEAH (IA), THF–C_6H_6, −70°, 7 h	$n\text{-}C_5H_{11}CHO$ (86)	68
C_8	3-Br-4-F-$C_6H_3CO_2CH_3$ (structure: Br, F, CO_2CH_3)	SMEAH + MPL, toluene, −50 to 25°, 16 h	3-Br-4-F-C_6H_3CHO (structure: Br, F, CHO) (53)	758
	$CH_3CO_2C_6H_4Cl\text{-}4$	LTBA, THF, −22°, 8 h	CH_3CHO (77)	64
	$ClCH_2CO_2C_6H_5$	LTBA (IA), THF, 0°, 30 min	$ClCH_2CHO$ (67)	64
	$CH_3CO_2C_6H_5$	LTBA (IA), THF, 0°, 4 h	CH_3CHO (71)	64
	$C_6H_5CO_2CH_3$	SMEAH, ether, −70°, 8 h	C_6H_5CHO (38)	68
		SMEAH + MHP (IA), toluene, −10°, 35 min	" (85)	73

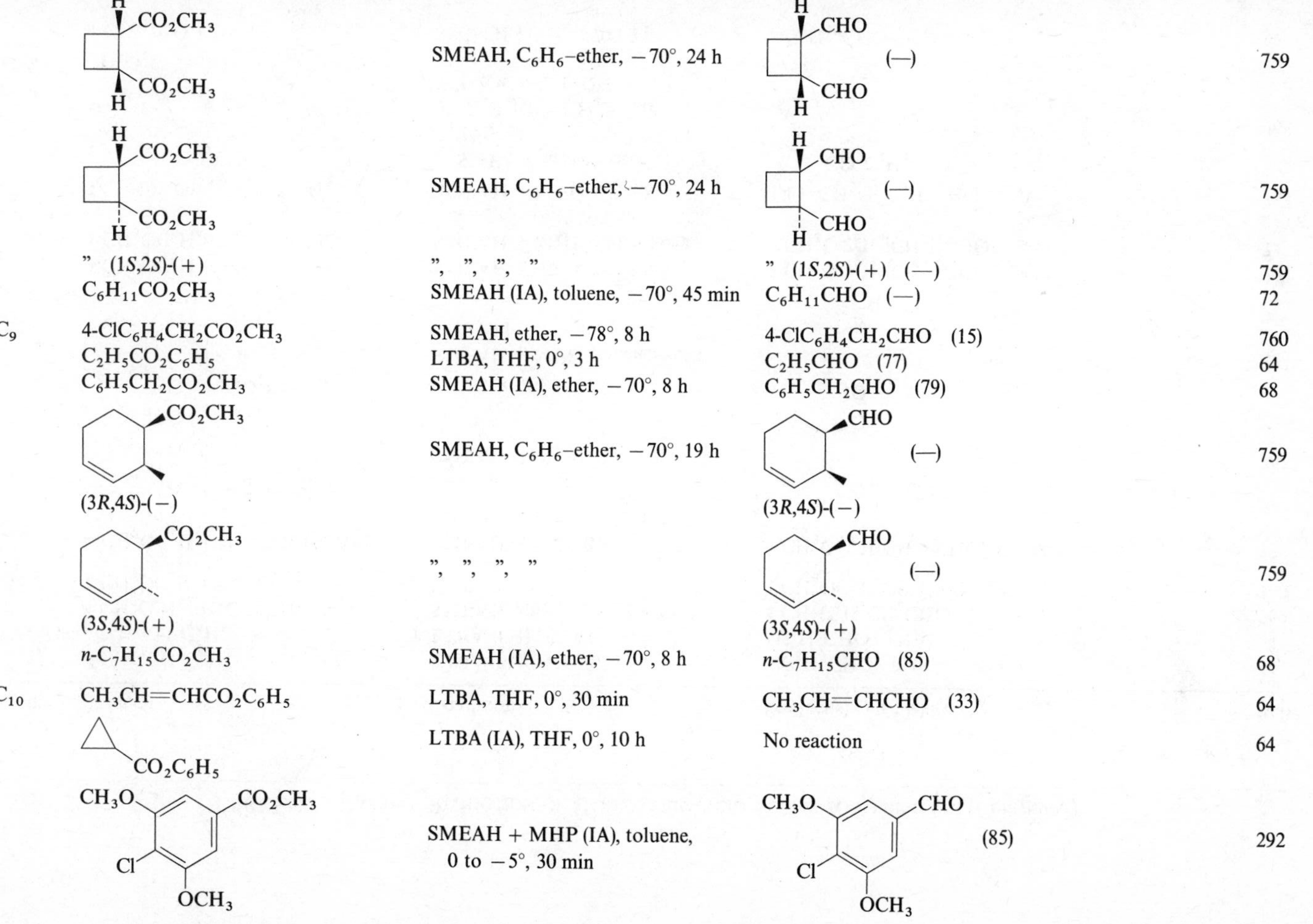

	Ester	Conditions	Product (Yield %)	Ref.
	cis-1,2-cyclobutane-$(CO_2CH_3)_2$ (structure)	SMEAH, C_6H_6–ether, $-70°$, 24 h	cis-1,2-cyclobutane-$(CHO)_2$ (structure) (—)	759
	trans-1,2-cyclobutane-$(CO_2CH_3)_2$ (structure)	SMEAH, C_6H_6–ether, $-70°$, 24 h	trans-1,2-cyclobutane-$(CHO)_2$ (structure) (—)	759
	" (1*S*,2*S*)-(+)	", ", ", "	" (1*S*,2*S*)-(+) (—)	759
	$C_6H_{11}CO_2CH_3$	SMEAH (IA), toluene, $-70°$, 45 min	$C_6H_{11}CHO$ (—)	72
C_9	$4\text{-}ClC_6H_4CH_2CO_2CH_3$	SMEAH, ether, $-78°$, 8 h	$4\text{-}ClC_6H_4CH_2CHO$ (15)	760
	$C_2H_5CO_2C_6H_5$	LTBA, THF, 0°, 3 h	C_2H_5CHO (77)	64
	$C_6H_5CH_2CO_2CH_3$	SMEAH (IA), ether, $-70°$, 8 h	$C_6H_5CH_2CHO$ (79)	68
	methyl 6-methylcyclohex-3-enecarboxylate (structure), (3*R*,4*S*)-(−)	SMEAH, C_6H_6–ether, $-70°$, 19 h	6-methylcyclohex-3-enecarbaldehyde (structure), (3*R*,4*S*)-(−) (—)	759
	methyl 6-methylcyclohex-3-enecarboxylate (structure), (3*S*,4*S*)-(+)	", ", ", "	6-methylcyclohex-3-enecarbaldehyde (structure), (3*S*,4*S*)-(+) (—)	759
	$n\text{-}C_7H_{15}CO_2CH_3$	SMEAH (IA), ether, $-70°$, 8 h	$n\text{-}C_7H_{15}CHO$ (85)	68
C_{10}	$CH_3CH{=}CHCO_2C_6H_5$	LTBA, THF, 0°, 30 min	$CH_3CH{=}CHCHO$ (33)	64
	cyclopropyl-$CO_2C_6H_5$ (structure)	LTBA (IA), THF, 0°, 10 h	No reaction	64
	3,5-(CH_3O)(Cl)(OCH_3)-$C_6H_2CO_2CH_3$ (structure: CH_3O, Cl, OCH_3, CO_2CH_3)	SMEAH + MHP (IA), toluene, 0 to $-5°$, 30 min	(CH_3O, Cl, OCH_3)-C_6H_2CHO (structure) (85)	292

TABLE X. Partial Reduction of Carboxylic Acid Esters to Aldehydes (*Continued*)

No. of Carbon Atoms	Reactant	Conditions	Product(s) and Yield(s) (%)	Refs.
C_{10} (*Contd.*)	n-$C_3H_7CO_2C_6H_5$	LTBA (IA), THF, 0°, 4 h	n-C_3H_7CHO (63)	64
	i-$C_3H_7CO_2C_6H_5$	LTBA, THF, 0°, 4 h	i-C_3H_7CHO (71)	64
	4-$CH_3C_6H_4CH_2CO_2CH_3$	SMEAH, ether, −78°, 8 h	4-$CH_3C_6H_4CH_2CHO$ (21)	760
	4-$CH_3OC_6H_4CH_2CO_2CH_3$	", ", ", "	4-$CH_3OC_6H_4CH_2CHO$ (30)	760
C_{11}	4-$CH_3OC_6H_4CH{=}CHCO_2CH_3$	SMEAH + MPL (IA), C_6H_6–toluene, 0°, 35 min	4-$CH_3OC_6H_4CH{=}CHCHO$ (84)	75
	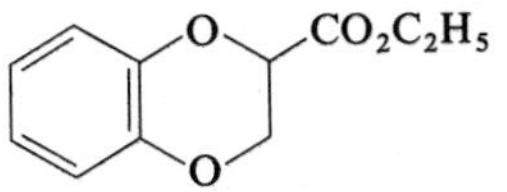 O, O, $CO_2C_2H_5$	SMEAH, THF–C_6H_6, −70°, 5 h	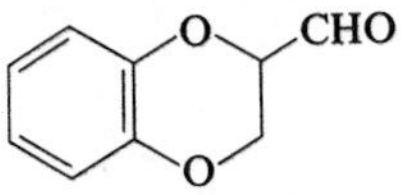 O, O, CHO (90)	761
	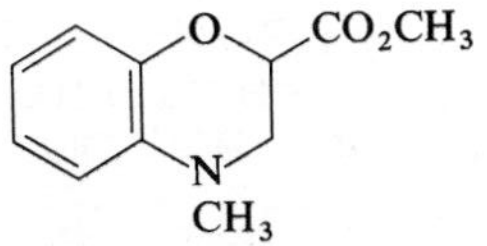 O, N, CH_3, CO_2CH_3	SMEAH + MPL, toluene, −40°, 20 min	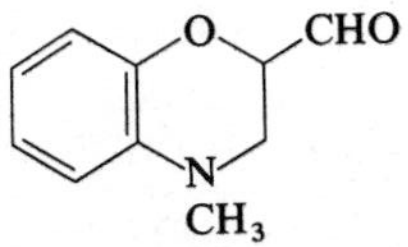 O, N, CH_3, CHO (19)	293
	3-$CH_3OC_6H_4(CH_2)_2CO_2CH_3$	SMEAH, THF, −75°, 3 h	3-$CH_3OC_6H_4(CH_2)_2CHO$ (90)	762
	4-$CH_3OC_6H_4(CH_2)_2CO_2CH_3$	SMEAH + MHP (IA), toluene, −55 to −40°, 70 min	4-$CH_3OC_6H_4(CH_2)_2CHO$ (88)	73
	3,4,5-$(CH_3O)_3C_6H_2CO_2CH_3$	SMEAH + MPL, toluene, −5°	3,4,5-$(CH_3O)_3C_6H_2CHO$ (94)	763
	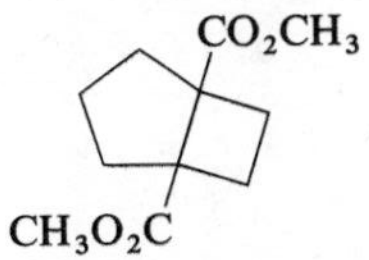 CO_2CH_3, CH_3O_2C	1. SMEAH, hexane–ether, −70°, 24 h 2. $(C_6H_5)_3P{=}CH_2$, −70° 3. $NaOC_2H_5$, EtOH	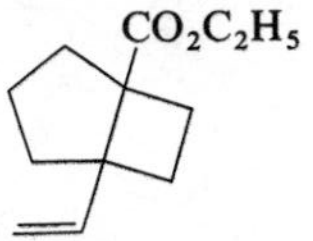 $CO_2C_2H_5$ (—)	764
	n-$C_9H_{19}CO_2CH_3$	SMEAH, ether, −70°, 2 h	n-$C_9H_{19}CHO$ (82)	68

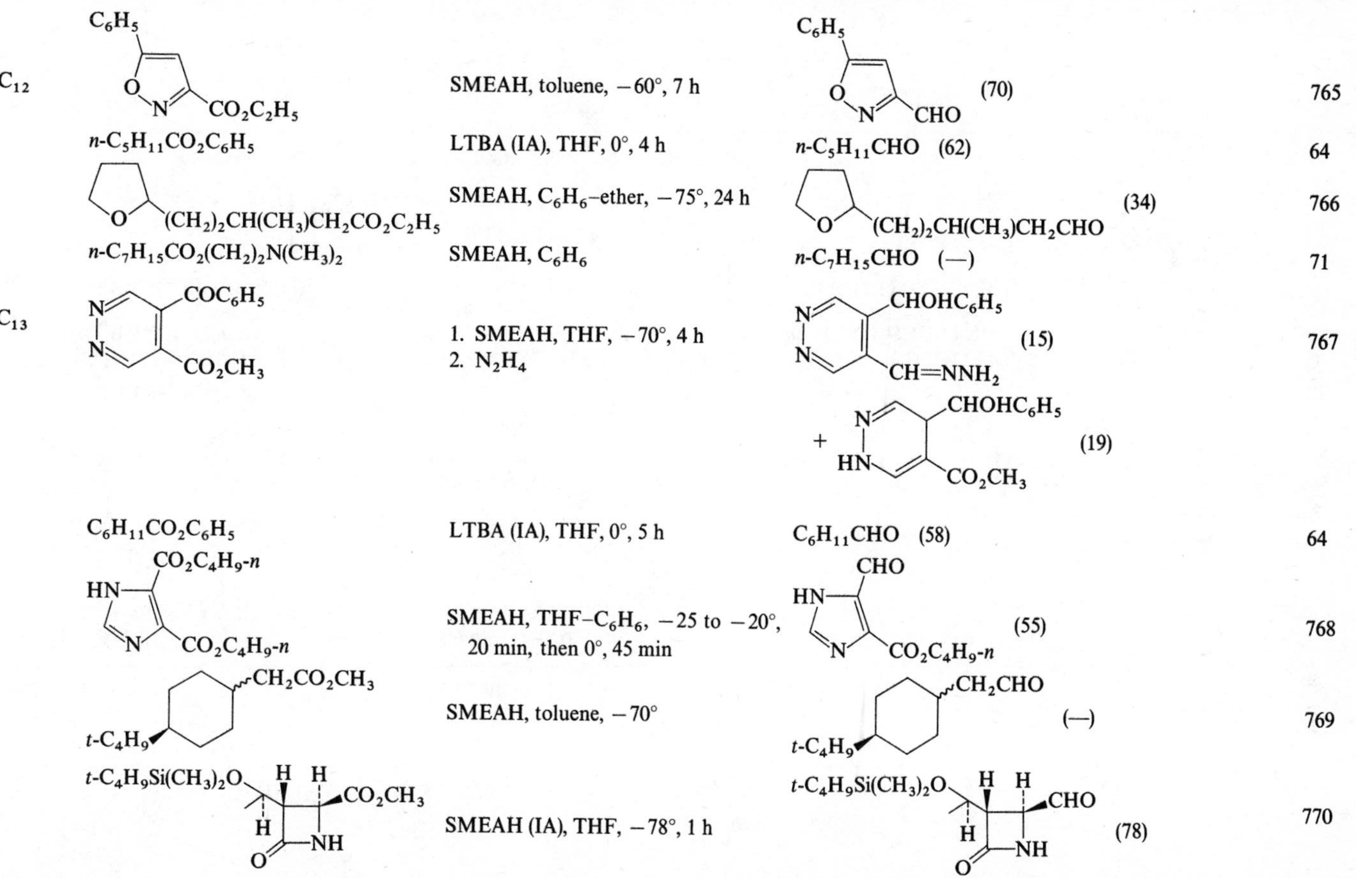

	Reactant	Conditions	Product(s) and Yield(s) (%)	Refs.
C_{12}	5-C_6H_5-isoxazole-3-$CO_2C_2H_5$ (structure)	SMEAH, toluene, −60°, 7 h	5-C_6H_5-isoxazole-3-CHO (structure) (70)	765
	n-$C_5H_{11}CO_2C_6H_5$	LTBA (IA), THF, 0°, 4 h	n-$C_5H_{11}CHO$ (62)	64
	2-tetrahydrofuryl-$(CH_2)_2CH(CH_3)CH_2CO_2C_2H_5$ (structure)	SMEAH, C_6H_6–ether, −75°, 24 h	2-tetrahydrofuryl-$(CH_2)_2CH(CH_3)CH_2CHO$ (structure) (34)	766
	n-$C_7H_{15}CO_2(CH_2)_2N(CH_3)_2$	SMEAH, C_6H_6	n-$C_7H_{15}CHO$ (—)	71
C_{13}	pyridazine, COC_6H_5, CO_2CH_3 (structure)	1. SMEAH, THF, −70°, 4 h 2. N_2H_4	pyridazine, $CHOHC_6H_5$, $CH{=}NNH_2$ (structure) (15) + dihydropyridazine (HN), $CHOHC_6H_5$, CO_2CH_3 (structure) (19)	767
	$C_6H_{11}CO_2C_6H_5$	LTBA (IA), THF, 0°, 5 h	$C_6H_{11}CHO$ (58)	64
	imidazole (HN), $CO_2C_4H_9$-n, $CO_2C_4H_9$-n (structure)	SMEAH, THF–C_6H_6, −25 to −20°, 20 min, then 0°, 45 min	imidazole (HN), CHO, $CO_2C_4H_9$-n (structure) (55)	768
	cyclohexane, t-C_4H_9, $CH_2CO_2CH_3$ (structure)	SMEAH, toluene, −70°	cyclohexane, t-C_4H_9, CH_2CHO (structure) (—)	769
	t-$C_4H_9Si(CH_3)_2O$, H, H, CO_2CH_3, azetidinone (O, NH) (structure)	SMEAH (IA), THF, −78°, 1 h	t-$C_4H_9Si(CH_3)_2O$, H, H, CHO, azetidinone (O, NH) (structure) (78)	770

TABLE X. Partial Reduction of Carboxylic Acid Esters to Aldehydes (*Continued*)

No. of Carbon Atoms	Reactant	Conditions	Product and Yield (%)	Refs.
C_{14}	$C_6H_5CH_2CO_2C_6H_5$	LTBA, THF, 0°, 1 h	$C_6H_5CH_2CHO$ (73)	64
	$C_6H_5OCH_2CO_2C_6H_5$	LTBA (IA), THF, 0°, 15 min	$C_6H_5OCH_2CHO$ (49)	64
C_{15}	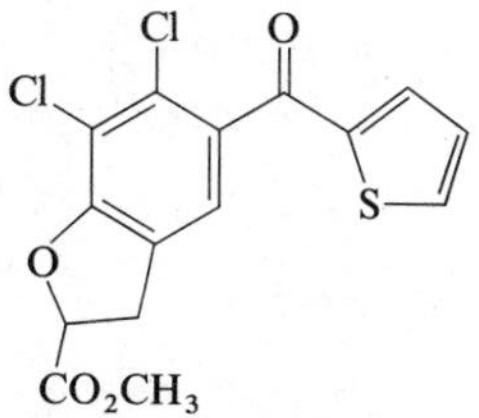 6,7-dichloro-5-(2-thienylcarbonyl)-2,3-dihydrobenzofuran-2-CO_2CH_3 (structure)	SMEAH, THF, −70°, 1 h	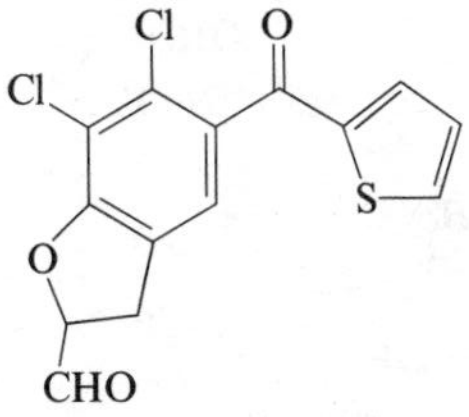 6,7-dichloro-5-(2-thienylcarbonyl)-2,3-dihydrobenzofuran-2-CHO (structure) (31)	771
	$C_6H_5CH{=}CHCO_2C_6H_5$	LTBA (IA), THF, 0°, 2 h	$C_6H_5CH{=}CHCHO$ (60)	64
	$(C_6H_5)_2CHCO_2CH_3$	SMEAH, ether, −70°, 36 h	$(C_6H_5)_2CHCHO$ (49)	68
C_{16}	3-[3,4,5-$(CH_3O)_3C_6H_2(CH_2)_2$]-isothiazole-4-CO_2CH_3 (structure) 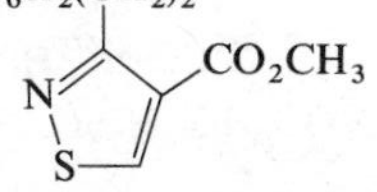	1. N_2H_4, O_2, Cu^{2+} 2. LTBA 3. $(C_6H_5)_3P{=}CHCH{=}CHCO_2CH_3$	3-[3,4,5-$(CH_3O)_3C_6H_2(CH_2)_2$]-4-($CH{=}CHCH{=}CHCO_2CH_3$)-isothiazole (structure) (—) 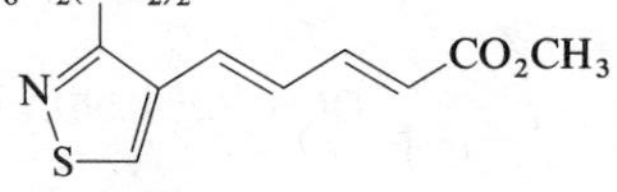	772, 773
C_{19}	n-$C_8H_{17}CH{=}CH(CH_2)_7CO_2CH_3$	SMEAH, ether, −70°, 24 h	n-$C_8H_{17}CH{=}CH(CH_2)_7CHO$ (70)	68

TABLE XI. Reduction of Acyl Chlorides

No. of Carbon Atoms	Reactant	Conditions	Product(s) and Yield(s) (%)	Refs.
C_4	n-C_3H_7COCl	$NaAlH(OC_2H_5)_3$, THF, 65°, 2 h	n-C_4H_9OH (85)	319
	i-C_3H_7COCl	SMEAH, toluene, 70–92°, 1 h	i-C_4H_9OH (99)	168
C_6	COCl, N, N, CH_3	LTBA, BME, −10 to 0°	CH_2O_2C, N, N, N, N, CH_3, CH_3 (27)	774
	n-$C_5H_{11}COCl$	SMEAH, C_6H_6, 70°, 1 h	n-$C_6H_{13}OH$ (99)	168
		$Ca[AlH_2(OC_4H_9$-$i)_2]_2 \cdot$THF, toluene, 80°, 30 min	" (~100)	165
C_7	C_6H_5COCl	$NaAlH_3(OCH_3)$, THF, 20–25°, 2 h	$C_6H_5CH_2OH$ (—)	775
		$NaAlH(OC_2H_5)_3$, THF, 65°, 4 h	" (87)	319
		SMEAH, C_6H_6, 70–90°, 4 h	" (94)	168, 68
		$NaAlH[OCH_2CH_2N(CH_3)_2]_3$, C_6H_6, reflux, 2.5 h	" (90)	238
		$NaAl_2H_4[OCH_2CH_2N(CH_3)_2]_3$, C_6H_6, 80°	" (97)	634
		$NaAlH_2(OCH_2CH_2OC_3H_7$-$i)_2$, C_6H_6, 25°, 2 h	" (80)	632
		$Ca[AlH_2(OC_4H_9$-$i)_2]_2 \cdot$THF, toluene, 25°, 1 h	" (~100)	165
C_8	ClOC, S, S, S, S, COCl	LTBA, THF, −78°, 1 h	$HOCH_2$, S, S, S, S, CH_2OH (16)	776

TABLE XI. REDUCTION OF ACYL CHLORIDES (*Continued*)

No. of Carbon Atoms	Reactant	Conditions	Product(s) and Yield(s)(%)	Refs.
C_8 (*Contd.*)	$C_6H_5CH_2COCl$	SMEAH, toluene, 70–102°, 1 h	$C_6H_5(CH_2)_2OH$ (98)	168
	n-$C_7H_{15}COCl$	SMEAH, C_6H_6, 70–85°, 1 h	n-$C_8H_{17}OH$ (88)	168
C_9	$C_6H_5CH{=}CHCOCl$	$NaAlH(OC_2H_5)_3$, THF, 65°, 4 h	$C_6H_5CH{=}CHCH_2OH$ (76)	319
		SMEAH (IA), C_6H_6, 0–18°, 1 h	" (52)	167, 168
		SMEAH, C_6H_6–toluene, 70–110°, 1 h	$C_6H_5(CH_2)_3OH$ (59)	167, 168
		$Ca[AlH_2(OC_4H_9\text{-}i)_2]_2 \cdot$THF, toluene, 100°, 5 h	" (62)	165
	(+)-$CH_3O_2C[CH(OAc)]_2COCl$	1. LTBA, THF, 0°, 1 h 2. Ac_2O	$CH_3O_2C[CH(OAc)]_2CH_2OAc$ (58)	776b
	n-$C'_8H_{17}COCl$	SMEAH, C_6H_6, 70–95°, 4 h	n-$C_9H_{19}OH$ (90)	168
C_{10}	$C_6H_5CH(OAc)COCl$	LTBA, BME, −10°	$C_6H_5CH(OAc)CH_2OH$ (54) + $C_6H_5CHOHCH_2OAc$ (—)	777
C_{11}	1-Adamantyl-COCl	SMEAH, C_6H_6	1-Adamantyl-CH_2OH (39)	778
C_{12}	n-$C_{11}H_{23}COCl$	SMEAH, toluene, 70–105°, 1 h	n-$C_{12}H_{25}OH$ (97)	168
C_{14}	$CH_3O_2CC(CH_3)_2CH(C_6H_5)CH_2COCl$	1. LTBA, THF, room temp, 2 h, then 60°, 40 min 2. NaOH, DME	$HO_2CC(CH_3)_2CH(C_6H_5)(CH_2)_2OH$ (46)	779
C_{15}	$C_6H_5CH(OBz)COCl$	LTBA, BME, −10°	$C_6H_5CHOHCH_2OBz$ (37)	777
C_{17}	1-CO_2H-2-(4-$CH_3OC_6H_4$)-3-C_6H_5-cyclopropene	1. $(COCl)_2$ 2. LTBA, BME, 0°, 2 h	1-CH_2OH-2-(4-$CH_3OC_6H_4$)-3-C_6H_5-cyclopropene (35)	294b

	Substrate	Conditions	Product(s) and Yield(s) (%)	Refs.
	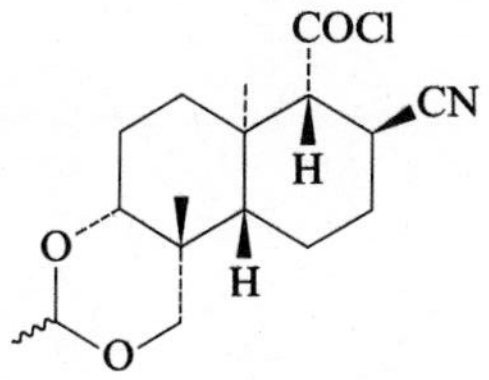 [structure: COCl, CN, H, H, cyclic acetal]	LTBA, THF, 0°, 1 h	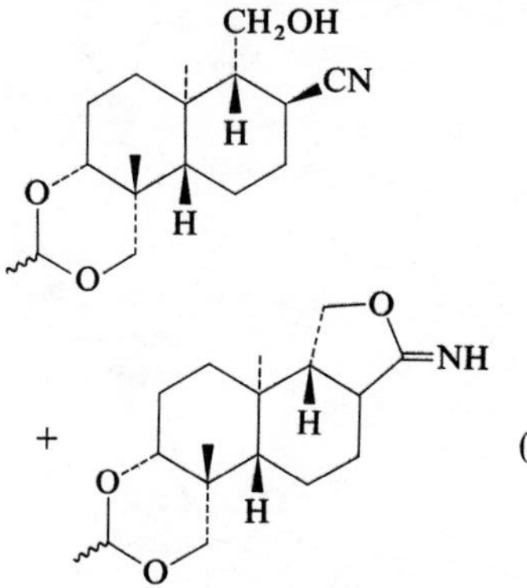 [structure: CH_2OH, CN, H, H, cyclic acetal] + [structure: =NH, O, H, H, cyclic acetal] (—)	780
C_{18}	$C_6H_5CH(COCl)O_2CC_6H_2(CH_3)_3$-2,4,6	LTBA, BME, −10°	$C_6H_5CH(CH_2OH)O_2CC_6H_2(CH_3)_3$-2,4,6 (77)	777
C_{19}	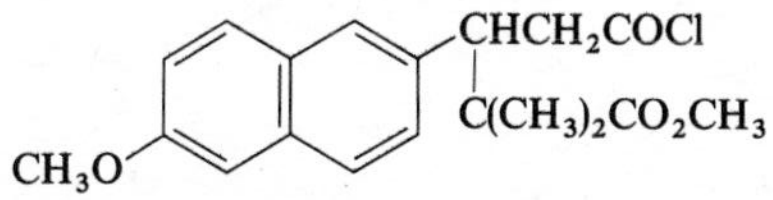 [structure: CH_3O-naphthyl-$CHCH_2COCl$ / $C(CH_3)_2CO_2CH_3$]	1. LTBA, THF, 0–60°, 40 min 2. NaOH, DME	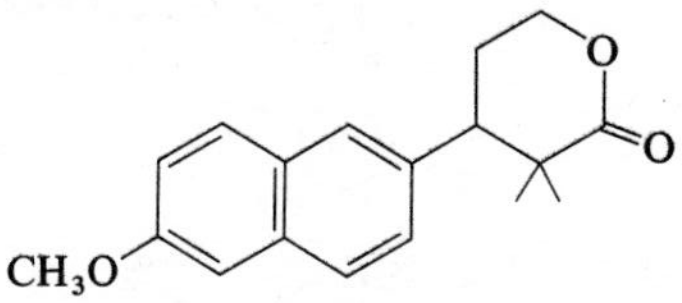 [structure: CH_3O-naphthyl lactone, O] (18)	779
C_{23}	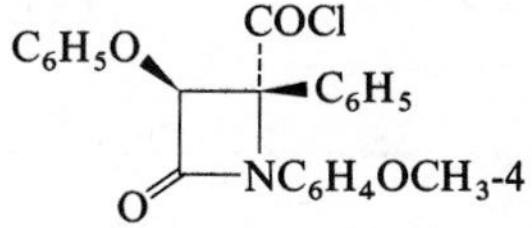 [structure: C_6H_5O, COCl, C_6H_5, O, $NC_6H_4OCH_3$-4]	LTBA, BME, −55°, 1.3 h	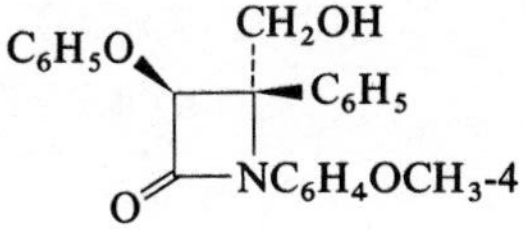 [structure: C_6H_5O, CH_2OH, C_6H_5, O, $NC_6H_4OCH_3$-4] (—)	190

TABLE XII. PARTIAL REDUCTION OF ACYL CHLORIDES TO ALDEHYDES

No. of Carbon Atoms	Reactant	Conditions	Products(s) and Yield(s) (%)	Refs.
C_2	CH_3COCl	Na saccharin, SMEAH (IA), C_6H_6, 0–5°, 2 h	CH_3CHO (76)	98
C_3	C_2H_5COCl	", ", ", ", "	C_2H_5CHO (78)	98
C_4	5-nitroisothiazole-3-COCl (O_2N, S, N ring, COCl)	LTBA, BME, −70°, 1 h, then room temp, 1 h	5-nitroisothiazole-3-CHO (O_2N, S, N ring, CHO) (48)	781
	$ClOC\text{–}CH{=}CH\text{–}COCl$	LTBA (IA), BME, −78°, 1 h; to room temp, 1 h	$OHC\text{–}CH{=}CH\text{–}CHO$ (59)	78
	$CH_3CH{=}CHCOCl$	", ", ", ", "	$CH_3CH{=}CHCHO$ (48)	78
	cyclopropyl-COCl	", ", ", ", "	cyclopropyl-CHO (42)	78
	$n\text{-}C_3H_7COCl$	", ", ", ", "	$n\text{-}C_3H_7CHO$ (37)	78
		$NaAlH(OC_4H_9\text{-}t)_3$, BME, −70°, 3 h	" (48)	80
		Na saccharin, SMEAH (IA), C_6H_6, 0–5°, 2 h	" (78)	98
	$i\text{-}C_3H_7COCl$	LTBA (IA), BME −78°, 1 h; to room temp, 1 h	$i\text{-}C_3H_7CHO$ (57)	78
		$NaAlH(OC_4H_9\text{-}t)_3$, BME, −70°, 3 h	" (34)	80
C_5	pyrazine-2-COCl (N, N ring, COCl)	LTBA, THF, −78°, 2 h	pyrazinyl-CO_2CH_2-pyrazinyl (55) + pyrazine-2-CHO (20)	302

Reactant	Conditions	Product(s) (Yield, %)	Refs.
3-Methyl-4-nitroisothiazole-5-COCl	LTBA, BME, −70°, 95 min	3-Methyl-4-nitroisothiazole-5-CHO + 3-methyl-4-nitroisothiazole-5-CH_2OH (—)	782
2-Thiophene-COCl	$NaAlH(OC_4H_9\text{-}t)_3$, BME, −70°, 3 h	2-Thiophene-CHO (63)	80
2-Furan-COCl	$NaAlH(OC_4H_9\text{-}t)_3$, BME, −70°, 3 h	2-Furan-CHO (85)	80
3-CH_3O-isoxazole-5-COCl	LTBA (IA), BME, −75°, 1 h; to room temp, 1 h	No reaction	783
3-O_2N-1-CH_3-pyrazole-4-COCl	LTBA	3-O_2N-1-CH_3-pyrazole-4-CHO (—)	784
ClOC-cyclopropane-COCl	LTBA	OHC-cyclopropane-CHO (43)	785
4-Br-3-methyl-4,5-dihydroisothiazole-5-COCl	LTBA (IA), BME–ether, −70°, 90 min	4-Br-3-methyl-4,5-dihydroisothiazole-5-CHO (44)	786
1-Methylcyclopropane-COCl	LTBA, BME −60°, 1 h; to room temp, 2 h	1-Methylcyclopropane-CHO (65)	787
Cyclobutane-COCl	LTBA (IA), BME −78°, 1 h; to room temp, 1 h	Cyclobutane-CHO (46)	78
$i\text{-}C_4H_9{}^{14}COCl$	$LiAlH(OC_2H_5)_3$, ether, −70°, 30 min	$i\text{-}C_4H_9{}^{14}CHO$ (—)	788
$t\text{-}C_4H_9COCl$	LTBA (IA), BME −78°, 1 h; to room temp, 1 h	$t\text{-}C_4H_9CHO$ (58)	78, 76, 77
	$NaAlH(OC_4H_9\text{-}t)_3$, BME, −70°, 3 h	" (45)	80

TABLE XII. PARTIAL REDUCTION OF ACYL CHLORIDES TO ALDEHYDES (*Continued*)

No. of Carbon Atoms	Reactant	Conditions	Products(s) and Yield(s) (%)	Refs.
C_6	2-oxo-2H-pyran-COCl	LTBA	No reaction	789
	ClOC, COCl (1H-pyrrole-3,4-diyl)	LTBA (IA), BME −78°, 1 h; to room temp, 1 h;	No reaction	790
	3-pyridyl-COCl	", ", ", ", ", "	3-pyridyl-CHO (69)	78
		$NaAlH(OC_4H_9\text{-}t)_3$, BME, −70°, 3 h	" (75)	80
	2-methyl-3-furyl-COCl	LTBA, BME	2-methyl-3-furyl-CHO (20)	791
	2-vinylcyclopropyl-COCl	LTBA	2-vinylcyclopropyl-CHO (—)	792
	$ClOC(CH_2)_4COCl$	LTBA (IA), BME −78°, 1 h; to room temp, 1 h	$OHC(CH_2)_4CHO$ (53)	78
	$n\text{-}C_5H_{11}COCl$	", ", ", ", ", ",	$n\text{-}C_5H_{11}CHO$ (41)	78, 77
		$NaAlH(OC_4H_9\text{-}t)_3$, BME, −70°, 3 h	" (45)	80
	$i\text{-}C_3H_7(CH_2)_2COCl$	LTBA, THF, −70°	$i\text{-}C_3H_7(CH_2)_2CHO$ (21)	793
	$C_2H_5C(CH_3)_2COCl$	LTBA, BME, −78°, 1.7 h	$C_2H_5C(CH_3)_2CHO$ (—)	794

	Reactant	Conditions	Product	(Yield, %)	Refs.
C_7	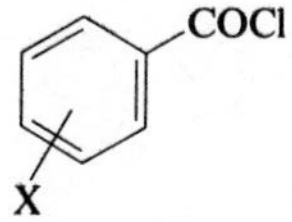		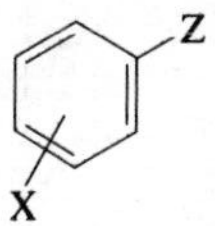		
	X				
	2-Br, 3-O_2N	LTBA (IA), BME, −78°, 1.5 h	Z = CHO	(20)	795
	3,5-$(O_2N)_2$	LTBA, BME, −78°, 1 h; to room temp, 1 h	"	(63)	796
		$NaAlH(OC_4H_9\text{-}t)_3$, THF, −70°, 3 h	"	(87)	80
	2-Br	" " " "	"	(61)	80
	4-Br	" " " "	"	(80)	80
	2-O_2N	LTBA (IA), BME, −78°, 1 h; to room temp, 1 h	"	(61)	78
	3-O_2N	", ", ", ", ", "	"	(88)	78
	4-O_2N		"	(67)	78, 77, 76
		$NaAlH(OC_4H_9\text{-}t)_3$, BME, −70°, 3 h	"	(78)	80
		Na saccharin, SMEAH, C_6H_6, 0–5°, 2 h	"	(75)	98
		$LiAlD(OC_4H_9\text{-}t)_3$ (IA), BME, −78°, 1 h, room temp, 1 h	Z = CDO	(57)	304
	2-Cl	LTBA (IA), BME, −78°, 1 h; to room temp, 1 h	Z = CHO	(20)	78, 76
		$NaAlH(OC_4H_9\text{-}t)_3$, BME, −70°, 2 h	"	(68)	80
		Na saccharin, SMEAH, C_6H_6, 0–5°, 2 h	"	(65)	98
	3-Cl	LTBA (IA), BME, −78°, 1 h; to room temp, 1 h	"	(76)	78, 77, 76
	4-Cl	", ", ", ", ", "	"	(70)	78, 77
		$NaAlH(OC_4H_9\text{-}t)_3$, BME, −70°, 3 h	"	(73)	80
		Na saccharin, SMEAH, C_6H_6, 0–5°, 2h	"	(72)	98
	H	LTBA (IA), BME, −78°, 1 h; to room temp, 1 h	"	(73)	78, 77
		$NaAlH(OC_4H_9\text{-}t)_3$, BME, −70°, 3 h	"	(80)	80
		Na saccharin, SMEAH, C_6H_6, 0–5°, 2 h	"	(80)	98
	$C_6H_5{}^{13}COCl$	LTBA, BME, −50°	$C_6H_5{}^{13}CHO$	(—)	797a
	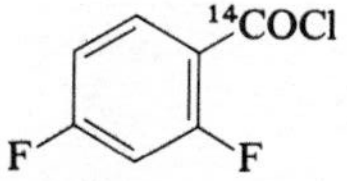	LTBA (IA), BME, −70 to −60°, 45 min, then −60 to 0°	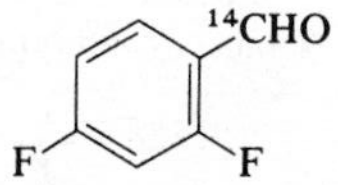	(—)	797b

TABLE XII. PARTIAL REDUCTION OF ACYL CHLORIDES TO ALDEHYDES (*Continued*)

No. of Carbon Atoms	Reactant	Conditions	Products(s) and Yield(s) (%)	Refs.
C_7 (*Contd.*)	2-furyl-CH=CHCOCl	$NaAlH(OC_4H_9\text{-}t)_3$, BME, −70°, 2 h	2-furyl-CH=CHCHO (72)	80
	NC-cyclopropyl-$CO_2C_2H_5$	1. H_3O^+, $SOCl_2$ 2. LTBA 3. H_3O^+, EtOH	NC-cyclopropyl-$CH(OC_2H_5)_2$ (—)	798
	$C_6H_{11}COCl$	LTBA (IA), BME, −78°, 1 h; to room temp, 1 h	$C_6H_{11}CHO$ (56)	78
		Na saccharin, SMEAH, C_6H_6, 0–5°, 2 h	" (75)	98
	$CH_3O_2CCH_2C(CH_3)_2COCl$	LTBA, THF, −70°, 30 min	$CH_3O_2CCH_2C(CH_3)_2CHO$ (—)	510
C_8	decafluorobicyclo[2.2.1]heptane-1-COCl	LTBA, BME, −78°, 1 h	decafluorobicyclo[2.2.1]heptyl-R, R = CHO, CH_2OH (—)	799
	X-C_6H_4-COCl, X		X-C_6H_4-CHO, X	
	2-CF_3	LTBA. BME, −70°	2-CF_3 (—)	800
	4-NC	", ", ", 1 h; to room temp, 1 h	4-NC (68)	78
	2-COCl	$NaAlH(OC_4H_9\text{-}t)_3$, BME, −70°, 2 h	2-CHO (55)	80
	3-COCl	LTBA, BME, −78°, 2 h; to room temp, 1 h	3-CHO (64)	78
		$NaAlH(OC_4H_9\text{-}t)_3$, BME, −70°, 2 h	3-CHO (62)	80
	4-COCl	LTBA, BME, −78°, 2 h; to room temp, 1 h	4-CHO (77)	78, 77, 76
		$NaAlH(OC_4H_9\text{-}t)_3$, BME, −70°, 2 h	4-CHO (78)	80
	2-CH_3	Na saccharin, SMEAH, C_6H_6, 0–5°, 2 h	2-CH_3 (63)	98

	Reactant	Conditions	Product(s) and Yield(s) (%)	Refs.
	4-CH_3	LTBA (IA), BME, $-78°$, 1 h; to room temp, 1 h	4-CH_3 (61)	78, 77, 76
		Na saccharin, SMEAH, C_6H_6, 0–5°, 2 h	4-CH_3 (70)	98
	2-CH_3S	LTBA, BME, $-70°$, 1.25 h	2-CH_3S (24)	801
	2-CH_3O	LTBA (IA), BME, $-78°$, 1 h; to room temp, 1 h	2-CH_3O (27)	78
	3-CH_3O	" " " " " " " " "	3-CH_3O (66)	78
	4-CH_3O	" " " " " " " " "	4-CH_3O (50)	78
	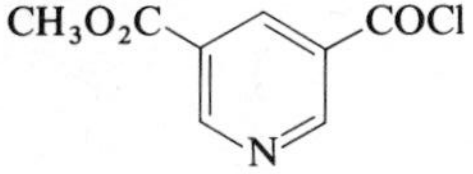 CH_3O_2C, COCl, N	LTBA (IA), THF, $-78°$, 30 min; to room temp, 10 h	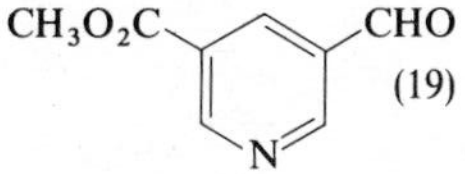 CH_3O_2C, CHO, N (19) 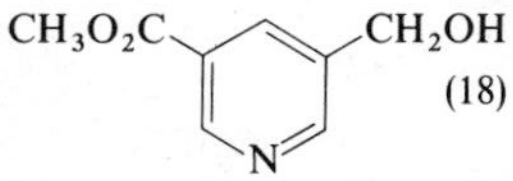 + CH_3O_2C, CH_2OH, N (18)	802
	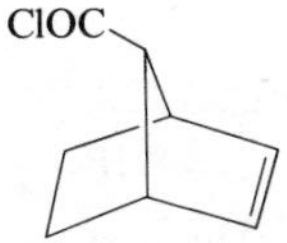 ClOC	LTBA (IA), BME, $-78°$	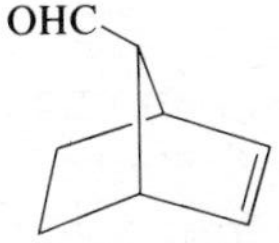 OHC (50)	803
	$C_2H_5O_2CCH_2C(CH_3)_2COCl$	LTBA (IA), DME, $-65°$, 2 h	$C_2H_5O_2CCH_2C(CH_3)_2CHO$ (56)	804
C_9	3,5-$(CF_3)_2C_6H_3COCl$	LTBA, BME, $-70°$, 1 h	3,5-$(CF_3)_2C_6H_3CHO$ (72)	805
	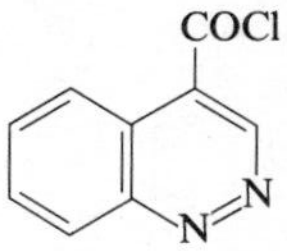 COCl, N, N	LTBA, BME, room temp, 1 h	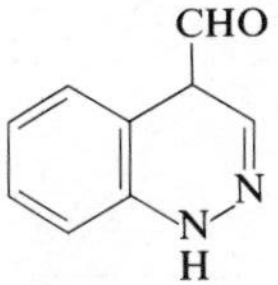 CHO, N, N, H (32)	806, 807
	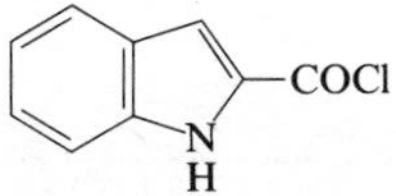 COCl, N, H	LTBA (IA), BME, $-70°$, then room temp, 1 h	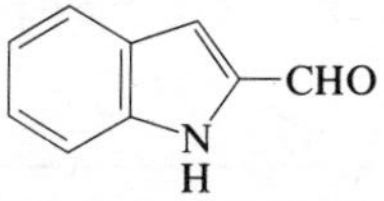 CHO, N, H (—)	808, 809

TABLE XII. Partial Reduction of Acyl Chlorides to Aldehydes (*Continued*)

No. of Carbon Atoms	Reactant	Conditions	Products(s) and Yield(s) (%)	Refs.
C_9 (*Contd.*)	ClOC, COCl, OCH_3 (benzene ring)	LTBA (IA), BME, −60 to −70°	OHC, CHO, OCH_3 (benzene ring) (61)	810
	COCl, NO_2, NHAc (benzene ring)	LTBA (IA), BME, −78°, 1 h; to room temp, 1 h	CHO, NO_2, NHAc (benzene ring) (50)	811
	$C_6H_5CH{=}CHCOCl$	" " " " " "	$C_6H_5CH{=}CHCHO$ (71)	78
		$NaAlH(OC_4H_9\text{-}t)_3$, BME, −70°, 2 h	" (69)	80
		Na saccharin, SMEAH, C_6H_6, 0–5°, 2 h	" (77)	98
	COCl (benzocyclobutene)	LTBA (IA), BME, −78°, 1 h	CHO (benzocyclobutene) (—)	812
	COCl, COCl (cycloheptadiene ring)	LTBA, THF, −70°	CHO, CHO (cycloheptadiene ring) (61)	813
	COCl (bicyclic cyclopropane ring)	LTBA, BME, −65°, 15 min	CHO (bicyclic cyclopropane ring) (50)	814

	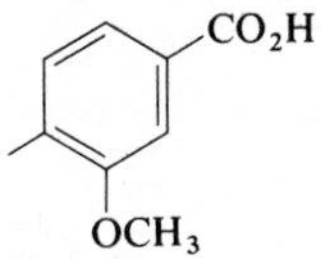	1. $SOCl_2$ 2. LTBA, BME, -78 to 20°, 3 h	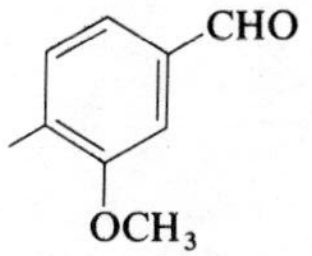 (62) + 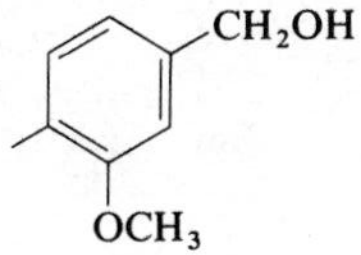 (13)	815
	 $R = CH_2COCl$	LTBA (IA), THF, $-40°$, 30 min; room temp, 30 min	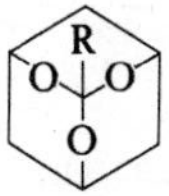 (27) $R = CH_2CHO$	816
	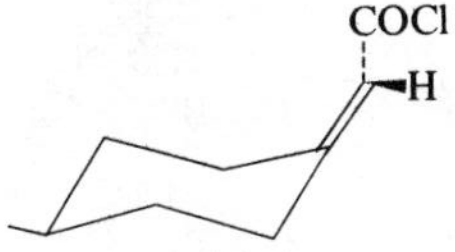 (*aR*)-(−)	LTBA, THF, $-78°$, then room temp, 45 min	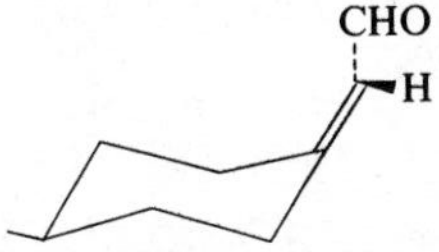 (75) (*aR*)-(−) + 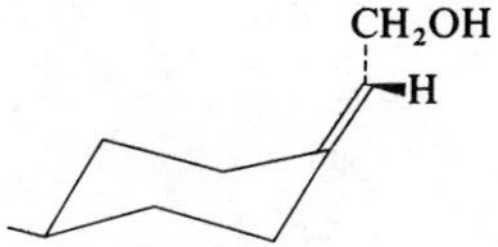 (~25%) (*aR*)-(−)	817
	n-$C_8H_{17}COCl$	$LiAlD(OC_4H_9$-$t)_3$, $-67°$, 1–2 h	n-$C_8H_{17}CDO$ (56)	307
C_{10}	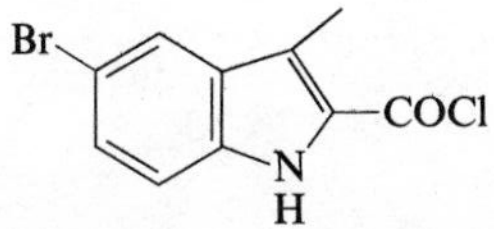	LTBA (IA), BME, -63 to $-67°$	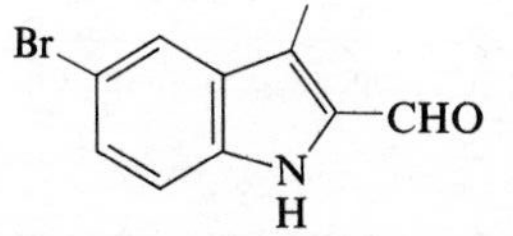 (—)	809

TABLE XII. PARTIAL REDUCTION OF ACYL CHLORIDES TO ALDEHYDES (*Continued*)

No. of Carbon Atoms	Reactant	Conditions	Products(s) and Yield(s) (%)	Refs.
C_{10} (*Contd.*)	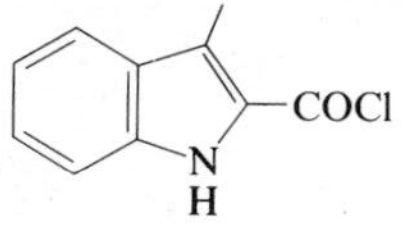 —COCl (N H)	LTBA (IA), BME, −70° to room temp, 1 h	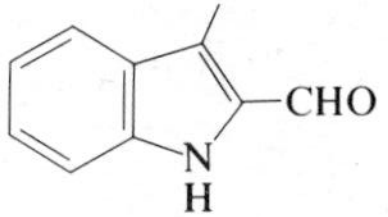 —CHO (N H) (—)	808
	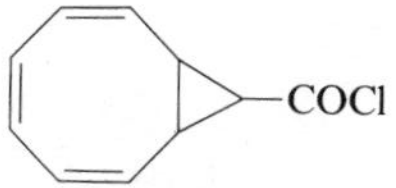 —COCl	LTBA, BME, −65°, 30 min	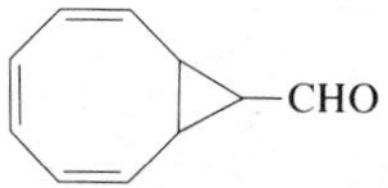 —CHO (39)	814
	$C_6H_5CH(OAc)COCl$	LTBA, BME, −70°	$C_6H_5CH(OAc)CHO$ (55)	818
	$4\text{-}(C_2H_5O_2C)C_6H_4COCl$	LTBA (IA), BME −78°, 1 h; to room temp, 1 h	$4\text{-}(C_2H_5O_2C)C_6H_4CHO$ (48)	78
	$CH_3C(=NOCH_2C_6H_5)COCl$	LTBA (IA), THF, −78°	$CH_3C(=NOCH_2C_6H_5)CHO$ (—)	819
	$C_6H_5C(CH_3)_2COCl$	LTBA, THF, −68°	$C_6H_5C(CH_3)_2CHO$ (50)	820
	$C_6H_5O(CH_2)_3COCl$	LTBA (IA), BME, −78°, 1.5 h; to room temp, 1 h	$C_6H_5O(CH_2)_3CHO$ (37)	821
	$2\text{-}CH_3OC_6H_4(CH_2)_2COCl$	LTBA (IA), BME, −78°, 1 h; to room temp, 1 h	$2\text{-}CH_3OC_6H_4(CH_2)_2CHO$ (8) + $2\text{-}CH_3OC_6H_4(CH_2)_3OH$ (40)	822
	$3,4,5\text{-}(CH_3O)_3C_6H_2COCl$	LTBA (IA), BME, −78°, 1 h	$3,4,5\text{-}(CH_3O)_3C_6H_2CHO$ (80)	823
	$CH_3O_2{}^{14}C(CH_2)_7{}^{14}CO_2H$	1. $SOCl_2$ 2. LTBA, THF–DME, −78°, 2 h	$CH_3O_2{}^{14}C(CH_2)_7{}^{14}CHO$ (55)	824
C_{11}	$1\text{-}C_{10}H_7COCl$	LTBA (IA), BME, −78°, 1 h; to room temp, 1 h	$1\text{-}C_{10}H_7CHO$ (68)	78
	$2\text{-}C_{10}H_7COCl$	LTBA (IA), BME, −78°, 1 h; to room temp, 1 h	$2\text{-}C_{10}H_7CHO$ (58)	78, 504
	$4\text{-}AcOC_6H_4CH{=}CHCOCl$	LTBA (IA), THF, −65° to room temp	$4\text{-}AcO(C_6H_4CH{=}CHCHO$ (74)	825
	$2,5\text{-}(AcO)_2C_6H_3COCl$	LTBA (IA), BME −70°; room temp, overnight	$2\text{-}AcO\text{-}5\text{-}HOC_6H_3CHO$ (20)	826
	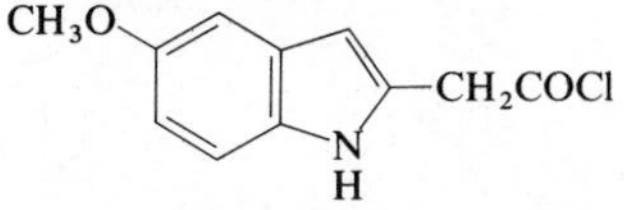 CH_3O —CH_2COCl (N H)	LTBA,THF, room temp, 4 h	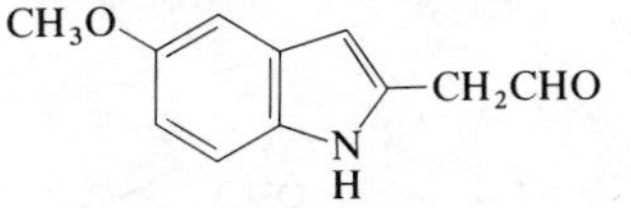 CH_3O —CH_2CHO (N H) (—)	827

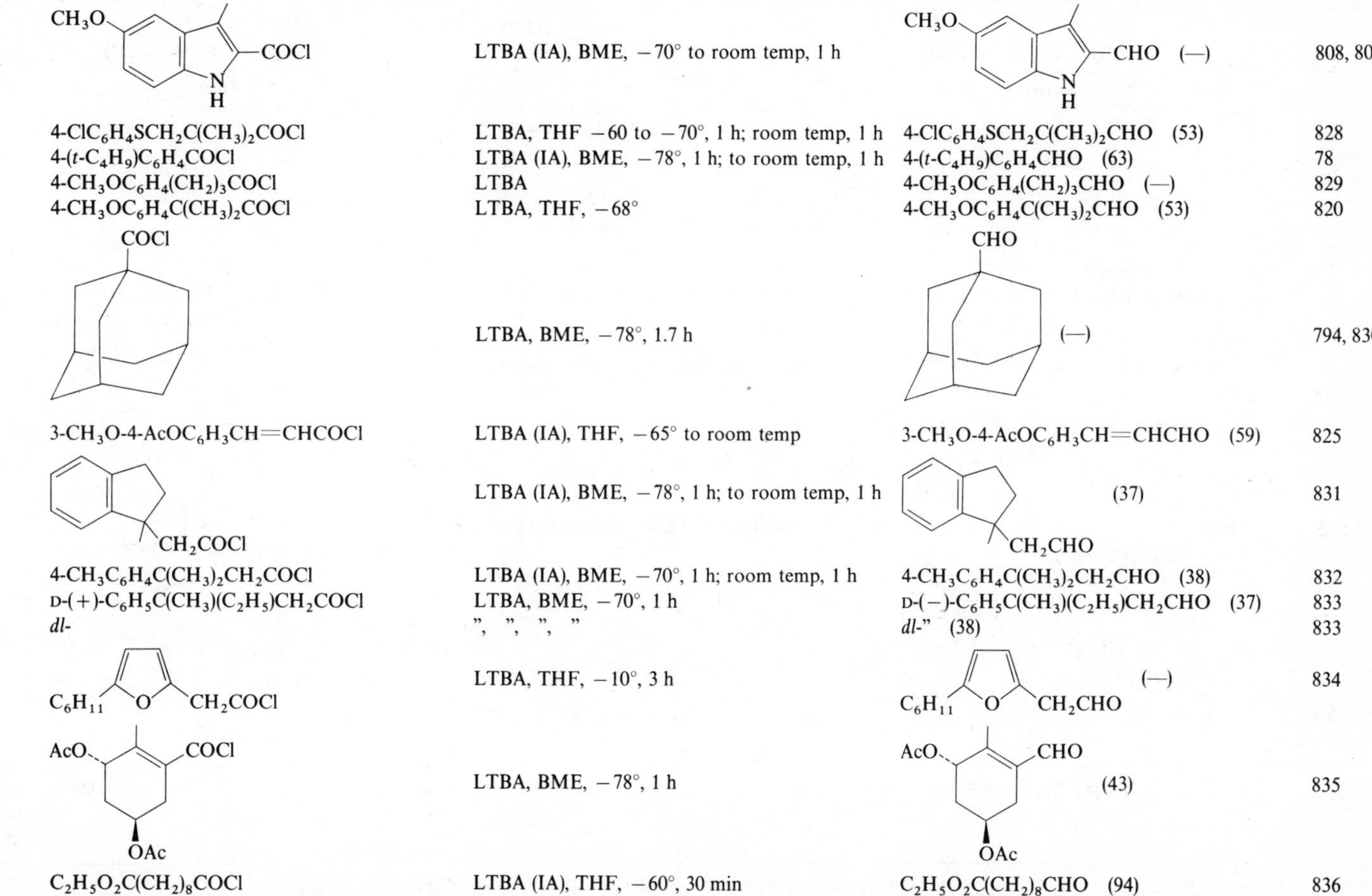

	Acid Chloride	Conditions	Product (Yield, %)	Refs.
	5-methoxy-3-methylindole-2-COCl (structure)	LTBA (IA), BME, −70° to room temp, 1 h	5-methoxy-3-methylindole-2-CHO (structure) (—)	808, 809
	4-$ClC_6H_4SCH_2C(CH_3)_2COCl$	LTBA, THF −60 to −70°, 1 h; room temp, 1 h	4-$ClC_6H_4SCH_2C(CH_3)_2CHO$ (53)	828
	4-(t-C_4H_9)C_6H_4COCl	LTBA (IA), BME, −78°, 1 h; to room temp, 1 h	4-(t-C_4H_9)C_6H_4CHO (63)	78
	4-$CH_3OC_6H_4(CH_2)_3COCl$	LTBA	4-$CH_3OC_6H_4(CH_2)_3CHO$ (—)	829
	4-$CH_3OC_6H_4C(CH_3)_2COCl$	LTBA, THF, −68°	4-$CH_3OC_6H_4C(CH_3)_2CHO$ (53)	820
	1-adamantyl-COCl (structure)	LTBA, BME, −78°, 1.7 h	1-adamantyl-CHO (structure) (—)	794, 830
C_{12}	3-CH_3O-4-$AcOC_6H_3CH{=}CHCOCl$	LTBA (IA), THF, −65° to room temp	3-CH_3O-4-$AcOC_6H_3CH{=}CHCHO$ (59)	825
	1-methylindan-1-yl-CH_2COCl (structure)	LTBA (IA), BME, −78°, 1 h; to room temp, 1 h	1-methylindan-1-yl-CH_2CHO (structure) (37)	831
	4-$CH_3C_6H_4C(CH_3)_2CH_2COCl$	LTBA (IA), BME, −70°, 1 h; room temp, 1 h	4-$CH_3C_6H_4C(CH_3)_2CH_2CHO$ (38)	832
	D-(+)-$C_6H_5C(CH_3)(C_2H_5)CH_2COCl$	LTBA, BME, −70°, 1 h	D-(−)-$C_6H_5C(CH_3)(C_2H_5)CH_2CHO$ (37)	833
	dl-	", ", ", "	*dl*-" (38)	833
	C_6H_{11}-furyl-CH_2COCl (structure)	LTBA, THF, −10°, 3 h	C_6H_{11}-furyl-CH_2CHO (structure) (—)	834
	AcO, OAc cyclohexenyl-COCl (structure)	LTBA, BME, −78°, 1 h	AcO, OAc cyclohexenyl-CHO (structure) (43)	835
	$C_2H_5O_2C(CH_2)_8COCl$	LTBA (IA), THF, −60°, 30 min	$C_2H_5O_2C(CH_2)_8CHO$ (94)	836

TABLE XII. PARTIAL REDUCTION OF ACYL CHLORIDES TO ALDEHYDES (*Continued*)

No. of Carbon Atoms	Reactant	Conditions	Products(s) and Yield(s) (%)	Refs.
C_{13}	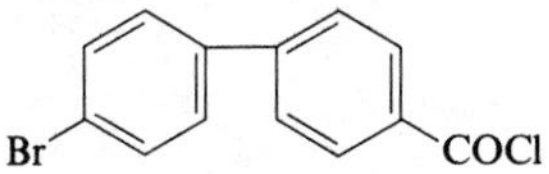 4'-Br-biphenyl-4-COCl (Br, COCl)	LTBA, BME, −60 to −70°	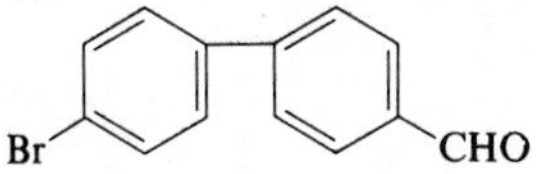 4'-Br-biphenyl-4-CHO (Br, CHO) (—)	837
	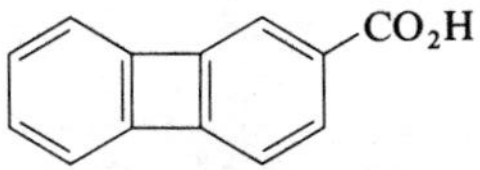 biphenylene-2-CO_2H	1. $SOCl_2$ 2. LTBA (IA), BME, −70°, 1 h	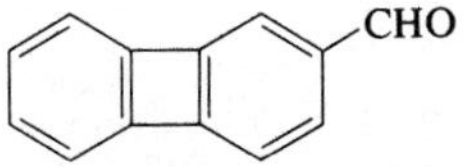 biphenylene-2-CHO (74)	838
	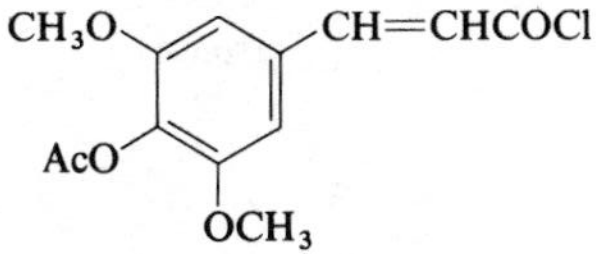 CH_3O, AcO, OCH_3-substituted phenyl–CH=CHCOCl	LTBA (IA), BME, −60° to room temp	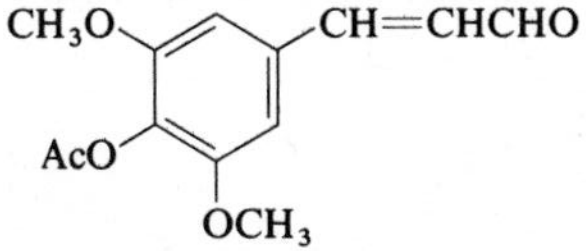 CH_3O, AcO, OCH_3-substituted phenyl–CH=CHCHO (36)	839
	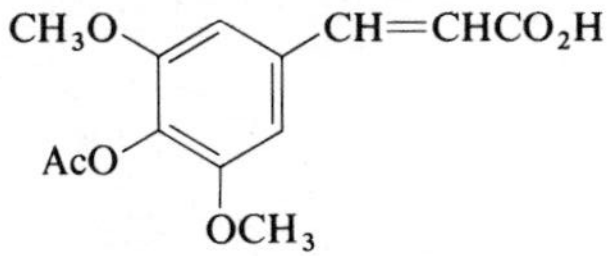 CH_3O, AcO, OCH_3-substituted phenyl–CH=CHCO_2H	1. $SOCl_2$ 2. LTBA, THF, room temp, 20°, 6 h	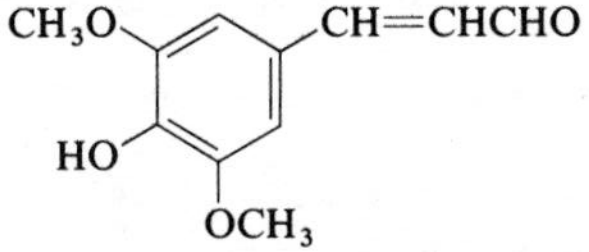 CH_3O, HO, OCH_3-substituted phenyl–CH=CHCHO (46)	839
	(+)-4-(i-C_3H_7)$C_6H_4CH_2CH(CH_3)COCl$	LTBA, THF −78°, 2 h, then room temp, 2 h	4-(i-C_3H_7)$C_6H_4CH_2CH(CH_3)CHO$ (51) (94% e.e.)	840
	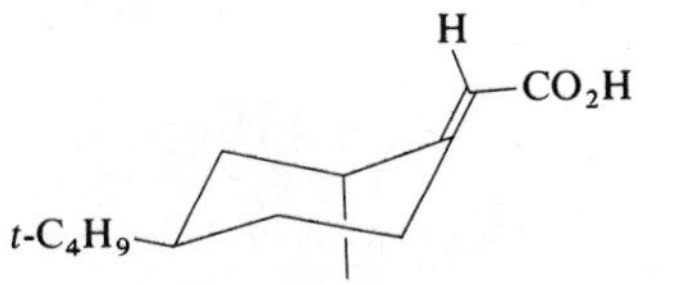 t-C_4H_9 bicyclic with =C(H)CO_2H, (2S,4R)-(+)	1. $SOCl_2$ 2. LTBA, THF, −78°, then room temp, 45 min	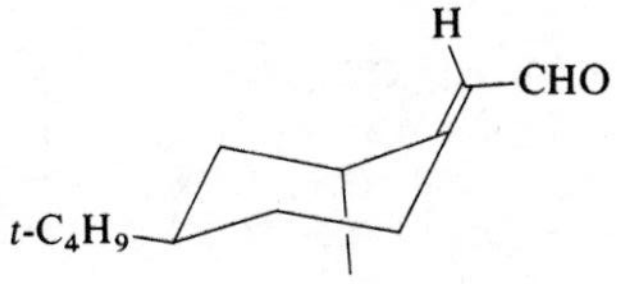 t-C_4H_9 bicyclic with =C(H)CHO, (2S,4R)-(+) (63)	817
C_{14}	4-($C_6H_5CH_2O$)$C_6H_4{}^{14}COCl$	1. LTBA (IA), BME, −75 to −10°, 1 h 2. H_3O^+	4-$HOC_6H_4{}^{14}CHO$ (41)	841

	Acid Chloride	Conditions	Product(s) and Yield(s) (%)	Refs.
	Cl, COCl, CH_3O, OCH_3 (structure)	LTBA (IA), THF, $-70°$, 2 h	Cl, CHO, CH_3O, OCH_3 (structure) (61)	842
	CH_3O, $CH(CH_3)COCl$ (naphthalene structure)	LTBA, THF, $-80°$, 1 h	CH_3O, $CH(CH_3)CHO$ (naphthalene structure) (—)	843, 844
	N_3, CH_2COCl, O, $NCH_2C_6H_3(OCH_3)_2$-2,4 (azetidinone structure)	LTBA	N_3, CH_2CHO, O, $NCH_2C_6H_3(OCH_3)_2$-2,4 (azetidinone structure) (—)	419
	N (piperidino), Cl, $CH(CH_3)COCl$ (structure)	LTBA, BME, -70 to $-80°$, 1 h	N (piperidino), Cl, $CH(CH_3)CHO$ (structure) (—)	845
	(+)-4-(t-C_4H_9)$C_6H_4CH_2CH(CH_3)COCl$	LTBA, THF	(+)-4-(t-C_4H_9)$C_6H_4CH_2CH(CH_3)CHO$ (61)	846
C_{15}	Anthracene-1-carbonyl chloride	LTBA (IA), BME, $-70°$ to room temp	1-Anthraldehyde (12)	847
	C_6H_5, CH_2COCl (cycloheptane structure)	LTBA, BME, $-78°$, 1 h	C_6H_5, CH_2CHO (cycloheptane structure) (31)	848
C_{16}	4-$C_6H_5C_6H_4C(CH_3)$=CHCOCl	LTBA	4-$C_6H_5C_6H_4C(CH_3)$=CHCHO (—)	849a
	COCl, C_6H_5, C_6H_5 (cyclopropane structure)	LTBA (IA), THF, $-78°$, 1.5 h, then to room temp	CHO, C_6H_5, C_6H_5 (cyclopropane structure) (45)	849b
	$C_6H_5CH_2C(=NOCH_2C_6H_5)COCl$	LTBA (IA), THF, $-78°$	$C_6H_5CH_2C(=NOCH_2C_6H_5)CHO$ (—)	819
	n-$C_{15}H_{31}COCl$	LTBA, BME $-70°$, 1 h; room temp, 1 h	n-$C_{15}H_{31}CHO$ (55)	850

TABLE XII. PARTIAL REDUCTION OF ACYL CHLORIDES TO ALDEHYDES (*Continued*)

No. of Carbon Atoms	Reactant	Conditions	Products(s) and Yield(s) (%)	Refs.
C_{17}	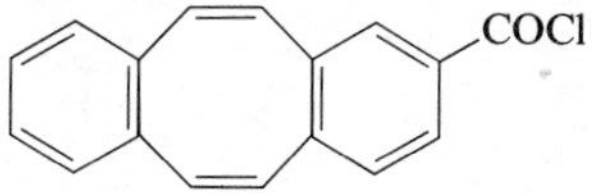 COCl	LTBA, THF	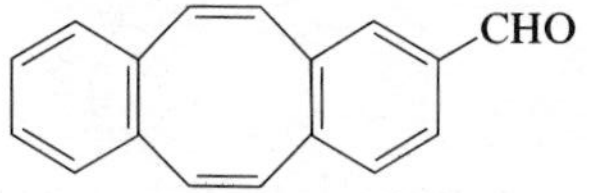 CHO (—)	851
	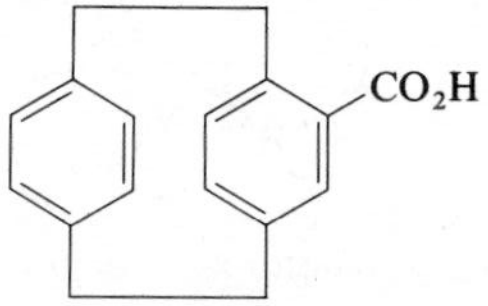 CO_2H	1. $SOCl_2$ 2. LTBA (IA), BME, −78°, 30 min; then 25°, 1 h	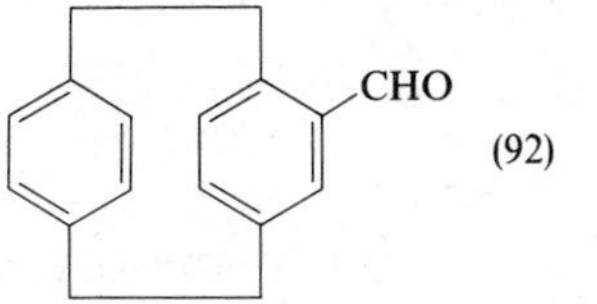 CHO (92)	852
	$(C_6H_5CH_2)_2CHCH_2COCl$	LTBA (IA), BME, −78°, 1 h; to room temp, 1 h	$(C_6H_5CH_2)_2CHCH_2CHO$ (68)	853
	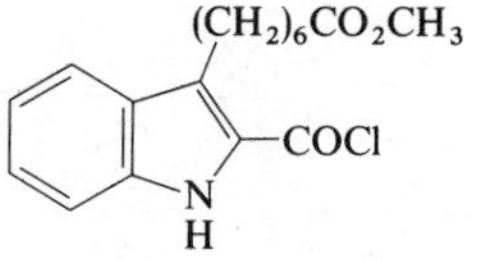 $(CH_2)_6CO_2CH_3$, COCl, N, H	LTBA, THF, −40°, 2 h	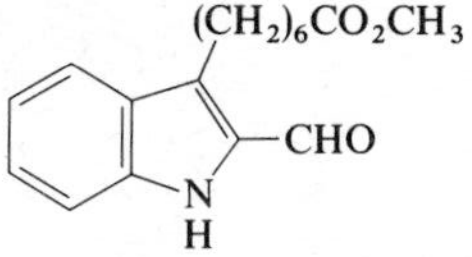 $(CH_2)_6CO_2CH_3$, CHO, N, H (64)	854
	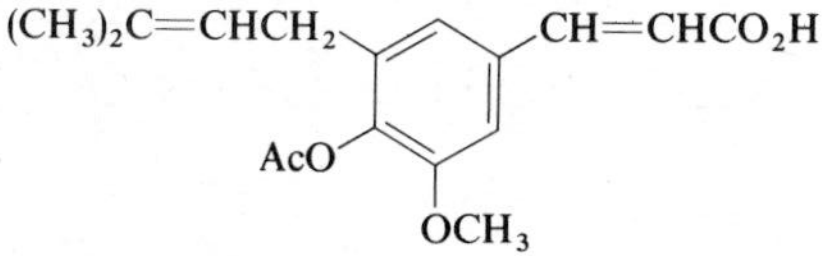 $(CH_3)_2C{=}CHCH_2$, $CH{=}CHCO_2H$, AcO, OCH_3	1. $(COCl)_2$ 2. LTBA (IA), THF, −65°	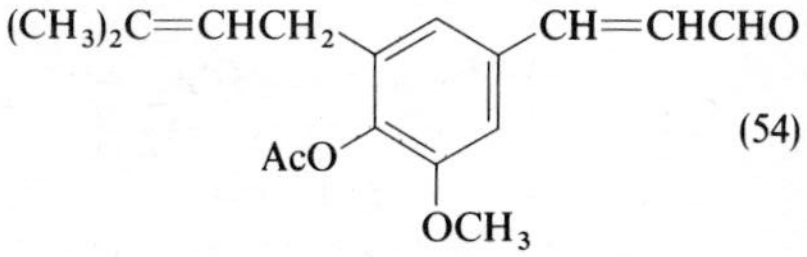 $(CH_3)_2C{=}CHCH_2$, $CH{=}CHCHO$, AcO, OCH_3 (54)	855
C_{18}	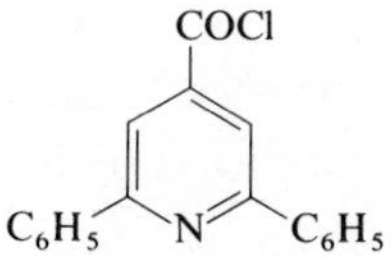 COCl, C_6H_5, N, C_6H_5	LTBA	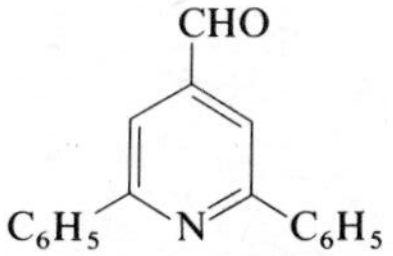 CHO, C_6H_5, N, C_6H_5 (64)	856
	$n\text{-}C_5H_{11}CH{=}CHCH_2CH{=}CH(CH_2)_7COCl$	LTBA, BME, −58 to −64°, 1 h	$n\text{-}C_5H_{11}CH{=}CHCH_2CH{=}CH(CH_2)_7CHO$ (50)	857
	$n\text{-}C_8H_{17}C{\equiv}C(CH_2)_7COCl$	", ", ", ", "	$n\text{-}C_8H_{17}C{\equiv}C(CH_2)_7CHO$ (59)	857
	$n\text{-}C_8H_{17}CH{=}CH(CH_2)_7COCl$	LTBA, BME, −58 to −64°, 1 h	$n\text{-}C_8H_{17}CH{=}CH(CH_2)_7CHO$ (45)	857

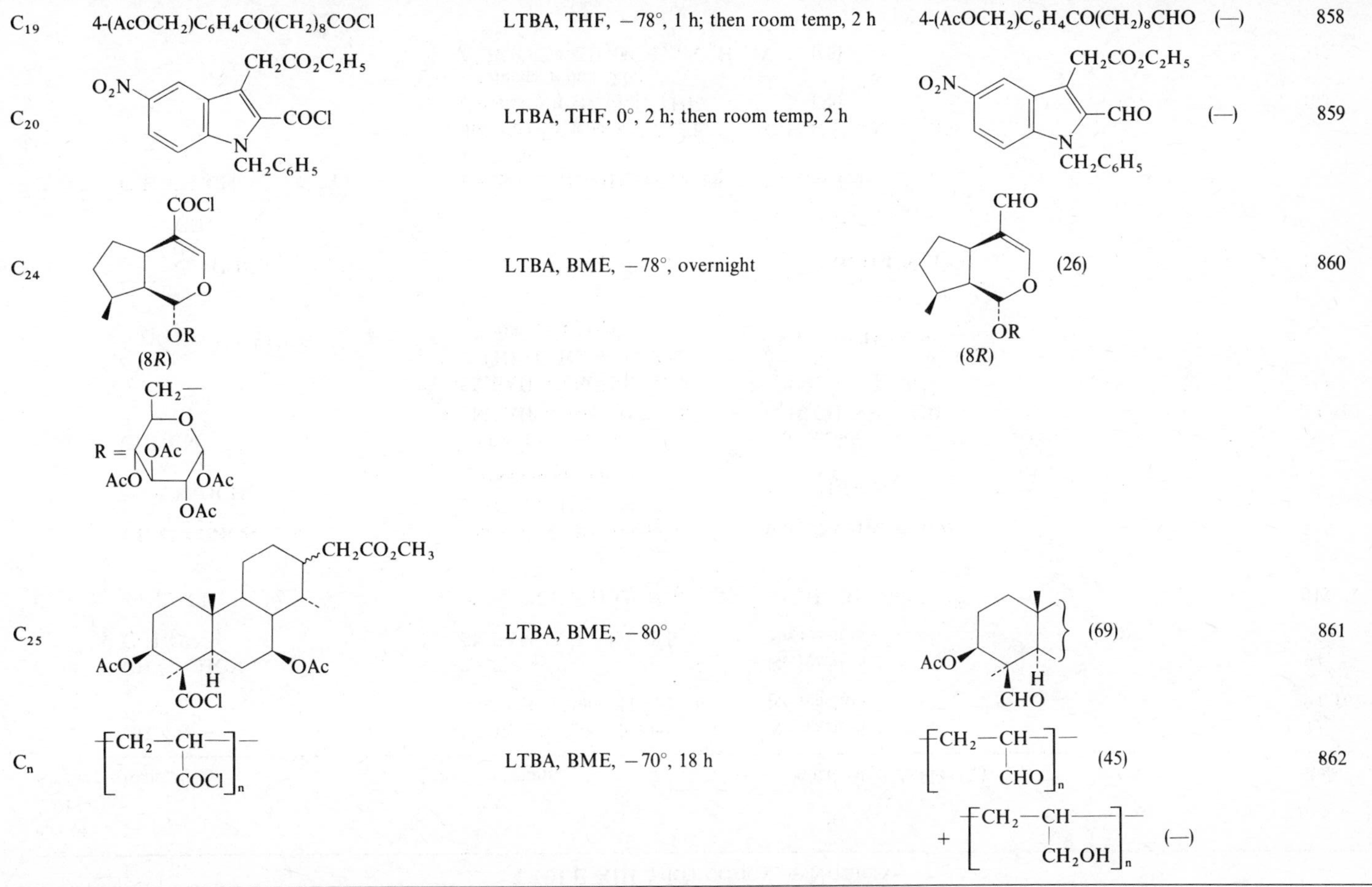

	Reactant	Conditions	Product(s) and Yield(s) (%)	Refs.
C_{19}	$4\text{-}(AcOCH_2)C_6H_4CO(CH_2)_8COCl$	LTBA, THF, −78°, 1 h; then room temp, 2 h	$4\text{-}(AcOCH_2)C_6H_4CO(CH_2)_8CHO$ (—)	858
C_{20}	O_2N, $CH_2CO_2C_2H_5$, COCl, N, $CH_2C_6H_5$	LTBA, THF, 0°, 2 h; then room temp, 2 h	O_2N, $CH_2CO_2C_2H_5$, CHO, N, $CH_2C_6H_5$ (—)	859
C_{24}	COCl, O, OR, (8*R*); R = CH_2—, O, OAc, AcO, OAc, OAc	LTBA, BME, −78°, overnight	CHO, O, OR, (8*R*) (26)	860
C_{25}	$CH_2CO_2CH_3$, AcO, H, COCl, OAc	LTBA, BME, −80°	AcO, H, CHO (69)	861
C_n	$-[CH_2-CH(COCl)]_n-$	LTBA, BME, −70°, 18 h	$-[CH_2-CH(CHO)]_n-$ (45) + $-[CH_2-CH(CH_2OH)]_n-$ (—)	862

TABLE XIII. REDUCTION OF NITRILES

No. of Carbon Atoms	Reactant	Conditions	Product(s) and Yield(s) (%)	Refs.
C_2	CH_3CN	LTBA, BME, 0°, 30 min	No reaction	77
		SMEAH, toluene, 115°, 3.5 h	No reaction	94, 167
C_3	$CH_2{=}CHCN$	", ", ", "	No reaction	94
	C_2H_5CN	SMEAH, C_6H_6, 20°, 18 h	No reaction	94
C_4	n-C_3H_7CN	$Al_2H_3(OCH_2CH_2OCH_3)_3$, C_6H_6, 80°, 2 h	n-$C_4H_9NH_2$ (61)	313
C_6	$CH_3C(CH_2OCH_3){=}CHCN$	SMEAH–CuBr–2-butanol, THF–C_6H_6, −78°, 2 h; room temp, 4 h	$CH_3CH(CH_2OCH_3)CH_2CN$ (29)	115
C_7	C_6H_5CN	LTBA, BME, 0°	No reaction	77
		SMEAH, toluene, 115°, 4 h	$C_6H_5CH_2NH_2$ (81)	94, 167
	2-furyl–CH=CHCN	SMEAH–CuBr–2-butanol, THF–C_6H_6, −78°, 2 h; room temp, 4 h	2-furyl–$(CH_2)_2CN$ (51)	115
	cyclopropyl–$C(CH_3){=}CHCN$	", ", ", ", ", "	cyclopropyl–$CH(CH_3)CH_2CN$ (70)	115
C_8	$C_6H_5CH_2CN$	$LiAlH(OC_2H_5)_3$ (IA), ether, 25°, 15 min	No reaction	81
		SMEAH, toluene, 115°, 3.5 h	$C_6H_5(CH_2)_2NH_2$ (13)	94
		$Ca[AlH_2(OC_4H_9\text{-}i)_2]_2\cdot THF$, toluene, reflux, 3 h	" (35)	165
		$Al_2H_3(OCH_2CH_2OCH_3)_3$, C_6H_6, 80°, 2 h	" (66)	313

	2-$CH_3C_6H_4CN$	SMEAH, toluene, 115°, 3.5 h	2-$CH_3C_6H_4CH_2NH_2$ (95)	94
	3-$CH_3C_6H_4CN$	", ", ", "	3-$CH_3C_6H_4CH_2NH_2$ (91)	94
	4-$CH_3C_6H_4CN$	$Al_2H_3(OCH_2CH_2OCH_3)_3$, C_6H_6, 80°, 2 h	4-$CH_3C_6H_4CH_2NH_2$ (97)	313
	4-$CH_3OC_6H_4CN$	SMEAH, toluene, reflux, 4 h	4-$CH_3OC_6H_4CH_2NH_2$ (95)	121
C_9	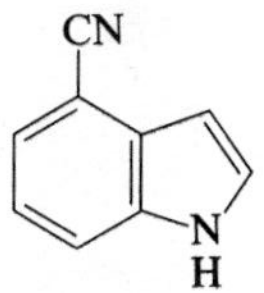 CN, N H	$LiAlH(OC_2H_5)_3$ (IA), THF, 20°, 15 h	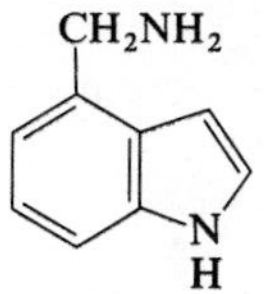 CH_2NH_2, N H (39) 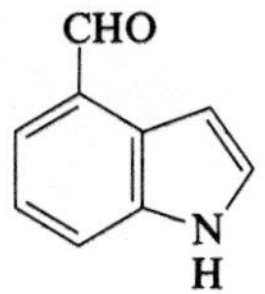 + CHO, N H (8)	863a
	$C_6H_5CH{=}CHCN$	SMEAH–CuBr–2-butanol, THF–C_6H_6, −78°, 2 h; room temp, 4 h	$C_6H_5(CH_2)_2CN$ (99)	115
	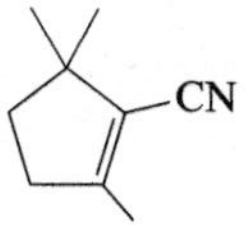 CN	SMEAH–CuBr–2-butanol, THF–C_6H_6, −78°, 2 h; room temp, 24 h	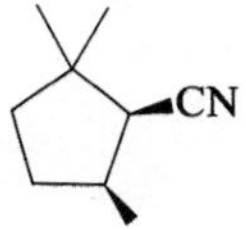 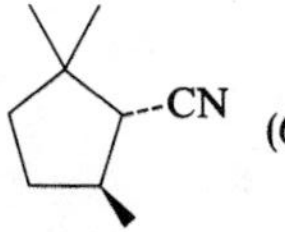 CN **I** + CN **II** (69) (**I**:**II** = 50:50)	117
	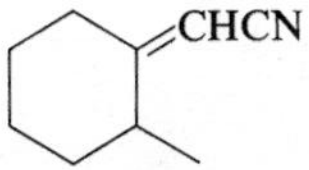 CHCN	SMEAH–CuBr–2-butanol, THF–C_6H_6, −78°, 2 h; room temp, 4 h	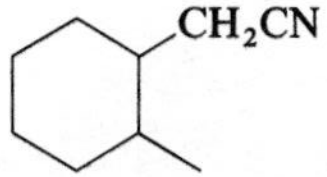 CH_2CN (92)	115
	$C_2H_5O_2C(CH_2)_2C(CH_3){=}CHCN$	", ", ", ", ", "	$C_2H_5O_2C(CH_2)_2CH(CH_3)CH_2CN$ (32)	115
C_{10}	2-$NC(CH_2)_2SC_6H_4CO_2H$	SMEAH, C_6H_6, room temp, 3 h	$(2\text{-}HOCH_2C_6H_4S)_2$ (40)	524
	$(n\text{-}C_3H_7)_2C{=}C(CH_3)CN$	SMEAH–CuBr–2-butanol, THF–C_6H_6, −78°, 2 h; room temp, 4 h	$(n\text{-}C_3H_7)_2CHCH(CH_3)CN$ (83)	115

TABLE XIII. REDUCTION OF NITRILES (*Continued*)

No. of Carbon Atoms	Reactant	Conditions	Product(s) and Yield(s)(%)	Ref.
C_{11}	$C_6H_5C(CH_3){=}C(CN)_2$	1. LAH + (−)-QN (1:1), ether 2. Aqueous KOH, heat	$C_6H_5CH(CH_3)CH_2CO_2H$ (*S*)-(+) (67) (6% e.e.)	735
		1. LAH + (+)-QND (1:1), ether 2. Aqueous KOH, heat	" (*R*)-(−) (69) (8% e.e.)	735
	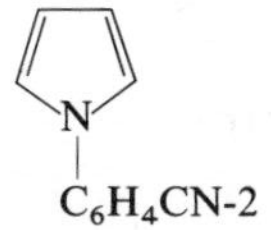 N-pyrrolyl-C_6H_4CN-2	SMEAH, toluene, 2.5 h	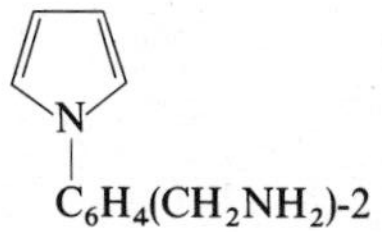 N-pyrrolyl-$C_6H_4(CH_2NH_2)$-2 (73)	863b
	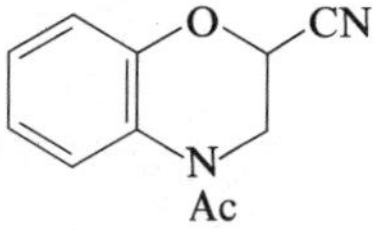 4-Ac-3,4-dihydro-2H-1,4-benzoxazine-2-CN	SMEAH–MPL, −40°, 2.5 h	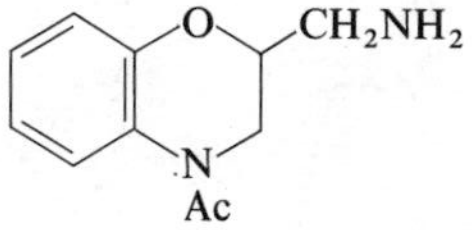 4-Ac-3,4-dihydro-2H-1,4-benzoxazine-2-CH_2NH_2 (56)	340
C_{13}	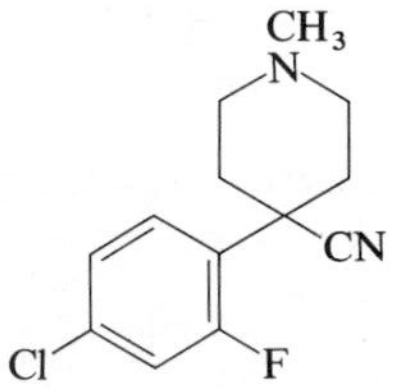 1-CH_3-4-(4-Cl-2-F-phenyl)piperidine-4-CN	$LiAlH(OC_2H_5)_3$, DME, reflux, 72 h	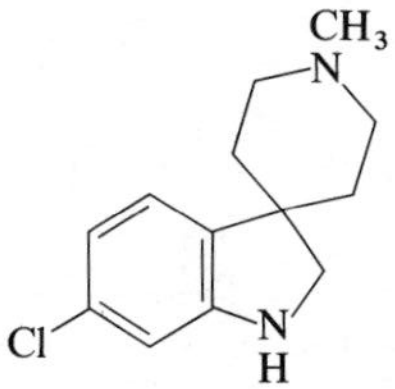 1'-CH_3-6-Cl-spiro[indoline-3,4'-piperidine] (93)	330
C_{17}	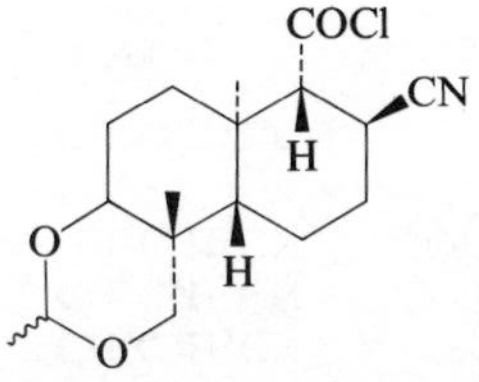 (structure: COCl, CN, H, H, O, O)	LTBA, THF, 0°, 1 h	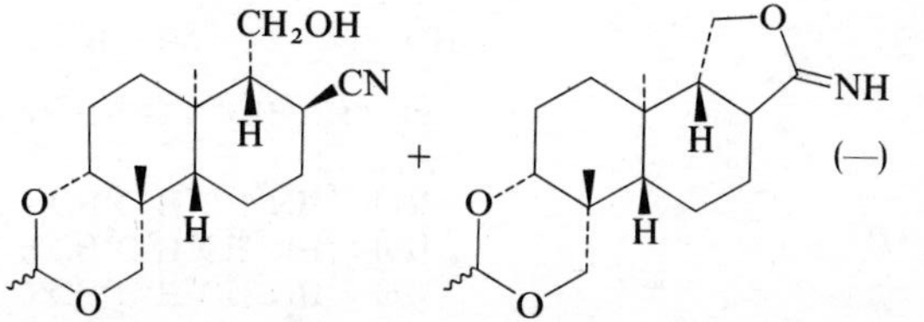 (structure: CH_2OH, CN, H, H, O, O) + (structure: O, =NH, H, H, O, O) (—)	780

	Reactant	Conditions	Product(s) and Yield(s) (%)	Refs.
C_{18}	(structure: OH, CN, NHTs)	SMEAH, C_6H_6, room temp, 18 h	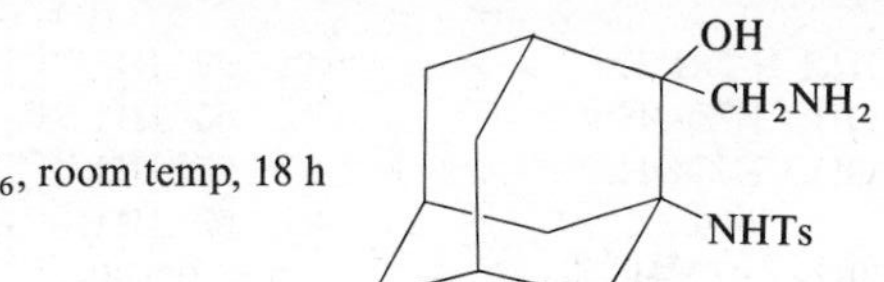 (OH, CH_2NH_2, NHTs) (45)	864
			+ 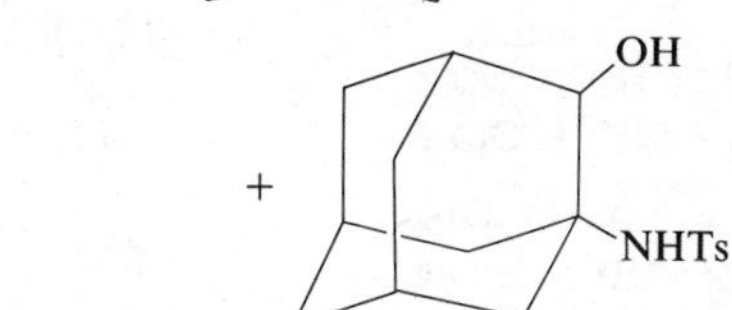 (OH, NHTs) (51)	
C_{20}	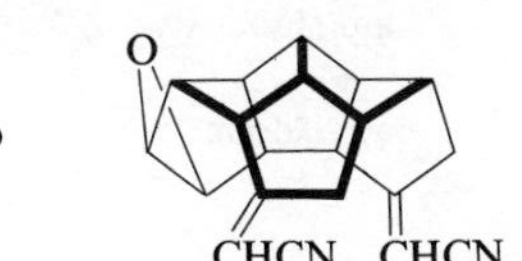 (O, CHCN, CHCN)	SMEAH–CuBr–2-butanol, THF, C_6H_6, −78°, 2 h; room temp, 4 h	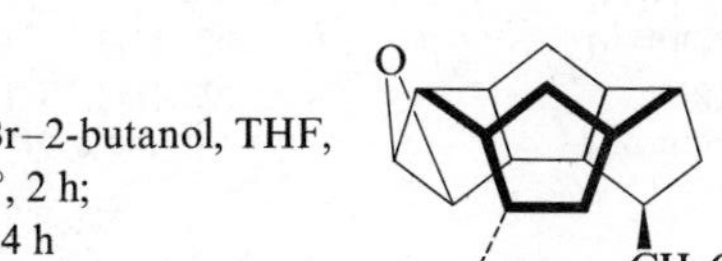 (O, CH_2CN, CH_2CN) (69)	116
C_n	$\left[-CH_2CH(C_6H_5)CH_2C(CH_3)(CN)-\right]_n$	SMEAH, *N*-methylmorpholine, 100°, 12 h	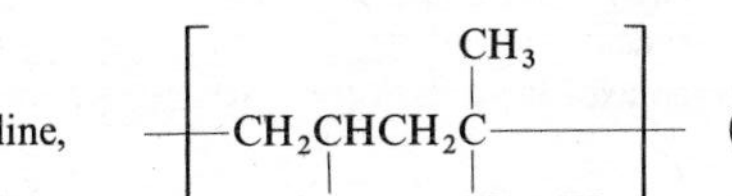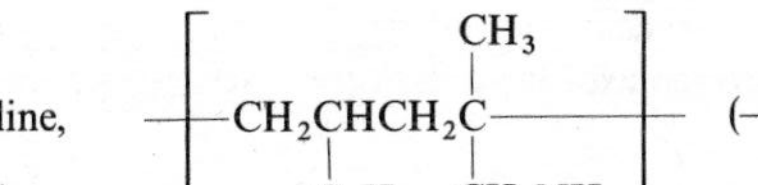 $\left[-CH_2CH(C_6H_5)CH_2C(CH_3)(CH_2NH_2)-\right]_n$ (—)	713

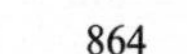

TABLE XIV. Partial Reduction of Nitriles to Aldehydes

No. of Carbon Atoms	Reactant	Conditions	Product(s) and Yield(s) (%)	Refs.
C_3	C_2H_5CN	$NaAlH(OC_2H_5)_3$, THF	C_2H_5CHO (30)	320, 319
C_4	cyclopropyl-CN	$LiAlH(OC_4H_9\text{-}n)_3$, ether, 0°, 1 h	cyclopropyl-CHO (66)	82
	$n\text{-}C_3H_7CN$	$LiAlH(OC_4H_9\text{-}n)_3$, ether, 0°, 1 h	$n\text{-}C_3H_7CHO$ (77)	82
	$i\text{-}C_3H_7CN$	$LiAlH(OC_2H_5)_3$, ether, 0°, 1 h	$i\text{-}C_3H_7CHO$ (81)	82
		$LiAlH_2(OC_2H_5)_2$, ether, 0°, 1 h	" (79)	82
C_5	3-Cyanofuran	SMEAH (IA), THF, 0°; reflux, 1 h	Furan-3-carboxaldehyde (55)	321
	3-Cyanothiophene	SMEAH (IA), THF, 0°; reflux, 1 h	Thiophene-3-carboxaldehyde (40)	321
	$t\text{-}C_4H_9CN$	$LiAlH(OC_2H_5)_3$, ether, 0°, 1 h	$t\text{-}C_4H_9CHO$ (89)	82
		$LiAlH(OC_4H_9\text{-}n)_3$, ether, 0°, 30 min; reflux, 30 min	" (79)	82
C_6	3-Cyanopyridine	$LiAlH(OC_2H_5)_3$, ether, 0°, 1 h	Pyridine-3-carboxaldehyde (58)	82
		$NaAlH(OC_2H_5)_3$, THF, 20°, 4 h	" (81)	319, 320
	4-Cyanopyridine	$NaAlH(OC_2H_5)_3$, THF, 20°, 4 h	Pyridine-4-carboxaldehyde (63)	319
	$NC(CH_2)_4CN$	$LiAlH(OC_2H_5)_3$, ether, 0°, 1 h	$OHC(CH_2)_4CHO$ (60)	82
	$n\text{-}C_5H_{11}CN$	", ", ", "	$n\text{-}C_5H_{11}CHO$ (55)	82, 81
		$NaAlH(OC_2H_5)_3$, THF	No reaction	319
C_7	$2\text{-}ClC_6H_4CN$	$LiAlH(OC_2H_5)_3$, ether, 0°, 1 h	$2\text{-}ClC_6H_4CHO$ (87)	82
	$4\text{-}ClC_6H_4CN$	", ", ", "	$4\text{-}ClC_6H_4CHO$ (84)	82
		$NaAlH(OC_2H_5)_3$, THF, 20°, 2 h	" (89)	319
	C_6H_5CN	$LiAlH(OC_2H_5)_3$, ether, 0°, 1 h	C_6H_5CHO (76)	82, 81
		$NaAlH(OC_2H_5)_3$, THF, 20°, 2 h	" (95)	319, 320
	$2\text{-}HOC_6H_4CN$	$NaAlH(OC_2H_5)_3$, THF, 20°, 2.5 h	$2\text{-}HOC_6H_4CHO$ (75)	319
	$4\text{-}HOC_6H_4CN$	$NaAlH(OC_2H_5)_3$, THF, 20°, 2 h	$4\text{-}HOC_6H_4CHO$ (70)	319
	$3\text{-}H_2NC_6H_4CN$	$NaAlH(OC_2H_5)_3$, THF, 65°, 30 min	$3\text{-}H_2NC_6H_4CHO$ (77)	319

	Nitrile	Reagent and conditions	Product (yield %)	Refs.
	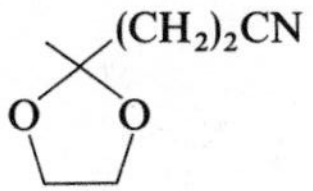 2-(1,3-dioxolan-2-yl, 2-methyl)-$(CH_2)_2CN$ [structure]	$LiAlH(OC_2H_5)_3$, ether, $-10°$; room temp, 1.5 h	No reaction	865
	$Cl(CH_2)_6CN$	$LiAlH(OC_2H_5)_3$, ether, 0°, 1 h	$Cl(CH_2)_6CHO$ (55)	866, 867
C_8	$2-NCC_6H_4CN$	$NaAlH(OC_2H_5)_3$, THF, 65°, 30 min	No reaction	319
	$4-NCC_6H_4CN$	", ", ", "	$4-OHCC_6H_4CHO$ (74)	319
	$C_6H_5CH_2CN$	$LiAlH(OC_2H_5)_3$, ether, 0°, 1 h	No reaction	81
		$NaAlH(OC_2H_5)_3$, THF	No reaction	319
	$2-CH_3C_6H_4CN$	$LiAlH(OC_2H_5)_3$, ether, 0°, 1 h	$2-CH_3C_6H_4CHO$ (87)	82
		$NaAlH(OC_2H_5)_3$, THF, 20°, 2 h	" (88)	319, 320
	$3-CH_3C_6H_4CN$	$NaAlH(OC_2H_5)_3$, THF, 65°, 3 h	$3-CH_3C_6H_4CHO$ (83)	319
	$4-CH_3C_6H_4CN$	$NaAlH(OC_2H_5)_3$, THF, 20°, 2 h	$4-CH_3C_6H_4CHO$ (88)	319
		$NaAlH[OCH_2CH_2N(CH_3)_2]_3$, C_6H_6, 25°, 2 h	" (75)	633
		$NaAl_2H_4[OCH_2CH_2N(CH_3)_2]_3$, C_6H_6, 25°, 1 h	" (73)	634
	$2-CH_3OC_6H_4CN$	$NaAlH(OC_2H_5)_3$, THF, 20°, 3.5 h	$2-CH_3OC_6H_4CHO$ (79)	319
	$4-CH_3OC_6H_4CN$	$LiAlH(OC_2H_5)_3$, ether, 0°, 1 h	$4-CH_3OC_6H_4CHO$ (81)	82
		$NaAlH(OC_2H_5)_3$, THF, 20°, 2 h	" (84)	319
	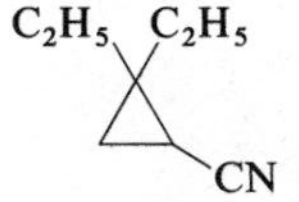 2,2-(C_2H_5, C_2H_5)-cyclopropyl-CN [structure]	$LiAlH(OC_2H_5)_3$, ether, 0°, 1 h	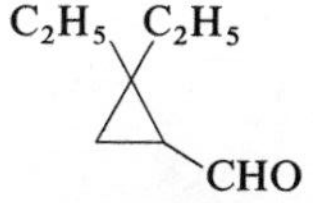 2,2-(C_2H_5, C_2H_5)-cyclopropyl-CHO [structure] (—)	868
	$(CH_3)_2C(CN)OCH(OC_2H_5)CH_3$	SMEAH (IA), ether–toluene, 0°, 3 h	$(CH_3)_2C(CHO)OCH(OC_2H_5)CH_3$ (57)	331
C_9	$4-(CH_3CF_2)C_6H_4CN$	$LiAlH(OC_2H_5)_3$, ether	$4-(CH_3CF_2)C_6H_4CHO$ (20)	869
	$C_6H_5CH{=}CHCN$	$LiAlH(OC_2H_5)_3$, ether, 0°, 1 h	$C_6H_5CH{=}CHCHO$ (61)	82
	$2,6-(CH_3)_2C_6H_3CN$	", ", ", "	$2,6-(CH_3)_2C_6H_3CHO$ (—)	870
	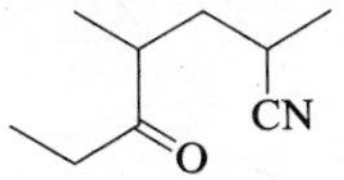 [structure: keto nitrile, CN, O]	$LiAlH_2(OC_2H_5)_2$	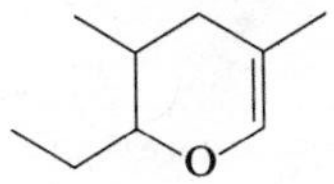 [structure: dihydropyran, O] (—)	871

TABLE XIV. PARTIAL REDUCTION OF NITRILES TO ALDEHYDES (*Continued*)

No. of Carbon Atoms	Reactant	Conditions	Product(s) and Yield(s) (%)	Refs.
C_9 (*Contd.*)	$(CH_3)_2C(OTHP)CN$	$LiAlH(OC_2H_5)_3$, ether, 0°, 45 min	$(CH_3)_2C(OTHP)CHO$ (—)	872
	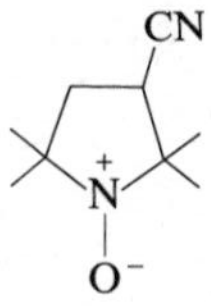 [structure: 2,2,5,5-tetramethylpyrrolidine N-oxide with CN]	$LiAlH(OC_2H_5)_3$, ether, 0°, 1 h	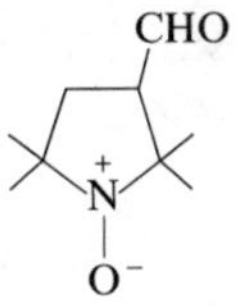 [structure: 2,2,5,5-tetramethylpyrrolidine N-oxide with CHO] (—)	444
C_{10}	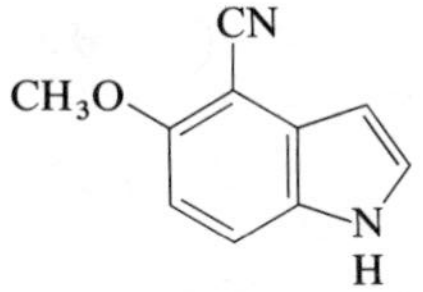 [structure: 5-CH_3O-4-CN-indole (N–H)]	$LiAlH(OC_2H_5)_3$	No reaction	873a
	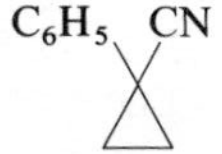 [structure: 1-C_6H_5-1-CN-cyclopropane]	$LiAlH_2(OC_2H_5)_2$, ether	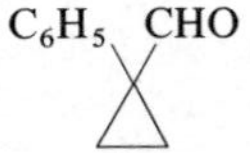 [structure: 1-C_6H_5-1-CHO-cyclopropane] (—)	873b
	$C_6H_5(CH_2)_3CN$	$LiAlH(OC_2H_5)_3$, ether, 0°, 1 h	$C_6H_5(CH_2)_3CHO$ (73)	82
	2,4,5-$(CH_3)_3C_6H_2CN$	$NaAlH(OC_2H_5)_3$, THF, 20°, 2.5 h	2,4,5-$(CH_3)_3C_6H_2CHO$ (77)	319
	$C_6H_5C(CH_3)_2CN$	$LiAlH_2(OC_2H_5)_2$, ether, 2 h	$C_6H_5C(CH_3)_2CHO$ (48)	325
	$C_6H_5O(CH_2)_3CN$	$LiAlH(OC_2H_5)_3$, ether, 0°, 1 h	$C_6H_5O(CH_2)_3CHO$ (66)	82
	 [structure: cyclopentane with $OCH(OC_2H_5)CH_3$ and CN]	SMEAH (IA), ether–toluene, 0°, 3 h	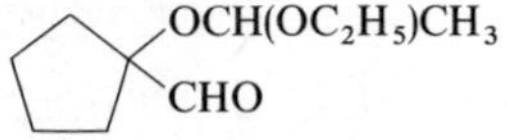 [structure: cyclopentane with $OCH(OC_2H_5)CH_3$ and CHO] (63)	331
C_{11}	1-$C_{10}H_7CN$	$LiAlH(OC_2H_5)_3$, ether, 0°, 1 h	1-$C_{10}H_7CHO$ (80)	82
	2-$C_{10}H_7CN$	$LiAlH(OC_2H_5)_3$, ether, 0°, 3 h	2-$C_{10}H_7CHO$ (50)	874
		$NaAlH(OC_2H_5)_3$, THF, 65°, 1 h	" (76)	319, 320

	Reactant	Conditions	Product(s) and Yield(s) (%)	Refs.
	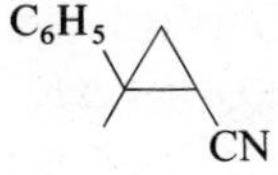 C_6H_5, CN cyclopropane	$LiAlH(OC_2H_5)_3$ (1 eq), ether, 0°, 1 h	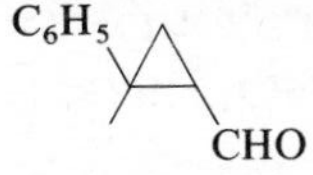 C_6H_5, CHO cyclopropane **I** (52) 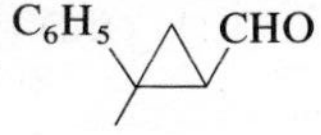 + C_6H_5, CHO cyclopropane **II** (3) 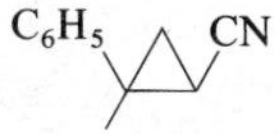 + C_6H_5, CN cyclopropane (2)	86
	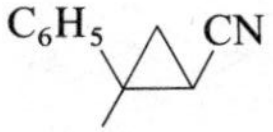 C_6H_5, CN cyclopropane	$LiAlH(OC_2H_5)_3$(1 eq), ether, 0°, 1 h	**II** (42) + **I** (18)	86
	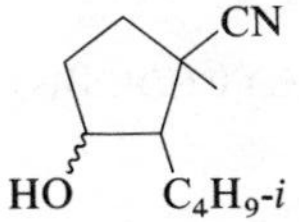 CN, HO, C_4H_9-*i* cyclopentane	$LiAlH(OC_2H_5)_3$, ether, 0°, 1 h	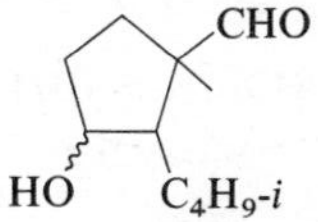 CHO, HO, C_4H_9-*i* cyclopentane (10)	875
	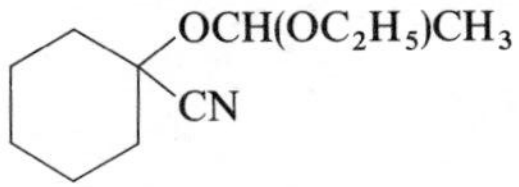 $OCH(OC_2H_5)CH_3$, CN cyclohexane	SMEAH (IA), ether–toluene, 0°, 3 h	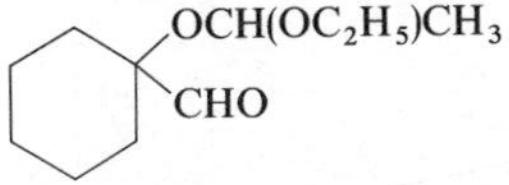 $OCH(OC_2H_5)CH_3$, CHO cyclohexane (53)	331
C_{12}	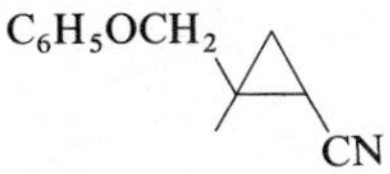 $C_6H_5OCH_2$, CN cyclopropane	$LiAlH(OC_2H_5)_3$ (1 eq), ether, 0°, 1 h; room temp, 1 h	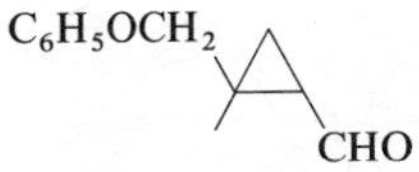 $C_6H_5OCH_2$, CHO cyclopropane (46) 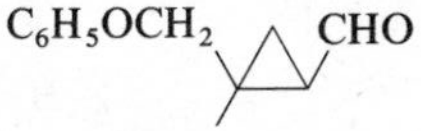 + $C_6H_5OCH_2$, CHO cyclopropane (24) 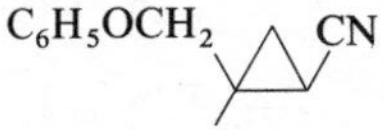 + $C_6H_5OCH_2$, CN cyclopropane (3)	86, 876
	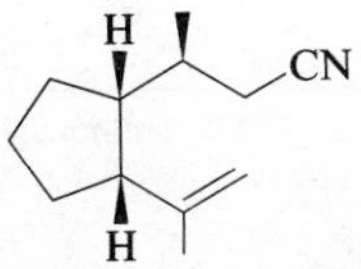 H, H, CN cyclopentane	$LiAlH(OC_2H_5)_3$, ether, −30°, 1 h	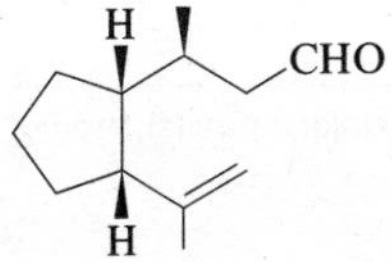 H, H, CHO cyclopentane (—)	877

TABLE XIV. PARTIAL REDUCTION OF NITRILES TO ALDEHYDES (*Continued*)

No. of Carbon Atoms	Reactant	Conditions	Product(s) and Yield(s) (%)	Refs.
C_{12} (*Contd.*)	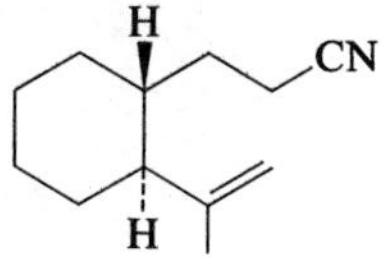 [structure] CN	$LiAlH(OC_2H_5)_3$, ether, −30°, 25 min; 0°, 15 min	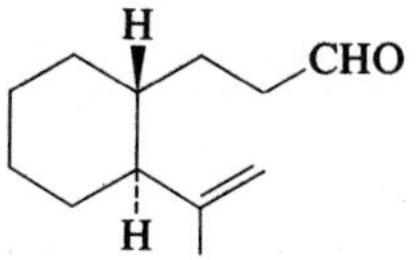 [structure] CHO (—)	878
	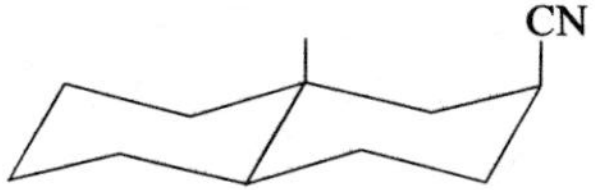 [structure] CN	$LiAlH(OC_2H_5)_3$ (1 eq), ether, 0°, 1 h; room temp, 1 h	 [structure] CHO (20)  + [structure] CHO (10) 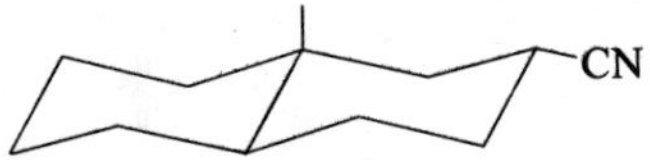 + [structure] CN (9)	86
	$THPO(CH_2)_6CN$	$LiAlH_2(OC_2H_5)_2$, ether, 0°, 1 h	$THPO(CH_2)_6CHO$ (45)	879
C_{13}	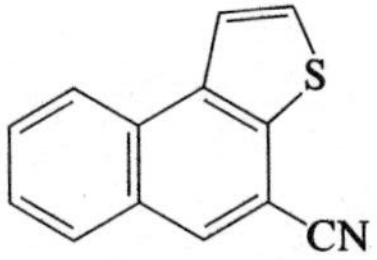 [structure] S, CN	$LiAlH(OC_2H_5)_3$, ether, 0°, 1 h	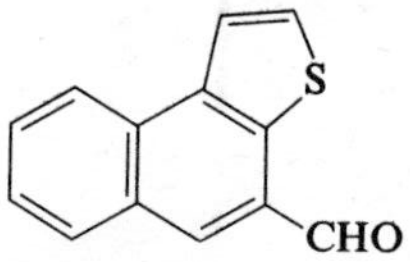 [structure] S, CHO (10)	880
	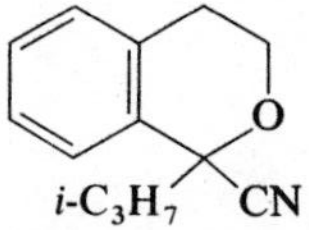 [structure] O, i-C_3H_7, CN	SMEAH, THF–toluene, −15°, 1 h; room temp, 1 h	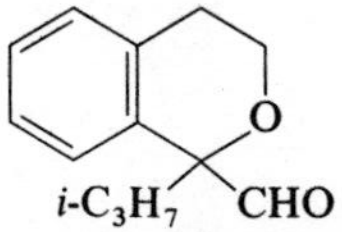 [structure] O, i-C_3H_7, CHO (78)	328
	3,4,5-$(CH_3O)_3C_6H_2C(CH_3)_2CN$	$LiAlH_2(OC_2H_5)_2$, ether, 3 h	3,4,5-$(CH_3O)_3C_6H_2C(CH_3)_2CHO$ (53) + 3,4,5-$(CH_3O)_3C_6H_2C(CH_3)_2CH_2NH_2$ (30)	325

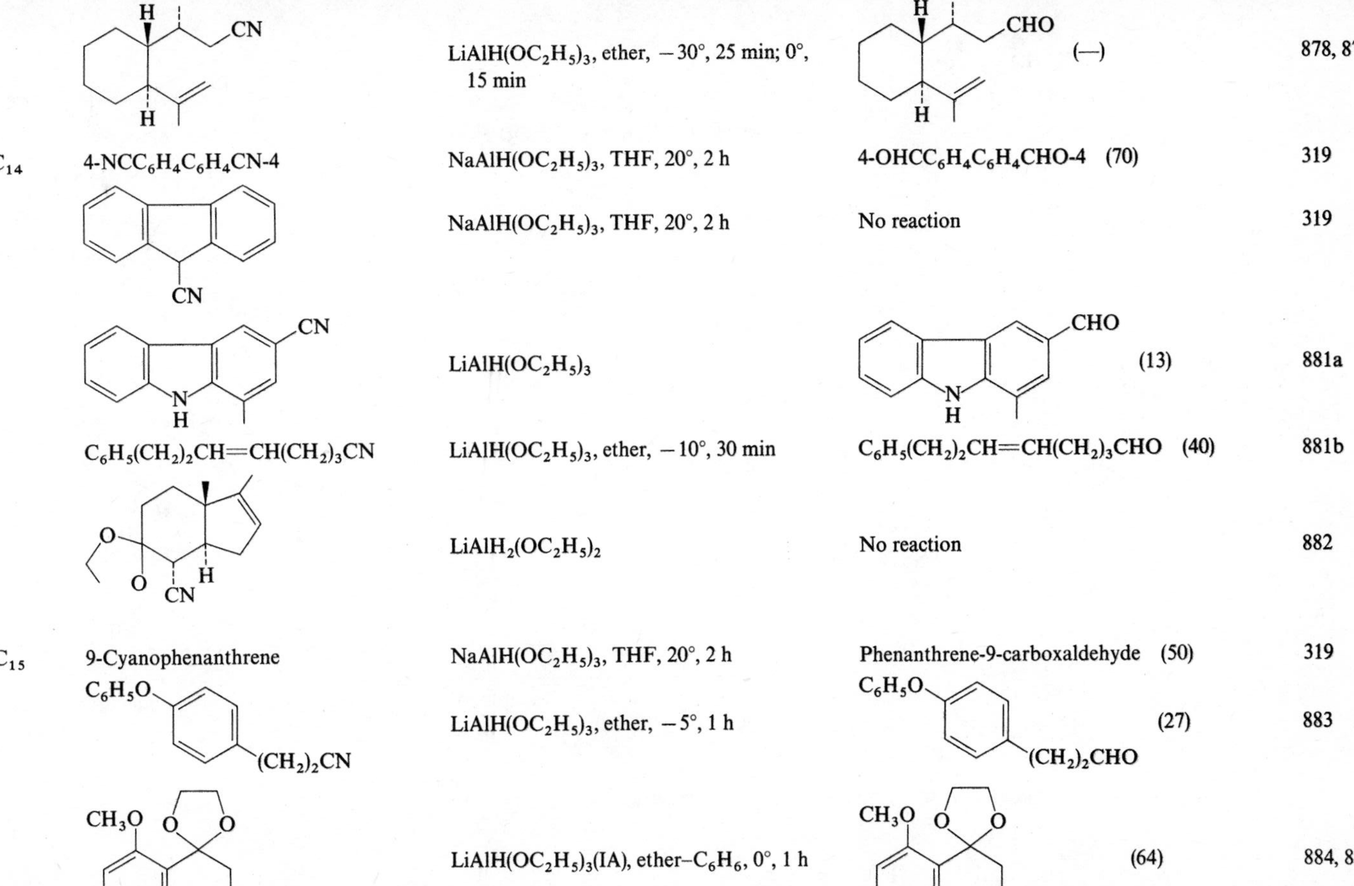

	Nitrile	Reagent and Conditions	Product (Yield, %)	Refs.
		$LiAlH(OC_2H_5)_3$, ether, $-30°$, 25 min; 0°, 15 min	(—)	878, 877
C_{14}	4-$NCC_6H_4C_6H_4CN$-4	$NaAlH(OC_2H_5)_3$, THF, 20°, 2 h	4-$OHCC_6H_4C_6H_4CHO$-4 (70)	319
		$NaAlH(OC_2H_5)_3$, THF, 20°, 2 h	No reaction	319
		$LiAlH(OC_2H_5)_3$	(13)	881a
	$C_6H_5(CH_2)_2CH{=}CH(CH_2)_3CN$	$LiAlH(OC_2H_5)_3$, ether, $-10°$, 30 min	$C_6H_5(CH_2)_2CH{=}CH(CH_2)_3CHO$ (40)	881b
		$LiAlH_2(OC_2H_5)_2$	No reaction	882
C_{15}	9-Cyanophenanthrene	$NaAlH(OC_2H_5)_3$, THF, 20°, 2 h	Phenanthrene-9-carboxaldehyde (50)	319
		$LiAlH(OC_2H_5)_3$, ether, $-5°$, 1 h	(27)	883
		$LiAlH(OC_2H_5)_3$(IA), ether–C_6H_6, 0°, 1 h	(64)	884, 885

TABLE XIV. PARTIAL REDUCTION OF NITRILES TO ALDEHYDES (*Continued*)

No. of Carbon Atoms	Reactant	Conditions	Product(s) and Yield(s) (%)	Refs.
C_{15} (*Contd.*)	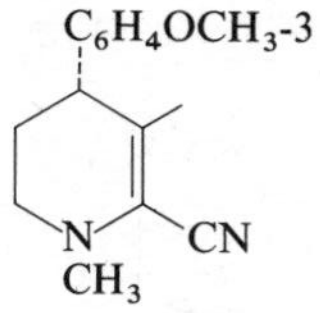 [structure: $C_6H_4OCH_3$-3, N–CH_3, CN]	$LiAlH_2(OC_2H_5)_2$, ether, 0° to room temp, 1.5 h	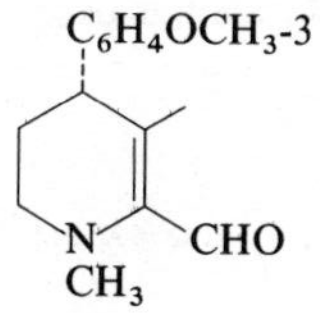 [structure: $C_6H_4OCH_3$-3, N–CH_3, CHO] (25)	327
C_{16}	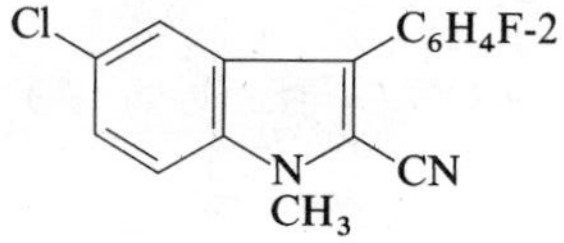 [structure: Cl, C_6H_4F-2, N–CH_3, CN]	SMEAH, toluene, reflux, 5 h	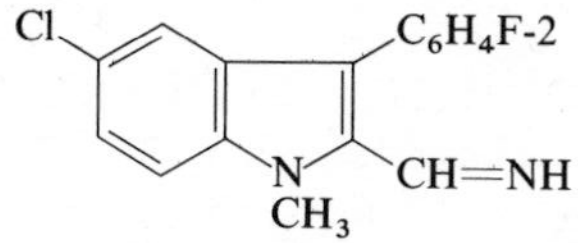 [structure: Cl, C_6H_4F-2, N–CH_3, CH=NH] (—)	329
	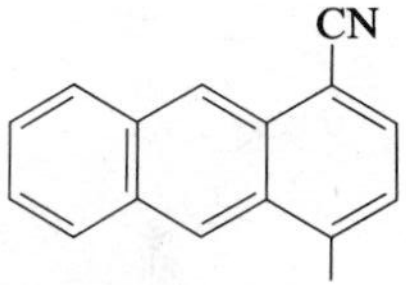 [structure: CN]	$NaAlH(OC_2H_5)_3$, THF, 20°, 3 h	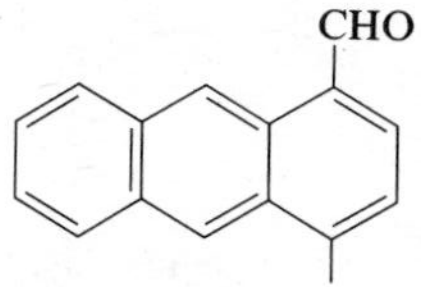 [structure: CHO] (70)	319
	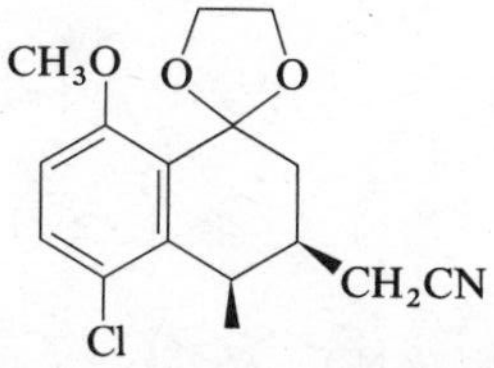 [structure: CH_3O, O, O, Cl, CH_2CN]	$LiAlH(OC_2H_5)_3$ (IA), C_6H_6–ether, 0°, 1 h	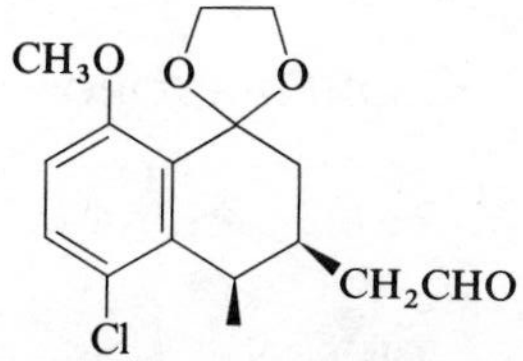 [structure: CH_3O, O, O, Cl, CH_2CHO] (63)	884

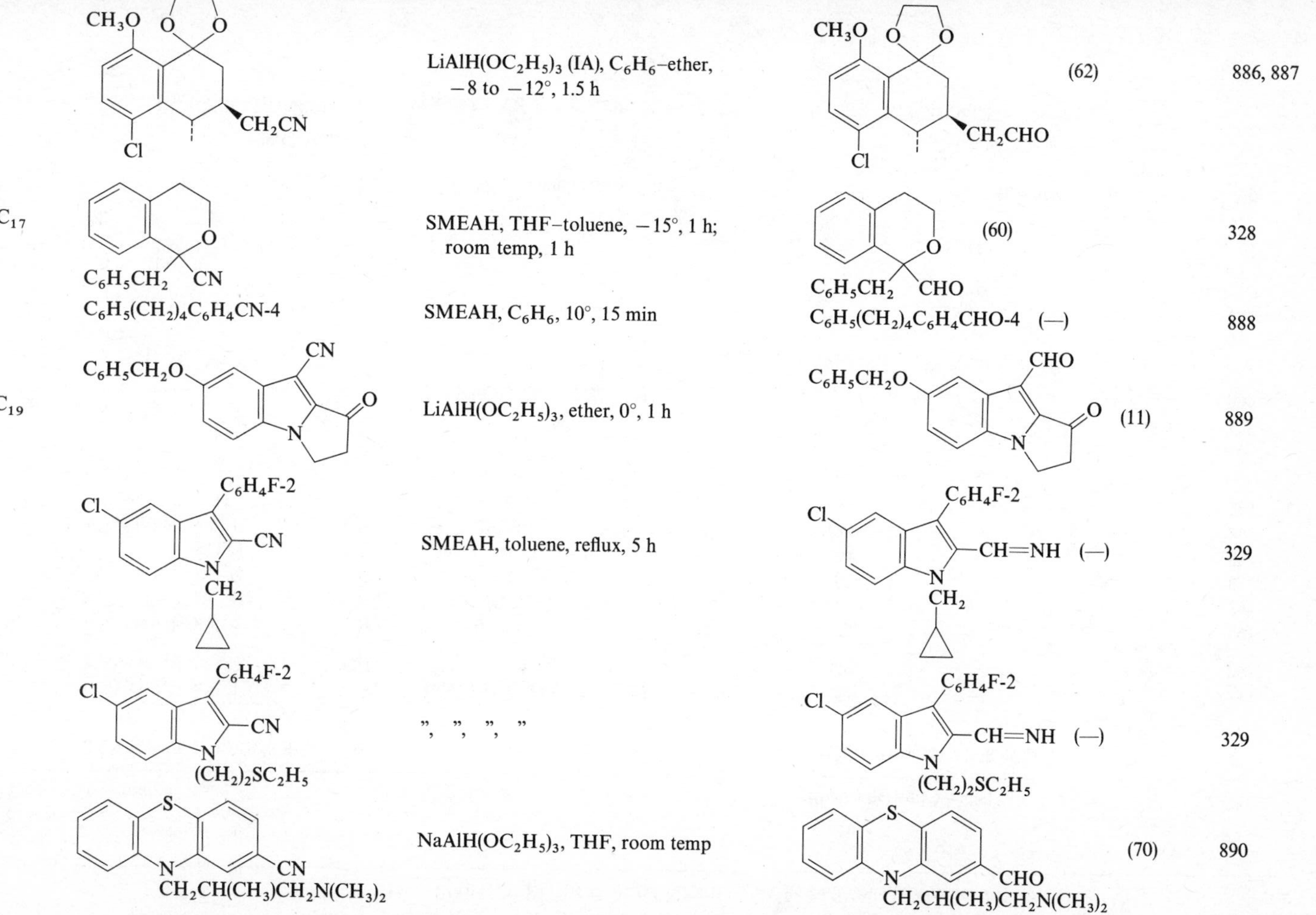

	Reactant	Reagents and conditions	Product (Yield, %)	Refs.
		$LiAlH(OC_2H_5)_3$ (IA), C_6H_6–ether, -8 to $-12°$, 1.5 h	(62)	886, 887
C_{17}		SMEAH, THF–toluene, $-15°$, 1 h; room temp, 1 h	(60)	328
	$C_6H_5(CH_2)_4C_6H_4CN$-4	SMEAH, C_6H_6, 10°, 15 min	$C_6H_5(CH_2)_4C_6H_4CHO$-4 (—)	888
C_{19}		$LiAlH(OC_2H_5)_3$, ether, 0°, 1 h	(11)	889
		SMEAH, toluene, reflux, 5 h	(—)	329
		", ", ", "	(—)	329
		$NaAlH(OC_2H_5)_3$, THF, room temp	(70)	890

TABLE XIV. PARTIAL REDUCTION OF NITRILES TO ALDEHYDES (*Continued*)

No. of Carbon Atoms	Reactant			Conditions	Product(s) and Yield(s) (%)	Refs.
	$RR^1C(CN)(CH_2)_2NR^2(CH_3)$				$RR^1C(CHO)(CH_2)_2NR^2(CH_3)$	
	R	R^1	R^2			
	1-Naphthyl	n-C_3H_7	CH_3	$LiAlH_2(OC_2H_5)_2$, ether, 25°, 2 h	(54)	891
	2-Naphthyl	i-C_3H_7	CH_3	", ", ", 5 h	(34)	891
C_{20}	1-Naphthyl	n-C_4H_9	CH_3	", ", 0°, 3 h	(73)	891
	"	s-C_4H_9	CH_3	$LiAlH_3(OC_2H_5)$, ", reflux, 3 h	(61)[a]	891
	"	i-C_4H_9	CH_3	$LiAlH_2(OC_2H_5)_2$, ", ", "	(67)	891
	"	i-C_3H_7	C_2H_5	$LiAlH_3(OC_2H_5)$, ", ", 8 h	(34)	891
	2-Naphthyl	s-C_4H_9	CH_3	", ", 25°, 3 h	(28)	891
	1-Naphthyl	$(CH_3)_2N(CH_2)_2$	CH_3	$LiAlH_2(OC_2H_5)_2$, ", ", 2 h	(56)	891
	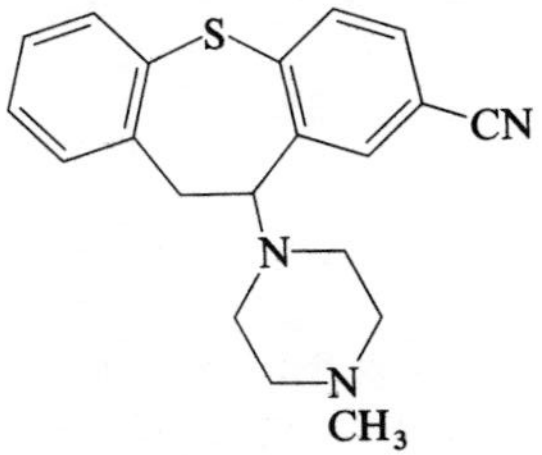 [structure: dibenzo[b,f]thiepin with CN, 10-(4-methylpiperazin-1-yl); labels S, CN, N, N, CH_3]			$LiAlH(OC_2H_5)_3$, THF, −10°	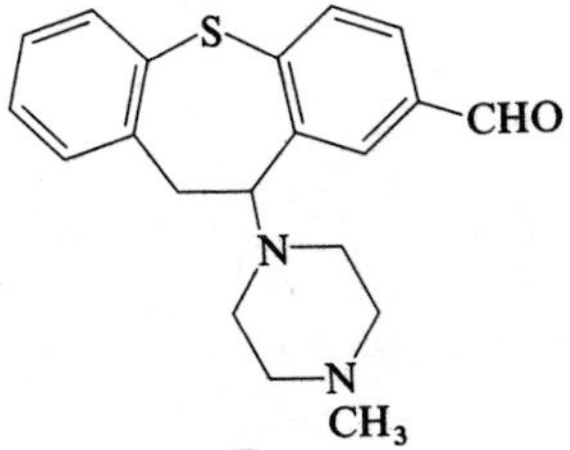 [structure: corresponding CHO; labels S, CHO, N, N, CH_3] (—)	892, 893
C_{21}	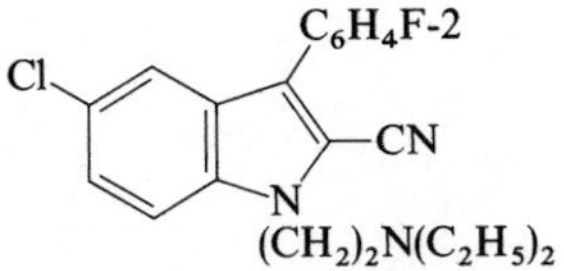 [structure: Cl-indole with C_6H_4F-2, CN, N-$(CH_2)_2N(C_2H_5)_2$]			SMEAH, toluene, reflux, 5 h	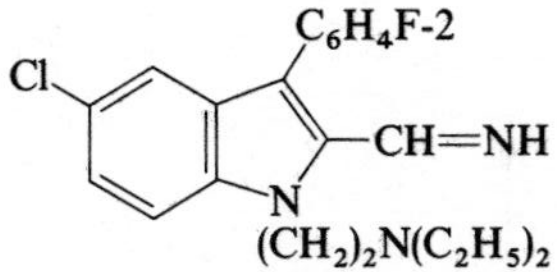 [structure: Cl-indole with C_6H_4F-2, $CH{=}NH$, N-$(CH_2)_2N(C_2H_5)_2$] (—)	329
	$RC(CN)(CH_2)_2NR^1R^2$ 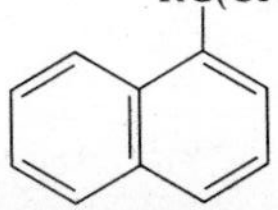 (R = 1-naphthyl, structure)				$RC(CHO)(CH_2)_2NR^1R^2$ 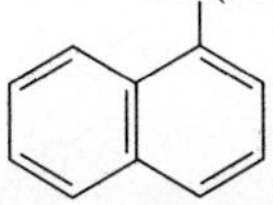 (R = 1-naphthyl, structure)	

	R	R^1	R^2			
	n-C_3H_7			$LiAlH_2(OC_2H_5)_2$, ether, reflux, 3 h	(53)	891
	n-C_3H_7		O	"	(57) + amine (5)	891
	i-C_3H_7		O	", reflux 60 h	(29)	891
	i-C_3H_7	C_2H_5	C_2H_5	", reflux 36 h	(32)	891
	s-C_4H_9	CH_3	C_2H_5	", reflux 10 h	(43)	891
C_{22}	n-C_3H_7			", reflux 3 h	(44)	891
	n-C_4H_9			", 25°, 4 h	(62)	891
	i-C_4H_9			", 25°, 4 h	(47)	891
	s-C_4H_9			", reflux 22 h	(43)	891
	n-C_4H_9		O	", 25°, 3 h	(58)	891
	i-C_4H_9		O	", reflux, 11 h	(16)	891
	s-C_4H_9		O	", reflux, 35 h	(33)	891
	s-C_4H_9	C_2H_5	C_2H_5	", reflux, 38 h	(28)	891
	$C_2H_5(CH_3)N(CH_2)_2$	CH_3	C_2H_5	", reflux, 3 h	(45)	891

TABLE XIV. Partial Reduction of Nitriles to Aldehydes (*Continued*)

No. of Carbon Atoms	Reactant R	R^1	R^2	Conditions	Product(s) and Yield(s) (%)	Refs.
C_{23}	$n\text{-}C_4H_9$	 [ring structure]		", reflux, 3 h	(47)	891
	$i\text{-}C_4H_9$	 [ring structure]		", reflux, 12 h	(37)	891
	$i\text{-}C_3H_7$	$i\text{-}C_3H_7$	$i\text{-}C_3H_7$	$LiAlH_3(OC_2H_5)$, ether, reflux, 93 h	(15)	891
C_{24}	 [ring structure] $N(CH_2)_2$	 [ring structure]		$LiAlH_2(OC_2H_5)_2$, ether, 25°, 2 h	(59)	891
	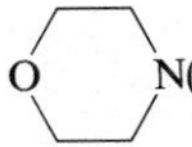 O [ring structure] $N(CH_2)_2$	 [ring structure] O		", 25°, 4 h	(64)	891
	$s\text{-}C_4H_9$	$i\text{-}C_3H_7$	$i\text{-}C_3H_7$	$LiAlH_3(OC_2H_5)$, ether, reflux, 90 h	(26)	891
	$(C_2H_5)_2N(CH_2)_2$	C_2H_5	C_2H_5	", reflux, 3 h	(34)	891
C_{26}	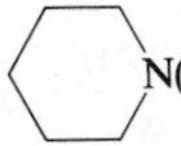 [ring structure] $N(CH_2)_2$	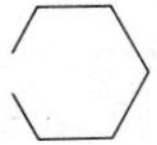 [ring structure]		", 25°, 4 h	(39)	891
C_{28}	$(i\text{-}C_3H_7)_2N(CH_2)_2$	$i\text{-}C_3H_7$	$i\text{-}C_3H_7$	$LiAlH_3(OC_2H_5)$, ether, reflux, 3 h	(46)	891

[a] Reduction of the nitrile to the aminomethyl compound occurred to the extent of 14%.

TABLE XV. REDUCTION OF CARBOXYLIC ACID AMIDES

No. of Carbon Atoms	Reactant	Conditions	Product(s) and Yield(s) (%)	Refs.
C_4	n-$C_3H_7CONH_2$	$Al_2H_3(OCH_2CH_2OCH_3)_3$, C_6H_6, 80°, 2 h	n-$C_4H_9NH_2$ (75)	313
		$NaAlH_2(OCH_2CH_2OC_4H_9$-$n)_2$, C_6H_6, 80°, 4 h	" (94)	632
C_6	n-$C_5H_{11}CONH_2$	LTMA, THF, 0°, 2 h; room temp, 2 h	n-$C_6H_{13}NH_2$ (86)	161
C_7	$C_6H_5CONH_2$	$Ca[AlH_2(OC_4H_9$-$i)_2]_2$·THF, C_6H_6, reflux, 3 h	$C_6H_5CH_2NH_2$ (62)	165
C_8	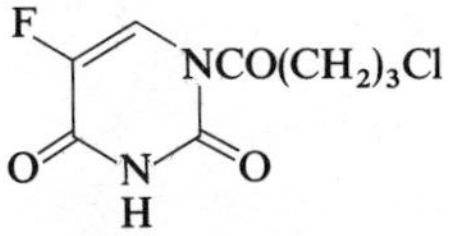 F; $NCO(CH_2)_3Cl$; O; N; O; H (structure)	SMEAH, toluene	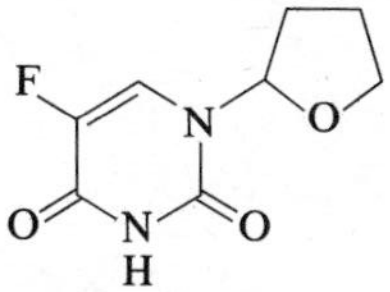 F; N; O; O; N; O; H (structure) (75)	894
	$CH_3CONHC_6H_5$	$Ca[AlH_2(OC_4H_9$-$i)_2]_2$·THF, C_6H_6, reflux, 4 h	$C_2H_5NHC_6H_5$ (93)	165
		SMEAH, C_6H_6, 80°, 1 h	" (85)	94, 167
	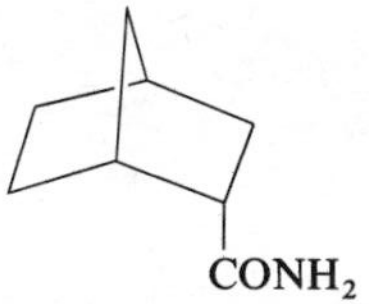 $CONH_2$ (structure)	1. SMEAH, C_6H_6, reflux, 2 h 2. HCl, EtOH	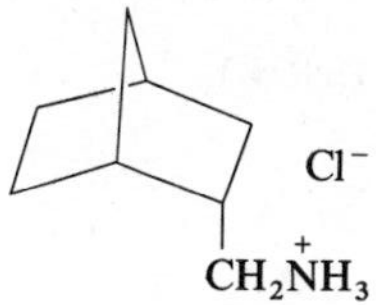 $CH_2\overset{+}{N}H_3$ Cl^- (structure) (69)	895
	$C_2H_5CO(CH_2)_3CONHCH_3$	SMEAH, C_6H_6, reflux, 2 h	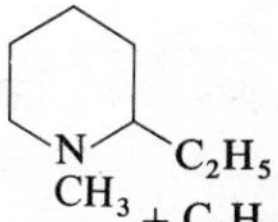 N; CH_3; C_2H_5 (structure) (49) + $C_2H_5CHOH(CH_2)_4NHCH_3$ (22)	94

TABLE XV. Reduction of Carboxylic Acid Amides (*Continued*)

No. of Carbon Atoms	Reactant	Conditions	Product(s) and Yield(s) (%)	Refs.
C_9	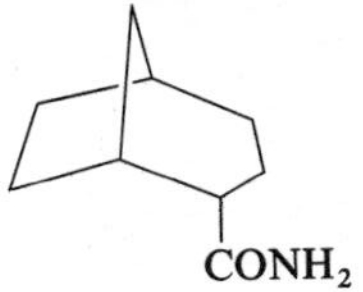 $CONH_2$	1. SMEAH, C_6H_6, reflux, 2 h 2. HCl, EtOH	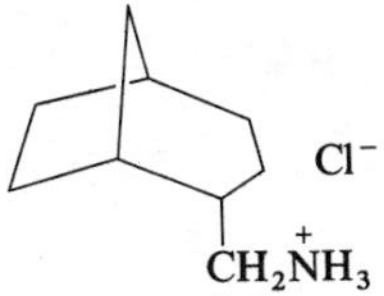 $CH_2\overset{+}{N}H_3$ Cl^- (77)	895
C_{10}	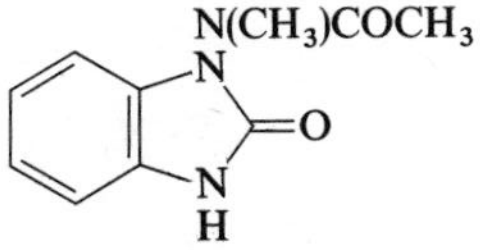 $N(CH_3)COCH_3$, N, =O, N, H	SMEAH, toluene–THF, reflux, 3 h	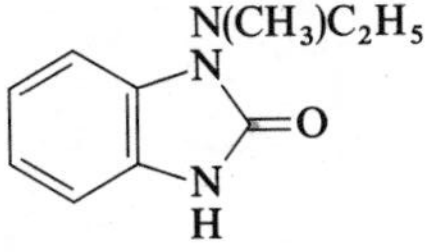 $N(CH_3)C_2H_5$, N, =O, N, H (52)	896
	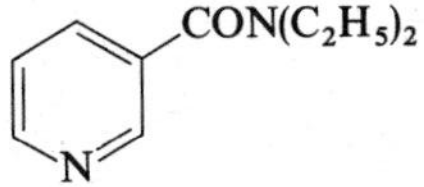 $CON(C_2H_5)_2$, N	SMEAH (IA), C_6H_6, 37°, 5 min	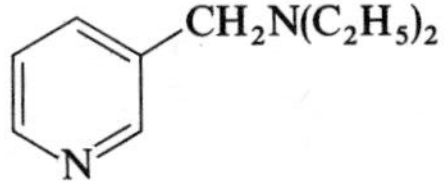 $CH_2N(C_2H_5)_2$, N (27)	94
C_{11}	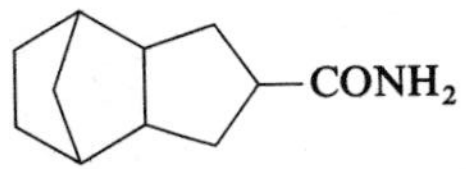 $CONH_2$	1. SMEAH, C_6H_6, reflux, 2 h 2. HCl, EtOH	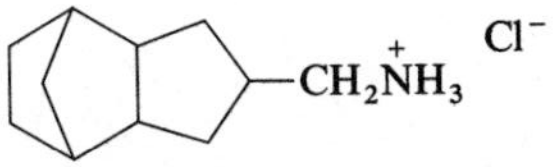 $CH_2\overset{+}{N}H_3$ Cl^- (44)	895
C_{12}	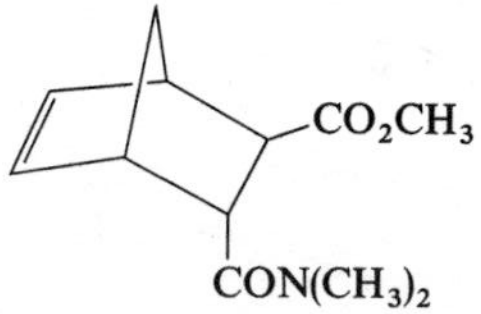 CO_2CH_3, $CON(CH_3)_2$	SMEAH, C_6H_6, reflux, 24 h	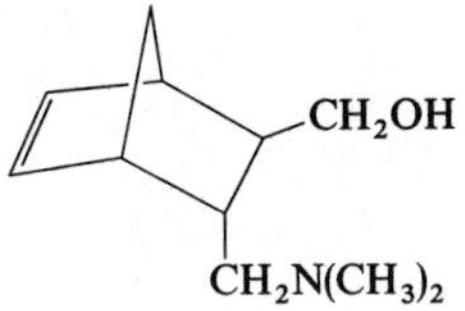 CH_2OH, $CH_2N(CH_3)_2$ (91)	650
	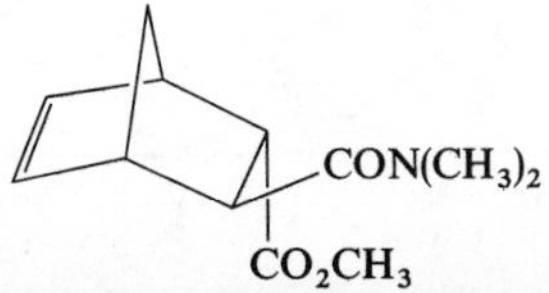 $CON(CH_3)_2$, CO_2CH_3	SMEAH, C_6H_6, reflux, 24 h	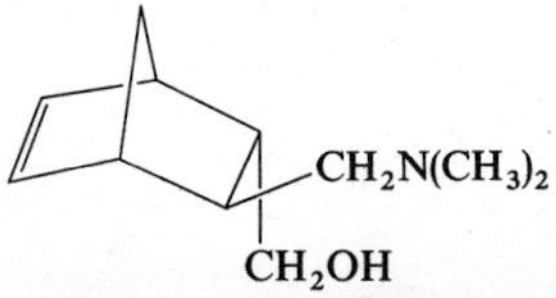 $CH_2N(CH_3)_2$, CH_2OH (74)	650

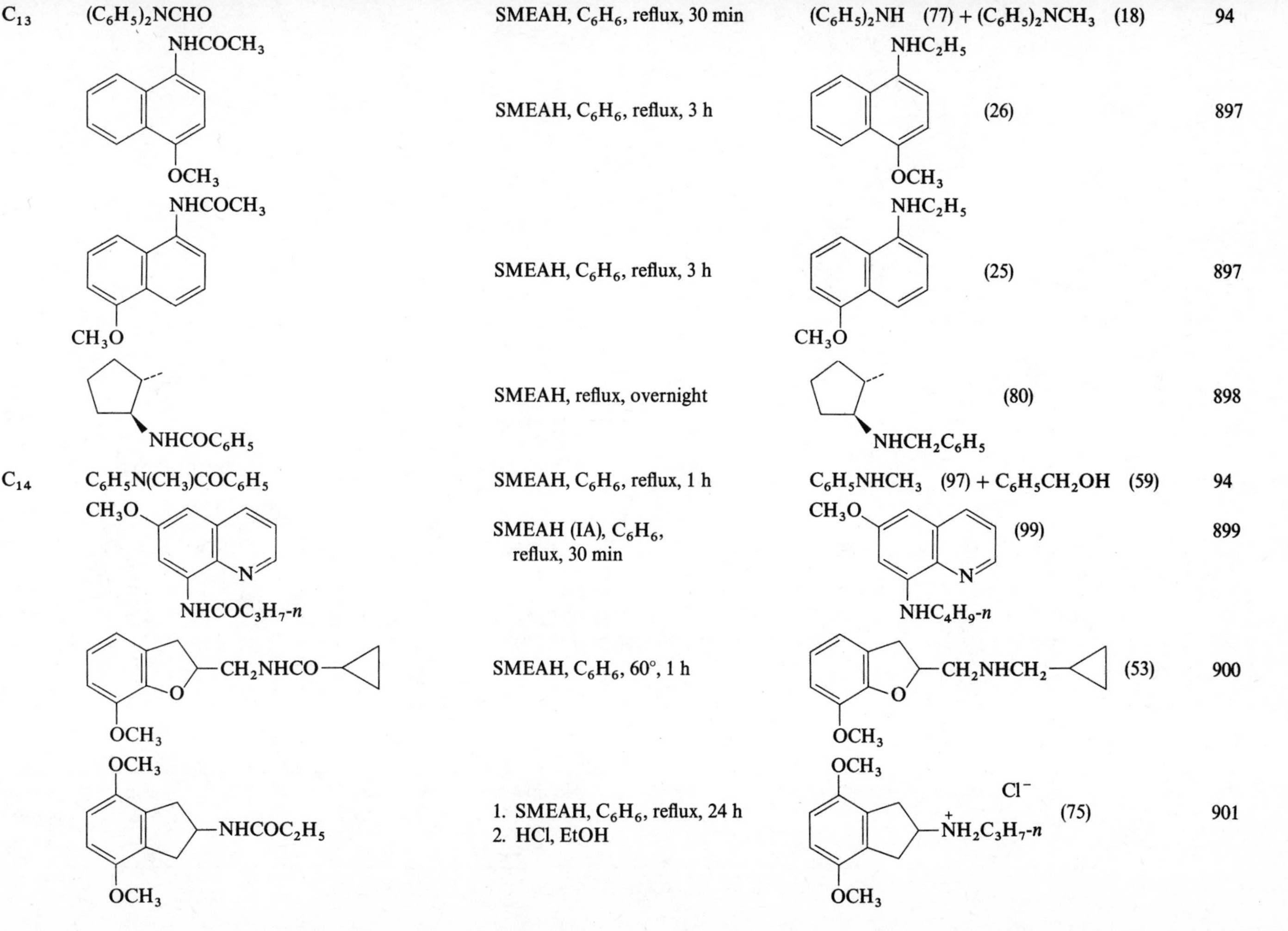

	Substrate	Conditions	Product(s) and Yield(s) (%)	Refs.
C_{13}	$(C_6H_5)_2NCHO$	SMEAH, C_6H_6, reflux, 30 min	$(C_6H_5)_2NH$ (77) + $(C_6H_5)_2NCH_3$ (18)	94
	4-methoxy-1-($NHCOCH_3$)naphthalene ($NHCOCH_3$, OCH_3)	SMEAH, C_6H_6, reflux, 3 h	4-methoxy-1-(NHC_2H_5)naphthalene (NHC_2H_5, OCH_3) (26)	897
	5-methoxy-1-($NHCOCH_3$)naphthalene ($NHCOCH_3$, CH_3O)	SMEAH, C_6H_6, reflux, 3 h	5-methoxy-1-(NHC_2H_5)naphthalene (NHC_2H_5, CH_3O) (25)	897
	cyclopentyl-$NHCOC_6H_5$	SMEAH, reflux, overnight	cyclopentyl-$NHCH_2C_6H_5$ (80)	898
C_{14}	$C_6H_5N(CH_3)COC_6H_5$	SMEAH, C_6H_6, reflux, 1 h	$C_6H_5NHCH_3$ (97) + $C_6H_5CH_2OH$ (59)	94
	6-CH_3O-8-($NHCOC_3H_7$-*n*)quinoline	SMEAH (IA), C_6H_6, reflux, 30 min	6-CH_3O-8-(NHC_4H_9-*n*)quinoline (99)	899
	7-OCH_3-dihydrobenzofuran-2-CH_2NHCO-cyclopropyl	SMEAH, C_6H_6, 60°, 1 h	7-OCH_3-dihydrobenzofuran-2-CH_2NHCH_2-cyclopropyl (53)	900
	4,7-di-OCH_3-indan-2-$NHCOC_2H_5$	1. SMEAH, C_6H_6, reflux, 24 h 2. HCl, EtOH	4,7-di-OCH_3-indan-2-$\overset{+}{N}H_2C_3H_7$-*n* Cl^- (75)	901

TABLE XV. REDUCTION OF CARBOXYLIC ACID AMIDES (*Continued*)

No. of Carbon Atoms	Reactant	Conditions	Product(s) and Yield(s) (%)	Refs.
C_{14} (*Contd.*)	$(CH_2)_3CONH_2$	1. SMEAH, C_6H_6, reflux, 24 h 2. HCl, EtOH	$(CH_2)_4\overset{+}{N}H_3$ Cl^- (49)	895
C_{15}	3-$CH_3C_6H_4OCH(CH_2)_2OH$ — $CONHC_4H_9$-*s*	SMEAH, C_6H_6	3-$CH_3C_6H_4OCH(CH_2)_2OH$ — $CH_2NHC_4H_9$-*s* (—)	902
C_{16}	$NHCOC_6H_5$, $CO_2C_2H_5$	SMEAH, C_6H_6, reflux, 2 h	$C_6H_5CH_2N$ (18) + $NHCOC_6H_5$, CH_2OH (9)	189
	$CH_2NHCOC_5H_9$, O, OCH_3	SMEAH, C_6H_6, 60°, 1 h	$CH_2NHCH_2C_5H_9$, O, OCH_3 (58)	900

Formula	Substrate	Conditions	Product (Yield, %)		Ref.
	$CON(CH_3)_2$; $[(CH_2)_2O]_3CH_2CH_2$ (pyridine crown)	SMEAH, C_6H_6, reflux, 30 min	$CH_2N(CH_3)_2$; $[(CH_2)_2O]_3CH_2CH_2$	(~100)	336
C_{17}	HCONH; OCH_3; OCH_3	SMEAH, C_6H_6, room temp, 1.5 h	CH_3NH; OCH_3; OCH_3	(86)	903
	$HOCH_2$; H; H; $NCOCH_3$; N H	SMEAH, THF, C_6H_6, 2 h	$HOCH_2$; H; H; NC_2H_5; N H	(58)	904
	$CH_2NHCOC_6H_{11}$; O; OCH_3	SMEAH, C_6H_6, 60°, 1 h	$CH_2NHCH_2C_6H_{11}$; O; OCH_3	(56)	900
C_{18}	$HOCH_2$; H; H; $NCOC_2H_5$; N H	SMEAH, THF, C_6H_6, reflux, 2 h	$HOCH_2$; H; H; NC_3H_7-*n*; N H	(65)	904

TABLE XV. REDUCTION OF CARBOXYLIC ACID AMIDES (*Continued*)

No. of Carbon Atoms	Reactant	Conditions	Product(s) and Yield(s) (%)	Refs.
C_{18} (*Contd.*)	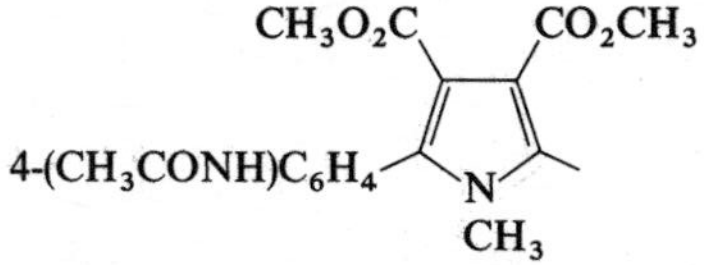 CH_3O_2C, CO_2CH_3, $4\text{-}(CH_3CONH)C_6H_4$, N, CH_3	$LiAlH_2(OC_2H_5)_2$	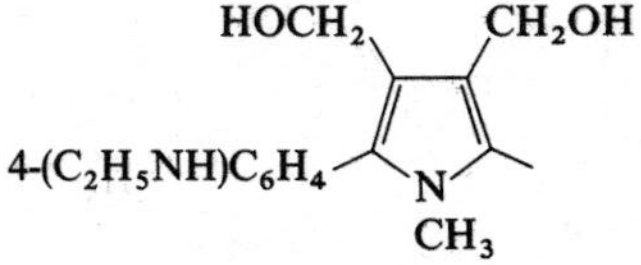 $HOCH_2$, CH_2OH, $4\text{-}(C_2H_5NH)C_6H_4$, N, CH_3 (—)	683
	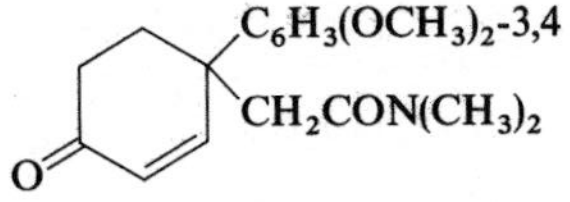 O, $C_6H_3(OCH_3)_2\text{-}3,4$, $CH_2CON(CH_3)_2$	$LiAlH(OC_2H_5)_3$, ether, 0°, 35 h	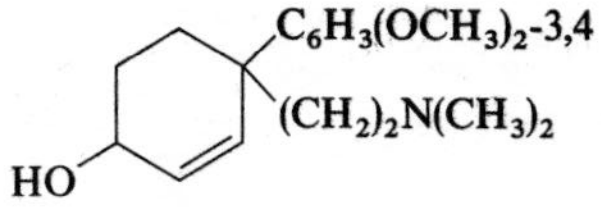 HO, $C_6H_3(OCH_3)_2\text{-}3,4$, $(CH_2)_2N(CH_3)_2$ (—)	905
C_{19}	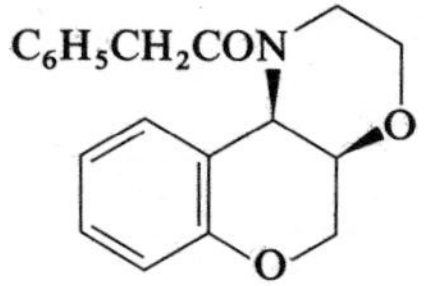 $C_6H_5CH_2CON$, O, O	SMEAH, THF, C_6H_6, reflux, 2 h	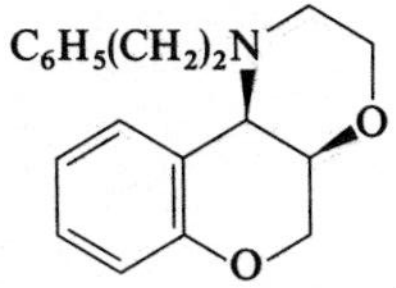 $C_6H_5(CH_2)_2N$, O, O (—)	906
	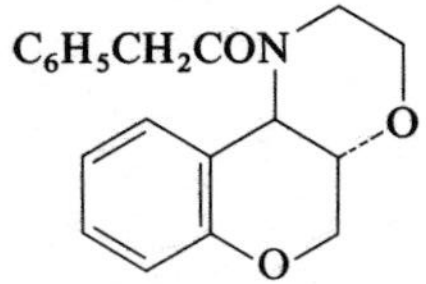 $C_6H_5CH_2CON$, O, O	″, ″, ″, ″, ″	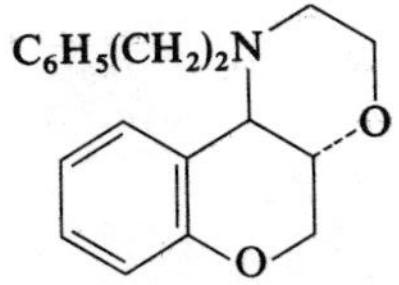 $C_6H_5(CH_2)_2N$, O, O (—)	906
C_{20}	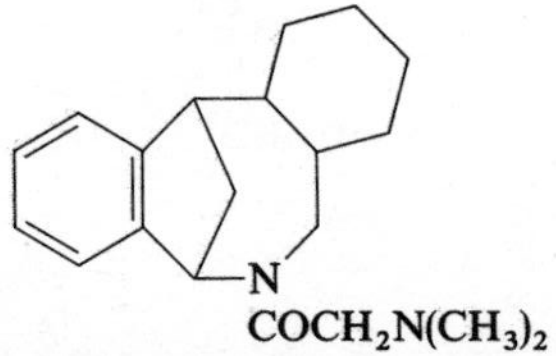 N, $COCH_2N(CH_3)_2$	SMEAH	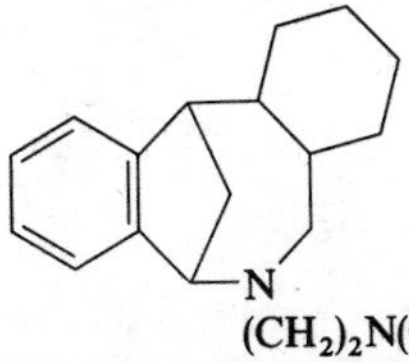 N, $(CH_2)_2N(CH_3)_2$ (—)	907

	Substrate	Conditions	Product (Yield, %)		Refs.
	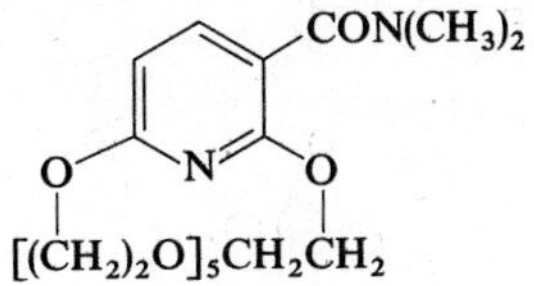 [structure: pyridine with 3-$CON(CH_3)_2$, 2,6-O-bridged by $[(CH_2)_2O]_5CH_2CH_2$]	SMEAH, C_6H_6, reflux, 30 min	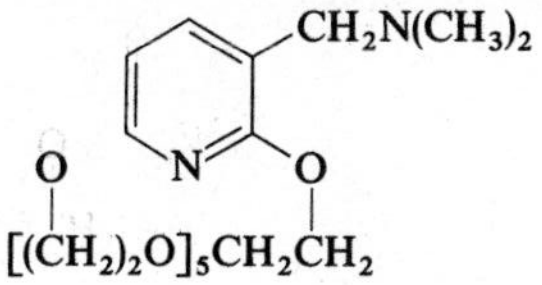 [structure: pyridine with 3-$CH_2N(CH_3)_2$, 2,6-O-bridged by $[(CH_2)_2O]_5CH_2CH_2$]	(~100)	336
C_{22}	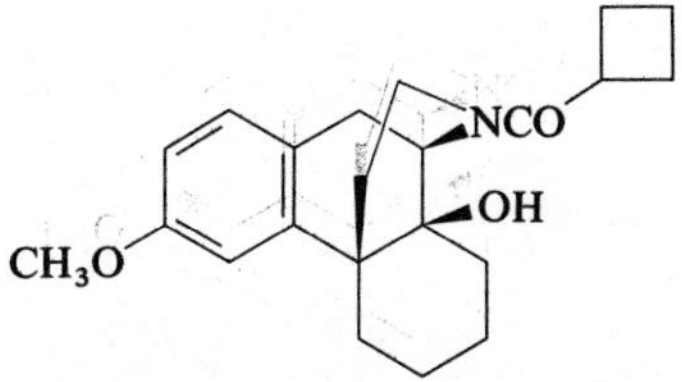 [structure: CH_3O-morphinan with NCO-cyclobutyl, OH]	SMEAH, C_6H_6, reflux, 3.5 h	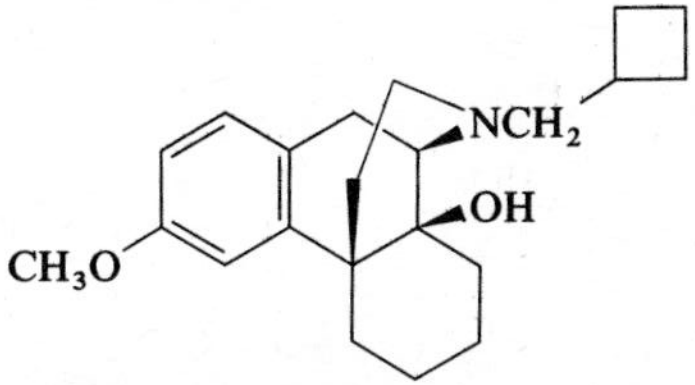 [structure: CH_3O-morphinan with NCH_2-cyclobutyl, OH]	(82)	908a
C_{23}	$4\text{-}FC_6H_4CH(CH_3)NHCOCH_2CH(C_6H_5)_2$	SMEAH, C_6H_6, reflux	$4\text{-}FC_6H_4CH(CH_3)NH(CH_2)_2CH(C_6H_5)_2$	(77)	908b
C_{24}	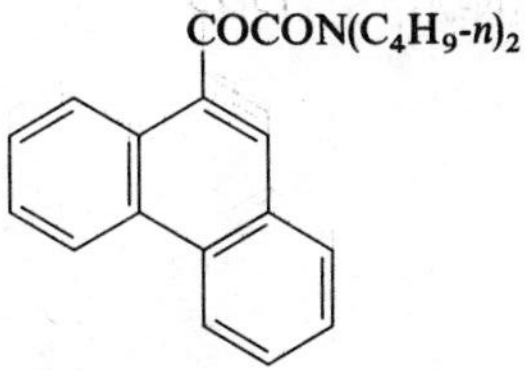 [structure: 9-phenanthryl-$COCON(C_4H_9\text{-}n)_2$]	SMEAH, C_6H_6, ether, room temp, 1 h	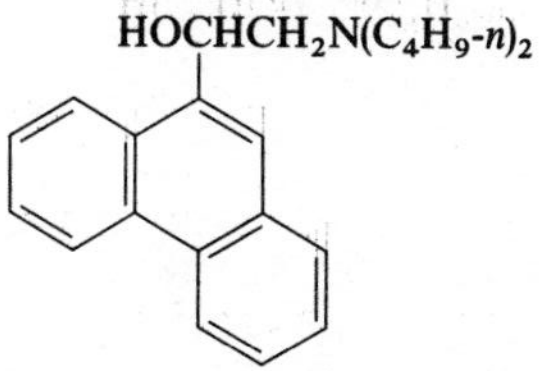 [structure: 9-phenanthryl-$HOCHCH_2N(C_4H_9\text{-}n)_2$]	(69)	909, 910
			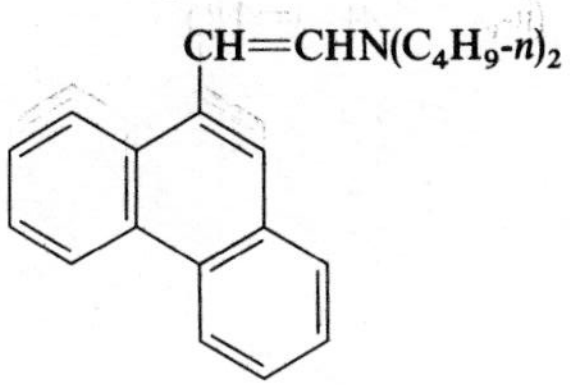 + [structure: 9-phenanthryl-$CH{=}CHN(C_4H_9\text{-}n)_2$]	(3)	

TABLE XV. REDUCTION OF CARBOXYLIC ACID AMIDES (*Continued*)

No. of Carbon Atoms	Reactant	Conditions	Product(s) and Yield(s) (%)	Refs.
C_{24} (*Contd.*)	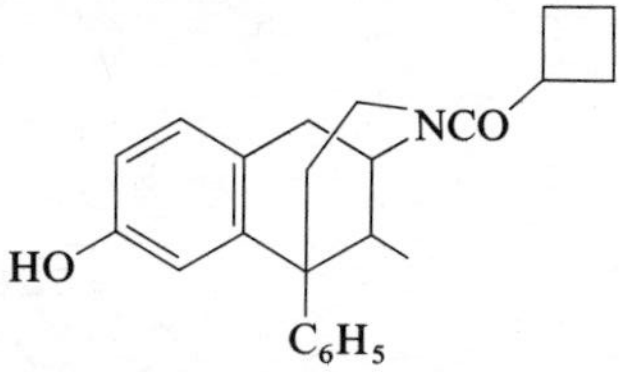 HO, NCO, C_6H_5, *dl-*	SMEAH, C_6H_6	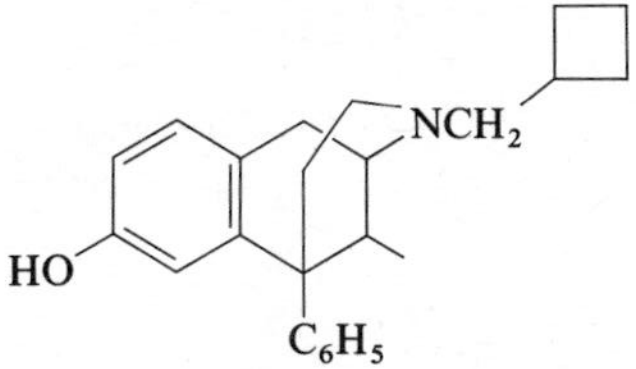 HO, NCH_2, C_6H_5, *dl-* (—)	911
	$C_6H_{11}CONHCH_2CH(CH_3)C(OCH_3)(C_6H_5)_2$	SMEAH, C_6H_6	$C_6H_{11}CH_2NHCH_2CH(CH_3)C(OCH_3)(C_6H_5)_2$ (—)	912
C_{25}	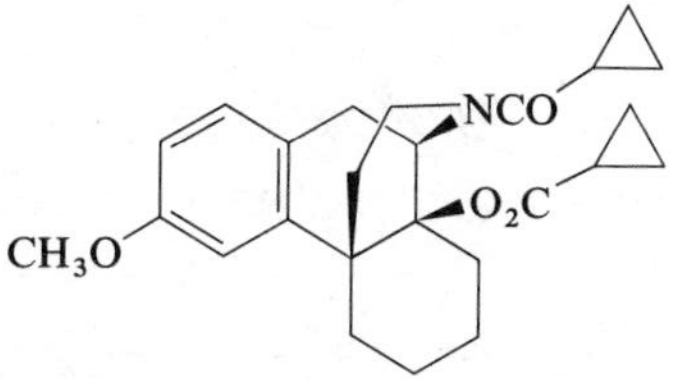 CH_3O, NCO, O_2C	SMEAH, C_6H_6, reflux	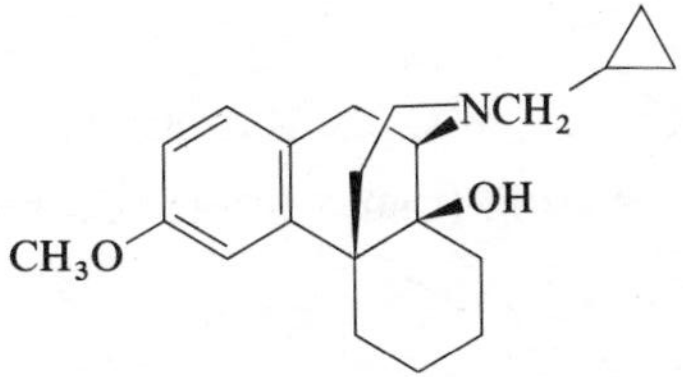 CH_3O, NCH_2, OH (—)	913
C_{26}	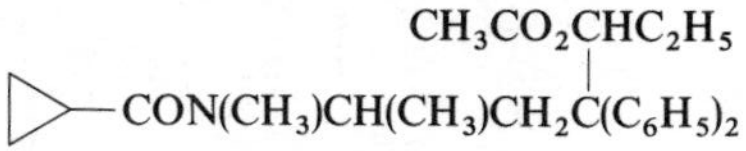 $CH_3CO_2CHC_2H_5$ / —$CON(CH_3)CH(CH_3)CH_2C(C_6H_5)_2$	SMEAH, C_6H_6, reflux, 30 min	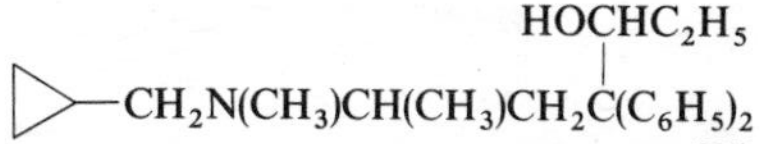 $HOCHC_2H_5$ / —$CH_2N(CH_3)CH(CH_3)CH_2C(C_6H_5)_2$ (78)	704
	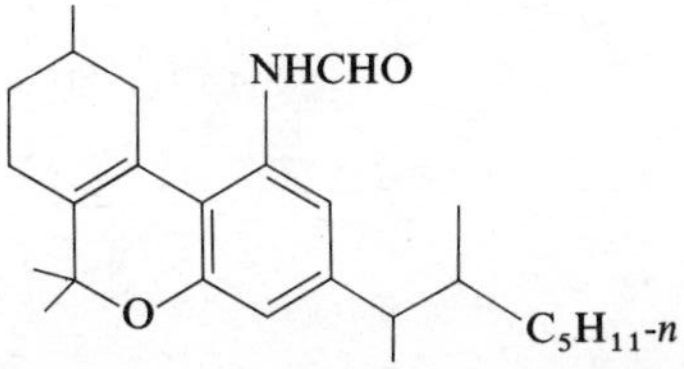 NHCHO, O, C_5H_{11}-*n*	SMEAH (IA), C_6H_6, room temp, 1 h	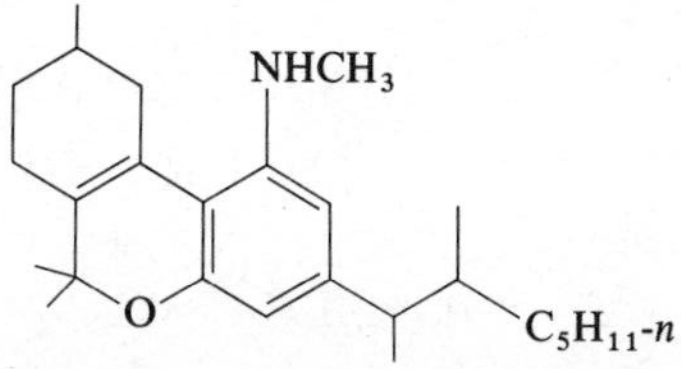 $NHCH_3$, O, C_5H_{11}-*n* (81)	914

	Reactant	Conditions	Product(s) and Yield(s) (%)		Refs.
C_{27}	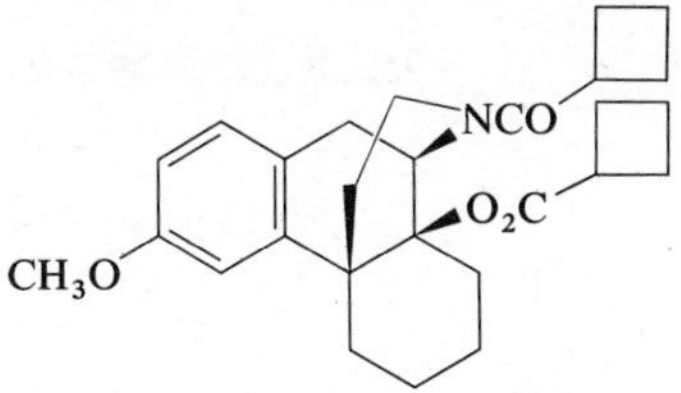 [structure: CH_3O-substituted morphinan with NCO-cyclobutyl and O_2C-cyclobutyl groups]	SMEAH, C_6H_6, reflux, 2 h	[structure: CH_3O-substituted morphinan with NCH_2-cyclobutyl and OH groups]	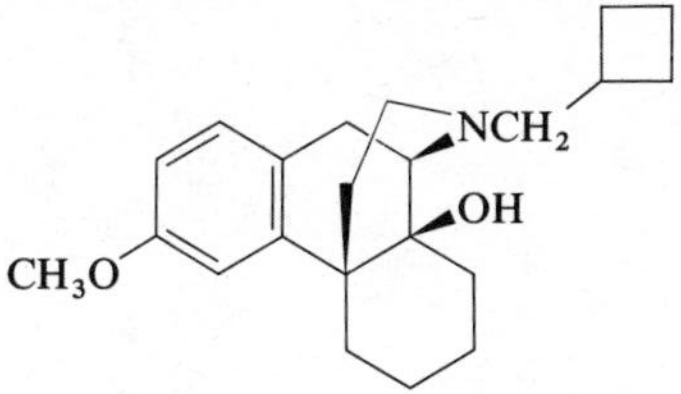 (85)	913
C_{31}	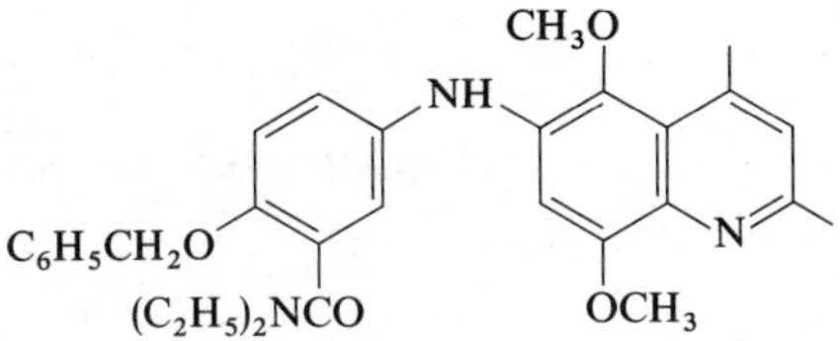 [structure: $C_6H_5CH_2O$, $(C_2H_5)_2NCO$-substituted phenyl–NH–quinoline bearing CH_3O, OCH_3]	SMEAH (IA), THF, reflux, 3 h	[structure: $C_6H_5CH_2O$, $(C_2H_5)_2NCH_2$-substituted phenyl–NH–quinoline bearing CH_3O, OCH_3]	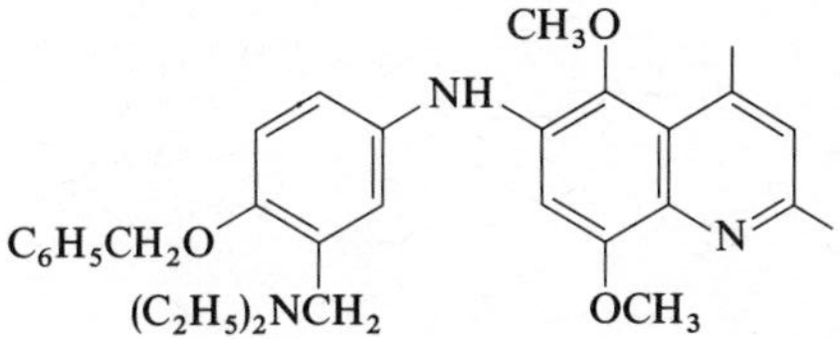 (57)	915
C_{32}	n-$C_{15}H_{31}CONHC_{16}H_{33}$-$n$	SMEAH, C_6H_6	$(n\text{-}C_{16}H_{33})_2NH$ (—)		916
C_{60}	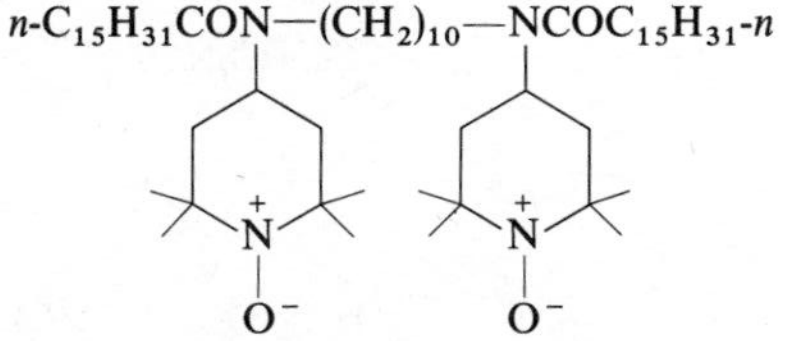 n-$C_{15}H_{31}CON$—$(CH_2)_{10}$—$NCOC_{15}H_{31}$-n [each N bearing a 2,2,6,6-tetramethylpiperidin-4-yl N^+—O^- group]	SMEAH, ether, 0°, 1 h	n-$C_{16}H_{33}N$—$(CH_2)_{10}$—$NC_{16}H_{33}$-n [each N bearing a 2,2,6,6-tetramethylpiperidin-4-yl N^+—O^- group]	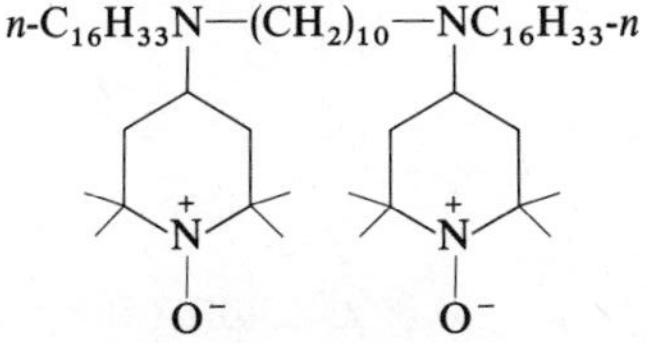 (47)	443
C_n	$[CH_2CH(CON(CH_3)_2)]_n$	SMEAH, N-methylmorpholine, 100°, 12 h	$[CH_2CH(CH_2N(CH_3)_2)]_n$	(—)	713

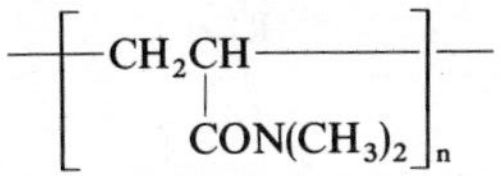

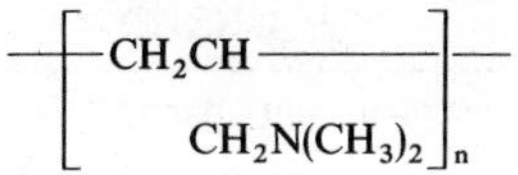

TABLE XVI. PARTIAL REDUCTION OF CARBOXYLIC ACID AMIDES TO ALDEHYDES

No. of Carbon Atoms	Reactant	Conditions	Product(s) and Yield(s) (%)	Refs.
C_4	$F_2CHCON(CH_3)_2$	$LiAlH_2(OC_2H_5)_2$, ether, 0°, 15 min	F_2CHCHO (—)	917
C_5	$CH_3CHClCON(CH_3)_2$	$LiAlH_2(OC_2H_5)_2$ (IA), ether, −20 to −30°, 1 h	$CH_3CHClCHO$ (84)	88
		$LiAlH(OC_2H_5)_3$, ether, −20 to −30°, 1 h	" (87)	88
C_6	▷—$CON(CH_3)_2$	$LiAlH_2(OC_2H_5)_2$ (IA), ether, 0°, 1 h	▷—CHO (78)	88, 89
		$LiAlH(OC_2H_5)_3$ (IA), ether, 0°, 1 h	" (67)	88
	n-C_3H_7CON◁	$LiAlH_2(OC_2H_5)_2$ (IA), ether, 0°, 1 h	n-C_3H_7CHO (83)	88
		$LiAlH(OC_2H_5)_3$ (IA), ether, 0°, 1 h	" (87)	88
		$NaAlH(OC_2H_5)_3$ (IA), THF, 0°, 1.5 h	" (82)	91
		$NaAlH(OCH_3)_3$ (IA), THF, 0°, 1.5 h	" (84)	91
	$CH_3CH{=}CHCON(CH_3)_2$	$LiAlH_2(OC_2H_5)_2$ (IA), ether, 0°, 1 h	No reaction	88
		$LiAlH(OC_2H_5)_3$, ether, 0°, 1 h	No reaction	88
	n-$C_3H_7CON(CH_3)_2$	$LiAlH_2(OC_2H_5)_2$ (IA), ether, 0°, 1 h	n-C_3H_7CHO (90)	88, 89
		$LiAlH(OC_2H_5)_3$, ether, 0°, 1 h	" (90)	88
		LTMA, ether, 0°, 1 h	" (59)	88
		$NaAlH(OCH_3)_3$ (IA), THF, 0°, 1.5 h	" (85)	91
		$NaAlH(OC_2H_5)_3$ (IA), THF, 0°, 1.5 h	" (91)	91
	i-$C_3H_7CON(CH_3)_2$	LTMA, ether, 0°, 1 h	i-C_3H_7CHO (70)	88
		$LiAlH(OC_2H_5)_3$, ether, 0°, 1 h	" (87)	88
		$LiAlH_2(OC_2H_5)_2$ (IA), ether, 0°, 1 h	" (89)	88, 89
	$C_2H_5SCH_2CON(CH_3)_2$	$LiAlH(OC_2H_5)_3$, ether, 0°, 1 h	$C_2H_5SCH_2CHO$ (69)	88, 89
		$LiAlH_2(OC_2H_5)_2$ (IA), ether, 0°, 1 h	" (70)	88

C_7	t-$C_4H_9CON(CH_3)_2$	$LiAlH_2(OC_2H_5)_2$ (IA), ether, 0°, 1 h	t-C_4H_9CHO (63)	88, 89
		$LiAlH(OC_2H_5)_3$, ether, 0°, 1 h	" (74)	88
	$(CH_3O)_3CCON(CH_3)_2$	$LiAlH_2(OCH_3)_2$ (IA), ether, reflux, 2 h	$(CH_3O)_3CCHO$ (28)	90
C_8	3-pyridyl-$CON(CH_3)_2$	$LiAlH_2(OC_2H_5)_2$ (IA), ether, 0°, 1 h	3-pyridyl-CHO (89)	89
	n-C_3H_7CON(pyrrolidine)	$LiAlH(OC_2H_5)_3$, ether, 0°, 1 h	n-C_3H_7CHO (70)	88
	n-$C_3H_7CON(C_2H_5)_2$	$LiAlH(OC_2H_5)_3$, ether, 0°, 1 h	" (59)	89
	n-$C_5H_{11}CON(CH_3)_2$	$LiAlH_2(OC_2H_5)_2$ (IA), ether, 0°, 1 h	n-$C_5H_{11}CHO$ (67)	88, 89
		$LiAlH(OC_2H_5)_3$, ether, 0°, 1 h	" (73)	88, 89
	$(C_2H_5)_2CHCON(CH_3)_2$	$LiAlH_2(OC_2H_5)_2$ (IA), ether, 0°, 1 h	$(C_2H_5)_2CHCHO$ (79)	88
C_9	2-$ClC_6H_4CON(CH_3)_2$	", ", "	2-ClC_6H_4CHO (77)	88, 89
		$LiAlH(OC_2H_5)_3$, ether, 0°, 1 h	" (80)	88
	4-$ClC_6H_4CON(CH_3)_2$	$LiAlH_2(OC_2H_5)_2$ (IA), ether, 0°, 1 h	4-ClC_6H_4CHO (78)	88
		$LiAlH(OC_2H_5)_3$, ether, 0°, 1 h	" (89)	88
	4-$O_2NC_6H_4CON(CH_3)_2$	", ", ", "	4-$O_2NC_6H_4CHO$ (60)	88
	$C_6H_5CON(CH_3)_2$	LTMA (IA), ether, 0°, 1 h	C_6H_5CHO (85)	88
		$LiAlH_2(OC_2H_5)_2$ (IA), ether, 0°, 1 h	" (83)	88
		$LiAlH(OC_2H_5)_3$, ether, 0°, 1 h	" (70)	88
	cyclopent-2-enyl-$CH_2CON(CH_3)_2$	$LiAlH_2(OC_2H_5)_2$ (IA), 0°; room temp, 12 h	cyclopent-2-enyl-CH_2CHO (23)	918
	$C_6H_{11}CON(CH_3)_2$	$LiAlH_2(OC_2H_5)_2$ (IA), ether, 0°, 1 h	$C_6H_{11}CHO$ (72)	88, 89
		$LiAlH(OC_2H_5)_3$, ether, 0°, 1 h	" (78)	88
C_{10}	4-ClC_6H_4CON(2-thioxothiazolidine)	LTBA (IA), THF, −20°; then 0°	4-ClC_6H_4CHO (86)	102

TABLE XVI. Partial Reduction of Carboxylic Acid Amides to Aldehydes (*Continued*)

No. of Carbon Atoms	Reactant	Conditions	Product(s) and Yield(s) (%)	Refs.
C_{10} (*Contd.*)	$4\text{-}O_2NC_6H_4CON$ (ring: S, =S)	LTBA (IA), THF, $-20°$; then $0°$	$4\text{-}O_2NC_6H_4CHO$ (88)	102
	C_6H_5CON (ring: O, =O)	LTBA (IA), THF, $-40°$; $-10°$, 1–2 h	C_6H_5CHO (60)	919
	C_6H_5CON (ring: S, =S)	LTBA (IA), THF, $-20°$; then $0°$	" (83)	102
	$C_6H_5CH_2CON(CH_3)_2$	$LiAlH_2(OC_2H_5)_2$ (IA), ether, 0°, 1 h	$C_6H_5CH_2CHO$ (79)	88
	$2\text{-}CH_3C_6H_4CON(CH_3)_2$	$LiAlH_2(OC_2H_5)_2$ (IA), ether, 0°, 1 h	$2\text{-}CH_3C_6H_4CHO$ (74)	920
	$2\text{-}CH_3OC_6H_4CON(CH_3)_2$	$LiAlH(OC_2H_5)_3$, ether, 0°, 1 h	$2\text{-}CH_3OC_6H_4CHO$ (74)	88
		$LiAlH_2(OC_2H_5)_2$ (IA), ether, 0°, 1 h	" (84)	88, 89
	(cyclohexenyl) $CH_2CON(CH_3)_2$	$LiAlH_2(OC_2H_5)_2$ (IA), 0°; room temp, 12 h	(cyclohexenyl) CH_2CHO (17)	918
	$CH_2CON(CH_3)_2$ (piperidine ring, N–CH_3)	$LiAlH(OC_2H_5)_3$, ether, 0–7°, 40 min; room temp, 20 min	CH_2CHO (piperidine ring, N–CH_3) (22)	921

C_{11}	$C_6H_5CH_2CON$ S S	LTBA (IA), THF, $-20°$, 5 h	$C_6H_5CH_2CHO$ (69)	102
	$C_6H_5CH{=}CHCON(CH_3)_2$	$LiAlH(OC_2H_5)_3$, ether, 0°, 1 h	$C_6H_5CH{=}CHCHO$ (7)	88
		$LiAlH_2(OC_2H_5)_2$ (IA), ether, 0°, 1 h	" (9)	88
	n-$C_3H_7CON(CH_3)C_6H_5$	$LiAlH(OC_2H_5)_3$ (IA), ether, 0°, 1 h	n-C_3H_7CHO (69)	88
		$LiAlH_2(OC_2H_5)_2$ (IA), ether, 0°, 1 h	" (82)	88
	$CON(CH_3)_2$	$LiAlH_2(OC_2H_5)_2$, ether, reflux, 1.5 h	CHO (43)	922
C_{12}	$C_6H_5(CH_2)_2CON$ S S	LTBA, THF, $-20°$; then 0°	$C_6H_5(CH_2)_2CHO$ (89)	102
	n-$C_3H_7CON(C_2H_5)C_6H_5$	$NaAlH(OCH_3)_3$ (IA), THF, 0°, 1.5 h	n-C_3H_7CHO (61)	91
	3,4,5-$(CH_3O)_3C_6H_2CON(CH_3)_2$	$LiAlH_2(OC_2H_5)_2$, ether, 0°, 2 h	3,4,5-$(CH_3O)_3C_6H_2CHO$ (60)	923
C_{13}	$C_6H_5(CH_2)_3CON$ S S	LTBA, THF, $-20°$; then 0°	$C_6H_5(CH_2)_3CHO$ (86)	102
	$C_6H_5CH(C_2H_5)CON$ S S	LTBA, THF, $-20°$; then 0°	$C_6H_5CH(C_2H_5)CHO$ (86)	102
	1-$C_{10}H_7CON(CH_3)_2$	$LiAlH_2(OC_2H_5)_2$ (IA), ether, 0°, 1 h	1-$C_{10}H_7CHO$ (81)	88, 89
	2-$C_{10}H_7CON(CH_3)_2$	$LiAlH(OC_2H_5)_3$, ether, 0°, 1 h	2-$C_{10}H_7CHO$ (81)	88
	4-(t-C_4H_9)$C_6H_4CON(CH_3)_2$	$LiAlH(OC_2H_5)_3$, ether, 0°, 2 h	4-(t-C_4H_9)C_6H_4CHO (61)	924
	4-$CH_3OC_6H_4CH(CH_3)CH_2CON(CH_3)_2$	$LiAlH(OC_2H_5)_3$, ether, 0°, 1 h	4-$CH_3OC_6H_4CH(CH_3)CH_2CHO$ (42)	925, 926

TABLE XVI. PARTIAL REDUCTION OF CARBOXYLIC ACID AMIDES TO ALDEHYDES (*Continued*)

No. of Carbon Atoms	Reactant	Conditions	Product(s) and Yield(s) (%)	Refs.
C_{13} (*Contd.*)	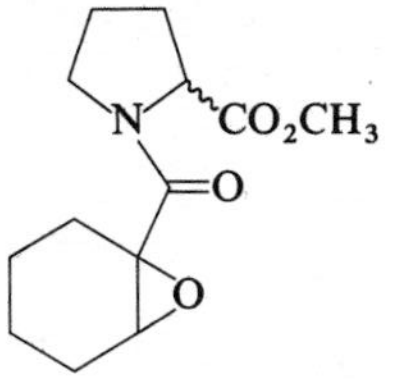 CO_2CH_3, O (1*R*,2*S*)-	SMEAH(IA), C_6H_6–ether, 0°, 1 h	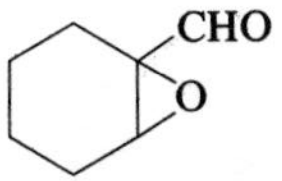 CHO, O (1*R*,2*S*)-(+) (58)(~100% e.e.)	103, 104
	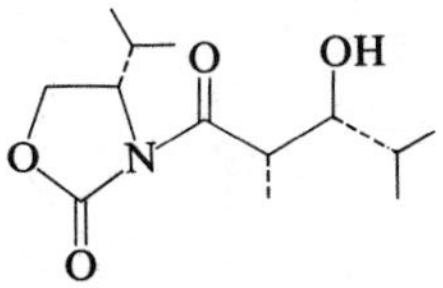 O, OH	1. SMEAH, THF, −78°, then −55 to −40°, 1 h 2. $[(C_2H_5O)_2P(O)\bar{C}HCO_2C_2H_5]K^+$	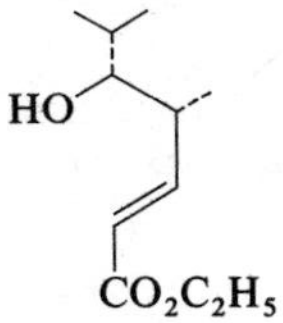 HO, $CO_2C_2H_5$ (45)	927a
	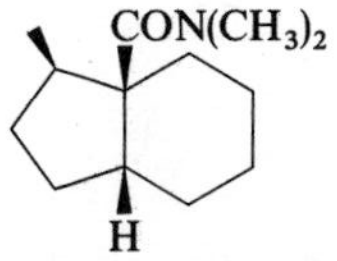 $CON(CH_3)_2$, H	$LiAlH_2(OC_2H_5)_2$, ether, 0°; reflux, 30 min	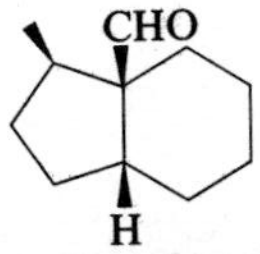 CHO, H (—)	927b
	$CH_2{=}CH(CH_2)_8CON(CH_3)_2$	$LiAlH_2(OC_2H_5)_2$ (IA), ether, 0°, 1 h	$CH_2{=}CH(CH_2)_8CHO$ (69)	88, 89
		$LiAlH(OC_2H_5)_3$, ether, 0°, 1 h	" (73)	88, 89
C_{14}	$C_6H_5CON(CH_3)C_6H_5$	SMEAH (IA), C_6H_6, 1–3°, 2 h 50 min	C_6H_5CHO (66) + $C_6H_5CH_2OH$ (11) + $C_6H_5NHCH_3$ (86)	94, 167
	$2\text{-}HSC_6H_4CON(CH_3)C_6H_5$	SMEAH, C_6H_6–THF, 0°, 21 h	$2\text{-}HSC_6H_4CHO$ (60)	928
	$1\text{-}C_{10}H_7CH_2CON(CH_3)_2$	$LiAlH_2(OC_2H_5)_2$ (IA), ether, 0°, 1 h	$1\text{-}C_{10}H_7CH_2CHO$ (72)	89

	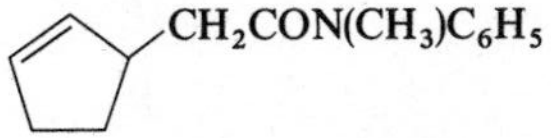 $CH_2CON(CH_3)C_6H_5$	$LiAlH_2(OC_2H_5)_2$ (IA), ether, 0–5°; then room temp, 12 h	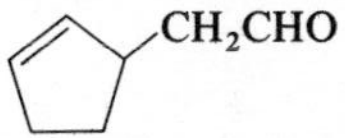 CH_2CHO (47)	918
	$C_5H_9CH_2CON(CH_3)C_6H_5$	", ", ", ", "	$C_5H_9CH_2CHO$ (55)	918
	n-$C_5H_{11}C{\equiv}CC{\equiv}C(CH_2)_2CON(CH_3)_2$	$LiAlH_2(OC_2H_5)_2$ (IA), ether, 0°, 30 min; reflux, 30 min	n-$C_5H_{11}C{\equiv}CC{\equiv}C(CH_2)_2CHO$ (44)	929
	trans-n-$C_5H_{11}C{\equiv}CCH{=}CH(CH_2)_2CON(CH_3)_2$	$LiAlH_2(OC_2H_5)_2$ (IA), ether, 0°, 40 min; reflux, 30 min	*trans*-n-$C_5H_{11}C{\equiv}CCH{=}CH(CH_2)_2CHO$ (76)	929
	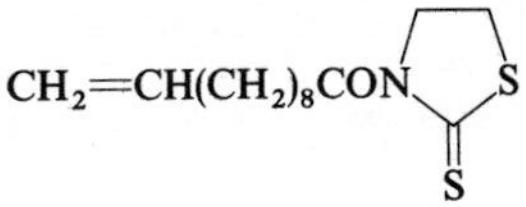 $CH_2{=}CH(CH_2)_8CON$ (S, S)	LTBA (IA), THF, −20°; then 0°	$CH_2{=}CH(CH_2)_8CHO$ (78)	102
	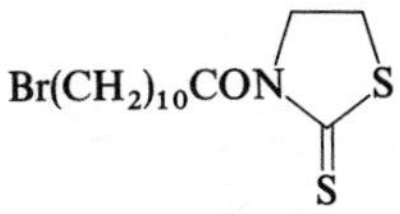 $Br(CH_2)_{10}CON$ (S, S)	", ", ", "	$Br(CH_2)_{10}CHO$ (80)	102
	n-$C_{11}H_{23}CON(CH_3)_2$	$LiAlH_2(OC_2H_5)_2$ (IA), ether, 0°, 1 h	n-$C_{11}H_{23}CHO$ (62)	88, 89
C_{15}	 $CH_2CON(CH_3)C_6H_5$	$LiAlH_2(OC_2H_5)_2$ (IA), ether, 0–5°; room temp, 12 h	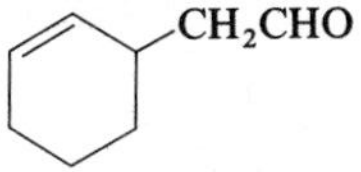 CH_2CHO (33)	918
	 $CH_2CON(CH_3)C_6H_5$	$LiAlH_2(OC_2H_5)_2$ (IA), ether, 0–5°; room temp, 12 h	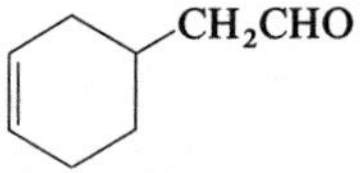 CH_2CHO (27)	918
	 $CH_2CON(CH_3)C_6H_5$	", ", ", ", "	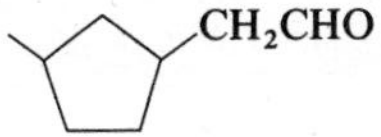 CH_2CHO (44)	918
	$C_6H_{11}CH_2CON(CH_3)C_6H_5$	", ", ", ", "	$C_6H_{11}CH_2CHO$ (45)	918
	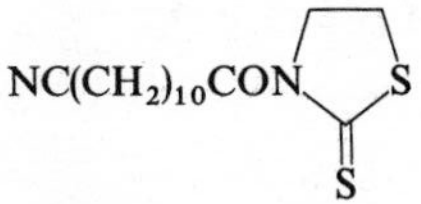 $NC(CH_2)_{10}CON$ (S, S)	−20°; then 0°	$NC(CH_2)_{10}CHO$ (71)	102

TABLE XVI. PARTIAL REDUCTION OF CARBOXYLIC ACID AMIDES TO ALDEHYDES (*Continued*)

No. of Carbon Atoms	Reactant	Conditions	Product(s) and Yield(s) (%)	Refs.
C_{16}	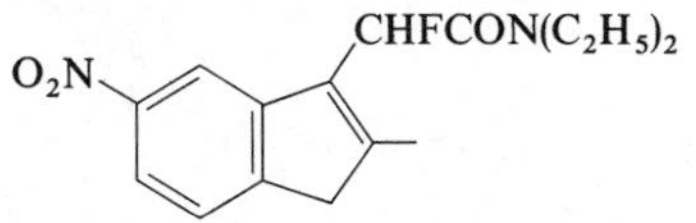 O_2N, $CHFCON(C_2H_5)_2$	LTBA, THF	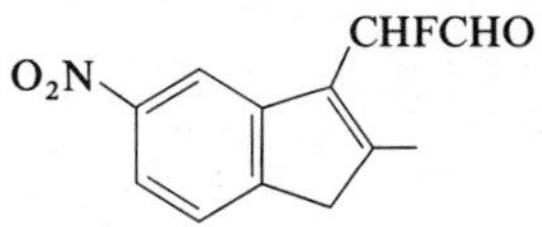 O_2N, CHFCHO (—)	930
	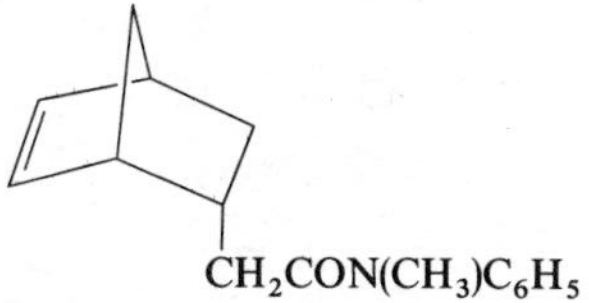 $CH_2CON(CH_3)C_6H_5$	$LiAlH_2(OC_2H_5)_2$ (IA), ether, 0–5°; room temp, 12 h	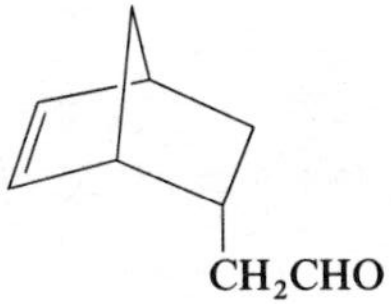 CH_2CHO (32)	918
	 $CH_2CON(CH_3)C_6H_5$	", ", ", ", "	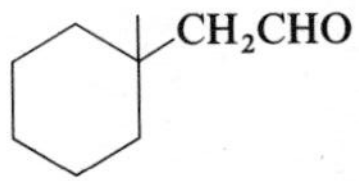 CH_2CHO (39)	918
	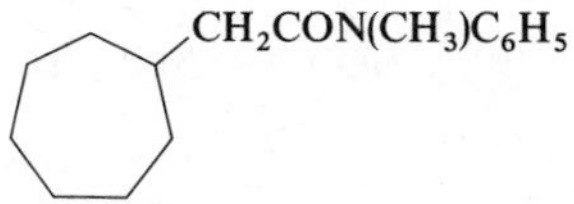 $CH_2CON(CH_3)C_6H_5$	", ", ", ", "	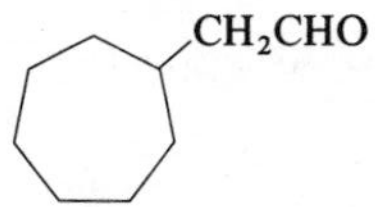 CH_2CHO (33)	918
	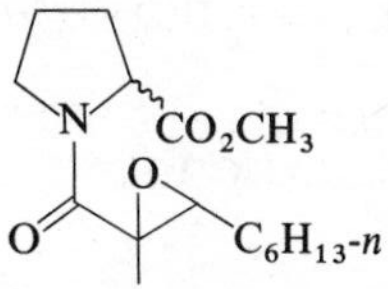 CO_2CH_3, O, C_6H_{13}-*n* (2*R*,3*S*)-	SMEAH (IA), C_6H_6–ether, 0°, 1 h	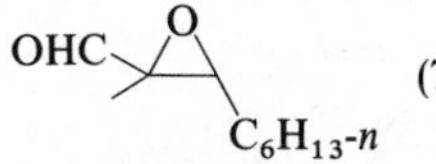 OHC, O, C_6H_{13}-*n* (2*R*,3*S*)-(+) (76) (~100% e.e.)	103, 104

	Substrate	Conditions	Product(s) and Yield(s) (%)	Refs.
C_{17}	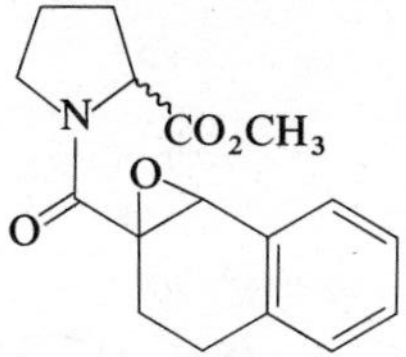 CO_2CH_3 (1*S*,2*R*)-	", ", ", "	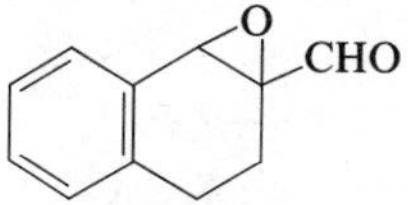 CHO (1*S*,2*R*)-(−) (76) (92% e.e.)	103, 104
	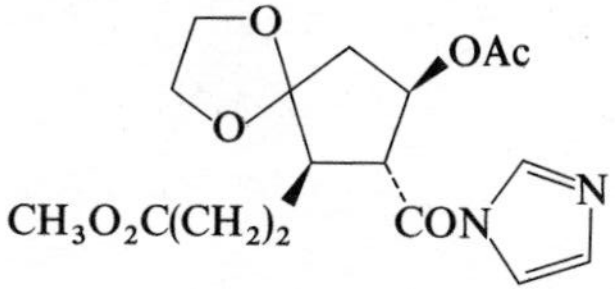 OAc $CH_3O_2C(CH_2)_2$ CON	LTBA, THF, 25°	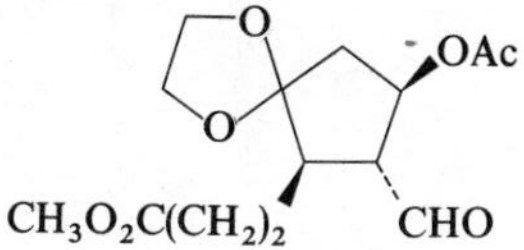 OAc $CH_3O_2C(CH_2)_2$ CHO (—)	931
	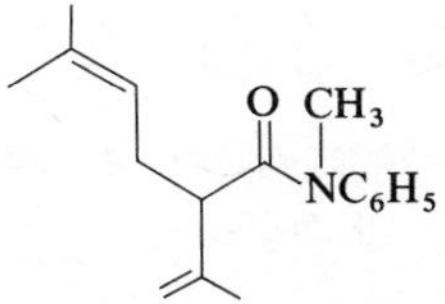 O CH_3 NC_6H_5	$LiAlH_2(OC_2H_5)_2$, ether, −78°	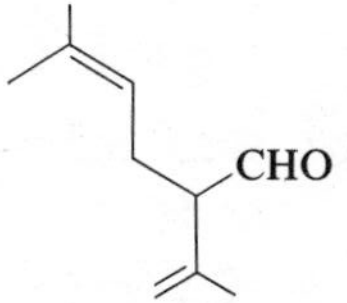 CHO (73)	932
	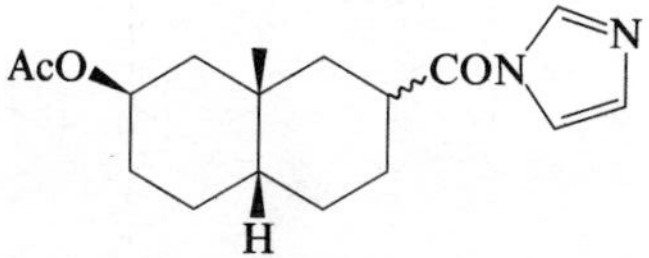 AcO CON H	LTBA	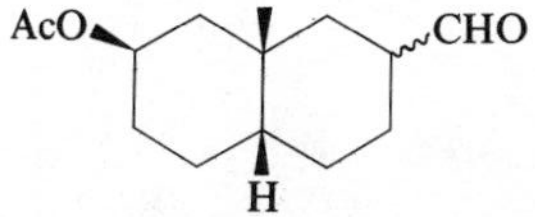 AcO CHO H (—)	933
C_{18}	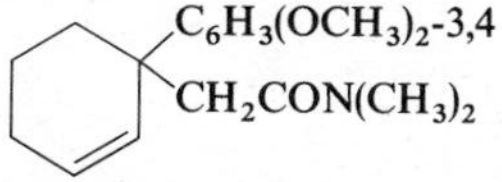 $C_6H_3(OCH_3)_2$-3,4 $CH_2CON(CH_3)_2$	$LiAlH(OC_2H_5)_3$, ether, 0°, 35 h	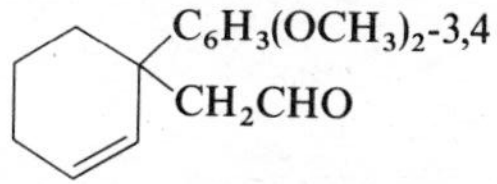 $C_6H_3(OCH_3)_2$-3,4 CH_2CHO (28) 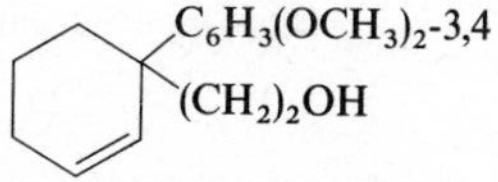 + $C_6H_3(OCH_3)_2$-3,4 $(CH_2)_2OH$ (36)	905

TABLE XVI. Partial Reduction of Carboxylic Acid Amides to Aldehydes (*Continued*)

No. of Carbon Atoms	Reactant	Conditions	Product(s) and Yield(s) (%)	Refs.
C_{18} (*Contd.*)	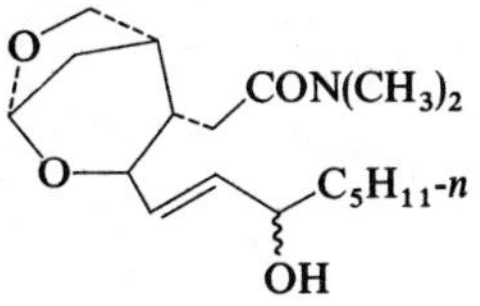 [structure] $CON(CH_3)_2$, C_5H_{11}-*n*, OH	$LiAlH(OC_2H_5)_3$, ether, −5°, 1 h	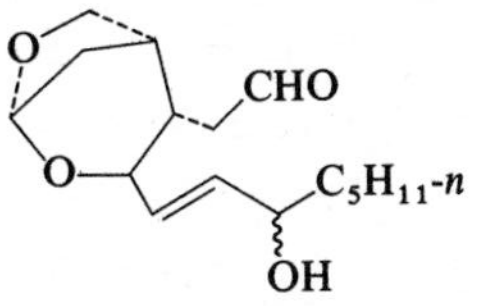 [structure] CHO, C_5H_{11}-*n*, OH (—)	934
	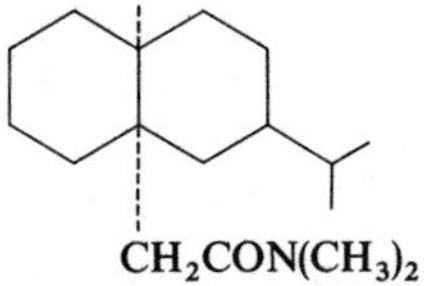 [structure] $CH_2CON(CH_3)_2$	$LiAlH_2(OC_2H_5)_2$	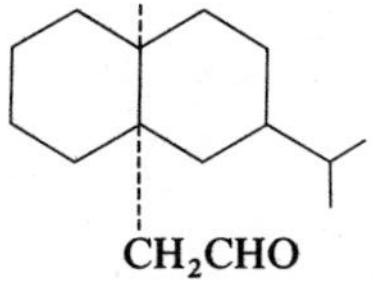 [structure] CH_2CHO (51)	935
	n-$C_{15}H_{31}CON(CH_3)_2$	$LiAlH_2(OCH_2CH_2OCH_3)_2$, toluene–ether, −10°, 2 h; −5°, 2 h	*n*-$C_{15}H_{31}CHO$ (41)	93
		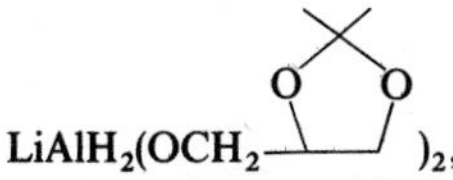 $LiAlH_2(OCH_2$–[structure]$)_2$, ether–toluene, −15°, 5 h	" (53)	93
C_{19}	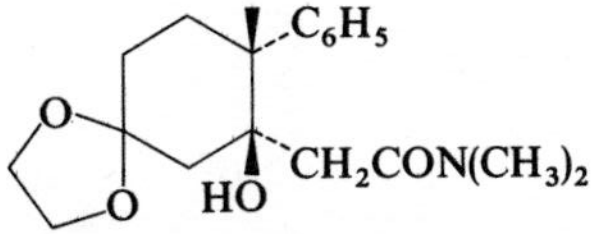 [structure] C_6H_5, $CH_2CON(CH_3)_2$, HO (*R*)-(−)-	SMEAH, C_6H_6–ether, −20°, 4 h	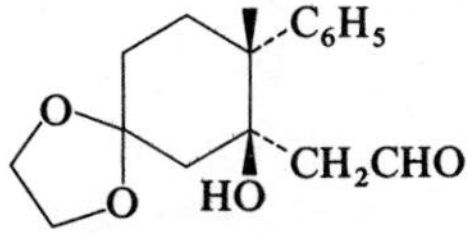 [structure] C_6H_5, CH_2CHO, HO (*R*)-(−)- (69)	936
C_{20}	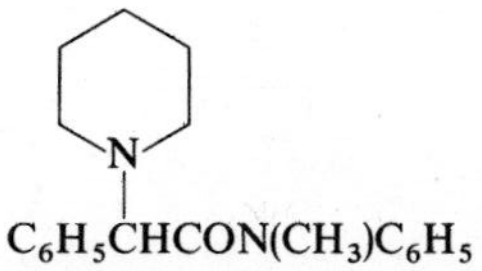 [structure] $C_6H_5CHCON(CH_3)C_6H_5$	$LiAlH(OC_2H_5)_3$, ether, 0°, 30 min	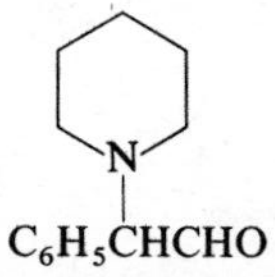 [structure] C_6H_5CHCHO (30)	95

	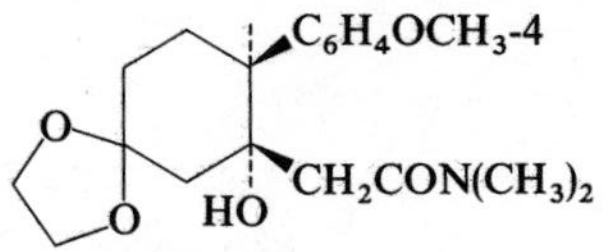	SMEAH, C_6H_6–ether, −20°, 4 h	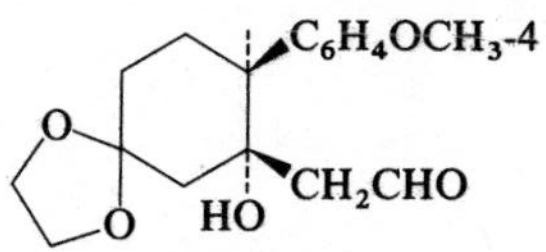 (65)	92
	CH_2(CH=CH$(CH_2)_7$CON$(CH_3)_2$)(CH=CHC_5H_{11}-*n*-)	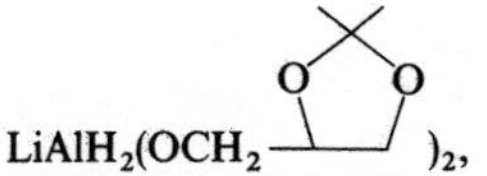 ether–toluene, −15°, 4 h	CH_2(CH=CH$(CH_2)_7$CHO)(CH=CHC_5H_{11}-*n*) (57)	93
	n-C_8H_{17}CH=CH$(CH_2)_7$CON$(CH_3)_2$	$LiAlH_2(OCH_2CH_2OCH_3)_2$, ether–toluene, −10 to −5°, 4 h	*n*-C_8H_{17}CH=CH$(CH_2)_7$CHO (44)	93
		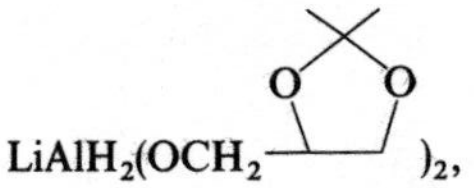 ether–toluene, −15°, 4 h	" (55)	93
C_{21}	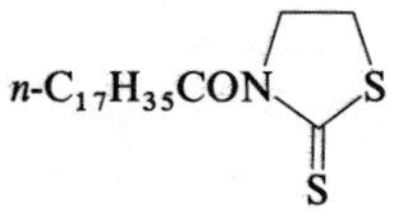	LTBA, THF, 0°, 4 h; room temp, 4 h	*n*-$C_{17}H_{35}$CHO (85)	102
C_{26}	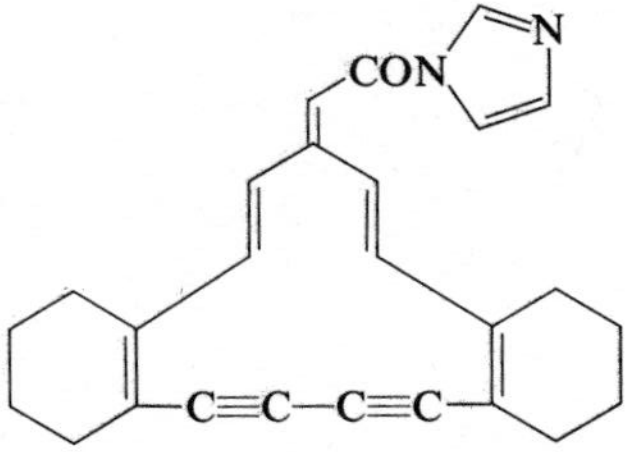	1. LTBA (IA), THF, room temp 2. H_3O^+ 3. MnO_2	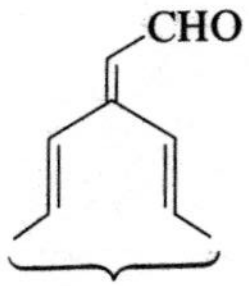 (61)	937

TABLE XVI. PARTIAL REDUCTION OF CARBOXYLIC ACID AMIDES TO ALDEHYDES (*Continued*)

No. of Carbon Atoms	Reactant	Conditions	Product(s) and Yield(s) (%)	Refs.
C_{36}	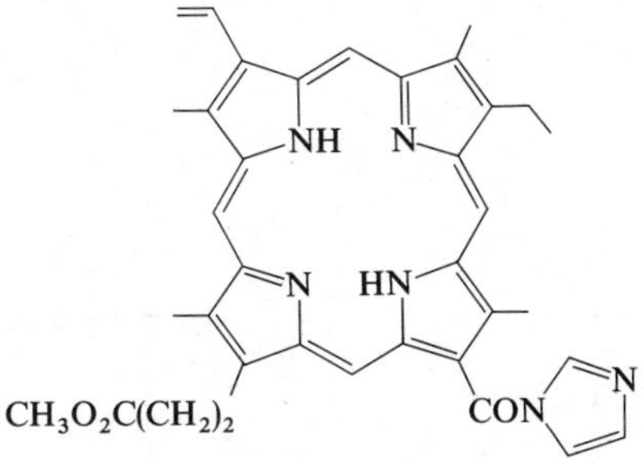 NH, N, N, HN; $CH_3O_2C(CH_2)_2$; CON (imidazolyl, N)	LTBA	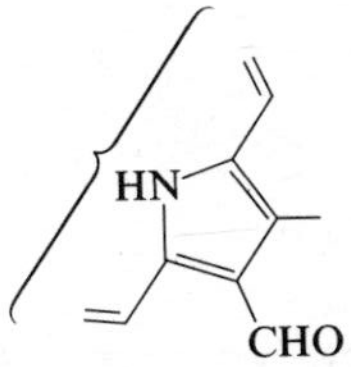 HN; CHO (—)	938
	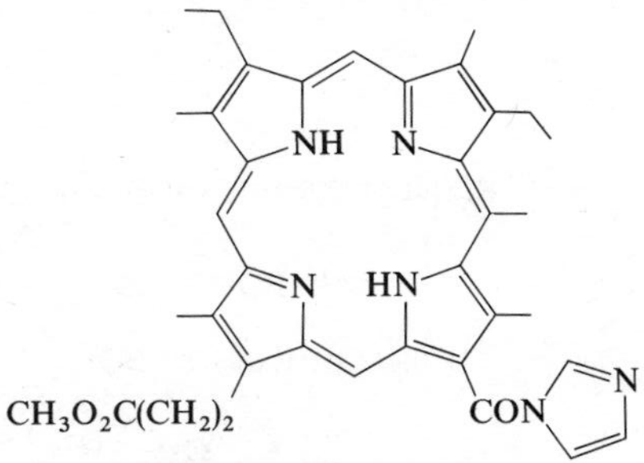 NH, N, N, HN; $CH_3O_2C(CH_2)_2$; CON (imidazolyl, N)	LTBA	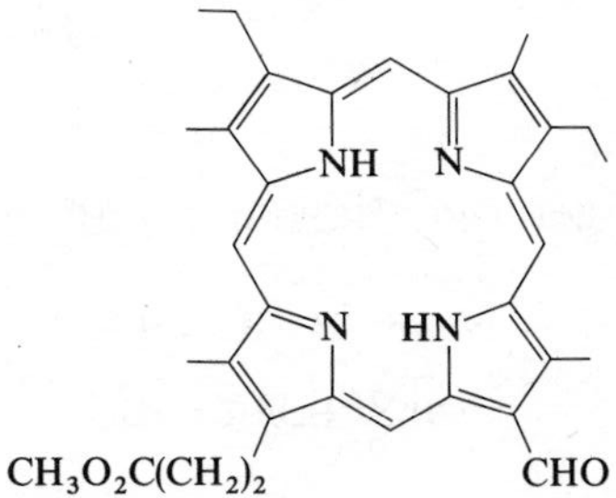 NH, N, N, HN; $CH_3O_2C(CH_2)_2$; CHO (—)	938

TABLE XVII. Reduction of Lactams

No. of Carbon Atoms	Reactant	Conditions	Product(s) and Yield(s) (%)	Refs.
C_4	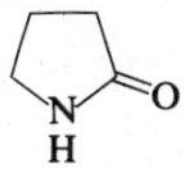 [structure]	SMEAH (IA), C_6H_6, room temp	NaAl([structure])$_2$$(OCH_2CH_2OCH_3)_2$ (—)	939
C_5	[structure]	1. SMEAH, C_6H_6 2. HCl, $CHCl_3$	[structure] $2Cl^-$ (—)	940, 941
	(S)-(−) [structure]	1. SMEAH (IA), THF, 0 to −10°, 2.5 h 2. KCN, room temp overnight; reflux, 30 min	(S)-(−) [structure] (46)	942, 943
		1. 1-Benzyl-2-pyrrolidinone (1 eq) 2. SMEAH, C_6H_6–ether, reflux, 30 min 3. $CO(CH_2CO_2H)_2$, NaOH 4. H_3O^+	[structure] (—)	944
	[structure]	SMEAH (IA), C_6H_6, room temp	NaAl([structure])$_2$$(OCH_2CH_2OCH_3)_2$ (—) 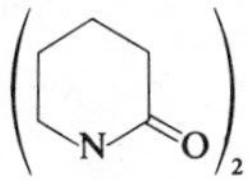	939
C_6	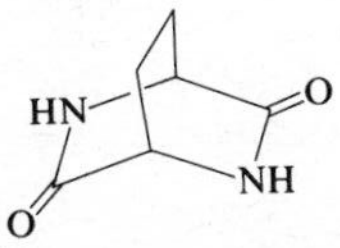 [structure]	SMEAH, C_6H_6, reflux, 47 h	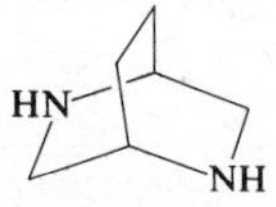 [structure] (58)	945

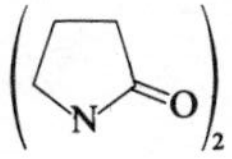

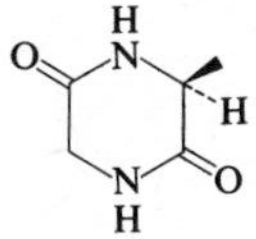

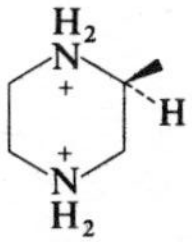

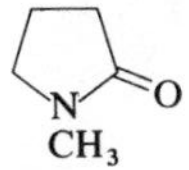

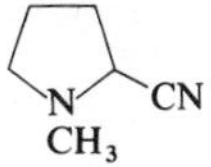

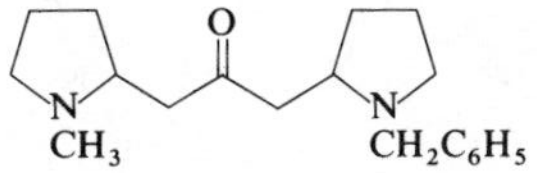

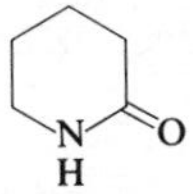

TABLE XVII. REDUCTION OF LACTAMS (*Continued*)

No. of Carbon Atoms	Reactant	Conditions	Product(s) and Yield(s) (%)	Refs.
C_6 (*Contd.*)	1-methyl-2-piperidone (N–CH_3)	SMEAH, C_6H_6, reflux, 2 h	1-methylpiperidine (N–CH_3) (86)	94
		$LiAlH_2(OC_2H_5)_2$	1-methyl-2-(1-methyl-1,4,5,6-tetrahydropyridin-3-yl)piperidine (N–CH_3, N–CH_3) (—)	349
	caprolactam (N–H)	SMEAH (IA), C_6H_6, room temp	$NaAl(\text{caprolactam-N})_2(OCH_2CH_2OCH_3)_2$ (—)	939
		SMEAH, C_6H_6, reflux, 1 h	hexamethyleneimine (N–H) (82)	94, 167
		$NaAlH_2(OCH_2CH_2OC_4H_9\text{-}n)_2$, C_6H_6, reflux, 4 h	" (89)	632
		$Al_2H_3(OCH_2CH_2OCH_3)_3$, C_6H_6, reflux, 2 h	" (97)	313
		$Ca[AlH_2(OC_4H_9\text{-}i)_2]_2 \cdot THF$, toluene, 80°, 2 h	" (70)	165
C_7	1,3-dimethyl-2-piperidone (N–CH_3)	1. $LiAlH(OC_2H_5)_3$, ether, 0°, 2 h 2. KCN, CH_3OH	1,3-dimethyl-2-cyanopiperidine (N–CH_3, CN) (80)	327

Formula	Substrate	Conditions	Product(s) and Yield(s) (%)	Refs.
C_8	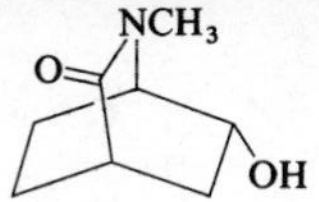 NCH_3, O, OH	SMEAH, C_6H_6, reflux, 4 h	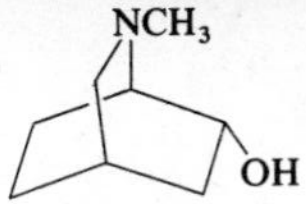 NCH_3, OH (67)	946, 387
C_9	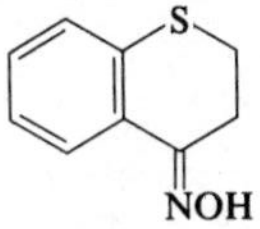 S, NOH	SMEAH, toluene, ether, reflux, 5 h	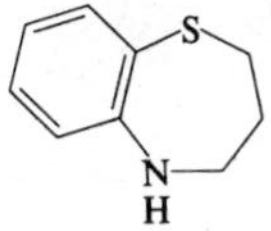 S, N H (~65)	392
	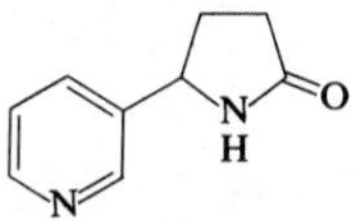 N, N H, O	SMEAH, C_6H_6, 15–20°	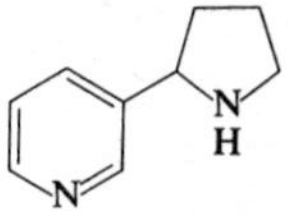 N, N H (90)	947
	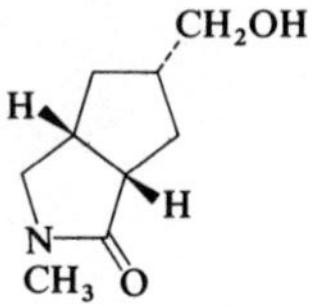 CH_2OH, H, H, N CH_3, O	SMEAH, C_6H_6, 50°, 2.5 h	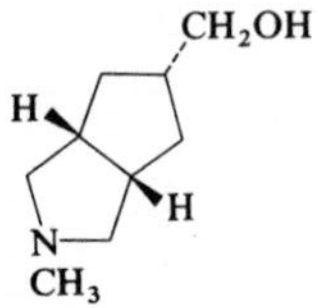 CH_2OH, H, H, N CH_3 (49)	948
C_{10}	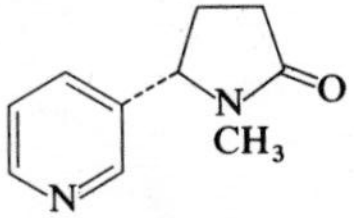 N, N CH_3, O (−)-	1. SMEAH (IA), C_6H_6, ether, 0°; then room temp, 19 h 2. 4-$CH_3C_6H_4SH$	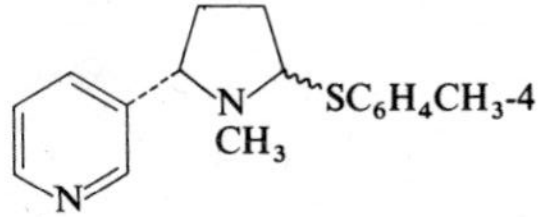 N, N CH_3, $SC_6H_4CH_3$-4 (—)	949
		SMEAH (IA), C_6H_6–ether, 0°; then room temp, 19 h	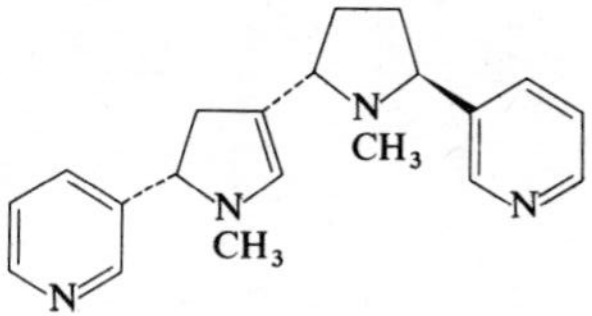 N, N CH_3, N CH_3, N (—)	949
	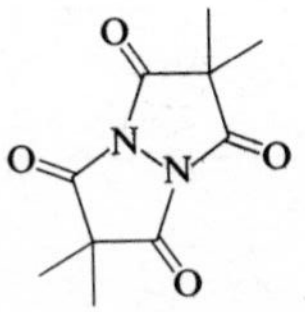 O, O, N, N, O, O	SMEAH, C_6H_6	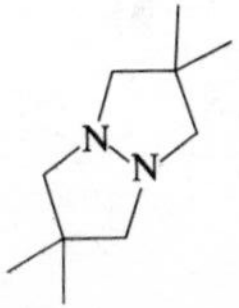 N, N (65)	950

TABLE XVII. REDUCTION OF LACTAMS (*Continued*)

No. of Carbon Atoms	Reactant	Conditions	Product(s) and Yield(s) (%)	Refs.
C_{11}	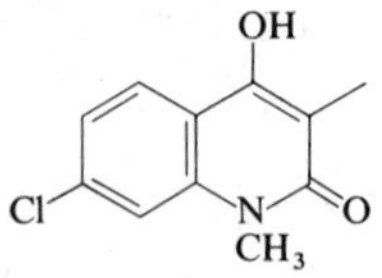 Cl, OH, N, CH_3, O	SMEAH (IA), C_6H_6–toluene, reflux, 1.5 h	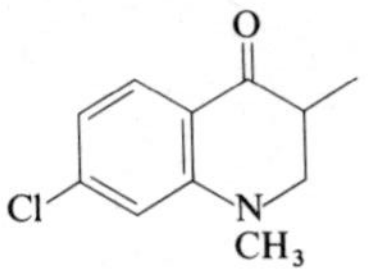 O, Cl, N, CH_3 (55)	951
	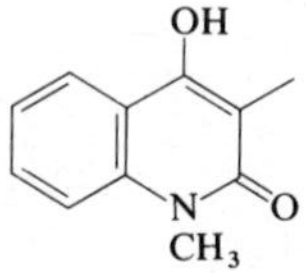 OH, N, CH_3, O	", ", ", ", "	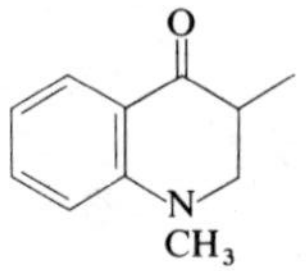 O, N, CH_3 (60)	951
	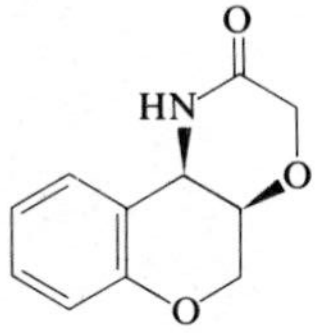 O, HN, O, O	SMEAH (IA), C_6H_6–THF, reflux, 3 h	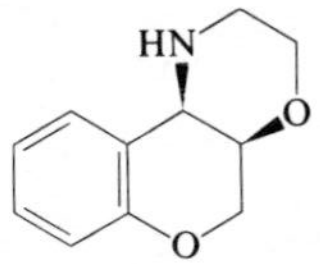 HN, O, O (45)	906
	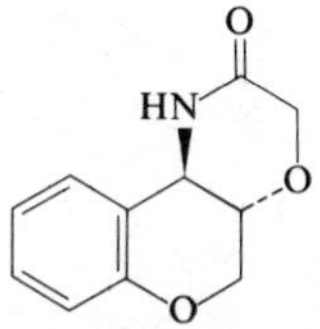 O, HN, O, O	", ", ", "	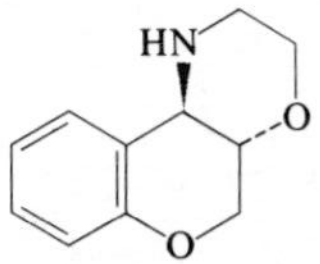 HN, O, O (52)	906
	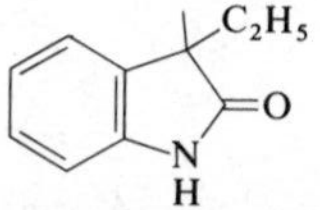 C_2H_5, O, N, H	SMEAH, C_6H_6, 5 h	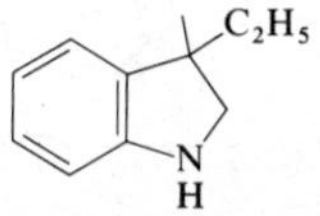 C_2H_5, N, H (57)	952, 953

	Substrate	Reagent and Conditions	Product (% Yield)	Refs.
	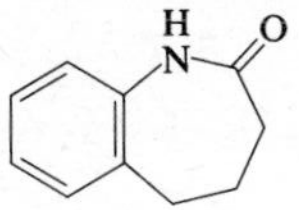 H, N, O	SMEAH, toluene, reflux 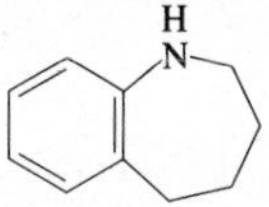	H, N (88)	954
	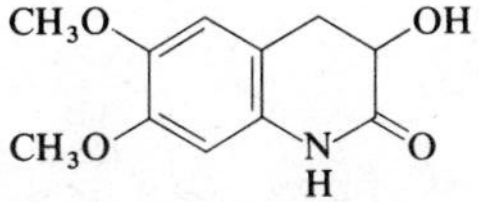 CH_3O, OH, CH_3O, N, H, O	SMEAH, C_6H_6, 60–70°, 1 h	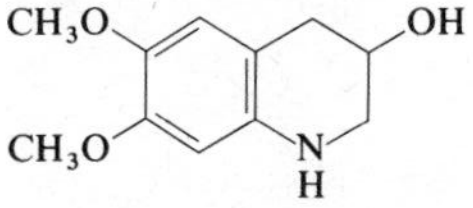 CH_3O, OH, CH_3O, N, H (44)	955
C_{12}	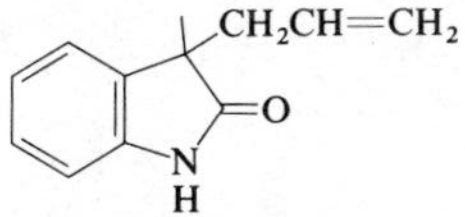 $CH_2CH{=}CH_2$, =O, N, H	SMEAH, C_6H_6, 5 h	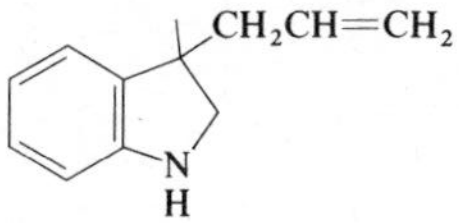 $CH_2CH{=}CH_2$, N, H (94)	952, 953
	NCH_3, O=	SMEAH	NCH_3 (—)	956
	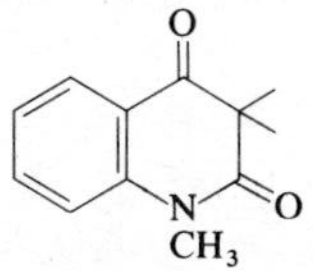 O, N, CH_3, O	SMEAH (IA), C_6H_6, reflux, 1.5 h 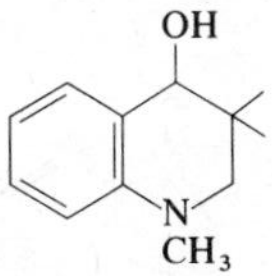	OH, N, CH_3 (68)	951
	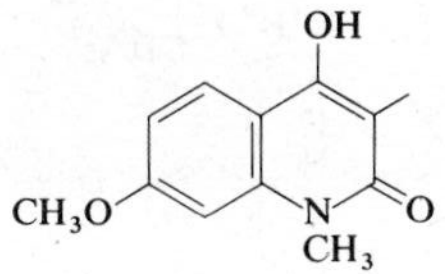 OH, CH_3O, N, CH_3, O	SMEAH (IA), C_6H_6–toluene, reflux, 1.5 h 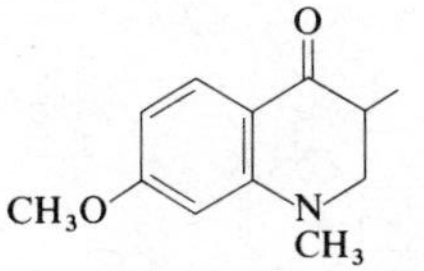	O, CH_3O, N, CH_3 (42)	951
	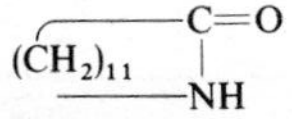 $(CH_2)_{11}$, C=O, NH	SMEAH (IA), C_6H_6, room temp	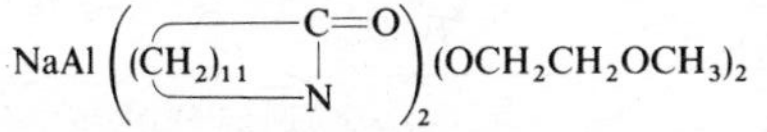 $NaAl\left((CH_2)_{11}, C{=}O, N\right)_2(OCH_2CH_2OCH_3)_2$ (—)	939
C_{13}	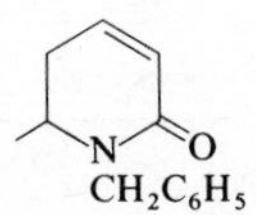 N, O, $CH_2C_6H_5$	$LiAlH_3(OC_2H_5)$, ether 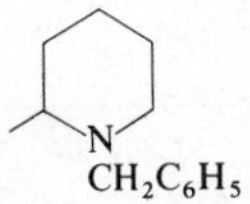	N, $CH_2C_6H_5$ (68)	364

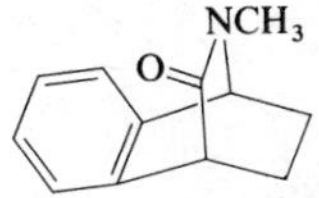

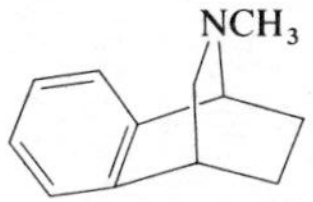

TABLE XVII. REDUCTION OF LACTAMS (*Continued*)

No. of Carbon Atoms	Reactant	Conditions	Product(s) and Yield(s) (%)	Refs.
C_{13} (*Contd.*)	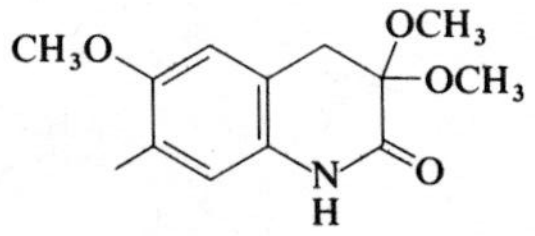 CH_3O, OCH_3, OCH_3, N H, O	SMEAH (IA), C_6H_6	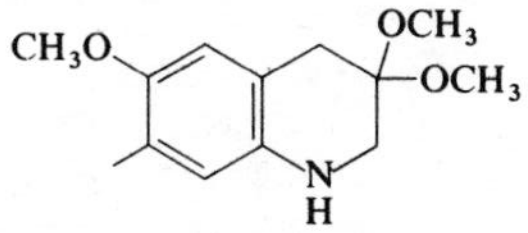 CH_3O, OCH_3, OCH_3, N H (57)	955
C_{14}	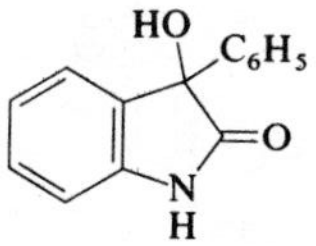 HO, C_6H_5, O, N H	SMEAH, C_6H_6, reflux, 2 h	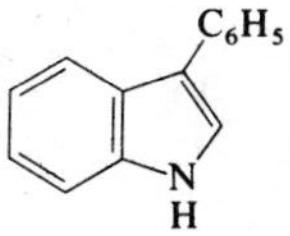 C_6H_5, N H (48)	360
	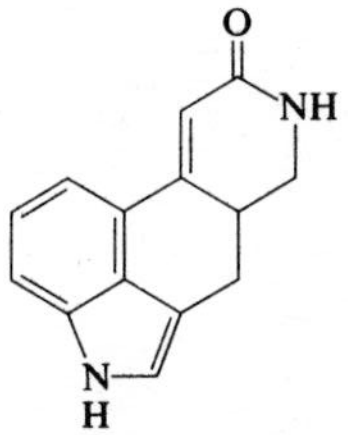 O, NH, N H	SMEAH, C_6H_6, room temp, 2 h	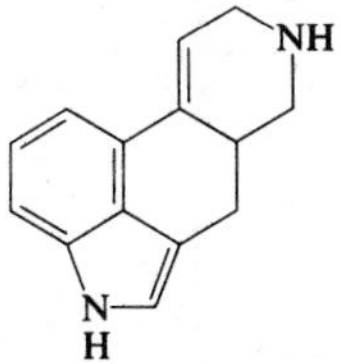 NH, N H (—)[a]	362
	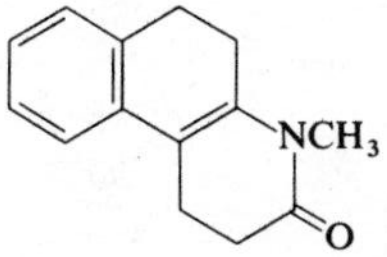 NCH_3, O	SMEAH (IA), C_6H_6, reflux, 6 h	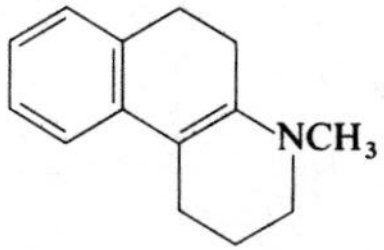 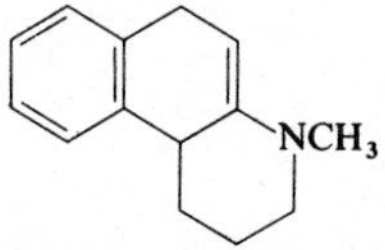 NCH_3 + NCH_3 (85)	957, 958

	Substrate	Conditions	Product (Yield, %)	Refs.
	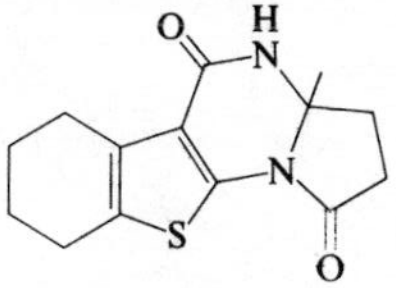	SMEAH (IA), C_6H_6, reflux, overnight	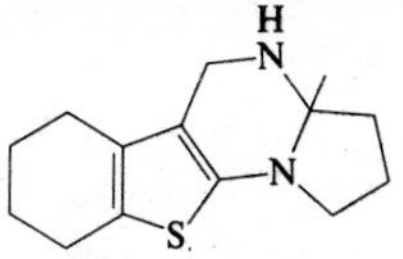 (86)	959
	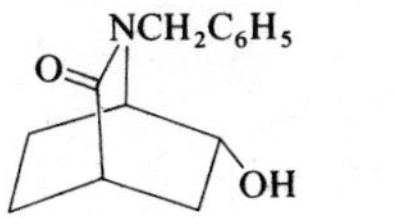	SMEAH, C_6H_6, reflux, 6 h	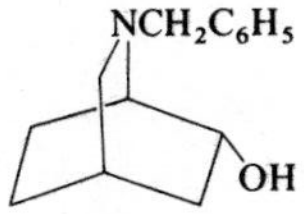 (80)	960
	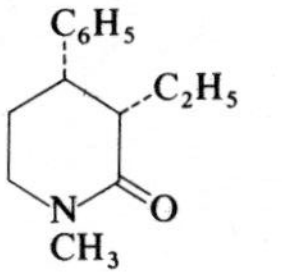	1. $LiAlH(OC_2H_5)_3$, ether, 0°, 5 min 2. KCN, CH_3OH	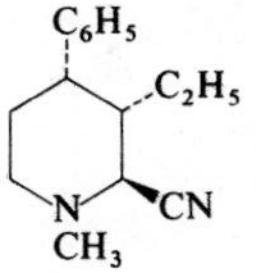 (89)	327
C_{15}	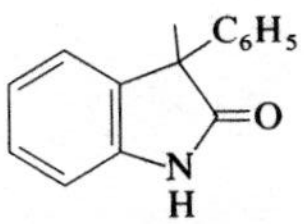	SMEAH, C_6H_6, reflux, 2 h	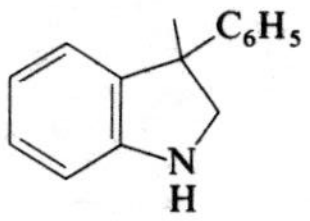 (88)	359, 360, 952, 953
	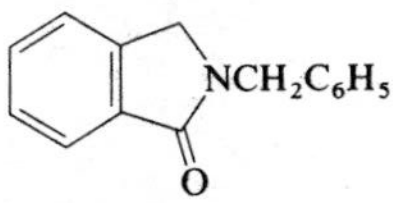	SMEAH, C_6H_6	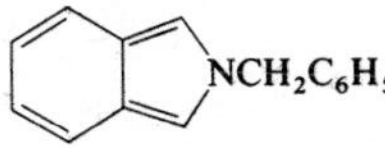 (60)	382
	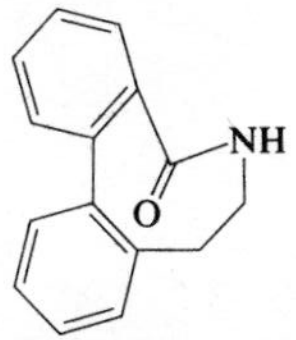	SMEAH, toluene	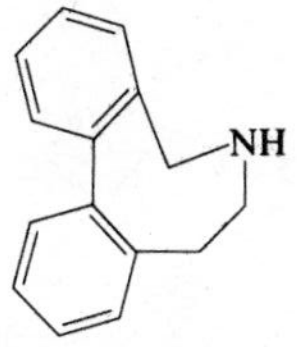 (92)	961
	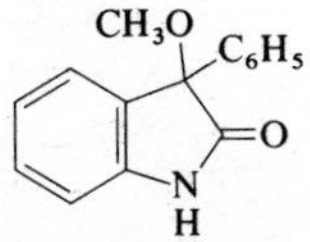	SMEAH, C_6H_6, reflux, 2 h	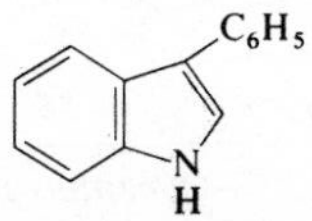 (63)	360

TABLE XVII. Reduction of Lactams (*Continued*)

No. of Carbon Atoms	Reactant	Conditions	Product(s) and Yield(s) (%)	Refs.
C_{15} (*Contd.*)	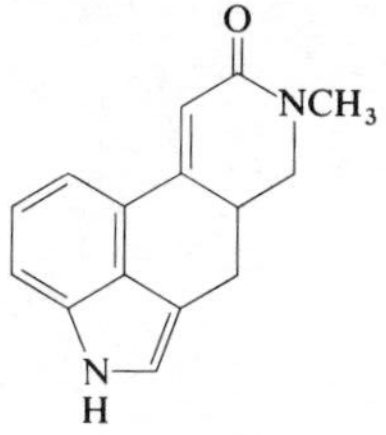 O, NCH_3, N H	SMEAH (IA), C_6H_6–toluene, 2 h	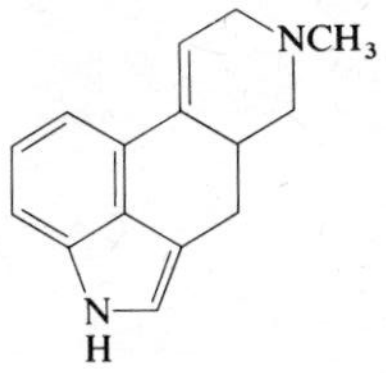 NCH_3, N H (91)	363, 362
	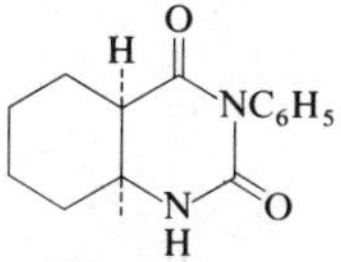 H, O, NC_6H_5, N H, O	SMEAH, C_6H_6, reflux, 36 h	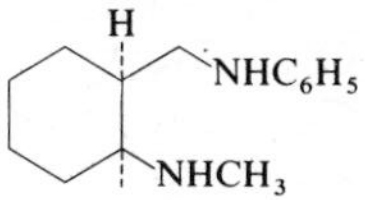 H, NHC_6H_5, $NHCH_3$ (65)	962
	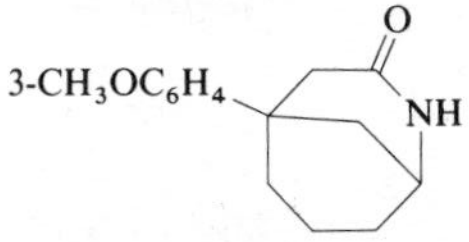 3-$CH_3OC_6H_4$, O, NH	SMEAH	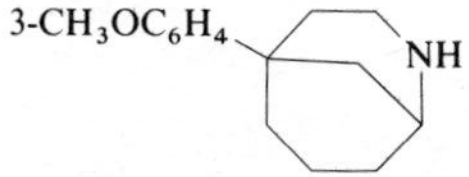 3-$CH_3OC_6H_4$, NH (—)	963
	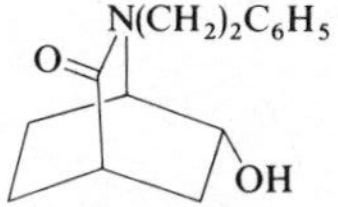 $N(CH_2)_2C_6H_5$, O, OH	SMEAH, C_6H_6, reflux, 4 h	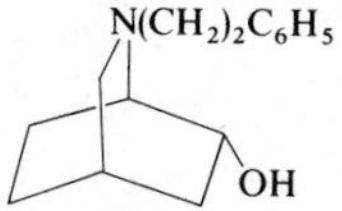 $N(CH_2)_2C_6H_5$, OH (71)	387
	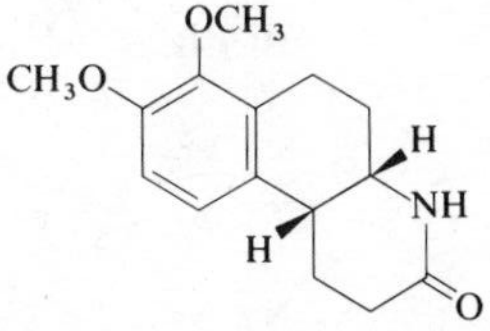 OCH_3, CH_3O, H, NH, H, O	SMEAH	No reaction	964

	Substrate	Conditions	Product(s) and Yield(s) (%)	Refs.
	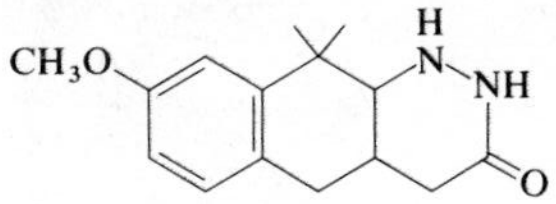 CH_3O; H, N–NH; =O	SMEAH, C_6H_6, 10 h	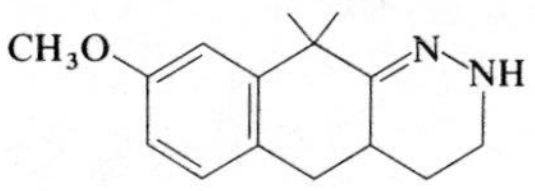 CH_3O; N–NH (24)	965
C_{16}	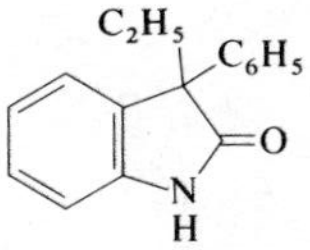 C_2H_5, C_6H_5; =O; N, H	SMEAH, C_6H_6, reflux, 2 h	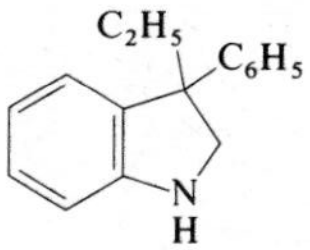 C_2H_5, C_6H_5; N, H (89)	359, 360
	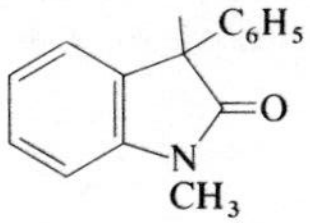 C_6H_5; =O; N, CH_3	SMEAH (1 eq), C_6H_6, reflux, 2 h	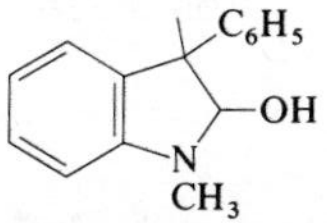 C_6H_5; –OH; N, CH_3 (75)	360
		SMEAH (excess), C_6H_6, reflux, 2 h	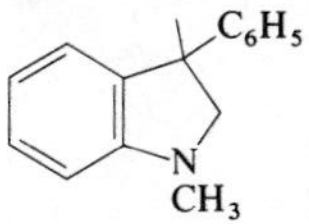 C_6H_5; N, CH_3 (72)	360
	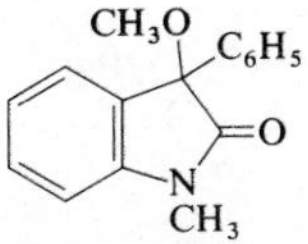 CH_3O, C_6H_5; =O; N, CH_3	SMEAH, C_6H_6, reflux, 10 h	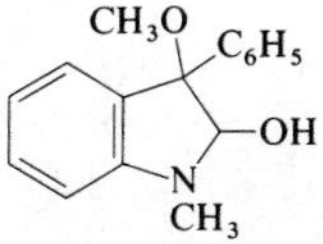 CH_3O, C_6H_5; –OH; N, CH_3 (57) 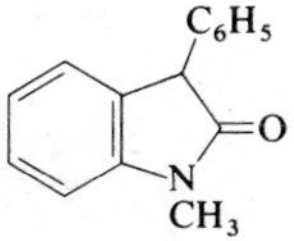 + C_6H_5; =O; N, CH_3 (12)	360
		SMEAH, C_6H_6, reflux, 2 h	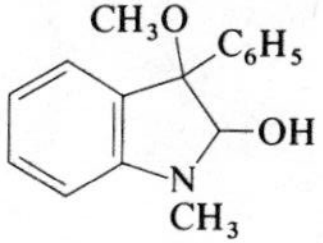 CH_3O, C_6H_5; –OH; N, CH_3 (72)	360
	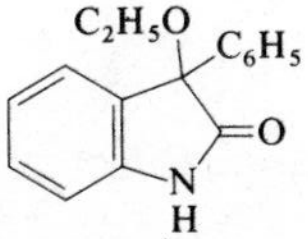 C_2H_5O, C_6H_5; =O; N, H	SMEAH, C_6H_6, reflux, 2 h	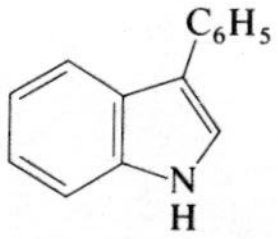 C_6H_5; N, H (59)	360

TABLE XVII. REDUCTION OF LACTAMS (*Continued*)

No. of Carbon Atoms	Reactant	Conditions	Product(s) and Yield(s) (%)	Refs.
C_{16} (*Contd.*)	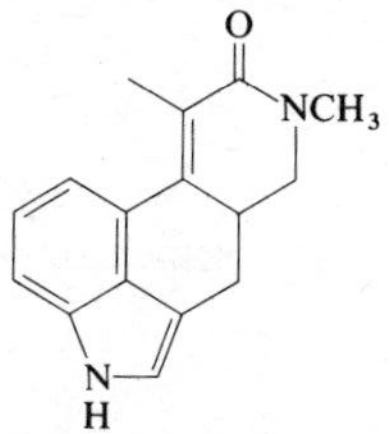 NCH$_3$	SMEAH (IA), C_6H_6–toluene, 2 h	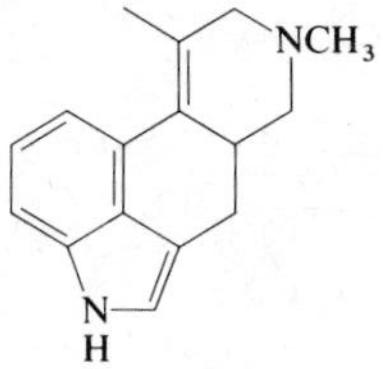 NCH$_3$ (85)	363, 362
	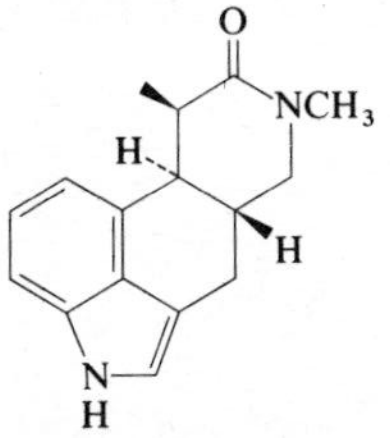 NCH$_3$ (6*aR*,10*R*,10*aR*)-	SMEAH (IA), C_6H_6–toluene, 2 h	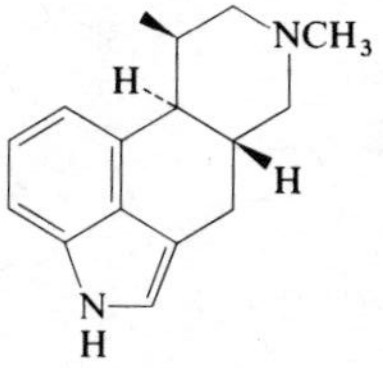 NCH$_3$ (64) (6*aR*,10*R*,10*aS*)-	363, 362
	NCH$_3$ (6*aR*,10*R*,10*aS*)-		NCH$_3$ (42) (6*aR*,10*R*,10*aR*)-	363

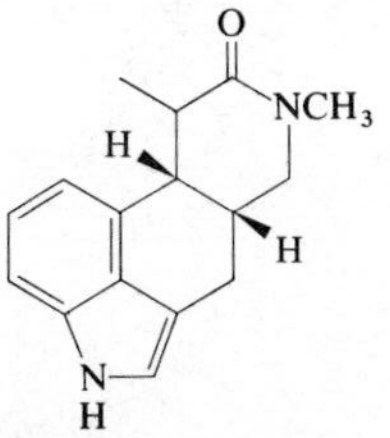

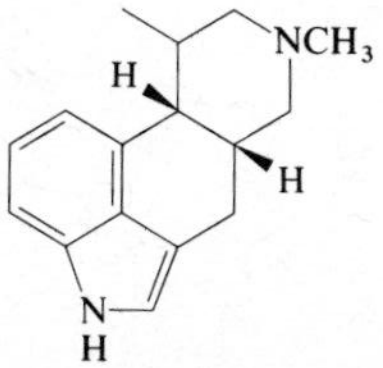

Reactant	Conditions	Product(s) and Yield(s) (%)	Refs.
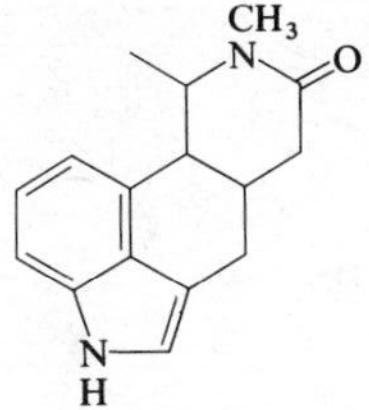	SMEAH, C_6H_6, room temp, 2 h	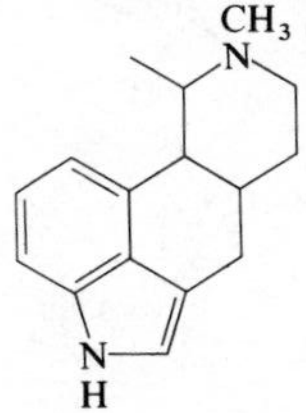 (—)	362
C_{17}	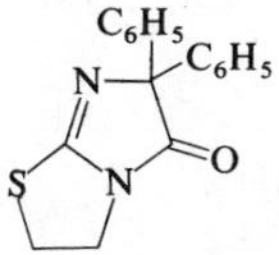 SMEAH, C_6H_6–THF, reflux, 1 h	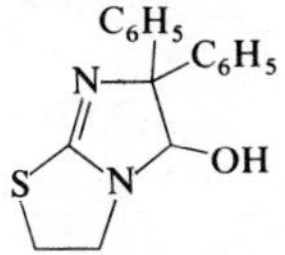 (59)	966
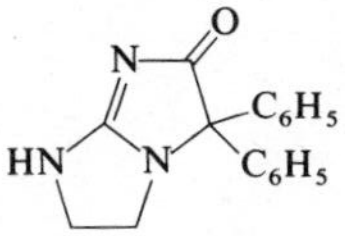	SMEAH, (IA), C_6H_6–THF, 0°, 1.5 h	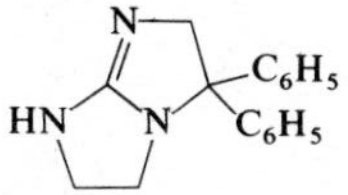 (57)	372, 967
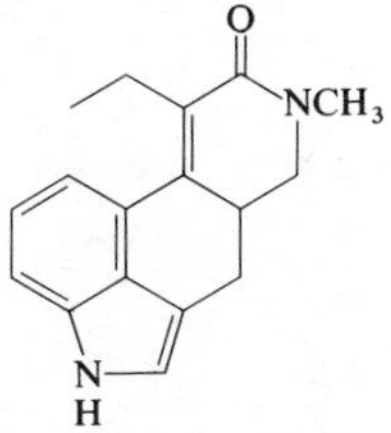	SMEAH, C_6H_6, room temp, 2 h	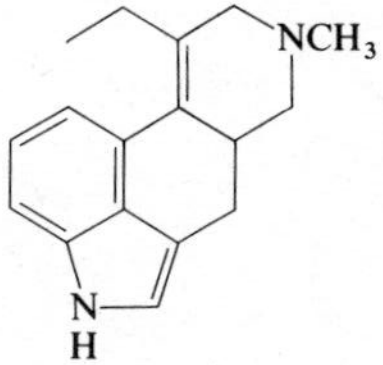 (80)[a]	362

TABLE XVII. Reduction of Lactams (*Continued*)

No. of Carbon Atoms	Reactant	Conditions	Product(s) and Yield(s) (%)	Refs.
C_{17} (*Contd.*)	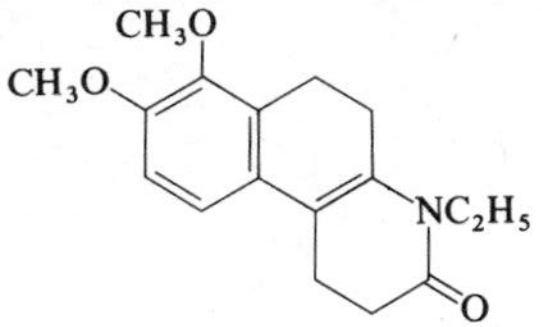 CH_3O, CH_3O, NC_2H_5, O	SMEAH (IA), C_6H_6, reflux, 6 h	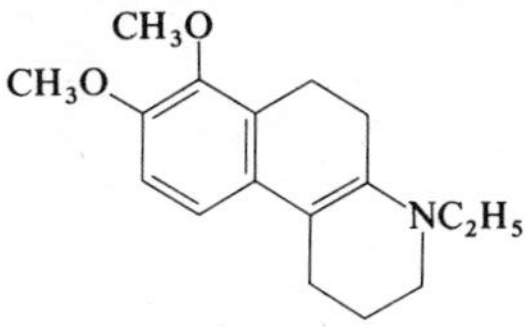 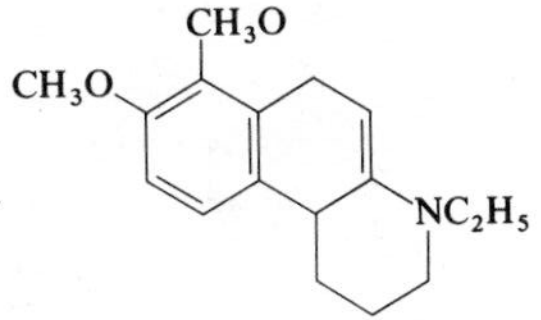 CH_3O, CH_3O, NC_2H_5 + CH_3O, CH_3O, NC_2H_5 (98)	964
	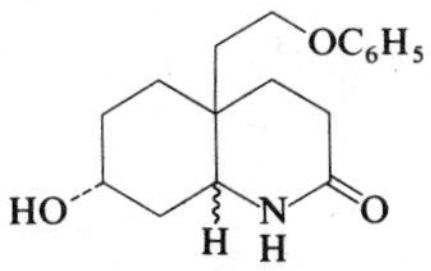 OC_6H_5, HO, N, H, H, O	$LiAlH_2(OCH_3)_2$	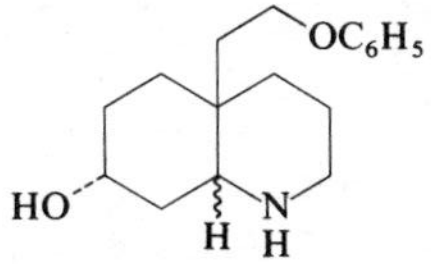 OC_6H_5, HO, N, H, H (—)	357
	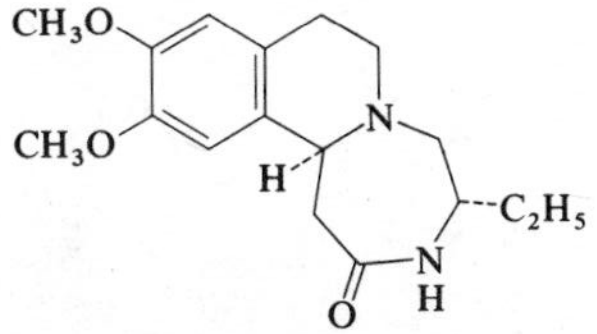 CH_3O, CH_3O, N, H, C_2H_5, N, H, O	SMEAH, THF—C_6H_6, reflux, 4 h	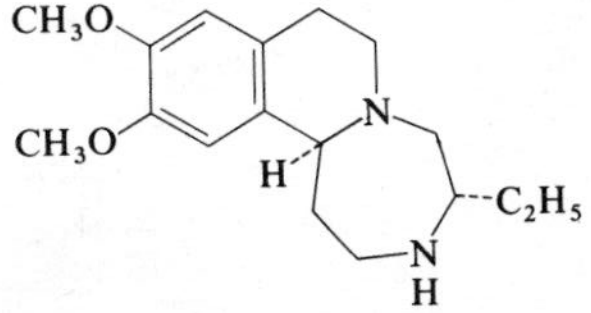 CH_3O, CH_3O, N, H, C_2H_5, N, H (90)	968
C_{18}	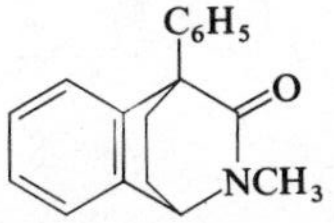 C_6H_5, O, NCH_3	$LiA1H_2(OC_2H_5)_2$, THF–toluene reflux, 2.75 h	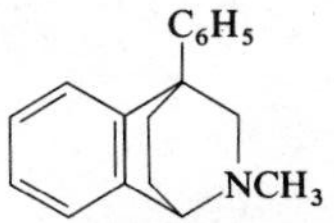 C_6H_5, NCH_3 (—)	969

Substrate	Conditions	Product(s) and Yield(s) (%)	Refs.
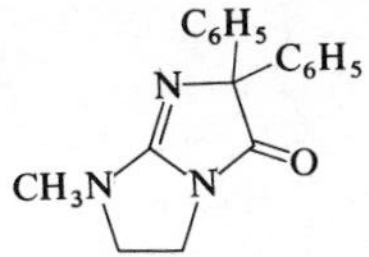	SMEAH (IA), THF–C_6H_6, 0°, 3 h	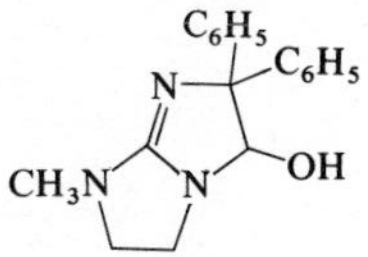 (75)	372
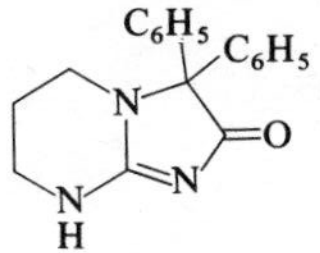	", ", ", "	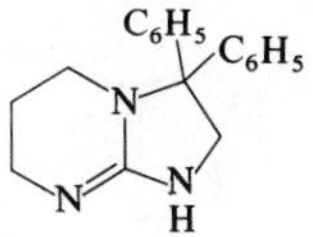 (62)	372
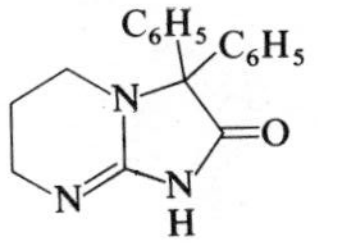	SMEAH, THF–C_6H_6, 0°, 1.5 h	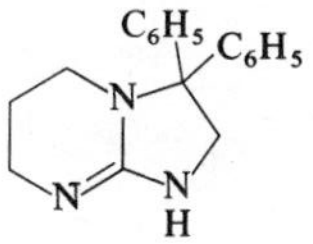 (81)	372
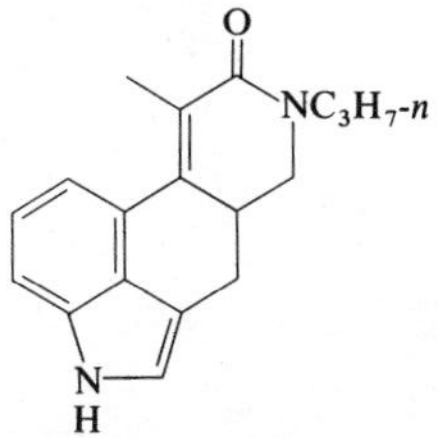	SMEAH, C_6H_6, room temp, 2 h	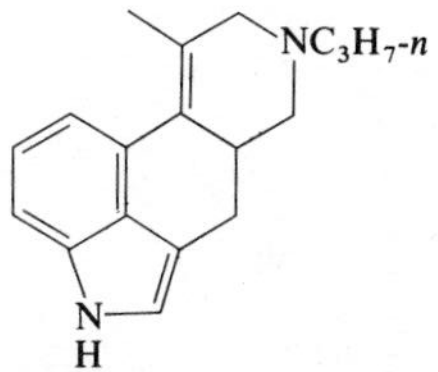 (76)[a]	362
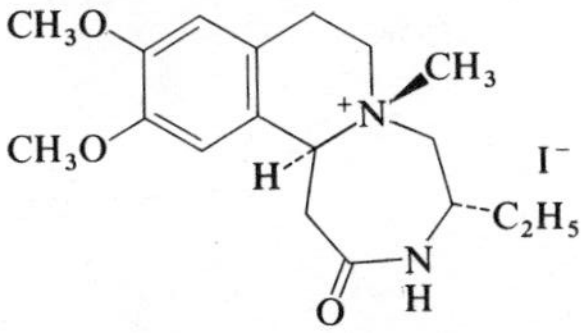	SMEAH, THF–C_6H_6, reflux, 2.5 h	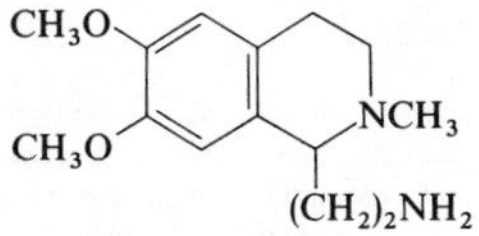 + 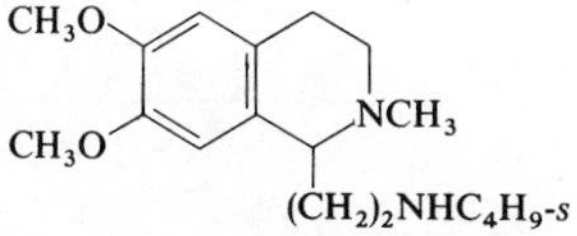 (—)	968

TABLE XVII. REDUCTION OF LACTAMS (*Continued*)

No. of Carbon Atoms	Reactant	Conditions	Product(s) and Yield(s) (%)	Refs.
C_{18} (*Contd.*)	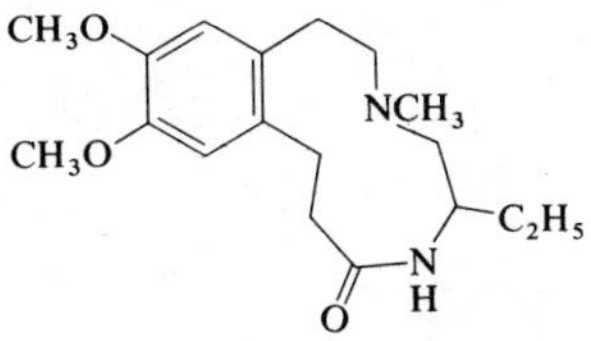 CH_3O, CH_3O, NCH_3, C_2H_5, N H, O	SMEAH, THF–C_6H_6, reflux, 2.5 h	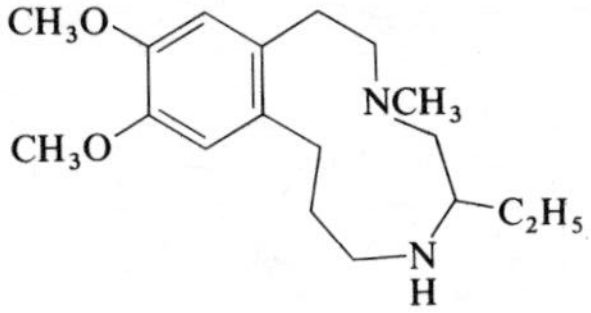 CH_3O, CH_3O, NCH_3, C_2H_5, N H (55)	968
C_{19}	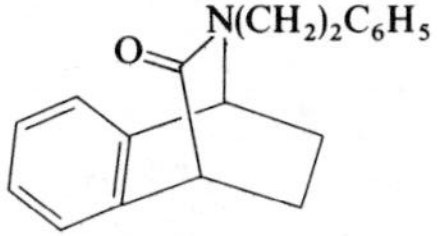 $N(CH_2)_2C_6H_5$, O	SMEAH	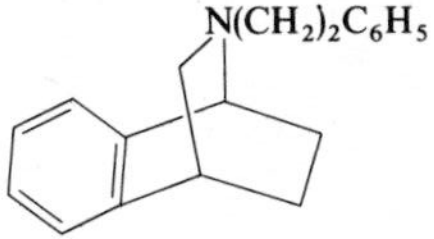 $N(CH_2)_2C_6H_5$ (—)	956
	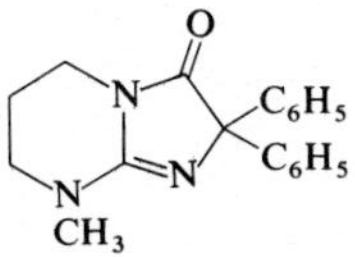 O, N, C_6H_5, C_6H_5, N, N, CH_3	SMEAH (IA), THF–C_6H_6, 0°, 3 h	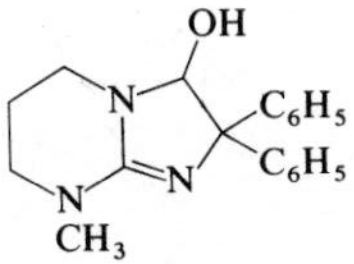 OH, N, C_6H_5, C_6H_5, N, N, CH_3 (80)	372
	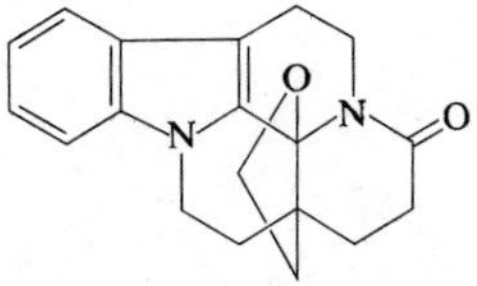 N, O, N, O	SMEAH, THF–C_6H_6, room temp, overnight	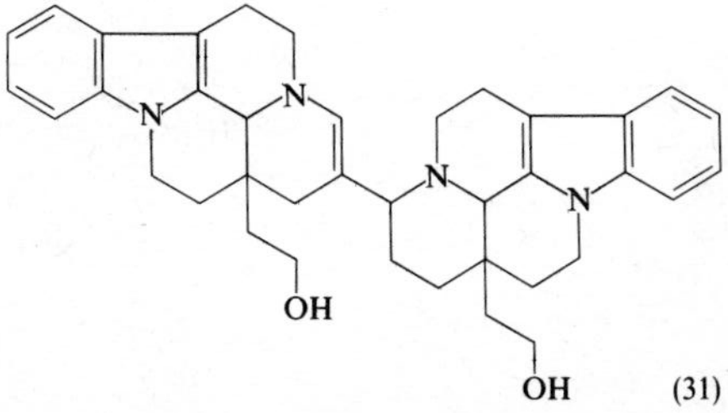 N, N, N, N, OH, OH (31)	970a

Formula	Substrate	Conditions	Product	Yield (%)	Ref.
	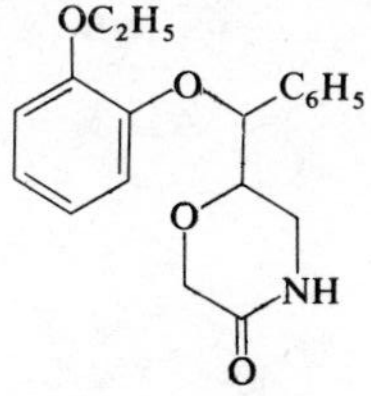 (2*RS*,3*RS*)	1. SMEAH, toluene, room temp, 4 h 2. MsOH	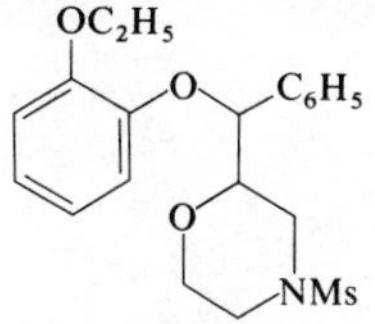 (2*RS*,3*RS*)	(72)	970b
	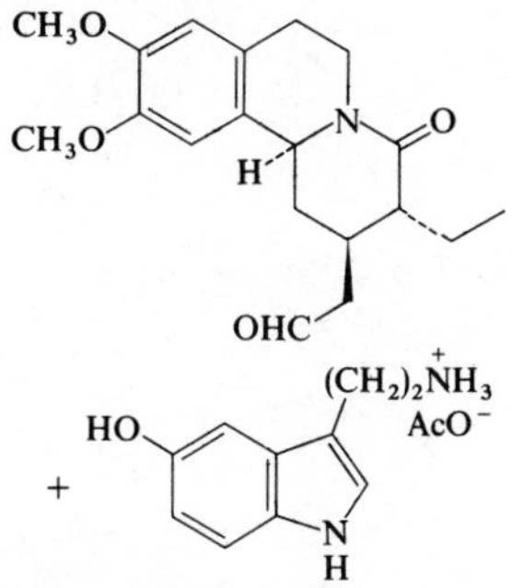 +	1. AcOH, room temp, 2 days 2. SMEAH, Py–toluene, room temp, 1 h	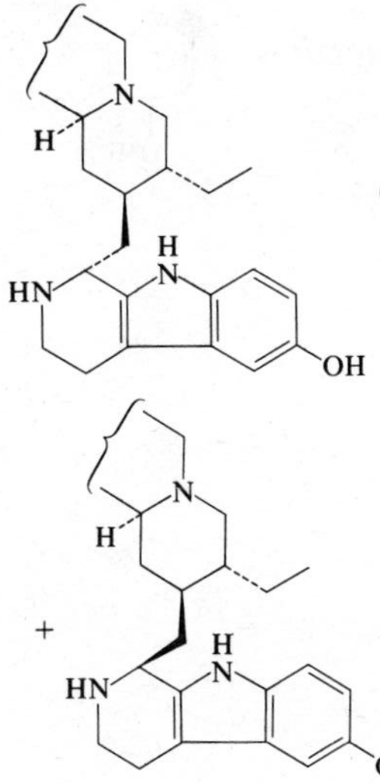	(21)	971
			+	(—)	
C_{20}	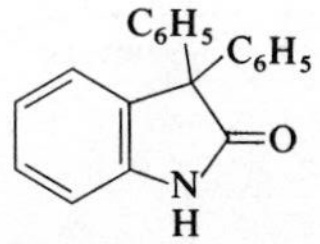	SMEAH, 5 h	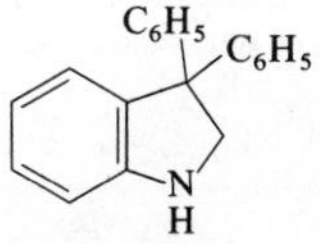	(90)	952, 953

TABLE XVII. REDUCTION OF LACTAMS (*Continued*)

No. of Carbon Atoms	Reactant	Conditions	Product(s) and Yield(s) (%)	Refs.
C_{20} (*Contd.*)	[structure] $C_6H_3(OCH_3)_2$-3,4	1. SMEAH (IA), THF–C_6H_6, 2 h 2. HCl	[structure] $C_6H_3(OCH_3)_2$-3,4, OH, Cl^- (70)	661
	[structure] Cl, $NHC_6H_3(OCH_3)_2$-3,4	SMEAH, THF, reflux, 2 h	No reaction	661
	[structure] C_2H_5O, $NCH_2C_6H_5$	SMEAH,	[structure] C_2H_5O, OH, $NCH_2C_6H_5$ (—)	972a
	[structure] $NC(=CH_2)C_3H_7$-*i*	SMEAH, C_6H_6, room temp, 2 h	[structure] $NC(=CH_2)C_3H_7$-*i* (80)[a]	362

	Substrate	Conditions	Product	(Yield, %)	Ref.
	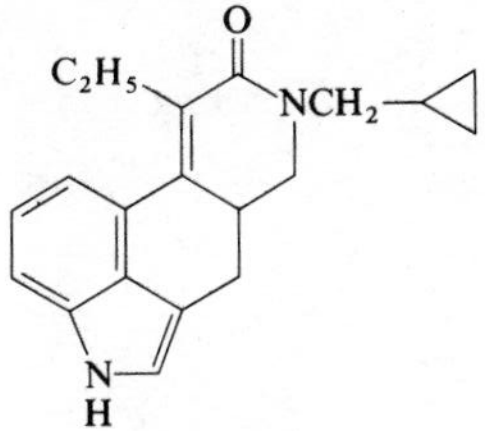 C_2H_5, O, NCH_2–cyclopropyl, NH	SMEAH, C_6H_6, room temp, 2 h	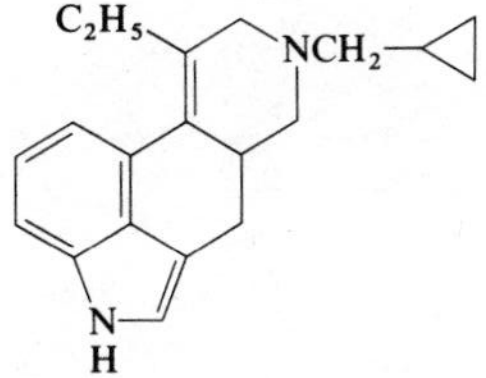 C_2H_5, NCH_2–cyclopropyl, NH	(—)[a]	362
	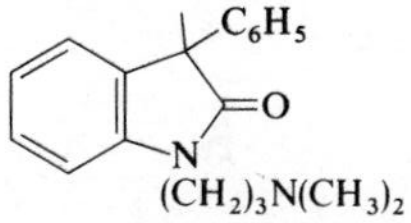 C_6H_5, =O, N, $(CH_2)_3N(CH_3)_2$	SMEAH, C_6H_6, reflux	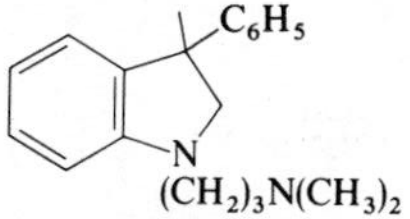 C_6H_5, N, $(CH_2)_3N(CH_3)_2$	(75)	359
	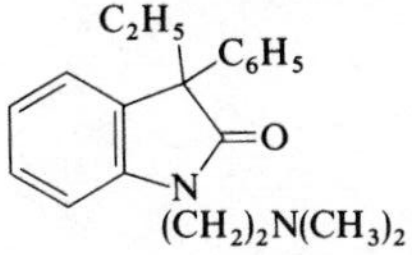 C_2H_5, C_6H_5, =O, N, $(CH_2)_2N(CH_3)_2$	", ", "	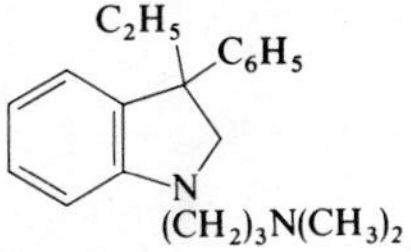 C_2H_5, C_6H_5, N, $(CH_2)_3N(CH_3)_2$	(77)	359
C_{21}	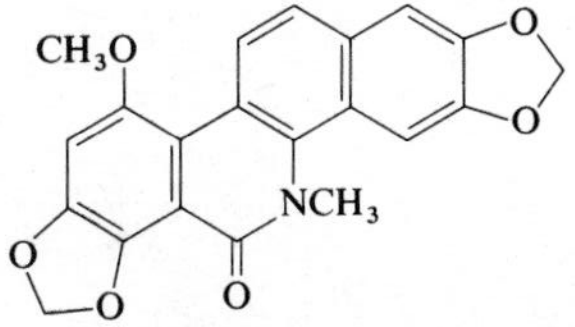 CH_3O, NCH_3, O	SMEAH (IA), C_6H_6–xylene, reflux, 30 min	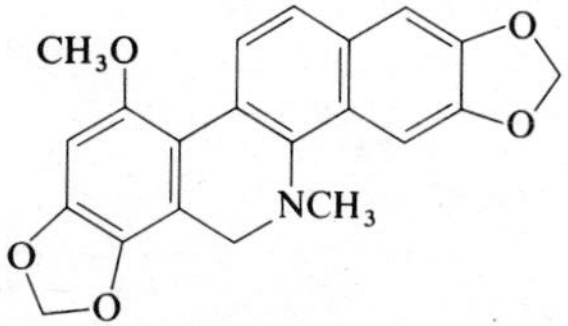 CH_3O, NCH_3	(79)	972b
	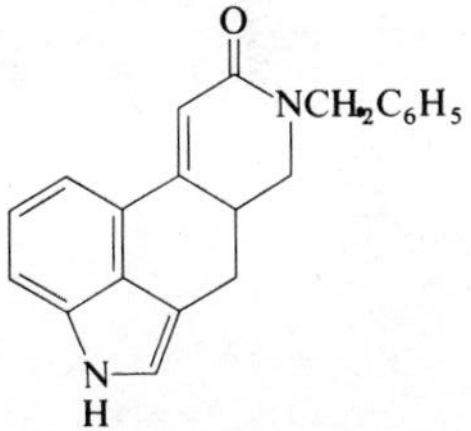 O, $NCH_2C_6H_5$, NH	SMEAH, C_6H_6, room temp. 2 h	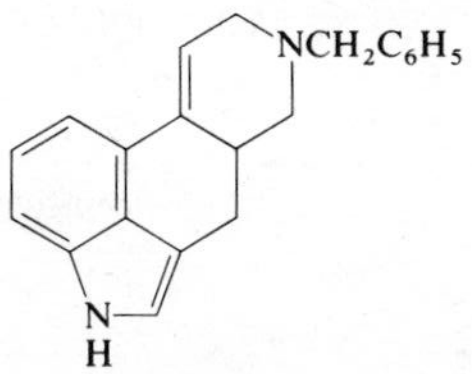 $NCH_2C_6H_5$, NH	(61)[a]	362

TABLE XVII. Reduction of Lactams (*Continued*)

No. of Carbon Atoms	Reactant	Conditions	Product(s) and Yield(s) (%)	Refs.
C_{21} (*Contd.*)	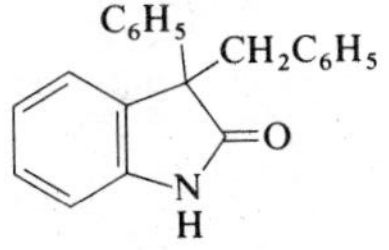 3-C_6H_5-3-$CH_2C_6H_5$-indolin-2-one	SMEAH, 5 h	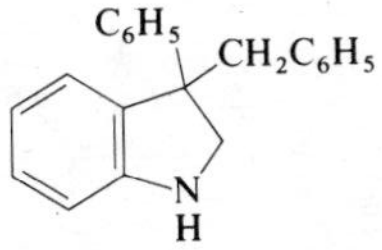 3-C_6H_5-3-$CH_2C_6H_5$-indoline (63)	952, 953
	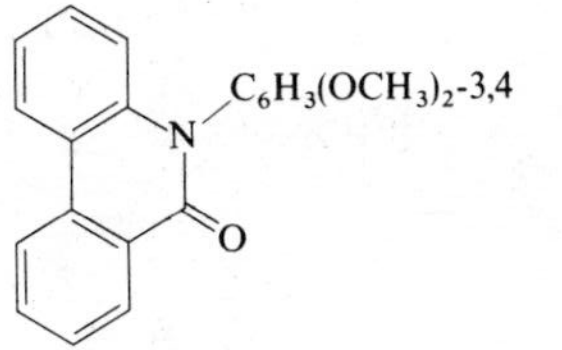 N-$C_6H_3(OCH_3)_2$-3,4 phenanthridinone	SMEAH (1A). (1 eq): C_6H_6, reflux, 42 h	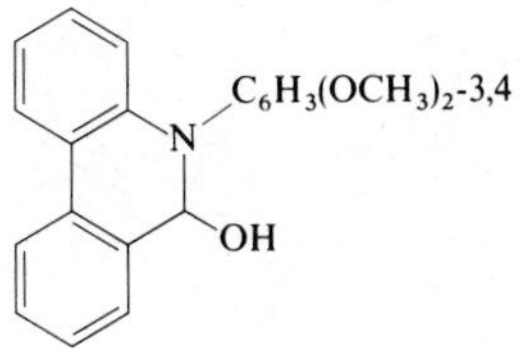 N-$C_6H_3(OCH_3)_2$-3,4, 6-OH dihydrophenanthridine (82)	973
		SMEAH (excess), C_6H_6, reflux	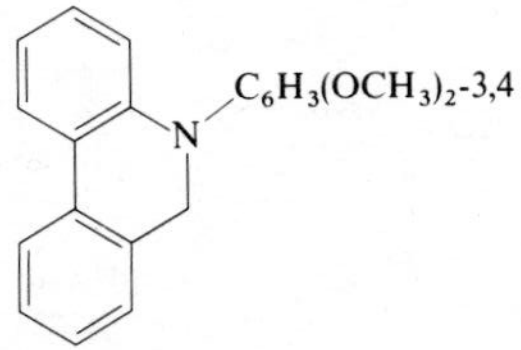 N-$C_6H_3(OCH_3)_2$-3,4 dihydrophenanthridine (—)	973
	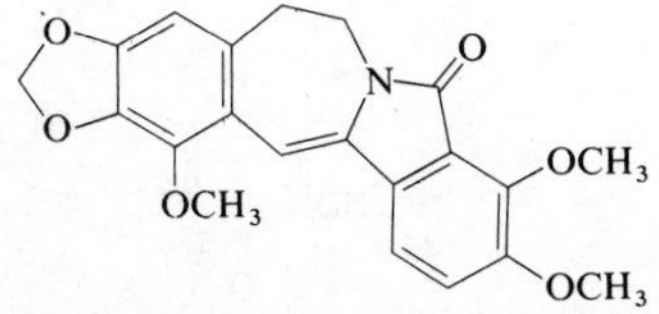 Methylenedioxy, OCH_3, OCH_3, OCH_3 lactam (N–C=O)	1. SMEAH, C_6H_6, reflux 2. HCl, C_2H_5OH	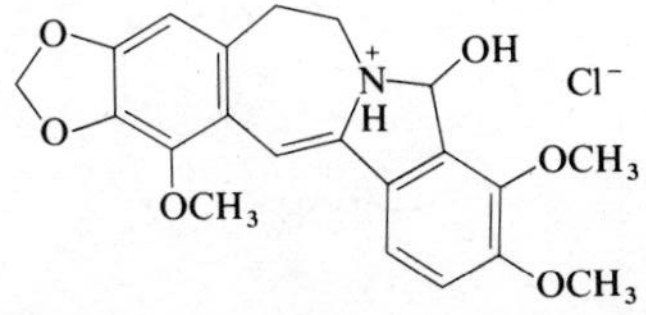 Corresponding $\overset{+}{N}$H, OH, Cl^- salt (—)[b]	974, 975

Substrate	Conditions	Product (yield %)	Ref.
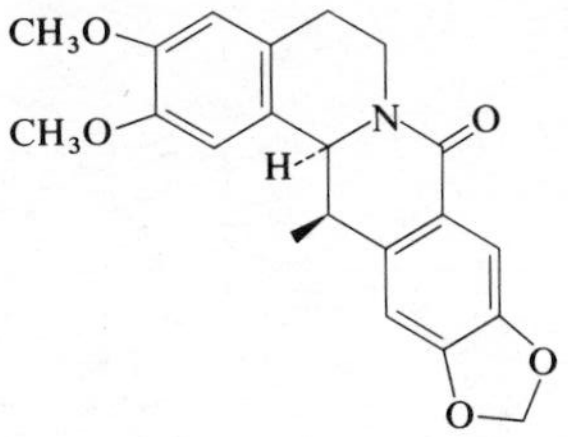 CH_3O, CH_3O, H, N, O, O, O	SMEAH, C_6H_6–toluene, 18 h 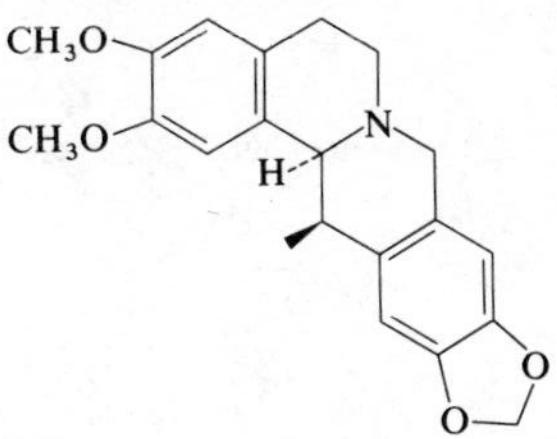	CH_3O, CH_3O, H, N, O, O (81)	976
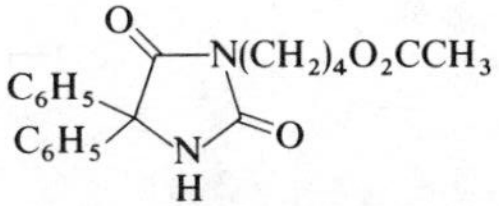 O, $N(CH_2)_4O_2CCH_3$, C_6H_5, C_6H_5, N, H, O	SMEAH (IA), THF–C_6H_6 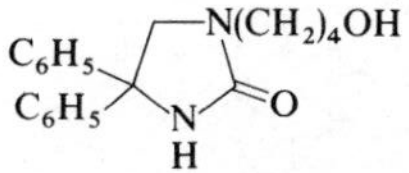	$N(CH_2)_4OH$, C_6H_5, C_6H_5, N, H, O (56)	372
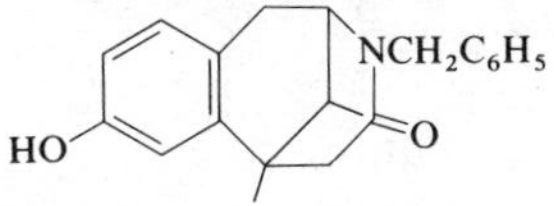 HO, $NCH_2C_6H_5$, O	SMEAH, xylene, reflux, 5 h	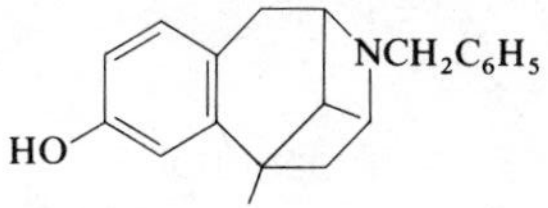 HO, $NCH_2C_6H_5$ (70)	977, 978
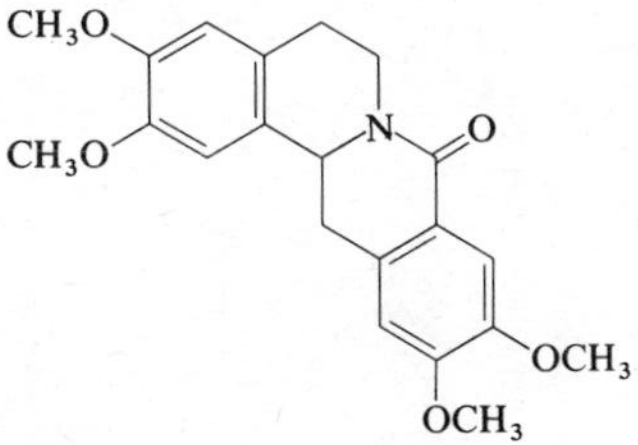 CH_3O, CH_3O, N, O, OCH_3, OCH_3	SMEAH, C_6H_6, 16 h 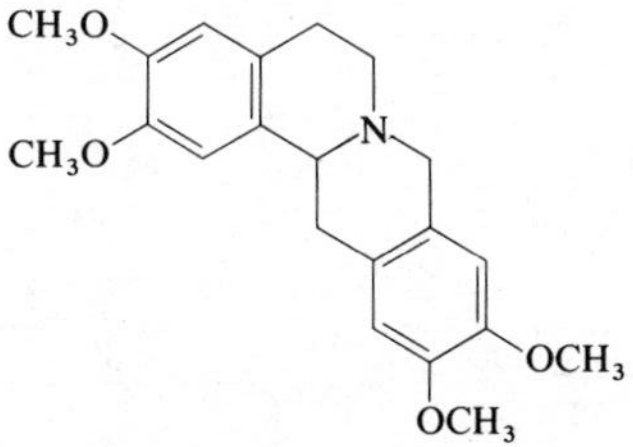	CH_3O, CH_3O, N, OCH_3, OCH_3 (80)	979
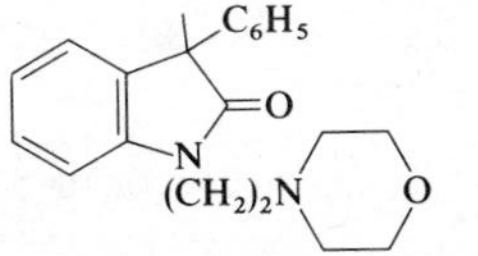 C_6H_5, O, N, $(CH_2)_2N$, O	SMEAH, C_6H_6, reflux 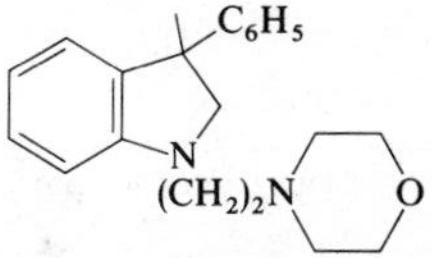	C_6H_5, N, $(CH_2)_2N$, O (77)	359

TABLE XVII. Reduction of Lactams (*Continued*)

No. of Carbon Atoms	Reactant	Conditions	Product(s) and Yield(s) (%)	Refs.
C_{21} (*Contd.*)	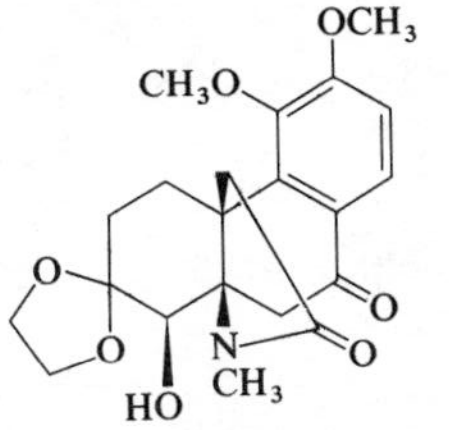 OCH_3, CH_3O, O, O, O, O, N, CH_3, HO	SMEAH, C_6H_6, reflux, 2 h	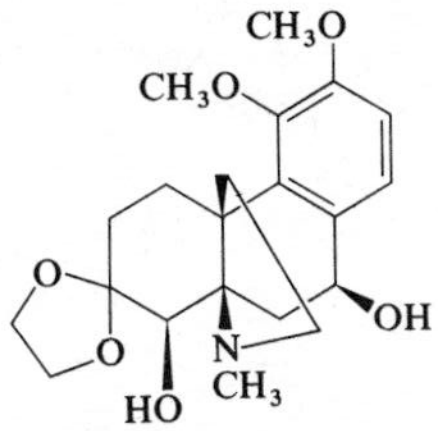 CH_3O, CH_3O, O, O, OH, N, CH_3, HO (36)	980, 981
	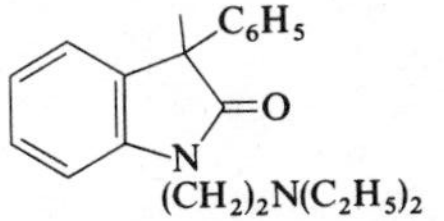 C_6H_5, O, N, $(CH_2)_2N(C_2H_5)_2$	SMEAH, C_6H_6, reflux	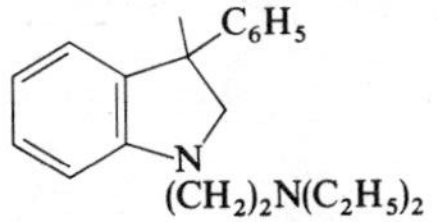 C_6H_5, N, $(CH_2)_2N(C_2H_5)_2$ (80)	359
	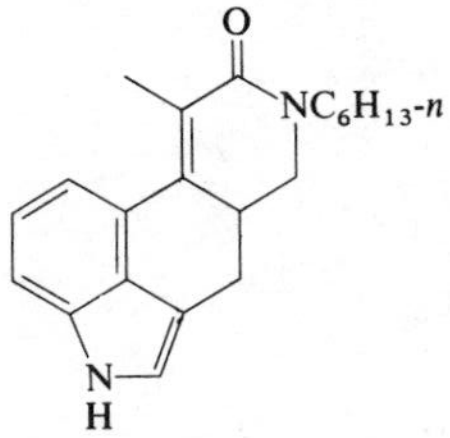 O, NC_6H_{13}-*n*, N, H	SMEAH, C_6H_6, room temp, 2 h	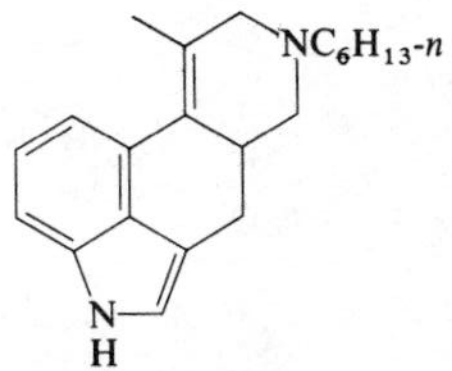 NC_6H_{13}-*n*, N, H (75)[a]	362
	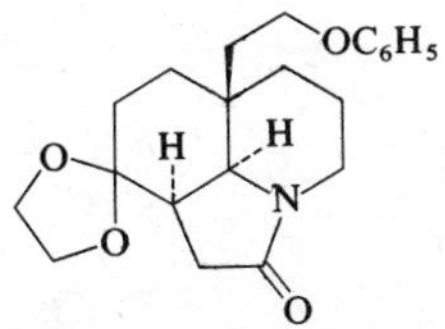 OC_6H_5, O, O, H, H, N, O	1. $LiAlH_2(OCH_3)_2$, THF, 0°, 30 min; reflux, 4 h 2. H_3O^+	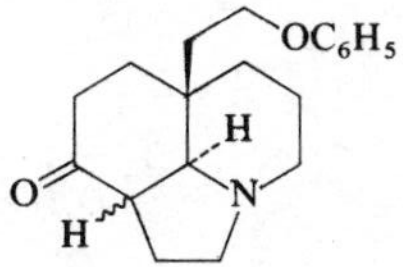 OC_6H_5, O, H, N, H (78)	358

	Substrate	Conditions	Product (Yield, %)	Refs.
C_{22}	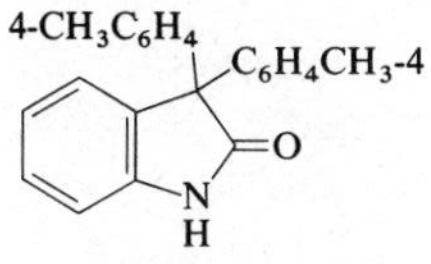 4-$CH_3C_6H_4$, $C_6H_4CH_3$-4, =O, N, H	SMEAH, 5 h	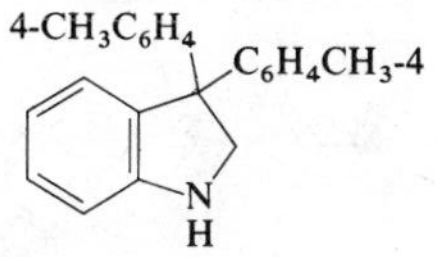 4-$CH_3C_6H_4$, $C_6H_4CH_3$-4, N, H (60)	952, 953
	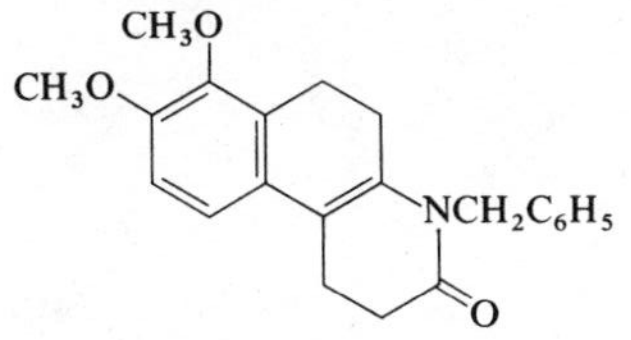 CH_3O, CH_3O, $NCH_2C_6H_5$, O	SMEAH, C_6H_6, reflux, 6 h	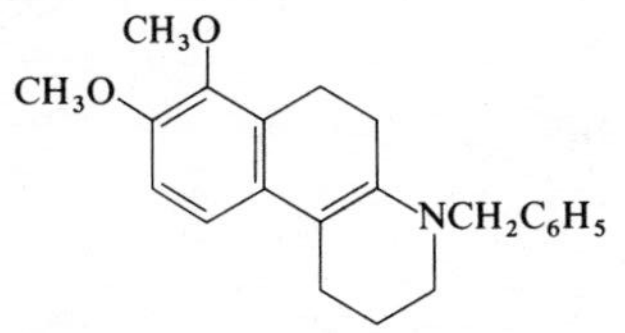 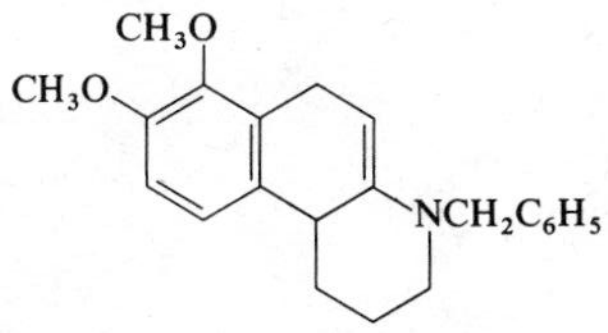 CH_3O, CH_3O, $NCH_2C_6H_5$ + CH_3O, CH_3O, $NCH_2C_6H_5$ (~100)	964
	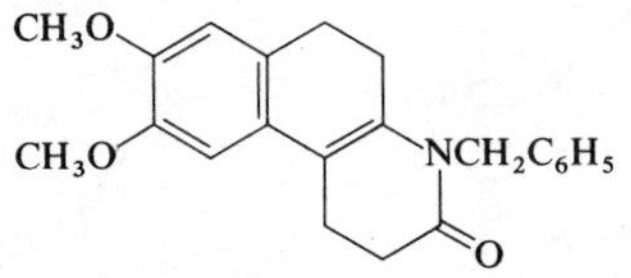 CH_3O, CH_3O, $NCH_2C_6H_5$, O	SMEAH, C_6H_6, reflux, 6 h; room temp, overnight	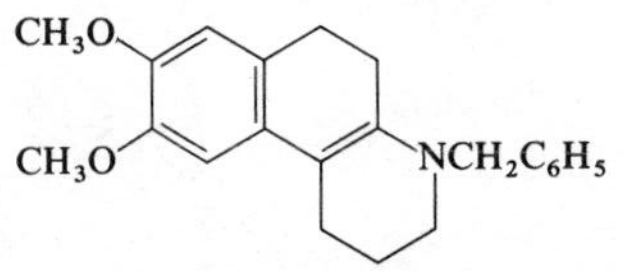 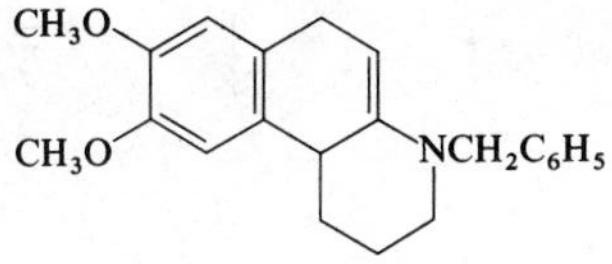 CH_3O, CH_3O, $NCH_2C_6H_5$ + CH_3O, CH_3O, $NCH_2C_6H_5$ (92)	982

TABLE XVII. REDUCTION OF LACTAMS (*Continued*)

No. of Carbon Atoms	Reactant	Conditions	Product(s) and Yield(s) (%)	Refs.
C_{22} (*Contd.*)	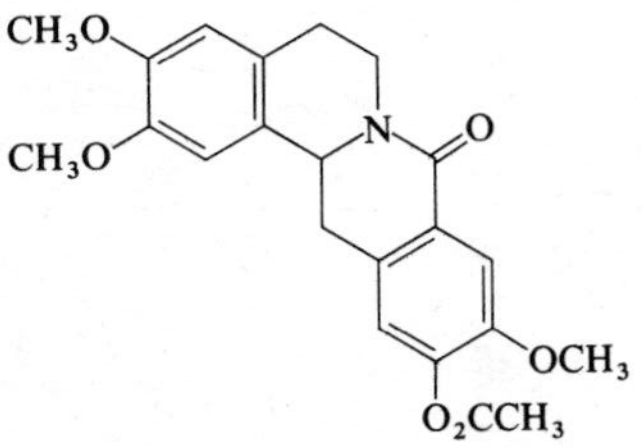 CH_3O, CH_3O, O, N, OCH_3, O_2CCH_3	SMEAH, C_6H_6, 16 h	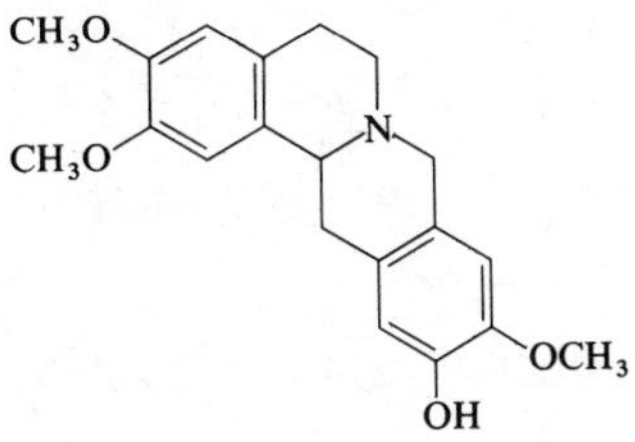 CH_3O, CH_3O, N, OCH_3, OH (62)	979
	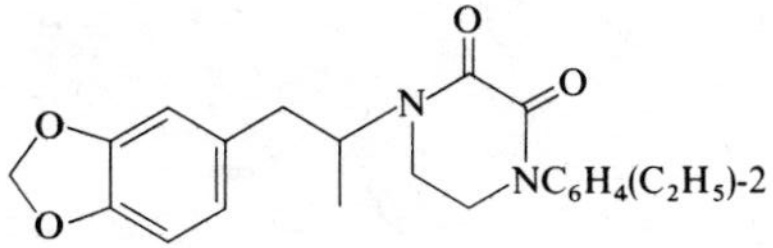 O, O, N, O, O, $NC_6H_4(C_2H_5)$-2	SMEAH, C_6H_6, reflux	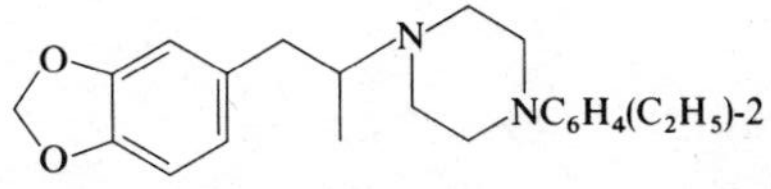 O, O, N, $NC_6H_4(C_2H_5)$-2 (—)	983
	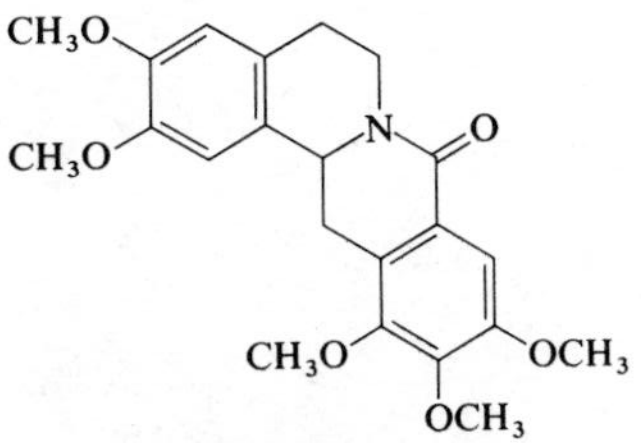 CH_3O, CH_3O, N, O, CH_3O, OCH_3, OCH_3	SMEAH, C_6H_6, 16 h	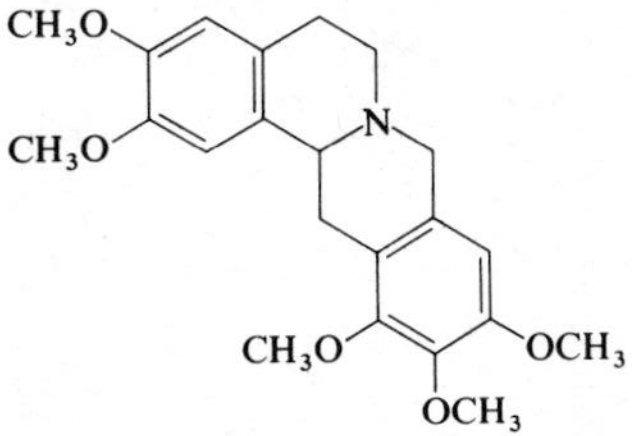 CH_3O, CH_3O, N, CH_3O, OCH_3, OCH_3 (90)	979
	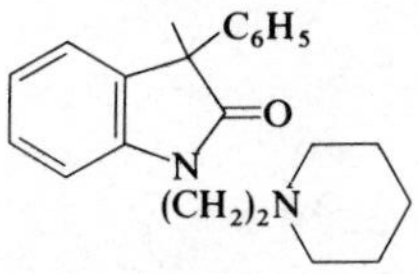 C_6H_5, O, N, $(CH_2)_2N$	SMEAH, C_6H_6, reflux, 2 h	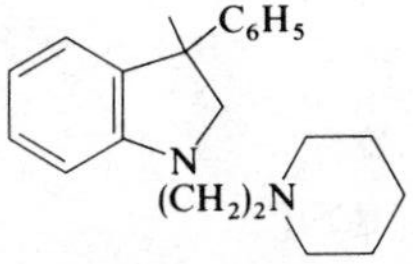 C_6H_5, N, $(CH_2)_2N$ (75)	359

	Reactant	Conditions	Product (Yield %)		Ref.
	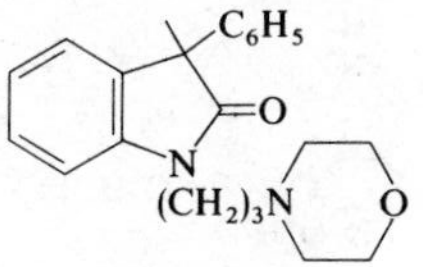 C_6H_5; O; N; $(CH_2)_3N$; O	", ", ", "	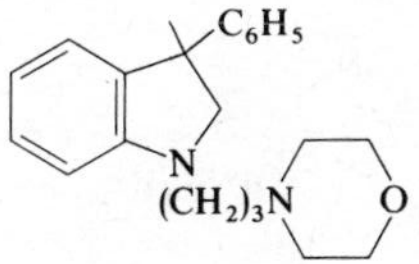 C_6H_5; N; $(CH_2)_3N$; O	(82)	359
	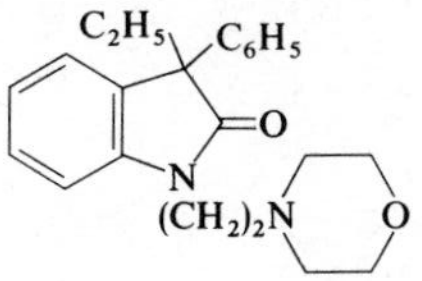 C_2H_5; C_6H_5; O; N; $(CH_2)_2N$; O	", ", ", "	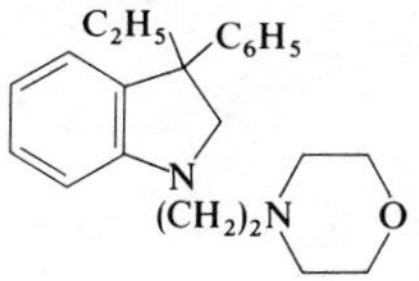 C_2H_5; C_6H_5; N; $(CH_2)_2N$; O	(81)	359
	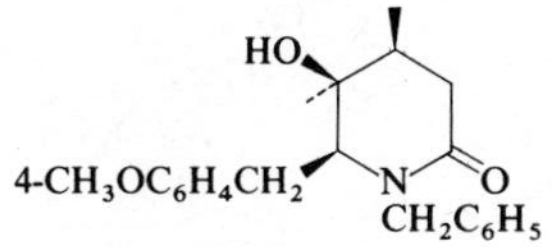 HO; 4-$CH_3OC_6H_4CH_2$; N; O; $CH_2C_6H_5$	SMEAH, xylene, reflux, 4 h	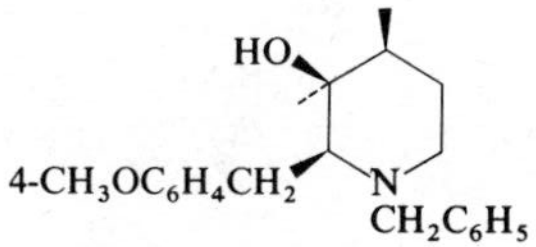 HO; 4-$CH_3OC_6H_4CH_2$; N; $CH_2C_6H_5$	(50)	984
	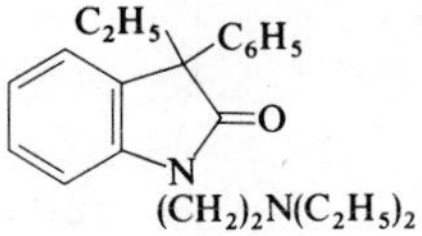 C_2H_5; C_6H_5; O; N; $(CH_2)_2N(C_2H_5)_2$	SMEAH, C_6H_6, reflux, 2 h	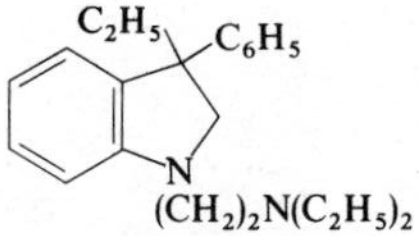 C_2H_5; C_6H_5; N; $(CH_2)_2N(C_2H_5)_2$	(81)	359
	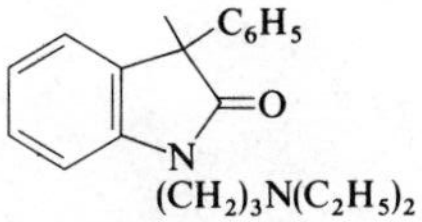 C_6H_5; O; N; $(CH_2)_3N(C_2H_5)_2$	", ", ", "	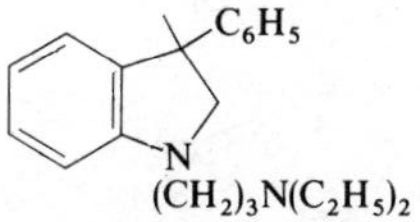 C_6H_5; N; $(CH_2)_3N(C_2H_5)_2$	(70)	359
C_{23}	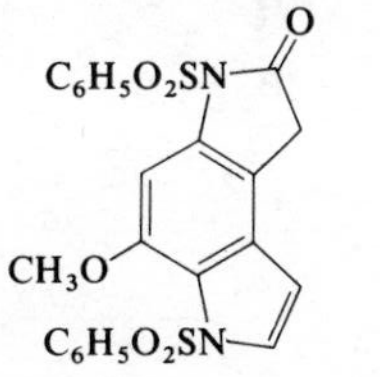 O; $C_6H_5O_2SN$; CH_3O; $C_6H_5O_2SN$	1. CH_3ONa 2. SMEAH	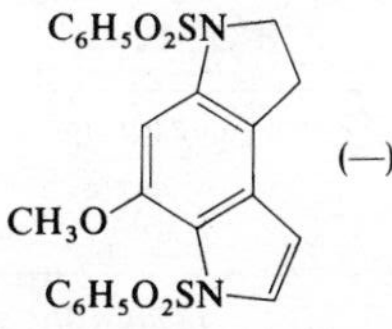 $C_6H_5O_2SN$; CH_3O; $C_6H_5O_2SN$	(—)	985a

TABLE XVII. REDUCTION OF LACTAMS (*Continued*)

No. of Carbon Atoms	Reactant	Conditions	Product(s) and Yield(s) (%)	Refs.
C_{23} (*Contd.*)	C_2H_5, C_6H_5, O, N, $(CH_2)_2N$	SMEAH, C_6H_6, reflux, 2 h	C_2H_5, C_6H_5, N, $(CH_2)_2N$ (76)	359
C_{24}	OCH_3, CH_3O, N, O, CH_3O, OCH_3	SMEAH, dioxane, reflux, 2 h	OCH_3, CH_3O, N, CH_3O, OCH_3 (87)	985b
	CH_3O, OCH_3, CH_3O, N, O, OCH_3, $CH_2C_6H_5$	SMEAH, C_6H_6, room temp, 15 h	CH_3O, OCH_3, CH_3O, N, OCH_3, $CH_2C_6H_5$ (65)	986, 987
	$C_2H_5O_2C$, 3,4-$(CH_3O)_2C_6H_3$, $CH_2C_6H_3(OCH_3)_2$-3,4, N, H, O	SMEAH, C_6H_6, 5°	$HOCH_2$, 3,4-$(CH_3O)_2C_6H_3$, $CH_2C_6H_3(OCH_3)_2$-3,4, N, H (—)	702

	Substrate	Conditions	Product (% yield)	Refs.
	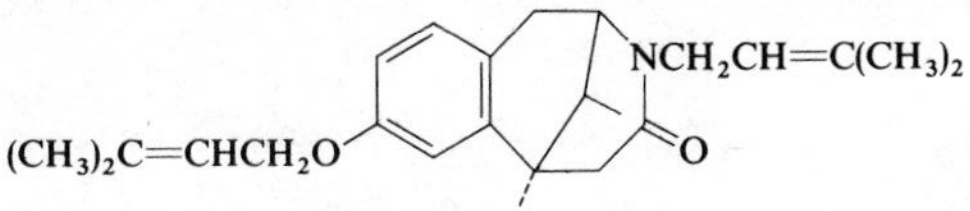 $(CH_3)_2C{=}CHCH_2O$, $NCH_2CH{=}C(CH_3)_2$, O	SMEAH, xylene, reflux, 42 h	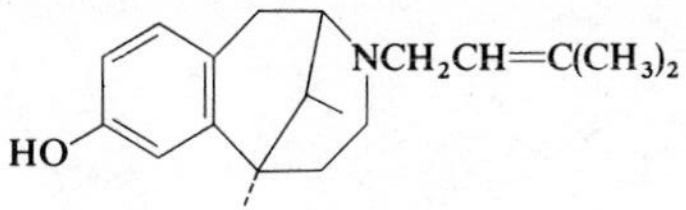 HO, $NCH_2CH{=}C(CH_3)_2$ (62)	988
C_{25}	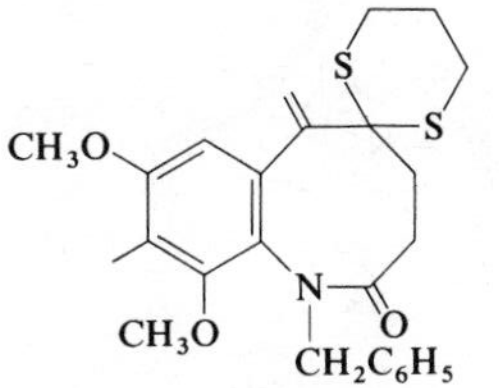 CH_3O, CH_3O, S, S, N, O, $CH_2C_6H_5$	SMEAH, C_6H_6, room temp, 6 h	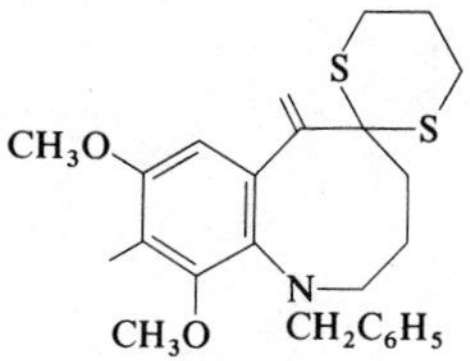 CH_3O, CH_3O, S, S, N, $CH_2C_6H_5$ (67)	986, 987
C_{26}	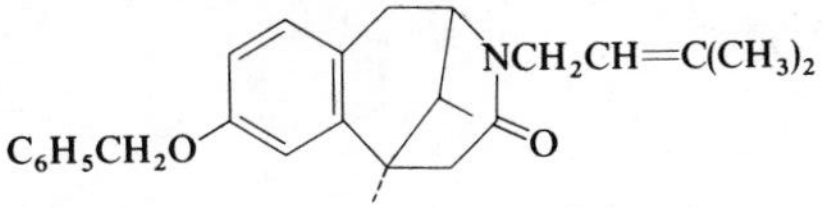 $C_6H_5CH_2O$, $NCH_2CH{=}C(CH_3)_2$, O	SMEAH, xylene, reflux, 60 h	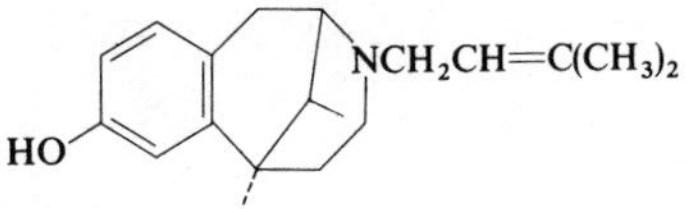 HO, $NCH_2CH{=}C(CH_3)_2$ (55)	988
C_{27}	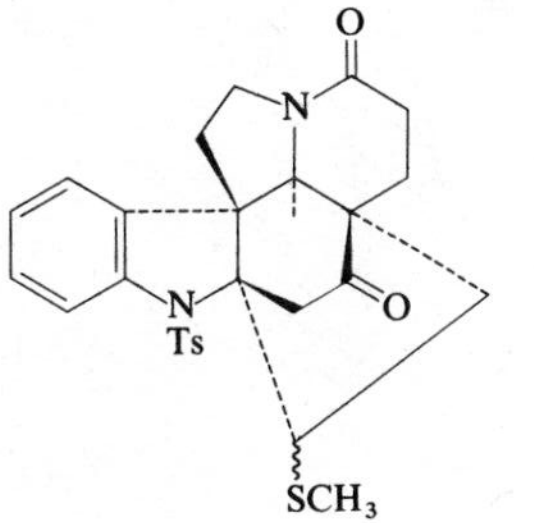 O, N, N, Ts, O, SCH_3	SMEAH, dioxane, reflux, 2.5 h	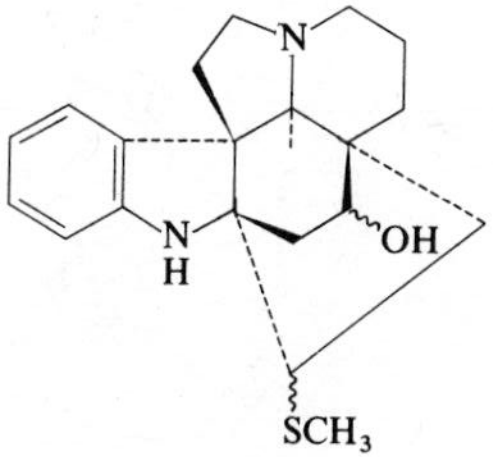 N, N, H, OH, SCH_3 (—)	989

TABLE XVII. REDUCTION OF LACTAMS (*Continued*)

No. of Carbon Atoms	Reactant	Conditions	Product(s) and Yield(s) (%)	Refs.
C_{28}	CH_3O, CH_3O, NH, O, N, O, $(CH_2)_2C_6H_3(OCH_3)_2$-3,4	LTBA (IA), THF, 5 h	CH_3O, CH_3O, NH, OH, N, O, $(CH_2)_2C_6H_3(OCH_3)_2$-3,4 (47)	990
C_{29}	CH_3O, CH_3O, N, O, H, H, N, HN	SMEAH, Py–toluene, room temp, 1 h	CH_3O, CH_3O, N, H, H, N, HN (41)	971
	CH_3O, CH_3O, N, O, H, H, N, N, OCH_3, OCH_3	SMEAH, C_6H_6–toluene, room temp, 1.5 h	CH_3O, CH_3O, N, H, H, N, N, OCH_3, OCH_3 (~100)	991, 992

Formula	Reactant	Reagent	Product(s) (Yield, %)	Refs.
C_{35}	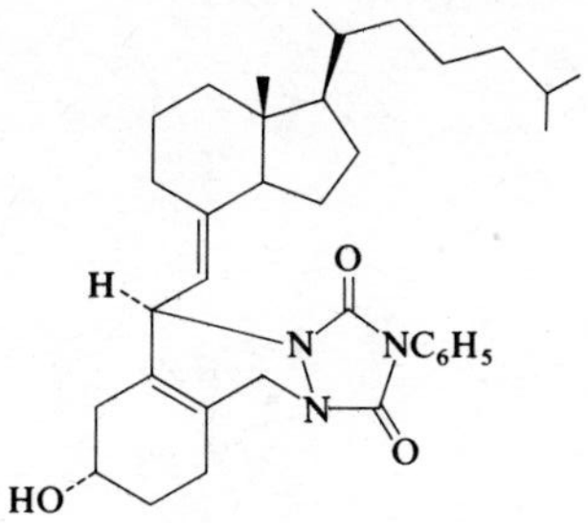	SMEAH	No reaction	993
C_{46}	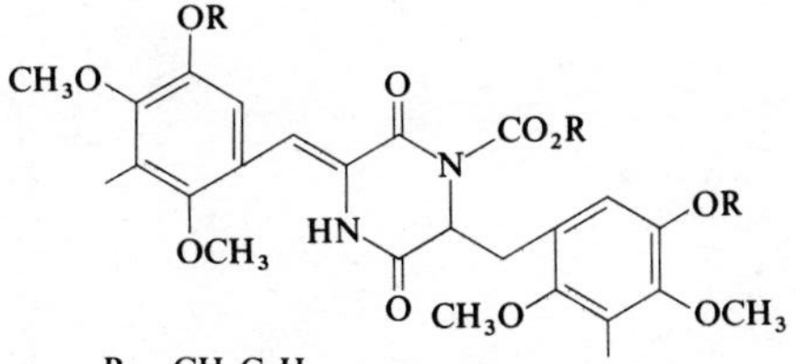 $R = CH_2C_6H_5$	1. LTBA, THF, room temp 2. HCO_2H, 60°	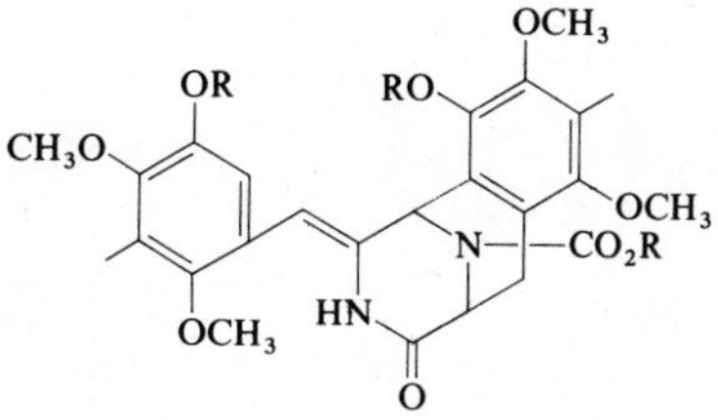 (—)	994

[a] The product was isolated as the maleate.

[b] The structure of the product given in the original paper[974] (a 7,8,13,13a-tetrahydro derivative) is a misprint.[995]

TABLE XVIII. REDUCTION OF CYCLIC IMIDES

Reactant	Conditions	Product(s) and Yield(s) (%)	Refs.
C_5			
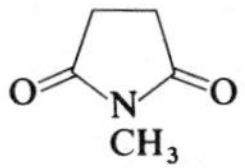 N-methylsuccinimide (CH_3 on N)	SMEAH, toluene, 100°, 20 min	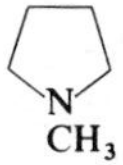 N-methylpyrrolidine (92)	94, 167
C_7			
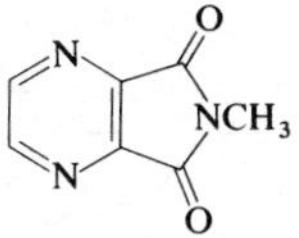 pyrazine-2,3-dicarboximide, NCH_3	SMEAH, THF–C_6H_6, 0°, room temp, 2 h	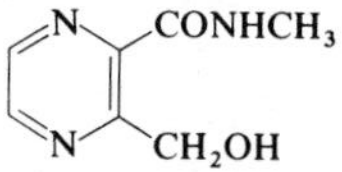 pyrazine with $CONHCH_3$ and CH_2OH (25)	996
C_8			
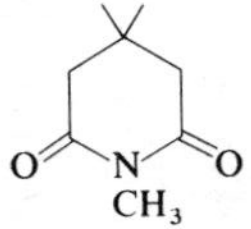 4,4-dimethyl-N-methylglutarimide (CH_3 on N)	SMEAH, C_6H_6, reflux, 6 h	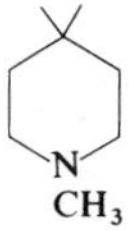 4,4-dimethyl-N-methylpiperidine (85)	378
C_9			
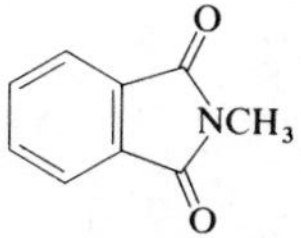 N-methylphthalimide (NCH_3)	SMEAH, toluene, reflux	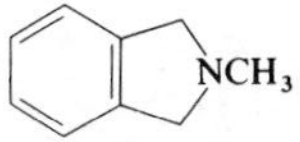 N-methylisoindoline, NCH_3 (76)	68
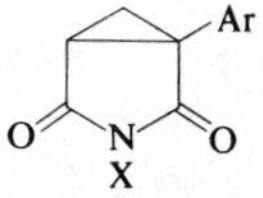 bicyclic imide with Ar, N–X (I)	1. SMEAH. C_6H_6, room temp, 30 min; reflux, 2 h 2. HCl	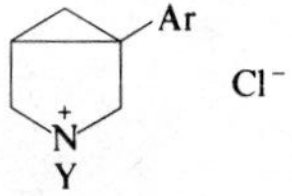 bicyclic ammonium salt with Ar, $\overset{+}{N}$–Y, Cl^- (II)	

Substrate	Conditions	Product(s) (Yield, %)	Refs.
C_{11}			
I, X = H, Ar = C_6H_4Cl-2		**II**, Y = H_2 (50)	997
I, X = H, Ar = C_6H_4Cl-3		**II**, Y = H_2 (70)	997
I, X = H, Ar = C_6H_4Cl-4		**II**, Y = H_2 (65)	997, 998
(+)-**I**, X = H, Ar = C_6H_4Cl-4		(+)-**II**, Y = H_2 (46)	997
(−)-**I**, X = H, Ar = C_6H_4Cl-4		(−)-**II**, Y = H_2 (37)	997
I, X = H, Ar = C_6H_4F-3		**II**, Y = H_2, (33)	997
I, X = H, Ar = C_6H_4F-4		**II**, Y = H_2 (81)	997
I, X = H, Ar = C_6H_5		**II**, Y = H_2 (34)	997, 999
(+)-**I**, X = H, Ar = C_6H_5		(+)-**II**, Y = H_2 (60)	997
(−)-**I**, X = H, Ar = C_6H_5		(−)-**II**, Y = H_2 (67)	997
C_{12}			
I, X = H, Ar = $C_6H_4CF_3$-3		**II**, Y = H_2 (47)	997
I, X = H, Ar = $C_6H_4CF_3$-4		**II**, Y = H_2 (56)	997
I, X = CH_3, Ar = C_6H_4Cl-4		**II**, Y = CH_3 (55)	997
I, X = H, Ar = $C_6H_4CH_3$-2		**II**, Y = H_2 (42)	997
I, X = H, Ar = $C_6H_4CH_3$-3		**II**, Y = H_2 (46)	997
I, X = H, Ar = $C_6H_4CH_3$-4		**II**, Y = H_2 (58)	997
(+)-**I**, X = H, Ar = $C_6H_4CH_3$-4		(+)-**II**, Y = H_2 (73)	997
(−)-**I**, X = H, Ar = $C_6H_4CH_3$-4		(−)-**II**, Y = H_2 (37)	997
I, X = H, Ar = $C_6H_4OCH_3$-3		**II**, Y = H_2 (24)	997
I, X = H, Ar = $C_6H_4OCH_3$-4		**II**, Y = H_2 (65)	997
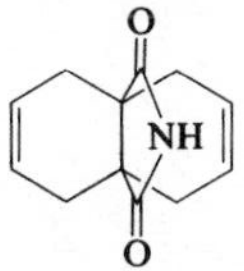 [structure: O, NH, O]	SMEAH, C_6H_6, reflux, 48 h	 [structure: NH] (80)	1000
C_{13}			
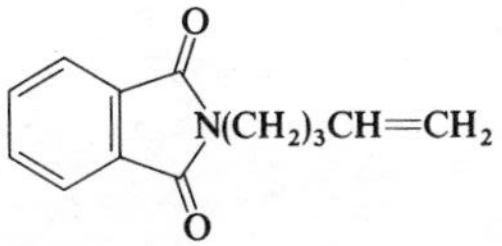 [structure: O, O] $N(CH_2)_3CH{=}CH_2$	SMEAH (IA), C_6H_6, room temp, 1 h	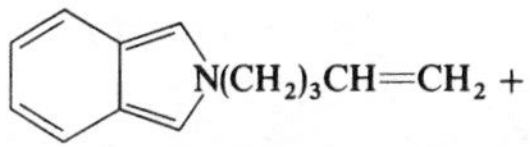 [structure] $N(CH_2)_3CH{=}CH_2$ (38) 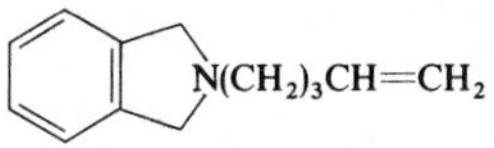 + [structure] $N(CH_2)_3CH{=}CH_2$ (37) 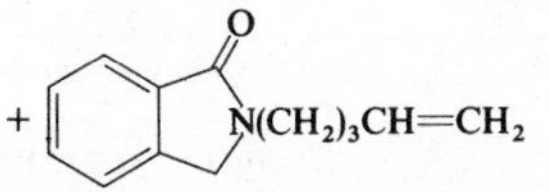 + [structure: O] $N(CH_2)_3CH{=}CH_2$ (15)	1001

TABLE XVIII. Reduction of Cyclic Imides (*Continued*)

Reactant	Conditions	Product(s) and Yield(s) (%)	Refs.
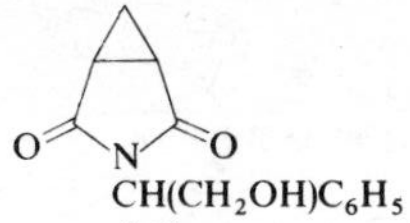 [structure] $CH(CH_2OH)C_6H_5$ -(+)	SMEAH, THF, −78°, 8 h	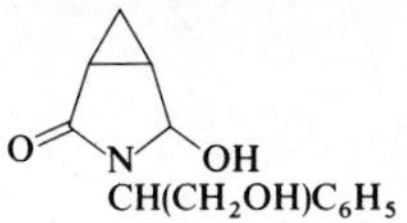 [structure] OH, $CH(CH_2OH)C_6H_5$ (85)	380, 381
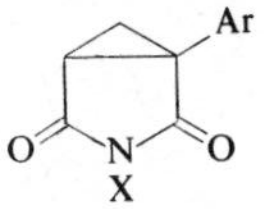 [structure] **I**	1. SMEAH, C_6H_6, room temp, 30 min; reflux, 2 h 2. HCl	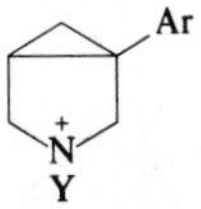 [structure] Cl^- **II**	
I, X = CH_3, Ar = $C_6H_4OCH_3$-3		II, Y = CH_3 (47)	997
I, X = H, Ar = $C_6H_4C_2H_5$-4		II, Y = H_2 (56)	997
C_{14}			
I, X = H, Ar = $C_6H_4(C_3H_7\text{-}i)$-4		II, Y = H_2 (71)	997
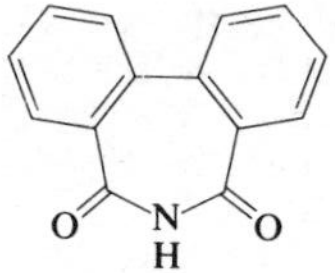 [structure]	1. LAH–$CH_3CO_2C_2H_5$ (excess) 2. Aqueous $HClO_4$	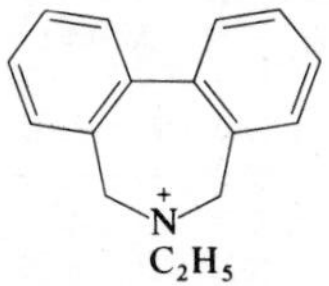 [structure] N^+–C_2H_5 ClO_4^- (73)	1002
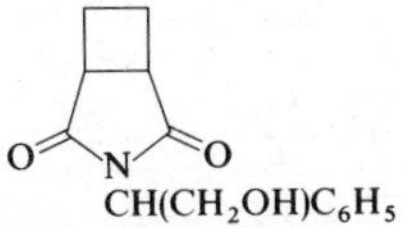 [structure] $CH(CH_2OH)C_6H_5$ (−)-	SMEAH, THF, −78°, 6 h	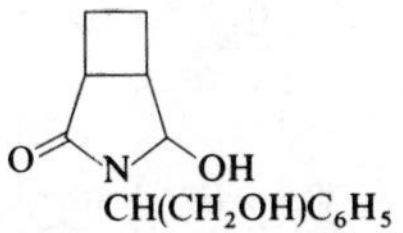 [structure] OH, $CH(CH_2OH)C_6H_5$ (89)	380, 381
C_{15}			
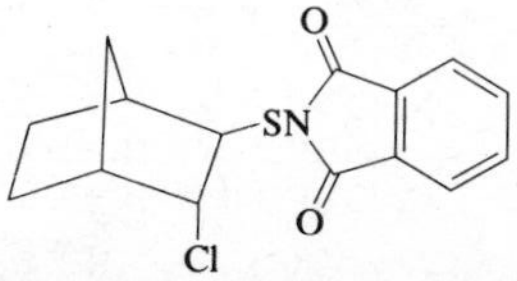 [structure] SN, Cl	$LiAlH_3(OC_2H_5)$, THF, −78°, 1.5 h	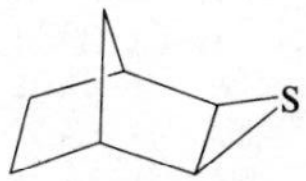 [structure] S (58)	495

Substrate	Conditions	Product (Yield %)		Refs.
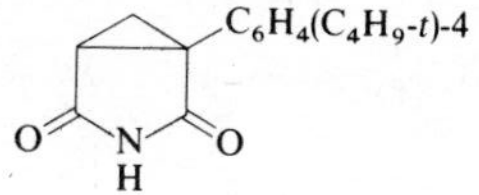 $C_6H_4(C_4H_9\text{-}t)\text{-}4$ (bicyclic imide, O, N–H, O)	1. SMEAH, C_6H_6, room temp, 30 min; reflux, 2 h 2. HCl	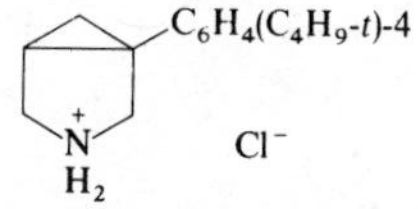 $C_6H_4(C_4H_9\text{-}t)\text{-}4$ ($\overset{+}{N}H_2$, Cl^-)	(45)	997
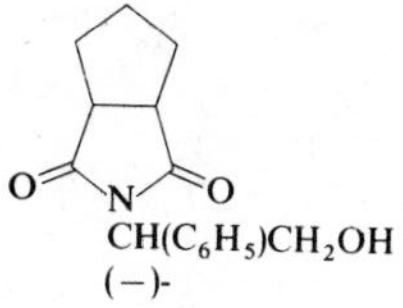 $CH(C_6H_5)CH_2OH$ (−)- (cyclopenta-fused imide)	SMEAH, THF, −78°, 4 h	OH, $CH(C_6H_5)CH_2OH$	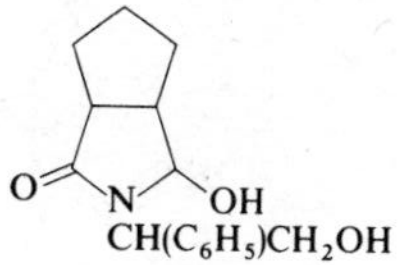 (94)	380, 381
C_{16}				
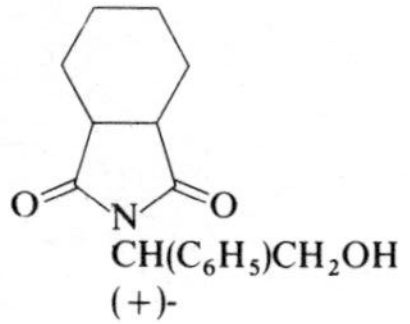 $CH(C_6H_5)CH_2OH$ (+)- (cyclohexa-fused imide)	SMEAH, THF, −78°, 2 h	OH, $CH(C_6H_5)CH_2OH$	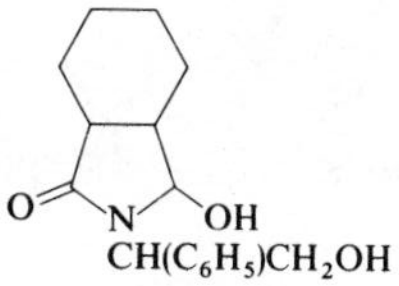 (95)	380, 381
C_{17}				
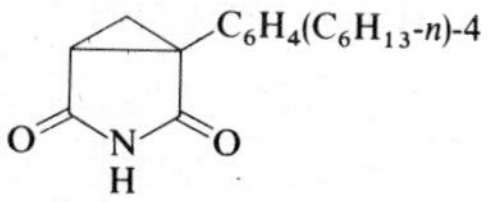 $C_6H_4(C_6H_{13}\text{-}n)\text{-}4$ (bicyclic imide, O, N–H, O)	1. SMEAH, C_6H_6, room temp, 30 min; reflux, 2 h 2. HCl	$C_6H_4(C_6H_{13}\text{-}n)\text{-}4$ ($\overset{+}{N}H_2$, Cl^-)	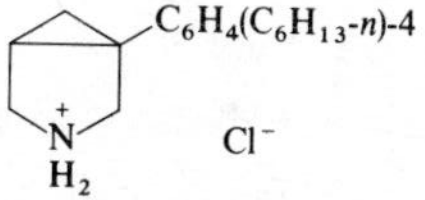 (56)	997
C_n				
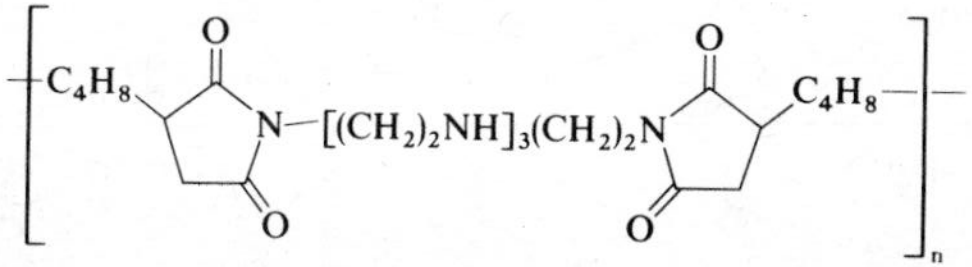 $[-C_4H_8\text{-succinimide-}N\text{—}[(CH_2)_2NH]_3(CH_2)_2N\text{-succinimide-}C_4H_8\text{—}]_n$	SMEAH (~7.5 eq), THF, 20–70°, 4 h	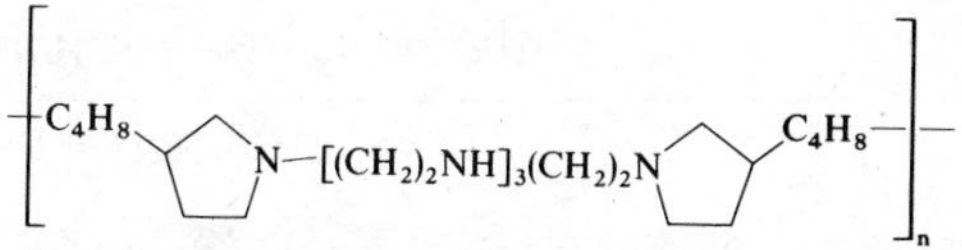 $[-C_4H_8\text{-pyrrolidine-}N\text{—}[(CH_2)_2NH]_3(CH_2)_2N\text{-pyrrolidine-}C_4H_8\text{—}]_n$	(—)	1003
	SMEAH (IA)(~3 eq), THF, 20–70°, 4 h	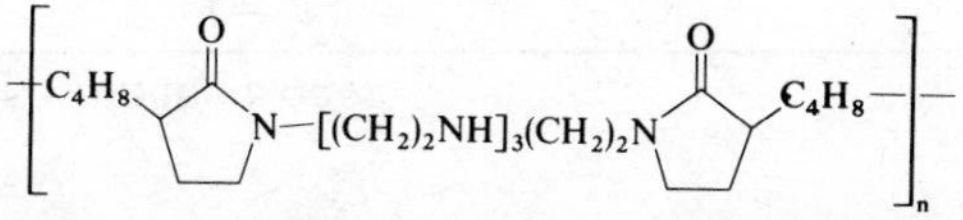 $[-C_4H_8\text{-pyrrolidinone-}N\text{—}[(CH_2)_2NH]_3(CH_2)_2N\text{-pyrrolidinone-}C_4H_8\text{—}]_n$	(—)	1003

TABLE XIX. REDUCTION OF IMINES AND IMINIUM SALTS

No. of Carbon Atoms	Reactant	Conditions	Product(s) and Yield(s) (%)	Refs.
C_8	$C_6H_5C(CH_3){=}NH$	LAH + (−)-MTH (1:1), ether, reflux, 4 h	$C_6H_5CHCH_3NH_2$ (43) (2% e.e.) (*R*)-(+)	388
		LAH + (−)-QN, ether, reflux, 4 h	" (45) (3% e.e.) (*R*)-(+)	388
		LAH + (+)-BN (1:1), ether, reflux, 4 h	" (35) (2% e.e.) (*S*)-(−)	388
C_9	$C_6H_5C(C_2H_5){=}NH$	LAH + (−)-MTH (1:1), ether, reflux, 4 h	$C_6H_5CH(C_2H_5)NH_2$ (33) (10% e.e.) (*R*)-(+)	388
		LAH + (+)-BN (1:1), ether, reflux, 4 h	" (44) (9% e.e.) (*S*)-(−)	388
C_{10}	$C_6H_5C(C_3H_7\text{-}n){=}NH$	LAH + (−)-MTH (1:1), ether, reflux, 4 h	$C_6H_5CH(C_3H_7\text{-}n)NH_2$ (38) (*R*)-(+)	388
		LAH + (+)-BN (1:1), ether, reflux, 4 h	" (37)	388
C_{11}	H, NCH_3	LTBA, −78° to room temp	H, $NHCH_3$ (−)	1004
	N^+, ClO_4^-	SMEAH, THF, reflux, 20 h	I + II (82) (I:II = 70:30)	134

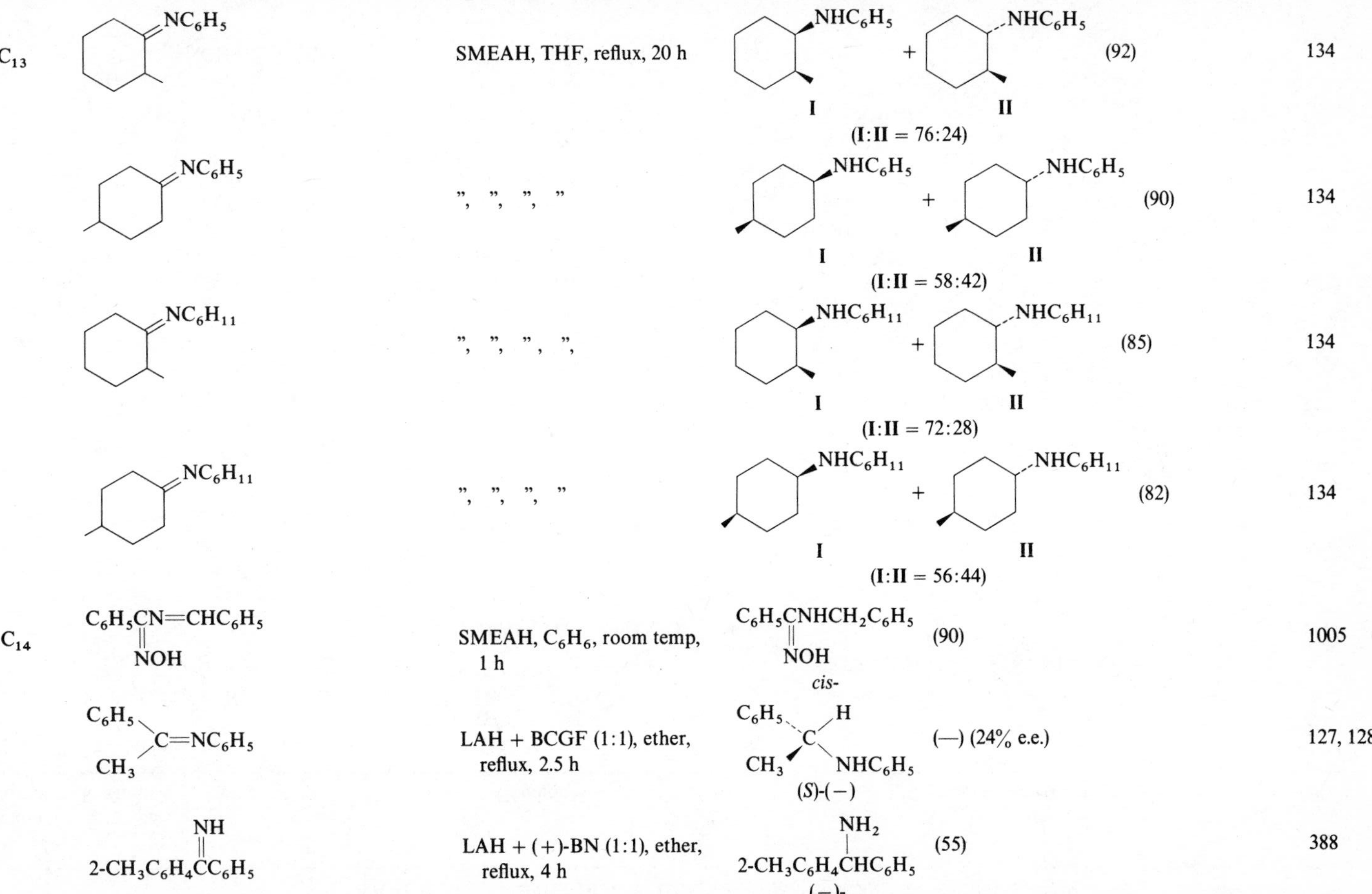

	Substrate	Reagent and Conditions	Product(s) and Yield(s) (%)	Refs.
C_{13}	2-methylcyclohexylidene-NC_6H_5	SMEAH, THF, reflux, 20 h	I (NHC_6H_5) + II (NHC_6H_5) (92) (I:II = 76:24)	134
	4-methylcyclohexylidene-NC_6H_5	", ", ", "	I (NHC_6H_5) + II (NHC_6H_5) (90) (I:II = 58:42)	134
	2-methylcyclohexylidene-NC_6H_{11}	", ", ", "	I (NHC_6H_{11}) + II (NHC_6H_{11}) (85) (I:II = 72:28)	134
	4-methylcyclohexylidene-NC_6H_{11}	", ", ", "	I (NHC_6H_{11}) + II (NHC_6H_{11}) (82) (I:II = 56:44)	134
C_{14}	$C_6H_5C(=NOH)N{=}CHC_6H_5$	SMEAH, C_6H_6, room temp, 1 h	$C_6H_5C(=NOH)NHCH_2C_6H_5$ *cis*- (90)	1005
	$(C_6H_5)(CH_3)C{=}NC_6H_5$	LAH + BCGF (1:1), ether, reflux, 2.5 h	$C_6H_5(CH_3)CH(NHC_6H_5)$ (*S*)-(−) (—) (24% e.e.)	127, 128
	$2\text{-}CH_3C_6H_4C(=NH)C_6H_5$	LAH + (+)-BN (1:1), ether, reflux, 4 h	$2\text{-}CH_3C_6H_4CH(NH_2)C_6H_5$ (−)- (55)	388

TABLE XIX. REDUCTION OF IMINES AND IMINIUM SALTS (*Continued*)

No. of Carbon Atoms	Reactant	Conditions	Products(s) and Yield(s) (%)	Refs.
C_{14} (*Contd.*)	$2\text{-}CH_3C_6H_4C(=\overset{+}{N}H_2)C_6H_5$ Cl^-	LAH + (−)-MTH (1:1), ether, reflux, 2 h	" (59) (+)-	388
	2-Methylcyclohexylidene-$NCH_2C_6H_5$	SMEAH, THF, reflux, 20 h	cis-2-Methyl-$NHCH_2C_6H_5$-cyclohexane (I) + trans-2-Methyl-$NHCH_2C_6H_5$-cyclohexane (II) (81) (I:II = 78:22)	134
	1-(4-t-C_4H_9-cyclohexylidene)pyrrolidinium ClO_4^-	", ", ", "	1-(4-t-C_4H_9-cyclohexyl)pyrrolidine I + II (92) (I:II = 70:30)	134
C_{15}	$(C_6H_5)(C_2H_5)C{=}NC_6H_5$	LAH + BCGF (1:1), ether, reflux, 2.5 h	$C_6H_5(C_2H_5)CH{-}NHC_6H_5$ (S)-(−) (—)	127, 128
C_{16}	3-C_6H_5-isoxazol-5-yl-$N{=}CHC_6H_5$	SMEAH, C_6H_6, 0°, 15 min	3-C_6H_5-isoxazol-5-yl-$NHCH_2C_6H_5$ (63)	1006
	$(C_6H_5)(CF_3)C{=}NCH(CH_3)C_6H_5$	LTBA, THF, 25°, 72 h	No reaction	386

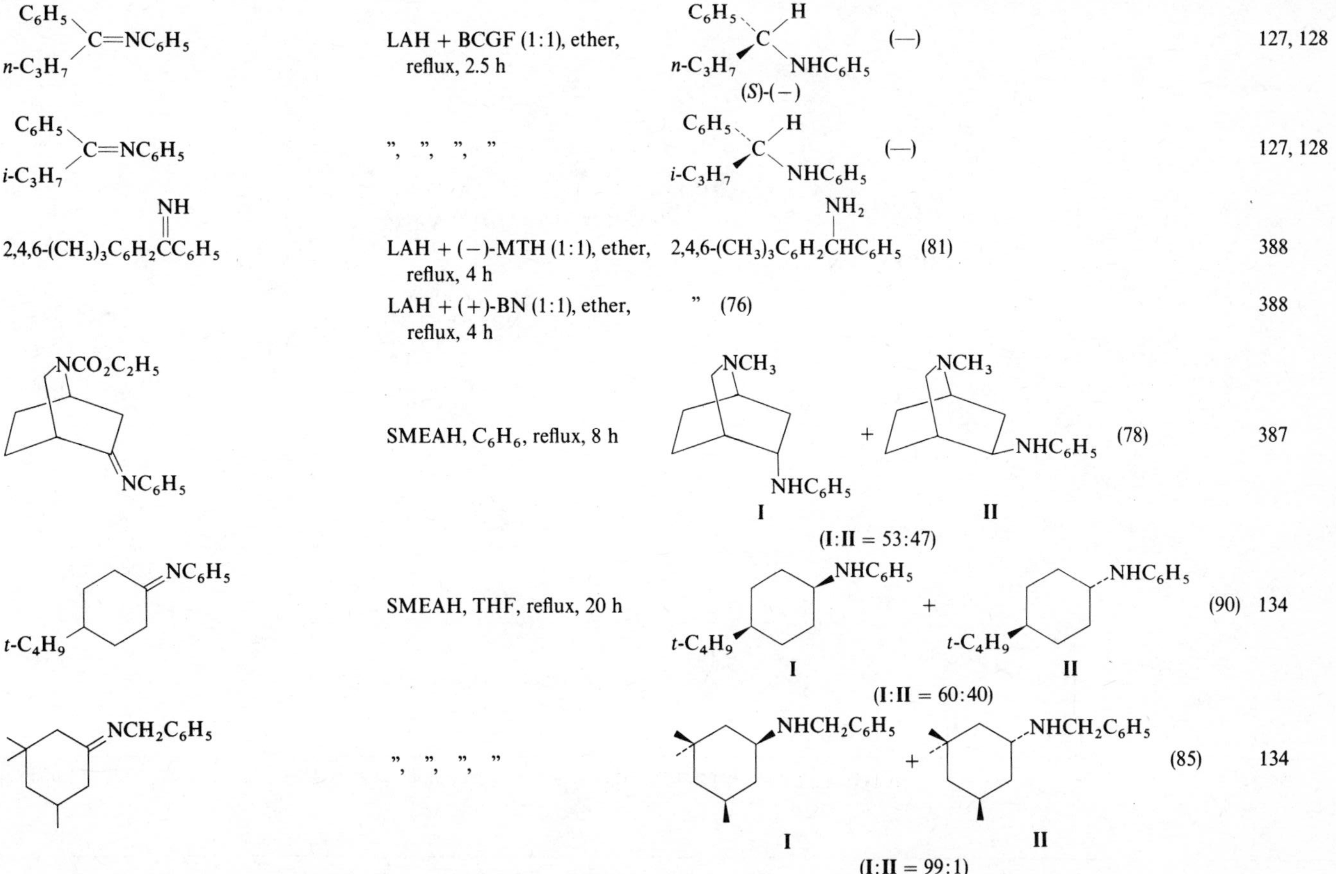

Substrate	Reagent and Conditions	Product(s) and Yield(s) (%)	Refs.
$C_6H_5(n\text{-}C_3H_7)C{=}NC_6H_5$	LAH + BCGF (1:1), ether, reflux, 2.5 h	$C_6H_5CH(n\text{-}C_3H_7)NHC_6H_5$ (S)-(−) (—)	127, 128
$C_6H_5(i\text{-}C_3H_7)C{=}NC_6H_5$	", ", ", "	$C_6H_5CH(i\text{-}C_3H_7)NHC_6H_5$ (—)	127, 128
2,4,6-$(CH_3)_3C_6H_2C({=}NH)C_6H_5$	LAH + (−)-MTH (1:1), ether, reflux, 4 h	2,4,6-$(CH_3)_3C_6H_2CH(NH_2)C_6H_5$ (81)	388
	LAH + (+)-BN (1:1), ether, reflux, 4 h	" (76)	388
$NCO_2C_2H_5$, NC_6H_5 (bicyclic)	SMEAH, C_6H_6, reflux, 8 h	NCH_3, NHC_6H_5 **I** + NCH_3, NHC_6H_5 **II** (78) (**I**:**II** = 53:47)	387
$t\text{-}C_4H_9$, NC_6H_5 (cyclohexylidene)	SMEAH, THF, reflux, 20 h	$t\text{-}C_4H_9$, NHC_6H_5 **I** + $t\text{-}C_4H_9$, NHC_6H_5 **II** (90) (**I**:**II** = 60:40)	134
$NCH_2C_6H_5$ (trimethylcyclohexylidene)	", ", ", "	$NHCH_2C_6H_5$ **I** + $NHCH_2C_6H_5$ **II** (85) (**I**:**II** = 99:1)	134

TABLE XIX. REDUCTION OF IMINES AND IMINIUM SALTS (*Continued*)

No. of Carbon Atoms	Reactant	Conditions	Product(s) and Yield(s) (%)	Refs.
C_{16} (*Contd.*)	4-t-C_4H_9-cyclohexylidene=NC_6H_{11}	SMEAH, THF, reflux, 20 h	t-C_4H_9-cyclohexyl-NHC_6H_{11} **I** + t-C_4H_9-cyclohexyl-NHC_6H_{11} **II** (89) (**I**:**II** = 58:42)	134
C_{17}	1-naphthyl-$C_6H_5C{=}NH$	LAH + (−)-MTH (1:1), ether, reflux, 4 h	1-naphthyl-$C_6H_5CHNH_2$ (+)- (32)	388
	1-naphthyl-$C_6H_5C{=}\overset{+}{N}H_2$ Cl^-	LAH + (+)-BN (1:1), ether, reflux, 2 h	" (−)- (16)	388
	4-t-C_4H_9-cyclohexylidene=$NCH_2C_6H_5$	SMEAH, THF, reflux, 20 h	t-C_4H_9-cyclohexyl-$NHCH_2C_6H_5$ **I** + t-C_4H_9-cyclohexyl-$NHCH_2C_6H_5$ **II** (94) (**I**:**II** = 64:36)	134

C_{18}	$C_6H_5C(C_3F_7){=}NCH(CH_3)C_6H_5$	SMEAH, THF, −78°, 72 h	$C_6H_5CH(C_3F_7)NHCH(CH_3)C_6H_5$ (80)	386
	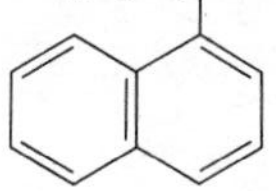 $2\text{-}CH_3C_6H_4\overset{+}{C}{=}NH_2$ (1-naphthyl) Cl^-	LAH + (−)-MTH (1:1), ether, reflux, 2 h	$2\text{-}CH_3C_6H_4CHNH_2$ 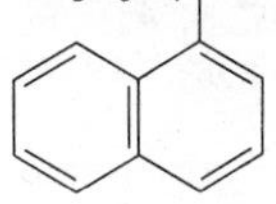 (1-naphthyl) (−)- (51)	388
	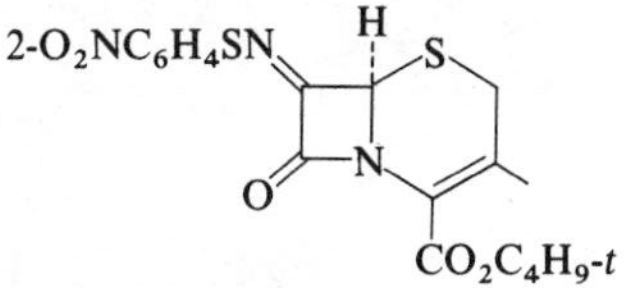 $2\text{-}O_2NC_6H_4SN{=}$ (cephem; H, S, N, O, $CO_2C_4H_9\text{-}t$)	LTBA, THF, −78°, 10 min; then 0°, 2 h	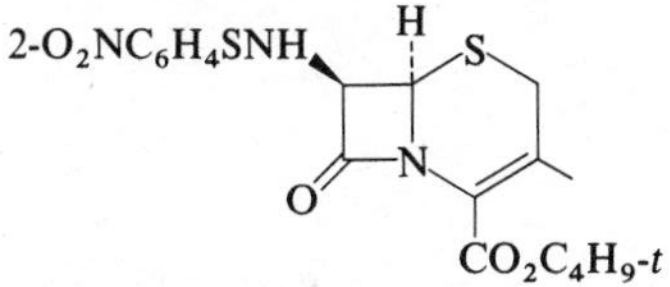 $2\text{-}O_2NC_6H_4SNH$ (cephem; H, S, N, O, $CO_2C_4H_9\text{-}t$) (50) 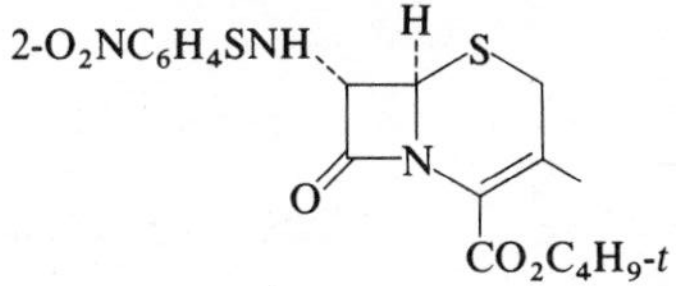 + $2\text{-}O_2NC_6H_4SNH$ (cephem; H, S, N, O, $CO_2C_4H_9\text{-}t$) (27)	1007
C_{19}	$(C_6H_5)_2C{=}NC_6H_5$	SMEAH, *p*-cymene, reflux, 5 h	$(C_6H_5)_2C(CH_3)_2 + C_6H_5NH_2$ (—)	1008–1010
C_{22}	*dl*-$C_6H_5COC(C_6H_5){=}NCH(CH_3)C_6H_5$	LTBA, THF	$C_6H_5COCH(C_6H_5)NHCH(CH_3)C_6H_5$ (81) [(2*R*,3*R*)/(2*S*,3*S*)]:[(2*R*,3*S*)/(2*S*,3*R*)] = 50:50	135
	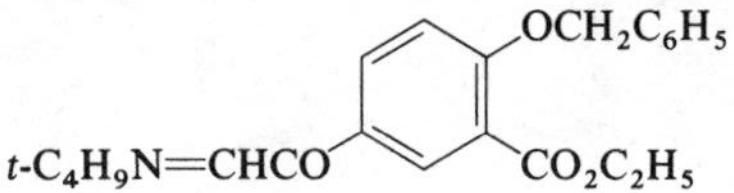 $t\text{-}C_4H_9N{=}CHCO$–C_6H_3($OCH_2C_6H_5$)($CO_2C_2H_5$)	SMEAH, C_6H_6, reflux, 1 h	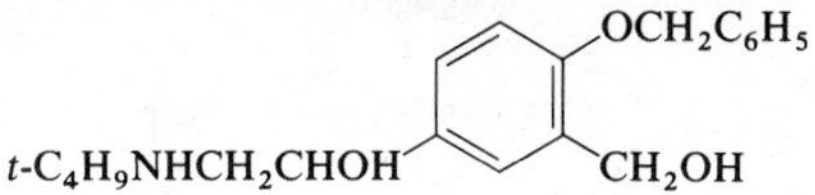 $t\text{-}C_4H_9NHCH_2CHOH$–$C_6H_3$($OCH_2C_6H_5$)($CH_2OH$) (—)	697

TABLE XX. Reduction of Oximes

No. of Carbon Atoms	Reactant	Conditions	Product(s) and Yield(s) (%)	Refs.
C_4	$C_2H_5C(CH_3)$=NOH	LAH + BCGF (1:1), ether, reflux, 2.5 h	$C_2H_5CH(CH_3)NH_2$ (*S*)-(+) (~70) (15% e.e.)	132
		LAH + BCGH + C_2H_5OH (1:1:1), ether, reflux, 2.5 h	" (*R*)-(−) (~55) (14% e.e.)	132
		LAH + (−)-MTH (1:1), ether, reflux, 30h	" (*S*)-(+) (7) (2% e.e.)	1011
C_5	*n*-$C_3H_7C(CH_3)$=NOH	LAH + BCGF (1:1), ether, reflux, 2.5 h	*n*-$C_3H_7CH(CH_3)NH_2$ (*S*)-(+) (~70)	132
	i-$C_3H_7C(CH_3)$=NOH	LAH + BCGF (1:1), ether, reflux, 2.5 h	*i*-$C_3H_7CH(CH_3)NH_2$ (*S*)-(+) (~70)	132
		LAH + (−)-MTH (1:1), ether, reflux, 33 h	" (*S*)-(+) (8) (5% e.e.)	1011
	$C_2H_5C(CH_3)$=NOCH$_3$	LAH + BCGF (1:1), ether, reflux, 2.5 h	$C_2H_5CH(CH_3)NH_2$ (*S*)-(+) (—) (18% e.e.)	132
C_6	(cyclohexane ring)=NOH	SMEAH, C_6H_6, reflux, 1 h	(cyclohexane ring)–NH_2 (89)	94, 167
		SMEAH, THF, 20°, 6 h	(cyclohexane ring fused)NH (25)	393
	n-$C_4H_9C(CH_3)$=NOH	LAH + BCGF (1:1), ether, reflux, 2.5 h	*n*-$C_4H_9CH(CH_3)NH_2$ (*S*) (~70) (22% e.e.)	132
	i-$C_4H_9C(CH_3)$=NOH	", ", ", "	*i*-$C_4H_9CH(CH_3)NH_2$ (*S*) (~70) (20% e.e.)	132
	t-$C_4H_9C(CH_3)$=NOH	LAH + BCGF (1:1), ether, reflux, 2.5 h	*t*-$C_4H_9CH(CH_3)NH_2$ (*S*)-(+) (~70)	132
		LAH + (−)-MTH (1:1), ether, reflux, 60 h	" (7)	1011

	Reactant	Conditions	Product(s) and Yield(s) (%)	Refs.
C_7	$C_6H_5CH{=}NOH$	$Ca[AlH_2(OC_4H_9\text{-}i)_2]_2 \cdot THF$, toluene, 90°, 3 h	$C_6H_5CH_2NH_2$ (43)	165
	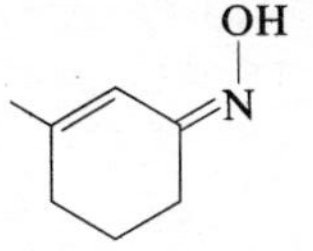 [structure] *syn-*	SMEAH	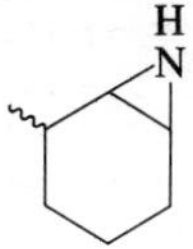 [structure] (65) 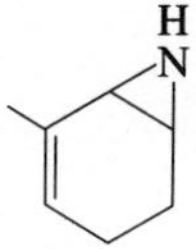 + [structure] (17) 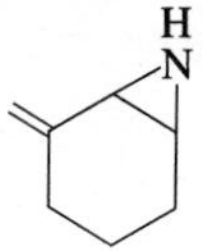 + [structure] (16)	154, 153
	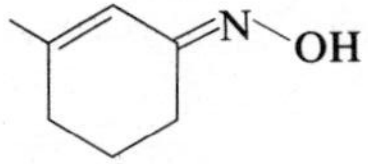 [structure] *anti-*	"	" (23) 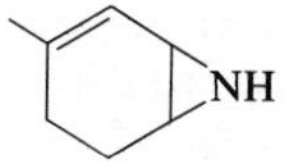 + [structure] NH (30)	154
	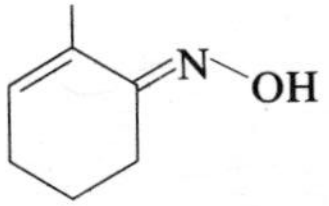 [structure] *anti-*	SMEAH, THF, reflux	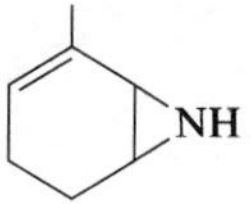 [structure] NH (26) 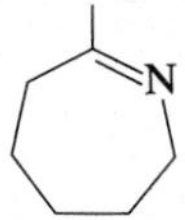 + [structure] (15) 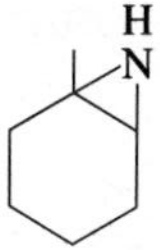 + [structure] (12) 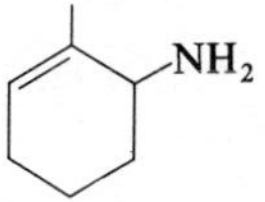 + [structure] NH_2 (12)	153, 154
	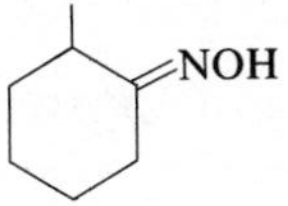 [structure] NOH	SMEAH, THF, 20°, 6 h	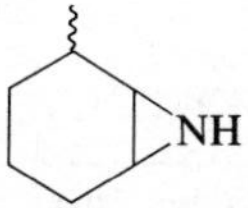 [structure] NH (17) (*cis*:*trans* = 46:54)	393
	$n\text{-}C_4H_9C(CH_3){=}NOCH_3$	LAH + BCGF (1:1), ether, reflux, 2.5 h	$n\text{-}C_4H_9CH(CH_3)NH_2$ (*S*)-(+) (—) (23% e.e.)	132
	$i\text{-}C_4H_9C(CH_3){=}NOCH_3$	", ", ", "	$i\text{-}C_4H_9CH(CH_3)NH_2$ (*S*)-(+) (—) (18% e.e.)	132

TABLE XX. REDUCTION OF OXIMES (*Continued*)

No. of Carbon Atoms	Reactant	Conditions	Product(s) and Yield(s) (%)	Refs.
C_8	4-$ClC_6H_4C(CH_3)$=NOH	SMEAH, THF, reflux, 2 h	4-$ClC_6H_4CH(CH_3)NH_2$ (**I**) + 4-ClC_6H_4-aziridine (N-H) (**II**) (—) (**I**:**II** = 55:45)	150
	$C_6H_5C(CH_3)$=NOH *syn,anti-*	LAH + BCGF (1:1), ether, reflux, 2.5 h	$C_6H_5CH(CH_3)NH_2$ (*S*)-(−) (~70) (11% e.e.)	132
		LAH + BCGF + C_2H_5OH (1:1:1), ether, reflux, 2.5 h	" (*R*)-(+) (~55) (18% e.e.)	132
		LAH + (−)-MTH (1:1), ether, reflux, 9 h	" (*S*)-(−) (13) (13% e.e)	1011
	anti-	SMEAH, THF, reflux, 2 h	C_6H_5-aziridine (N-H) (**I**) + $C_6H_5CH(CH_3)NH_2$ (**II**) + $C_6H_5NHC_2H_5$ (**III**) (—) (**I**:**II**:**III** = 52:43:5)	150
	n-$C_6H_{13}C(CH_3)$=NOH	LAH + BCGF (1:1), ether, reflux, 2.5 h	n-$C_6H_{13}CH(CH_3)NH_2$ (*S*)-(+) (~70) (24% e.e)	132
C_9	thiochroman-4-one oxime (=NOH)	SMEAH, toluene–ether, reflux, 5 h	2,3,4,5-tetrahydro-1,5-benzothiazepine (N-H) (~65)	392
	$C_6H_5C(CH_3)$=$NOCH_3$ *syn,anti-*	LAH + BCGF (1:1), ether, reflux, 2.5 h	$C_6H_5CH(CH_3)NH_2$ (**I**) (*S*)-(−) (—) (13% e.e.)	132

anti-	SMEAH, THF, reflux, 2 h	**I** + C_6H_5-aziridine (NH) (—) **II** (**I**:**II** = 66:32)	150
$C_6H_5CH(CH_3)CH{=}NOH$	SMEAH, C_6H_6, reflux, 1 h	$C_6H_5CH(CH_3)CH_2NH_2$ (60)	94
$C_6H_5CH_2C(CH_3){=}NOH$	LAH + BCGF (1:1), ether, reflux, 2.5 h	$C_6H_5CH_2CH(CH_3)NH_2$ **I** (*S*)-(+) (~70) (21% e.e.)	132
	LAH + BCGF + C_2H_5OH (1:1:1), ether, reflux, 2.5 h	**I** (*R*)-(−) (~55) (14% e.e.)	132
	SMEAH, THF, reflux, 2 h	**I** + cis-2-C_6H_5-3-methylaziridine (NH) **II** + trans-2-C_6H_5-3-methylaziridine (NH) **III** + 2-$C_6H_5CH_2$-aziridine (NH) (—) **IV** (**I**:**II**:**III**:**IV** = 18:55:24:3)	150
$C_6H_5C(C_2H_5){=}NOH$ *syn,anti*-	LAH + BCGF (1:1), ether, reflux, 2.5 h	$C_6H_5CH(C_2H_5)NH_2$ **I** (*S*)-(−) (~70) (14% e.e.)	132
anti-	SMEAH, THF, reflux, 2 h	**I** + cis-2-C_6H_5-3-methylaziridine (NH) **II** + $C_6H_5NHC_3H_7$-*n* **III** (—) (**I**:**II**:**III** = 70:20:10)	150

TABLE XX. REDUCTION OF OXIMES (*Continued*)

No. of Carbon Atoms	Reactant	Conditions	Product(s) and Yield(s) (%)	Refs.
C_9 (*Contd.*)	[structure] *syn-*	SMEAH	[structure] (52) + [structure] (23) + [structure] (14)	154
	$C_2H_5C(CH_3){=}NOTHP$	LAH + BCGF (1:1), ether, reflux, 2.5 h	$C_2H_5CH(CH_3)NH_2$ (*S*)-(+) (—) (21% e.e.)	132
	$C_6H_{11}C(CH_3){=}NOCH_3$	", ", ", "	$C_6H_{11}CH(CH_3)NH_2$ (*S*)-(+) (—) (44% e.e.)	132
	n-$C_6H_{13}C(CH_3){=}NOCH_3$	LAH + BCGF (1:1), ether, reflux, 2.5 h	n-$C_6H_{13}CH(CH_3)NH_2$ (*S*)-(+) (—) (23% e.e.)	132
C_{10}	$C_6H_5CH_2C(CH_3){=}NOCH_3$ *syn,anti-*	", ", ", "	$C_6H_5CH_2CH(CH_3)NH_2$ (*S*)-(+) (—) (18% e.e.) **I**	132
		SMEAH, THF, reflux, 2 h	**I** + [structure] **II** + [structure] **III** + [structure] (—) **IV** (**I**:**II**:**III**:**IV** = 21:56:20:2)	150
	anti-	SMEAH, THF, reflux, 2 h	(**I** + **II** + **IV** (—) (**I**:**II**:**IV** = 43:53:3)	150

Substrate	Conditions	Product(s) (Yield, %)	Refs.
$C_6H_5C(C_2H_5){=}NOCH_3$	LAH + BCGF (1:1), ether, reflux, 2.5 h	$C_6H_5CH(C_2H_5)NH_2$ (S)-(−) (—) (18% e.e.)	132
$C_6H_5C(C_3H_7\text{-}n){=}NOH$	SMEAH, THF, reflux, 2 h	$C_6H_5CH(C_3H_7\text{-}n)NH_2$ (I) + $C_6H_5NHC_4H_9\text{-}n$ (II) 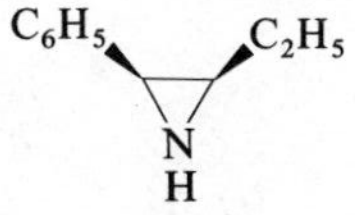 + cis-2-C_6H_5-3-C_2H_5-aziridine (III) (—) (I:II:III = 60:25:15)	150
$C_6H_5C(C_3H_7\text{-}i){=}NOH$	", ", ", "	$C_6H_5CH(C_3H_7\text{-}i)NH_2$ (I) + $C_6H_5NHC_4H_9\text{-}i$ (II) (—) (I:II = 62:38)	150
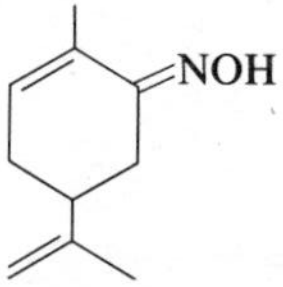 carvone oxime, anti-	", ", ", "	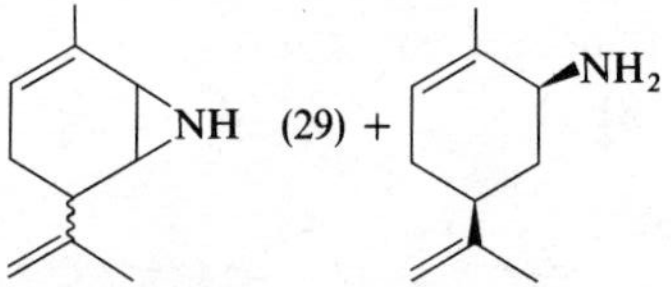 aziridine NH (29) + amine NH_2 (25) 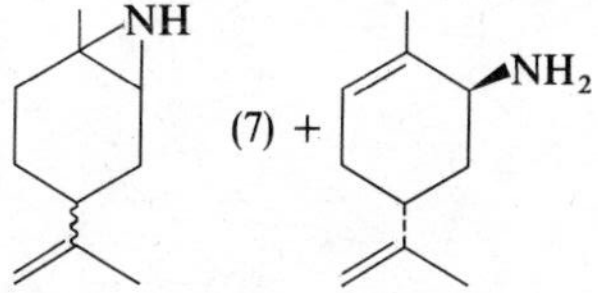 + aziridine NH (7) + amine NH_2 (6)	153, 154

TABLE XX. REDUCTION OF OXIMES (*Continued*)

No. of Carbon Atoms	Reactant	Conditions	Product(s) and Yield(s) (%)	Refs.
C_{10} (*Contd.*)	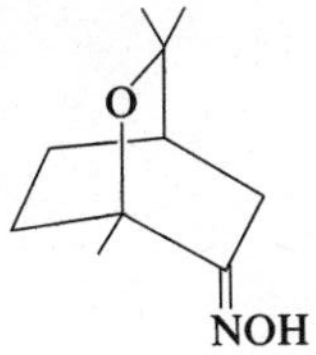 NOH	SMEAH, C_6H_6, reflux, 48 h	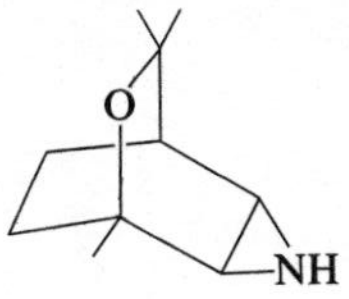 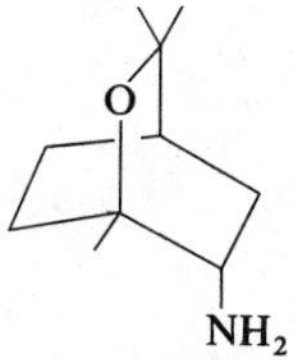 NH (50) 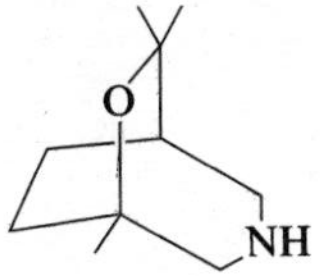 + NH **I** **II** + NH_2 + NH_2 (—) **III** 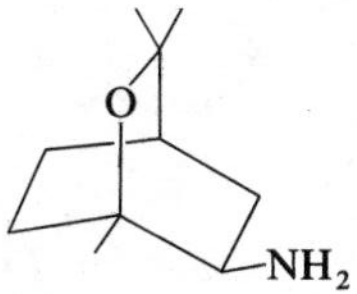 **IV** [**I**:**II**:(**III** + **IV**) = 78:20:2]	1012
	$n\text{-}C_3H_7C(CH_3)=NOTHP$	LAH + BCGF (1:1), ether, reflux, 2.5 h	$n\text{-}C_3H_7CH(CH_3)NH_2$ (*S*)-(+) (—)	132
	$i\text{-}C_3H_7C(CH_3)=NOTHP$	", ", ", "	$i\text{-}C_3H_7CH(CH_3)NH_2$ (*S*)-(+) (—)	132
	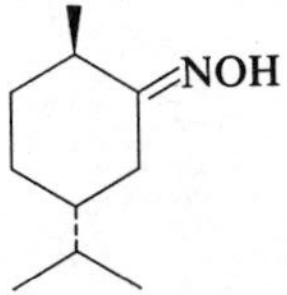 NOH	SMEAH, THF, 20°, 6 h	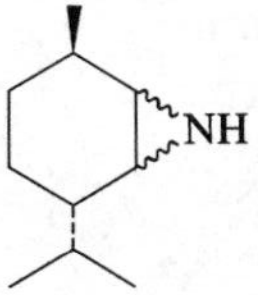 NH (7)	393
	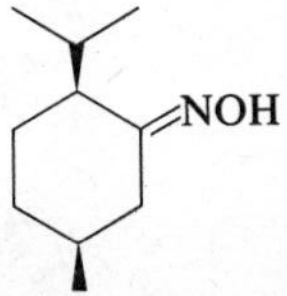 NOH	", ", ", "	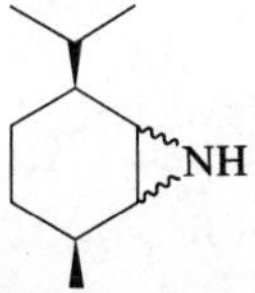 NH (56) (*trans*:*cis* = 73:27)	393

C_{11}	*n*-$C_4H_9C(CH_3)$=NOTHP	LAH + BCGF (1:1), ether, reflux, 2.5 h	*n*-$C_4H_9CH(CH_3)NH_2$ (*S*)-(+) (—) (24% e.e.)	132
	i-$C_4H_9C(CH_3)$=NOTHP	" " " "	*i*-$C_4H_9CH(CH_3)NH_2$ (*S*)-(+) (—) (19% e.e.)	132
	t-$C_4H_9C(CH_3)$=NOTHP	" " " "	*t*-$C_4H_9CH(CH_3)NH_2$ (*S*)-(+) (—)	132
C_{12}	1-$C_{10}H_7C(CH_3)$=NOH	SMEAH, THF, reflux, 2 h	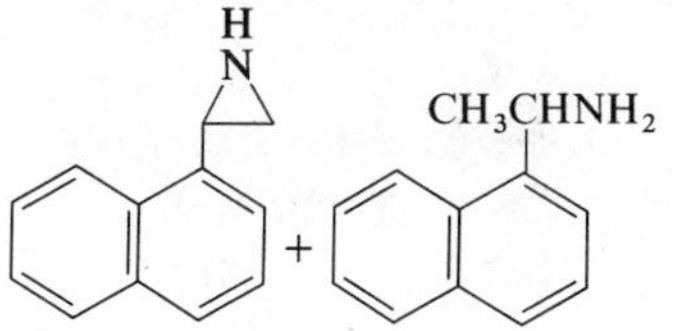 1-naphthyl-aziridine (**I**) + 1-$C_{10}H_7$-CH_3CHNH_2 (**II**) (—) (**I**:**II** = 75:25)	150
		LAH + BCGF (1:1), ether, reflux, 2.5 h	**II** (*S*)-(−) (~60) (10% e.e.)	132
C_{13}	1-$C_{10}H_7C(CH_3)$=NOCH$_3$	SMEAH, THF, reflux, 2 h	**I** + **II** (—) (**I**:**II** = 63:36)	150
		LAH + BCGF (1:1), ether, reflux, 2.5 h	**II** (*S*)-(—) (—) (14% e.e.)	132
	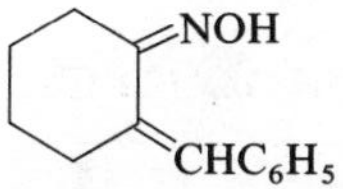 2-benzylidenecyclohexanone oxime (=NOH, =CHC$_6$H$_5$)	LTBA, THF, reflux, 8 h	No reaction	1013
	$C_6H_5C(CH_3)$=NOTHP *syn,anti-*	LAH + BCGF (1:1), reflux, 2.5 h	$C_6H_5CH(CH_3)NH_2$ (**I**) (*S*)-(−) (—) (4% e.e.)	132
		LAH + BCGF + C_2H_5OH (1:1:1), ether, reflux, 2.5 h	**I** (*R*)-(+) (~55) (13% e.e.)	132
	anti-	SMEAH, THF, reflux, 2 h	**I** 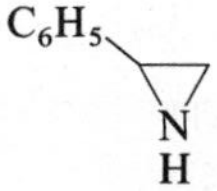 + 2-C_6H_5-aziridine (N–H) (**II**) + $C_6H_5NHC_2H_5$ (**III**) (—) (**I**:**II**:**III** = 56:38:6)	150

TABLE XX. REDUCTION OF OXIMES (*Continued*)

No. of Carbon Atoms	Reactant	Conditions	Product(s) and Yield(s) (%)	Refs.
C_{13} (*Contd.*)	$C_6H_{11}C(CH_3)$=NOTHP	LAH + BCGF (1:1), ether, reflux, 2.5 h	$C_6H_{11}CH(CH_3)NH_2$ (*S*)-(+) (—) (50% e.e.)	132
	n-$C_6H_{13}C(CH_3)$=NOTHP	LAH + BCGF (1:1), reflux, 2.5 h	*n*-$C_6H_{13}CH(CH_3)NH_2$ (*S*)-(+) (—) (27% e.e.)	132
C_{14}	$C_6H_5CH_2C(C_6H_5)$=NOH	SMEAH, THF, reflux, 2 h	2,3-(C_6H_5)$_2$-aziridine (NH) **I** + $C_6H_5CH_2CH(C_6H_5)NH_2$ **II** (—) (**I**:**II** = 55:45)	150
		LAH + BCGF (1:1), ether, reflux, 2.5 h	**II** (*S*)-(−) (~60) (25% e.e.)	132
	$C_6H_5CH_2C(CH_3)$=NOTHP *syn,anti-*	" " " "	$C_6H_5CH_2CH(CH_3)NH_2$ **I** (*S*)-(+) (—) (22% e.e.)	132
	anti-	SMEAH, THF, reflux, 2 h	**I** + 2-C_6H_5-3-methylaziridine (NH) **II** + 2-$C_6H_5CH_2$-aziridine (NH) **III** (—) (**I**:**II**:**III** = 35:60:5)	150
	$C_6H_5C(C_2H_5)$=NOTHP	LAH + BCGF (1:1), ether, reflux, 2.5 h	$C_6H_5CH(C_2H_5)NH_2$ **I** (*S*)-(−) (—) (17% e.e.)	132
		SMEAH, THF, reflux, 2 h	**I** + 2-C_6H_5-3-methylaziridine (NH) **II** + $C_6H_5NHC_3H_7$-*n* **III** (—) (**I**:**II**:**III** = 67:23:10)	150

	Substrate	Reagents and conditions	Product(s) and yield(s) (%)	Refs.
	3,4,5-$(CH_3O)_3C_6H_2CH_2C(=NOH)CO_2C_2H_5$	1. SMEAH, C_6H_6, −50°, 40 min 2. HCl, CH_2Cl_2	3,4,5-$(CH_3O)_3C_6H_2CH_2CH(\overset{+}{N}H_3)CH_2OH$ Cl^- (44)	667
C_{15}	$C_6H_5CH_2C(C_6H_5){=}NOTHP$	LAH + BCGF (1:1), ether, reflux, 2.5 h	$C_6H_5CH_2CH(C_6H_5)NH_2$ (S)-(−) (—) (27% e.e.)	132
	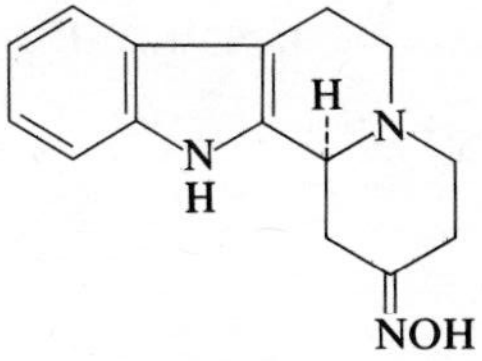 [indoloquinolizidine oxime; NOH]	SMEAH, $THF–C_6H_6$, reflux, overnight	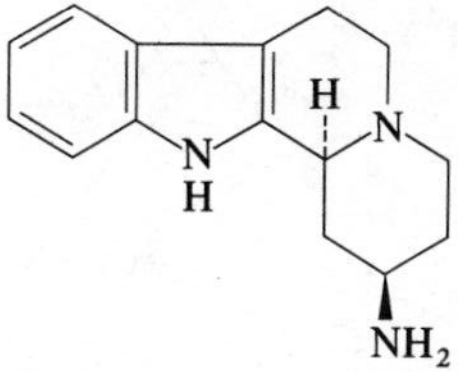 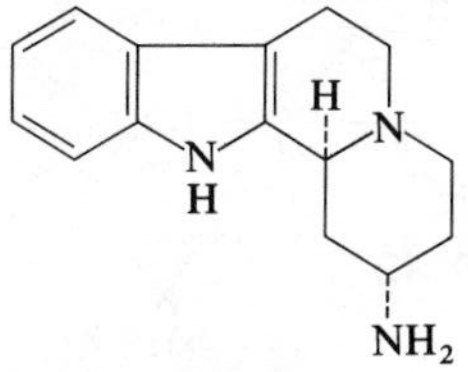 [indoloquinolizidine amine; NH_2] (39) + [indoloquinolizidine amine, epimer; NH_2] (16)	1014
	$C_6H_5C(C_3H_7\text{-}n){=}NOTHP$	SMEAH, THF, reflux, 2 h	$C_6H_5CH(C_3H_7\text{-}n)NH_2$ **I** 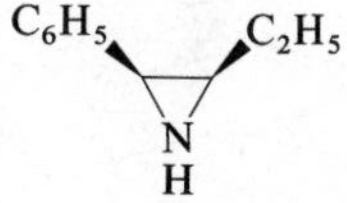 + [aziridine: C_6H_5, C_2H_5, NH] **II** + $C_6H_5NHC_4H_9\text{-}n$ **III** (—) (**I**:**II**:**III** = 73:18:9)	150
C_{16}	$(C_6H_5CH_2)_2C{=}NOCH_3$	", ", ", "	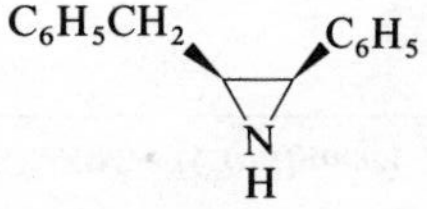 [aziridine: $C_6H_5CH_2$, C_6H_5, NH] **I** + $(C_6H_5CH_2)_2CHNH_2$ **II** (—) (**I**:**II** = 55:45)	150

TABLE XX. Reduction of Oximes (*Continued*)

No. of Carbon Atoms	Reactant	Conditions	Product(s) and Yield(s) (%)	Refs.
C_{16} (*Contd.*)	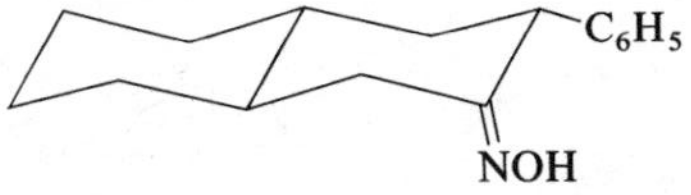 (structure: C_6H_5, NOH)	SMEAH	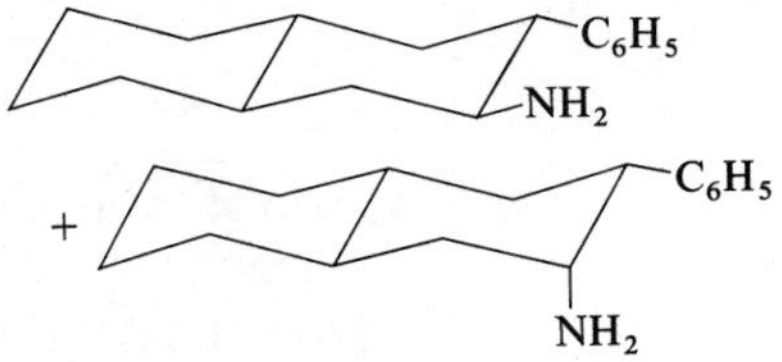 (structure: C_6H_5, NH_2) + (structure: C_6H_5, NH_2) (—)	1015
C_{17}	1-$C_{10}H_7C(CH_3)$=NOTHP	SMEAH, THF, reflux, 2 h	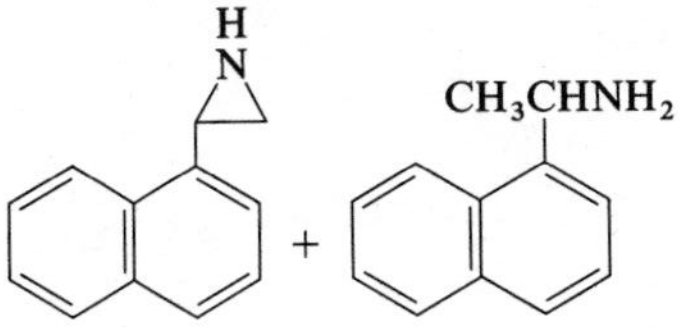 (structure: H, N) **I** + CH_3CHNH_2 (structure) **II** (—) (**I**:**II** = 71:28)	150
		LAH + BCGF (1:1), ether, reflux, 2.5 h	**II** (*S*)-(−) (—) (11% e.e.)	132
C_{19}	$C_6H_5CH_2C(C_6H_5)$=NOTHP	SMEAH, THF, reflux, 2 h	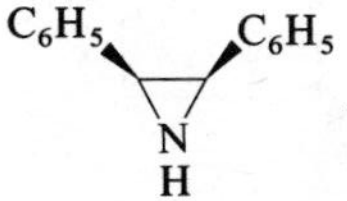 C_6H_5, C_6H_5, N, H (structure) **I** + $C_6H_5CH_2CH(C_6H_5)NH_2$ **II** (—) (**I**:**II** = 48:52)	150
		LAH + BCGF (1:1), ether, reflux, 2.5 h	**II** (*S*)-(−) (—) (26% e.e.)	132

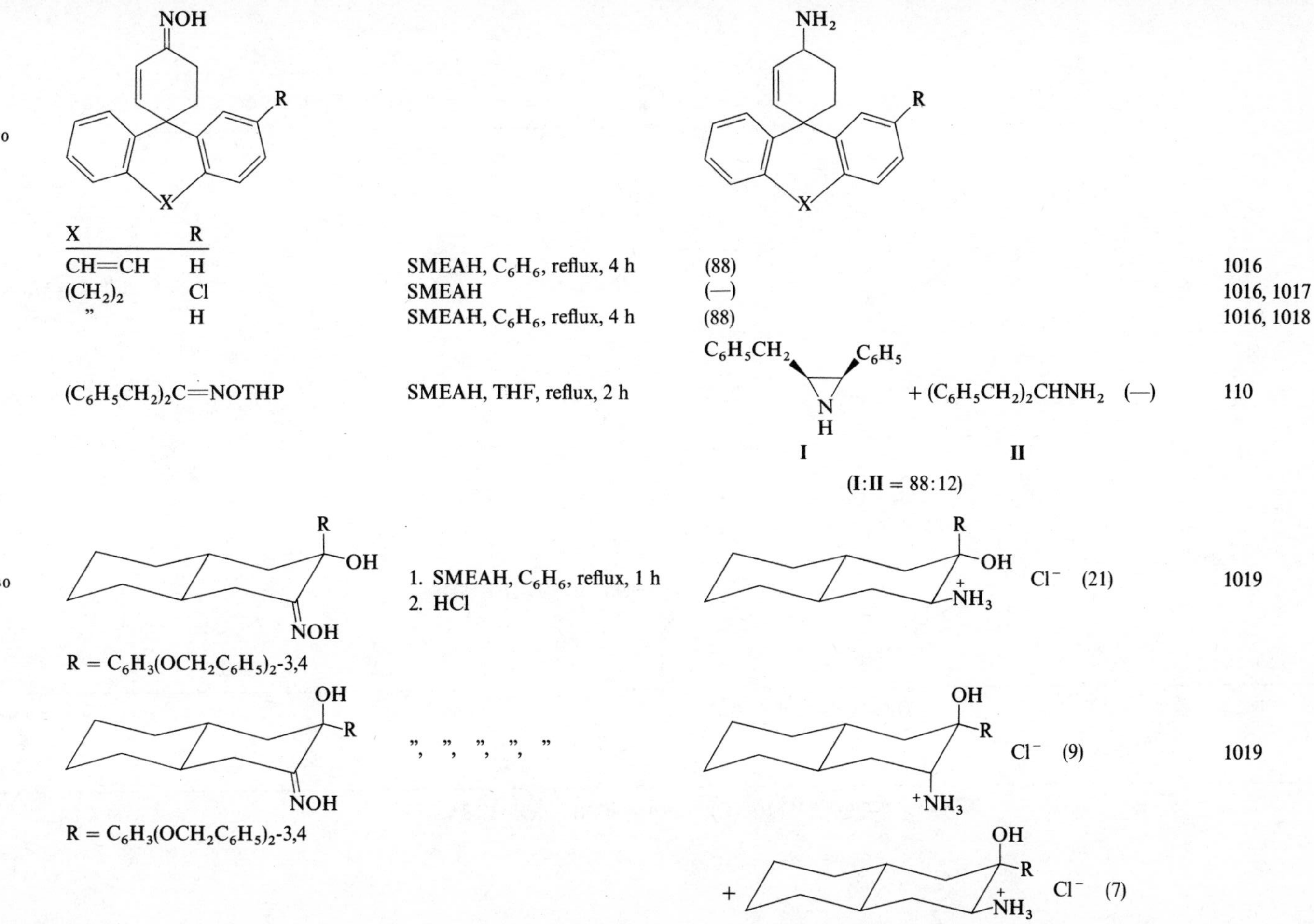

Formula	Substrate	Conditions	Product(s) and Yield(s) (%)	Refs.
C_{20}	(spiro oxime, X, R)		(spiro amine, NH_2)	
	X: CH=CH, R: H	SMEAH, C_6H_6, reflux, 4 h	(88)	1016
	X: $(CH_2)_2$, R: Cl	SMEAH	(—)	1016, 1017
	X: ", R: H	SMEAH, C_6H_6, reflux, 4 h	(88)	1016, 1018
	$(C_6H_5CH_2)_2C{=}NOTHP$	SMEAH, THF, reflux, 2 h	$C_6H_5CH_2$, C_6H_5 aziridine (N–H) **I** + $(C_6H_5CH_2)_2CHNH_2$ **II** (—) (**I**:**II** = 88:12)	110
C_{30}	(oxime, R, OH, NOH) $R = C_6H_3(OCH_2C_6H_5)_2$-3,4	1. SMEAH, C_6H_6, reflux, 1 h 2. HCl	(R, OH, $\overset{+}{N}H_3$) Cl^- (21)	1019
	(oxime, OH, R, NOH) $R = C_6H_3(OCH_2C_6H_5)_2$-3,4	", ", ", ", "	(OH, R, $^+NH_3$) Cl^- (9) + (OH, R, $\overset{+}{N}H_3$) Cl^- (7)	1019

TABLE XX. REDUCTION OF OXIMES (*Continued*)

No. of Carbon Atoms	Reactant	Conditions	Product(s) and Yield(s) (%)	Refs.
C_n	Ⓟ-C_6H_4-C(=NOH)-C_6H_5 [a]	SMEAH, C_6H_6, 60°, 12 h	Ⓟ-C_6H_4-CH(NH_2)-C_6H_5 + Ⓟ-C_6H_4-CH_2NH-C_6H_5 (—)	149, 1020
	Ⓟ-C_6H_4-C(=NOH)-C_6H_4-OCH_3-4 [a]	", ", ", "	Ⓟ-C_6H_4-CH(NH_2)-C_6H_4-OCH_3-4 **I** + Ⓟ-C_6H_4-CH_2NH-C_6H_4-OCH_3-4 **II** (—) (**I**:**II** = 58:42)	149

[a] P = Copolymer of styrene–divinylbenzene (1–2%).

TABLE XXI. REDUCTION OF *N*-ALKYLIDENEPHOSPHINIC AMIDES

No. of Carbon Atoms	Reactant	Conditions	Product(s) and Yield(s) (%)	Ref.
C_{20}	$C_6H_5C(CH_3){=}NP(O)(C_6H_5)_2$	LAH + (−)-QN (1:1), THF–ether, room temp, 24 h	$C_6H_5\overset{\star}{C}H(CH_3)NHP(O)(C_6H_5)_2$ (*R*)-(+) (75) (29% e.e.)	395
C_{24}	$1\text{-}C_{10}H_7C(CH_3){=}NP(O)(C_6H_5)_2$	", ", ", "	$1\text{-}C_{10}H_7\overset{\star}{C}H(CH_3)NHP(O)(C_6H_5)_2$ (*R*)-(−) (54) (10% e.e.)	395
	$2\text{-}C_{10}H_7C(CH_3){=}NP(O)(C_6H_5)_2$	", ", ", "	$2\text{-}C_{10}H_7\overset{\star}{C}H(CH_3)NHP(O)(C_6H_5)_2$ (*R*)-(+) (41) (21% e.e.)	395

TABLE XXII. REDUCTION OF ISOCYANATES AND THIOCYANATES

No. of Carbon Atoms	Reactant	Conditions	Product(s) and Yield(s) (%)	Refs.
C_5	$n\text{-}C_4H_9NCO$	SMEAH	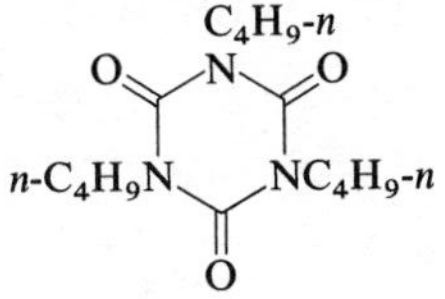 1,3,5-tris(C_4H_9-n)-triazinetrione: $n\text{-}C_4H_9N$, $NC_4H_9\text{-}n$, $C_4H_9\text{-}n$, O, O, O (96)	401
C_6	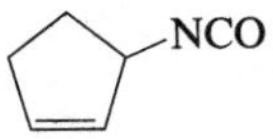 cyclopent-2-enyl–NCO	SMEAH (IA), C_6H_6, reflux, 12 h	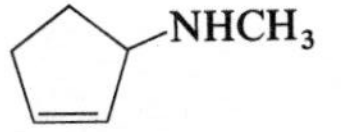 cyclopent-2-enyl–$NHCH_3$ (65)[a]	397
C_7	C_6H_5NCO	SMEAH	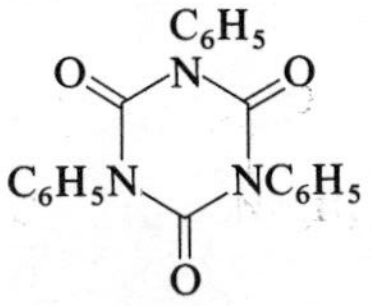 1,3,5-tris(C_6H_5)-triazinetrione: C_6H_5N, NC_6H_5, C_6H_5, O, O, O (94)	401
		LTBA	" (89)	400
	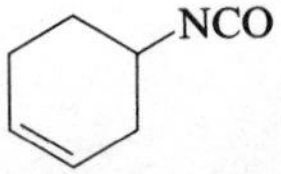 cyclohex-3-enyl–NCO	SMEAH (IA), C_6H_6, reflux, 12 h	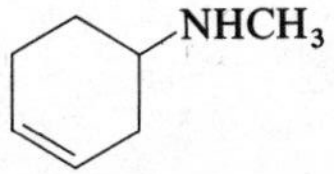 cyclohex-3-enyl–$NHCH_3$ (55)[a]	397

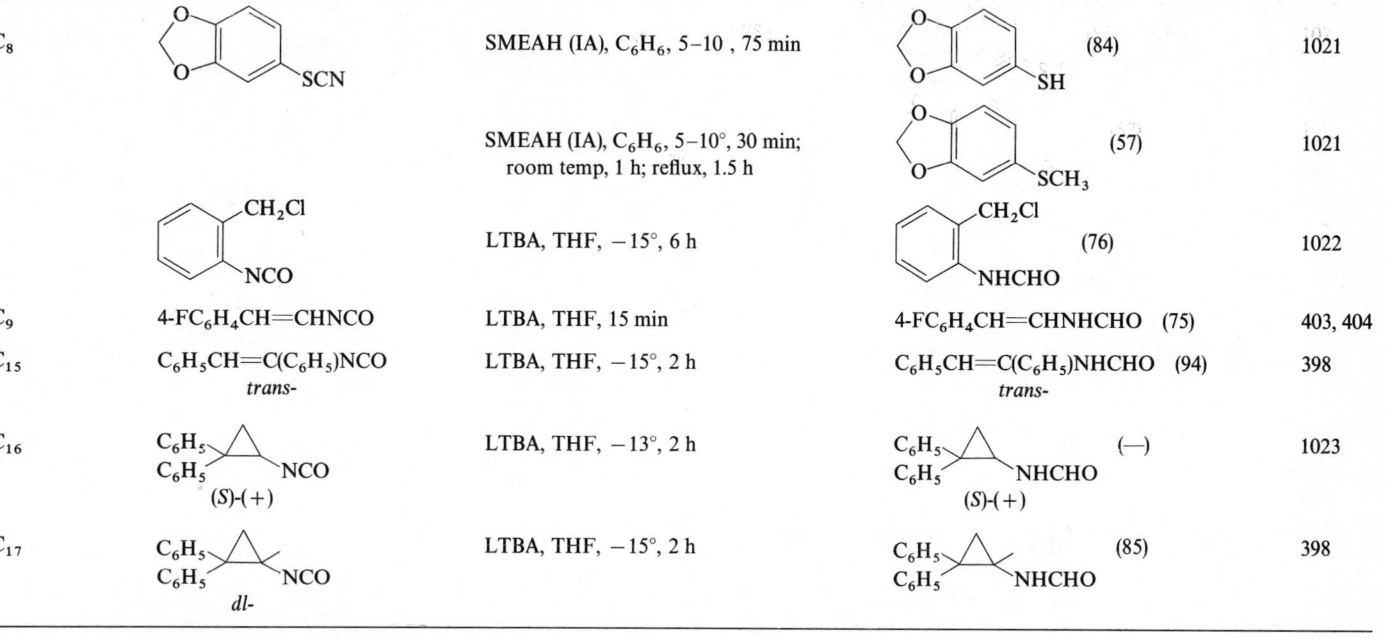

C_8	SCN	SMEAH (IA), C_6H_6, 5–10 , 75 min	SH (84)	1021
		SMEAH (IA), C_6H_6, 5–10°, 30 min; room temp, 1 h; reflux, 1.5 h	SCH_3 (57)	1021
	CH_2Cl, NCO	LTBA, THF, −15°, 6 h	CH_2Cl, NHCHO (76)	1022
C_9	$4\text{-}FC_6H_4CH{=}CHNCO$	LTBA, THF, 15 min	$4\text{-}FC_6H_4CH{=}CHNHCHO$ (75)	403, 404
C_{15}	$C_6H_5CH{=}C(C_6H_5)NCO$ *trans-*	LTBA, THF, −15°, 2 h	$C_6H_5CH{=}C(C_6H_5)NHCHO$ (94) *trans-*	398
C_{16}	C_6H_5, C_6H_5, NCO (*S*)-(+)	LTBA, THF, −13°, 2 h	C_6H_5, C_6H_5, NHCHO (—) (*S*)-(+)	1023
C_{17}	C_6H_5, C_6H_5, NCO *dl-*	LTBA, THF, −15°, 2 h	C_6H_5, C_6H_5, NHCHO (85)	398

[a] The yield is based on the starting carboxylic acid.

TABLE XXIII. Reduction of Urethanes, Ureas, and Cyanamides

No. of Carbon Atoms	Reactant	Conditions	Product(s) and Yield(s) (%)	Refs
C_9	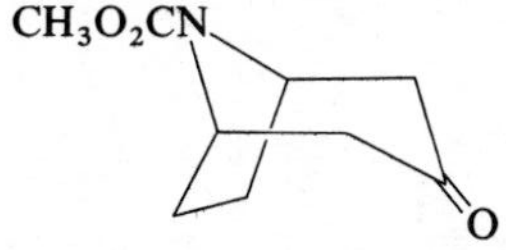 CH_3O_2CN [bicyclic ketone, =O]	SMEAH, C_6H_6, 25°, 22 h	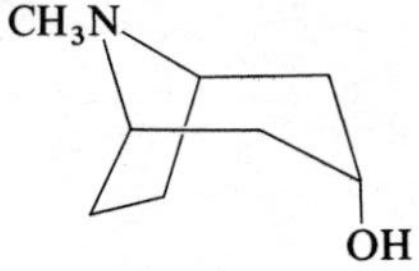 CH_3N [bicyclic, OH] (54) 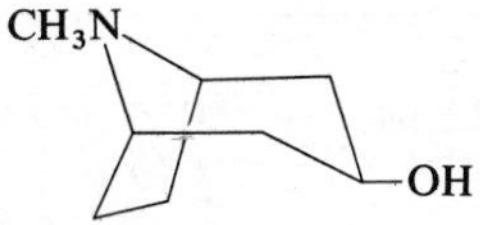 + CH_3N [bicyclic, —OH] (36)	406
		LTBA, THF, 25°, 18 h	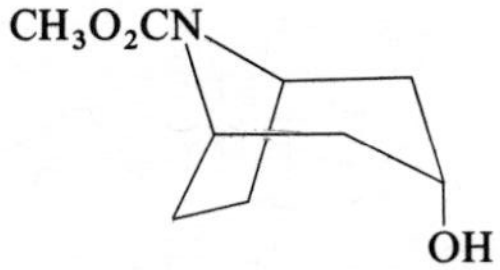 CH_3O_2CN [bicyclic, OH] (54) 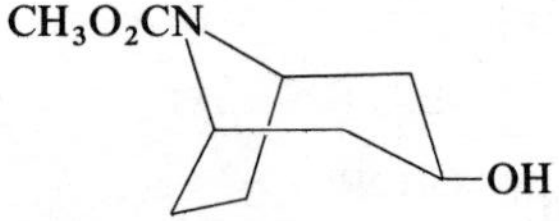 + CH_3O_2CN [bicyclic, —OH] (36)	406
C_{10}	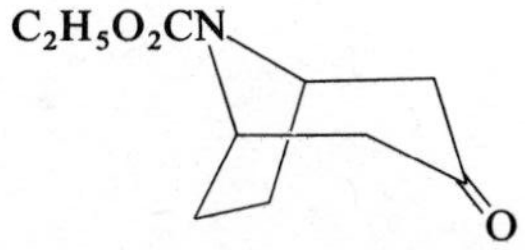 $C_2H_5O_2CN$ [bicyclic ketone, =O]	LTBA, THF, room temp, 3 h	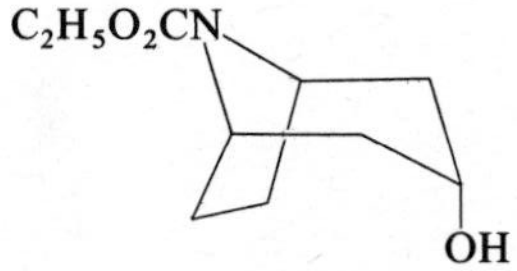 $C_2H_5O_2CN$ [bicyclic, OH] (68) 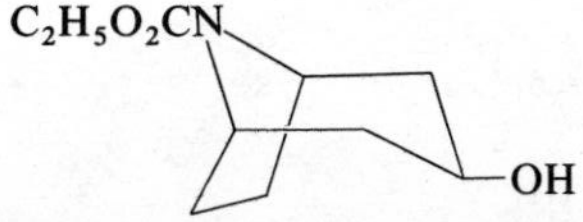 + $C_2H_5O_2CN$ [bicyclic, —OH] (23)	409

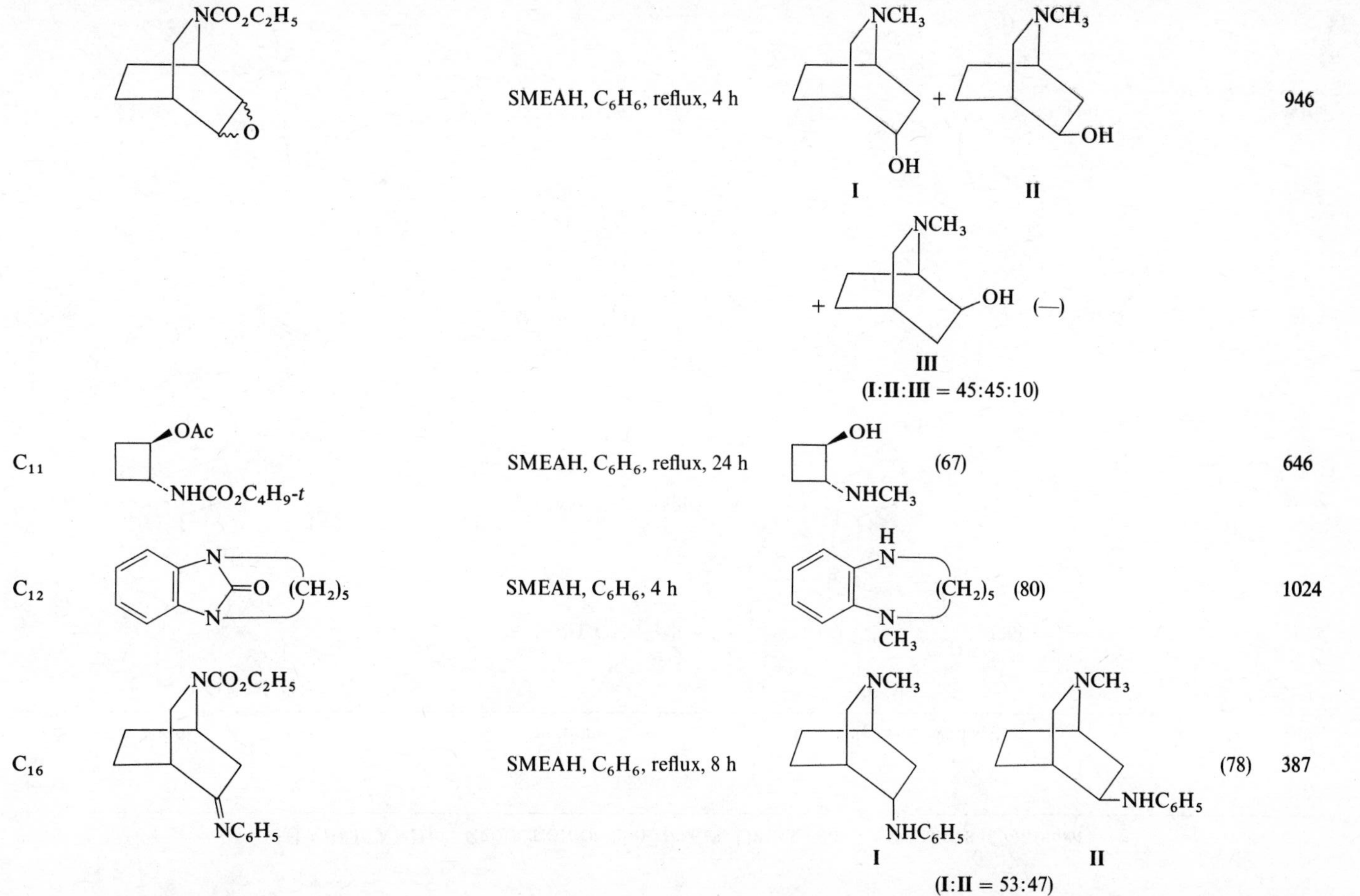
NCO$_2$C$_2$H$_5$
O
SMEAH, C$_6$H$_6$, reflux, 4 h
NCH$_3$
OH
I
+
NCH$_3$
OH
II
946
+
NCH$_3$
OH
(—)
III
(I:II:III = 45:45:10)
C$_{11}$
OAc
NHCO$_2$C$_4$H$_9$-t
SMEAH, C$_6$H$_6$, reflux, 24 h
OH
NHCH$_3$
(67)
646
C$_{12}$
N
O
N
(CH$_2$)$_5$
SMEAH, C$_6$H$_6$, 4 h
H
N
N
CH$_3$
(CH$_2$)$_5$
(80)
1024
C$_{16}$
NCO$_2$C$_2$H$_5$
NC$_6$H$_5$
SMEAH, C$_6$H$_6$, reflux, 8 h
NCH$_3$
NHC$_6$H$_5$
I
NCH$_3$
NHC$_6$H$_5$
II
(I:II = 53:47)
(78)
387

TABLE XXIII. REDUCTION OF URETHANES, UREAS, AND CYANAMIDES (*Continued*)

No. of Carbon Atoms	Reactant	Conditions	Product(s) and Yield(s) (%)	Refs.
C_{16} (*Contd.*)	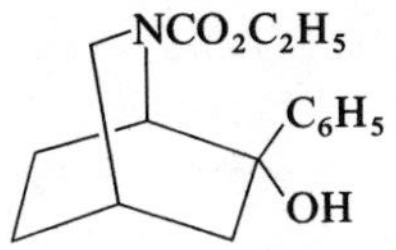 $NCO_2C_2H_5$, C_6H_5, OH	SMEAH, C_6H_6, reflux, 3 h	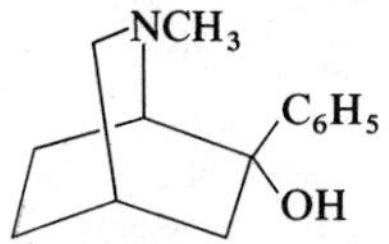 NCH_3, C_6H_5, OH (36)	1025
	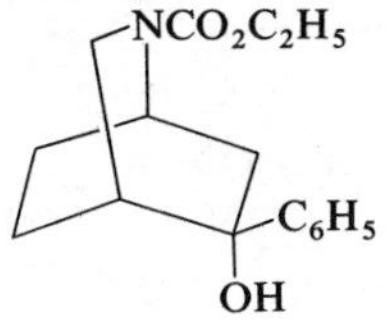 $NCO_2C_2H_5$, C_6H_5, OH	SMEAH, C_6H_6, reflux, 2 h	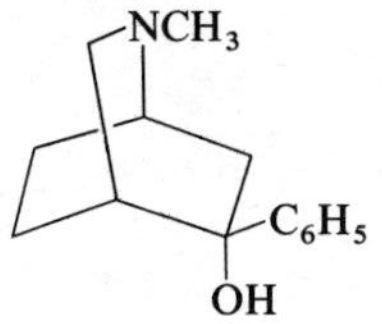 NCH_3, C_6H_5, OH (26)	1025
C_{20}	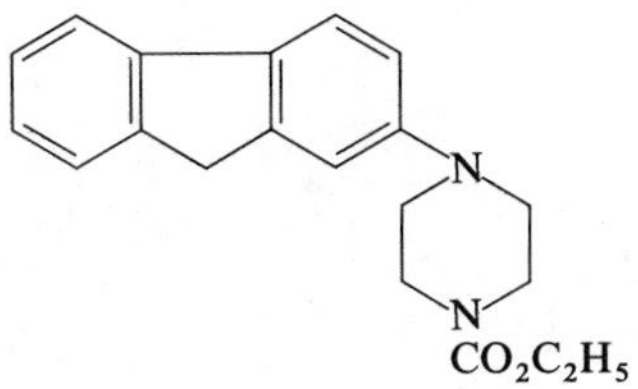 N, N, $CO_2C_2H_5$	SMEAH, C_6H_6	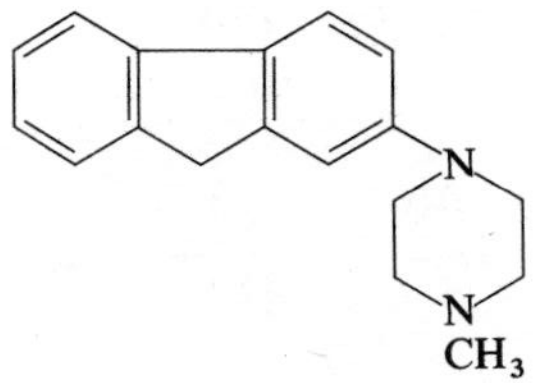 N, N, CH_3 (~100)	1026, 1027
	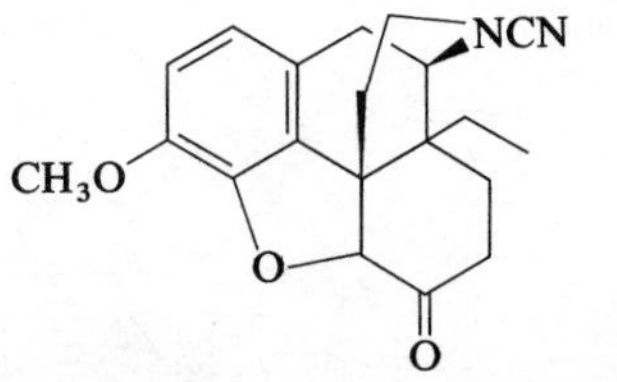 CH_3O, O, NCN, O	SMEAH, C_6H_6, 15 h	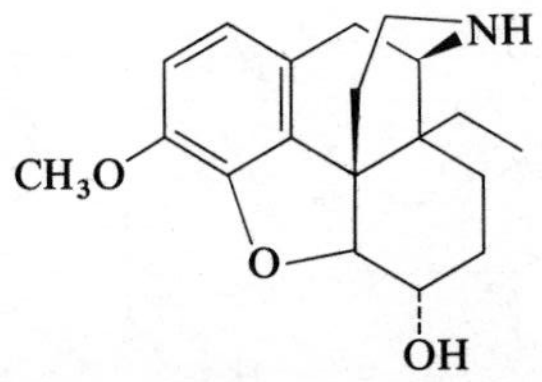 CH_3O, O, NH, OH (44)	410a

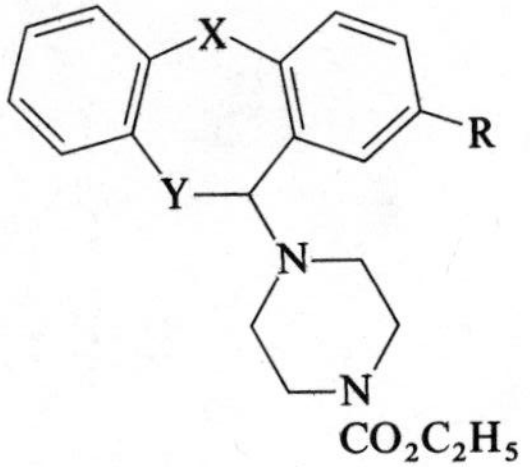

	Substrate	Reagent	Product (Yield, %)	Refs.
	X Y R (structure with piperazine N-$CO_2C_2H_5$)	$SMEAH, C_6H_6$	(structure with piperazine N-CH_3)	
C_{20}	S O H		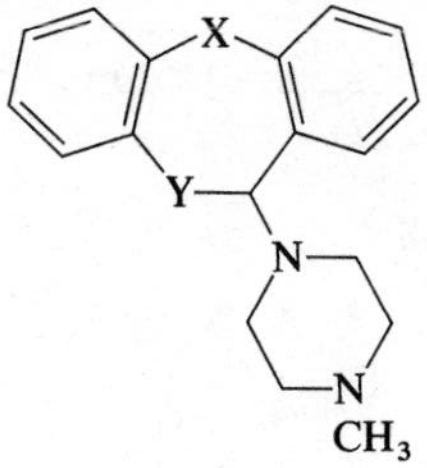 (77)	1028a
C_{21}	S CH_2 H		(—)	1028b
C_{22}	CH_2 CH_2 Cl		(92)	1029,1030
	S CH_2 CH_3O		(—)	1030
C_{24} 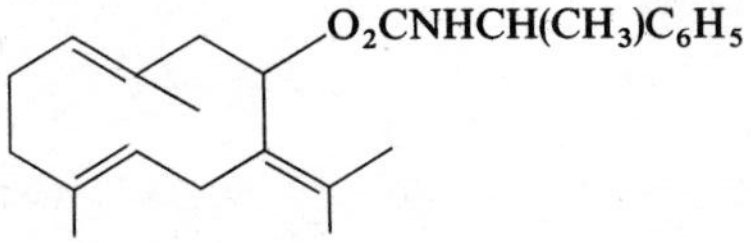 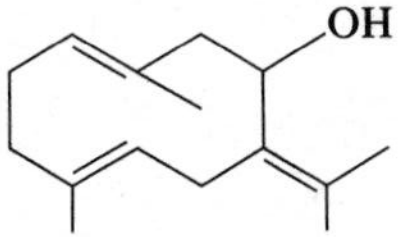	$O_2CNHCH(CH_3)C_6H_5$ (structure)	SMEAH	OH (structure) (—)	1031

TABLE XXIV. REDUCTION OF NITRO COMPOUNDS

No. of Carbon Atoms	Reactant	Conditions	Product(s) and Yield(s) (%)	Refs.
C_1	CH_3NO_2	$NaAlH(OC_2H_5)_3$, THF, 65°, 30 min	CH_3NH_2 (65)	319
		SMEAH, C_6H_6, 20–40°, 6 min	" (45)	167, 411
C_3	i-$C_3H_7NO_2$	$NaAlH(OC_2H_5)_3$, THF, 65°, 2 h	i-$C_3H_7NH_2$ (72)	319
C_6	2,5-$Cl_2C_6H_3NO_2$	SMEAH (IA), (1.5 eq) C_6H_6, 0°, then room temp, 20 min	2,5-$Cl_2C_6H_3N{=}\overset{+}{N}(O^-)C_6H_3Cl_2$-2,5 (35)	413
	3,4-$Cl_2C_6H_3NO_2$	SMEAH (IA), (2 eq) C_6H_6, 0°, then room temp, 20 min	3,4-$Cl_2C_6H_3N{=}NC_6H_3Cl_2$-3,4 (15) + 3,4-$Cl_2C_6H_3N{=}\overset{+}{N}(O^-)C_6H_3Cl_2$-3,4 (40)	413
	2-$BrC_6H_4NO_2$	SMEAH (IA) (2 eq), C_6H_6, 0°, then room temp, 20 min	$C_6H_5N{=}\overset{+}{N}(O^-)C_6H_5$ (62)	413
	4-$BrC_6H_4NO_2$	SMEAH (IA) (2.5 eq), C_6H_6, 0°, then room temp, 20 min	" (75)	413
	3-$BrC_6H_4NO_2$	SMEAH (IA) (2 eq), C_6H_6, 0°, then room temp, 20 min	$C_6H_5N{=}NC_6H_5$ (84)	413
	2-$ClC_6H_4NO_2$	SMEAH (IA) (2 eq), C_6H_6, 0°, then reflux, 30 min	2-$ClC_6H_4N{=}\overset{+}{N}(O^-)C_6H_4Cl$-2 (38)	413
	3-$ClC_6H_4NO_2$	", ", ", ", "	3-$ClC_6H_4N{=}NC_6H_4Cl$-3 (85)	413
	4-$ClC_6H_4NO_2$	SMEAH (IA)(1.5 eq), C_6H_6, 0°, then room temp, 20 min	4-$ClC_6H_4N{=}\overset{+}{N}(O^-)C_6H_4Cl$-4 (78)	413
		SMEAH (IA) (2 eq), C_6H_6, 0°, then reflux, 30 min	4-$ClC_6H_4N{=}NC_6H_4Cl$-4 (76)	413

	$2\text{-}IC_6H_4NO_2$	SMEAH (IA) (2.5 eq), C_6H_6, 0°, then room temp, 20 min	$C_6H_5N{=}NC_6H_5$ (65)	413
	$3\text{-}IC_6H_4NO_2$	" " " " "	" (65)	413
	$4\text{-}IC_6H_4NO_2$	" " " " "	" (65)	413
	$C_6H_5NO_2$	$NaAlH(OC_2H_5)_3$, THF, 65°, 4 h	" (68)	319
		SMEAH (IA) (2 eq), C_6H_6, 0°, then room temp, 20 min	" (92)	413, 411, 167
		$NaAlH[OCH_2CH_2N(CH_3)_2]_3$, C_6H_6, 23°, 30 min	" (46)	633
		$NaAl_2H_4[OCH_2CH_2N(CH_3)_2]_3$, C_6H_6, 25°, 2 h	" (5) + $C_6H_5N{=}\overset{+}{N}(O^-)C_6H_5$ (43)	634
		$NaAlH[O(CH_2)_3OCH_3]_3$, C_6H_6, 25°, 30 min	" (57) + $C_6H_5NHNHC_6H_5$ (33)	632
C_7	NO_2, Cl, Br	SMEAH (IA) (2.5 eq), C_6H_6, 0°, then reflux, 30 min	N=N, Cl, Cl (65)	413
	$4\text{-}CH_3C_6H_4NO_2$	SMEAH (IA) (2 eq), C_6H_6, 0°, then room temp, 20 min	$4\text{-}CH_3C_6H_4N{=}NC_6H_4CH_3\text{-}4$ (95)	413, 238a
	$4\text{-}CH_3OC_6H_4NO_2$	" " " " "	$4\text{-}CH_3OC_6H_4N{=}NC_6H_4OCH_3\text{-}4$ (83)	413
C_8	$4\text{-}[(CH_3)_2N]C_6H_4NO_2$	" " " " "	$4\text{-}[(CH_3)_2N]C_6H_4N{=}\overset{+}{N}(O^-)C_6H_4[N(CH_3)_2]\text{-}4$ (80)	413
		SMEAH (IA) (2 eq), C_6H_6, 0°, then reflux, 30 min	$4\text{-}[(CH_3)_2N]C_6H_4N{=}NC_6H_4[N(CH_3)_2]\text{-}4$ (85)	413
C_9	$3\text{-}(C_2H_5O_2C)C_6H_4NO_2$	SMEAH (IA) (4 eq), C_6H_6, 0°, then reflux, 30 min	$3\text{-}(HOCH_2)C_6H_4N{=}NC_6H_4(CH_2OH)\text{-}3$ (68)	413
C_{10}	$1\text{-}C_{10}H_7NO_2$	SMEAH (IA) (2 eq), C_6H_6, 0°, then room temp, 20 min	No reaction	413

TABLE XXIV. REDUCTION OF NITRO COMPOUNDS (*Continued*)

No. of Carbon Atoms	Reactant	Conditions	Product(s) and Yield(s) (%)	Refs.
C_{10} (*Contd.*)	2-$C_{10}H_7NO_2$	SMEAH (IA) (2 eq), C_6H_6, 0°, then room temp, 20 min	2-$C_{10}H_7N{=}NC_{10}H_7$-2 (78)	413
	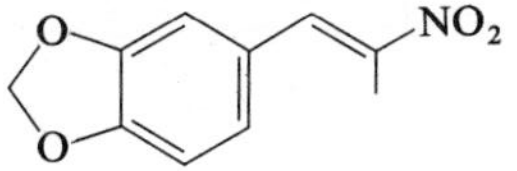 O, O, NO_2	SMEAH, C_6H_6, reflux	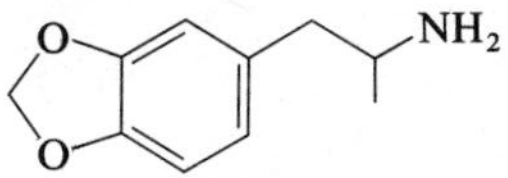 O, O, NH_2 (85)	415
C_{11}	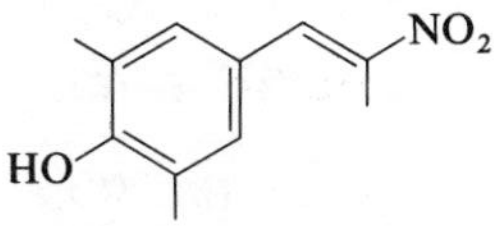 HO, NO_2	", ", "	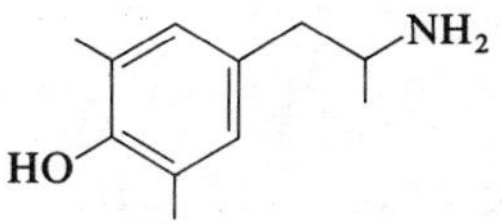 HO, NH_2 (75)	415
C_{12}	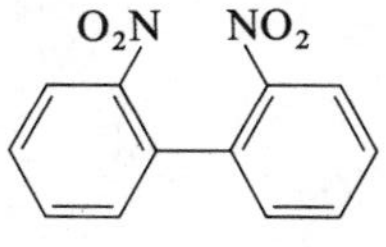 O_2N, NO_2	SMEAH (IA) (2 eq), C_6H_6, 0°, then room temp, 20 min	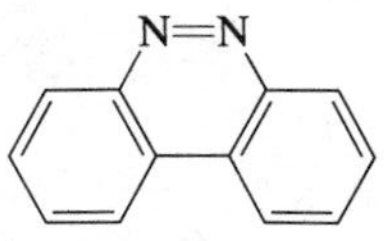 N=N (76)	413
	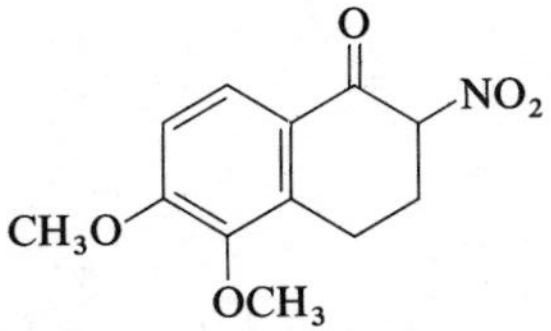 CH_3O, OCH_3, O, NO_2	1. Pd/C, H_2 2. SMEAH	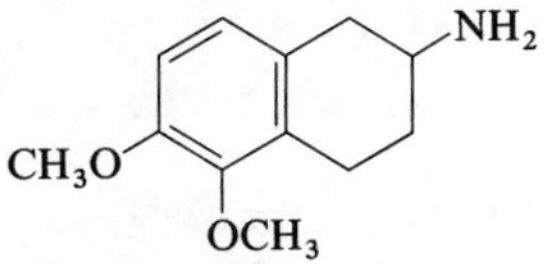 CH_3O, OCH_3, NH_2 (—)	1032
	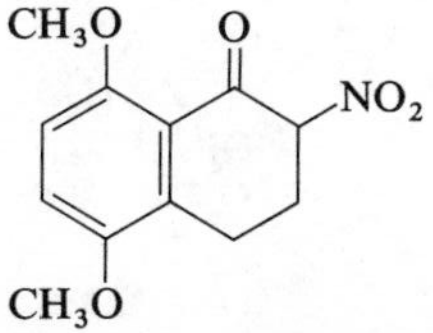 CH_3O, CH_3O, O, NO_2	1. SMEAH, C_6H_6, reflux, 6 h 2. HCl, C_6H_6	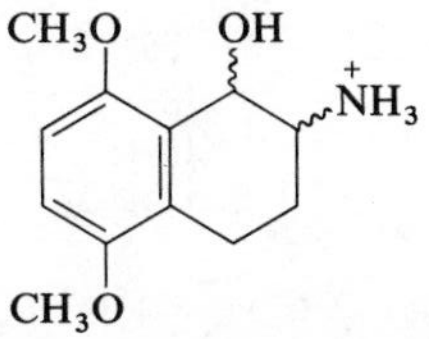 CH_3O, CH_3O, OH, $\overset{+}{N}H_3$ Cl^- (39)	1033

C_{13}	CH_3O, O, NO_2, OCH_3	1. SMEAH, C_6H_6, reflux, 6 h 2. HCl, $CHCl_3$–ether	CH_3O, OH, $\overset{+}{N}H_3$, OCH_3 Cl^- (62)	1034
C_{15}	R, NO_2, R, R R = i-C_3H_7	SMEAH, C_6H_6, 25°, 48 h	R, R, $N=\overset{+}{N}$, O^-, R, R, R, R (—)	414
C_{19}	CH_3O, CH_3O, $NCH_2C_6H_5$, CH_2NO_2	1. SMEAH, C_6H_6, reflux, 1 h 2. HCl	CH_3O, CH_3O, $NCH_2C_6H_5$, $CH_2\overset{+}{N}H_3$ Cl^- (25)	1035

TABLE XXV. Reduction of *N*-Nitroso and Azido Compounds

No. of Carbon Atoms	Reactant	Conditions	Product(s) and Yield(s) (%)	Ref.
C_4	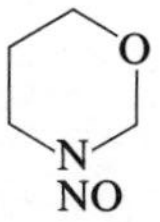 O, N, NO	SMEAH, C_6H_6, 10°, 1 h	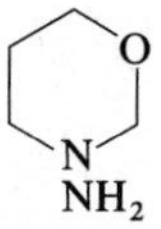 O, N, NH_2 (67)	416
C_5	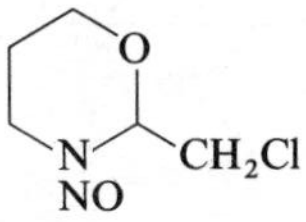 O, N, CH_2Cl, NO	" " " "	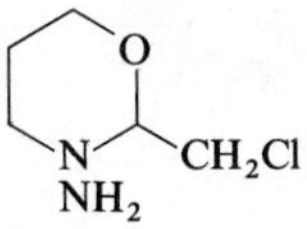 O, N, CH_2Cl, NH_2 (53)	416
C_6	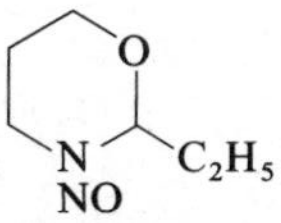 O, N, C_2H_5, NO	" " " "	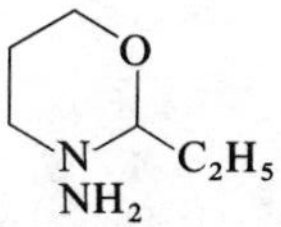 O, N, C_2H_5, NH_2 (41)	416
C_{11}	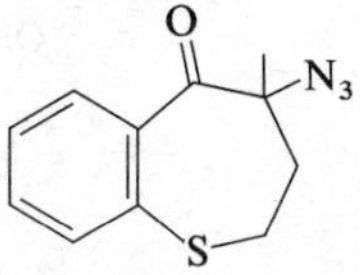 O, N_3, S	SMEAH, C_6H_6, 16 h	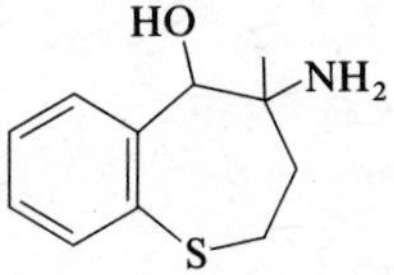 HO, NH_2, S (65)	421a

TABLE XXVI. REDUCTION OF HYDRAZONIUM AND GUANIDINIUM SALTS

No. of Carbon Atoms	Reactant	Conditions	Product(s) and Yield(s) (%)	Ref.
C_7	$(CH_3)_2N$, $(CH_3)_2N$ $\overset{+}{C}$ $N(CH_3)_2$ Cl^-	SMEAH (IA), C_6H_6, 60°, 1 h	$HC[N(CH_3)_2]_3$ (55)	1036
C_9	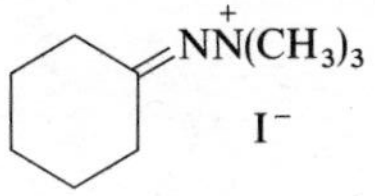 $N\overset{+}{N}(CH_3)_3$ I^-	SMEAH, THF, 70°, 2 d	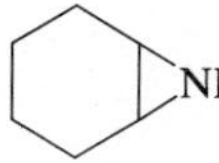 NH (55)	393
C_{10}	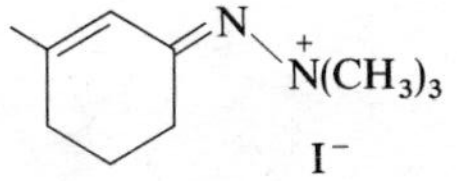 N $\overset{+}{N}(CH_3)_3$ I^- *anti-*	SMEAH	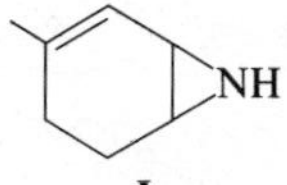 NH (60) 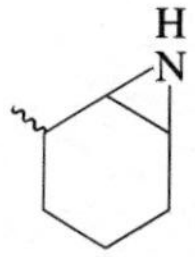 + H N (6) I II	154
		SMEAH–H_2O (7:3)	" (68) + " (2)	154
	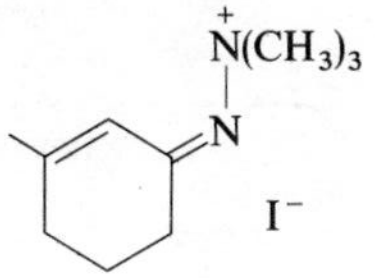 $\overset{+}{N}(CH_3)_3$ N I^- *syn-*	SMEAH	**II** (92) 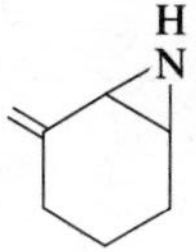 + H N (7) 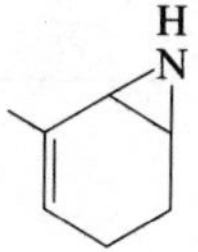 + H N (1) III IV	154
		SMEAH–H_2O (7:3)	**III** (72) + **IV** (16) + **II** (12)	154
	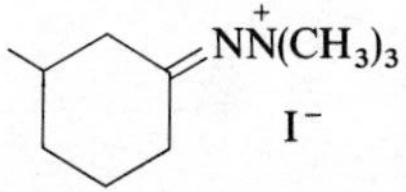 $N\overset{+}{N}(CH_3)_3$ I^-	SMEAH, THF, 70°, 2 d	**II** (55) (*cis*:*trans* = 52:48)	393

TABLE XXVI. REDUCTION OF HYDRAZONIUM AND GUANIDINIUM SALTS (*Continued*)

No. of Carbon Atoms	Reactant	Conditions	Product(s) and Yield(s) (%)	Ref.
C_{12}	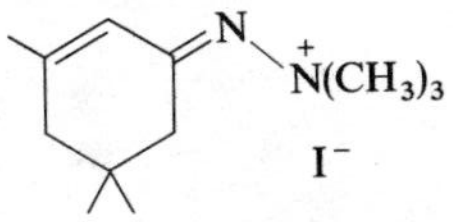 $\overset{+}{N}(CH_3)_3$ I^- *anti-*	SMEAH	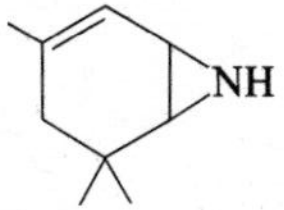 NH (98) 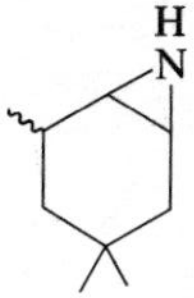 + H N (2)	154
		$SMEAH-H_2O$ (7:3)	" (96)	154
C_{13}	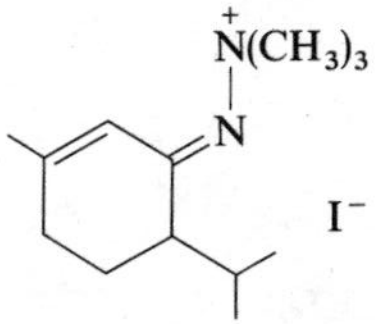 $\overset{+}{N}(CH_3)_3$ N I^- *syn-*	SMEAH	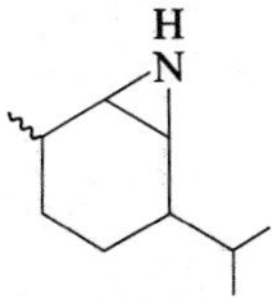 H N (75) 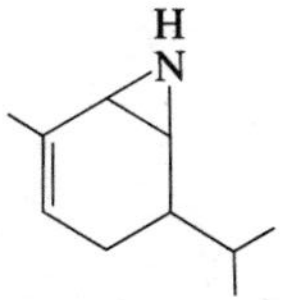 + H N (17) 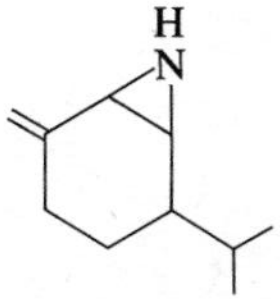 + H N (8) **I** **II** **III**	154
		$SMEAH-H_2O$ (7:3)	**I** (25) + **II** (51) + **III** (17)	154
	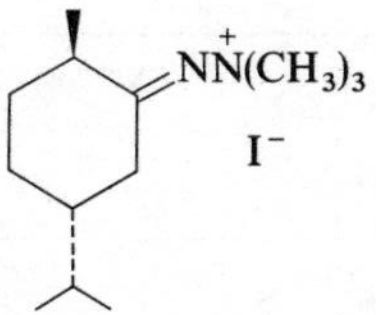 $N\overset{+}{N}(CH_3)_3$ I^-	SMEAH, THF, 70°, 2 d	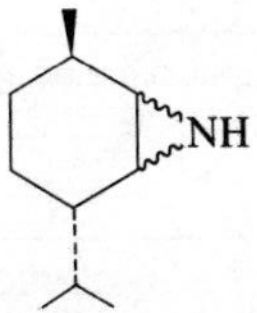 NH (50) (*cis*:*trans* = 50:50)	393

Substrate	Conditions	Product(s) and Yield(s) (%)	Refs.
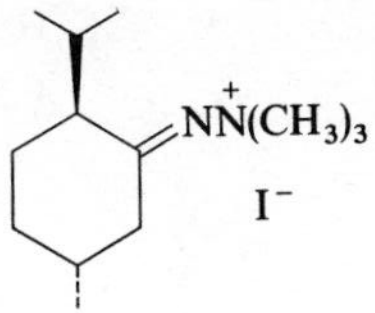 $=N\overset{+}{N}(CH_3)_3$ I^-	SMEAH, THF, 70°, 2 d	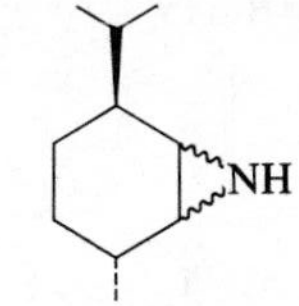 NH (46) (*cis*:*trans* = 50:50)	393
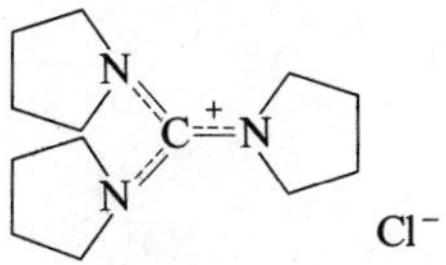 $C\overset{+}{=}N$ Cl^-	SMEAH (IA), C_6H_6, 60°, 1 h	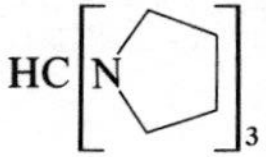 $HC[N]_3$ (33)	1036
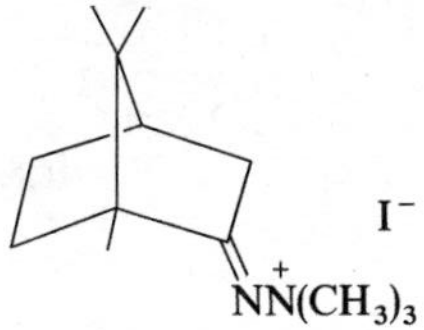 I^- $\overset{+}{N}N(CH_3)_3$	SMEAH, THF, 70°, 2 d	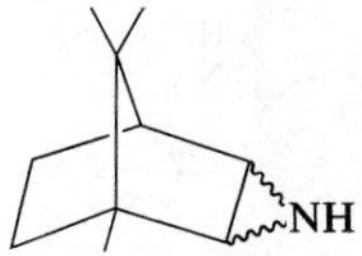 NH (70) (*endo*:*exo* = 81:19)	393
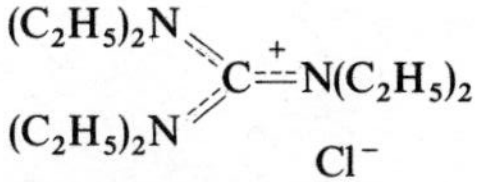 $(C_2H_5)_2N$, $(C_2H_5)_2N$ $\overset{+}{C}=N(C_2H_5)_2$ Cl^-	SMEAH (IA), C_6H_6, 60°, 1 h	$HC[N(C_2H_5)_2]_3$ (42)	1036

TABLE XXVII. Reduction of Heterocyclic Bases and Their Quaternary Salts

No. of Carbon Atoms	Reactant	Conditions	Product(s) and Yield(s) (%)	Refs.
C_6	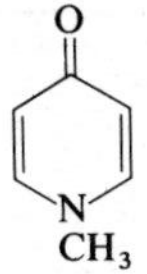 1-methyl-4-pyridone (CH_3 on N)	LTBA, THF, 15°, 3 h	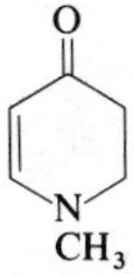 1-methyl-2,3-dihydro-4-pyridone (87) 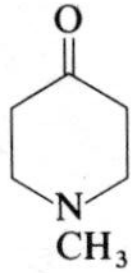 + 1-methyl-4-piperidone (5)	447
		SMEAH (IA), THF, 0°, 4 h	" (76)	448
	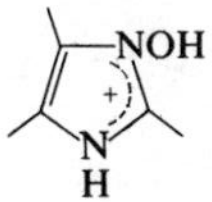 1-hydroxy-2,4,5-trimethylimidazolium BF_4^- (NOH, N–H)	SMEAH	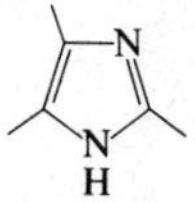 2,4,5-trimethylimidazole (N–H) (80)	449
C_7	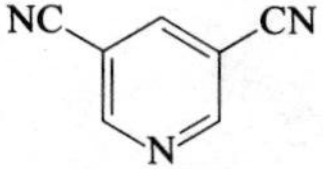 3,5-dicyanopyridine (NC, CN)	$NaAlH_2(OC_2H_5)_2$ (IA), THF, room temp, 30 min	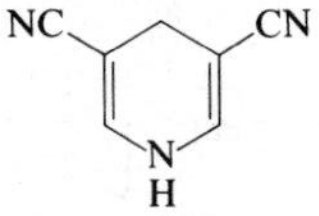 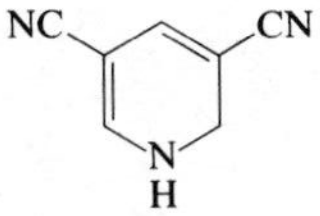 3,5-dicyano-1,4-dihydropyridine (N–H) **I** + 3,5-dicyano-1,2-dihydropyridine (N–H) **II** (48) (**I**:**II** = 55:45)	315
		SMEAH (IA), THF, room temp, 30 min	**I** (12)	315
	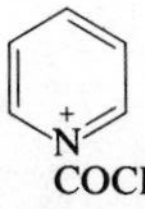 1-acetylpyridinium Cl^- ($COCH_3$ on N^+)	LTBA–CuBr, THF, −23°, 1.5 h	 1-acetyl-1,2,3,4-tetrahydropyridine ($COCH_3$ on N) (36)	439

	Substrate	Reagent and conditions	Product (Yield, %)	Refs.
	[structure] N–CH_2OCH_3	LTBA, THF, 15°, 2.5 h	[structure] N–CH_2OCH_3 (78)	447
		SMEAH (IA), THF, 0°, 4 h	" (75)	448
C_8	[structure] CH_3N	SMEAH, C_6H_6, reflux, 2 h	[structure] CH_3N, NH (66)	1037
	[structure] Cl^-, $CO_2C_2H_5$	LTBA–CuBr, THF, −23°, 1.5 h	[structure] $CO_2C_2H_5$ (20)	439
	[structure] OH, BF_4^-, H	SMEAH	[structure] H (50)	449
C_9	[structure]	1. $BH_3 \cdot THF$ (IA), THF, −70°, 30 min 2. SMEAH (IA), C_6H_6, −78°, 30 min 3. $ClCO_2CH_3$, room temp, 18 h	[structure] CO_2CH_3 (90)	445, 446
	[structure] Cl, C_6H_5 + [structure] C_6H_5, Cl	SMEAH, C_6H_6, 1 h	[structure] C_6H_5, H (~100)	426
	[structure]	LTBA, BME, room temp, 1 h	[structure] H (—)	807, 806

TABLE XXVII. REDUCTION OF HETEROCYCLIC BASES AND THEIR QUATERNARY SALTS (*Continued*)

No. of Carbon Atoms	Reactant	Conditions	Product(s) and Yield(s) (%)	Refs.
C_9 (Contd.)	3-(2-methyl-1,3-dioxolan-2-yl)pyridine	1. $ClCO_2CH_2C_6H_5$ 2. SMEAH, THF, −70°, 30 min	N-$CO_2CH_2C_6H_5$ dihydropyridine (58) + I ($CO_2CH_2C_6H_5$) + II ($CO_2CH_2C_6H_5$) (I + II:16)	1038a–c, 1039
	4-pyridone, N-$(CH_2)_2CH{=}CH_2$	SMEAH (IA), THF, 0°, 4 h	2,3-dihydro-4-pyridone, N-$(CH_2)_2CH{=}CH_2$ (87)	448
	pyridinium, N-COC_3H_7-*n*, Cl^-	LTBA–CuBr, THF, −23°, 1.5 h	1,4-dihydropyridine, N-COC_3H_7-*n* (32)	439

	Substrate	Conditions	Product(s) and Yield(s) (%)	Refs.
C_{10}	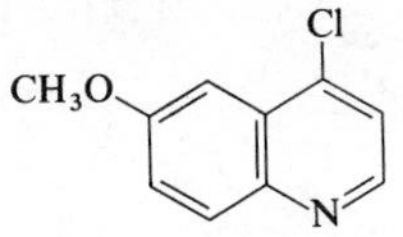 CH_3O, Cl, N	1. $BH_3 \cdot THF$ (IA), THF, $-78°$, 30 min 2. SMEAH (IA), C_6H_6, 25°, 18 h 3. $ClCO_2CH_3$, 25°	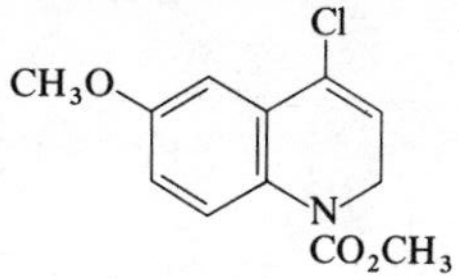 CH_3O, Cl, N, CO_2CH_3 (86)	445, 446
	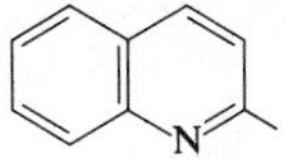 N	1. $BH_3 \cdot THF$ (IA), THF, $-78°$, 30 min 2. SMEAH, C_6H_6, 25°, 18 h 3. $ClCO_2CH_3$, 25°	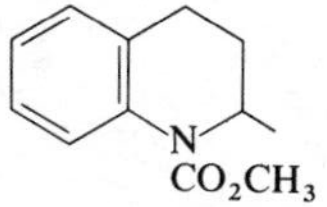 N, CO_2CH_3 (75)	446
	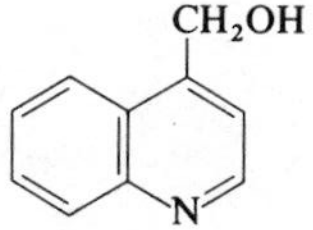 CH_2OH, N	1. $BH_3 \cdot THF$ (IA), THF, $-78°$, 30 min 2. SMEAH (IA), C_6H_6, $-78°$, 45 min 3. $ClCO_2CH_3$, 25°	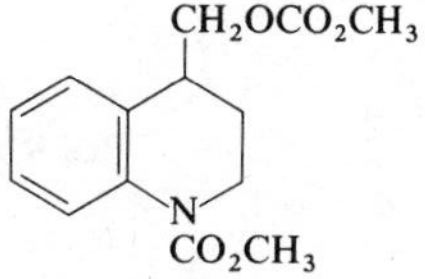 $CH_2OCO_2CH_3$, N, CO_2CH_3 (—)	446
	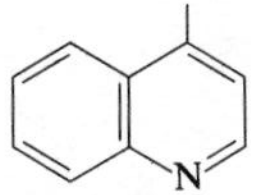 N	1. $BH_3 \cdot THF$ (IA), THF, $-78°$, 30 min 2. SMEAH (IA), C_6H_6, $-78°$, 30 min 3. $ClCO_2CH_3$, room temp, 18 h	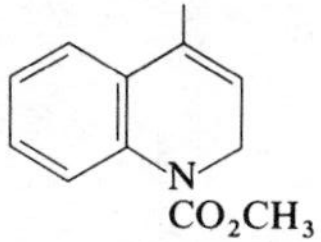 N, CO_2CH_3 (87)	445, 446
	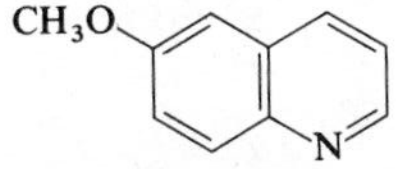 CH_3O, N		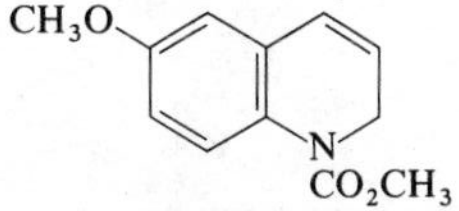 CH_3O, N, CO_2CH_3 (89)	445
	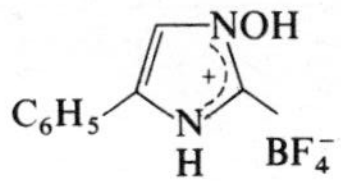 C_6H_5, NOH, +, N, H, BF_4^-	SMEAH	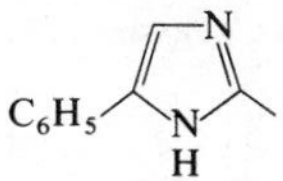 C_6H_5, N, N, H (50)	449
	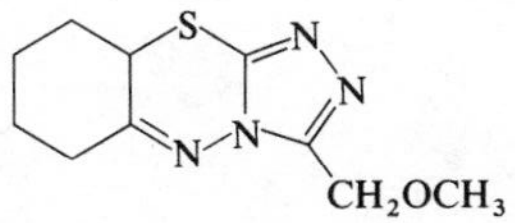 S, N, N, N, N, CH_2OCH_3	SMEAH (IA), THF, room temp, 3 h	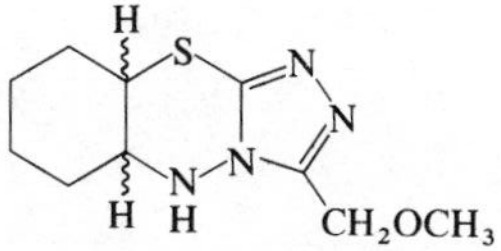 H, S, N, N, N, N, H, H, CH_2OCH_3 (15)	1040

TABLE XXVII. REDUCTION OF HETEROCYCLIC BASES AND THEIR QUATERNARY SALTS (*Continued*)

No. of Carbon Atoms	Reactant	Conditions	Product(s) and Yield(s) (%)	Refs.
C_{11}	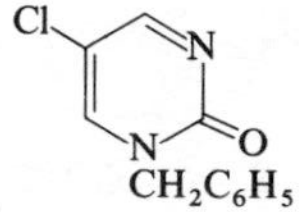 Cl, N, N, O, $CH_2C_6H_5$	LTBA (IA), THF, room temp, 30 min	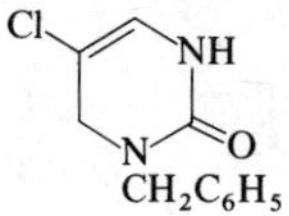 Cl, NH, N, O, $CH_2C_6H_5$ (76) 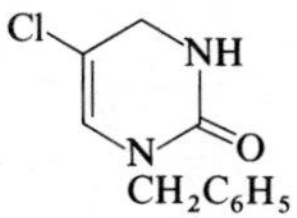 + Cl, NH, N, O, $CH_2C_6H_5$ (15)	1041a
	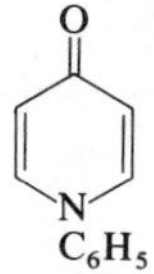 O, N, C_6H_5	LTBA, THF, 15°, 2 h	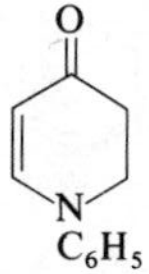 O, N, C_6H_5 (88)	447
	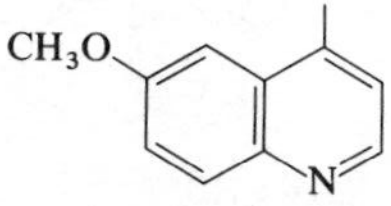 CH_3O, N	1. $BH_3 \cdot THF$ (IA), THF, −78°, 30 min 2. SMEAH (IA), C_6H_6, −78°, 30 min 3. $ClCO_2CH_3$, room temp, 18 h	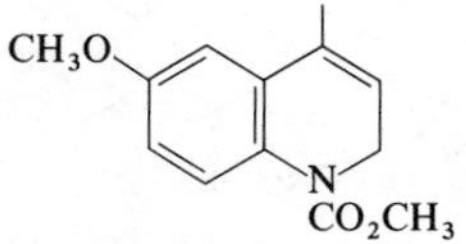 CH_3O, N, CO_2CH_3 (95)	445
	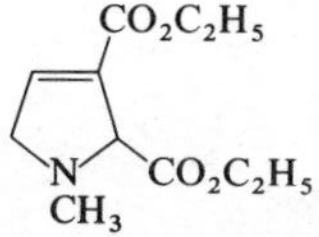 $CO_2C_2H_5$, N, $CO_2C_2H_5$, CH_3	SMEAH, C_6H_6–ether, room temp, 15 min	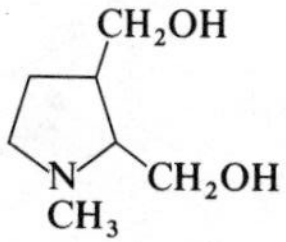 CH_2OH, N, CH_2OH, CH_3 (56)	726
C_{12}	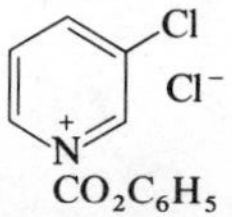 Cl, Cl^-, $\overset{+}{N}$, $CO_2C_6H_5$	LTBA–CuBr, THF, −23°, 1.5 h	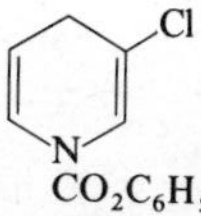 Cl, N, $CO_2C_6H_5$ (40)	439

	Reactant	Conditions	Product (Yield, %)	Refs.
	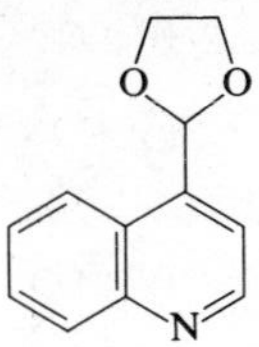	1. $BH_3 \cdot THF$ (IA), THF, $-78°$, 30 min 2. SMEAH (IA), C_6H_6, $-78°$, 30 min 3. $ClCO_2CH_3$, room temp, 18 h	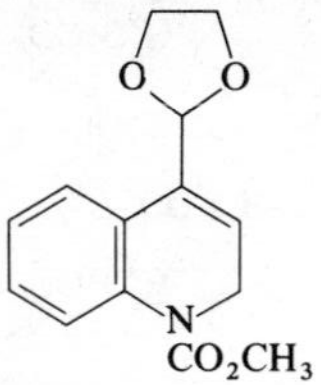 CO_2CH_3 (85)	445
	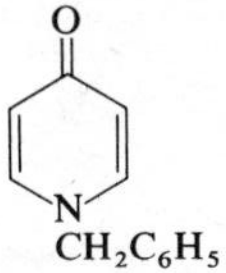 $CH_2C_6H_5$	LTBA, THF, 15°, 2.5 h	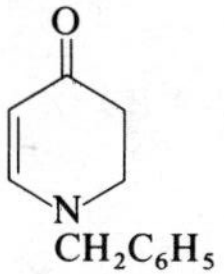 $CH_2C_6H_5$ (85)	448, 447
	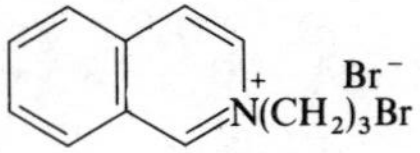 Br^- $N(CH_2)_3Br$	LTBA, THF, 0°, 45 min	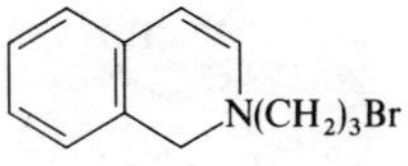 $N(CH_2)_3Br$ (68)	1041b
	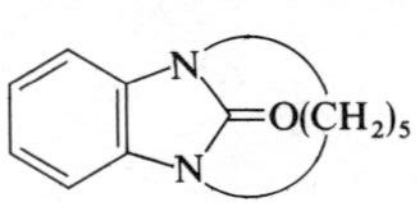 $O(CH_2)_5$	SMEAH, C_6H_6, 4 h	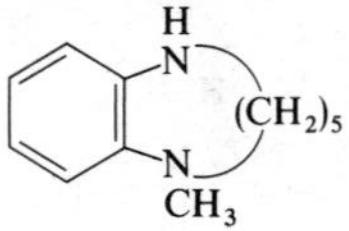 $(CH_2)_5$, CH_3 (80)	1024
	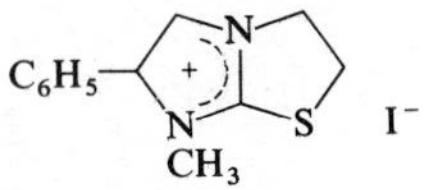 C_6H_5, CH_3, I^-	SMEAH, C_6H_6–toluene, room temp, 18 h	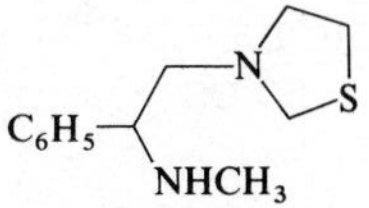 C_6H_5, $NHCH_3$ (44)	1042
	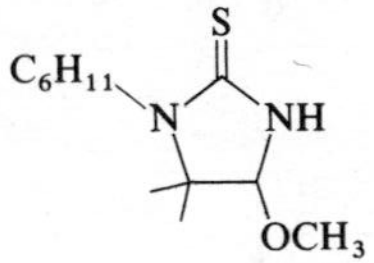 C_6H_{11}, OCH_3	SMEAH (IA), C_6H_6, room temp, 2 h	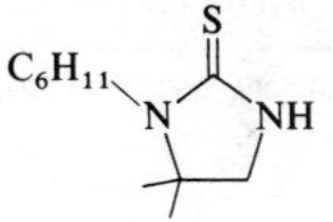 C_6H_{11} (89)	1043
C_{13}	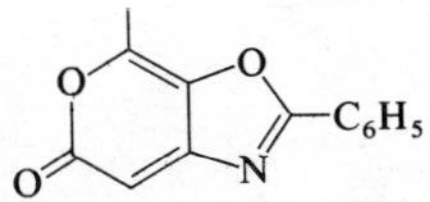 C_6H_5	SMEAH	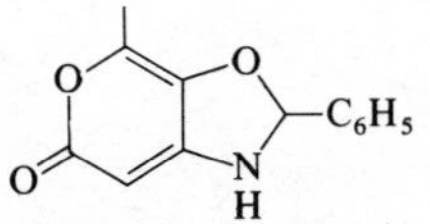 C_6H_5 (—)	1044

TABLE XXVII. Reduction of Heterocyclic Bases and Their Quaternary Salts (*Continued*)

No. of Carbon Atoms	Reactant	Conditions	Product(s) and Yield(s) (%)	Refs.
C_{13} (*Contd.*)	4-COC_6H_5-5-CO_2CH_3-pyridazine	1. SMEAH, THF, −70°, 4 h 2. N_2H_4	4-$CHOHC_6H_5$-5-CO_2CH_3-1,4-dihydropyridazine (19) + 4-$CHOHC_6H_5$-5-$CH{=}NNH_2$-pyridazine (15)	767
	5-Cl-4-CO_2CH_3-1-$CH_2C_6H_5$-pyrimidin-2-one	LTBA (IA), THF, −78°, 30 min	5-Cl-4-CO_2CH_3-1-$CH_2C_6H_5$-3,6-dihydropyrimidin-2-one (87)	1041a
	1-$CO_2CH_2C_6H_5$-pyridinium Cl^-	LTBA–CuBr, THF, −23°, 1.5 h	1-$CO_2CH_2C_6H_5$-1,2,3,4-tetrahydropyridine (35)	439
	3-CH_3-1-$CO_2C_6H_5$-pyridinium Cl^-	" " " "	5-CH_3-1-$CO_2C_6H_5$-1,2,3,4-tetrahydropyridine (51)	439
C_{14}	3-CO_2CH_3-1-$CO_2C_6H_5$-pyridinium Cl^-	" " " "	5-CO_2CH_3-1-$CO_2C_6H_5$-1,2,3,4-tetrahydropyridine (41)	439

	Reactant	Conditions	Product(s) and Yield(s) (%)	Refs.
	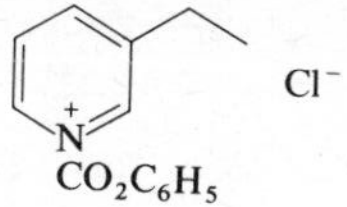 $CO_2C_6H_5$, Cl^-	" " " "	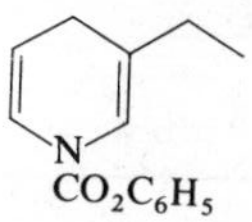 $CO_2C_6H_5$ (39)	439
	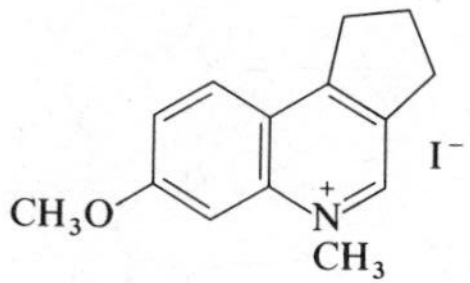 CH_3O, CH_3, I^-	LTBA, ether, room temp, 4 h	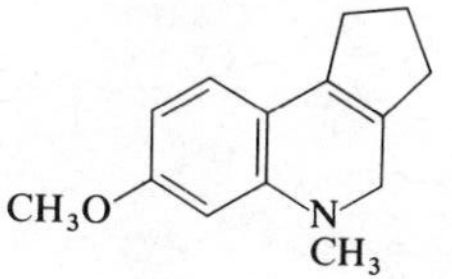 CH_3O, CH_3 (48)	1045
	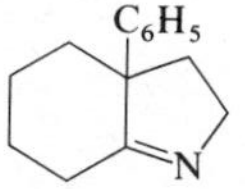 C_6H_5	1. LTBA 2. Ac_2O, Py	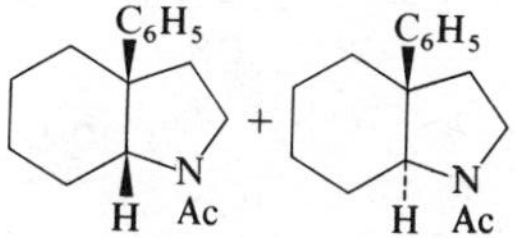 C_6H_5, H, Ac, **I** + C_6H_5, H, Ac, **II** (—) (**I**:**II** = 55:45)	1046
C_{15}	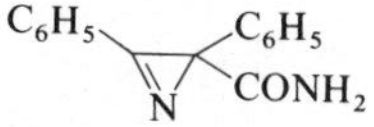 C_6H_5, C_6H_5, $CONH_2$	SMEAH, C_6H_6, 0°	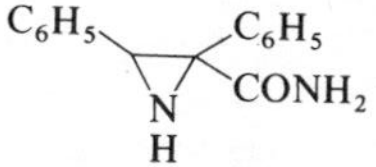 C_6H_5, C_6H_5, $CONH_2$, H (71)	151
	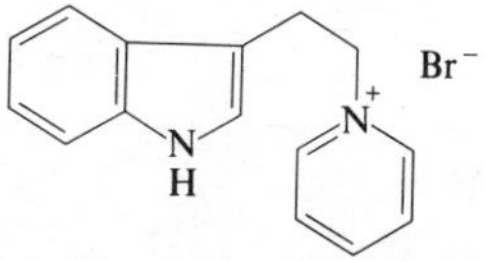 H, Br^-	LTBA, THF, room temp, 25 h	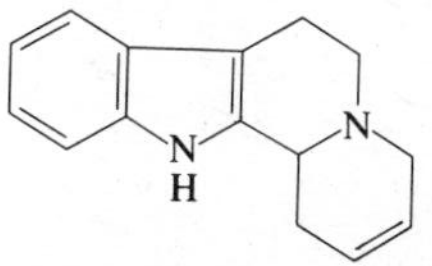 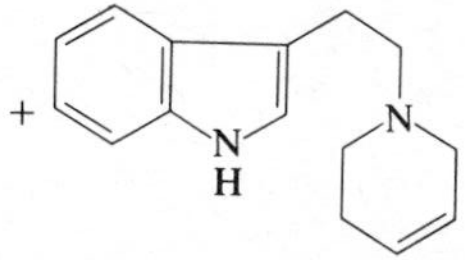 H (30) + H (5)	1047

No. of Carbon Atoms	Reactant	Conditions	Product(s) and Yield(s) (%)	Refs.
C_{16}	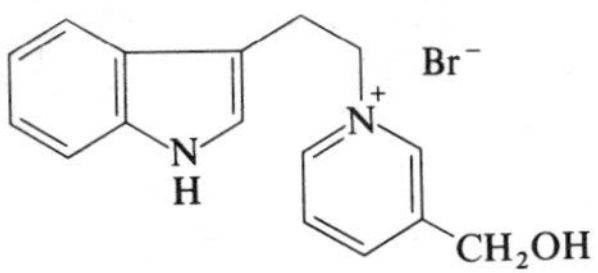 Br^- CH_2OH (N-H indole / pyridinium structure)	LTBA, THF, reflux, 2 h	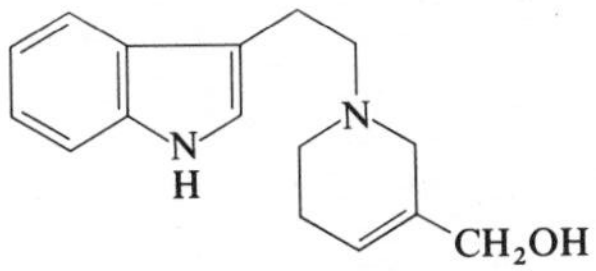 CH_2OH (28) 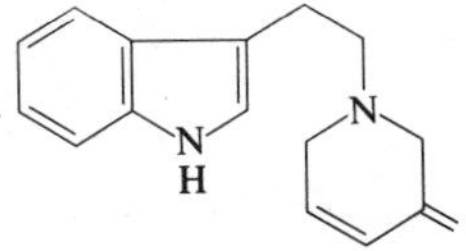 + (17)	682
	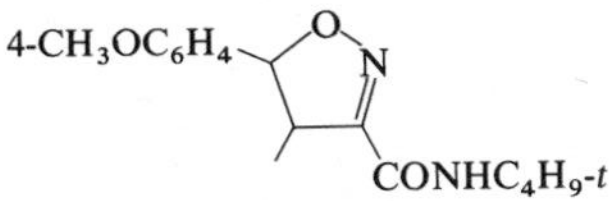 4-$CH_3OC_6H_4$ … $CONHC_4H_9$-*t*	1. SMEAH, THF, −78 to 25° 2. HCl, C_2H_5OH	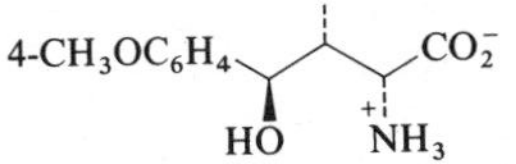 4-$CH_3OC_6H_4$ … CO_2^-, HO, $\overset{+}{N}H_3$ (62)	451
		1. SMEAH, THF, −78 to 25° 2. HBr, AcOH	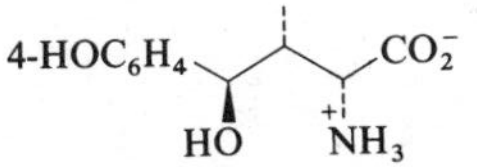 4-HOC_6H_4 … CO_2^-, HO, $\overset{+}{N}H_3$ (56)	451
C_{17}	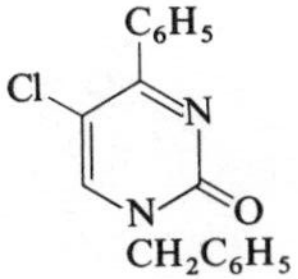 Cl, C_6H_5, $CH_2C_6H_5$	LTBA (IA), THF, room temp, 30 min	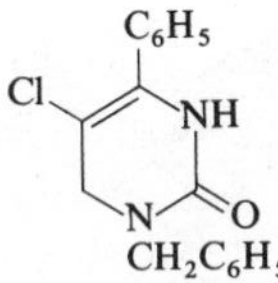 Cl, C_6H_5, NH, $CH_2C_6H_5$ (79)	1041a
	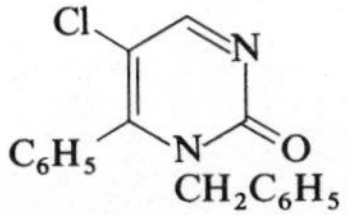 Cl, C_6H_5, $CH_2C_6H_5$	" " " " "	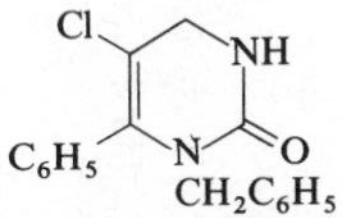 Cl, NH, C_6H_5, $CH_2C_6H_5$ (69)	1041a

Substrate	Conditions	Product(s) (Yield, %)	Refs.
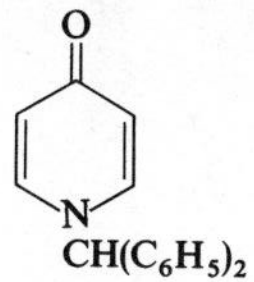 $CH(C_6H_5)_2$	SMEAH (IA), THF, 0°, 4 h	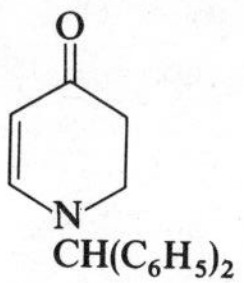 $CH(C_6H_5)_2$ (60)	448
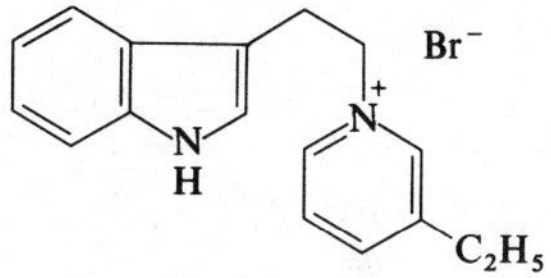 Br^-, C_2H_5	LTBA, THF	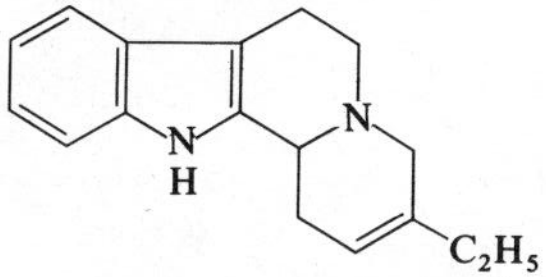 C_2H_5 (36)	1047
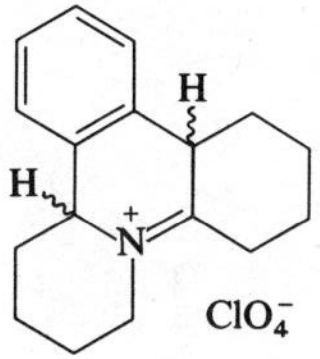 ClO_4^-	LTBA, THF, reflux, 4 h	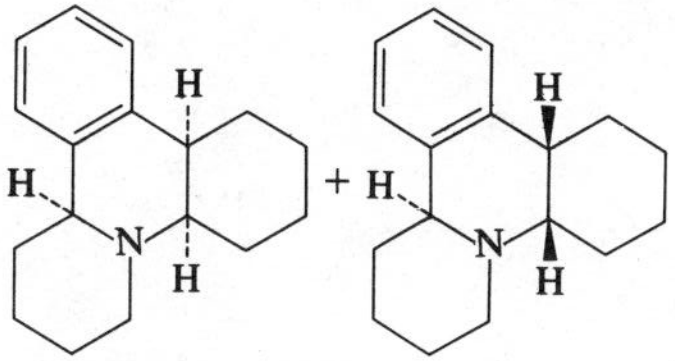 I + II 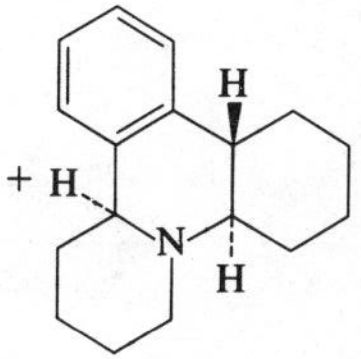 + III (—) (I:II:III = 79:19:2)	1048
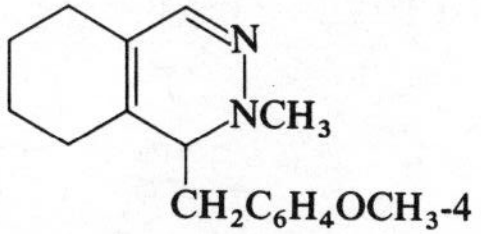 NCH_3, $CH_2C_6H_4OCH_3$-4	SMEAH, C_6H_6, reflux, 18 h	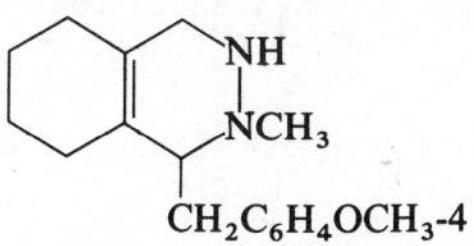 NCH_3, $CH_2C_6H_4OCH_3$-4 (21)	1049

TABLE XXVII. Reduction of Heterocyclic Bases and Their Quaternary Salts (*Continued*)

No. of Carbon Atoms	Reactant	Conditions	Product(s) and Yield(s) (%)	Refs.
C_{18}	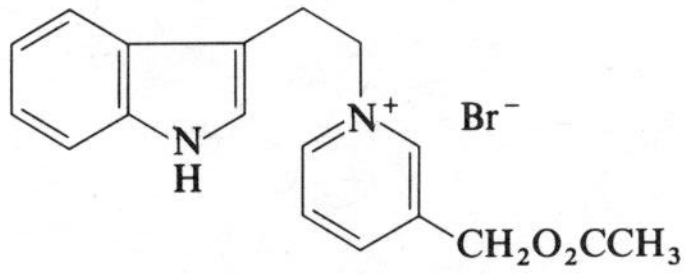 [structure] Br^-, $CH_2O_2CCH_3$	LTBA, THF, reflux, 2 h	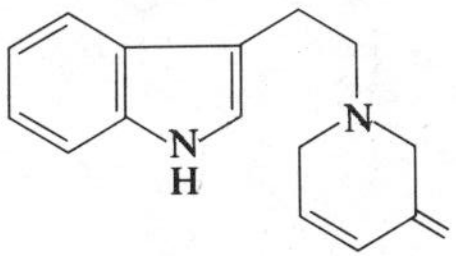 [structure] (31)	682
	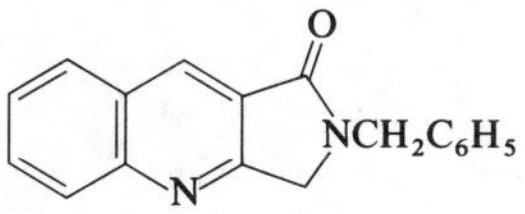 [structure] $NCH_2C_6H_5$	SMEAH	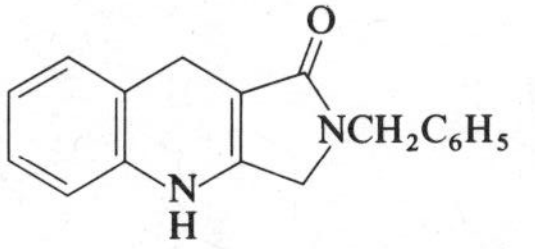 [structure] $NCH_2C_6H_5$ (52)	972a
	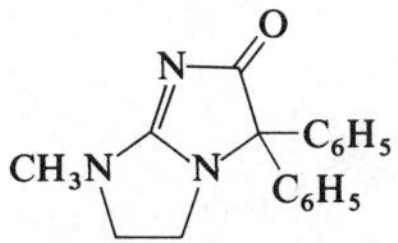 [structure] CH_3N, C_6H_5, C_6H_5	SMEAH (IA), THF, C_6H_6, 0°, 4 h	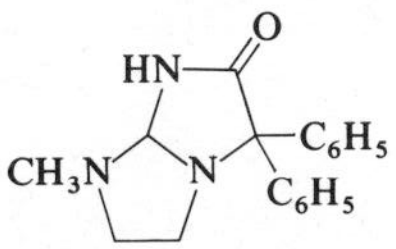 [structure] CH_3N, C_6H_5, C_6H_5 (86)	372
C_{19}	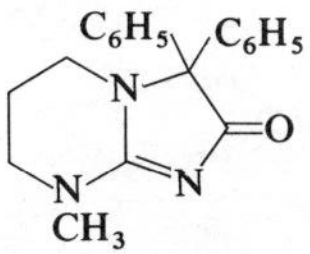 [structure] C_6H_5, C_6H_5, CH_3	SMEAH (IA), THF, C_6H_6, 0°, 5 h	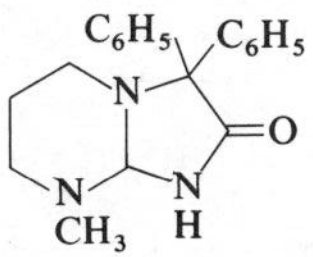 [structure] C_6H_5, C_6H_5, CH_3 (85)	372
	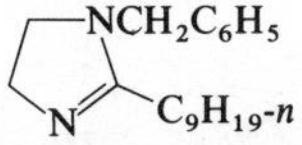 [structure] $NCH_2C_6H_5$, C_9H_{19}-*n*	LTBA	No reaction	1050a

	Substrate	Reagent/Conditions	Product (Yield %)	Refs.
C_{20}	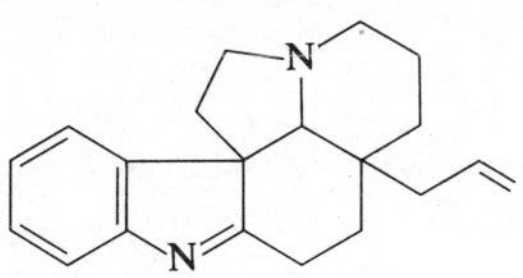	SMEAH, C_6H_6	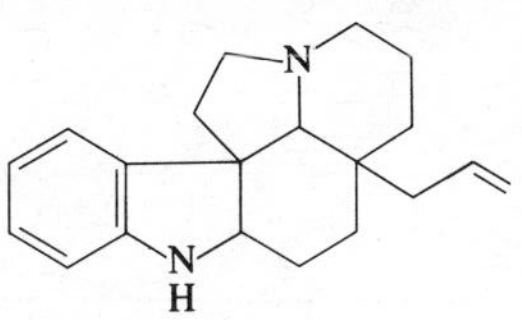 (—)	1050b
C_{21}	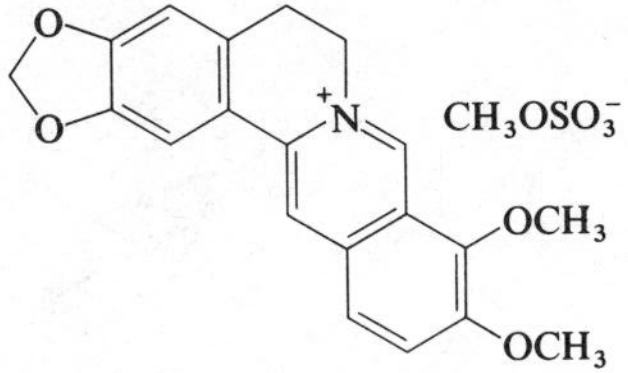 $CH_3OSO_3^-$	SMEAH, C_6H_6–THF	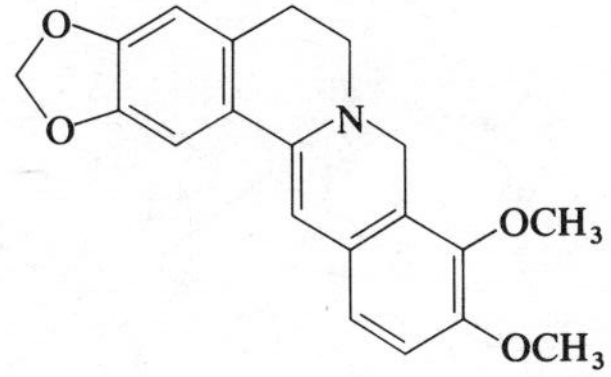 (—)	440
C_{22}	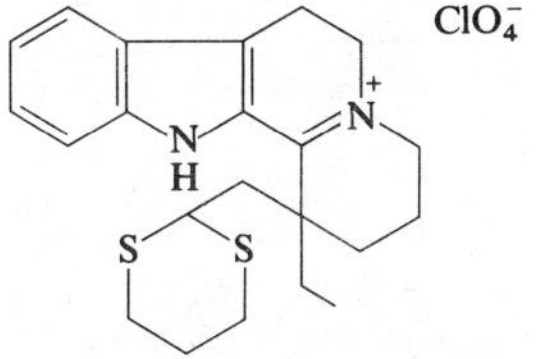 ClO_4^-	LTBA, THF, 0°, 3 h	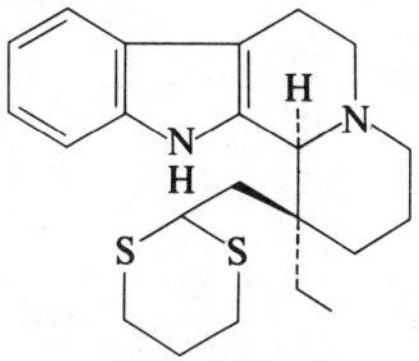 (–)- (79)	1051a–c
C_{23}	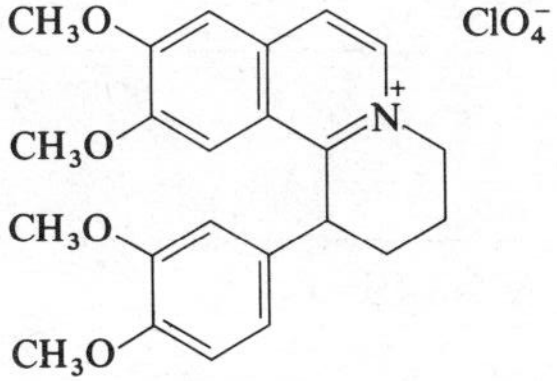 ClO_4^-	SMEAH, ether–toluene, reflux, 30 min	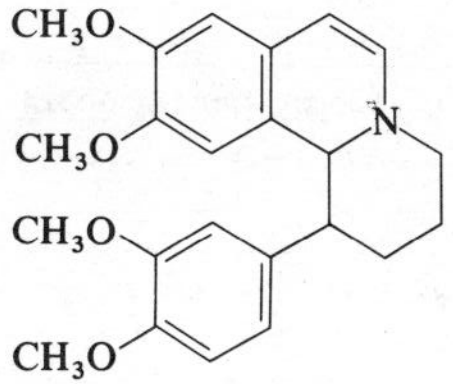 (79)	1052

TABLE XXVII. REDUCTION OF HETEROCYCLIC BASES AND THEIR QUATERNARY SALTS (*Continued*)

No. of Carbon Atoms	Reactant	Conditions	Product(s) and Yield(s) (%)	Refs.
C_{26}	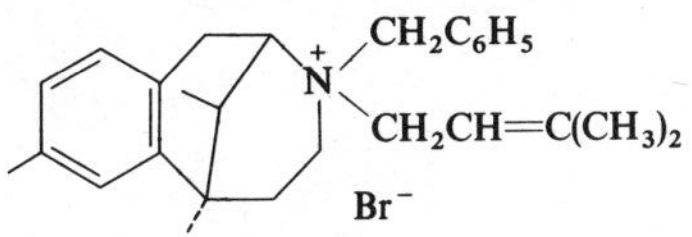 $\overset{+}{N}(CH_2C_6H_5)CH_2CH{=}C(CH_3)_2$ Br^-	SMEAH, xylene, reflux, 17 h	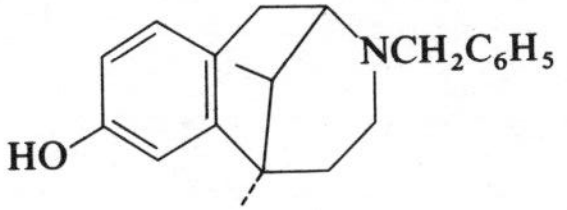 HO, $NCH_2C_6H_5$ (43) 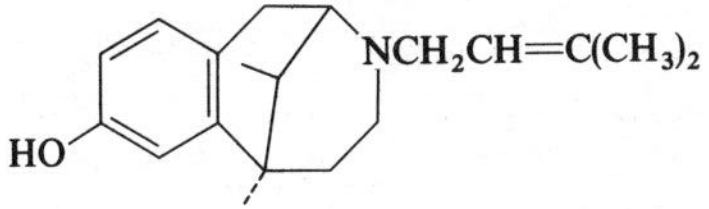 + HO, $NCH_2CH{=}C(CH_3)_2$ (11)	988
C_{36}	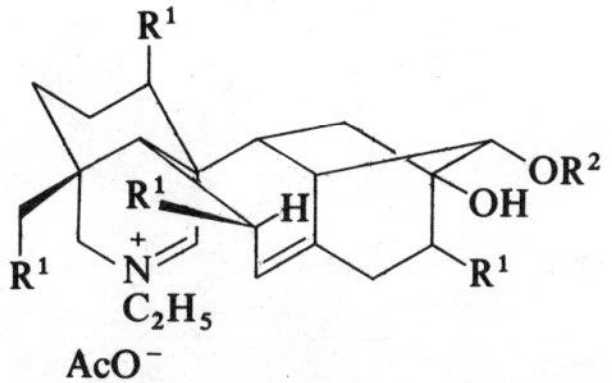 R^1, R^1, H, OR^2, OH, R^1, R^1, $\overset{+}{N}$, C_2H_5 AcO^- $R^1 = OCH_3$; R^2 = veratryl	LTBA	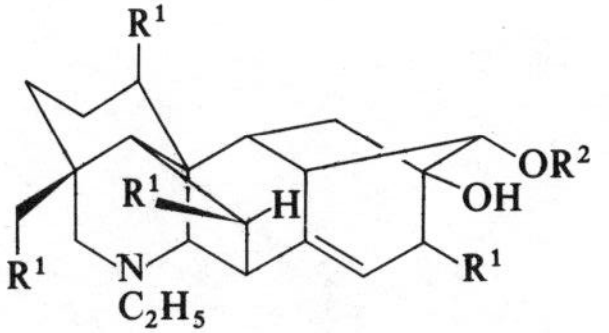 R^1, R^1, H, OR^2, OH, R^1, R^1, N, C_2H_5 (—)	1053

TABLE XXVIII. REDUCTION OF SULFOXIDES

No. of Carbon Atoms	Reactant	Conditions	Product(s) and Yield(s) (%)	Ref.
C_2	$(CH_3)_2SO$	SMEAH (IA), C_6H_6, 65°	$(CH_3)_2S$ (70)	457a
C_3	*dl*-$CH_3S(O)C_2H_5$	LAH + (+)-QND (1:1), ether, reflux, 40 h	$CH_3SC_2H_5 + CH_3\overset{*}{S}(O)C_2H_5$ (24)	461
C_4	*dl*-$CH_3S(O)C_3H_7$-*n*	LAH + (+)-QND (1:1), ether, reflux, 40 h	$CH_3SC_3H_7$-*n* + (*R*)-(−)-$CH_3\overset{*}{S}(O)C_3H_7$-*n* (40)	461
	dl-$CH_3S(O)C_3H_7$-*i*	LAH + (+)-QND (1:1), ether, reflux, 40 h	$CH_3SC_3H_7$-*i* + (*R*)-(−)-$CH_3\overset{*}{S}(O)C_3H_7$-*i* (40)	461
C_5	*dl*-$CH_3S(O)C_4H_9$-*n*	LAH + (−)-QN (1:1), ether	$CH_3SC_4H_9$-*n* + (*S*)-(+)-$CH_3\overset{*}{S}(O)C_4H_9$-*n* (54)	461
		LAH + (−)-(1*R*,2*S*)-ephedrine (1:1), ether	" " (55)	461
		LAH + (−)-MTH (1:1), ether	" " (42)	461
		LAH + (−)-CND (1:1), ether	" + (*R*)-(−)-$CH_3\overset{*}{S}(O)C_4H_9$-*n* (67)	461
		LAH + (−)-MTH (1:2), ether	" + " (—)	461
	dl-$CH_3S(O)C_4H_9$-*i*	LAH + (+)-QND (1:1), ether, reflux, 40 h	$CH_3SC_4H_9$-*i* + (*R*)-(−)-$CH_3\overset{*}{S}(O)C_4H_9$-*i* (35)	461
C_8	*dl*-$CH_3S(O)CH_2C_6H_5$	LAH + (+)-QND (1:1), ether, reflux, 40 h	$CH_3SCH_2C_6H_5 + CH_3\overset{*}{S}(O)CH_2C_6H_5$ (72)	461
	dl-$CH_3S(O)C_6H_4CH_3$-4	", ", ", "	$CH_3SC_6H_4CH_3$-4 + (*S*)-(−)-$CH_3\overset{*}{S}(O)C_6H_4CH_3$-4 (45)	461
	$(n\text{-}C_4H_9)_2SO$	SMEAH (IA), C_6H_6, reflux, 30 min	$(n\text{-}C_4H_9)_2S$ (98)	457a
C_{12}	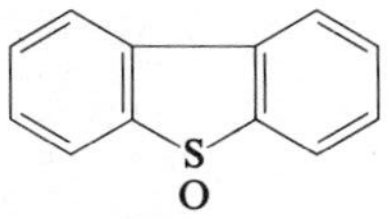 [dibenzothiophene S-oxide structure]	SMEAH (IA), C_6H_6, reflux, 30 min	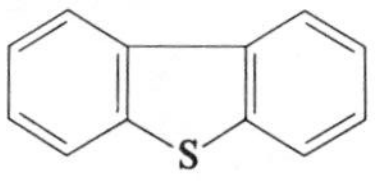 [dibenzothiophene structure] (95)	457a
	$(C_6H_5)_2SO$	SMEAH (IA), C_6H_6, reflux, 30 min	$(C_6H_5)_2S$ (97)	457a
C_{15}	$C_6H_5S(O)(=NCH_3)CH_2COC_6H_5$	LTBA	No reaction	1054
C_{18}	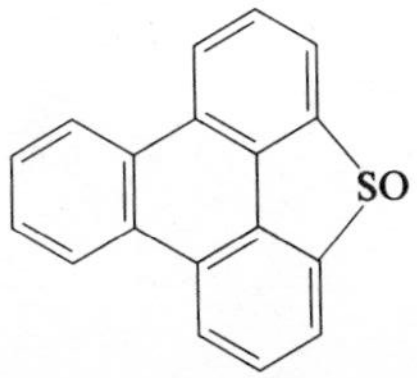 [structure]-SO	SMEAH, C_6H_6, reflux, 1 h	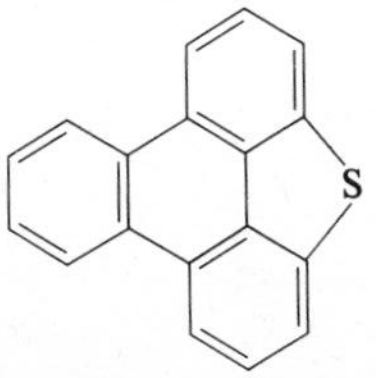 [structure]-S (82)	1055

TABLE XXIX. Reduction of Alkanesulfonic and Arenesulfonic Acid Esters

No. of Carbon Atoms	Reactant	Conditions	Product(s) and Yield(s) (%)	Ref.
C_8	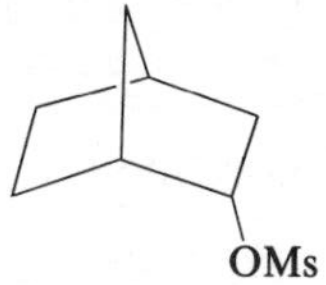 OMs	LTMA–CuI, THF, room temp, 5 h	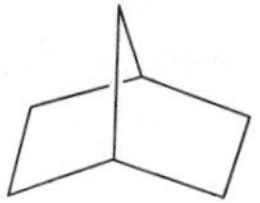 (75)	472
		$LiAlD(OCH_3)_3$–CuI, THF, room temp	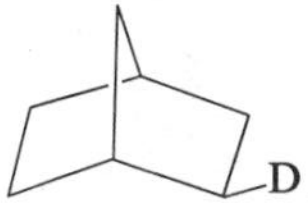 D (—)	472
	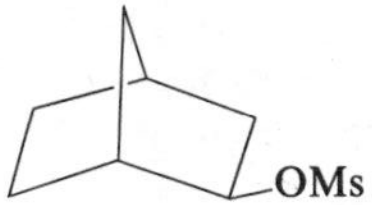 OMs	LTMA–CuI, THF, room temp, 2 h	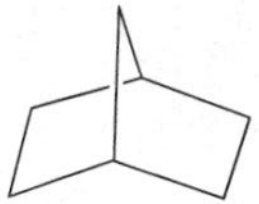 (99)	472
		$LiAlD(OCH_3)_3$–CuI, THF, room temp	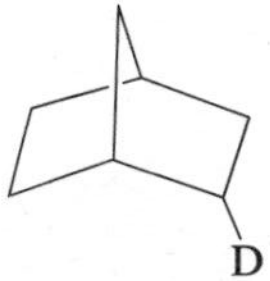 D (—)	472
C_{10}	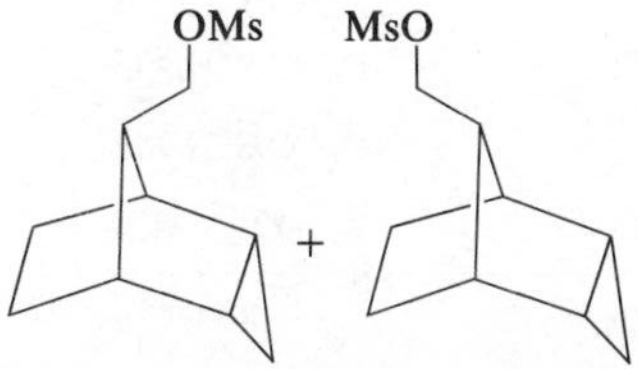 OMs MsO **I** + **II** (**I**:**II** = 78:22)	LTMA–CuI, THF, room temp, overnight	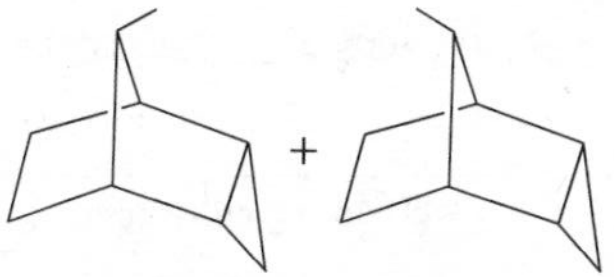 **III** + **IV** (—) (**III**:**IV** = 80:20)	1056

	Substrate	Reagent and Conditions	Product(s) and Yield(s) (%)	Refs.
	n-$C_9H_{19}OMs$	LTMA–CuI, THF, room temp, 1 h	n-C_9H_{20} (99)	472
C_{16}	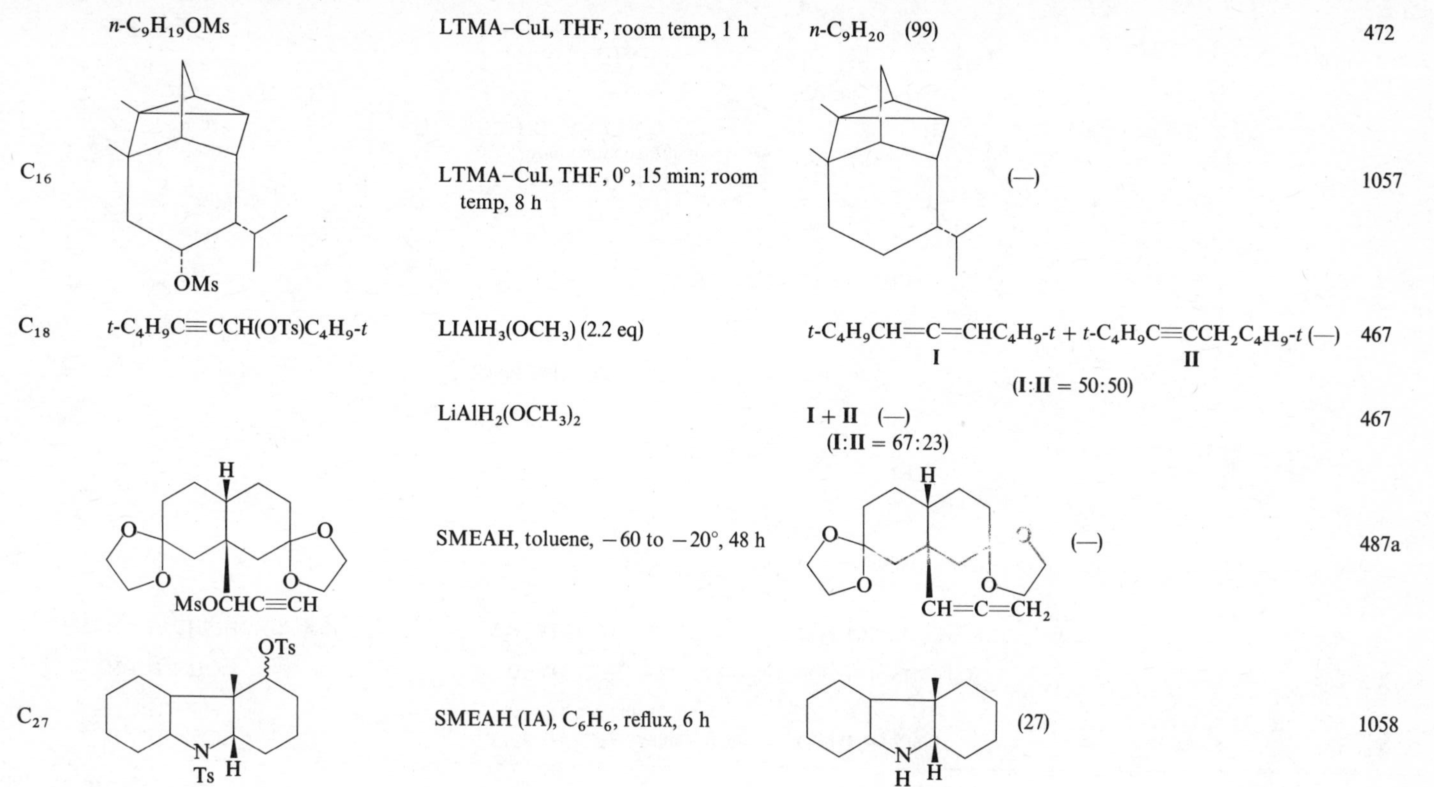	LTMA–CuI, THF, 0°, 15 min; room temp, 8 h	(—)	1057
C_{18}	t-$C_4H_9C{\equiv}CCH(OTs)C_4H_9$-$t$	$LIAlH_3(OCH_3)$ (2.2 eq)	t-$C_4H_9CH{=}C{=}CHC_4H_9$-t (**I**) + t-$C_4H_9C{\equiv}CCH_2C_4H_9$-$t$ (**II**) (—) (**I**:**II** = 50:50)	467
		$LiAlH_2(OCH_3)_2$	**I** + **II** (—) (**I**:**II** = 67:23)	467
		SMEAH, toluene, −60 to −20°, 48 h	(—)	487a
C_{27}		SMEAH (IA), C_6H_6, reflux, 6 h	(27)	1058

TABLE XXX. REDUCTION OF SULFONAMIDES

No. of Carbon Atoms	Reactant	Conditions	Product(s) and Yield(s) (%)	Refs.
C_7	$C_6H_5NHO_2SCF_3$	SMEAH, C_6H_6, reflux, 10 min	$C_6H_5NH_2$ (94)	492, 494, 493
	C_6H_5NHMs	SMEAH, toluene, reflux	" (63)	488
C_8	$C_6H_5CH_2NHO_2SCF_3$	SMEAH, C_6H_6, reflux, <10 min	$C_6H_5CH_2NH_2$ (95)	492, 494, 493
C_{11}	$C_6H_5CH_2CH(CH_3)N(Ms)CH_3$	SMEAH, C_6H_6, reflux	$C_6H_5CH_2CH(CH_3)NHCH_3$ (67)	488
C_{12}	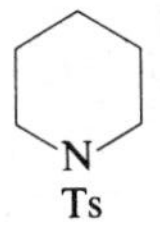 N–Ts (piperidine ring)	" " "	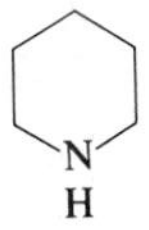 N–H (piperidine ring) (75)	488
	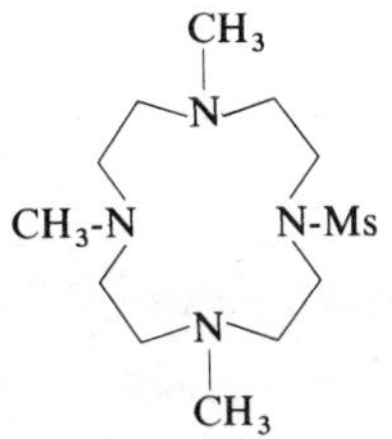 CH_3-N, N-Ms, N–CH_3, N–CH_3 (macrocycle)	SMEAH (IA), toluene, 100°, 25 h	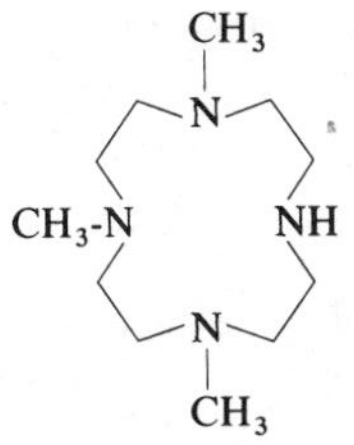 CH_3-N, NH, N–CH_3, N–CH_3 (macrocycle) (50)	1059a
C_{13}	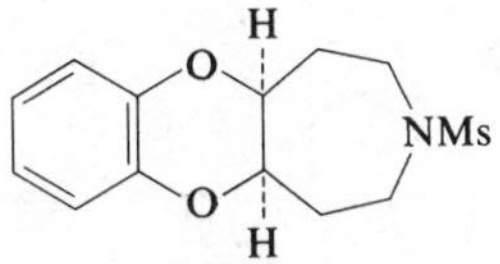 O, O, H, H, NMs (fused ring system)	1. SMEAH, C_6H_6, reflux, 4 h; room temp, overnight 2. HCl, $AcOC_2H_5$	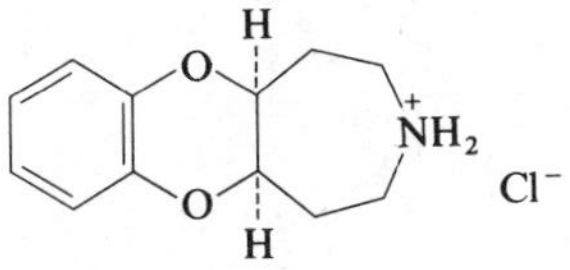 O, O, H, H, $\overset{+}{N}H_2$ Cl^- (fused ring system) (69)	1059b
	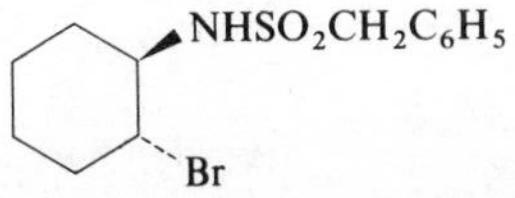 $NHSO_2CH_2C_6H_5$, Br (cyclohexane)	SMEAH, C_6H_6, reflux, 2 h	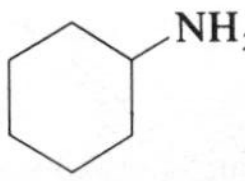 NH_2 (cyclohexane) (60)	490

C_{15}	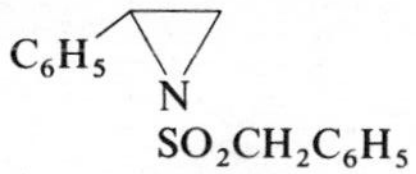 1-($SO_2CH_2C_6H_5$)-2-C_6H_5-aziridine [structure]	" " " "	$C_6H_5CH(CH_3)NH_2$ + $C_6H_5(CH_2)_2NH_2$ (20) I + II (**I**:**II** = 90:10)	490
	$C_6H_5CH(CH_2Br)NHSO_2CH_2C_6H_5$	SMEAH, C_6H_6, reflux, 2 h	**I** + **II** (46) (**I**:**II** = 90:10)	490
	$C_6H_5CHBrCH_2NHSO_2CH_2C_6H_5$	" " " "	**I** + **II** (22) (**I**:**II** = 90:10)	490
	$C_6H_5(CH_2)_2NHSO_2CH_2C_6H_5$	" " " "	**II** (98)	490
	[structure: N-Ts bicyclic enamine, H, H]	" " " "	[structure: bicyclic NH amine, H, H] (46)	1060
	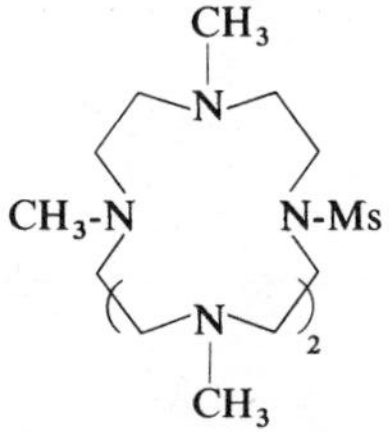 [structure: macrocycle CH_3-N, N-Ms, CH_3]	SMEAH (IA), toluene, reflux, 23 h	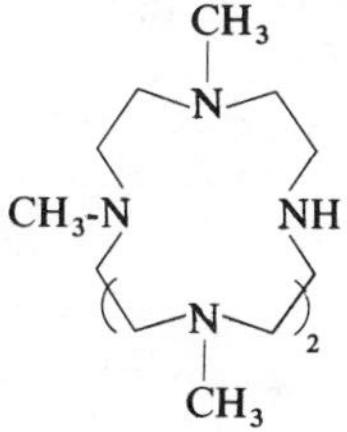 [structure: macrocycle CH_3-N, NH, CH_3] (65)	1059a
C_{16}	$C_6H_5CH_2CH(CH_3)NHTs$ (*S*)-	SMEAH, toluene, reflux, 72 h	$C_6H_5CH_2CH(CH_3)NH_2$ (27) (*S*)-	488
C_{17}	$C_6H_5CH_2CH(CH_3)N(Ts)CH_3$	SMEAH, C_6H_6, reflux	$C_6H_5CH_2CH(CH_3)NHCH_3$ (77)	488
	$C_6H_5CH(OC_2H_5)CH_2NHO_2SCH_2C_6H_5$	" " "	$C_6H_5CH(OC_2H_5)CH_2NH_2$ (22)	490
C_{18}	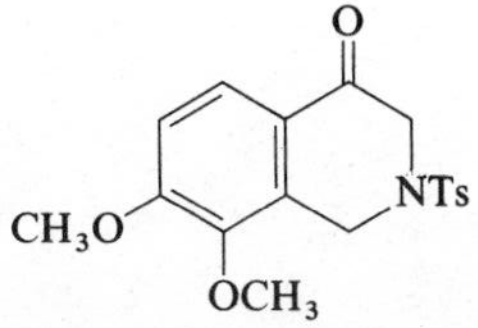 [structure: 7,8-(CH_3O)$_2$-2-Ts-tetrahydroisoquinolin-4-one; O, NTs, CH_3O, OCH_3]	SMEAH, C_6H_6, reflux, 12 h	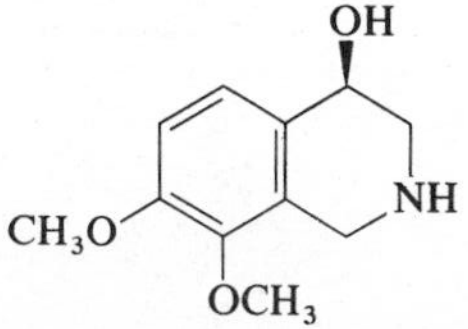 [structure: OH, NH, CH_3O, OCH_3] (62)	1061a, 1061b

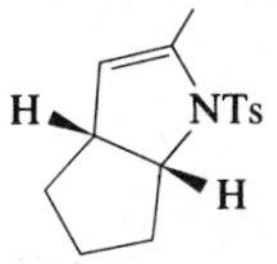

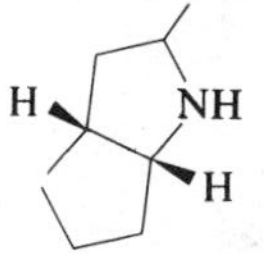

TABLE XXX. REDUCTION OF SULFONAMIDES (*Continued*)

No. of Carbon Atoms	Reactant	Conditions	Product(s) and Yield(s) (%)	Refs.
C_{18} (*Contd.*)	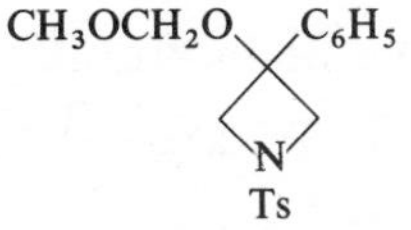 CH_3OCH_2O, C_6H_5 (azetidine, N-Ts)	SMEAH, toluene, reflux	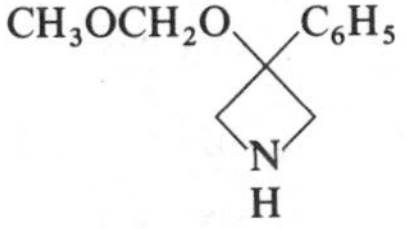 CH_3OCH_2O, C_6H_5 (azetidine, N-H) (69)	488
	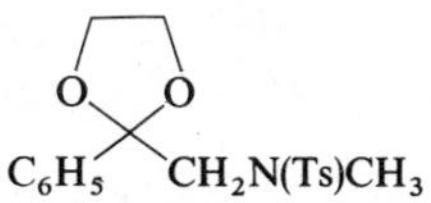 O O (dioxolane) C_6H_5 $CH_2N(Ts)CH_3$	SMEAH, toluene, reflux, 22 h	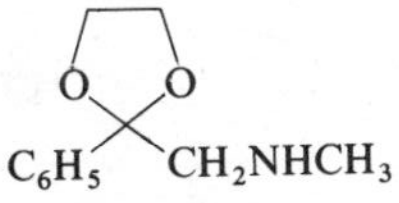 O O (dioxolane) C_6H_5 CH_2NHCH_3 (56)	488
C_{19}	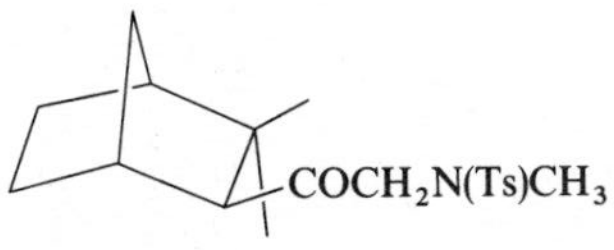 $COCH_2N(Ts)CH_3$ (bicyclic)	" " " "	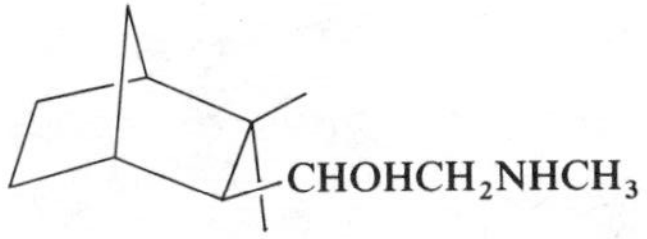 $CHOHCH_2NHCH_3$ (bicyclic) (60)	1062
C_{20}	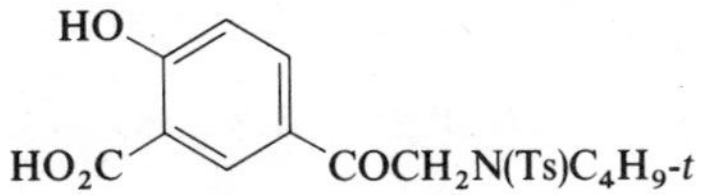 HO, HO_2C, $COCH_2N(Ts)C_4H_9$-*t* (benzene ring)	SMEAH, C_6H_6, reflux, 12 h	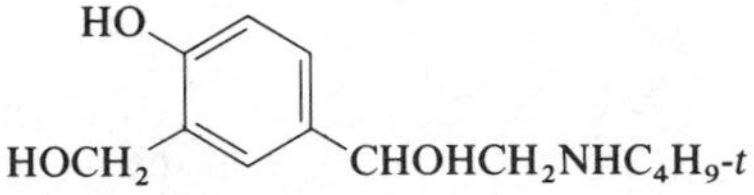 HO, $HOCH_2$, $CHOHCH_2NHC_4H_9$-*t* (benzene ring) (—)	489
	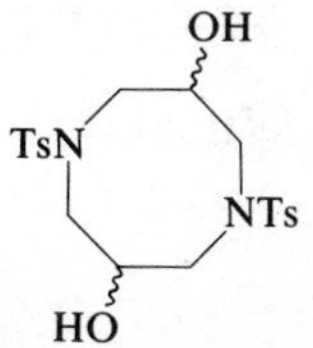 OH, TsN, NTs, HO (8-membered ring)	SMEAH, C_6H_6, reflux, 20 h	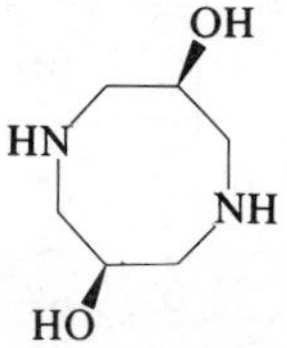 OH, HN, NH, HO (8-membered ring) (33)	488

	Reactant	Reagents and conditions	Product(s) (Yield, %)	Refs.
	2-benzyloxyphenoxy, H, OH, NC_3H_7-i, Ms; $OCH_2C_6H_5$; (2*S*)-(−)	SMEAH, toluene	2-benzyloxyphenoxy, H, OH, NC_3H_7-i, H; $OCH_2C_6H_5$; (2*S*)-(−) (—) + H, OH, NC_3H_7-i, CH_3; $OCH_2C_6H_5$ (~5)	1063
		1. $CH_2{=}C(CH_3)OCH_3$, $POCl_3$, toluene 2. $(C_2H_5)_3N$ 3. SMEAH, toluene, 80–82°, 2 h	" (—)	1063
	$C_6H_5CH_2O$-phenoxy, H, OH, NC_3H_7-i, Ms; (2*S*)-(−)	SMEAH, toluene	$C_6H_5CH_2O$-phenoxy, H, OH, NC_3H_7-i, H; (2*S*)-(−) (—) + $C_6H_5CH_2O$-phenoxy, H, OH, NC_3H_7-i, CH_3 (~5)	1063
		1. $CH_2{=}C(CH_3)OCH_3$ $POCl_3$, toluene 2. $(C_2H_5)_3N$ 3. SMEAH, toluene, 80–82°, 2 h	" (87)	1063
C_{22}	$C_6H_5SO_2HN$-phenyl-pyrrole, CH_3O_2C, CO_2CH_3, N-CH_3	$LiAlH_2(OC_4H_9$-$t)_2$	H_2N-phenyl-pyrrole, $HOCH_2$, CH_2OH, N-CH_3 (—)	683

TABLE XXX. REDUCTION OF SULFONAMIDES (*Continued*)

No. of Carbon Atoms	Reactant	Conditions	Product(s) and Yield(s) (%)	Refs.
C_{22} (*Contd.*)	[structure: NSO$_2$CH$_2$ (camphorsulfonyl) azepine, O, O]	SMEAH, C_6H_6, reflux, 4.5 h	[structure: NH, O, O] (—)	1059b
	[structure: TsN, Ts-N, O, O macrocycle]	SMEAH, toluene, reflux, 14.5 h	[structure: HN, H-N, O, O macrocycle] (37)	1064
	$[HO(CH_2)_2NTs(CH_2)_2]_2O$	SMEAH, C_6H_6, reflux, 1 d	$[HO(CH_2)_2NH(CH_2)_2]_2O$ (62)	1065
C_{27}	[structure: N-Ts, H, O, O, SCH_3]	SMEAH	[structure: N-H, H, OH, SCH_3] (—)	989
	[structure: OTs, N-Ts, H]	SMEAH (IA), C_6H_6, reflux, 6 h	[structure: N-H, H] (27)	1058

	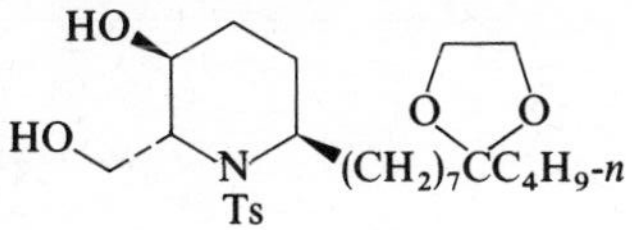	SMEAH, C_6H_6, reflux, 24 h	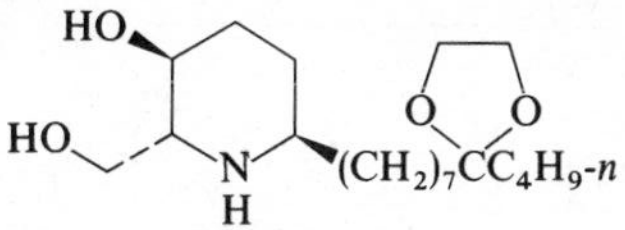 (—)	1066
C_{28}	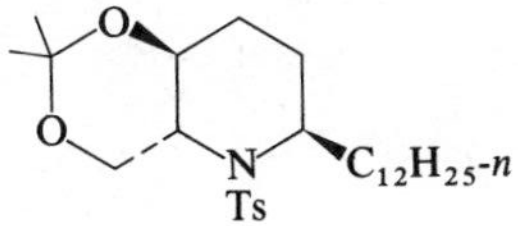	"	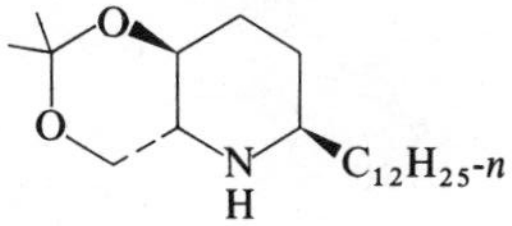 (—)	1066
	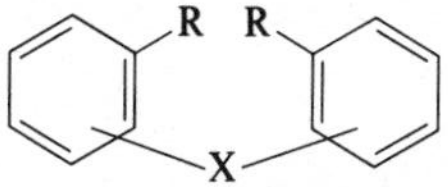 R = $OCHOHCH_2N(Ms)C_3H_7$-*i* (*S*,*S*)-(−)		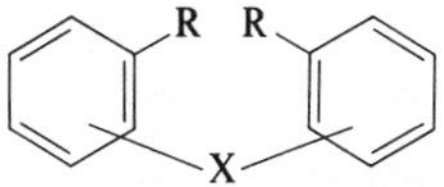 R = $OCHOHCH_2NHC_3H_7$-*i* (*S*,*S*)-(−)	
	X			
C_{27}	$O(CH_2)_3O$ (*o*,*o*)	1. $CH_2{=}C(CH_3)OCH_3$, $POCl_3$, toluene 2. $(C_2H_5)_3N$ 3. SMEAH, toluene, 80–85°, 1.5 h	(82)[a]	1063
	" (*m*,*m*)	", ", "	(85)[b]	1063
	" (*p*,*p*)	SMEAH (IA), toluene, 80°, 3.5 h	(56)[c]	1063
	$OCH_2CHOHCH_2O$ (*o*,*o*)	1. $CH_2{=}C(CH_3)OCH_3$, $POCl_3$, toluene 2. $(C_2H_5)_3N$ 3. SMEAH, toluene, 80–85°, 1.5 h	(72)[c]	1063

TABLE XXX. Reduction of Sulfonamides (*Continued*)

No. of Carbon Atoms	Reactant		Conditions	Product(s) and Yield(s) (%)	Refs.
C_{28}	$O(CH_2)_2O(CH_2)_2O$	(*o,o*)	SMEAH, toluene, 80–85°, 1.5 h	(86)[c]	1063
	"	(*m,m*)	" " "	(85)	1063
	"	(*p,p*)	SMEAH (IA), toluene, 80°, 3.5 h	(47)[c]	1063
C_{30}	$O(CH_2)_6O$	(*o,o*)	"	(78)[c]	1063
	"	(*p,p*)	"	(48)[c]	1063
	$O[(CH_2)_2O]_3$	(*o,o*)	1. $CH_2{=}C(CH_3)OCH_3$, $POCl_3$, toluene 2. $(C_2H_5)_3N$ 3. SMEAH, toluene, 80–85°, 1.5 h	(82)[b]	1063
C_{32}	$O(CH_2)_8O$	(*o,o*)	SMEAH (IA), toluene, 80°, 3.5 h	(51)[c]	1063
	$O[(CH_2)_2O]_4$	(*o,o*)	1. $CH_2{=}C(CH_3)OCH_3$, $POCl_3$, toluene 2. $(C_2H_5)_3N$ 3. SMEAH, toluene, 80–85°, 1.5 h	(76)	1063
C_{34}	$O(CH_2)_{10}O$	(*o,o*)	SMEAH (IA), toluene, 80°, 3.5 h	(52)[b]	1063
	"	(*p,p*)	"	(68)[c]	1063
C_{36}	$O(CH_2)_{12}O$	(*o,o*)	"	(45)[c]	1063
C_{38}	$O(CH_2)_{14}O$	(*o,o*)	"	(61)[c]	1063
C_{44}	$O(CH_2)_{20}O$	(*o,o*)	"	(37)[c]	1063
C_{29}	Ts-N, TsN, O, N-Ts (ring)		SMEAH, toluene, reflux, 14.5 h	H-N, HN, O, N-H (ring) (17)	1064

[a] The product was isolated as the maleate. [b] The product was isolated as the fumarate. [c] The product was isolated as the hydrochloride.

TABLE XXXI. REDUCTION OF SULFUR HETEROCYCLIC COMPOUNDS

No. of Carbon Atoms	Reactant	Conditions	Product(s) and Yield(s) (%)	Refs.
C_5	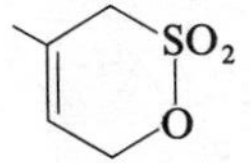 SO_2, O	SMEAH, C_6H_6-ether, reflux, 20 h	$(CH_3)_2C{=}CHCH_2OH$ (—)	1067
	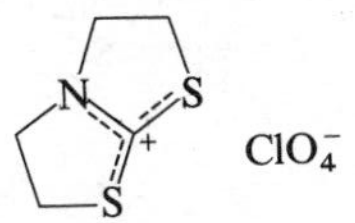 N, S, +, S ClO_4^-	SMEAH, THF, −5°, 4 h	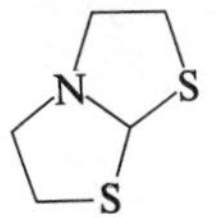 N, S, S (57)	425
	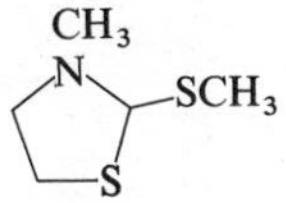 CH_3, N, SCH_3, S	SMEAH, C_6H_6, room temp, 24 h	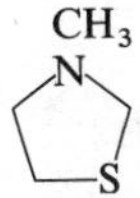 CH_3, N, S (—)	425
	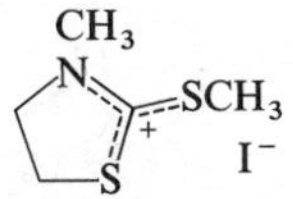 CH_3, N, SCH_3, +, S I^-	", ", ", "	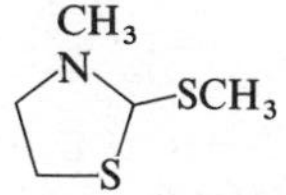 CH_3, N, SCH_3, S (36)	425
C_6	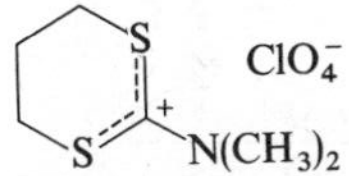 S, +, S, $N(CH_3)_2$ ClO_4^-	", ", ", "	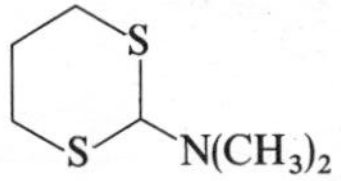 S, S, $N(CH_3)_2$ (34)	425
C_{10}	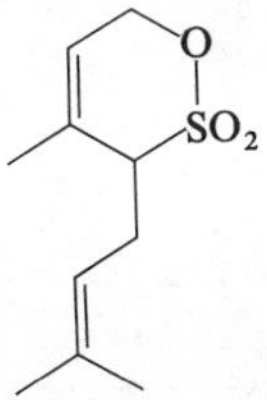 O, SO_2	SMEAH, C_6H_6–ether, 25°, 5 h	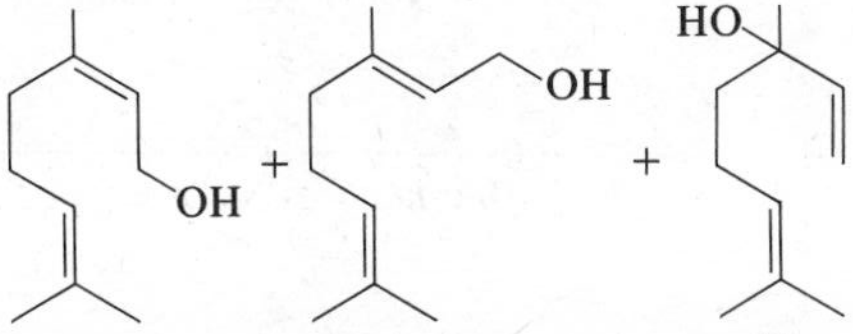 OH **I** + OH **II** + HO **III** (**I**:**II**:**III** = 8:62:30) (12)	1067

TABLE XXXI. REDUCTION OF SULFUR HETEROCYCLIC COMPOUNDS (*Continued*)

No. of Carbon Atoms	Reactant	Conditions	Product(s) and Yield(s) (%)	Refs.
C_{10} (*Contd.*)	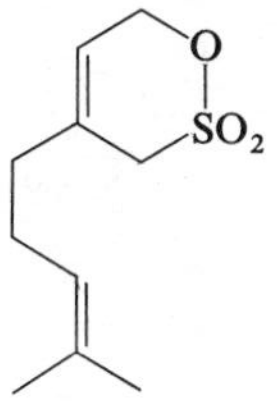 SO_2, O	SMEAH, C_6H_6–ether, 25°, 5 h	**I + II + III** (**I**:**II**:**III** = 8:62:30) (10)	1067
	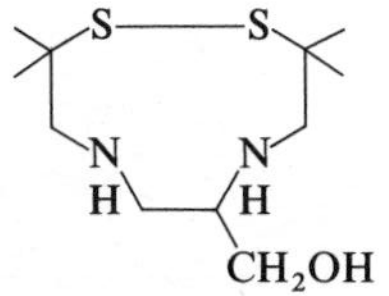 S, S, N H, N H, CH_2OH	1. SMEAH, C_6H_6–toluene, reflux, 30 min, then room temp, 2 h 2. HCl, ether	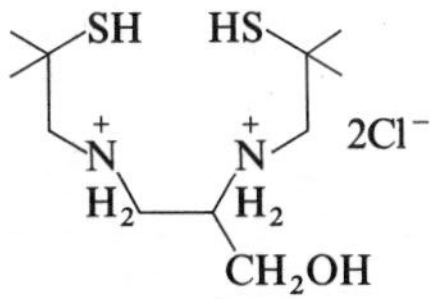 SH, HS, $\overset{+}{N}H_2$, $\overset{+}{N}H_2$, CH_2OH, $2Cl^-$ (18)	1068
	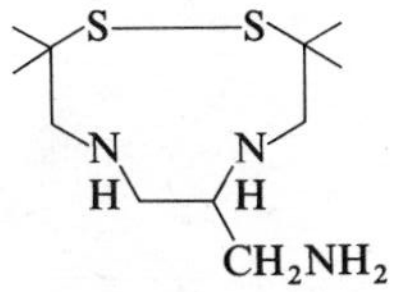 S, S, N H, N H, CH_2NH_2	1. SMEAH, C_6H_6–toluene, reflux, 2 h 2. HCl, ether	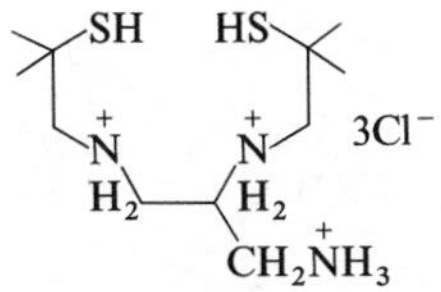 SH, HS, $\overset{+}{N}H_2$, $\overset{+}{N}H_2$, $CH_2\overset{+}{N}H_3$, $3Cl^-$ (43)	1068
	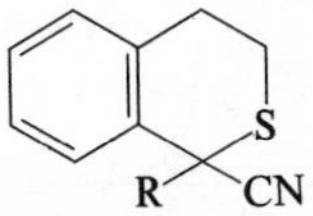 S, R, CN	SMEAH, THF, toluene, −15°, 1 h; room temp, 1 h	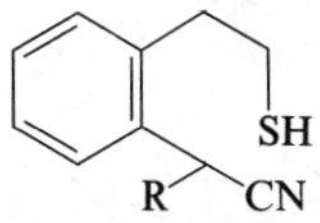 SH, R, CN	

	Reactant	Reagents	Product (Yield, %)		Refs.
	R				
C_{11}	CH_3			(45)	328
C_{13}	i-C_3H_7			(40)	328
C_{17}	[cyclic bis-imine disulfide: S—S, N, N, $C_6H_4OCH_3$-2]	1. SMEAH, C_6H_6 2. HCl	[SH, HS, $\overset{+}{N}H_2$, $\overset{+}{N}H_2$, $C_6H_4OCH_3$-2] $2Cl^-$	(69)	1068, 1069
C_{19}	[cyclic bis-imine disulfide: C_2H_5, C_2H_5, S—S, C_2H_5, C_2H_5, N, N, cyclohexane]	" " "	[C_2H_5, C_2H_5, SH, HS, C_2H_5, C_2H_5, $\overset{+}{N}H_2$, $\overset{+}{N}H_2$, cyclohexane] $2Cl^-$		1069

ADDENDA TO THE TABLES

Listed below are references to recent papers that are not cited in the corresponding tables. The cut-off date is December 1985.

Table I.	Ref. 531a
Table IV.	Refs. 590a, 590b
Table VI.	Refs. 601a, 601b, 606a
Table VIII.	Refs. 637b, 658a, 675a, 696a
Table IX.	Refs. 637b, 741a, 743c
Table XVII.	Refs. 531a, 960a
Table XX.	Ref. 1014a
Table XXVII.	Ref. 1041c
Table XXX.	Ref. 1059c

REFERENCES

[1] J. Málek, *Org. React.* **34**, 1 (1985).

[2] S. C. Stinson, *Chem. Eng. News*, **1980** (November 3), 18.

[3] H. J. Sanders, *Chem. Eng. News*, **1972** (June 19), 29.

[4] C. F. Lane, *Aldrichimica Acta*, **8**, 20 (1975).

[5] H. C. Brown, Y. M. Choi, and S. Narasimhan, *J. Org. Chem.*, **47**, 3153 (1982).

[6] H. C. Brown, *Government Reports and Announcements Index* (*U.S.*), **82**, 3196 (1982) [*C.A.*, **97**, 143899p (1982)].

[7] H. C. Brown and S. Krishnamurthy, *Tetrahedron*, **35**, 567 (1979).

[8] R. O. Hutchins, K. Learn, B. Nazer, D. Pytlewski, and A. Pelter, *Org. Prep. Proced. Int.*, **16**, 335 (1984) [*C.A.*, **101**, 211204m (1984)].

[9] T. Onak, *Organoborane Chemistry*, Academic Press, New York, 1975.

[10] J. H. Babler, *Synth. Commun.*, **12**, 839 (1982).

[11] N. M. Yoon, B. T. Cho, U. J. Yoo, and G. P. Kim, *Teahan Hwahakhoe Chi*, **27**, 434 (1983) [*C.A.*, **100**, 138654f (1984)].

[12] S. K. Chung and G. Han, *Synth. Commun.*, **12**, 903 (1982).

[13] S. W. Heinzman and B. Ganem, *J. Am. Chem. Soc.*, **104**, 6801 (1982).

[14a] H. C. Brown, J. S. Cha, B. Nazer, C. S. Kim, S. Krishnamurthy, and C. A. Brown, *J. Org. Chem.*, **49**, 885 (1984).

[14b] G. W. Gribble and C. F. Nutaitis, *Org. Prep. Proced. Int.*, **17**, 317 (1985).

[14c] M. Hudlicky, *Reductions in Organic Chemistry*, Ellis Horwood Ltd., New York, 1984.

[14d] O. Štrouf, B. Čásenský, and V. Kubánek, *Sodium Dihydrido-bis*(*2-methoxyethoxy*)*aluminate. A Versatile Organometallic Hydride*, Elsevier Science Publishers, Amsterdam, 1985.

[15] E. R. Grandbois, S. I. Howard, and J. D. Morrison, in *Asymmetric Synthesis*, J.D. Morrison, Ed., Vol. 2, Academic Press, New York, 1983, pp. 71–90.

[16] H. B. Kagan and C. J. Fiaud, *Top. Stereochem.*, **10**, 1975 (1979).

[17] M. Nishizawa and R. Noyori, *Kagaku, Zokan*, **91**, 181 (1981) [*C.A.*, **95**, 96306q (1981)].

[18] H. Haubenstock, *Top. Stereochem.*, **14**, 231 (1983).

[19] J. D. Morrison, E. R. Grandbois, and G. R. Weisman, *American Chemical Society Symposium Series, 1982*, 185 (Asymmetric Reactions and Processes in Chemistry, 278) [*C.A.*, **97**, 162272v (1982)].

[20] H. C. Brown, P. K. Jadhav, and A. K. Mandel, *Tetrahedron*, **37**, 3547 (1981).

[21] H. C. Brown and B. Singaram, *J. Am. Chem. Soc.*, **106**, 1797 (1984).

[22] D. A. Evans, J. V. Nelson, and T. R. Taber, *Top. Stereochem.*, **13**, 1 (1982).

[23] H. S. Mosher and J. D. Morrison, *Science*, **221**, 1013 (1983).

[24] H. C. Brown and P. K. Jadhav, in *Asymmetric Synthesis*, J. D. Morrison, Ed., Vol. 2, Academic Press, New York, 1983, pp. 1–43.

[25] M. M. Midland, in *Asymmetric Synthesis*, J. D. Morrison, Ed., Vol. 2, Academic Press, New York, 1983, pp. 45–69.

[26a] R. Noyori, *New Frontiers in Organometallic and Inorganic Chemistry, Proceedings of the 2nd China–Japan–U.S.A. Trilateral Seminar, 1982* (published in 1984), 159 [*C.A.*, **102**, 203440r (1985)].

[26b] M. Ishiguro, N. Koizumi, M. Yasuda, and N. Ikekawa, *J. Chem. Soc., Chem. Commun.*, **1981**, 115.

[26c] N. Koizumi, M. Ishiguro, M. Yasuda, and N. Ikekawa, *J. Chem. Soc., Perkin Trans. 1*, **1983**, 1401.

[26d] T. Tanaka, N. Okamura, K. Bannai, A. Hazato, S. Sugiura, K. Manabe, F. Kamimoto, and S. Kurozumi, *Chem. Pharm. Bull.* **33**, 2359 (1985).

[26e] H. Suemune, A. Akashi, and K. Sakai, *Chem. Pharm. Bull.*, **33**, 1055 (1985).

[26f] P. Koch, Y. Nakatani, B. Luu, and G. Ourisson, *Bull. Soc. Chim. Fr.*, Part 2, **1983**, 189.

[26g] F. J. Sardina, A. Mouriño, and L. Castedo, *Tetrahedron Lett.*, **24**, 4477 (1983).

[26h] J. P. Vigneron, R. Méric, M. Larchevêque, A. Debal, J. Y. Lallemand, G. Kunesch, P. Zagatti, and M. Gallois, *Tetrahedron*, **40**, 3521 (1984).

[26i] T. G. Schenck and B. Bosnich, *J. Am. Chem. Soc.*, **107**, 2058 (1985).

[26j] M. Kawasaki and S. Terashima, *Chem. Pharm. Bull.*, **33**, 347 (1985).

[26k] R. A. Russell, A. S. Kraus, R. W. Irwine, and R. N. Warrener, *Aust. J. Chem.*, **38**, 179 (1985).

[26m] V. M. Potapov, V. M. Demyanovich, and V. I. Maleev, *Zh. Org. Khim.*, **21**, 1758 (1985); *Engl. Transl.*, p. 1606 [*C.A.*, **105**, 114661h (1986)].

[27] T. Sato, Y. Goto, and T. Fujisawa, *Tetrahedron Lett.*, **23**, 4111 (1982).

[28] T. Sato, Y. Gotoh, Y. Wakabayashi, and T. Fujisawa, *Tetrahedron Lett.*, **24**, 4123 (1983).

[29] T. Sato, M. Watanabe, N. Honda, and T. Fujisawa, *Chem. Lett.*, **1984**, 1175.

[30] K. Hiroi, R. Kitayama, and S. Sato, *Chem. Pharm. Bull.*, **32**, 2628 (1984).

[31] J. W. Harris and W. M. Jones, *J. Am. Chem. Soc.*, **104**, 7329 (1982).

[32] S. Kiyooka, T. Adachi, H. Kojo, and K. Suzuki, *Kochi Daigaku Rigakubu Kiyo, Kagaku*, **4**, 1 (1983) [*C.A.*, **100**, 120599g (1984)].

[33] T. Sato, Y. Gotoh, M. Watanabe, and T. Fujisawa, *Chem. Lett.*, **1983**, 1533.

[34] J. Yamashita, S. Tomiyama, H. Hashimoto, K. Kitahara, and H. Sato, *Chem. Lett.*, **1984**, 749.

[35] J. Yamashita, H. Kawahara, S. Ohashi, Y. Honda, T. Kenmotsu, and H. Hashimoto, *Technol. Rep. Tohoku Univ.*, **48**, 211 (1983) [*C.A.*, **101**, 6722n (1984)].

[36a] M. Kawasaki, Y. Suzuki, and S. Terashima, *Chem. Lett.*, **1984**, 239.

[36b] M. Kawasaki, Y. Suzuki, and S. Terashima, *Chem. Pharm. Bull.*, **33**, 52 (1985).

[36c] Sagami Chemical Research Center, Japanese Patent (Tokkyo Koho) 60 72,976 (85 72,976) (1985) [*C.A.*, **103**, 178052b (1985)].

[37] B. L. Allwood, H. Shahriari-Zavareh, J. F. Stoddart, and D. J. Williams, *J. Chem. Soc., Chem. Commun.*, **1984**, 1461.

[38a] S. Itsuno and K. Ito, *J. Org. Chem.*, **49**, 555 (1984).

[38b] S. Itsuno, M. Nakano, K. Miyazaki, H. Masuda, K. Ito, A. Hirao, and S. Nakahama, *J. Chem. Soc., Perkin Trans. 1*, **1985**, 2039.

[39] S. Krishnamurthy, F. Vogel, and H. C. Brown, *J. Org. Chem.*, **42**, 2534 (1977).

[40] H. C. Brown and G. G. Pai, *J. Org. Chem.*, **47**, 1606 (1982).

[41] M. M. Midland and J. I. McLoughlin, *J. Org. Chem.*, **49**, 1316 (1984).

[42] S. Itsuno, K. Ito, A. Hirao, and S. Nakahama, *J. Chem. Soc., Chem. Commun.*, **1983**, 469.

[43] M. M. Midland and A. Kazubski, *J. Org. Chem.*, **47**, 2495 (1982).

[44] K. Soai, T. Yamanoi, and H. Oyamada, *Chem. Lett.*, **1984**, 251.

[45] K. Soai, H. Oyamada, and T. Yamanoi, *J. Chem. Soc., Chem. Commun.*, **1984**, 413.

[46] J. J. Bloomfield and S. L. Lee, *J. Org. Chem.*, **32**, 3919 (1967).

[47] M. E. Birckelbaw, P. W. Le Quesne, and C. K. Wocholski, *J. Org. Chem.*, **35**, 558 (1970).

[48] D. E. Burke and P. W. Le Quesne, *J. Org. Chem.*, **36**, 2397 (1971).

[49] M. J. Begley, D. W. Knight, and G. Pattenden, *Tetrahedron Lett.*, **1975**, 4279.

[50] D. W. Knight and G. Pattenden, *J. Chem. Soc., Chem. Commun.*, **1975**, 876.

[51] M. M. Kayser and P. Morand, *Can. J. Chem.*, **56**, 1524 (1978).

[52] D. W. Knight and G. Pattenden, *J. Chem. Soc., Perkin Trans. 1*, **1979**, 62.

[53] M. M. Kayser and P. Morand, *Tetrahedron Lett.*, **1979**, 695.

[54] S. Krishnamurthy and W. B. Vreeland, *Heterocycles*, **18** (Special Issue), 265 (1982).

[55] M. M. Kayser, J. Salvador, and M. Morand, *Can. J. Chem.*, **61**, 439 (1983).

[56] M. A. Makhlouf and B. Rickborn, *J.* Org. *Chem.*, **46**, 4810 (1981).

[57] P. Morand, J. Salvator, and M. M. Kayser, *J. Chem. Soc., Chem. Commun.*, **1982**, 458.

[58] G. K. Cooper and L. J. Dolby, *J. Org. Chem.*, **44**, 3414 (1979).

[59] H. O. House, *Modern Synthetic Reactions*, 2nd ed., Benjamin, Menlo Park, CA, 1972.

[60] M. M. Kayser and O. Eisenstein, *Can. J. Chem.*, **59**, 2457 (1981).

[61] M. M. Kayser and P. Morand, *Can. J. Chem.*, **58**, 2484 (1980).

[62] M. M. Kayser, J. Salvador, and P. Morand, *Can. J. Chem.*, **60**, 1199 (1982).

[63] H. C. Brown and P. M. Weissman, *Isr. J. Chem.*, **1**, 430 (1963) [*C.A.*, **60**, 11923e (1964)].

[64] P. M. Weissman and H. C. Brown, *J. Org. Chem.*, **31**, 283 (1966).

[65] H. C. Brown and N. M. Yoon, *J. Am. Chem. Soc.*, **88**, 1464 (1966).

[66] H. C. Brown, *U.S. Clearinghouse Fed. Sci. Tech. Inform.*, AD 645581 [*C.*A., **67**, 99306x (1967)].

[67] J. Málek and M. Černý, *Synthesis*, **1972**, 217.

[68] J. Vít, *Eastman Org. Chem. Bull.*, **42**, 1 (1970) [*C.A.*, **74**, 99073p (1971)].

[69] J. Vít, C. Papaionnou, H. Cohen, and D. Batesky, *Eastman Org. Chem. Bull.*, **46**, 1 (1974) [*C. A.*, **80**, 120098m (1974)].

[70a] L. Li and Y. Zheng, *Huaxue Shiji*, **5**, 182 (1983) [*C. A.*, **99**, 105303y (1983)].

[70b] E. I. Edwards, R. Epton, and G. Marr, *J. Organomet. Chem.*, **122**, C49 (1976).

[71] J. Vít, U.S. Patent 3,660,416 (1972) [*C.A.*, **77**, 100787v (1972)].

[72] J. Vít, U.S. Patent 3,839,367 (1974) [*C.A.*, **81**, 169132c (1974)].

[73] R. Kanazawa and T. Tokoroyama, *Synthesis*, **1976**, 526.

[74] R. B. Greenwald and D. H. Evans, *J. Org. Chem.*, **41**, 1470 (1976).

[75] D. H. Evans and R. B. Greenwald, U.S. Patent 4,302,612 (1981) [*C.A.*, **96**, 103653b (1982)].

[76] H. C. Brown and R. F. McFarlin, *J. Am. Chem. Soc.*, **78**, 252 (1956).

[77] H. C. Brown and R. F. McFarlin, *J. Am. Chem. Soc.*, **80**, 5372 (1958).

[78] H. C. Brown and B. C. Subba Rao, *J. Am. Chem. Soc.*, **80**, 5377 (1958).

[79] H. C. Brown, R. F. McFarlin, and B. C. Subba Rao, U.S. Patent 3,147,272 (1964) [*C.A.*, **62**, 11689c (1965)].

[80] L. I. Zakharkin, D. N. Maslin, and V. V. Gavrilenko, *Zh. Org. Khim.*, **2**, 2197 (1966); *Engl. Transl., p.* 2153 [*C.A.*, **66**, 85572y (1967)].

[81] H. C. Brown and C. J. Shoaf, *J. Am. Chem. Soc.*, **86**, 1079 (1964).

[82] H. C. Brown and C. P. Garg, *J. Am. Chem. Soc.*, **86**, 1085 (1964).

[83] H. C. Brown, C. J. Shoaf, and C. P. Garg., *Tetrahedron Lett.*, **1959**, 9.

[84] C. P. Garg, Ph.D. Dissertation, Purdue University, Lafayette, Indiana; *Diss. Abstr.*, **23**, 830 (1962) [*C.A.*, **58**, 2329g (1963)].

[85a] H. Haubenstock and T. Mester, Jr., *J. Org. Chem.*, **48**, 945 (1983).

[85b] A. I. Belokon, V. N. Bochkarev, V. V. Gavrilenko, T. D. Danina, and G. I. Belik, *Zh. Obshch. Khim.*, **53**, 2042 (1983); *Engl. Transl.*, p. 1843 [*C.A.*, **100**, 22683k (1984)].

[86] D. de Peretti, T. Strzalko-Bottin, and J. Seyden-Penne, *Bull. Soc. Chim. Fr.*, **1974**, 2925.

[87] F. Weygand, G. Eberhardt, H. Linden, F. Schäfer, and I. Eigen, *Angew. Chem.*, **65**, 525 (1953).

[88] H. C. Brown and A. Tsukamoto, *J. Am. Chem. Soc.*, **86**, 1089 (1964).

[89] H. C. Brown and A. Tsukamoto, *J. Am. Chem. Soc.*, **81**, 502 (1959).

[90] R. Nicoletti and L. Baiocchi, *Ann. Chim.* (*Rome*), **50**, 1502 (1960) [*C.A.*, **56**, 9937h (1962)].

[91] L. I. Zakharkin, D. N. Maslin, and V. V. Gavrilenko, *Tetrahedron*, **25**, 5555 (1969).

[92] T. Sone, S. Terashima, and S. I. Yamada, *Chem. Pharm. Bull.*, **24**, 1273 (1976).

[93] A. V. Chebyshev, N. G. Evstratova, G. A. Serebrennikova, and R. P. Evstigneeva, *Zh. Org. Khim.*, **13**, 1175 (1977); *Engl. Transl.*, p. 1081 [*C.A.*, **87**, 84458n (1977)].

[94] M. Černý, J. Málek, M. Čapka, and V. Chvalovský, *Collect. Czech. Chem. Commun.*, **34**, 1033 (1969).

[95] P. Duhamel, L. Duhamel, and P. Siret, *C. R. Acad. Sci., Ser. C*, **270**, 1750 (1970).

[96a] T. C. McMorris, *J. Org. Chem.*, **35**, 458 (1970).

[96b] F. Sondheimer, W. McCrae, and W. G. Salmond, *J. Am. Chem. Soc.*, **91**, 1228 (1969).

[96c] Schering A.-G., French Patent 1,508,947 (1968) [*C.A.*, **70**, 47719b (1969)].

[96d] Schering A.-G., French Patent 1,509,563 (1968) [*C.A.*, **70**, 88132y (1969)].

[96e] U. Kerb, G. Schulz, P. Hocks, R. Wiechert, A. Furlenmeier, A. Fürst, A. Langemann, and G. Waldvogel, *Helv. Chim. Acta*, **49**, 1601 (1966).

[96f] V. V. K. Prasad and S. C. Franklin, *J. Labelled Compd. Radiopharm.*, **22**, 353 (1985).

[97] C. H. Kuo, D. Taub, and N. L. Wendler, *Tetrahedron Lett.*, **1972**, 5317.

[98] N. S. Ramegowda, M. N. Modi, A. K. Koul, J. M. Bora, C. K. Narang, and N. K. Mathur, *Tetrahedron*, **29**, 3985 (1973).

[99] A. I. Meyers and D. L. Comins, *Tetrahedron Lett.*, **1978**, 5179.

[100] T. Izawa and T. Mukaiyama, *Chem. Lett.*, **1977**, 1443.

[101] T. Izawa and T. Mukaiyama, *Chem. Lett.*, **1978**, 409.

[102] T. Izawa and T. Mukaiyama, *Bull. Chem. Soc. Jpn.*, **52**, 555 (1979).

[103] M. Hayashi, S. Terashima, and K. Koga, *Tetrahedron*, **37**, 2797 (1981).

[104] S. Terashima, M. Hayashi, and K. Koga, *Tetrahedron Lett.*, **21**, 2733 (1980).

[105] J. P. Morizur, G. Muzard, J. J. Basselier, and J. Kossanyi, *Bull. Soc. Chim. Fr.*, **1975**, 257.

[106] R. S. Davidson, W. H. H. Günther, S. M. Waddington-Feather, and B. Lythgoe, *J. Chem. Soc.*, **1964**, 4907.

[107] O. Eisenstein, J. M. Lefour, C. Minot, N. T. Anh, and G. Soussan, *C. R. Acad. Sci., Ser. C*, **274**, 1310 (1972).

[108] J. Bottin, O. Eisenstein, C. Minot, and N. T. Anh, *Tetrahedron Lett.*, **1972**, 3015.

[109] J. Durand, N. T. Anh, and J. Huet, *Tetrahedron Lett.*, **1974**, 2397.

[110] D. H. Miles, P. Loew, W. S. Johnson, A. F. Kluge, and J. Meinwald, *Tetrahedron Lett.*, **1972**, 3019.

[111] J. A. Marshall and R. D. Carroll, *J. Org. Chem.*, **30**, 2748 (1965).

[112] D. Nasipury, A. Sarkar, and S. K. Konar, *J. Org. Chem.*, **47**, 2840 (1982).

[113] M. F. Semmelhack and R. D. Stauffer, *J. Org. Chem.*, **40**, 3619 (1975).

[114] M. F. Semmelhack, R. D. Stauffer, and A. Yamashita, *J. Org. Chem.*, **42**, 3180 (1977).

[115] M. E. Osborn, J. F. Pegues, and L. A. Paquette, *J. Org. Chem.*, **45**, 167 (1980).

[116] M. E. Osborn, S. Kuroda, J. L. Muthard, J. D. Kramer, P. Engel, and L. A. Paquette, *J. Org. Chem.*, **46**, 3379 (1981).

[117] J. Huguet, M. Karpf, and A. S. Dreiding, *Helv. Chim. Acta*, **65**, 2413 (1982).

[118] P. Vermeer, J. Meijer, C. Eylander, and L. Brandsma, *Recl., J. Roy. Neth. Chem. Soc.*, **95**, 25 (1976).

[119] M. Černý and J. Málek, *Tetrahedron Lett.*, **1969**, 1739.

[120] M. Černý and J. Málek, *Collect. Czech. Chem. Commun.*, **35**, 1216 (1970).

[121] M. Černý and J. Málek, *Collect. Czech. Chem. Commun.*, **35**, 2030 (1970).

[122] M. Černý and J. Málek, *Collect. Czech. Chem. Commun.*, **35**, 3079 (1970).

[123] M. Černý and J. Málek, *Collect. Czech. Chem. Commun.*, **36**, 2394 (1971).

[124] L. H. Conover and D. S. Tarbell, *J. Am. Chem. Soc.*, **72**, 3586 (1950).

[125] D. O. Cheng, T. L. Bowman, and E. LeGoff, *J. Heterocycl. Chem.*, **13**, 1145 (1976).

[126] N. G. Gaylord, *Experientia*, **10**, 166 (1954).

[127] S. R. Landor, O. O. Sonola, and A. R. Tatchell, *J. Chem. Soc., Perkin Trans. 1*, **1978**, 605.

[128] S. R. Landor, O. O. Sonola, and A. R. Tatchell, *Bull. Chem. Soc. Jpn.*, **57**, 1658 (1984).

[129] S. R. Landor, B. J. Miller, and A. R. Tatchell, *J. Chem. Soc., C*, **1966**, 1822.

[130] S. R. Landor, B. J. Miller, and A. R. Tatchell, *J. Chem. Soc., C*, **1966**, 2280.

[131] S. R. Landor, B. J. Miller, and A. R. Tatchell, *J. Chem. Soc., C*, **1967**, 197.

[132] S. R. Landor, O. O. Sonola, and A. R. Tatchell, *J. Chem. Soc., Perkin Trans. 1*, **1974**, 1902.

[133] F. Cistone, Ph.D. Dissertation, Drexel University, Philadelphia, PA, 1980; *Diss. Abstr. Int. B*, **41**, 3784 (1981) [*C.A.*, **95**, 24155s (1981)].

[134] R. O. Hutchins, W. Y. Su, R. Sivakumar, F. Cistone, and Y. P. Stercho, *J. Org. Chem.*, **48**, 3412 (1983).

[135] B. Alcaide, C. Lopez Mardomingo, R. Perez-Ossorio, and J. Plumet, *An. Quím.*, *Sec. C*, **78**, 278 (1982) [*C.A.*, **97**, 181486j (1982)].

[136] J. Martin, W. Parker, and R. A. Raphael, *J. Chem. Soc.*, **1964**, 289.

[137] J. Martin, W. Parker, B. Shroot, and T. Stewart, *J. Chem. Soc.*, *C*, **1967**, 101.

[138] S. C. Chen, *Synthesis*, **1974**, 691.

[139] W. Carruthers and M. I. Qureshi, *J. Chem. Soc.*, *C*, **1970**, 2238.

[140] E. W. Colvin, R. A. Raphael, and J. S. Roberts, *J. Chem. Soc.*, *D*, **1971**, 858.

[141] E. W. Colvin, S. Malchenko, R. A. Raphael, and J. S. Roberts, *J. Chem. Soc.*, *Perkin Trans. 1*, **1973**, 1989.

[142] E. W. Colvin, S. Malchenko, R. A. Raphael, and J. S. Roberts, *J. Chem. Soc.*, *Perkin Trans. 1*, **1978**, 658.

[143a] A. Kasal, *Collect. Czech. Chem. Commun.*, **43**, 1778 (1978).

[143b] A. Kasal, *Collect. Czech. Chem. Commun.*, **48**, 1489 (1983).

[144] A. T. Blomquist, B. F. Hallam, and A. D. Josey, *J. Am. Chem. Soc.*, **81**, 678 (1959).

[145] S. H. Graham and A. J. S. Williams, *Tetrahedron*, **21**, 3263 (1965).

[146] K. Kotera and K. Kitahonoki, *Org. Prep. Proced. Int.*, **1**, 305 (1969).

[147] H. Tanida, T. Okada, and K. Kotera, *Bull. Chem. Soc. Jpn.*, **46**, 934 (1973).

[148] J. P. Freeman, *Chem. Rev.*, **73**, 283 (1973).

[149] R. C. Orlowski, R. Walter, and DeLoss Winkler, *J. Org. Chem.*, **41**, 3701 (1976).

[150] S. R. Landor, O. O. Sonola, and A. R. Tatchell, *J. Chem. Soc.*, *Perkin Trans. 1*, **1974**, 1294.

[151] T. Nishiwaki and F. Fujiyama, *Synthesis*, **1972**, 569.

[152] L. Ferrero, M. Rouillard, M. Decouzon, and M. Azzaro, *Tetrahedron Lett.*, **1974**, 131.

[153] L. Ferrero, S. Geribaldi, M. Rouillard, and M. Azzaro, *Can. J. Chem.*, **53**, 3227 (1975).

[154] Y. Girault, M. Decouzon, and M. Azzaro, *Tetrahedron Lett.*, **25**, 2763 (1984).

[155] T. Kametani, S. P. Huang, A. Ujiie, M. Ihara, and K. Fukumoto, *Heterocycles*, **4**, 1223 (1976).

[156] T. Kametani, A. Ujiie, S. P. Huang, M. Ihara, and K. Fukumoto, *J. Chem. Soc.*, *Perkin Trans. 1*, **1977**, 394.

[157] T. Kametani, S. P. Huang, C. Koseki, M. Ihara, and K. Fukumoto, *J. Org. Chem.*, **42**, 3040 (1977).

[158] K. Fukumoto, *Heterocycles*, **15**, 9 (1981).

[159] S. H. Pine, *Org. React.* **18**, 403 (1970).

[160] J. Kuszmann, P. Sohár, G. Horváth, and Z. Méhesfalvi-Vajna, *Tetrahedron*, **30**, 3905 (1974).

[161] H. C. Brown and P. M. Weissman, *J. Am. Chem. Soc.*, **87**, 5614 (1965).

[162] S. Cucinella, German Offen. 2,849,767 (1979) [*C.A.*, **91**, 174814n (1979)].

[163] S. Cucinella, G. Dozzi, and G. Del Piero Assoreni, *J. Organomet. Chem.*, **224**, 1 (1982).

[164] G. Dozzi, S. Cucinella, and M. Bruzzone, *J. Organomet. Chem.*, **224**, 13 (1982).

[165] S. Cucinella, G. Dozzi, and M. Bruzzone, *J. Organomet. Chem.*, **224**, 21 (1982).

[166] N. K. Efimov, S. N. Tsiomo, A. I. Gorbunov, G. I. Samokhvalov, L. P. Davydova, A. V. Bogatskii, and N. G. Luk'yanenko, *Nov. Obl. Primeneniya Metallorgan. Soedin., M.*, **1983**, 112 [*C.A.*, **101**, 151362z (1984)].

[167] V. Bažant, M. Čapka, M. Černý, V. Chvalovský, K. Kochloefl, M. Kraus, and J. Málek, *Tetrahedron Lett.*, **1968**, 3303.

[168] M. Černý, J. Málek, M. Čapka, and V. Chvalovský, *Collect. Czech. Chem. Commun.*, **34**, 1025 (1969).

[169] A. Hajós, *Complex Hydrides and Related Reducing Agents in Organic Synthesis*, Elsevier, Amsterdam, 1979.

[170] K. K. Chan, N. Cohen, J. P. De Noble, A. C. Specian, Jr., and G. Saucy, *J. Org. Chem.*, **41**, 3497 (1976).

[171] N. Cohen, W. F. Eichel, R. J. Lopresti, C. Neukom, and G. Saucy, *J. Org. Chem.*, **41**, 3505 (1976).

[172] K. K. Chan and G. Saucy, U.S. Patent 4,045,475 (1977) [*C.A.*, **87**, 183979q (1977)].

[173] W. G. Brown, *Org. React.* **6**, 469 (1951).
[174] H. C. Brown, P. M. Weissman, and N. M. Yoon, *J. Am. Chem. Soc.* **88**, 1458 (1966).
[175] N. M. Yoon and B. T. Cho, *Tetrahedron Lett.*, **23**, 2475 (1982).
[176] V. Prelog and D. Bedekovič, *Helv. Chim. Acta*, **62**, 2285 (1979).
[177] N. Pourahmady and E. J. Eisenbraun, *J. Org. Chem.*, **48**, 3067 (1983).
[178] R. Pappo and C. J. Jung, German Offen. 2,321,984 (1973) [*C.A.*, **80**, 26827b (1974)].
[179] R. Pappo and C. J. Jung, U.S. Patent 3,969,391 (1976) [*C.A.*, **86**, 55057e (1977)].
[180] J. Nickel, E. Seeger, W. Engel, G. Engelhardt, and A. Eckenfels, German Offen. 2,047,804 (1972) [*C.A.*, **77**, 34157m (1972)].
[181] J. Kraemer, H. E. Radunz, D. Orth, M. Baumgarth, and P. Bruneau, German Offen. 2,513,371 (1976) [*C.A.*, **86**, 55069k (1977)].
[182] H. E. Radunz, J. Kraemer, M. Baumgarth, and D. Orth, German Offen. 2,422,924 (1975) [*C.A.*, **84**, 121284p (1976)].
[183] S. W. Chaikin and W. G. Brown, *J. Am. Chem. Soc.*, **71**, 122 (1949).
[184] H. C. Brown, S. C. Kim, and S. Krishnamurthy, *J. Org. Chem.*, **45**, 1 (1980).
[185] W. R. Roush, *J. Am. Chem. Soc.*, **100**, 3599 (1978).
[186] N. R. A. Beeley, R. Peel, J. K. Sutherland, J. J. Holohan, K. B. Mallion, and G. J. Sependa, *Tetrahedron*, **37** (Suppl. No. 1), 411 (1981).
[187] H. Heimgartner, L. Ulrich, H. J. Hansen, and H. Schmid, *Helv. Chim. Acta*, **54**, 2313 (1971).
[188] E. Ade, G. Helmchen, and G. Heiligenmann, *Tetrahedron Lett.*, **21**, 1137 (1980).
[189] T. P. Johnston, G. S. McCaleb, S. D. Clayton, J. L. Frye, C. A. Krauth, and J. A. Montgomery, *J. Med. Chem.*, **20**, 279 (1977).
[190] A. K. Bose, M. Tsai, J. C. Kapur, and M. S. Manhas, *Tetrahedron*, **29**, 2355 (1973).
[191] E. Stephen and H. Stephen, *J. Chem. Soc.*, **1957**, 490.
[192] T. Fujisawa, T. Mori, S. Tsuge, and T. Sato, *Tetrahedron Lett.*, **24**, 1543 (1983).
[193] T. D. Hubert, D. P. Eyman, and D. F. Wiemer, *J. Org. Chem.*, **49**, 2279 (1984).
[194] W. Kreiser, L. Janitschke, W. Voss, L. Ernst, and W. S. Sheldrick, *Chem. Ber.*, **112**, 397 (1979).
[195] D. Taub, N. N. Girotra, R. D. Hoffsommer, C. H. Kuo, H. L. Slate, S. Weber, and N. L. Wendler, *Tetrahedron*, **24**, 2443 (1968).
[196] D. Gravel, R. Deziel, F. Brisse, and L. Hechler, *Can. J. Chem.*, **59**, 2997 (1981).
[197] D. M. Bailey and R. E. Johnson, *J. Org. Chem.*, **35**, 3574 (1970).
[198] S. Narasimhan, *Heterocycles*, **18**, 131 (1982).
[199a] K. Miyano, Y. Ohfune, S. Azuma, and T. Matsumoto, *Tetrahedron Lett.*, **1974**, 1545.
[199b] G. R. Lenz, *J. Org. Chem.*, **41**, 3532 (1976).
[199c] V. Pouzar, P. Drašar, I. Černý, and M. Havel, *Collect. Czech. Chem. Commum.*, **50**, 869 (1985).
[199d] M. Harnik and Y. Aharonovitz, Israeli Patent 62, 501 (1985) [*C.A.*, **103**, 160768e (1985)].
[200] M. Cushman and J. Mathew, *J. Org. Chem.*, **46**, 4921 (1981).
[201] D. O. Shah and K. N. Trivedi, *Indian J. Chem.*, **15B**, 599 (1977).
[202] B. J. Clark, R. Grayshan, and D. D. Miller, *J. Chem. Soc., Perkin Trans. 1*, **1983**, 2237.
[203] E. Gellert, *J. Nat. Prod.*, **45**, 50 (1982) [*C.A.*, **96**, 117643k (1982)].
[204] T. F. Buckley III and H. Rapoport, *J. Org. Chem.*, **48**, 4222 (1983).
[205a] C. Tamm, *Helv. Chim. Acta*, **43**, 338 (1960); German Offen. 1,134,668 (1962) [*C.A.*, **58**, 3496f (1963)].
[205b] U. Stache, W. Fritsch, W. Haede, K. Radscheit, and K. Fachinger, *Justus Liebigs Ann. Chem.*, **726**, 136 (1969).
[205c] W. Fritsch, U. Stache, W. Haede, K. Radscheit, and H. Ruschig, *Justus Liebigs Ann. Chem.*, **721**, 168 (1969).
[205d] U. Stache, K. Radscheit, W. Fritsch, H. Kohl, W. Haede, and H. Ruschig, *Tetrahedron Lett.*, **1969**, 3033.
[205e] U. Stache, K. Radscheit, W. Fritsch, and W. Haede, *Justus Liebigs Ann. Chem.*, **1974**, 608.
[205f] Y. Kamano and G. R. Pettit, *J. Org. Chem.*, **39**, 2629 (1974).
[205g] Y. Kamano, G. R. Pettit, and M. Tozawa, *J. Chem. Soc., Perkin Trans. 1*, **1975**, 1972.

[206a] L. Lábler, *Collect. Czech. Chem. Commun.*, **26**, 724 (1961).

[206b] M. S. Ragab, H. Linde, and K. Meyer, *Helv. Chim. Acta*, **45**, 1794 (1962).

[206c] U. Kerb, H. D. Berndt, U. Eder, R. Wiechert, P. Buchschacher, A. Furlenmeier, A. Fürst, and M. Müller, *Experientia*, **27**, 759 (1971).

[206d] U. Valcavi, B. Corsi, R. Caponi, S. Innocenti, and P. Martelli, *J. Med. Chem.*, **18**, 1258 (1975).

[206e] R. M. Weier and L. M. Hofmann, *J. Med. Chem.*, **20**, 1304 (1977).

[206f] H. P. Albrecht, *Justus Liebigs Ann. Chem.*, **1980**, 886.

[206g] U. Stache, K. Radscheit, W. Haede, and W. Fritsch, *Justus Liebigs Ann. Chem.*, **1982**, 342.

[206h] R. Dhal, E. Brown, and J. P. Robin, *Tetrahedron*, **39**, 2787 (1983).

[207a] L. Sawlewicz and K. Meyer, *Pharm. Acta Helv.*, **45**, 261 (1970) [*C.A.*, **72**, 133061d (1970)]

[207b] M. E. Wolff and S. Y. Cheng, *J. Org. Chem.*, **32**, 1029 (1967).

[207c] P. Wieland, *Helv. Chim. Acta*, **62**, 2276 (1979).

[207d] P. Welzel, H. Stein, and T. Milkova, *Justus Liebigs Ann. Chem.*, **1982**, 2119.

[207e] M. Ando, A. Ono, and K. Takase, *Chem. Lett.*, **1984**, 493.

[207f] D. W. Brooks, H. S. Bevinakatti, E. Kennedy, and J. Hathaway, *J. Org. Chem.*, **50**, 628 (1985).

[208a] A. Lardon, K. Stöckel, and T. Reichstein, *Helv. Chim. Acta*, **53**, 167 (1970).

[208b] U. Stache, K. Radscheit, W. Fritsch, W. Haede, H. Kohl, and H. Ruschig, *Justus Liebigs Ann. Chem.*, **750**, 149 (1971).

[208c] C. R. Lenz and J. A. Schulz, *J. Org. Chem.*, **43**, 2334 (1978).

[208d] K. Nickisch, H. Laurent, and R. Wiechert, *Tetrahedron Lett.*, **22**, 3833 (1981).

[208e] J. Engel, O. Isaac, K. Posselt, K. Thiemer, and H. Uthemann, *Arzneim.-Forsch.*, **33**, 1215 (1983) [*C.A.*, **100**, 34743t (1984)].

[208f] H. Asada, T. Miyase, and S. Fukushima, *Chem. Pharm. Bull.*, **32**, 3403 (1984).

[208g] R. Kwok and M. E. Wolff, *J. Org. Chem.*, **28**, 423 (1963).

[208h] F. S. Alvarez and A. N. Watt, *J. Org. Chem.*, **37**, 3725 (1972).

[209] R. B. Woodward, E. Logusch, K. P. Nambiar, K. Sakan, D. E. Ward, B. W. Au-Yeung, P. Balaram, L. J. Browne, P. J. Card, C. H. Chen, R. B. Chênevert, A. Fliri, K. Frobel, H. J. Gais, D. G. Garratt, K. Hayakawa, W. Heggie, D. P. Hesson, D. Hoppe, I. Hoppe, J. A. Hyatt, D. Ikeda, P. A. Jacobi, K. S. Kim, Y. Kobuke, K. Kojima, K. Krowicki, V. J. Lee, T. Leutert, S. Malchenko, J. Martens, R. S. Matthews, B. S. Ong, J. B. Press, T. V. Rajan Babu, G. Rousseau, H. M. Sauter, M. Suzuki, K. Tatsuta, L. M. Tolbert, E. A. Truesdale, I. Uchida, Y. Ueda, T. Uyehara, A. T. Vasella, W. C. Vladuchick, P. A. Wade, R. M. Williams, and H. N. C. Wong, *J. Am. Chem. Soc.*, **103**, 3213 (1981).

[210] R. Noyori, *Kankyo Kagaku Tokubetsu Kenkyu Kenkyu Hokokushu*, **1980**, B43-R-32-6, Kagakuteki Kankyo Kaizen Gijutsu, 33 [*C.A.*, **93**, 166992v (1980)].

[211] R. Noyori, I. Tomino, and M. Nishizawa, *J. Am. Chem. Soc.*, **101**, 5843 (1979).

[212] R. Noyori, *Pure Appl. Chem.*, **53**, 2315 (1981).

[213] R. Noyori, I. Tomino, M. Yamada, and M. Nishizawa, *J. Am. Chem. Soc.*, **106**, 6717 (1984).

[214] R. Noyori, I. Tomino, Y. Tanimoto, and M. Nishizawa, *J. Am. Chem. Soc.*, **106**, 6709 (1984).

[215] R. Noyori and M. Suzuki, *Angew. Chem.*, **96**, 854 (1984); *Angew. Chem. Int. Ed., Engl.*, **23**, 847 (1984).

[216] E. J. Corey, K. Shimoji, and C. Shih, *J. Am. Chem. Soc.*, **106**, 6425 (1984).

[217] K. Yamamoto, H. Fukushima, and M. Nakazaki, *J. Chem. Soc., Chem. Commun.*, **1984**, 1490.

[218] H. Suda, S. Kanoh, N. Umeda, M. Ikka, and M. Toto, *Chem. Lett.*, **1984**, 899.

[219] J. A. Marshall and R. D. Royce, Jr., *J. Org. Chem.*, **47**, 693 (1982).

[220] J. W. Clark-Lewis and D. C. Skingle, *Aust. J. Chem.*, **20**, 2169 (1967).

[221a] J. Kitchin and R. J. Stoodley, *J. Chem. Soc., Perkin Trans. 1*, **1973**, 22.

[221b] K. Kefurt, Z. Kefurtová, and J. Jarý, *Collect. Czech. Chem. Commun.*, **41**, 1791 (1976).

[221c] K. Kefurt, K. Čapek, J. Čapková, Z. Kefurtová, and J. Jarý, *Collect. Czech. Chem. Commun.*, **37**, 2985 (1972).

[221d] V. N. Shibaev, V. A. Petrenko, L. L. Danilov, N. S. Utkina, M. I. Struchkova, and N. K. Kochetkov, *Izv. Akad. Nauk SSSR, Ser. Khim.*, **1980**, 158; *Engl. Transl.*, p. 143 [*C.A.*, **92**, 215658 (1980)].

[221e] H. Wehrli, *Chimia*, **23**, 403 (1969) [*C.A.*, **72**, 43977c (1970)].

[222] E. Winterfeldt, *Synthesis*, **1975**, 617.

[223a] W. E. Parham and L. D. Huestis, *J. Am. Chem. Soc.*, **84**, 813 (1962).

[223b] L. Eignerová and A. Kasal, *Collect. Czech. Chem. Commun.*, **41**, 1056 (1976).

[223c] A. Kasal and A. Trka, *Collect, Czech. Chem. Commun.*, **42**, 1389 (1977).

[224] N. N. Girotra and N. L. Wendler, *Tetrahedron Lett.*, **1975**, 227.

[225] F. J. McQuillin and R. B. Yeats, *J. Chem. Soc.*, **1965**, 4273.

[226] R. Kanazawa, H. Kotsuki, and T. Tokoroyama, *Tetrahedron Lett.*, **1975**, 3651.

[227] T. Tokoroyama, K. Matsuo, H. Kotsuki, and R. Kanazawa, *Tetrahedron*, **36**, 3377 (1980).

[228] H. K. Hung, H. Y. Lam, W. Niemczura, M. C. Wang, and C. M. Wong, *Can. J. Chem.*, **56**, 638 (1978).

[229] M. Shamma and L. Töke, *Tetrahedron*, **31**, 1991 (1975).

[230] H. Disselnkötter, F. Lieb, H. Oedinger, and D. Wendisch, *Justus Liebigs Ann. Chem.*, **1982**, 150.

[231] I. Nagakura, S. Maeda, M. Ueno, M. Funamizu, and Y. Kitahara, *Chem. Lett.*, **1975**, 1143.

[232] S. Tsuboi, K. Shimozuma, and A. Takeda, *J. Org. Chem.*, **45**, 1517 (1980).

[233] M. M. Midland and A. Tramontano, *Tetrahedron Lett.*, **21**, 3549 (1980).

[234] K. Matsumoto, P. Stark, and R. G. Meister, *J. Med. Chem.*, **20**, 25 (1977).

[235] M. K. Rastogi, C. Kamla, R. P. Kapoor, and C. P. Garg, *Indian J. Chem.*, **17B**, 34 (1979).

[236a] A. Yogev and Y. Mazur, *J. Am. Chem. Soc.*, **87**, 3520 (1965).

[236b] A. Kasal, *Collect. Czech. Chem. Commun.*, **44**, 1619 (1979).

[236c] E. W. Colvin, J. Martin, W. Parker, R. A. Raphael, B. Shroot, and M. Doyle, *J. Chem. Soc., Perkin Trans. 1*, **1972**, 860.

[236d] A. Kasal, *Collect. Czech. Chem. Commun.*, **43**, 498 (1978).

[237] W. R. Roush and T. E. D'Ambra, *J. Org. Chem.*, **46**, 5045 (1981).

[238a] J. Vit, B. Čásenský, M. Mamula, and J. Macháček, U.S. Patent, 3,829,449 (1974).

[238b] E. T. Kaiser, U.S. Patent 4,163,744 (1979) [*C.A.*, **92**, 59114t (1980)].

[238c] V. Kumar and W. A. Remers, *J. Med. Chem.*, **22**, 432 (1979).

[238d] K. L. Bhat, Shin-Yih Chen, and M. M. Joullié, *Heterocycles*, **23**, 691 (1985).

[238e] E. T. Kaiser, U.S. Patent 4,354,972 (1982) [*C.A.*, **98**, 72568n (1983)].

[239] A. Holý, *Collect. Czech. Chem. Commun.*, **49**, 2148 (1984).

[240] N. Cohen, R. J. Lopresti, and G. Saucy, *J. Am. Chem. Soc.*, **101**, 6710 (1979).

[241] D. M. Gash and P. D. Woodgate, *Aust. J. Chem.*, **32**, 1863 (1979).

[242a] S. Torii and T. Inokuchi, *Bull. Chem. Soc. Jpn.*, **53**, 2642 (1980).

[242b] C. Iwata, Y. Takemoto, H. Kubota, T. Kuroda, and T. Imanishi, *Tetrahedron Lett.*, **26**, 3231 (1985).

[242c] C. Iwata, H. Kubota, M. Yamada, Y. Takemoto, S. Uchida, T. Tanaka, and T. Imanishi, *Tetrahedron Lett.*, **25**, 3339 (1984).

[243] J. A. Marshall and H. Cohen, *J. Org. Chem.*, **30**, 3475 (1965).

[244] J. A. Marshall, M. T. Pike, and R. D. Carroll, *J. Org. Chem.*, **31**, 2933 (1966).

[245] J. A. Marshall, N. H. Andersen, and R. A. Hochstetler, *J. Org. Chem.*, **32**, 113 (1967).

[246] B. Weiss, *Chem. Phys. Lipids*, **19**, 347 (1977).

[247] S. Takano, M. Morimoto, and K. Ogasawara, *Yakugaku Zasshi*, **103**, 1257 (1983) [*C.A.*, **101**, 7444k (1984)].

[248] J. K. Whitesell, D. Deyo, and A. Bhattacharya, *J. Chem. Soc., Chem. Commun.*, **1983**, 802.

[249] K. Yamakawa, K. Nishitani, A. Murakami, and A. Yamamoto, *Chem. Pharm. Bull.*, **31**, 3397 (1983).

[250] D. D. Weller and E. P. Stirchak, *J. Org. Chem.*, **48**, 4873 (1983).

[251] J. D. Elliot, A. B. Kelson, N. Purcell, R. J. Stoodley, and M. N. Palfreyman, *J. Chem. Soc., Perkin Trans. 1*, **1983**, 2441.

[252] N. Mongelli, A. Andreoni, L. Zuliani, and C. A. Gandolfi, *Tetrahedron Lett.*, **24**, 3527 (1983).
[253] G. Schulz and H. Berner, *Tetrahedron*, **40**, 905 (1984).
[254a] B. Douglas and J. A. Weisbach, U.S. Patent, 3,455,936 (1969) [*C.A.*, **71**, 91450b (1969)].
[254b] B. Douglas and J. A. Weisbach, U.S. Patent 3,507,877 (1970) [*C.A.*, **73**, 14825s (1970)].
[254c] J. Fajkoš and F. Šorm, *Chem. Listy*, **52**, 505 (1958) [*C.A.*, **53**, 4349b (1959)].
[254d] J. Joska, J. Fajkoš, and F. Šorm, *Chem. Ind.* (*London*), **1958**, 1665.
[254e] J. Fajkoš and F. Šorm, *Collect. Czech. Chem. Commun.*, **24**, 766 (1959).
[254f] J. Fajkoš and F. Šorm, *Chem. Listy*, **52**, 2115 (1958) [*C.A.*, **53**, 5342d (1959)].
[255] M. Arnó, M. Carda, J. A. Marco, and E. Seoane, *Chem. Lett.*, **1984**, 1021.
[256] M. Kocór and W. Wojciechowska, *Pol. J. Chem.*, **54**, 1907 (1980) [*C.A.*, **95**, 98131c (1981)].
[257] S. Torii, H. Tanaka, T. Inokuchi, and K. Tomozane, *Bull. Chem. Soc. Jpn.*, **55**, 3947 (1982).
[258] A. Horeau, H. B. Kagan, and J. P. Vigneron, *Bull. Soc. Chim. Fr.*, **1968**, 3795.
[259] Y. Konishi, M. Kawamura, Y. Iguchi, Y. Arai, and M. Hayashi, *Tetrahedron*, **37**, 4391 (1981).
[260a] Nitto Electric Industrial Co., Ltd., Japanese Patent (Tokkyo Koho) 59,157,055 (84,157,055) (1984) [*C.A.*, **102**, 61994y (1985)].
[260b] W. H. Okamura, R. Peter, and W. Reischl, *J. Am. Chem. Soc.*, **107**, 1034 (1985).
[260c] K. H. Marx, P. Raddatz, and E. Winterfeldt, *Justus Liebigs Ann. Chem.*, **1984**, 474.
[261] T. Takahashi, H. Okumoto, and J. Tsuji, *Tetrahedron Lett.*, **25**, 1925 (1984).
[262] E. J. Corey and N. W. Boaz, *Tetrahedron Lett.*, **25**, 3055 (1984).
[263] F. Yasuhara and S. Yamaguchi, *Tetrahedron Lett.*, **21**, 2827 (1980).
[264] K. Kabuto and H. Ziffer, *J. Org. Chem.*, **40**, 3467 (1975).
[265] K. Kabuto, H. Shindo, and H. Ziffer, *J. Org. Chem.*, **42**, 1742 (1977).
[266] D. Valentine, Jr. and J. W. Scott, *Synthesis*, **1978**, 329.
[267] P. W. Collins, E. Z. Dajani, R. Pappo, A. F. Gasiecki, R. G. Bianchi, and E. M. Woods, *J. Med. Chem.*, **26**, 786 (1983).
[268] C. J. Sih, J. B. Heather, G. P. Peruzzotti, P. Price, R. Sood, and L. F. Hsu Lee, *J. Am. Chem. Soc.*, **95**, 1676 (1973).
[269] C. J. Sih, J. B. Heather, R. Sood, P. Price, G. P. Peruzzotti, L. F. Hsu Lee, and S. S. Lee, *J. Am. Chem. Soc.*, **97**, 865 (1975).
[270] J. G. Murphy, *J. Med. Chem.*, **8**, 267 (1965).
[271] O. E. Edwards, L. Fonzes, and L. Marion, *Can. J. Chem.*, **44**, 583 (1966).
[272] M. D. Higgs and D. J. Faulkner, *J. Org. Chem.*, **43**, 3454 (1978).
[273] D. B. Stierle and D. J. Faulkner, *J. Org. Chem.*, **45**, 3396 (1980).
[274] S. W. Pelletier, J. Finer-Moore, R. C. Desai, N. V. Mody, and H. K. Desai, *J. Org. Chem.*, **47**, 5290 (1982).
[275] R. van der Linde, L. van der Wolf, H. J. J. Pabon, and D. A. van Dorp, *Recl., J. Roy. Neth. Chem. Soc.*, **94**, 257 (1975).
[276] B. Lythgoe, R. Manwaring, J. R. Milner, T. A. Moran, M. E. N. Nambudiry, and J. Tideswell, *J. Chem. Soc., Perkin Trans. 1*, **1978**, 387.
[277a] M. Karpf and C. Djerassi, *Tetrahedron Lett.*, **21**, 1603 (1980).
[277b] P. Drašar, V. Pouzar, I. Černý, M. Havel, S. N. Ananchenko, and I. V. Torgov, *Collect. Czech. Chem. Commun.*, **47**, 1240 (1982).
[277c] L. Nedelec and V. Torelli, European Patent Application 23,856 (1981) [*C.A.*, **95**, 98147n (1981)].
[277d] L. Nedelec, V. Torelli, and M. Hardy, French Demande 2,498,607 (1982) [*C.A.*, **98**, 72560d (1983)].
[277e] L. Nedelec and V. Torelli, French Demande 2,498,608 (1982) [*C.A.*, **98**, 72562f (1983)].
[277f] J. Žemlička, J. V. Freisler, R. Gasser, and J. P. Horwitz, *J. Org, Chem.*, **38**, 990, 1973.
[277g] C. A. Henrick, F. Schaub, and J. B. Siddall, *J. Am. Chem. Soc.*, **94**, 5374 (1972).
[278] C. A. Henrick, F. Schaub, and J. B. Siddall, *J. Am. Chem. Soc.*, **94**, 8647 (1972).
[279] T. Sugiyama, A. Kobayashi, K. Yamashita, and T. Suzuki, *Agric. Biol. Chem.*, **36**, 2275 (1972) [*C.A.*, **78**, 110644f (1973)].
[280] S. Uesato, K. Kobayashi, and H. Inouye, *Chem. Pharm. Bull.*, **30**, 3942 (1982).
[281] R. Noyori, I. Umeda, and T. Ishigami, *J. Org. Chem.*, **37**, 1542 (1972).

[282] B. R. Snider and D. Rodini, *Tetrahedron Lett.*, **1978**, 1399.
[283] J. M. Fortunato and B. Ganem, *J. Org. Chem.*, **41**, 2194 (1976).
[284] L. I. Zakharkin and I. M. Khorlina, *Tetrahedron Lett.*, **1962**, 619.
[285] L. I. Zakharkin and I. M. Khorlina, *Izv. Akad. Nauk SSSR, Ser. Khim.*, **1962**, 538; *Engl. Transl.*, p. 497 [*C.A.*, **57**, 14924d (1962)].
[286] L. I. Zakharkin and I. M. Khorlina, *Izv. Akad. Nauk SSSR, Ser. Khim.*, **1963**, 316; *Engl. Transl.*, p. 288.
[287] L. I. Zakharkin, V. V. Gavrilenko, D. N. Maslin, and I. M. Khorlina, *Tetrahedron Lett.*, **1963**, 2087.
[288] L. I. Zakharkin, V. V. Gavrilenko, and D. N. Maslin, *Izv. Akad. Nauk SSSR, Ser. Khim.*, **1964**, 926; *Engl. Transl.*, p. 867. [*C.A.*, **61**, 5505c (1964)].
[289] L. I. Zakharkin and I. M. Khorlina, *Izv. Akad. Nauk SSSR, Ser. Khim.*, **1964**, 465; *Engl. Transl.*, p. 435 [*C.A.*, **60**, 15765d (1964)].
[290] C. L. J. Wang, *Tetrahedron Lett.*, **23**, 1067 (1982).
[291] S. Ayral-Kaloustian and W. C. Agosta, *J. Org, Chem.*, **46**, 4880 (1981).
[292] I. Kompis and A. Wick, *Helv. Chim. Acta*, **60**, 3025 (1977).
[293] H. Bartsch and O. Schwarz, *J. Heterocycl. Chem.*, **19**, 1189 (1982).
[294a] C. H. DePuy, R. L. Parton, and T. Jones, *J. Am. Chem. Soc.*, **99**, 4070 (1977).
[294b] R. Breslow, J. Lockhart, and A. Small, *J. Am. Chem. Soc.*, **84**, 2793 (1962).
[295] E. Schenker, *Angew. Chem.*, **73**, 81 (1961).
[296] M. N. Rerick, in *Reduction*, R. L. Augustine, Ed., Marcel Dekker, New York, 1968, pp. 1–94.
[297] W. H. Tamblyn, D. H. Weingold, E. D. Snell, and R. E. Waltermire *Tetrahedron Lett.*, **23**, 3337 (1982).
[298a] W. H. White and W. K. Fife, *J. Am. Chem. Soc.*, **83**, 3846 (1961).
[298b] S. I. Goldberg, *J. Org. Chem.*, **25**, 482 (1960).
[298c] M. Cais and A. Modiano, *Chem. Ind.* (*London*), **1960**, 202.
[298d] T. H. Coffield, K. G. Ihrman, and W. Burns, *J. Am. Chem. Soc.*, **82**, 1251 (1960).
[298e] A. Bowers and J. A. Edwards, U.S. Patent 3,155,695 (1964) [*C.A.*, **62**, 10488a (1965)].
[298f] E. Ritchie, R. G. Senior, and W. C. Taylor, *Aust. J. Chem.*, **22**, 2371 (1969).
[298g] T. H. Coffield, K. G. Ihrman, and W. Burns, *J. Am. Chem. Soc.*, **82**, 4209 (1960).
[299] B. H. Lee and M. J. Miller, *Tetrahedron Lett.*, **25**, 927 (1984).
[300] M. Pailer and H. Gutwillinger, *Monatsh. Chem.*, **108**, 653 (1977).
[301] E. H. White, R. E. K. Winter, R. Graeve, U. Zirngibl, E. W. Friend, H. Maskill, U. Mende, G. Kreiling, H. P, Reisenauer, and G. Maier, *Chem. Ber.*, **114**, 3906 (1981).
[302] H. Rutner and P. E. Spoerri, *J. Org. Chem.*, **28**, 1898 (1963).
[303] E. D. Bergmann and A. Cohen, *Tetrahedron Lett.*, **1965**, 1151.
[304] G. E. Lewis, *J. Org. Chem.*, **30**, 2433 (1965).
[305] D. Blackburn and G. Burguard, *J. Pharm. Sci.*, **54**, 1586 (1965).
[306] W. G. Filby and K. Günther, *J. Labelled Compd. Radiopharm.*, **9**, 321 (1973).
[307] H. Rakoff, *J. Labelled Compd. Radiopharm.*, **12**, 473 (1976).
[308] S. E. Najjar, M. I. Blake, and M. C. Lu, *J. Labelled Compd. Radiopharm.*, **15** (Suppl. Vol.), 71 (1978).
[309] G. W. J. Fleet and P. J. C. Harding, *Tetrahedron Lett.*, **1979**, 975.
[310] R. O. Hutchins and M. Markowitz, *Tetrahedron Lett.*, **21**, 813 (1980).
[311] P. Four and F. Guibe, *J. Org. Chem.*, **46**, 4439 (1981).
[312] E. Mosettig, *Org. React*, **4**, 362 (1948).
[313] O. Kříž, B. Čásenský, and O. Štrouf, *Collect. Czech. Chem. Commun.*, **38**, 842 (1973).
[314] F. E. Ziegler and H. Lim, *J. Org. Chem.*, **49**, 3278 (1984).
[315] J. Kuthan, J. Procházková, and E. Janečková, *Collect. Czech. Chem. Commun.*, **33**, 3558 (1968).
[316] G. R. Newkome and T. Kawato, *J. Org. Chem.*, **44**, 2693 (1979).
[317] E. Auderhaar, J. E. Baldwin, D. H. R. Barton, D. J. Faulkner, and M. Slaytor, *J. Chem. Soc., C*, **1971**, 2175.

[318] J. A. Profitt, D. S. Watt, and E. J. Corey, *J. Org. Chem.*, **40**, 127 (1975).

[319] G. Hesse and R. Schrödel, *Justus Liebigs Ann. Chem.*, **607**, 24 (1957).

[320] G. Hesse and R. Schrödel, *Angew. Chem.*, **68**, 438 (1956).

[321] I. Stibor, M. Janda, and J. Šrogl, *Z. Chem.*, **10**, 342 (1970).

[322] L. I. Zakharkin and I. M. Khorlina, *Dokl. Akad. Nauk SSSR*, **116**, 422 (1957) [*C.A.*, **52**, 8040f (1958)].

[323] E. Mosettig, *Org. React.* **8**, 218 (1954).

[324] K. R. Varma and E. Caspi, *J. Org. Chem.*, **34**, 2489 (1969).

[324a] B. Gadsby, M. R. G. Leeming, G. Greenspan, and H. Smith, *J. Chem. Soc., C*, **1968**, 2647.

[324b] H. Saeki, T. Iwashige, E. Ohki, K. Furuya, and M. Shirasaka, *Ann. Sankyo Res. Lab.*, **19**, 137 (1967) [*C.A.*, **68**, 96075f (1968)].

[324c] J. A. Montgomery and K. Hewson, *J. Org. Chem.*, **29**, 3436 (1964).

[325] N. M. Waldron, *J. Chem. Soc., C*, **1968**, 1914.

[326] W. H. Moos, R. D. Gless, and H. Rapoport, *J. Org. Chem.*, **48**, 227 (1983).

[327] R. D. Gless and H. Rapoport, *J. Org. Chem.*, **44**, 1324 (1979).

[328] H. Böhme and P. N. Sutoyo, *Justus Liebigs Ann. Chem.*, **1982**, 1643.

[329] A. Matsuura, M. Akatsu, M. Sunagawa, K. Ishizumi, and J. Katsube, Japanese Patent 77 91,893 [*C.A.*, **88**, 22874g (1978)].

[330] H. H. Ong, J. A. Profitt, J. Fortunato, E. J. Glamkowski, D. B. Ellis, H. M. Geyer III, J. C. Wilker, and H. Burghard, *J. Med. Chem.*, **26**, 981 (1983).

[331] M. Schlosser and Z. Brich, *Helv. Chim. Acta*, **61**, 1903 (1978).

[332] K. Kato, T. Takita, and H. Umezawa, *Tetrahedron Lett.*, **21**, 4925 (1980).

[333a] P. G. Baraldi, A. Barco, S. Benetti, F. Moroder, G. P. Pollini, and D. Simoni, *J. Org. Chem.*, **48**, 1297 (1983).

[333b] R. E. Ireland and F. R. Brown, Jr., *J. Org. Chem.*, **45**, 1868 (1980).

[333c] G. J. Bird, D. J. Collins, F. W. Eastwood, R. H. Exner, M. L. Romanelli, and D. D. Small, *Aust, J. Chem.*, **32**, 783 (1979).

[334] Continental Oil Co., British Patent 938,044 (1964) [*C.A.*, **60**, 2764e (1964)].

[335] K. Kefurt, Z. Kefurtová, and J. Jarý, *Collect. Czech. Chem. Commun.*, **37**, 1035 (1972).

[336] G. R. Newkome, T. Kawato, and A. Nayak, *J. Org. Chem.*, **44**, 2697 (1979).

[337] D. Berney and K. Schuh, *Helv. Chim. Acta*, **63**, 918 (1980).

[338] J. P. Kutney, R. A. Badger, J. F. Beck, H. Bosshardt, F. S. Matough, V. E. Ridaura-Sanz, Y. H. So, R. S. Sood, and B. R. Worth, *Can. J. Chem.*, **57**, 289 (1979).

[339] A. Brossi and S. Teitel, *Helv. Chim. Acta*, **53**, 1779 (1970).

[340] H. Bartsch and O. Schwarz, *Arch. Pharm.* (*Weinheim*, W. Germany), **315**, 538 (1982) [*C.A.*, **97**, 92217e (1982)].

[341] A. Hajós, *Komplexe Hydride*, VEB Deutscher Verlag der Wissenschaften, East Berlin, 1966.

[342] H. C. Brown and A. Tsukamoto, *J. Am. Chem. Soc.*, **83**, 4549 (1961).

[343] L. I. Zakharkin and I. M. Khorlina, *Izv. Akad. Nauk SSSR, Ser. Khim.*, **1959**, 2146; *Engl. Transl.*, p. 1865 [*C.A.*, **54**, 10932b (1960)].

[344a] D. B. Tulshian and B. Fraser-Reid, *J. Org. Chem.*, **49**, 518 (1984).

[344b] K. Šindelář, J. Holubek, and M. Protiva, *Heterocycles*, **9**, 1498 (1978).

[345] E. Renk, P. R. Shafer, W. H. Graham, R. H. Mazur, and J. D. Roberts, *J. Am. Chem. Soc.*, **83**, 1987 (1961).

[346] M. Grdinic, D. A. Nelson, and V. Boekelheide, *J. Am. Chem. Soc.*, **86**, 3357 (1964).

[347] L. Ruzicka, M. Kobelt, O. Häflinger, and V. Prelog, *Helv. Chim. Acta*, **32**, 544 (1949).

[348] M. Ferles, *Chem. Listy*, **52**, 2184 (1958) [*C.A.*, **53**, 6225h (1959)].

[349] G. A. Swan and J. D. Wilcock, *J. Chem. Soc., Perkin Trans. 1*, **1974**, 885.

[350] V. Kubánek, J. Králíček, J. Mařík, and J. Kondelíková, *Chem. Prům.*, **25**, 628 (1975) [*C.A.*, **85**, 6120r (1976)].

[351] V. Kubánek, J. Králíček, and J. Kondelíková, Czech. Patent 166,482 (1976) [*C.A.*, **87**, 40072c (1977)].

[352] B. Čásenský, J. Macháček, O. Kříž, and V. Kubánek, German Offen. 2,655,889 (1977) [*C.A.*, **87**, 138040r (1977)].

[353] O. Kříž, B. Čásenský, J. Macháček, and P. Sochor, *Chem. Prům.*, **28**, 76 (1978) [*C.A.*, **89**, 110523t (1978)].

[354] V. Kubánek, J. Králíček, J. Šejba, and J. Mařík, *Chem. Prům.*, **28**, 412 (1978) [*C.A.*, **89**, 163993c (1978)].

[355] B. Čásenský and J. Mařík, *Chem. Prům.*, **29**, 651 (1979) [*C.A.*, **92**, 94929j (1980)].

[356a] H. Takahata, H. Okajima, and T. Yamazaki, *Chem. Pharm. Bull.*, **28**, 3632 (1980).

[356b] A. I. Meyers and K. T. Wanner, *Tetrahedron Lett.*, **26**, 2047 (1985).

[357] Y. Ban, I. Iijima, I. Inoue, M. Akagi, and T. Oishi, *Tetrahedron Lett.*, **1969**, 2067.

[358] I. Inoue and Y. Ban, *J. Chem. Soc., C*, **1970**, 602.

[359] N. Hirose, S. Sohda, S. Kuriyama, and S. Toyoshima, *Chem. Pharm. Bull.*, **20**, 1669 (1972).

[360] N. Hirose, S. Sohda, S. Kuriyama, and S. Toyoshima, *Chem. Pharm. Bull.*, **21**, 960 (1973).

[361] R. V. Stevens, R. K. Mehra, and R. L. Zimmermann, *J. Chem. Soc., Chem. Commun.*, **1969**, 877.

[362] D. C. Horwell and D. E. Tupper, British Patent 2,031,409 B (1980) [*C.A.*, **94**, 83967k (1981)].

[363] D. C. Horwell, D. E. Tupper, and W. H. Hunter, *J. Chem. Soc., Perkin Trans. 1*, **1983**, 1545.

[364] M. Shamma and P. D. Rosenstock, *J. Org. Chem.*, **26**, 718 (1961).

[365] R. J. Stoodley and N. S. Watson, *J. Chem. Soc., Perkin Trans. 1*, **1973**, 2105.

[366] P. L. Southwick, N. Latif, B. M. Fitzgerald, and N. M. Zaczek, *J. Org. Chem.*, **31**, 1 (1966).

[367] M. J. Schneider and T. M. Harris, *J. Org. Chem.*, **49**, 3681 (1984).

[368] R. M. Williams, O. P. Anderson, R. W. Armstrong, J. Josey, H. Meyers, and C. Eriksson, *J. Am. Chem. Soc.*, **104**, 6092 (1982).

[369] M. Yamasaki, *Osaka Daigaku Igaku Zasshi*, **25**, 159 (1973) [*C.A.*, **83**, 43270u (1975)].

[370] Chau-der Li, M. H. Lee, and A. C. Sartorelli, *J. Med. Chem.*, **22**, 1030 (1979).

[371] V. E. Marques, L. M. Twanmoh, H. B. Wood, Jr., and J. S. Driscoll, *J. Org. Chem.*, **37**, 2558 (1972).

[372] A. Kosasayama, T. Konno, K. Higashi, and F. Ishikawa, *Chem. Pharm. Bull.*, **27**, 848 (1979).

[373] T. Kametani, N. Tagaki, M. Toyota, T. Honda, and K. Fukumoto, *Heterocycles*, **16**, 591 (1981).

[374] T. Naito, Y. Tada, Y. Nishiguchi, C. Hashimoto, T. Kiguchi, and I. Ninomiya, *Heterocycles*, **19**, 163 (1982).

[375] T. Naito, Y. Tada, C. Hashimoto, K. Katsumi, T. Kiguchi, and I. Ninomiya, *Tennen Yuki Kagobutsu Toronkai Koen Yoshishu*, **24**, 46C (1981) [*C.A.*, **96**, 143132n (1982)].

[376] T. Naito, Y. Tada, and I. Ninomiya, *Heterocycles*, **16**, 1141 (1981).

[377a] T. Naito, K. Katsumi, Y. Tada, and I. Ninomiya, *Heterocycles*, **20**, 779 (1983).

[377b] I. Ninomiya, C. Hashimoto, T. Kiguchi, and T. Naito, *J. Chem. Soc., Perkin Trans. 1*, **1984**, 2911.

[378] H. Posvic and J. C. de Meireles, *J. Heterocycl. Chem.*, **17**, 1241 (1980).

[379] W. L. Meyer, T. E. Goodwin, R. J. Hoff, and C. W. Sigel, *J. Org. Chem.*, **42**, 2761 (1977).

[380] T. Mukaiyama, H. Yamashita, and M. Asami, *Chem. Lett.*, **1983**, 385.

[381] Y. Nagao and E. Fujita, *J. Synth. Org. Chem. Jpn.*, **42**, 622 (1984); W. N. Speckamp and H. Hiemstra, *Tetrahedron*, **41**, 4367 (1985); Mitsui Toatsu Chemicals, Inc., Japanese Patent (Tokkyo Koho) 59,161,345 (84,161,345) (1984) [C.A., 102, 149102s (1985)].

[382] D. L. Garmaise and A. Ryan, *J. Heterocycl. Chem.*, **7**, 413 (1970).

[383] J. Altman, E. Babad, J. Itzchaki, and D. Ginsburg, *Tetrahedron* (*Suppl. 8*), **22**, 279 (1966).

[384] R. A. Partyka, R. T. Strandridge, H. G. Howell, and A. T. Shulgin, U.S. Patent 4,105,695 (1978) [*C.A.*, **90**, 151799b (1979)].

[385] H. Newman, *J. Org. Chem.*, **39**, 100 (1974).

[386] W. H. Pirkle and J. R. Hauske, *J. Org. Chem.*, **42**, 2436 (1977).

[387] S. J. Law, D. H. Lewis, and R. F. Borne, *J. Heterocycl. Chem.*, **15**, 273 (1978).

[388] O. Červinka, V. Suchan, O. Kotýnek, and V. Dudek, *Collect. Czech. Chem. Commun.*, **30**, 2484 (1965).

[389] S. Karady, J. S. Amato, L. M. Weinstock, and M. Sletzinger, *Tetrahedron Lett.*, **1978**, 403.

[390a] H. Feuer and B. F. Vincent, Jr., *J. Am. Chem. Soc.*, **84**, 3771 (1962).

[390b] H. Feuer, B. F. Vincent, Jr., and R. S. Bartlett, *J. Org. Chem.*, **30**, 2877 (1965).

[391] S. Allenmark, *Tetrahedron Lett.*, **1972**, 2885.
[392] L. M. Meshcheryakova, V. A. Zagorevskii, and E. K. Orlova, *Khim. Geterotsikl. Soedin.*, **1980**, 853 [*C.A.*, **93**, 186299c (1980)].
[393] Y. Girault, M. Decouzon, and M. Azzaro, *Tetrahedron Lett.*, **1976**, 1175.
[394] S. R. Landor, Y. M. Chan, O. O. Sonola, and A. R. Tatchell, *J. Chem. Soc., Perkin Trans. 1*, **1984**, 493.
[395] B. Krzyzanowska and W. J. Stec, *Synthesis*, **1982**, 270.
[396] R. Annunziata, M. Cinquini, and F. Cozzi, *J. Chem. Soc., Perkin Trans. 1*, **1982**, 339.
[397] R. A. Perry, S. C. Chen, B. C. Menon, K. Hanaya, and Y. L. Chow, *Can. J. Chem.*, **54**, 2385 (1976).
[398] H. M. Walborsky and G. E. Niznik, *J. Org. Chem.*, **37**, 187 (1972).
[399] P. A. S. Smith, *The Organic Chemistry of Open-Chain Nitrogen Compounds*, Benjamin, New York, 1965.
[400] S. E. Ellzey, Jr. and C. H. Mack, *J. Org. Chem.*, **28**, 1600 (1963).
[401] Z. Bukač and J. Šebenda, U.S. Patent 3,962,239 (1976) [*C.A.*, **85**, 177495w (1976)].
[402a] Z. Bukač, Czech. Patent 202,899 (1982) [*C.A.*, **98**, 198280f (1983)].
[402b] Z. Bukač, R. Puffr, and J. Šebenda, *Chem. Prům.*, **34**, 374 (1984) [*C.A.*, **102**, 5864u (1985)].
[403] I. J. Massey and I. T. Harrison, *Chem. Ind. (London)*, **1977**, 920.
[404] I. T. Harrison, W. Kurz, I. J. Massey, and S. H. Unger, *J. Med. Chem.*, **21**, 588 (1978).
[405a] P. D. Klimstra, E. F. Nutting, and R. E. Counsell, *J. Med. Chem.*, **9**, 693 (1966).
[405b] G. D. Searle & Co., British Patent 1,038,349 (1966) [*C.A.*, **65**, 13795b (1966)].
[405c] P. D. Klimstra, U.S. Patent 3,301,850 (1967) [*C.A.*, **66**, 85954z (1967)].
[405d] W. Merkel and W. Ried, *Chem. Ber.*, **106**, 471 (1973).
[406] Y. Hayakawa, Y. Baba, S. Makino, and R. Noyori, *J. Am. Chem. Soc.*, **100**, 1786 (1978).
[407a] A. Klemer, B. Brandt, U. Hofmeister, and E. R. Rüter, *Justus Liebigs Ann. Chem.*, **1983**, 1920.
[407b] G. L. Olson, H. C. Cheung, K. D. Morgan, J. F. Blount, L. Todaro, L. Berger, A. B. Davidson, and E. Boff, *J. Med. Chem.*, **24**, 1026 (1981).
[408] A. P. Marchand and R. W. Allen, *Tetrahedron Lett.*, **1975**, 67.
[409] S. J. Daum, C. M. Martini, R. K. Kullnig, and R. L. Clarke, *J. Org. Chem.*, **37**, 1665 (1972).
[410a] W. Fleischhacker and B. Richter, *Chem. Ber.*, **112**, 2539 (1979).
[410b] G. Schramm and H. Riedl, U.S. Patent 3,673,175 (1972) [*C.A.*, **77**, 102032u (1972)].
[410c] G. Schramm and H. Riedl, German Offen. 2,021,761 (1971) [*C.A.*, **76**, 59892q (1972)].
[411] M. Kraus and K. Kochloefl, *Collect. Czech. Chem. Commun.*, **34**, 1823 (1969).
[412] Y. Kamitori, M. Hojo, R. Masuda, T. Inoue, and T. Izumi, *Tetrahedron Lett.*, **24**, 2575 (1983).
[413] J. F. Corbett, *J. Chem. Soc., Chem. Commun.*, **1968**, 1257.
[414] L. R. C. Barclay, I. T. McMaster, and J. K. Burgess, *Tetrahedron Lett.*, **1973**, 3947.
[415] J. R. Butterick and A. M. Unrau, *J. Chem. Soc., Chem. Commun.*, **1974**, 307.
[416] K. Eiter, K. F. Hebenbrock, and H. J. Kabbe, *Justus Liebigs Ann. Chem.*, **765**, 55 (1972).
[417] G. Snatzke, H. Laurent, and R. Wiechert, *Tetrahedron*, **25**, 761 (1969).
[418a] P. D. Klimstra, U.S. Patent 3,350,424 (1967) [*C.A.*, **68**, 114848s (1968)].
[418b] K. J. Rorig and H. A. Wagner, U.S. Patent 3,412,094 (1968) [*C.A.*, **70**, 68410j (1969)].
[418c] P. D. Klimstra, U.S. Patent 3,388,123 (1968) [*C.A.*, **70**, 4414f (1969)].
[418d] P. D. Klimstra, U.S. Patent, 3,511,859 (1970) [*C.A.*, **73**, 15112u (1970)].
[419] T. F. Buckley and J. G. Gleason, German Offen. 2,740,280 (1978) [*C.A.*, **89**, 6351t (1978)].
[420] G. Just and T. J. Liak, *Can. J. Chem.*, **56**, 211 (1978).
[421a] D. Huckle, I. M. Lockhart, and N. E. Webb, *J. Chem. Soc., C*, **1971**, 2252.
[421b] G. Eberle, I. Lagerlund, I. Ugi, and R. Urban, *Tetrahedron*, **34**, 977 (1978).
[422] G. A. Boswell, Jr., *Chem. Ind. (London)*, **1965**, 1929.
[423] G. A. Boswell, Jr., *J. Org. Chem.*, **33**, 3699 (1968).
[424a] E. Galantay, I. Bacso, and R. V. Coombs, *Synthesis*, **1974**, 344.
[424b] I. Bacso, German Offen. 2,301,911 (1973) [*C.A.*, **79**, 105468t (1973)].
[424c] I. Bacso, U.S. Patent 3,803,183 (1974) [*C.A.*, **81**, 13708j (1974)].
[424d] I. Bacso and E. E. Galantay, U.S. Patent 3,907,845 (1975) [*C.A.*, **84**, 31303q (1976)].

[424e] I. Bacso, U.S. Patent 3,978,048 (1976) [*C.A.*, **86**, 29998j (1977)].

[425] K. Hiratani, T. Nakai, and M. Okawara, *Bull. Chem. Soc. Jpn.*, **46**, 3872 (1973).

[426] A. Padwa, T. J. Blacklock, P. H. J. Carlsen, and M. Pulwer, *J. Org. Chem.*, **44**, 3281 (1979).

[427] O. Červinka, *Chimia* (*Aarau*), **13**, 332 (1959) [*C.A.*, **54**, 9711b (1960)].

[428] O. Červinka, *Collect. Czech. Chem. Commun.*, **30**, 2403 (1965).

[429] O. Červinka, *Collect. Czech. Chem. Commun.*, **26**, 673 (1961).

[430] J. L. Herrmann, R. J. Cregge, J. E. Richman, C. L. Semmelhack, and R. H. Schlessinger, *J. Am. Chem. Soc.*, **96**, 3702 (1974).

[431] J. L. Herrmann, R. J. Cregge, J. E. Richman, G. R. Kieczykowski, S. N. Normandin, M. L. Quesada, C. L. Semmelhack, A. J. Poss, and R. H. Schlessinger, *J. Am. Chem. Soc.*, **101**, 1540 (1979).

[432] M. Ferles, *Collect. Czech. Chem. Commun.*, **24**, 2221 (1959).

[433] M. Ferles, J. Janoušková, and O. Fuchs, *Collect. Czech. Chem. Commun.*, **36**, 2389 (1971).

[434] M. Holík and M. Ferles, *Collect. Czech. Chem. Commun.*, **32**, 3067 (1967).

[435] A. Šilhánková, M. Holík, and M. Ferles, *Collect. Czech. Chem. Commun.*, **33**, 2494 (1968).

[436] M. Ferles, *Collect. Czech. Chem. Commun.*, **23**, 479 (1958).

[437] M. Ferles, M. Kovařík, and Z. Vondráčková, *Collect. Czech. Chem. Commun.*, **31**, 1348 (1966).

[438] M. Ferles, A. Attia, and A. Šilhánková, *Collect. Czech. Chem. Commun.*, **38**, 615 (1973).

[439] D. L. Comins and A. H. Abdullah, *J.* Org. Chem., **49**, 3392 (1984).

[440] H. Nishimura, S. Naruto, and H. Mizuta, Japanese Patent 77 07,998 [*C.A.*, **87**, 23588h (1977)].

[441] M. Jankovský and M. Ferles, *Collect. Czech. Chem. Commun.*, **35**, 2802 (1970).

[442] D. J. Kosman and L. H. Piette, *J. Chem. Soc., Chem. Commun.*, **1969**, 926.

[443] P. Rey and H. M. McConnell, *J. Am. Chem. Soc.*, **99**, 1637 (1977).

[444] K. Hideg, H. O. Hankovszky, L. Lex, and G. Kulcsár, *Synthesis*, **1980**, 911.

[445] D. E. Minter and P. L. Stotter, *J. Org. Chem.*, **46**, 3965 (1981).

[446] B. K. Blackburn, J. F. Frysinger, and D. E. Minter, *Tetrahedron Lett.*, **25**, 4913 (1984).

[447] Y. Tamura, M. Kumitomo, T. Masui, and M. Terashima, *Chem. Ind.* (*London*), **1972**, 168.

[448] P. Guerry and R. Neier, *Synthesis*, **1984**, 485.

[449] M. L. Scheinbaum and M. B. Dines, *Tetrahedron Lett.*, **1971**, 2205.

[450] A. Alberola, A. M. Gonzáles, M. A. Laguna, and F. J. Pulido, *Synthesis*, **1984**, 510.

[451] B. J. Banks, A. G. M. Barrett, M. A. Russell, and D. J. Williams, *J. Chem. Soc., Chem. Commun.*, **1983**, 873.

[452] S. Oae and H. Togo, *Kagaku* (*Kyoto*), **38**, 48 (1983) [*C.A.*, **98**, 197543g (1983)].

[453a] J. H. Rigby, J. M. Sage, and J. Raggon, *J. Org. Chem.*, **47**, 4815 (1982).

[453b] M. Kishi, S. Ishihara, and T. Komeno, *Tetrahedron*, **30**, 2135 (1974).

[454] F. G. Bordwell and W. H. McKellin, *J. Am. Chem. Soc.*, **73**, 2251 (1951).

[455] E. N. Karaulova and G. D. Galpern, *Zh. Obshch. Khim.*, **29**, 3033 (1959); *Engl. Transl.*, p. 2998 [*C.A.*, **54**, 12096d (1960)].

[456] S. Oae and H. Togo, *Kagaku* (*Kyoto*), **37**, 812 (1982) [*C.A.*, **98**, 16023u (1983)].

[457a] T. L. Ho and C. M. Wong, *Org. Prep. Proced. Int.*, **7**, 163 (1975) [C.A., **84**, 30328q (1976)].

[457b] L. Weber, *Chem. Ber.*, **116**, 2022 (1983).

[457c] L. Weber and R. Boese, *Chem. Ber.*, **118**, 1545 (1985).

[458] J. Drabowitz and M. Mikolajczyk, *Synthesis*, **1976**, 527.

[459] T. L. Ho, *Synthesis*, **1979**, 1.

[460] V. Ceré, C. Paolucci, S. Pollicino, E. Sandri, and A. Fava, *J. Org. Chem.*, **44**, 4128 (1979).

[461] M. Mikolajczyk and J. Drabowicz, *Phosphorus Sulfur*, **1**, 301 (1976) [*C.A.*, **86**, 170427n (1977)].

[462] T. J. Barton and R. C. Kippenhan, Jr., *J. Org. Chem.*, **37**, 4194 (1972).

[463] W. P. Weber, P. Stromquist, and T. I. Ito, *Tetrahedron Lett.*, **1974**, 2595.

[464] S. Oae and H. Togo, *Kagaku* (*Kyoto*), **37**, 909 (1982) [*C.A.*, **98**, 160185r (1983)].

[465a] B. Koutek, L. Pavlíčková, J. Velek, and M. Souček, *Collect. Czech. Chem. Commun.*, **41**, 2250 (1976).

[465b] H. Velgová and V. Černý, *Collect. Czech. Chem. Commun.*, **38**, 2976 (1973).

[466a] J. A. Marshall and J. A. Ruth, *J. Org. Chem.*, **39**, 1971 (1974).
[466b] A. Wettstein, K. Heusler, and P. Wieland, U.S. Patent 3,085,089 (1963) [*C.A.*, **59**, 10210g (1963)].
[466c] K. Prezewowsky, H. Laurent, H. Hofmeister, R. Wiechert, F. Neumann, and Y. Nishino, German Offen. 2,426,777 (1975) [*C.A.*, **84**, 122139g (1976)].
[466d] H. Breuer and K. Engel, *Justus Liebigs Ann. Chem.*, **1978**, 580.
[466e] M. N. Galbraith, D. H. S. Horn, E. J. Middleton, and R. J. Hackney, *Aust. J. Chem.*, **22**, 1059 (1969).
[467] W. T. Borden and E. J. Corey, *Tetrahedron Lett.*, **1969**, 313.
[468] H. C. Brown, P. Heim, and N. M. Yoon, *J. Am. Chem. Soc.*, **92**, 1637 (1970).
[469] S. Krishnamurthy and H. C. Brown, *J. Org. Chem.*, **41**, 3064 (1976).
[470] S. Krishnamurthy, *J. Organomet. Chem.*, **156**, 171 (1978).
[471] S. Masamune, G. S. Bates, and P. E. Georghiou, *J. Am. Chem. Soc.*, **96**, 3686 (1974).
[472a] S. Masamune, P. A. Rossy, and G. S. Bates, *J. Am. Chem. Soc.*, **95**, 6452 (1973).
[472b] T. C. Chang, J. F. Gadberry, and D. A. Lightner, *Synth. Commun.*, **14**, 1321 (1984).
[473a] A. Zobáčová, V. Heřmánková, and J. Jarý, *Collect. Czech. Chem. Commun.*, **42**, 2540 (1977).
[473b] S. Krishnamurthy, *J. Org. Chem.*, **45**, 2550 (1980).
[474] F. Nerdel, H. Kaminski, and D. Frank, *Tetrahedron Lett.*, **1967**, 4973.
[475] J. V. Paukstelis and Jar-lin Kao, *J. Am. Chem. Soc.*, **94**, 4783 (1972).
[476] P. Ma, V. S. Martin, S. Masamune, K. B. Sharpless, and S. M. Viti, *J. Org. Chem.*, **47**, 1378 (1982).
[477] S. M. Viti, *Tetrahedron Lett.*, **23**, 4541 (1982).
[478] N. Minami, S. S. Ko, and Y. Kishi, *J. Am. Chem. Soc.*, **104**, 1109 (1982).
[479] J. M. Finan and Y. Kishi, *Tetrahedron Lett.*, **23**, 2719 (1982).
[480] K. C. Nicolaou and J. Uenishi, *J. Chem. Soc., Chem. Commun.*, **1982**, 1292
[481a] S. Takano, S. Otaki, and K. Ogasawara, *J. Chem. Soc., Chem. Commun.*, **1983**, 1172.
[481b] P. C. B. Page, J. E. Carefull, L. H. Powel, and I. O. Sutherland, *J. Chem. Soc., Chem. Commun.*, **1985**, 882.
[481c] S. Hashimoto, Y. Arai, and N. Hamanaka, *Tetrahedron Lett.*, **26**, 2679 (1985).
[481d] M. Marek, K. Kefurt, J. Staněk, Jr., and J. Jarý, *Collect. Czech. Chem. Commun.*, **41**, 2596 (1976).
[481e] L. Pettersson, T. Frejd, and G. Madnusson, *J. Org. Chem.*, **49**, 4540 (1984).
[482] A. M. Mubarak and D. M. Brown, *Tetrahedron Lett.*, **21**, 2453 (1980).
[483] A. M. Mubarak and D. M. Brown, *J. Chem. Soc., Perkin Trans. 1*, **1982**, 809.
[484] R. O. Hutchins, D. Kandasamy, C. A. Maryanoff, D. Masilamani, and B. E. Maryanoff, *J. Org. Chem.*, **42**, 82 (1977).
[485] C. F. Lane, *Synthesis*, **1975**, 135.
[486] A. Claesson and L. I. Olsson, *J. Am. Chem. Soc.*, **101**, 7302 (1979).
[487a] G. Stork, R. K. Boeckmann, Jr., D. F. Taber, W. C. Still, and J. Singh, *J. Am. Chem. Soc.*, **101**, 7107 (1979).
[487b] J. J. Edwards, U.S. Patent 3,904,611 (1975) [*C.A.*, **84**, 44543t (1976)].
[487c] B. W. Metcalf and J. O'Neal Johnston, U.S. Patent 4,289,762 (1981) [*C.A.*, **96**, 104601v (1982)].
[488] E. H. Gold and E. Babad, *J. Org. Chem.*, **37**, 2208 (1972).
[489] E. H. Gold and E. Babad, U.S. Patent 3,956,390 (1976) [*C.A.*, **85**, 62788b (1976)].
[490] H. Terauchi, S. Takemura, and Y. Ueno, *Chem. Pharm. Bull.*, **23**, 640 (1975).
[491] W. Wierenga, *J. Am. Chem. Soc.*, **103**, 5621 (1981).
[492] J. B. Hendrickson and R. Bergeron, *Tetrahedron Lett.*, **1973**, 3839.
[493] J. B. Hendrickson, R. Bergeron, A. Giga, and D. Sternbach, *J. Am. Chem. Soc.*, **95**, 3412 (1973).
[494] J. B. Hendrickson, R. Bergeron, and D. Sternbach, *Tetrahedron*, **31**, 2517 (1975).
[495] M. U. Bombala and S. V. Ley, *J. Chem. Soc., Perkin Trans. 1*, **1979**, 3013.
[496] S. Masamune, P. Ma, H. Okumoto, J. W. Ellingboe, and Y. Ito, *J. Org. Chem.*, **49**, 2834 (1984).

[497] K. Prasad and O. Repič, *Tetrahedron Lett.*, **25**, 3391 (1984).

[498] M. Honda, T. Katsuki, and M. Yamaguchi, *Tetrahedron Lett.*, **25**, 3857 (1984).

[499a] G. J. McGarvey, M. Kimura, T. Oh, and J. M. Williams, *J. Carbohydr. Chem.*, **3**, 125 (1984).

[499b] S. Masamune, W. Choy, J. S. Petersen, and L. R. Sita, *Angew. Chem.*, **97**, 1 (1985); *Angew. Chem., Int. Ed. Engl.*, **24**, 1 (1985).

[500] W. Pickenhagen and H. Brönner-Schindler, *Helv. Chim. Acta*, **67**, 947 (1984).

[501] R. J. Stoodley, *Tetrahedron*, **31**, 2321 (1975).

[502] J. C. Sheehan and C. A. Panetta, *J. Org. Chem.*, **38**, 940 (1973).

[503] D. R. Burfield and R. H. Smithers, *J. Org. Chem.*, **48**, 2420 (1983).

[504] D. A. Lightner, D. T. Hefelfinger, T. W. Powers, G. W. Frank, and K. N. Trueblood, *J. Am. Chem. Soc.*, **94**, 3492 (1972).

[505] S. Auricchio, A. Ricca, and O. V. De Pava, *J. Org. Chem.*, **48**, 602 (1983).

[506] P. R. O. de Montellano, J. S. Wei, R. Castillo, C. K. Hsu, and A. Boparai, *J. Med. Chem.*, **20**, 243 (1977).

[507] R. Baudouy and J. Gore, *Bull. Soc. Chim. Fr.*, Part 2, **1975**, 2159.

[508] R. J. McMahon, K. E. Wiegers, and S. G. Smith, *J. Org. Chem.*, **46**, 99 (1981).

[509] S. Ställberg-Stenhagen, *J. Am. Chem. Soc.*, **69**, 2568 (1947).

[510] R. F. C. Brown, V. M. Clark, and A. Todd, *Tetrahedron* (Suppl. No. 8), Part I, 15 (1966).

[511] K. Šindelář, J. Holubek, E. Svátek, M. Ryska, A. Dlabač, and M. Protiva, *Collect. Czech. Chem. Commun.*, **47**, 1367 (1982).

[512] J. O. Jílek, K. Šindelář, J. Pomykáček, C. Horešovský, K. Pelz, E. Svátek, B. Kakáč, J. Holubek, J. Metyšová, and M. Protiva, *Collect. Czech. Chem. Commun.*, **38**, 115 (1973).

[513] B. Prasad, A. K. Saund, J. M. Bora, and N. K. Mathur, *Indian J. Chem.*, **12**, 290 (1974).

[514] J. Bruhn, J. Zsindely, H. Schmid, and G. Fráter, *Helv. Chim. Acta*, **61**, 2542 (1978).

[515] J. M. Bachhawat, C. K. Narang, and N. K. Mathur, *Indian J. Chem. Educ.*, **3**, 23 (1972) [*C.A.*, **77**, 48016b (1972)].

[516] J. Šrogl, M. Janda, I. Stibor, V. Skála, P. Trška, and M. Ryska, *Collect. Czech. Chem. Commun.*, **39**, 3109 (1974).

[517] S. Brandänge, S. Josephson, and S. Vallén, *Acta Chem. Scand.*, **27**, 3668 (1973).

[518] J. R. DeMember, R. B. Greenwald, and D. H. Evans, *J. Org. Chem.*, **42**, 3518 (1977).

[519] J. O. Jílek, J. Metyšová, J. Němec, Z. Šedivý, J. Pomykáček, and M. Protiva, *Collect. Czech. Chem. Commun.*, **40**, 3386 (1975).

[520] M. Protiva, M. Rajšner, V. Trčka, M. Vaněček, J. Němec, and Z. Šedivý, *Collect. Czech. Chem. Commun.*, **40**, 3904 (1975).

[521] R. L. Cargill and B. W. Wright, *J. Org. Chem.*, **40**, 120 (1975).

[522] O. Červinka and O. Kříž, *Z. Chem.*, **11**, 63 (1971).

[523] O. Červinka and O. Kříž, *Collect. Czech. Chem. Commun.*, **38**, 938 (1973).

[524] R. Smrž, J. O. Jílek, K. Šindelář, B. Kakáč, E. Svátek, J. Holubek, J. Grimová, and M. Protiva, *Collect. Czech. Chem. Commun.*, **41**, 2771 (1976).

[525] R. Zurflüh, L. L. Dunham, V. L. Spain, and J. B. Siddall, *J. Am. Chem. Soc.*, **92**, 425 (1970).

[526] S. G. Spanton and G. D. Prestwich, *Tetrahedron*, **38**, 1921 (1982).

[527] P. Zeltner, G. A. Huber, R. Peters, F. Tátrai, L. Boksányi, and E. Kováts, *Helv. Chim. Acta*, **62**, 2495 (1979).

[528] V. Bártl, J. Metyšová, and M. Protiva, *Collect. Czech. Chem. Commun.*, **38**, 2778 (1973).

[529] M. Protiva, V. Bártl, and J. Metyšová, Czech. Patent 159,522 (1975) [*C.A.*, **85**, 21451t (1976)].

[530] V. Bártl, J. Metyšová, and M. Protiva, *Collect. Czech. Chem. Commun.*, **38**, 1693 (1973).

[531] M. Protiva, V. Bártl, and J. Metyšová, Czech. Patent 153,855 (1974) [*C.A.*, **84**, 44146r (1976)].

[531a] G. R. Brown, A. J. Foubister, and B. Wright, *J. Chem. Soc., Perkin Trans. 1*, **1985**, 2577.

[532] J. D. McChesney and R. A. Swanson, *J. Org. Chem.*, **47**, 5201 (1982).

[533] R. L. Markezich, W. E. Willy, B. E. McCarry, and W. S. Johnson, *J. Am. Chem. Soc.*, **95**, 4414 (1973).

[534] W. S. Johnson, B. .E McCarry, R. L. Markezich, and S. G. Boots, *J. Am. Chem. Soc.*, **102**, 352 (1980).

[535] W. Roelofs, A. Comeau, A. Hill, and G. Milicevic, *Science*, **174**, 297 (1971).

[536] W. Roelofs, J. Kochansky, R. Cardé, H. Arn, and S. Rauscher, *Mitt. Schweiz, Entomol. Ges.* (*Bull. Soc. Entomol. Suisse*), **46**, 71 (1973).

[537] W. Roelofs, J. Kochansky, and R. Cardé, U.S. Patent 3,845,108 (1974) [*C.A.*, **82**, 111610m (1975)].

[538] G. Holan and D. F. O'Keefe, *Tetrahedron Lett.*, **1973**, 673.

[539] I. Červená, K. Šindelář, Z. Kopicová, J. Holubek, E. Svátek, J. Metyšová, M. Hrubantová, and M. Protiva, *Collect. Czech. Chem. Commun.*, **42**, 2001 (1977).

[540] J. O. Jílek, K. Šindelář, M. Rajšner, A. Dlabač, J. Metyšová, Z. Votava, J. Pomykáček, and M. Protiva, *Collect. Czech. Chem. Commun.*, **40**, 2887 (1975).

[541] J. O. Jílek, J. Holubek, E. Svátek, M. Ryska, J. Pomykáček, and M. Protiva, *Collect. Czech. Chem. Commun.*, **44**, 2124 (1979).

[542] J. Jílek, J. Pomykáček, J. Methyšová, and M. Protiva, *Collect. Czech. Chem. Commun.*, **46**, 1607 (1981).

[543] K. Šindelář, B. Kakáč, E. Svátek, J. Holubek, J. Metyšová, M. Hrubantová, and M. Protiva, *Collect. Czech. Chem. Commun.*, **38**, 3321 (1973).

[544] I. Červená, J. Metyšová, E. Svátek, B. Kakáč, J. Holubek, M. Hrubantová, and M. Protiva, *Collect. Czech. Chem. Commun.*, **41**, 881 (1976).

[545] J. Jílek, M. Protiva, and J. Vít, Czech. Patent 131, 742 (1969) [*C.A.*, **73**, 45099n (1970)].

[546] K. Šindelář, J. Metyšová, and M. Protiva, *Collect. Czech. Chem. Commun.*, **34**, 3801 (1969).

[547] H. H. Ong, J. A. Profitt, T. L. Spaulding, and J. C. Wilker, *J. Med. Chem.*, **22**, 834 (1979).

[548] K. Šindelář, J. Metyšová, J. Holubek, Z. Šedivý, and M. Protiva, *Collect. Czech. Chem. Commun.*,**42**, 1179 (1977).

[549] K. Šindelář, J. O. Jílek, J. Metyšová, J. Pomykáček, and M. Protiva, *Collect. Czech. Chem. Commun.*, **39**, 3548 (1974).

[550] J. O. Jílek, J. Pomykáček, J. Metyšová, M. Bartošová, and M. Protiva, *Collect. Czech. Chem. Commun.*, **43**, 1747 (1978).

[551] I. Červená, K. Šindelář, J. Metyšová, E. Svátek, M. Ryska, M. Hrubantová, and M. Protiva, *Collect. Czech. Chem. Commun.*, **42**, 1705 (1977).

[552] M. Gerecke, E. Kyburz, and J. P. Kaplan, German Offen. 2,412,520 (1974) [*C.A.*, **82**, 43466s (1975)].

[553] Z. Kopicová, J. Metyšová, and M. Protiva, *Collect. Czech. Chem. Commun.*, **40**, 3519 (1975).

[554] Z. Kopicová, Z. Šedivý, F. Hradil, and M. Protiva, *Collect. Czech. Chem. Commun.*, **37**, 1371 (1972).

[555] K. Šindelář, J. Metyšová, and M. Protiva, *Collect. Czech. Chem. Commun.*, **37**, 1734 (1972).

[556] K. Šindelář, A. Dlabač, J. Metyšová, B. Kakáč, J. Holubek, E. Svátek, Z. Šedivý, and M. Protiva, *Collect. Czech. Chem. Commun.*, **40**, 1940 (1975).

[557] V. Bártl, J. Metyšová, J. Metyš, J. Němec, and M. Protiva, *Collect. Czech. Chem. Commun.*, **38**, 2301 (1973).

[558] M. Protiva. V. Bártl, J. Metyš. and J. Němec. Czech. Patent 160,402 (1975) [*C.A.*, **86,** 5493y (1977)].

[559] K. Šindelář, B. Kakáč, E. Svátek, J. Metyšová, and M. Protiva, *Collect. Czech. Chem. Commun.*, **38**, 1579 (1973).

[560] M. Protiva, K. Šindelář, Z. Šedivý, and J. Metyšová, *Collect. Czech. Chem. Commun.*, **44**, 2108 (1979).

[561] M. Protiva, K. Šindelář, A. Dlabač, J. Metyšová, and Z. Šedivý, Czech. Patent 193,369 (1981) [*C.A.*, **96**, 217877h (1982)].

[562] T. Takahashi, Japanese Patent (Tokkyo Koho) 79 106, 463 [*C.A.*, **92**, 94233c (1980)].

[563] N. L. Allinger, T. J. Walter, and M. G. Newton, *J. Am. Chem. Soc.*, **96**, 4588 (1974).

[564] N. L. Allinger and T. J. Walter, *J. Am. Chem. Soc.*, **94**, 9267 (1972).

[565] L. Colombo, C. Gennari, C. Scolastico, F. Aragozzini, and C. Merendi, *J. Chem. Soc., Perkin Trans. 1*, **1980**, 2549.

[566] D. O. Shah and K. N. Trivedi, *Curr. Sci.*, **43**, 278 (1974) [*C.A.*, **81**, 77724u (1974)].

[567] V. Valenta, J. Metyšová, Z. Šedivý, and M. Protiva, *Collect. Czech. Chem. Commun.*, **39**, 783 (1974).

[568] K. Šindelář, Z. Kopicová, J. Metyšová, and M. Protiva, *Collect. Czech. Chem. Commun.*, **40**, 3530 (1975).

[569] M. Protiva, K. Šindelář, Z. Šedivý, J. Holubek, and M. Bartošová, *Collect. Czech. Chem. Commun.*, **46**, 1808 (1981).

[570] M. Protiva, K. Šindelář, Z. Šedivý, and J. Pomykáček, *Collect. Czech. Chem. Commun.*, **44**, 2987 (1979).

[571] M. Rajšner, Z. Kopicová, J. Holubek, E. Svátek, J. Metyš, M. Bartošová, F. Mikšík, and M. Protiva, *Collect. Czech. Chem. Commun.*, **43**, 1760 (1978).

[572] M. Rajšner and M. Protiva, Czech. Patent 169,331 (1977) [*C.A.*, **88**, 120770q (1978)].

[573] M. Rajšner, M. Protiva, and F. Mikšík, Czech. Patent 189,164 (1981) [*C.A.*, **96**, 142717v (1982)].

[574] I. Červená, E. Svátek, J. Metyšová, and M. Protiva, *Collect. Czech. Chem. Commun.*, **39**, 3733 (1974).

[575] J. O. Jílek, J. Metyšová, J. Pomykáček, and M. Protiva, *Collect. Czech. Chem. Commun.*, **39**, 3338 (1974).

[576] Z. Kopicová and M. Protiva, *Collect. Czech. Chem. Commun.*, **39**, 3147 (1974).

[577] W. J. Hammar, U.S. Patent 3,960,944 (1976) [*C.A.*, **85**, 108458z (1976)].

[578] J. W. Clark-Lewis and D. P. Cox, *Aust. J. Chem.*, **29**, 191 (1976).

[579] N. Cohen, J. W. Scott, F. T. Bizzaro, R. J. Lopresti, W. F. Eichel, G. Saucy, and H. Mayer, *Helv. Chim. Acta*, **61**, 837 (1978).

[580] K. Takahashi, S. Shibata, S. Yano, M. Harada, H. Saito, Y. Tamura, and A. Kumagai, *Chem. Pharm. Bull.*, **28**, 3449 (1980).

[581] S. Shibata, A. Kumagai, M. Harada, S. Yano, H. Saito, and K. Takahashi, British Patent Application 2,075,835 (1981) [*C.A.*, **96**, 143114h (1982)].

[582] R. B. Silverman and M. A. Levy, *J. Org. Chem.*, **45**, 815 (1980).

[583] E. Duran, L. Gorrichon, L. Cazaux, and P. Tisnes, *Tetrahedron Lett.*, **25**, 2755 (1984).

[584] A. Vystrčil, V. Pouzar, and V. Křeček, *Collect. Czech. Chem. Commun.*, **38**, 3902 (1973).

[585] A. Fischli, M. Klaus, H. Mayer, P. Schönholzer, and R. Rüegg, *Helv. Chim. Acta*, 58, 564 (1975).

[586] H. Upadek and K. Bruns, European Patent Application 53,717 (1982) [*C.A.*, **97**, 198414g (1982)].

[587a] A. W. Johnson, G. Gowda, A. Hassanali, J. Knox, S. Monaco, Z. Razavi, and G. Rosebery, *J. Chem. Soc., Perkin Trans. 1*, **1981**, 1734.

[587b] P. Cannone and M. Akssira, *Tetrahedron*, **41**, 3695 (1985).

[588] I. D. Rae and A. M. Redwood, *Aust, J. Chem.*, **27**, 1143 (1974).

[589] A. Sugimoto, H. Tanaka, Y. Eguchi, S. Ito, Y. Takashima, and M. Ishikawa, *J. Med. Chem.*, **27**, 1300 (1984).

[590] E. E. Smissman, R. J. Murray, and J. D. McChesney, *J. Med. Chem.*, **19**, 148 (1976).

[590a] N. Bhaduri and A. P. Bhaduri, *Indian J. Chem.*, **23B**, 209 (1984).

[590b] Y. Takei, K. Mori, and M. Matsui, *Agr. Biol. Chem.*, **37**, 637 (1973).

[591] J. K. Whitesell and R. S. Matthews, *J. Org. Chem.*, **43**, 1650 (1978).

[592] P. D. Hobbs and P. D. Magnus, *J. Am. Chem. Soc.*, **98**, 4594 (1976).

[593] P. A. Grieco, J. Inanaga, N. H. Lin, and T. Yanami, *J. Am. Chem. Soc.*, **104**, 5781 (1982).

[594] G. R. Newkome, T. Kawato, and W. H. Benton, *J. Org. Chem.*, **45**, 626 (1980).

[595] J. Hrbek, Jr., L. Hruban, V. Šimánek, F. Šantavý, and G. Snatzke, *Collect. Czech. Chem. Commun.*, **38**, 2799 (1973).

[596] V. Šimánek and A. Klásek, *Tetrahedron Lett.*, **1971**, 4133.

[597] J. Rebek, Jr., D. F. Tai, and Y. K. Shue, *J. Am. Chem. Soc.*, **106**, 1813 (1984).

[598] J. Rebek, Jr., and Y. K. Shue, *Tetrahedron Lett.*, **23**, 279 (1982).

[599] J. Colonge, M. Constantini, and M. Ducloux, *Bull. Soc. Chim. Fr.*, **1966**, 2005.
[600] P. Dufêÿ, *Bull. Soc. Chim. Fr.*, **1968**, 4653.
[601] P. D. Hobbs and P. D. Magnus, *J. Chem. Soc., Chem. Commun.*, **1974**, 856.
[601a] L. Skattebøl and Y. Stenstrom, *Acta Chem. Scand.*, **B39**, 291 (1985).
[601b] C. A. Elliger, *Org. Prep. Proced. Int.*, **17**, 419 (1985).
[602] R. T. Borchardt and L. A. Cohen, *J. Am. Chem. Soc.*, **95**, 8308 (1973).
[603] Y. Tsuda, K. Yoshimoto, T. Yamashita, and M. Kaneda, *Chem. Pharm. Bull.*, **29**, 3238 (1981).
[604] F. Johnson, K. G. Paul, D. Favara, R. Ciabatti, and U. Guzzi, *J. Am. Chem. Soc.*, **104**, 2190 (1982).
[605] D. J. Brecknell and R. M. Carman, *Aust. J. Chem.*, **32**, 2455 (1979).
[606] T. Oritani, K. Yamashita, and M. Matsui, *Agric. Biol. Chem.*, **34**, 1244 (1970) [*C.A.*, **74**, 3750h (1971)].
[606a] R. V. Stevens and A. P. Vinogradoff, *J. Org. Chem.*, **50**, 4056 (1985).
[607a] W. Klötzer, S. Teitel, and A. Brossi, *Helv. Chim. Acta*, **55**, 2228 (1972).
[607b] W. Klötzer, S. Teitel, and A. Brossi, *Helv. Chim. Acta*, **54**, 2057 (1971).
[607c] R. B. Yeats, *Tetrahedron Lett.*, **24**, 3423 (1983).
[608] G. Doria, P. Gaio, and C. Gandolfi, *Tetrahedron Lett.*, **1972**, 4307.
[609] H. Schmidhammer, *Sci. Pharm.*, **49**, 304 (1981) [*C.A.*, **96**, 143124m (1982)].
[610] W. Klötzer, S. Teitel, J. F. Blount, and A. Brossi, *J. Am. Chem. Soc.*, **93**, 4321 (1971).
[611] F. Cassidy, R. W. Moore, and G. Wootton, *Tetrahedron Lett.*, **22**, 253 (1981).
[612] W. Klötzer, S. Teitel, J. F. Blount, and A. Brossi, *Monatsh. Chem.*, **103**, 435 (1972).
[613] A. Brossi, W. Klötzer, and S. Teitel, U.S. Patent 3,946,041 (1976) [*C.A.*, **85**, 33246t (1976)].
[614] K. C. Nicolaou, S. P. Seitz, and M. R. Pavia, *J. Am. Chem. Soc.*, **103**, 1222 (1981).
[615] K. Čapek, V. Kubelka, J. Paleček, J. Staněk, I. Stibor, I. Veselý, V. Dudek, Z. Havel, and M. Janda, Czech. Patent 204,595 (1983) [*C.A.*, **101**, 72513w (1984)].
[616] C. Gandolfi, G. Doria, and P. Gaio, *Farmaco, Ed. Sci.*, **27**, 1125 (1972) [*C.A.*, **78**, 97160y (1973)].
[617] R. Hohlbrugger and W. Klötzer, *Chem. Ber.*, **112**, 849 (1979).
[618] R. Hohlbrugger and W. Klötzer, *Chem. Ber.*, **107**, 3457 (1974).
[619] C. Kamla, M. K. Rastogi, R. P. Kapoor, and C. P. Garg, *Indian J. Chem.*, **16B**, 417 (1978).
[620] W. Carruthers and M. I. Qureshi, *J. Chem. Soc., Chem. Commun.*, **1969**, 832.
[621] S. C. Welch and R. L. Walters, *J. Org. Chem.*, **39**, 2665 (1974).
[622] A. G. Schultz and J. D. Godfrey, *J. Am. Chem. Soc.*, **102**, 2414 (1980).
[623] K. Ito, T. Iida, and T. Funatani, *Yakugaku Zasshi*, **99**, 349 (1979) [*C.A.*, **91**, 52694c (1979)].
[624] J. A. Elix and U. Engkaninan, *Aust. J. Chem.*, **29**, 2693 (1976).
[625] O. Paleta, L. Štěpán, J. Svoboda, and V. Dědek, Czech. Patent, 204,892 (1982) [*C.A.*, **99**, 70203m (1983)].
[626] J. G. Cannon, D. M. Crockatt, J. P. Long, and W. Maixner, *J. Med. Chem.*, **25**, 1091 (1982).
[627] R. B. Silverman and M. A. Levy, *J. Org. Chem.*, **45**, 815 (1980).
[628] A. I. Meyers and C. E. Whitten, *Heterocycles*, **4**, 1687 (1976).
[629] C. G. Pitt, H. H. Seltzman, Y. Sayed, C. E. Twine, Jr., and D. L. Williams, *J. Org. Chem.*, **44**, 677 (1979).
[630] L. K. Sydnes and L. Skattebøl, *Acta Chem. Scand., Ser. B.*, **32**, 632 (1978).
[631] J. A. Benson and P. B. Dervan, *J. Am. Chem. Soc.*, **94**, 7597 (1972)
[632] O. Kříž, B. Čásenský, and O. Štrouf, *Collect. Czech. Chem. Commun.*, **38**, 2076 (1973).
[633] O. Kříž and *J*, Macháček, *Collect. Czech. Chem. Commun.*, **37**, 2175 (1972).
[634] O. Kříž, J. Macháček, and O. Štrouf, *Collect. Czech. Chem. Commun.*, **38**, 2072 (1973).
[635a] R. A. Pascal, Jr. and Y. C. J. Chen, *J. Org. Chem.*, **50**, 408 (1985).
[635b] W. L. F. Armarego, B. A. Milloy, and S. C. Sharma, *J. Chem. Soc., Perkin Trans. 1*, **1972**, 2485.
[636a] Z. J. Vejdělek, J, Němec, Z. Šedivý, L. Tůma, and M. Protiva, *Collect. Czech. Chem. Commun.*, **39**, 2276 (1974).
[636b] E. J. Corey and J. P. Dittami, *J. Am. Chem. Soc.*, **107**, 256 (1985).

[637a] J. Hebký and J. Poláček, *Collect. Czech. Chem Commun.*, **35**, 667 (1970).

[637b] O. P. Vig, M. L. Sharma, and R. Gauba, *Indian J. Chem.*, **24B**, 313 (1985) [*C.A.*, **103**, 178089u (1985)].

[638] J. G. Cannon and B. J. Demopoulos, *J. Heterocycl. Chem.*, **19**, 1195 (1982).

[639a] G. Stork, J. D. Melton, and D. Kim, Columbia University, New York, unpublished results; quoted by I. Paterson and M. M. Mansuri, *Tetrahedron*, **41**, 3569 (1985).

[639b] J. G. Cannon, D. L. Kolbe, J. P. Long, and T. Verimer, *J. Med. Chem.*, **23**, 750 (1980).

[640] M. Julia and F. Le Goffic, *C. R. Acad. Sci., Ser. C*, **255**, 714 (1962).

[641] A. H. Dekmezian and M. K. Kalouistan, *Synth. Commun.*, **9**, 431 (1979).

[642] W. S. Johnson, Y. Q. Chen, and M. S. Kellog, *J. Am. Chem. Soc.*, **105**, 6653 (1983).

[643a] E. Götschi, F. Schneider, H. Wagner, and K. Bernauer, *Helv. Chim. Acta*, **60**, 1416 (1977).

[643b] E. G. Lewars and G. Morrison, *Can. J. Chem.*, **55**, 975 (1977).

[644] H. Ishii, T. Ichikawa, T. Deushi, K. I. Harada, T. Watanabe, E. Ueda, T. Ishida, M. Sakamoto, E. Kawanabe, T. Takahashi, Y. I. Ishikawa, K. Takizawa, T. Matsuda, and I. S. Chen, *Chem. Pharm. Bull.*, **31**, 3024 (1983).

[645] Y. Asaka, T. Kamikawa, and T. Kubota, *Tetrahedron Lett.*, **1972**, 1597.

[646] J. G. Cannon, T. Lee, and V. Sankaran, *J. Med. Chem.*, **18**, 1027 (1975).

[647] T. Taguchi, T. Takigawa, Y. Tawara and Y. Kobayashi, *Tetrahedron Lett.*, **25**, 5689 (1984).

[648] N. Cohen and G. Saucy, U.S. Patent 4,041,058 (1977) [*C.A.*, **87**, 184731q (1977)].

[649] P. A. Marshall and R. H. Prager, *Aust. J. Chem.*, **30**, 151 (1977)

[650] W. L. Nelson, D. S. Freeman, and R. Sankar, *J. Org. Chem.*, **40**, 3658 (1975).

[651] M. S. Allen, N. Darby, P. Salisbury, E. R. Sigurdson, and T. Money, *Can. J. Chem.*, **57**, 733 (1979).

[652] A. Ichihara, R. Kimura, S. Yamada, and S. Sakamura, *J. Am. Chem. Soc.*, **102**, 6353 (1980).

[653] R. C. Zurflueh and J. B. Siddall, German Offen. 2,050,343 (1971) [*C.A.*, **75**, 151423q (1971)].

[654] R. W. Thies and J. E. Billigmeier, *J. Org. Chem.*, **38**, 1758 (1973).

[655] Mitsubishi Chemical Industries Co., Ltd., Japanese Patent (Tokkyo Koho) 58 04,786 (83 04,786) [*C.A.*, **98**, 179434t (1983)].

[656a] R. Murphy and R. H. Prager, *J. Organomet. Chem.*, **156**, 133 (1978).

[656b] G. W. H. Cheeseman, S. A. Eccleshall, and T. Thornton, *J. Heterocycl. Chem.*, **22**, 809 (1985).

[657] R. Baudouy and J. Gore, *Tetrahedron Lett.*, **1974**, 1593.

[658] R. Baudouy and J. Gore, *Bull. Soc. Chim. Fr.*, Part 2, **1975**, 2166.

[658a] H. J. Liu and T. K. Ngooi, *Can. J. Chem.*, **62**, 2676 (1984).

[659] R. Murphy and R. H. Prager, *Aust. J. Chem.*, **31**, 1629 (1978).

[660] C. Jutz and H. G. Peuker, *Synthesis*, **1975**, 431.

[661] R. H. Prager and S. T. Were, *Aust. J. Chem.*, **36**, 1441 (1983).

[662] J. Rebek, Jr., J. E. Trend, R. V. Wattley, and S. Chakravorti, *J. Am. Chem. Soc.*, **101**, 4333 (1979).

[663a] G. B. Trimitsis, A. Tuncay, R. D. Beyer, and K. J. Ketterman, *J. Org. Chem.*, **38**, 1491 (1973).

[663b] G. R. Brown and A. J. Foubister, *J. Chem. Soc., Chem. Commun.*, **1985**, 455.

[664] B. Glatz, G. Helmchen, H. Muxfeldt, H. Porcher, R. Prewo, J. Senn, J. J. Stezowski, R. J. Stojda, and D. R. White, *J. Am. Chem. Soc.*, **101**, 2171 (1979).

[665] T. H. Smith, A. N. Fujiwara, W. W. Lee, H. Y. Wu, and D. W. Henry, *J. Org. Chem.*, **42**, 3653 (1977).

[666] H. Ishii, T. Ishikawa, I. S. Chen, and S. T. Lu, *Tetrahedron Lett.*, **23**, 4345 (1982).

[667a] P. D. Cooper, *J. Med. Chem.*, **16**, 1057 (1973).

[667b] K. R. Scott, J. A. Moore, T. B. Zalucky, J. M. Nicholson, J. A. M. Lee, and C. N. Hinko, *J. Med. Chem.*, **28**, 413 (1985).

[668] H. C. Kluender and G. P. Peruzzotti, *Tetrahedron Lett.*, **1977**, 2063.

[669] Hisamitsu Pharmaceutical Co., Inc., Japanese Patent (Tokkyo Koho) 82 32,238 [*C.A.*, **97**, 6012e (1982)].

[670] E. Campaigne and G. M. Shutske, *J. Heterocycl. Chem.*, **12**, 317 (1975).

[671] P. Anastasis and P. E. Brown, *J. Chem. Soc., Perkin Trans. 1*, **1983**, 197.

[672] T. A. Hicks, C. E. Smith, W. R. N. Williamson, and E. H. Day, *J. Med. Chem.*, **22**, 1460 (1979).

[673] R. N. Booher, German Offen. 2,361,340 (1974) [*C.A.*, **81**, 77935p (1974)].

[674] R. N. Booher, S. E. Smits, W. W. Turner, Jr., and A. Pohland, *J. Med. Chem.*, **20**, 885 (1977).

[675] O. D. Dailey, Jr., and P. L. Fuchs, *J. Org. Chem.*, **45**, 216 (1980).

[675a] O. P. Vig, M. L. Sharma, S. Kumari, and Veena Rani, *Indian J. Chem.*, **24B**, 675 (1985).

[676] T. S. Lin, R. E. Harmon, W. Pierantoni, and J. C. W. Su, *Org. Prep. Proced. Int.*, **6**, 185 (1974) [*C.A.*, **81**, 169678k (1974)].

[677] I. Lantos and D. Ginsburg, *Tetrahedron*, **28**, 2507 (1972).

[678] M. Ito, R. Masahara, and K. Tsukida, *Tetrahedron Lett.*, **1977**, 2767.

[679] J. Ohishi, K. Tsuneoka, S. Ikegami, and S. Akaboshi, *J. Org. Chem.*, **43**, 4013 (1978).

[680] J. W. Clark-Lewis and M. M. Mahandru, *Aust. J. Chem.*, **24**, 563 (1971).

[681] D. N. Jones, T. P. Kogan, and R. F. Newton, *J. Chem. Soc., Perkin Trans. 1*, **1982**, 1333.

[682] F. E. Ziegler and J. G. Sweeny, *J. Org. Chem.*, **32**, 3216 (1967).

[683] W. K. Anderson and M. J. Halat, *J. Med. Chem.*, **22**, 977 (1979).

[684] M. Yoshimoto, H. Miyzawa, H. Nakao, K. Shinkai, and M. Arakawa, *J. Med. Chem.*, **22**, 491 (1979).

[685] D. B. Rusterholz, C. F. Barfknecht, and J. A. Clemens, *J. Med. Chem.*, **20**, 85 (1977).

[686] P. Y. Johnson, I. Jacobs, and D. J. Kerkman, *J. Org. Chem.*, **40**, 2710 (1975).

[687] D. D. Weller and D. W. Ford, *Tetrahedron Lett.*, **25**, 2105 (1984).

[688] O. P. Vig, M. L. Sharma, S. Kiran, and J. Singh, *Indian J. Chem.*, **22B**, 746 (1983).

[689] T. R. Govindachary, K. R. Ravindranath, and N. Viswanathan, *J. Chem. Soc., Perkin Trans. 1*, **1974**, 1215.

[690] N. Harada, Y. Takuma, and H. Uda, *Bull. Chem. Soc. Jpn.*, **50**, 2033 (1977).

[691] N. Harada, Y. Takuma, and H. Uda, *J. Am. Chem. Soc.*, **98**, 5408 (1976).

[692] W. S. Johnson, U.S. Patent 4,032,579 (1977) [*C.A.*, **87**, 201884m (1977)].

[693] K. Okamoto, M. Watanabe, M. Kawada, G. Goto, Y. Ashida, K. Oda, A. Yajima, I. Imada, and H. Morimoto, *Chem. Pharm. Bull.*, **30**, 2797 (1982).

[694] M. J. Wanner, G. J. Koomen, and U. K. Pandit, *Heterocycles*, **19**, 2295 (1982).

[695] M. J. Wanner, G. J. Koomen, and U. K. Pandit, *Tetrahedron*, **39**, 3673 (1983).

[696] J. P. Kutney and G. B. Fuller, *Heterocycles*, **3**, 197 (1975).

[696a] A. Abad, M. Arno, L. R. Domingo, and R. J. Zaragoza, *Tetrahedron*, **41**, 4937 (1985).

[697] Anon., Japanese Patent 75 130,732 [*C.A.*, **85**, 142808y (1976)].

[698] I. Kubo, T. Kamikawa, and T. Kubota, *Tetrahedron*, **30**, 615 (1974).

[699] J. W. Huffman and J. J. Gibbs, *J. Org. Chem.*, **39**, 2501 (1974).

[700a] H. J. Liu and M. G. Kulkarni, *Tetrahedron Lett.*, **26**, 4847 (1985).

[700b] N. Cohen, G. Saucy, and K. K. Chan, German Offen. 2,602,509 (1976) [*C.A.*, **86**, 55601c (1977)].

[701] W. A. Ayer, L. M. Browne, J. R. Mercer, D. R. Taylor, and D. E. Ward, *Can. J. Chem.*, **56**, 717 (1978).

[702] M. Nakanishi and H. Yuki, German Offen. 2,409,646 (1974) [*C.A.*, **81**, 169433b (1974)].

[703a] J. Goto, N. Goto, and T. Nambara, *Chem. Pharm. Bull.*, **30**, 4597 (1982).

[703b] Shimadzu Corp., Japanese Patent (Tokkyo Koho) 59,175,462 (84,175,462) (1984) [*C.A.*, **102**, 78602s (1985)].

[704a] D. L. Lattin, B. Caviness, R. G. Hudson, D. L. Greene, P. K. Raible, and J. B. Richardson, *J. Med. Chem.*, **24**, 903 (1981).

[704b] S. Sakai, H. Takayama, K. Yamaguchi, N. Ide, and T. Okamoto, *Yakugaku Zasshi*, **104**, 731 (1984) [*C.A.*, **102**, 95873y (1985)].

[705] H. de Koning, A. Springer-Fidder, M. J. Moolenaar, and H. O. Huisman, *Recl. Trav. Chim. Pays-Bas*, **92**, 237 (1973).

[706] K. K. Chan, A. C. Specian, Jr., and G. Saucy, *J. Org. Chem.*, **43**, 3435 (1978).

[707] K. K. Chan and G. Saucy, U.S. Patent 4,029,678 (1977) [*C.A.*, **88**, 38000e (1978)].

[708] P. Capdevieille and J. Rigaudy, *Tetrahedron*, **35**, 2101 (1979).

[709] J. W. Scott, F. T. Bizzarro, D. R.. Parrish, and G. Saucy, *Helv. Chim. Acta*, **59**, 290 (1976).

[710] J. W. Scott, D. R. Parrish, and G. Saucy, U.S. Patent 4,026,907 (1977) [*C.A.*, **87**, 184732r (1977)].

[711] G. L. Olson, H. C. Cheung, K. Morgan, and G. Saucy, *J. Org. Chem.*, **45**, 803 (1980).

[712] G. L. Olson and G. Saucy, German Offen. 2,820,863 (1978) [*C.A.*, **90**, 103617u (1979)].

[713] H. L. Cohen, *J. Polym. Sci., Polym. Chem. Ed.*, **13**, 745 (1975).

[714a] F. Boyer-Kawenoki, *C. R. Acad. Sci., Ser. C*, **252**, 1792 (1961).

[714b] I. Artaud and P. Viout, *J. Chem. Soc., Perkin Trans. 1*, **1985**, 1257.

[715a] J. Thiem, H. Mohn, and A. Heesing, *Synthesis*, **1985**, 775.

[715b] D. Miller, L. Mandel, and R. A. Day, Jr., *J. Org. Chem.*, **36**, 1683 (1971).

[716] R. F. Childs, E. F. Lund, A. G. Marshall, W. J. Morrisey, and C. V. Rogerson, *J. Am. Chem. Soc.*, **98**, 5924 (1976).

[717] M. Kraus, *Collect. Czech. Chem. Commun.*, **37**, 460 (1972).

[718] E. Piers and E. H. Ruediger, *J. Org. Chem.*, **45**, 1725 (1980).

[719] G. I. Myagkova, V. P. Nechiporenko, and R. P. Evstigneeva, *Dokl. Vses. Konf. Khim. Atsetilena, 4th*, **1**, 233 (1972) [*C.A.*, **79**, 31433d (1973)].

[720] R. G. Salomon, S. Ghosh, M. G. Zagorski, and M. Reitz, *J. Org. Chem.*, **47**, 829 (1982).

[721] J. J. McCullough, W. K. MacInnis, C. J. L. Lock, and R. Faggiani, *J. Am. Chem. Soc.*, **104**, 4644 (1982).

[722] A. J. Lovey and B. A. Pawson, *J. Med. Chem.*, **25**, 71 (1982).

[723] H. Kotsuki, A. Kawamura, M. Ochi, and T. Tokoroyama, *Chem. Lett.*, **1981**, 917.

[724a] J. V. Frosch, I. T. Harrison, B. Lythgoe, and A. K. Saksena, *J. Chem. Soc., Perkin Trans. 1*, **1974**, 2005.

[724b] J. Csekö, H. O. Hankovszky, and K. Hideg, *Can. J. Chem.*, **63**, 940 (1985).

[724c] S. S. Al-Hassan, R. J. Cameron, A. W. C. Curran, W. J. S. Lyall, S. H. Nicholson, D. R. Robinson, A. Stuart, C. J. Suckling, I. Stirling, and H. C. S. Wood, *J. Chem. Soc., Perkin Trans. 1*, **1985**, 1645.

[725] F. Orsini and F. Pelizzoni, *J. Org. Chem.*, **45**, 4726 (1980).

[726] A. R. Mattocks, *J. Chem. Soc., Perkin Trans. 1*, **1974**, 707.

[727] C. P. Casey and D. F. Marten, *Synth. Commun.*, **3**, 321 (1973).

[728] G. Stork, R. L. Danheiser, and B. Ganem, *J. Am. Chem. Soc.*, **95**, 3414 (1973).

[729] L. A. Paquette, P. Charumilind, T. M. Kravetz, M. C. Böhm, and R. Gleiter, *J. Am. Chem. Soc.*, **105**, 3126 (1983).

[730] D. L. Dare, I. D. Entwistle, and R. A. W. Johnstone, *J. Chem. Soc., Perkin Trans. 1*, **1973**, 1130.

[731] H. Ishii, T. Ishikawa, T. Tohojoh, K. Murakami, E. Kawanabe, S. T. Lu, and I. S. Chen, *J. Chem. Soc., Perkin Trans. 1*, **1982**, 2051.

[732] J. Wolinsky and H. S. Hauer, *J. Org. Chem.*, **34**, 3169 (1969).

[733] B. A. McAndrew, *J. Chem. Soc., Perkin Trans. 1*, **1979**, 1837.

[734] K. A. Parker and W. S. Johnson, *J. Am. Chem. Soc.*, **96**, 2556 (1974).

[735] D. Cabaret and Z. Welvart, *J. Organomet. Chem.*, **78**, 295 (1974).

[736] H. M. Balba and G. G. Still, *J. Labelled Compd. Radiopharm.*, **15** (Suppl. Vol.), 309 (1978).

[737] P. M. McCurry, Jr., *Tetrahedron Lett.*, **1971**, 1845.

[738] J. P. Morizur, G. Bidan, and J. Kossanyi, *Tetrahedron Lett.*, **1975**, 4167.

[739] G. Stork, Yo. Nakahara, Yy. Nakahara, and W. J. Greenlee, *J. Am. Chem. Soc.*, **100**, 7775 (1978).

[740] D. J. Field and D. W. Jones, *J. Chem. Soc., Perkin Trans. 1*, **1980**, 714.

[741] M. Ito, K. Hirata, A. Kodama, K. Tsukida, H. Matsumoto, K. Horiuchi, and T. Yoshizawa, *Chem. Pharm. Bull.*, **26**, 925 (1978).

[741a] O. P. Vig, R. Vig, D. M. Dua, and Veena Rani, *Indian J. Chem.*, **23B**, 274 (1984).

[742] E. P. Brody and C. D. Gutsche, *Tetrahedron*, **33**, 723 (1977).

[743a] N. Ya. Grigor'eva, I. M. Avrutov, A. V. Semenovskii, V. N. Odinokov, V. R. Akhunova, and G. A. Tolstikov, *Izv. Akad, Nauk SSSR, Ser. Khim.*, **1979**, 382; *Engl. Transl.*, p. 353 [*C.A.*, **91**, 74720k (1979)].

[743b] V. N. Odinokov, O. S. Kukovinets, N. I. Sakharova, and G. A. Tolstikov, *Zh. Org. Khim.*, **20**, 1866 (1984); *Engl. Transl.*, p. 1702 [*C.A.*, **102**, 95842n (1985)].

[743c] O. P. Vig, R. Nanda, R. Gauba, and S. K. Puri, *Indian J. Chem.*, **24B**, 918 (1985).

[744] L. E. Friedrich and R. A. Cormier, *J. Org. Chem.*, **36**, 3011 (1971).

[745] E. C. Kornfeld and N. J. Bach, U.S. Patent 3, 929, 796 (1975) [*C.A.*, **84**, 135920e (1976)].

[746] N. J. Bach and E. C. Kornfeld, *Tetrahedron Lett.*, **1974**, 3225.

[747] B. A. Pawson, H. C. Cheung, R. J. L. Han, P. W. Trown, M. Buck, R. Hansen, W. Bollag, U. Ineichen, H. Pleil, R. Rüegg, N. M. Dunlop, D. L. Newton, and M. B. Sporn, *J. Med. Chem.*, **20**, 918 (1977).

[748] G. Ohloff and W. Giersch, *Helv. Chim. Acta*, **63**, 76 (1980).

[749] R. Baker, P. Bevan, and R. C. Cookson, *J. Chem. Soc., Chem. Commun.*, **1975**, 752.

[750] T. Matsumoto, K. Miyano, S. Kagawa, S. Yu, J. Ogawa, and A. Ichihara, *Tetrahedron Lett.*, **1971**, 3521.

[751] M. F. Semmelhack, S. Tomoda, and K. M. Hurst, *J. Am. Chem. Soc.*, **102**, 7567 (1980).

[752] R. M. Coates, D. A. Ley, and P. L. Cavender, *J. Org. Chem.*, **43**, 4915 (1978).

[753] H. H. Inhoffen, W. Kreiser, and M. Nazir, *Justus Liebigs Ann. Chem.*, **755**, 1 (1972).

[754] M. Zajer, C. Szantay, L. Szabo, G. Kalaus, E. Karpati, G. Kiraly, A. Kiraly, B. Rosdy, L. Forgach, and L. Szporny, French Demande 2,499,573 (1982) [*C.A.*, **98**, 16917p (1983)].

[755a] Z. Horii, K. Ohkawa, and C. Iwata, *Chem. Pharm. Bull.*, **20**, 624 (1972).

[755b] L. N. Polyachenko, L. P. Davidova, E. N. Darskaya, T. M. Filippova, and G. I. Samokhvalov, *Zh. Org. Khim.*, **21**, 756 (1985); *Engl. Transl.*, p. 685.

[756a] F. Ü. Afifi-Yazar, O. Sticher, S. Uesato, K. Nagajima, and H. Inouye, *Helv. Chim. Acta*, **64**, 16 (1981).

[756b] M. J. Hughes, E. J. Thomas, M. D. Turnbull, R. H. Jones, and R. E. Warner, *J. Chem. Soc. Chem. Commun.*, **1985**, 755.

[757] M. B. Gravestock, W. S. Johnson, B. E. McCarry, R. J. Parry, and B. E. Ratcliffe, *J. Am. Chem. Soc.*, **100**, 4274 (1978).

[758] B. Gallenkamp, German Offen. 3,129,651 (1983) [*C.A.*, **98**, 178948b (1983)].

[759] J. A. Berson, P. B. Dervan, R. Malherbe, and J. A. Jenkins, *J. Am. Chem. Soc.*, **98**, 5937 (1976).

[760] J. B. Lambert, H. W. Mark, and E. S. Magyar, *J. Am. Chem. Soc.*, **99**, 3059 (1977).

[761] N. Petragnani, T. Brocksom, and A. Moro, *Farmaco, Ed. Sci.*, **32**, 512 (1977) [*C.A.*, **87**, 135220p (1977)].

[762] P. A. Bartlett and W. S. Johnson, *J. Am. Chem. Soc.*, **95**, 7501 (1973).

[763] I. Simonyi, L. Ladanyi, G. Racz, A. Mandi, and S. Pataki, Hungarian Patent 23,998 (1982) [*C.A.*, **99**, 53359z (1983)].

[764] K. J. Shea, S. Wise, L. D. Burke, P. D. Davis, J. W. Gilman, and A. C. Greeley, *J. Am. Chem. Soc.*, **104**, 5708 (1982).

[765] Y. Hamashima and K. Minami, Japanese Patent 75 131,965 [*C.A.*, **85**, 78107q (1976)].

[766] K. Hejno and F. Šorm, *Collect. Czech. Chem. Commun.*, **41**, 1225 (1976).

[767] S. Chimichi and R. Nesi, *J. Chem. Soc., Perkin Trans. 1*, **1979**, 2215.

[768] A. G. Caldwell, C. J. Harris, R. Stepney, and N. Whittaker, *J. Chem. Soc., Perkin Trans. 1*, **1980**, 495.

[769] G. Ohloff. W. Giersch, W. Thommen, and B. Willhalm, *Helv. Chim. Acta*, **66**, 1343 (1983).

[770] M. Shiozaki, N. Ishida, T. Hiraoka, and H. Maruyama, *Tetrahedron*, **40**, 1795 (1984).

[771] W. F. Hoffman, O. W. Woltersdorf, Jr., F. C. Novello, E. J. Cragoe, Jr., J. P. Springer, L. S. Watson, and G. M. Fanelli, Jr., *J. Med. Chem.*, **24**, 865 (1981).

[772] R. B. Woodward, *The Harvey Lecture Ser.*, **59**, 31 (1963–64; published 1965) [*C.A.*, **63**, 8422f (1965)].

[773] T. Kametani and K. Fukumoto, *Heterocycles* (Special Issue), **10**, 542 (1978).

[774] C. Wijnberger and C. L. Habraken, *J. Heterocycl. Chem.*, **6**, 545 (1969).

[775] M. Cole, U.S. Patent 3,184,492 (1965) [*C.A.*, **63**, 7933h (1965)].

[776a] W. R. Hertler, *J. Org. Chem.*, **41**, 1412 (1976).

[776b] H. J. Bestmann and R. Schmiechen, *Chem. Ber.*, **94**, 751 (1961).

[777] G. Berti, F. Bottari, and B. Macchia, *Ann. Chim.* (*Rome*), **52**, 1101 (1962) [*C.A.*, **59**, 3818e (1963)].
[778] D. D. Roberts and C. S. Wu, *J. Org. Chem.*, **39**, 1570 (1974).
[779] A. Horeau, J. Jacques, and R. Emiliozzi, *Bull. Soc. Chim. Fr.*, **1959**, 1854.
[780] S. W. Pelletier and S. Prabhakar, *J. Am. Chem. Soc.*, **90**, 5318 (1968).
[781] R. J. A. Walsh and K. R. H. Wooldridge, *J. Chem. Soc., Perkin Trans. 1*, **1972**, 1247.
[782] M. P. L. Caton, D. H. Jones, R. Slack, S. Squires, and K. R. H. Wooldridge, *J. Med. Chem.*, **8**, 680 (1965).
[783] P. Krogsgaard-Larsen and S. B. Christensen, *Acta Chem. Scand.*, **B30**, 281 (1976).
[784] H. Berger, R. Gall, K. Stach, M. Thiel, and W. Voemel, German Offen. 2,558,117 (1977) [*C.A.*, **87**, 135320w (1977)].
[785] T. Shono, M. Kimura, A. Oku, and R. Oda, *Kogyo Kagaku Zasshi*, **69**, 2147 (1966) [*C.A.*, **66**, 94708z (1967)].
[786] D. Buttimore, D. H. Jones, R. Slack, and K. R. H. Wooldridge, *J. Chem. Soc.*, **1963**, 2032.
[787] K. Rajeswari and T. S. Sorensen, *J. Am. Chem. Soc.*, **95**, 1239 (1973).
[788] D. G. O'Donovan and E. Barry, *J. Chem. Soc., Perkin Trans. 1*, **1974**, 2528.
[789] J. T. Kurek and G. Vogel, and *J. Heterocycl. Chem.*, **5**, 275 (1968).
[790] C. V. Greco and F. Pellegrini, *J. Chem. Soc., Perkin Trans. 1*, **1972**, 720.
[791] E. Bisagni, J. M. Lhoste, and W. C. Hung, *J. Heterocycl. Chem.*, **18**, 755 (1981).
[792] K. C. Das and B. Weinstein, *Tetrahedron Lett.*, **1969**, 3459.
[793] M. Julia and G. Le Thuillier, *Bull. Soc. Chim. Fr.*, **1966**, 717.
[794] D. E. Applequist and L. Kaplan, *J. Am. Chem. Soc.*, **87**, 2194 (1965).
[795] L. K. A. Rahman and R. M. Scrowston, *J. Chem. Soc., Perkin Trans. 1*, **1984**, 385.
[796] J. E. Siggins, A. A. Larsen, J. H. Ackerman, and C. D. Carabateas, *Org. Synth.*, **53**, 52 (1973).
[797a] J. E. Parente, J. M. Risley, and R. L. Van Etten, *J. Am. Chem. Soc.*, **106**, 8156 (1984).
[797b] P. W. Scott and D. R. Hawkins, *J. Labelled Compd. Radiopharm.*, **20**, 575 (1983) [*C.A.*, **99**, 194872y (1983)].
[798] J. A. Day, B. R. J. Devlin, and R. J. G. Searle, European Patent Application 7,652 (1980) [*C.A.*, **93**, 239213d (1980)].
[799] P. L. Coe, J. H. Sleigh, and J. C. Tatlow, *J. Chem. Soc., Perkin Trans. 1*, **1980**, 217.
[800] L. G. Humber, *J. Med. Chem.*, **7**, 826 (1964).
[801] R. P. Dickinson and B. Iddon, *J. Chem. Soc., C*, **1971**, 3447.
[802] B. Wenkert, K. G. Dave, I. Dainis, and G. D. Reynolds, *Aust. J. Chem.*, **23**, 73 (1970).
[803] J. W. Wilt and A. A. Levin, *J. Org. Chem.*, **27**, 2319 (1962).
[804] M. Julia, S. Julia, and B. Cochet, *Bull. Soc. Chim. Fr.*, **1964**, 1487.
[805] E. A. Nodiff, A. J. Saggiomo, M. Shimbo, E. H. Chen, H. Otomasu, Y. Kondo, T. Kikuchi, B. L. Verma, S. Matsuura, K. Tanabe, M. P. Tyagi, and S. Morosawa, *J. Med. Chem.*, **15**, 775 (1972).
[806] R. N. Castle and M. Onda, *J. Org. Chem.*, **26**, 4465 (1961).
[807] L. S. Besford, C. Allen, and J. M. Bruce, *J. Chem. Soc.*, **1963**, 2867.
[808] Y. Sato and Y. Matsumoto, *Takamine Kenkyusho Nempo*, **11**, 33 (1959) [*C.A.*, **55**, 5456d (1961)].
[809] G. Sunagawa, H. Sato, and Y. Matsumoto, Japanese Patent 7, 480 (1962) [*C.A.*, **59**, 1596d (1963)].
[810] K. Tsuda, S. Ikuma, M. Kawamura, R. Tachikawa, and T. Miyadera, *Chem. Pharm. Bull.*, **10**, 865 (1962).
[811] M. F. Zipplies, C. Krieger, and H. A. Staab, *Tetrahedron Lett.*, **24**, 1925 (1983).
[812] C. Kaiser and C. L. Zirkle, U.S. Patent 3,149,159 (1964) [*C.A.*, **62**, 14530d (1965)].
[813] E. Vogel, H. M. Deger, J. Sombroek, J. Palm, A. Wagner, and J. Lex, *Angew. Chem.*, **92**, 43 (1980); *Angew. Chem., Int. Ed. Engl.*, **19**, 41 (1980).
[814] M. Jones, Jr., S. D. Reich, and L. T. Scott, *J. Am. Chem. Soc.*, **92**, 3118 (1970).
[815] B. T. Ho, W. M. McIsaac, R. An, L. W. Tansey, K. E. Walker, L. F. Englert, Jr., and M. B. Noel, *J. Med. Chem.*, **13**, 26 (1970).
[816] F. Bohlmann and W. Sucrow, *Chem. Ber.*, **97**, 1839 (1964).

[817] M. Duraisamy and H. M. Walborsky, *J. Am. Chem. Soc.*, **105**, 3252 (1983).
[818] J. J. Riehl and A. Fougerousse, *Bull. Soc. Chim. Fr.*, **1968**, 4083.
[819] D. A. Coviello and W. H. Hartung, *J. Org. Chem.*, **24**, 1611 (1959).
[820] A. L. Ternay, Jr., D. Davenport, and G. Bledsoe, *J. Org. Chem.*, **39**, 3268 (1974).
[821] L. H. Slaugh, *J. Am. Chem. Soc.*, **83**, 2737 (1961).
[822] J. R. L. Smith, R. O. C. Norman, M. E. Rose, and A. C. W. Curran, *J. Chem. Soc., Perkin Trans. 1*, **1979**, 2863.
[823] D. C. Ayres, B. G. Carpenter, and R. C. Denney, *J. Chem. Soc.*, **1965**, 3578.
[824] G. C. Barley, E. R. H. Jones, V. Thaller, and R. A. Vere Hodge, *J. Chem. Soc., Perkin Trans. 1*, **1973**, 151.
[825] I. A. Pearl and S. F. Darling, *J. Org. Chem.*, **22**, 1266 (1957).
[826] H. Yee and A. J. Boyle, *J. Org. Chem.*, **27**, 2929 (1962).
[827] Merck & Co., Inc., Belgian Patent 649,169 (1964) [*C.A.*, **64**, 6622b (1966)].
[828] R. H. Rynbrandt and F. E. Dutton, *J. Org. Chem.*, **40**, 2282 (1975).
[829] T. Okuno, Y. Ishita, A. Sugawara, Y. Mori, K. Sawai, and T. Matsumoto, *Tetrahedron Lett.*, **1975**, 335.
[830] M. J. Perkins and B. P. Roberts, *J. Chem. Soc., Perkin Trans. 1*, **1974**, 297.
[831] J. W. Wilt and C. A. Schneider, *J. Org. Chem.*, **26**, 4196 (1961).
[832] C. Rüchardt, *Chem. Ber.*, **94**, 2609 (1961).
[833] C. Rüchardt and H. Trautwein, *Chem. Ber.*, **98**, 2478 (1965).
[834] Merck & Co., Inc., Dutch Patent Application 6,517,090 (1966) [*C.A.*, **66**, 2467f (1967)].
[835] R. G. Harrison, B. Lythgoe, and P. W. Wright, *J. Chem. Soc., Perkin Trans. 1*, **1974**, 2654.
[836] F. Bohlmann, M. Ganzer, M. Krüger, and E. Nordhoff, *Tetrahedron*, **39**, 123 (1983).
[837] Suwa Seikosha Co., Ltd., Japanese Patent (Tokkyo Koho) 57 118,581 (1982) [*C.A.*, **98**, 16703r (1983)].
[838] A. Burger and S. S. Hillery, *J. Med. Chem.*, **13**, 1232 (1970).
[839] I. A. Pearl, *J. Org. Chem.*, **24**, 736 (1959).
[840] K. Hamada, K. Nishizawa, Y. Hazama, and T. Akitani, Japanese Patent (Tokkyo Koho) 80 27,166 [*C.A.*, **94**, 65323t (1981)].
[841] H. Grisebach and L. Patschke, *Chem. Ber.*, **93**, 2326 (1960).
[842] H. Muxfeldt, *Chem. Ber.*, **92**, 3122 (1959).
[843] J. H. Fried and I. T. Harrison, German Offen. 2,043,048 (1971) [*C.A.*, **75**, 20035s (1971)].
[844] J. H. Fried and I. T. Harrison, U.S. Patent 3,958,012 (1976) [*C.A.*, **85**, 46259z (1976)].
[845] J. Borck, J. Dahm, V. Koppe, J. Krämer, G. Schorre, J. W. Hovy, and E. Schorscher, German Offen. 1,900,585 (1970) [*C.A.*, **74**, 53554t (1971)].
[846] K. Hamada, K. Nishizawa, S. Hazama, and T. Akitani, Japanese Patent (Tokkyo Koho) 80 36,459 [*C.A.*, **93**, 185962b (1980)].
[847] G. Culbertson and R. Pettit, *J. Am. Chem. Soc.*, **85**, 741 (1963).
[848] J. W. Wilt, J. F. Zawadzki, and D. G. Schultenover, S. J., *J. Org. Chem.*, **31**, 876 (1966).
[849a] P. LeRoy Anderson and D. A. Brittain, German Offen. 2,439,294 (1975) [*C.A.*, **82**, 170341x (1975)].
[849b] M. I. Komendantov and T. B. Panosiuk, *Zh. Org. Khim.*, 21, 1437 (1985); *Engl. Transl.*, p. 1308.
[850] P. V. Rao, S. Ramachandran, and D. G. Cornwell, *J. Lipid Res.*, **8**, 380 (1967).
[851] G. H. Senkler, Jr., D. Gust, P. X. Riccobono, and K. Mislow, *J. Am. Chem. Soc.*, **94**, 8626 (1972).
[852] A. Burger, D. J. Abraham, J. P. Buckley, and W. J. Kinnard, *Monatsh. Chem.*, **95**, 1721 (1964).
[853] L. H. Slaugh, E. F. Magoon, and V. P. Guinn, *J. Org. Chem.*, **28**, 2643 (1963).
[854] A. Barco, S. Benetti, G. P. Pollini, P. G. Baraldi, M. Guarneri, D. Simoni, C. B. Vicentini, P. G. Borasio, and A. Capuzzo, *J. Med. Chem.*, **21**, 988 (1978).
[855] J. A. Diment, E. Ritchie, and W. C. Taylor, *Aust. J. Chem.*, **20**, 565 (1967).
[856] S. Hünig and G. Ruider, *Tetrahedron Lett.*, **1968**, 773.
[857] H. Rakoff, *J. Am. Oil Chem. Soc.*, **46**, 277 (1969) [*C.A.*, **71**, 38264t (1969)].

[858] P. R. Jones, M. D. Saltzman, and R. J. Panicci, *J. Org. Chem.*, **35**, 1200 (1970).

[859] T. Y. Shen and L. H. Sarett, U.S. Patent 3,271,416 (1966) [*C.A.*, **66**, 18668w (1967)].

[860] F. Murai and M. Tagawa, *Chem. Pharm. Bull.*, **28**, 1730 (1980).

[861] V. P. Arya and B. G. Engel, *Helv. Chim. Acta*, **44**, 1650 (1961).

[862] R. C. Schulz, P. Elzer, and W. Kern, *Chimia* (*Aarau*), **13**, 237 (1959) [*C.A.*, **54**, 1281f (1960)].

[863a] H. Plieninger, M. Höbel, and V. Liede, *Chem. Ber.*, **96**, 1618 (1963).

[863b] G. W. H. Cheeseman and S. G. Greenberg, *J. Heterocycl. Chem.*, **16**, 241 (1979).

[864] W. L. F. Armarego and P. G. Tucker, *Aust. J. Chem.*, **32**, 1805 (1979).

[865] E. Brown and R. Dhal, *Bull. Soc. Chim. Fr.*, **1972**, 4292.

[866] E. Crundwell and A. L. Cripps, *J. Med. Chem.*, **15**, 754 (1972).

[867] E. Crundwell and A. L. Cripps, *Chem. Ind.* (*London*), **1971**, 767.

[868] H. M. Frey and R. K. Solly, *J. Chem. Soc.*, *B*, **1970**, 996.

[869] F. Mathey and J. Bensoam, *Tetrahedron*, **31**, 391 (1975).

[870] J. M. Ribó and F. R. Trull, *Monatsh. Chem.*, **114**, 1087 (1983).

[871] W. Francke and W. Mackenroth, *Angew. Chem.*, **94**, 704 (1982); *Angew. Chem., Int. Ed. Engl.*, **21**, 698 (1982).

[872] J. E. Baldwin, R. C. Thomas, L. I. Kruse, and L. Silberman, *J. Org. Chem.*, **42**, 3846 (1977).

[873a] L. I. Kruse and M. D. Meyer, *J. Org. Chem.*, **49**, 4761 (1984).

[873b] K. W. Blake, I. Gillies, and R. C. Denney, *J. Chem. Soc.*, *Perkin Trans. 1*, **1981**, 700.

[874] J. N. Chatterjea and K. Prasad, *J. Indian Chem. Soc.*, **37**, 357 (1960) [*C.A.*, **55**, 533h (1961)].

[875] J. Kossanyi, B. Guiard, and B. Furth, *Bull. Soc. Chim. Fr.* **1974**, 305.

[876] T. Strzalko and J. Seyden-Penne, *C. R. Acad. Sci., Ser. C*, **269**, 604 (1969).

[877] J. A. Marshall and N. H. Andersen, *Tetrahedron Lett.*, **1967**, 1219.

[878] J. A. Marshall, N. H. Andersen, and J. W. Schlicher, *J. Org. Chem.*, **35**, 858 (1970).

[879] M. Schwarz, J. E. Oliver, and P. E. Sonnet, *J. Org. Chem.*, **40**, 2410 (1975).

[880] B. P. Das, J. A. Campbell, F. B. Samples, R. A. Wallace, L. K. Whisenant, R. W. Woodard, and D. W. Boykin, Jr., *J. Med. Chem.*, **15**, 370 (1972).

[881a] N. S. Narasimhan and S. M. Gokhale, *J. Chem. Soc., Chem. Commun.*, **1985**, 86.

[881b] J. C. Chottard and M. Julia, *Bull. Soc. Chim. Fr.*, **1968**, 3700.

[882] R. V. Stevens, F. C. A. Gaeta, and D. S. Lawrence, *J. Am. Chem. Soc.*, **105**, 7713 (1983).

[883] A. Franke, G. Mattern, and W. Traber, *Helv. Chim. Acta*, **58**, 283 (1975).

[884] H. Muxfeldt, E. Jacobs, and K. Uhlig, *Chem. Ber.*, **95**, 2901 (1962).

[885] H. Muxfeldt, German Offen. 1,147,212 (1963) [*C.A.*, **59**, 9933h (1963)].

[886] H. Muxfeldt, W. Rogalski, and K. Striegler, *Chem. Ber.*, **95**, 2581 (1962).

[887] H. Muxfeldt, German Offen. 1,143,192 (1963) [*C.A.*, **58**, 12484g (1963)].

[888] B. R. Baker and R. E. Gibson, *J. Med. Chem.*, **14**, 315 (1971).

[889] G. R. Allen, Jr., and M. J. Weiss, *J. Org. Chem.*, **30**, 2904 (1965).

[890] Société des Usines Chimiques Rhône-Poulenc, French Patent, 1,276,217 (1962) [*C.A.*, **57**, 13771h (1962)].

[891] G. Pala, A. Donetti, A. Mantegani, E. Crescenzi, B. Lumachi, and G. Coppi, *J. Med. Chem.*, **13**, 1089 (1970).

[892] K. Šindelář, J. Metyšová, B. Kakáč, J. Holubek, B. Svátek, J. O. Jílek, J. Pomykáček, and M. Protiva, *Collect. Czech. Chem. Commun.*, **39**, 2099 (1974).

[893] M. Protiva, K. Šindelář, and J. Metyšová, Czech. Patent 155,091 (1974) [*C.A.*, **83**, 164232s (1975)].

[894] N. Suzuki, T. Sone, S. Tagaki, E. Muramatsu, Y. Kobayashi, M. Wakabayashi, and T. Sowa, Japanese Patent 77 59,168 [*C.A.*, **87**, 168079c (1977)].

[895] W. A. Bolhofer, C. N. Habecker, A. M. Pietruszkiewicz, M. L. Torchiana, H. I. Jacoby, and C. A. Stone, *J. Med. Chem.*, **22**, 295 (1979).

[896] M. J. Kornet, W. Beaven, and T. Varia, *J. Heterocycl. Chem.*, **22**, 1089 (1985).

[897] D. G. Markees, L. S. Schwab, and A. Vegotsky, *J. Med. Chem.*, **17**, 137 (1974).

[898] D. L. Venton, S. E. Enke, and G. C. Le Breton, *J. Med. Chem.*, **22**, 824 (1979).

[899] F. I. Carroll, J. T. Blackwell, A Philip, and C. E. Twine, *J. Med. Chem.*, **19**, 1111 (1976).

[900] N. Hirose, S. Kuriyama, and S. Toyoshima, *Chem. Pharm. Bull.*, **24**, 2661 (1976).

[901] R. D. Sindelar, J. Mott, C. F. Barfknecht, S. P. Arneric, J. R. Flynn, J. P. Long, and R. K. Bhatnagar, *J. Med. Chem.*, **25** 858 (1982).

[902] A. Kotani and M. Shiroki, Japanese Patent 77 153,922 [*C.A.*, **88**, 169760a (1978)].

[903] K. Šindelář, B. Kakáč, J. Holubek, B. Svátek, M. Ryska, J. Metyšová, and M. Protiva, *Collect. Czech. Chem. Commun.*, **41**, 1396 (1976).

[904] A. M. Crider, J. M. Robinson, H. G. Floss, J. M. Cassady, and J. A. Clemens, *J. Med. Chem.*, **20**, 1473 (1977).

[905] K. Psotta and A. Wiechers, *Tetrahedron*, **35**, 255 (1979).

[906] D. R. Julian and Z. S. Matusiak, *J. Heterocycl. Chem.*, **12**, 1179 (1975).

[907] R. P. Stein and D. J. Delecki, U.S. Patent 4,129,561 (1978) [*C.A.*, **90**, 152035m (1979)].

[908a] J. Mouralová, J. Hájíček, and J. Trojánek, *Česk. Farm.*, **32**, 23 (1983) [*C.A.*, **98**, 179710e (1983)].

[908b] M. Rajsner, L. Bláha, J. Pirková, V. Trčka, J. Muratová, and M. Vaněček, Czech. Patent 217,009 (1984) [*C.A.*, **102**, 220550u (1985)].

[909] C. Temple, Jr., J. D. Rose, and J. A. Montgomery, *Chem. Ind.* (*London*), **1971**, 883.

[910] C. Temple, Jr., J. D. Rose, and J. A. Montgomery, *J. Pharm. Sci.*, **61**, 1297 (1972) [*C.A.*, **77**, 101279t (1972)].

[911] N. Yokoyama, P. I. Almaula, F. B. Block, F. R. Granat, N. Gottfried, R. T. Hill, E. H. McMahon, W. F. Munch, H. Rachlin, J. K. Saelens, M. G. Siegel, H. C. Tomaselli, and F. H. Clarke, *J. Med. Chem.*, **22**, 537 (1979).

[912] R. Hoellinger, W. Wendtland, and G. Schneider, Canadian Patent 986,539 (1976) [*C.A.*, **85**, 159639s (1976)].

[913] J. Mouralová, Z. Veselý, J. Hodková, and J. Trojánek, Czech. Patent 196,935 (1981) [*C.A.*, **97**, 92609j (1982)].

[914] K. Matsumoto, P. Stark, and R. G. Meister, *J. Med. Chem.*, **20**, 17 (1977).

[915] C. Temple, Jr., J. D. Rose, and J. A. Montgomery, *J. Med. Chem.*, **17**, 972 (1974).

[916] A. R. Kraska and J. J. Plattner., German Offen. 2,528,731 (1976) [*C.A.*, **84**, 164386r (1976)].

[917] J. G. Topliss, M. H. Sherlock, F. H. Clarke, M. C. Daly, B. W. Petersen, J. Lipski, and N. Sperber, *J. Org. Chem.*, **26**, 3842 (1961).

[918] C. W. Whitehead, J. J. Traverso, H. R. Sullivan, and F. J. Marshall, *J. Org. Chem.*, **26**, 2814 (1961).

[919] T. Kunieda, T. Higuchi, Y. Abe, and M. Hirobe, *Tetrahedron*, **39**, 3253 (1983).

[920] H. O. House, F. J. Sauter, W. G. Kenyon, and J. J. Riehl, *J. Org. Chem.*, **33**, 957 (1968).

[921] P. G. Gassman and B. L. Fox, *J. Org. Chem.*, **32**, 480 (1967).

[922] R. R. Sauers, S. B. Schlosberg, and P. E. Pfeffer, *J. Org. Chem.*, **33**, 2175 (1968).

[923] T. J. Perun, L. Zeftel, R. G. Nelb, and D. S. Tarbell, *J. Org. Chem.*, **28**, 2937 (1963).

[924] R. A. Abramovitch and D. L. Struble, *Tetrahedron*, **24**, 705 (1968).

[925] K. Mori and M. Matsui, *Tetrahedron*, **24**, 3127 (1968).

[926] K. Mori and M. Matsui, *Tetrahedron Lett.*, **1967**, 2515.

[927a] A. I. Meyers, R. F. Spohn, and R. J. Lindermann, *J. Org. Chem.*, **50**, 3633 (1985).

[927b] R. Bloch, J. L. Bouket, and J. M. Conia, *Bull. Soc. Chim. Fr.*, **1969**, 489.

[928] M. F. Corrigan and B. O. West, *Aust. J. Chem.*, **29**, 1413 (1976).

[929] A. Meisters and P. C. Wailes, *Aust. J. Chem.*, **13**, 347 (1960).

[930] T. Y. Shen and H. Jones, German Offen. 2,321,151 (1973) [*C.A.*, **80**, 27016y (1974)].

[931] D. Taub, S. Zelawski, and N. L. Wendler, *Tetrahedron Lett.*, **1975**, 3667.

[932] J. A. Oakleaf, M. T. Thomas, A. Wu, and V. Snieckus, *Tetrahedron Lett.*, **1978**, 1645.

[933] K. Naya, Y. Makiyama, T. Matsuura, N. Ii, H. Nagano, and T. Takahashi, *Chem. Lett.*, **1978**, 301.

[934] T. K. Schaaf, D. L. Bussolotti, and M. J. Parry, *J. Am. Chem. Soc.*, **103**, 6502 (1981).

[935] D. J. Dawson and R. E. Ireland, *Tetrahedron Lett.*, **1968**, 1899.

[936] T. Sone, S. Terashima, and S. Yamada, *Chem. Pharm. Bull.*, **24**, 1288 (1976).

[937] P. D. Howes and F. Sondheimer, *J. Org. Chem.*, **43**, 2158 (1978).

[938] G. F. Griffiths, G. W. Kenner, S. W. McCombie, K. M. Smith, and M. J. Sutton, *Tetrahedron*, **32**, 275 (1976).

[939] O. Kříž, J. Fusek, and B. Čásenský, *Collect. Czech. Chem. Commun.*, **48**, 2188 (1983).

[940] W. L. F. Armarego, H. Schou, and P. Waring, *J. Chem. Res.* (*S*), **1980**, 133.

[941] W. L. F. Armarego, H. Schou, and P. Waring, *J. Chem. Res.* (*M*), **1980**, 1951.

[942] E. B. Sanders, H. V. Secor, and J. I. Seeman, *J. Org. Chem.*, **43**, 324 (1978).

[943] T. S. Osdene and E. B. Sanders, U.S. Patent 4,093,620 (1978) [*C.A.*, **89**, 179851r (1978)].

[944] U. Ekevåg, M. Elander, L. Gawell, K. Leander, and B. Lüning, *Acta Chem. Scand., Ser. B*, **27**, 1982 (1973).

[945] H. Newman, *J. Heterocycl. Chem.*, **11**, 449 (1974).

[946] R. F. Borne, C. R. Clark, and J. M. Holbrook, *J. Med. Chem.*, **16**, 853 (1973).

[947] T. Kakizawa, K. Sakai, Y. Naoi, and M. Onaya, Japanese Patent 76 29,484 [*C.A.*, **85**, 94232g (1976)].

[948] R. W. Lockhart, M. Kitadani, F. W. B. Einstein, and Y. L. Chow, *Can. J. Chem.*, **56**, 2897 (1978).

[949] S. Brandänge and L. Lindblom, *Acta Chem. Scand., Ser. B*, **33**, 187 (1979).

[950] D. S. Kemp, J. C. Chabala, and S. A. Marson, *Tetrahedron Lett.*, **1978**, 543.

[951] A. B. Daruwala, J. E. Gearien, W. J. Dunn, III, P. S. Benoit, and L. Bauer, *J. Med. Chem.*, **17**, 819 (1974).

[952] K. Takayama, M. Isobe, K. Harano, and T. Taguchi, *Tetrahedron Lett.*, **1973**, 365.

[953] M. Isobe, K. Harano, and T. Taguchi, *Yakugaku Zasshi*, **94**, 343 (1974) [*C.A.*, **81**, 77766j (1974)].

[954] J. Jindřichovský, A. Rybář, and R. Frimm, Czech. Patent 210,993 (1984) [*C.A.*, **101**, 130603p (1984)].

[955] N. Kawahara, T. Nakajima, T. Itoh, and H. Ogura, *Heterocycles*, **16**, 729 (1981).

[956] R. F. Borne, S. J. Law, P. W. With, and J. C. Murphy, *J. Pharm. Sci.*, **66**, 594 (1977) [*C.A.*, **87**, 102139r (1977)].

[957] J. G. Cannon, G. J. Hatheway, J. P. Long, and F. M. Sharabi, *J. Med. Chem.*, **19**, 987 (1976).

[958] F. M. Sharabi, J. P. Long, J. G. Cannon, and G. J. Hatheway, *J. Pharmacol. Exp. Ther.*, **199**, 630 (1976) [*C.A.*, **86**, 37497u (1977)].

[959] M. S. Manhas, M. Sugiura, and H. P. S. Chawla, *J. Heterocycl. Chem.*, **15**, 949 (1978).

[960] R. F. Borne, C. R. Clark, and R. L. Peden, *J. Heterocycl. Chem.*, **10**, 241 (1973).

[960a] A. I. Meyers, R. Hanreich, and K. T. Wanner, *J. Am. Chem. Soc.*, **107**, 7776 (1985).

[961] F. Szemes and A. Rybář, Czech. Patent, 209,400 (1983) [*C.A.*, **100**, 68192c (1984)].

[962] W. L. F. Armarego and T. Kobayashi, *J. Chem. Soc., C*, **1971**, 3222.

[963] M. E. Rogers and E. L. May, *J. Med. Chem.*, **17**, 1328 (1974).

[964] J. G. Cannon, C. Suarez-Gutierrez, T. Lee, J. P. Long, B. Costall, D. H. Fortune, and R. J. Naylor, *J. Med. Chem.*, **22**, 341 (1979).

[965] T. Tanaka, T. Hayasaka, K. Saito, and T. Goto, *Yakugaku Zasshi*, **99**, 335 (1979) [*C.A.*, **91**, 211355m (1979)].

[966] K. Okada, J. A. Kelley, and J. S. Driscoll, *J. Org. Chem.*, **42**, 2594 (1977).

[967] F. Ishikawa, T. Imano, K. Higashi, and Y. Abiko, Japanese Patent (Tokkyo Koho) 78 79,888 [*C.A.*, **90**, 23095g (1979)].

[968] M. J. Begley and N. Whittaker, *J. Chem. Soc., Perkin Trans. 1*, **1973**, 2830.

[969] CIBA Ltd., French Patent 1,449,604 (1966) [*C.A.*, **67**, 21847e (1967)].

[970a] D. L. Coffen, D. A. Katonak, and F. Wong, *J. Am. Chem. Soc.*, **96**, 3966 (1974).

[970b] P. Melloni, A. Della Torre, E. Lazzari, G. Mazzini, and M. Meroni, *Tetrahedron*, **41**, 1393 (1985).

[971] T. Kametani, Y. Suzuki, and M. Ihara, *Can. J. Chem.*, **57**, 1679 (1979).

[972a] T. Tanaka, K. Mashimo, and M. Wagatsuma, *Tetrahedron Lett.*, **1971**, 2803.

[972b] H. Ishii, E. Ueda, K. Nakajima, T. Ishida, T. Ishikawa, K. I. Harada, I. Ninomiya, T. Naito, and T. Kiguchi, *Chem. Pharm. Bull.*, **26**, 864 (1978).

[973] S. I. Clarke and R. H. Prager, *Aust. J. Chem.*, **35**, 1645 (1982).

[974] Z. Veselý, J. Holubek, and J. Trojánek, *Chem. Ind.* (*London*), **1973**, 478.

[975] Z. Veseleý and J. Trojánek, Czech. Patent 154,973 (1974) [*C.A.*, **82**, 98001s (1975)].

[976] G. R. Lenz, *J. Org. Chem.*, **41**, 2201 (1976).

[977] T. Kametani, S. P. Huang, M. Ihara, and K. Fukumoto, *Chem. Pharm. Bull.*, **23**, 2010 (1975).
[978] T. Kametani, Japanese Patent 76 80,860 [*C.A.*, **86**, 72474f (1977)].
[979] G. R. Lenz, *J. Org. Chem.*, **39**, 2846 (1974).
[980] T. Ibuka, K. Tanaka, and Y. Inubushi, *Chem. Pharm. Bull.*, **22**, 907 (1974).
[981] T. Ibuka, K. Tanaka, and Y. Inubushi, *Tetrahedron Lett.*, **1972**, 1393.
[982] J. G. Cannon, T. Lee, H. D. Goldman, J. P. Long, J. R. Flynn, T. Verimer, B. Costall, and R. J. Naylor, *J. Med. Chem.*, **23**, 1 (1980).
[983] E. O. Renth, A. Mentrup, and K. Schromm, German Offen. 2,050,684 (1972) [*C.A.*, **77**, 34573u (1972)].
[984] T. Kametani, H. Seto, H. Nemoto, and K. Fukumoto, *J. Org. Chem.*, **42**, 3605 (1977).
[985a] G. A. Kraus and S. Yue, *J. Chem. Soc., Chem. Commun.*, **1983**, 1198.
[985b] M. Ihara, M. Tsuruta, K. Fukumoto, and T. Kametani, *J. Chem. Soc., Chem. Commun.*, **1985**, 1159.
[986] T. Kametani, T. Ohsawa, and M. Ihara, *J. Chem. Res.* (*S*), **1979**, 364.
[987] T. Kametani, T. Ohsawa, and M. Ihara, *J. Chem. Res.* (*M*), **1979**, 4438.
[988] T. Kametani, S. P. Huang, M. Ihara, and K. Fukumoto, *J. Org. Chem.* **41**, 2545 (1976).
[989] T. Ohnuma, T. Oishi, and Y. Ban, *J. Chem. Soc., Chem. Commun.*, **1973**, 301.
[990] A. Mondon and H. Witt, *Chem. Ber.*, **103**, 1522 (1970)
[991] T. Kametani, S. A. Surgenor, and K. Fukumoto, *J. Chem. Soc., Perkin Trans. 1*, **1981**, 920.
[992] Sendai Institute of Heterocyclic Chemistry, Japanese Patent (Tokkyo Koho) 82 38,780 [*C.A.*, **97**, 145104z (1982)].
[993] D. J. Aberhart and C. T. Hsu, *J. Org. Chem.*, **41**, 2098 (1976).
[994] T. Fukuyama and R. A. Sachleben, *J. Am. Chem. Soc.*, **104**, 4957 (1982).
[995] J. Trojánek, Research Institute for Pharmacy and Biochemistry, Prague, personal communication.
[996] H. von Dobeneck and B. Hansen, *Chem. Ber.*, **105**, 3611 (1972).
[997] J. W Epstein, H. J. Brabander, W. J. Fanshawe, C. M. Hofmann, T. C. McKenzie, S. R. Safir, A. C. Osterberg, D. B. Cosulich, and F. M. Lovell, *J. Med. Chem.*, **24**, 481 (1981).
[998] W. J. Fanshawe, J. W. Epstein, L. S. Crawley, C. M. Hofmann, and S. R. Safir, German Offen. 2,740,562 (1978) [*C.A.*, **89**, 6216j (1978)].
[999] W. J. Fanshawe, J. W. Epstein, L. S. Crawley, C. M. Hofmann, and S. R. Safir, U.S. Patent 4,088,652 (1978) [*C. A.*, **89**, 129383u (1978)].
[1000] J. Kalo, D Ginsburg, and E. Vogel, *Tetrahedron*, **33**, 1177 (1977).
[1001] E. Ciganek, *J. Org. Chem.*, **45**, 1512 (1980).
[1002] H. G. Thomas and P. C. Ruenitz, *J. Heterocycl. Chem.*, **21**, 1057 (1984).
[1003] J. C. Nnadi and P. S. Landis, U.S. Patent 3,799,877 (1974) [*C.A.*, **81**, 153475f (1974)].
[1004] A. P. Kozikowski and R. J. Schmiesing, *J. Org. Chem.*, **48**, 1000 (1983).
[1005] A. Gieren, B. Dederer, I. Ugi, and S. Stüber, *Tetrahedron Lett.*, **1977**, 1507.
[1006] T. Nishiwaki and F. Fujiyama, *J. Chem. Soc., Perkin Trans. 1*, **1972**, 1456.
[1007] T. Kobayashi, K. Iino, and T. Hiraoka, *Chem. Pharm. Bull.*, **27**, 2727 (1979).
[1008] J. Málek and M. Černý, unpublished results.
[1009] M. Černý and J. Málek, *Tetrahedron Lett.*, **1972**, 691.
[1010] J. Málek and M. Černý, *J. Organomet. Chem.*, **84**, 139 (1975).
[1011] G. Bellucci, F. Macchia, and M. Poggianti, *Chim. Ind.* (*Milan*), **50**, 1324 (1968).
[1012] F. Bondavalli, A. Ranise, P. Schenone, and S. Lanteri, *J. Chem. Soc., Perkin Trans. 1*, **1978**, 804.
[1013] J. R. Dimmock, W. A. Turner, P. J. Smith, and R. G. Sutherland *Can. J. Chem.*, **51**, 427 (1973).
[1014] S. S. Klioze, F. J. Ehrgott, Jr., J. C Wilker, and D. L. Woodward, *J. Med. Chem.*, **22**, 1497 (1979).
[1014a] Sagami Chemical Research Center, Japanese Patent (Tokkyo Koho) 60 23,393 (85 23,393) (1985) [*C.A.* **103**, 215719x (1985)].
[1015] E. E. Smissman and T. L. Pazdernik, *J. Med. Chem.*, **16**, 18 (1973).
[1016] B. S. E. Carnmalm, T. De Paulis, S. B. Ross, S. I. Rämsby, N. E. Stjernström, and S. O. Ogren, German Offen. 2,227,922 (1973) [*C.A.*, **78**, 159333t (1973)].

[1017] B. S. E. Carnmalm, S. I. Rämsby, N. E. Stjernström, and S. O. Ogren, European Patent Application 16,746 (1980) [*C.A.*, **94**, 156419v (1981)].

[1018] B. Carnmalm, E. Jacupovic, L. Johansson, T. de Paulis, S. Rämsby, N. E. Stjernström, A. L. Renyi, S. B. Ross, and S. O. Ögren, *J. Med. Chem.*, **17**, 65 (1974).

[1019] E. E. Smissman and R. T. Borchardt, *J. Med. Chem.*, **14**, 377 (1971).

[1020] P. G. Pietta, P. F. Cavallo, K. Takahashi, and G. R. Marshall, *J. Org. Chem.*, **39**, 44 (1974).

[1021] L. D. Ruffin and S. B. Bowlus, *J. Heterocycl. Chem.*, **20**, 461 (1983).

[1022] R. A. Michelin, G. Facchin, and P. Uguagliati, *Inorg. Chem.*, **23**, 961 (1984).

[1023] M. P. Periasamy and H. M. Walborsky, *J. Am. Chem. Soc.*, **99**, 2631 (1977).

[1024] R. Fielden, O. Meth-Cohn, and H. Suschitzky, *J. Chem. Soc., Perkin Trans. 1*, **1973**, 705.

[1025] R. F. Borne, C. R. Clark, and N. A. Wade, *J. Heterocycl. Chem.*, **11**, 311 (1974).

[1026] V. Valenta and M. Protiva, *Collect. Czech. Chem. Commun.*, **41**, 906 (1976).

[1027] M. Protiva, V. Valenta, and A. Čapek, Czech. Patent 183,126 (1980) [*C.A.*, **94**, 192373j (1981)].

[1028a] K. Šindelář, J. Holubek, J. Schlanger, A. Dlabač, M. Valchář, and M. Protiva, *Collect. Czech. Chem. Commun.*, **50**, 503 (1985).

[1028b] M. Protiva, K. Pelz, J. Jílek, V. Seidlová, and J. Metyšová, Czech. Patent 131,189 (1969) [*C.A.*, **73**, 56127w (1970)]; French Patent 1,566,933 (1969) [*C.A.*, **72**, 90336c (1970)].

[1029] M. Protiva and J. Jílek, South African Patent 71 04,647 (1972) [*C.A.*, **78**, 72209h (1973)].

[1030] M. Protiva, J. Jílek, and J. Vít, Czech. Patent 149,277 (1973) [*C.A.*, **80**, 70836c (1974)].

[1031] R. K. Hill, M. G. Fracheboud, S. Sawada, R. M. Carlson, and S. J. Yan, *Tetrahedron Lett.*, **1978**, 945.

[1032] J. C. Kim, S. K. Lee, C. B. Kim, S. K. Han, S. K. Choi, and K. H. Lee, *Teahan Hwahak* Hoechi, **21**, 187 (1977) [*C.A.*, **87**, 184254m (1977)].

[1033] C. F. Barfknecht, D. E. Nichols, D. B. Busterholz, J. P. Long, J. A. Engelbrecht, J. M. Beaton, R. J. Bradley, and D. C. Dyer, *J. Med. Chem.*, **16**, 804 (1973).

[1034] D. E. Nichols, C. F. Barfknecht, J.P. Long, R. T. Standridge, H. G. Howell, R. A. Partyka, and D. C. Dyer, *J. Med. Chem.*, **17**, 161 (1974).

[1035] D. Beaumont, R. D. Waigh, M. Sunbhanich, and M. W. Nott, *J. Med. Chem.*, **26**, 507 (1983).

[1036] W. Kantlehner, P. Speh, and H. J. Bräuner, *Synthesis*, **1983**, 905.

[1037] I. Ahmed, G. W. H. Cheeseman, and B. Jaques, *Tetrahedron*, **35**, 1145 (1979).

[1038a] M. Natsume and I. Utsunomiya, *Chem. Pharm. Bull.*, **32**, 2477 (1984).

[1038b] I. Utsunomiya and M. Natsume, *Heterocycles*, **23**, 223 (1985).

[1038c] M. Natsume, I. Utsunomiya, K. Yamaguchi, and Shin-Ichiro Sakai, *Tetrahedron*, **41**, 2115 (1985).

[1039] I. Utsunomiya and M. Natsume, *Heterocycles*, **21** (No. 2, Special Issue), 726 (1984).

[1040] W. L. Albrecht, W. D. Jones, Jr., and F. W. Sweet, *J. Heterocycl. Chem.*, **15**, 209 (1978).

[1041a] F. Rise and K. Undheim, *Acta Chem. Scand.*, **B39**, 195 (1985).

[1041b] N. Finch and C. W. Gemenden, *J. Org. Chem.*, **38**, 437 (1973).

[1041c] A. Rasmussen, F. Rise, and K. Undheim, *Acta Chem. Scand.*, **B39**, 235 (1985).

[1042] D. C. H. Bigg, A. W. Faull, and S. R. Purvis, *J. Heterocycl. Chem.*, **14**, 603 (1977).

[1043] N. De Kimpe, R. Verhe, L. De Buyck, N. Schamp, J. P. Declercq, G. Germain, and M. Van Meersche, *J. Org. Chem.*, **42**, 3704 (1977).

[1044] C. M. Wong, T. L. Ho, and W. P. Niemczura, *Can. J. Chem.*, **53**, 3144 (1975).

[1045] V. N. Gogte, M. A. Salama, and B. D. Tilak, *Tetrahedron*, **26**, 173 (1970).

[1046] H. W. Whitlock, Jr., and G. L. Smith, *Tetrahedron Lett.*, **1965**, 1389.

[1047] E. Wenkert, R. A. Massy-Westropp, and R. G. Lewis, *J. Am. Chem. Soc.*, **84**, 3732 (1962).

[1048] G. Laus and G. Van Binst, *Tetrahedron*, **35**, 849 (1979).

[1049] T. Kametani, T. Kohno, K. Kigasawa, M. Hiiragi, K. Wakisaka, and T. Uryu, *Chem. Pharm. Bull.*, **19**, 1794 (1971).

[1050a] M. W. Anderson, R. C. F. Jones, and J. Saunders, *J. Chem. Soc., Chem. Commun.*, **1982**, 282.

[1050b] J. Hájíček and J. Trojánek, Czech. Patent 214,361 (1984) [*C.A.*, **102**, 221069f (1985)].

[1051a] S. Takano, S. Hatakeyama, and K. Ogasawara, *J. Chem. Soc., Perkin Trans. 1*, **1980**, 457.

[1051b] S. Takano, S. Hatakeyama, and K. Ogasawara, *J. Chem. Soc., Chem. Commun.*, **1977**, 68.

[1051c] S. Takano, M. Yonaga, M. Morimoto, and K. Ogasawara, *J. Chem. Soc., Perkin Trans. 1*, **1985**, 305.

[1052] J. Knabe and P. Dorfmüller, *Arch. Pharm.* (Weinheim, W. Germany), **318**, 531 (1985).

[1053] O.E. Edwards, *J. Chem. Soc., Chem. Commun.*, **1965**, 318.

[1054] C. R. Johnson and C. J. Stark, Jr., *J. Org. Chem.*, **47**, 1196 (1982).

[1055] L. H. Klemm and R. F. Lawrence, *J. Heterocycl. Chem.*, **16**, 599 (1979).

[1056] R. W. Hoffmann, N. Hauel, and B. Landmann, *Chem. Ber.*, **116**, 389 (1983).

[1057] S. W. Baldwin and J. C. Tomesch, *J. Org. Chem.*, **45**, 1455 (1980).

[1058] Y. Ban, K. Yoshida, J. Goto, T. Oishi, and E. Takeda, *Tetrahedron*, **39**, 3657 (1983).

[1059a] J. F. Pilichowski, J. M. Lehn, J. P. Sauvage, and J. C. Gramain, *Tetrahedron*, **41**, 1959 (1985).

[1059b] N. Finch, L. Blanchard, and L. H. Werner, *J. Org. Chem.*, **42**, 3933 (1977).

[1059c] D. Fréhel, A. Badorc, J. M. Pereillo, and J. P. Maffrand, *J. Heterocycl. Chem.*, **22**, 1011 (1985).

[1060] L. S. Hegedus and J. M. McKearin, *J. Am. Chem. Soc.*, **104**, 2444 (1982).

[1061a] W. Nagata, H. Itazaki, K. Okada, T. Wakabayashi, K. Shibata, and N. Tokutake, *Chem. Pharm. Bull.*, **23**, 2867 (1975).

[1061b] W. Nagata, Japanese Patent 75 19,756 [*C.A.*, **83**, 114234y (1975)].

[1062] G. Buchbauer, W. Pernold, M. Ittner, M. T. Ahmadi, R. Dobner, and R. Reidiger, *Monatsh. Chem.*, **116**, 1209 (1985).

[1063] R. W. Kierstead, A. Faraone, F. Mennona, J. Mullin, R. W. Guthrie, H. Crowley, B. Simko, and L. C. Blaber, *J. Med. Chem.*, **26**, 1561 (1983).

[1064] E. Buhleier, W. Rasshofer, W. Wehner, F. Luppertz, and F. Vögtle, *Justus Liebigs Ann. Chem.*, **1977**, 1344.

[1065] A. P. King and C. G. Krespan, *J. Org. Chem.*, **39**, 1315 (1974).

[1066] A. B. Holmes, J. Thompson, A. J. G. Baxter, and J. Dixon, *J. Chem. Soc., Chem. Commun.*, **1985**, 37.

[1067] A. V. Semenovskii, E. V. Polunin, I. M. Zaks, and A. M. Moiseenkov, *Izv. Akad. Nauk SSSR, Ser. Khim.*, **1979**, 1327 [*C.A.*, **91**, 123689t (1979)].

[1068] H. F. Kung, C. C. Yu, J. Billings, M. Molnar, and M. Blau, *J. Med. Chem.*, **28**, 1280 (1985).

[1069] H. F. Kung, C. C. Yu, J. Billings, M. Molnar, and M. Blau, *J. Labelled Compd. Radiopharm.*, **21**, 1016 (1984).

AUTHOR INDEX, VOLUMES 1–36

CHAPTER AND TOPIC INDEX, VOLUMES 1–36

Many chapters contain brief discussions of reactions and comparisons of alternative synthetic methods related to the reaction that is the subject of the chapter. These related reactions and alternative methods are not usually listed in this index. In this index, the volume number is in **BOLDFACE**, the chapter number is in ordinary type.

Organic Reactions